The Freshwater Budget of the Arctic Ocean

NATO Science Series

A Series presenting the results of activities sponsored by the NATO Science Committee. The Series is published by IOS Press and Kluwer Academic Publishers, in conjunction with the NATO Scientific Affairs Division.

A. Life Sciences	IOS Press
B. Physics	Kluwer Academic Publishers
C. Mathematical and Physical Sciences	Kluwer Academic Publishers
D. Behavioural and Social Sciences	Kluwer Academic Publishers
E. Applied Sciences	Kluwer Academic Publishers
F. Computer and Systems Sciences	IOS Press

1. Disarmament Technologies	Kluwer Academic Publishers
2. Environmental Security	Kluwer Academic Publishers
3. High Technology	Kluwer Academic Publishers
4. Science and Technology Policy	IOS Press
5. Computer Networking	IOS Press

NATO-PCO-DATA BASE

The NATO Science Series continues the series of books published formerly in the NATO ASI Series. An electronic index to the NATO ASI Series provides full bibliographical references (with keywords and/or abstracts) to more than 50000 contributions from international scientists published in all sections of the NATO ASI Series.
Access to the NATO-PCO-DATA BASE is possible via CD-ROM "NATO-PCO-DATA BASE" with user-friendly retrieval software in English, French and German (WTV GmbH and DATAWARE Technologies Inc. 1989).

The CD-ROM of the NATO ASI Series can be ordered from: PCO, Overijse, Belgium

Series 2. Environment Security – Vol. 70

The Freshwater Budget of the Arctic Ocean

edited by

Edward Lyn Lewis

Emeritus, Institute of Ocean Sciences,
Sidney, B.C., Canada

Associate Editors:

E. Peter Jones

Bedford Institute of Oceanography,
Dartmouth, N.S., Canada

Peter Lemke

Professor of Meteorology,
Institute of Marine Research,
Kiel, Germany

Terry D. Prowse

National Water Research Institute,
Saskatoon, Sk., Canada

and

Peter Wadhams

Reader in Polar Studies,
Scott Polar Research Institute,
University of Cambridge, U.K.

Springer Science+Business Media, B.V.

Proceedings of the NATO Advanced Research Workshop on
The Freshwater Budget of the Arctic Ocean
Tallinn, Estonia
27 April-1 May 1998

A C.I.P. Catalogue record for this book is available from the Library of Congress.

ISBN 978-0-7923-6439-9   ISBN 978-94-011-4132-1 (eBook)
DOI 10.1007/978-94-011-4132-1

Printed on acid-free paper

TABLE OF CONTENTS

PREFACE

Following a decision by the Arctic Ocean Sciences Board (AOSB) in July 1996 the then chairman, Geoffrey Holland, wrote a letter of invitation to a meeting to plan a "Symposium on the Freshwater Balance of the Arctic". The meeting was held in Ottawa on November 12-13 1996 and was attended by representatives of various organisations, including the U.S. National Science Foundation (NSF), as well as individual scientists. Results of this meeting included:

- Co-sponsorship with AOSB by the Scientific Committee on Ocean Research (SCOR), the Arctic Climate System Study (ACSYS) and the Global Energy and Water Cycle Experiment (GEWEX).
- A decision to apply for funding as a Advanced Research Workshop (ARW) of the North Atlantic Treaty Organisation (NATO) Scientific Affairs Division.
- That expenses would be covered in part by funds available through an existing NSF grant to the SCOR Executive offices in Baltimore, MD.
- The appointment of myself to be Chairman/Manager for the Symposium.
- Provision of a recommended list of Scientific Advisors to assist the Chairman in selecting key speakers.

NATO ARWs have to follow clear guidelines. They require a balance between numbers of attendees from NATO countries and Cooperation Partner (CP) countries, which consist mainly of states of the Russian Federation and Eastern Europe. The key speakers at the ARW must be identified in the application for funding and the total attendance is limited and by invitation only. Normally it must be held in a CP country and co-directors are required, one each from a NATO and a CP country: Dr. Igor Shiklomanov, Director of the State Hydrological Institute, St. Petersburg. Russian Federation, agreed to act with me in this role. The Proceedings must be published in book form.

Application for NATO funding was made in May 1997 under the heading "The Freshwater Budget of the Arctic Ocean" with the meeting to be held in Tallinn, the capital of Estonia. At the same time SCOR agreed to provide administrative support, especially for making speakers' travel arrangements. Approval was received from NATO in September 1997. With funding assured, definite invitations went out to potential key speakers and speakers. It was decided to try and make the book not merely a conference "Proceedings", but a text on the topic, a status report for 1998. To ensure coverage, invited attendees were carefully selected by area of expertise as well as reputation and, as NATO funds covered all a speaker's expenses, it was thought appropriate to propose a tentative title for the paper. To provide as wide a view as possible, it was suggested that principal authors should incorporate contributions from their colleagues into the final text. A glance at the authors' listing at the head of many of the papers will show the success of this policy.

In October 1997 I traveled to Tallinn. An introduction to the Estonian Marine Institute (EMI) and its staff was provided through the good offices of Dr.Penti Malkki, Director of the Finnish Institute of Marine Research, Helsinki. From that time on my main contact in Estonia was Dr.Jüri Elken of EMI who accepted the position of Chairman of the local arrangements committee. His help was invaluable. Hotels with facilities adequate for the meeting were visited and an agreement entered into with the Hotel Viru. It was becoming clear that the Editor needed help to provide adequate assessments and criticisms of the papers. Subjects ranged from river hydrology through meteorology and ocean physics/chemistry to modelling of air-ice-ocean interface processes: data collection methods and their limitations were also discussed. It was indeed fortunate that the four Associate Editors listed on the title page agreed to help. Between us, we possessed adequate knowledge and each paper was assigned to a suitable Associate Editor for assessment before coming on to me for final comments.

NATO ARW No. 971307, "The Freshwater Budget of the Arctic Ocean", was held at the Hotel Viru, Viru Valjak 4, EE0001, Tallinn, Estonia, on the five days April 27 - May 1, 1998 inclusive. Most attendees arrived on the evening of Saturday, 25 April, which gave the North Americans a day to recover from the time shift before commencing work. This meeting was not advertised, but knowledge of it spread by mouth, and the topic and its timeliness was such that it proved impossible to invite all those who wished to come. In addition to the forty-three scientists who attended with all expenses paid, an additional seven came as auditors, covering all their own costs. This brought the total up to fifty, the maximum allowed by the NATO guidelines. An ARW office was established in the Hotel Viru, which served for typing, photocopying, travel arrangements and all liason with the Hotel staff. Presentations were organised so that on Monday, after welcoming addresses, key speakers gave some of the more general background papers. In the following four days topics were arranged to progress from studies of the freshwater inputs to the Arctic ocean to descriptions of its distribution and redistribution within the ocean to estimates of oceanic import/export of water and ice through the straits joining the Arctic to the World ocean. The Editors led discussions after each related group of papers had been presented.

It was some twenty months after Tallinn that the final manuscripts came to hand, a time of frustration to those who completed their manuscripts before the end of 1998. In the November of that year all five Editors met for four days in Cambridge U.K. when papers to hand were read, discussed, and an initial response made to each principal author. Revised versions came in to individual Editors, who made further comment, but by the summer of 1999 most of the papers were in near-final form. Some papers were still absent at the end of the year and it was decided that only twenty-three of the twenty-seven papers given in Tallinn would be published. Attempts to cope with the inevitable gap are the short paper by an Associate Editor (Prowse) and Flegg and reference to recently published Journal papers in the Introduction. Very many authors, informed of the closing date, provided final updated versions of their work. As a result, it is thought that the present volume describes of our state of knowledge in 1999 on the Freshwater Budget of the Arctic Ocean.

Victoria B.C. Feb. 2000   E. Lyn Lewis

ACKNOWLEGDEMENTS

This book originated in a decision by Mr. G. Holland and his colleagues in the AOSB at their meeting in Helsinki in April 1996. Since that time the importance of freshwater in the Arctic Ocean to climate has become more and more apparent and the perspicacity of the AOSB choice a matter for congratulation.

My debt to my four Associates is profound and difficult to exaggerate. Dr. E. Peter Jones of the Ocean Science Division, Bedford Institute of Oceanography, Dartmouth N.S. Canada; Prof. Dr. Peter Lemke, Director, Institut für Meereskunde, Kiel, Germany; Dr. Terry D. Prowse, National Water Research Institute, Saskatoon, SK, Canada; Dr. Peter Wadhams, Reader in Polar studies, Scott Polar Research Institute, University of Cambridge U.K. Most polar scientists reading this book will be familiar with these men and their work. In addition to their direct influence on authors, it was of great value to have such knowledgeable people as consultants when initial, tentative conclusions needed confirmation.

In Tallinn, the Chairman of the Local Arrangements Committee, Dr. Jüri Elken gave much of his time to organising the meeting. Quite literally all communications within Estonia had to pass through his hands as Estonian is a non - Indo-European language and was opaque to all other organisers. In addition to thanking him for a first class job I wish to apologise; he suffered from having far to many E-mails from this office. Sometimes I think that his accomplishment was in spite of me and my detailed requests.

Dr. L. Veiga da Cunha of the NATO Scientific and Environmental Affairs Division was my correspondent in all matters dealing with this ARW and was in Tallinn for the greater part of our meeting. He has now left that office, but leaves a memory of the efficient way in which he dealt with administrative problems.

It is very pleasant to recall my contacts with the Executive Director of SCOR, Ms. E Gross. She offered encouragement and understanding, coped with my rudimentary accountancy, and advised on how to serve the interests of science and administration simultaneously. Her assistant Ms. W. A. Ross dealt effectively with the myriad practical details.

I want to thank individuals whose help at the Viru Hotel had a material affect on our meeting. Dr Vladimir Kattsov used his fluent knowledge of English to provide a simultaneous translation of some of his compatriots' presentations and so made them immediately available to the rest of the audience. It has been said that the basis of a good conference is well-housed and fed delegates and a smoothly running administration. My wife, Kládía, was responsible for all the day to day "housekeeping" details, from meeting people at the airport, to running the office, to coping with any and all problems within the hotel, to making and altering travel arrangements. My sister Susan did all the necessary typing.

Financial support was supplied by the U.S. National Science Foundation under its Grant No.OCE - 9422144 to SCOR. Environ. ARW 971307 was the designation of the support received from the NATO Scientific and Environmental Affairs Division.

THE FRESHWATER BUDGET OF THE ARCTIC OCEAN

Workshop Summary by Mark C. Serreze

The water that runs to the Arctic sea,
Ought to be balancing P minus E.
While all of us know this has to be true,
Why then, we ask, are our budgets askew?

Its tough to assess the land-ocean link
When data on precip. and snowfall just stink.
It's also hard to attain our goals
From streamflow records with numerous holes.

Rawinsonde data can help in this matter,
But estimates also contain ample scatter.
Is numerical weather prediction a cure?
Some of us think so, others aren't sure.

Similar problems abound in the sea,
Where fluxes of fresh and of salt don't agree.
A Sverdrup here and a Sverdrup there,
Add it all up and it's really a scare.

How do we measure the average rate
Of freshwater export out of Fram Strait?
And how do the volumes of sea ice and brine
Affect the strength of the halocline?

The Atlantic inflow, where does it go?
Does the import have links with the NAO?
This is an issue that needs lots of thinking.
Likely combined with needed drinking.

To conclude, it seems that we still don't know
Just how to balance the H20.
But as the modeling crowd might say with inflection,
Lets try and do it without flux correction.

INTRODUCTION

E. L. Lewis
Institute of Ocean Sciences
Sidney B. C. Canada

This Advanced Research Workshop joined together meteorologists, hydrologists, oceanographers and sea-ice specialists in an attempt to present the latest information about the Arctic Ocean Freshwater Budget, a snapshot of our state of knowledge. Modellers informed us of the capability of their representations and major needs for more information. Twenty-seven papers were presented over five days. Twenty-three are published here, plus a contribution by Prowse & Flegg which attempts to fill the gap left by at least two of the four papers that were not available in time. All the principal authors have responded to detailed criticisms by the Editors, who themselves act as referees for papers in well known geophysical journals. This book is thus considered to be a compilation of "refereed" papers.

Important interpretive notes in this Introduction are in black italics.

Changes in the freshwater flow from the Arctic Ocean may have very significant consequences for living human beings. Year by year the observational evidence makes predictions of consequent global climatic effects more likely to be true. We must become able to predict, with far greater certainty, the magnitude of such effects and the probability of their occurrence within a given time scale. This will require internationally organised commitments of scientific labour and money. This Introduction, primarily written to provide a framework in which to place the chapters of this book, also outlines the reasons for this present, major concern.

Due to the latitudinal variation in the sun's incoming radiation, there is a net evaporation of fresh water from the surface in equatorial regions and a net precipitation in polar regions. In the Northern hemisphere, this produces an overall southward flow of freshwater. The details of this freshwater circulation and local mixing are subject to weather, topography, permafrost, sea-ice growth/decay etc., and on longer time scales and distances, to climate. There is certainly some connection between the ENSO oscillation in the Pacific (giving rise to El Niño) and precipitation in the Arctic, though few details of the linkage are known. Better documented is the connection to the North Atlantic Oscillation (NAO) index, defined in terms of the variations in the atmospheric pressure difference between the Azores high and the Icelandic low. An excellent introduction to global factors is given in Chapters 1 & 2, which provide a setting for the following papers. It is this picture of the world heat engine as a whole that must provide the background to Arctic studies. The hydrological cycle of the Arctic Ocean has intimate linkages to global climate. Changes in one affect the other, often with a feedback.

The combined effects of large river runoff, advection of meteoric water, low evaporation rates and distillation by freezing, contribute to the formation of a strong halocline in the upper Arctic ocean, which limits thermal communication between the sea ice and the warmer waters of Atlantic origin below. Sea ice and freshened surface waters are transported from the marginal seas by winds and currents, ultimately exiting the Arctic Ocean through Fram and Davis Straits. Variations in the freshwater outflow from these regions, which appear to be influenced by the NAO, effect the density structure of the Arctic Ocean itself and so the surface heat balance, Another feedback is the effect these variations have on the density profile of the water column in the Greenland and Labrador seas, where, at present, convection takes place mixing surface waters downwards with those at greater depth. In descending these waters remove carbon dioxide (and pollutants) from the surface and thus variations in the local exchange of surface with deeper waters play a role in climate change and environmental quality. On a much larger scale this downward convection motion produces dense deep waters that flow outwards from these two centres and can influence the entire North Atlantic. Such deep outflows must be replaced by inward flowing waters at lesser depths. One such inward flow is the so - called North Atlantic Drift, which brings warm surface water to Northern Europe from the Gulf of Mexico. It is possible that enhanced outflow of fresher waters from the Arctic Ocean may diminish this heat source, and cause a major change in the climate of European maritime states. .

The primary pathways for the influx of freshwater to the Arctic Ocean are precipitation, solid and liquid, less evaporation, runoff from Arctic terrestrial watersheds and oceanic advection, which includes the advection of sea ice. The large-scale atmospheric circulation affects all three of these pathways, either through wind forcing in the case of the ocean or directly through the atmospheric-surface exchanges, which are largely driven by atmospheric advection and moisture flux convergence. Chapter 3 concentrates on the effects of the atmospheric moisture transport and problems in its measurement and estimation.

Despite the importance of freshwater fluxes to the thermohaline circulation of the world ocean and the global heat balance, the hydrology of the Arctic basin land surface is not well understood. Chapter 4 explores the factors controlling river discharge. Space-time analysis of historical streamflow and climatological observations, along with new data sets such as reanalysis data from the global weather forecast centers, can offer insights into some of the controlling processes and their characteristics. The distribution of river water inflow around the Arctic basin is considered, also problems with measurement techniques.

There is a desirable overlap between Chapters 3 & 4, which extends into Chapter 5. This chapter looks at the same components of freshwater flow but from other side, that of the receiving ocean. The halocline of the Arctic Ocean is part of a much larger, salinity-stratified system that covers the high-latitude oceans of the Northern Hemisphere between the subarctic fronts of the Pacific and Atlantic oceans. This larger system exists simply because of the already noted need of the Earth's climate system to redistribute heat poleward. Fresh water, delivered to the polar regions by atmospheric transport and by ocean and river inflows, enters the Arctic Ocean and is stored within the various layers of the halocline, itself serving as an extremely complex and poorly understood reservoir. Further distillation of fresh water may occur during the melt/freeze cycle of sea ice, provided that the ice and resulting excess brine formed by freezing in winter can be separated and exported before they are reunited by melting and wind mixing during the following summer. The ultimate sink of fresh water is its export southwards into the North Atlantic, for it must ultimately replace the fresh water evaporating from low latitude oceans.

Thus Chapters 3,4 &5 give the basic physical background for reading the more specialised papers, and are joined by Chapter 6, focused on the capability of a particular climate General Circulation Model (the U.K. Meteorological Office HADCM3), hereinafter GCM, to represent the components of the Arctic freshwater budget. They form an extended and detailed "Introduction" to all the following chapters which are ordered in a "atmospheric", "hydrologic" and "oceanographic" sequence.

Atmospheric contributions to the freshwater budget are all dependent on precipitation minus evaporation (P-E) and the geographical distribution of this net deposition of water may, in turn, influence sea ice distribution and thickness with consequent feedback into the atmosphere. Direct measurements of precipitation are sparse and often of uncertain error bar. Two techniques are used to estimate P and E. One is based on vertical profiles of atmospheric humidity etc. measured by instruments dependent from a rising balloon, with its transit velocity determined by a tracking station. This is the "Rawinsonde" data, available from widely spaced stations around the Arctic. It cannot be properly used to interpolate on desirable smaller intervals of distance or time; there are also has some problems in standardisation of sensors and reporting schedules. Chapter 7 discusses these matters and considers the annual and interannual variability in P-E. The second technique is termed "Reanalysis" and is the subject of Chapter 8. Predictions of weather conditions have been available globally for many years from operational weather forecasting centres. These may be compared to values of the variables measured subsequently and corrections made. Data may then be corrected and recalculated as necessary to produce values of P-E on a fixed grid of time and distance intervals. The chapters overlap and there is a very useful comparison of the two systems and their limitations. Both chapters are describing response to climate, whereas Chapter 9 describes the reaction of the northward atmospheric moisture flux to events in the weather, especially to cyclogenesis. The pattern of flow and the origin of the water evaporated may change on these short time scales.

Chapter10 compares values for Arctic P, E, and P-E obtained from a variety of GCMs and compares these to the experimental data available. There is a considerable scatter in these predictions, which tend to give too high values for P, E, and P-E. The ECHAM-4 model of the Max Planck Institute for Meteorology has the best agreement with observations, but it is uncertain if this improvement is due to a difference in the physical formulation of this model or other causes. This raises once again the inadequacy of the observational data, especially those of river discharge, which leads in to Chapter 11, which discusses the detail of flow gauging, its computation and analysis. These discharge measurements suffer from a variety of problem ranging from lack of coverage to instrumentation difficulties and some of the causes for error are not independent. An uncertainty of around 30% is estimated for near coastal stations.

At the Workshop it was found that there were a number of definitions extant of the area constituting the Arctic drainage basin. For example probably most of the water flow from the islands of the Canadian Arctic Archepeligo drains into Baffin Bay and the fresh water flowing into Hudson's Bay enters the Labrador Sea. These flows certainly affect the North Atlantic but are not part of the Arctic Ocean's freshwater budget. After much discussion, and with the idea of making the chapters of this book easily comparable, it was decided that only those areas that actually drained into the Arctic Ocean should be included. However differences still remain in this book and *it is most important that the reader should consult Chapter 12, (especially Figure 1) before reading the other studies of river discharge that follow.* The definition of Arctic Ocean drainage basin recommended at the ARW is illustrated there, designated AORB. It is thought that, in future, reference should be made to this chapter in order to relate new information to the existing body. A remarkable conclusion is that the annual drainage per square kilometer is nearly independent of the basin areas presupposed.

Chapter 13 considers river discharge into the Arctic Ocean from all the circumpolar catchment areas and gives a time series analysis of flow variations over the 75 year interval, 1921 - 1996. Also considered are the effects of human activities during this period. Do the discharges reflect changing annual mean air temperatures in Arctic regions over selected intervals? The possible effects of climatic change are estimated on the basis of existing model studies. Once again absence of data or uncertainties in error bars is a limitation. The following two chapters are focused on the freshwater flow from Siberia and, it is thought, contain much material that has not been available in English previously. In Chapter 14 the riverine discharge from the entire Eurasian watershed is divided into sections, each listing the flow from major rivers entering a named sea along the Siberian coast. A chemical analysis of the water from each river is presented together with the weight of the total suspended sediment per cubic meter. Seasonal variations are considered. Chapter 15 studies cause and effect linkages between atmosphere, the Siberian rivers and conditions in the Russian shelf seas. Regimes of global atmospheric circulation are correlated to changes in precipitation and hence river runoff, both seasonal and interannual. With time lags these are related to sea ice conditions offshore of the estuary and then to exchanges through Fram Strait. The effects of climatic change are noted. This is an inadequate précis of the chapter's content: it is thought to be the most detailed description available of Siberian contributions to the Arctic Ocean freshwater budget.

Chapter 16 studies the Mackenzie River discharge, which constitutes about 11% of the total gauged flow of surface water into the Arctic Ocean. It considers all the factors contributing to the discharge on a monthly, annual and interannual bases and is probably the most detailed study of physical processes available for any of the four major Arctic rivers. The importance of the local topography is underlined and the general applicability of the approach presented to other river basins is inferred.

In Chapter 17 the viewpoint is changed to that of the estuary. An Arctic estuary viewed on the scale of the shelf is, in winter, "negative" in the usual nomenclature, which is that the removal of freshwater exceeds that supplied by runoff. This occurs because comparatively fresh sea ice is produced in open areas and the salt is rejected into the underlying waters. This ice, of course, may melt in other seasons, but in the interval it has often been moved elsewhere, while the salt remains within the water mass. Changes in estuarine type are explored over four naturally occurring seasons, defined by the runoff volume and air temperature. Possible changes due to climatic warming include a reduction in the shelves' contribution to the Arctic Ocean halocline. The analysis is conducted for the Mackenzie River and Beaufort Sea shelf as the specific oceanographic data necessary is easily available, however the discussion should be applicable to all the major Arctic rivers.

The next two chapters consider the whole Arctic Ocean as an estuary, as seen from satellite borne sensors and from the chemical analysis of tracers in seawater respectively. Both are written for readers without specialised knowledge of the techniques involved and discuss problems as well as results. They are excellent surveys of what is known and the limitations of the methods. Chapter18 describes what may be learned from satellite sensors alone and in conjunction with other measurements or models. As the atmosphere is the medium for transmission of signals between satellite and surface, quantities such as moisture content may be calculated. Information on the sea-ice extent, concentration and motion is also available directly. Sea-ice type is more difficult to determine, while for volume fluxes (such as the export through Fram Strait), the thickness must presently be found from other sources. Techniques under development are described, such as determination of ice thickness from radar altimeter readings.

Chapter 19 reports on the use of tracers, especially the stable isotopes of water, to describe the distribution of runoff, sea-ice melt and freshwater from the Pacific, within the Arctic Ocean. Residence times are estimated using other tracers. These data constitute a valuable method for the calibration of GCMs.

The only significant net flow of oceanic "fresh" water (i.e.defined as having a salinity below some standard value) into the Arctic Ocean is through the Bering Straits. Most of this flow exits between the islands of the Canadian Arctic Archipelago and though Kennedy Channel, between the Archipelago and Greenland. Counter flows complicate the issue at both locations. The net flow of fresh water is joined by the runoff from the islands themselves though it is probable that some of the summer flow from the icecap of northern Ellesmere Island may not leave the Arctic Ocean immediately. These flows are the subject of Chapter 20 with the overall conclusion that the exit flows are very poorly known and that the common methods of flux calculation are suspect. The outflow of freshwater through Fram Strait in both liquid and solid forms has already been considered in Chapter 18, where estimates using satellite records have been compared to those available from other sources. The inflow of the warmer, deeper, so-called Atlantic water, through Fram Strait and the Barents sea has a profound effect on the internal structure ot the Arctic Ocean. So does the seasonal freezing of sea ice, which has about 10% of the salinity of the water from which it was formed, and usually will have been moved before melting. The last four chapters are concerned with these internal processes that, together with the wind field, modify the input/output relationship for freshwater in the Arctic Ocean.

Chapter 21 is concerned with the modification of the Atlantic water within the Ocean. . Given sufficient wind, inputs of freshwater, both from atmosphere directly and from rivers, may mix downward and the heat in the Atlantic water used to melt ice which was produced elsewhere. The freezing of surface waters over the immense Siberian shelves produce pools of rich dense brine which flow downwards. An extremely complex situation is described, where major efforts are being undertaken presently in the hope of producing a better understanding. There has been a rapid increase in the heat content of Atlantic water within the Arctic Ocean basin since 1990 and the reasons are still unclear, though a change in atmospheric circulation patterns is the likely cause, the NAO being the prime candidate for further study. It is very important to be able to ascribe the recent changes to climatic change or to accept them as a natural long-term variability and Chapter 22 focuses on the modelling of exchanges between the Nordic

Seas and the Arctic Ocean both for Atlantic and fresh water. Experimental observations are compared to the results of modelling with an accent on the NAO to explain the changes. Once again this chapter is a report from a very active area of research.

Chapter 23 presents the theory, data, and models that describe the solid/liquid phase change for seawater in the Arctic Ocean; it then does the same with the wind-driven dynamic regime included. On an annual basis there are regions of net ice growth and net melt; the predictions of various models are compared. This paper gives the present understanding of the annual redistribution of the freshwater input to the Arctic Ocean, a redistribution that plays a fundamental role in controlling the output. Interannual variations are also considered. Chapter 24 deals with the same topic, but almost entirely from an observational viewpoint. Probably the data utilised are the most extensive set of direct measurements in existence. The main cause of interannual variations in freshwater content of the Arctic Ocean is attributed to changes in the summer input. Chapters 23 &24 are complementary, together giving a very good view of the present state of knowledge.

There are a number of different "reference salinities" used in this book to define what is the "freshwater" component in flows of diluted seawater.. Difference in this reference reflect directly into volume calculations in the budget and must be considered in any comparison between chapters.

A chapter was planned to describe the major effort by Norwegian investigators to measure the flux of ice out through Fram Strait. This work did not materialise, but some of the results are discussed, in association with satellite measurements, in Chapter 18. However a recent reanalysis of data from Fram Strait [1] has suggested that systematic differences in calculated area flux occur, depending on whether the ice velocities are calculated from pressure differences (as was done by the Norwegians, Vinje et al [2]) or from two different SSM/I image-matching techniques (Kwok and Rothrock, [3] ; Emery et al., [4]). Comparisons show that Vinje's values are substantially (up to 50%) higher than Kwok's, which are slightly higher than Emery's. An additional factor enters when area fluxes are converted into volume fluxes: Mooney [1] re-calculated mean ice thickness from moored upward sonar data used by Vinje et al., finding lower mean values owing to certain zero depth readings having apparently been given an erroneous finite depth value by the ULS internal processor. This again implies that volume fluxes calculated from satellite velocity data and reanalysed thickness data are smaller than those reported by in [2]. An example is the winter of 1994-5, where the approximately 4000 cu km flux described by Vinje et al. becomes 2800 using Kwok's method, 1800 by Emery's and 1500 by Emery's with thickness recalculation. *Clearly these issues need to be resolved because of the important influence of Fram Strait volume flux on the oceanographic structure of the Greenland Sea.* This paragraph follows information provided by Associate Editor Wadhams.

Towards the end of each chapter the authors have listed the urgent research needs of their field. The common denominator of all these studies is the requirement for better data, especially in regards to meteorological and runoff measurements. Precipitation values are gauged by sensors whose response under the varying conditions of Arctic weather leaves much to be desired. River flow measurements appear crude to an outsider and lack of funds allows only inadequate coverage. At present modelling is utilised to fill some of the gaps, but necessity drives researchers to use this expedient. A better understanding needs better data as a prerequisite.

References.

1. A.T. Mooney, *Sea ice flux estimates through the Fram Strait.* Transfer thesis from MPhil to PhD, Univ. College London 1999.
2. Vinje, T., N. Nordlund and A. Kvambekk (1998). *Monitoring ice thickness in Fram Strait.* J. Geophys. Res., **103**, 10437-10449
3. Kwok, R. and D.A. Rothrock (1999). *Variability of Fram Strait ice flux and North Atlantic Oscillation.,* J. Geophys. Res., **104**, 5177-5189.
4. Emery, W.J., C.W. Fowler and J.A. Maslanik (1997). Satellite-derived maps of arctic and antarctic ice motion: 1988 to 1994. Geophys. Res. Lett., **24**, 897-900.

OCEANIC FRESHWATER FLUXES IN THE CLIMATE SYSTEM

ANDERS STIGEBRANDT
Dept. of Oceanography
Earth Sciences Centre
Göteborg University
P.O. Box 460
S-40530 Göteborg
Sweden

1. Introduction

Atmospheric transports of freshwater, within and between ocean basins and between oceans and continents, are integrated parts of the climate system of the earth of decisive importance for the distribution of salinity in the oceans. The latter enforces restrictions on the oceans regarding, for instance, possible locations of areas where deepwater and sea ice may be formed. Such locations strongly influence the regional climate and are of great importance for the appearance of the large-scale thermohaline circulation of the oceans. However, the long-term oceanic response to atmospheric freshwater forcing is complicated because it also involves thermally and wind forced circulation as well as astronomically forced diapycnal mixing and modification by basin topography. The main purpose of this paper is to describe and discuss major features of the oceanic response to present day's systematic atmospheric transports of freshwater.

The distribution of salinity in the sea obviously tends to change due to atmospheric transports of freshwater from one part of the sea to another. In a long-term steady state situation, however, systematic freshwater transports by the atmosphere, e.g. from subtropical to tropical and subpolar areas, between oceans as well as from sea to land, are counteracted by equally large freshwater transports in the oceans restoring the freshwater and salt balances.

Transport mechanisms maintaining the oceanic salt and freshwater balances under atmospheric freshwater forcing may be illustrated by examples from less complicated estuaries and coastal seas. For a simple example, we consider atmospheric transports of freshwater to a deep estuary and its watershed (Fig. 1). If diapycnal mixing processes do not occur in the estuary, the resulting density distribution in the estuary would be a layer of freshwater on top of seawater. Horizontal pressure differences, due to the difference in vertical density distribution between the estuary and the sea outside, would force a surface current transporting the freshwater out of the estuary. In this idealised case without mixing, no salt transport is induced by the freshwater supply (Fig. 1a). The response to freshwater supply in real estuaries, however, is complicated by local mixing of seawater and freshwater, usually brought about mainly by the action of local winds and tides. The mixing induces a superposed circulation of sea water in the estuary that may

E.L. Lewis et al. (eds.), The Freshwater Budget of the Arctic Ocean, 1-20.
© 2000 *Kluwer Academic Publishers.*

multiply the rate of outflow from the surface layer although, of course, the net flow out of the estuary still equals the net supply of freshwater (Fig. 1b). The response of the salinity distribution in an estuary may be further complicated by so-called remote forcing, driven by sea level and density variations in the coastal sea outside the estuary. Obviously, a certain freshwater forcing does not give the same response in all estuaries since the flows restoring the freshwater and salt balances are influenced by basin topography and, in several different ways, by winds and tides.

Freshwater circulation in an estuary, however, may be somewhat misleading as a model for the freshwater circulation in the oceans because the dominating restoring flow in the estuary increases with the content of freshwater (positive feedback). This transport mechanism certainly operates also in the ocean but other transport mechanisms independent of the freshwater content (lacking feedback to freshwater content) might be equally strong and even stronger. Such independent transports may be performed by wind-forced Ekman currents and thermally dominated baroclinic currents. However, if such completely independent transport mechanisms tend to accumulate freshwater in an area of the sea this will lead to increasingly strong opposing transports by freshwater gradients.

Sea salt tends to stay in the liquid phase, not only when moisture is evaporated from the sea surface by heating but also when water is transferred to solid phase by cooling because much of the salt initially enclosed in the ice is soon ejected as brine to the underlying water. When sea ice subsequently melts, almost fresh liquid water is added to the sea surface. Thus, the distribution of salt in the oceans may be influenced by not only evaporation and precipitation but also by freezing and melting of sea ice. As an example of the latter it may be mentioned that a lot of sea ice produced in the Arctic Ocean is exported to and melts in the Nordic Seas and some of the salt ejected on shallow Arctic shelves by ice production may eventually contribute to increase the salinity of the deep water in the Nordic Seas, a major deepwater source for the oceans. Accordingly, ice-related changes of the salinity in high-latitude seas may confuse the picture of the ocean response to atmospheric freshwater transports.

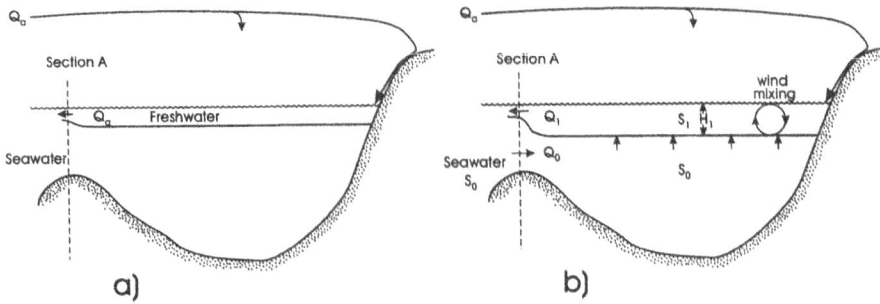

Figure 1. Circulation and stratification in an estuary driven by atmospheric freshwater supply Q_a; a) without mixing, the surface layer is pure freshwater and the outflow $Q_1 = Q_a$ b) with wind forced local mixing entraining sea water into the surface layer which has the salinity S_1. The outflow is $Q_1 = Q_0 + Q_a$ and the salinity of seawater is S_0.

Two different definitions of freshwater in the oceans, the 'absolute' and the 'relative', are used in this paper. According to the absolute definition, seawater (of salinity S) is composed of freshwater, i.e. pure water (H_2O), and sea salt. Thus the fraction S of seawater is sea salt and the fraction (1-S) is freshwater. This absolute definition of freshwater has been used in studies of net (barotropic) mass flows through vertical cross-sections of the oceans, e.g. in Wijffels et al. [1]. In the relative definition freshwater is treated like a tracer and the amount of freshwater is related to the salinity of a reference seawater. Thus, the relative definition of freshwater reads Seawater = Freshwater + Reference Seawater. If the reference seawater has salinity S_{ref} and seawater salinity S, the fraction of freshwater equals $(S_{ref}-S)/S_{ref}$ and the fraction of reference water equals S/S_{ref}. With the relative definition, the amount of freshwater depends on the chosen value of S_{ref}. The relative definition is frequently used in studies of e.g. local and regional freshwater budgets, estuarine circulation, diapycnal mixing and freshwater as a tracer. If S_{ref} is taken equal to 1 the two definitions fall together and the relative amount of freshwater becomes equal to the absolute amount, meaning that also all the water (H_2O) in the reference sea water is considered to belong to the freshwater pool and the reference seawater is just the salt.

The salinity distribution in the oceans obviously plays a major role for the spatial and temporal shape of large-scale thermohaline circulation and sea ice coverage. Pre-historic oceanic response, in terms of salinity distribution, thermohaline circulation and sea ice coverage, to guessed pre-historic atmospheric freshwater fluxes and continental runoff has been the subject for much speculation, often based on loosely formulated mechanisms and models. Such paleoceanographic applications make it still more important to identify major processes for the freshwater circulation in the sea and develop models that can be used to quantify ocean response. It will be shown in this paper that the 'super-fjord heat pump' mechanism, introduced by Berger and Jansen [2] to explain temperature variations in the Nordic Seas during the last deglaciation, does not work as supposed by these authors.

Below I will first, in section 2, present estimates of the meridional heat transports in the climate system. Meridional freshwater transports by the atmosphere and oceans as determined by two independent methods are described in section 3. Estimates of interoceanic atmospheric freshwater transports are described in section 4. Large-scale oceanic response in salinity and sea level due to atmospheric freshwater transports is discussed in section 5. Among other things, I will discuss how the net flow of freshwater by the atmosphere to the Pacific causes both lower salinity and higher sea level in the Pacific than in the Atlantic Ocean. This in turn causes an oceanic net flow toward the Atlantic Ocean through the Bering Strait and a net circulation of salt in the oceans. In a steady state, the ocean must set up transports restoring the spatial freshwater and salinity distributions. To some extent, these restoring ocean flows might be driven by salinity gradients but also thermally and wind driven circulation and tidal mixing may provide important forcing for the resulting flows as discussed in section 6. Different modes of salt and freshwater transports in the ocean are discussed in section 7 where particular attention is paid to baroclinic transports due to horizontal salinity (freshwater) gradients and possible topographic control of such transports. Furthermore, horizontal dispersion by fluctuating currents is also discussed based on observations of summertime dispersion of 'juvenile' freshwater in the seasonal thermal stratification of the Baltic Sea.

Consequences of the atmospheric freshwater transports for the vertical circulation of the oceans are discussed in Section 8. The paper is concluded in section 9 by a few remarks and a list of questions to answer by future research.

2. Meridional Energy Transports by Atmosphere and Ocean in the Global Climate System

The climate system receives energy from the sun and loses energy to space. The meridional distributions of incoming and outgoing energy fluxes to the earth have been determined from satellite radiation measurements in the Earth Radiation Budget Experiment (ERBE). The annual mean value of absorbed short wave radiation ($0.2\mu m < \lambda < 4\mu m$) has a maximum at the equator of about 320 W m^{-2} and minima at the poles of about 60 W m^{-2}. The latitudinal distribution of emitted long wave radiation from the Earth ($4\mu m < \lambda < 100\mu m$), however, is more evenly distributed due to the small temperature difference of about 30 K between low and high latitudes. It has a broad maximum of about 270 W m^{-2} in tropical regions and a minimum of about 160 W m^{-2} at high latitudes, cf. Fig. 2.

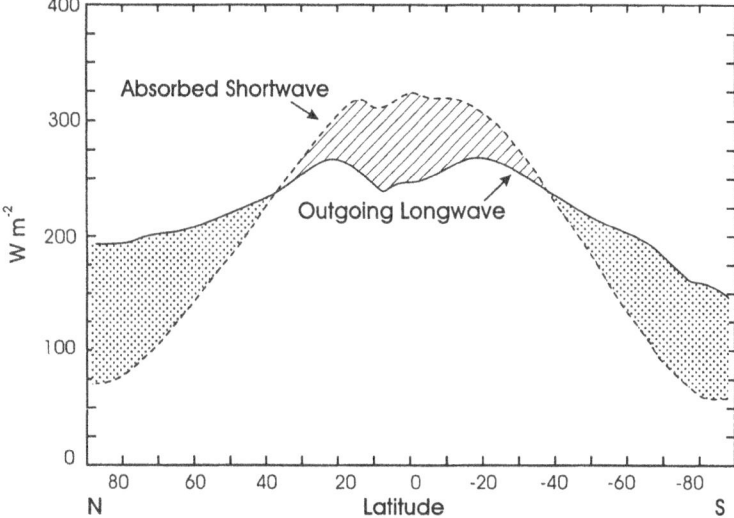

Figure 2. The meridional distribution of incoming and outgoing energy fluxes to the earth and the net top-of-the-atmosphere radiative imbalance. (Redrawn from Trenberth and Solomon [3]).

The net top-of-the-atmosphere radiative imbalance, with an energy excess between 35°S and 35°N and an energy deficit at higher latitudes, drives the general circulation of the atmosphere and the oceans that has to bring about energy transports balancing the radiative imbalance. The total poleward meridional energy transport by atmosphere and oceans has maxima of 5.5-6.0 PW (1 PW = 10^{15} W) at about 35°S and 35°N. However, the partitioning of meridional energy transport between atmosphere and ocean is rather uncertain and has therefore so far been dismissed as a tool for validation of climate models (Keith [4]). Recent estimates, also shown in Fig. 3, suggest that the oceanic energy transport dominates at latitudes lower than 20° while the atmospheric transport is dominating at higher latitudes.

3.   Meridional Freshwater Transports by the Atmosphere and the Oceans

Poleward atmospheric moisture transport Q_a (kg s^{-1}) carries part of the meridional atmospheric energy transport H_{am} (W). The relationship between these quantities is $H_{am} = Q_a L$ where L ($\sim 2.3 \; 10^6$ J kg^{-1}) is the latent heat of moisture. An energy transport of 1 PW thus corresponds to a moisture (freshwater) transport of about 0.44 10^6 ton s^{-1}. Moisture is transported both polewards and equatorwards from the subtropics which have greater evaporation E than precipitation P, i.e. a positive E-P. The equatorward transport

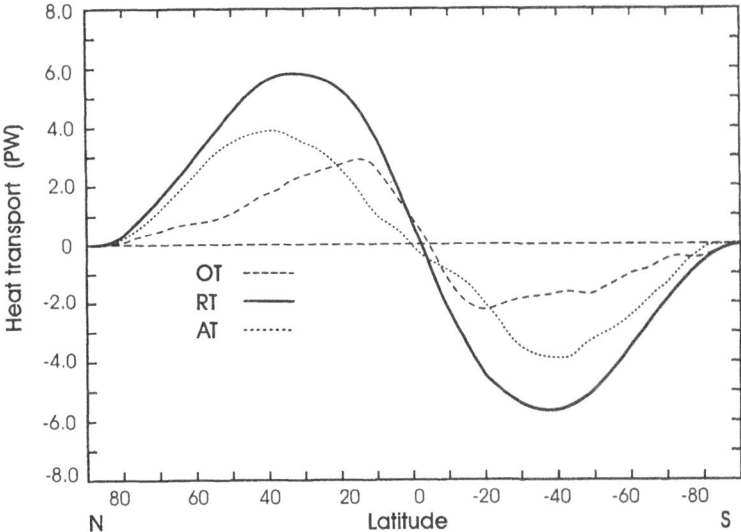

Figure 3. Meridional energy transport by atmosphere (AT) and oceans (OT) as required by the radiation imbalance (RT). (Redrawn from Trenberth and Solomon [3]).

of moisture, mainly due to the Hadley circulation, causes a net equatorward transport of latent heat. The convergence of moisture transports near the equator causes large precipitation and a negative E-P. The moisture transport across mid latitudes is mainly due to transient eddies. The northward atmospheric freshwater transport as estimated by Keith [4] from 3 years of filed analyses of atmospheric data from an operational weather prediction centre (ECMWF) is shown in Fig. 4.

   For an ocean basin, changes in the zonally integrated meridional mass transport are related to surface fluxes in the following way

$$\frac{d}{dy} \int \int \rho \, v \, dx \, dz = \int F(x,y) \, dx \qquad (1)$$

Here x,y and z are the zonal, meridional and vertical coordinates, respectively, v is the meridional velocity and ρ the density. F(x,y) is the net gain of mass at the ocean surface due to precipitation P, evaporation E, and land runoff R. Hence F(x,y)=P-E+R

There is no significant atmospheric pathway for salt, thus

$$\frac{d}{dy}\int\int \rho v S \, dx \, dz = 0 \qquad (2)$$

where S is salinity of seawater. The freshwater balance (freshwater in the absolute sense) is the difference between (1) and (2):

$$\frac{d}{dy}\int\int \rho v (1-S) \, dx \, dz = \int F(x,y) \, dx \qquad (3)$$

Wijffels et al. [1] computed the net gain or loss of freshwater by the ocean in 5° latitude bands using the air-sea freshwater exchange (P-E) and runoff (R) at the coasts estimated by Baumgartner and Reichel [5]. For the North Atlantic, they used the estimates of Schmitt et al. [6]. Wijffels et al. [1] obtained the meridional freshwater flow for each ocean through integration of Eq. (3) from sections with known transports which for the Atlantic and Pacific is the flow through the Bering Strait (Roach et al. [7]) and for the Indian Ocean the vanishing flow through the northern boundary. The meridional transport through each latitude can be summed over the three oceans to get the global freshwater transport as a function of latitude and the result is shown in Fig. 4.

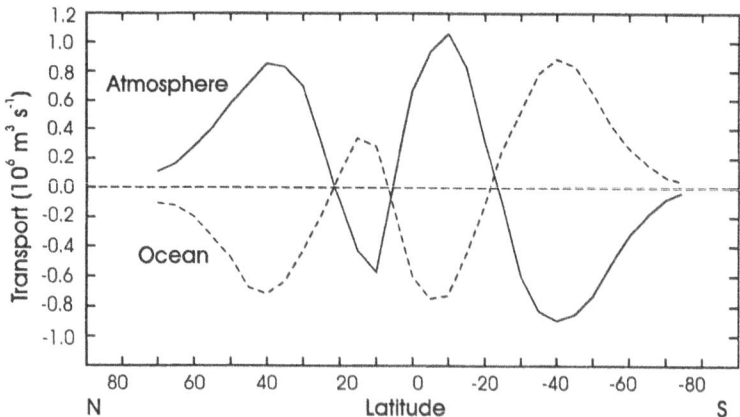

Figure 4. The northward transport of freshwater in the ocean (broken line) and in the atmosphere (solid line). (From Wijffels et al. [1] and Keith [4], respectively).

The sum of the meridional ocean and atmosphere freshwater transports must equal the meridional freshwater flow over land due to river and ground water flow. The latter flows are generally believed to be quite small compared to flows in the ocean and atmosphere. Fig. 4 illustrates the complementary transports of freshwater by the ocean and atmosphere. The estimated freshwater transports in the atmosphere and ocean are counter-directional and quite similar in magnitude and maxima and minima appear at about the same latitudes. The estimated atmospheric freshwater transport is somewhat greater than the estimated ocean freshwater transport. The discrepancy should be a result of inaccuracies in data and in methods used to compute the transports.

4. Interocean Freshwater Transports by the Atmosphere

The atmospheric mean (14 years) flow of freshwater was estimated by Chen et al. [8] using upper air data by the Global Data Assimilation System (GDAS) at the US National Meteorological Center. Convergence-divergence of atmospheric freshwater flows are shown in Fig. 5 together with precipitation estimates for 1979-1980 derived from space-based data by Susskind and Pfaendtner [9]. It can be seen that water vapour diverges out of the subtropical highs over the three oceans and converges toward the rainfall belts of the tropics and midlatitudes of both hemispheres.

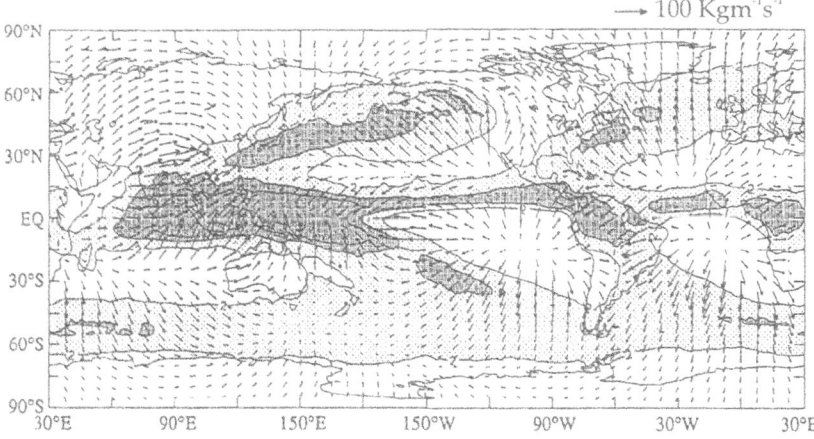

Figure 5. Estimates of atmospheric average divergent water vapour transport and precipitation. Precipitation larger than 2 mm d^{-1} and smaller than 4 mm d^{-1} is lightly stippled. Precipitation larger than 4 mm d^{-1} is heavily stippled. (From Chen et al., [8]).

Due to the general zonal character of the atmospheric circulation, the presence of high meridional mountain ridges tends to create net moisture fluxes between ocean basins. Both the Atlantic and Indian Oceans have positive basin averages of E-P and the excess moisture is exported to the Pacific that accordingly has a negative basin average E-P. For a steady state ocean the imbalances in E-P between different ocean basins, brought about by atmospheric transports have to be compensated by restoring freshwater transports in the oceans. These transports have to be such that the salinity distribution is preserved as discussed in section 5 below.

Inter-oceanic atmospheric freshwater flows were estimated in an indirect way by

Wijffels et al. [1] who calculated the net freshwater loss to the atmosphere in the Pacific, Atlantic and Indian Oceans using compiled data of the E-P field and of the runoff from Baumgartner and Reichel [5] but with modification for the North Atlantic from Schmitt et al. [6]. Recently Jourdan et al. [10] presented similar computations based on blended satellite and ship evaporation and SSM/I retrieved precipitation fields and Baumgartner and Reichel [5] runoff estimates. The results of these two estimates of inter-oceanic atmospheric freshwater flows generally match except for in the South Pacific where Jourdan et al. [10] obtained a much larger freshwater transport which has no counterpart in Fig. 4. It may be of interest to quote the sums of net atmospheric fluxes and river runoff (flow unit 1 Sv (Sverdrup) $= 10^6$ m^3 s^{-1}) for different parts of the oceans according to Baumgartner and Reichel [5]: North Pacific 0.75, South Pacific 0.13, North Atlantic -0.12, South Atlantic -0.33, Indian Ocean -0.44.

5. Oceanic Response in Salinity and Sea Level to Atmospheric Freshwater Transports

Atmospheric moisture transports from arid to humid areas as well as transports from the oceans to continents (cf. Fig. 5) effect the surface salinity of the ocean. There is no simple generally applicable model relating values of E-P and surface salinity because surface salinity also depends on vertical mixing and circulation in the oceans. This is of course true in particular in coastal areas with large runoff from land. The spatial distribution of surface salinity in the oceans (Fig. 6) shows some regular features with maxima in the subtropics where E-P is positive and lower salinity in tropic and subpolar areas where E-P is negative. The quite evident interoceanic differences, with highest salinity in the Atlantic Ocean and lowest in the Pacific, are correlated to ocean means of E-P but are also, of course, influenced by large-scale ocean circulation. It should be noted that there is no feedback mechanism between surface salinity and air-sea water exchange. Evaporation and precipitation are insensitive to variations in surface salinity and the direction of air-sea water exchange depends on other variables than salinity. This fundamental property of air-sea water exchange is often violated in ocean circulation modelling by the introduction of a salinity restoring boundary condition at the air-sea interface. Model studies of the oceanic response to atmospheric freshwater forcing of course require models free from violating boundary conditions on the air-sea water exchange.

Spatial salinity variations in the upper ocean cause corresponding sea level variations by the so-called steric effect. Assuming a level of no motion at 1200 metres depth, present day sea level in the N. Pacific stands about 0.65 m higher than in the N. Atlantic. However, this sea level difference will vary on long time-scales in response to the salinity difference between the N. Pacific and the N. Atlantic. The sea level difference between the N. Pacific and the N. Atlantic explains the northward barotropic flow through the Bering Strait (Stigebrandt [12, 13]). This author also suggested that the salinity difference between the N. Pacific and the N. Atlantic is adjusted in such a way that the flow through the Bering Strait closes the global freshwater budget (Stigebrandt [12, 13]). Present day mean flow through the Bering Strait seems to be about 0.8 Sv of water with mean salinity 32.5 (Roach et al. [7]).

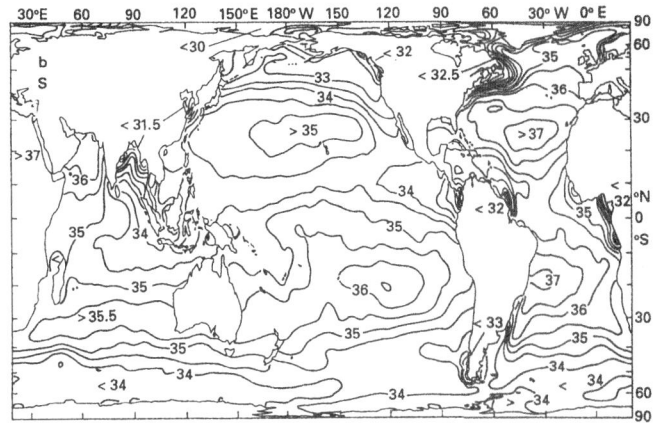

Figure 6. Spatial distribution of surface salinity in the oceans. (From [11]).

Some of the consequences of present day atmospheric freshwater transports are summarized in Fig. 7. Due to atmospheric interocean freshwater transports, N. Atlantic has high and N. Pacific has low salinity. This leads to 1) higher sea level in N. Pacific than in N. Atlantic which implies 2) barotropic flow from the Pacific to Atlantic Ocean through Bering Strait, 3) deepwater production in the salty Atlantic Ocean and 4) an accompanying "conveyor belt" circulation with surface flow toward the Atlantic. Note that the rate of vertical circulation of the ocean (conveyor belt) is controlled by the rate of diapycnal mixing in the abyss (probably mainly forced by tides) which means that it does not depend on specific locations for deepwater formation, see section 8 below.

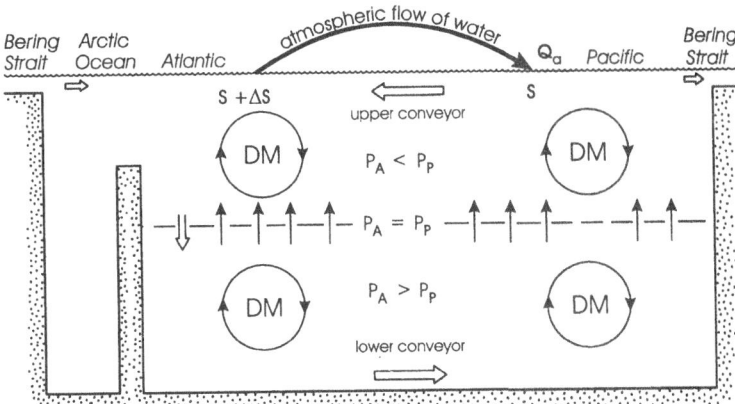

Figure 7. Sketch of present day's atmospheric freshwater transports and some consequences of it. DM is diapycnal mixing, P_A (P_P) pressure in the Atlantic (Pacific) and S salinity. (Modified after Stigebrandt [12]).

6. Barotropic Transport of Freshwater by the Oceans

The global distribution of (absolute) freshwater transport in the ocean presented by Wijffels et al. [1] is based on an integration point at Bering Strait, which connects the Pacific and Atlantic Oceans via the Arctic Oceans. As mentioned in the previous section, about $0.8 \cdot 10^6$ m^{3}s^{-1} of relatively fresh water flows through the Bering Strait from the Pacific into the Arctic Ocean. Baumgartner and Reichel's tabulation of the net gain of freshwater by 5° latitude intervals is then integrated from the reference location at Bering Strait to yield the meridional freshwater transport in each ocean. Freshwater transport in the Pacific is directed northward at nearly all latitudes. In the Atlantic, the freshwater transport is directed southward at all latitudes, with a small southward freshwater transport out of the Atlantic across 35°S (Fig. 8). Salt transport, which must be considered jointly with the freshwater transport, is northward throughout the Pacific and southward throughout the Atlantic (in the same direction as the freshwater flux) and is equal to the salt transport through the Bering Strait, see [1]. The circulation around Australasia associated with the poorly known Pacific-Indian through-flow modifies the above scenario only in the South Pacific and Indian Oceans. A moderate choice for the through-flow indicates that it dominates the absolute meridional fluxes of freshwater and salt in these oceans. The global freshwater scheme presented by Wijffels et al. [1] suggests the need for a careful assessment of the treatment of ocean freshwater and salt transports in inverse, numerical, and climate models.

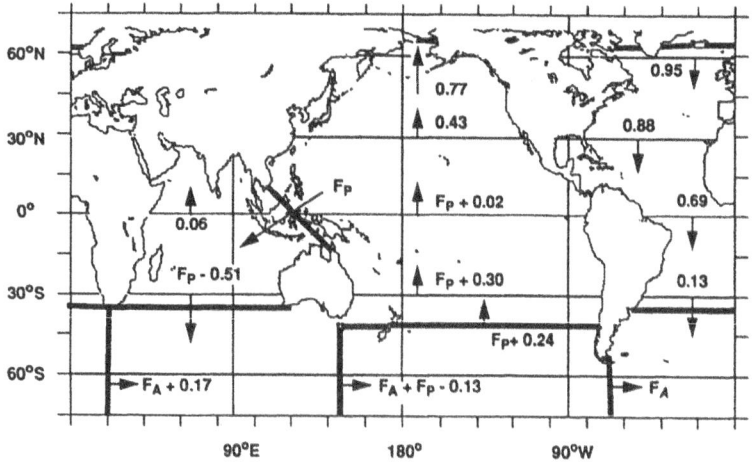

Figure 8. Freshwater (absolute) transports (x10^9 kg s^{-1}) in the World ocean as estimated by Wijffels et al. [1]. Here F_P and F_A refer to the freshwater fluxes of the Pacific-Indian through-flow and the Antarctic Circumpolar Current in Drake Passage, respectively.

It should be emphasised that the transport through the oceans estimated in [1] is the net (barotropic) transport due to the atmospheric freshwater transport. This should not be confused with the vertically stratified so-called thermohaline circulation often termed the "conveyor belt" which essentially is a baroclinic transport with about equally large counter-directed flows. The difference between these two types of circulation can be demonstrated by an estuarine analogue to the oceanic freshwater circulation (Fig. 1). Atmospheric fluxes Q_a bring freshwater to the estuary. Together with mixing, this leads to a surface layer of salinity S_1 and thickness H_1 (Fig. 1b). The net flow of freshwater (in the absolute sense) through the mouth (section A) is given by

$$\iint_A \rho v (1 - S) \, dA = Q_a \tag{4}$$

Equation (4) explains the mean freshwater transport through section A in the same way as the freshwater transports through the oceans were explained using Equation (3) by Wijffels et al. [1]. However, Eq. (4) cannot explain the baroclinic response to Q_a, i.e. the salinity difference S_0-S_1 and the thickness H_1 of the surface layer. Thus, the supply of freshwater in combination with the action of mixing processes induces also a baroclinic circulation, a conveyor belt circulation, and the mixing tends to increase the salinity S_1 and the thickness H_1 of the surface layer. To predict estuary, and ocean, salinity distributions in response to atmospheric freshwater forcing one obviously needs to consider vertical stratification and account for diapycnic mixing processes and baroclinic and possibly other kinds of transport processes. This is discussed in section 7.1 below.

7.   Modes of Freshwater Transport in the Ocean

Atmospheric freshwater transports tend to change the distribution of salinity in the ocean. This is restored by a complex system of transports including 1) barotropic transports due to unbalanced P-E+R as described in sections 3 and 6 above, 2a) baroclinic transports due to horizontal salinity (freshwater) gradients, 2b) baroclinic transports due to horizontal temperature gradients, 3) Ekman transports by steady winds in the upper layers and 4) dispersive transports by unsteady motions, e.g. due to unsteady winds and eddies shed by large-scale currents.

Ocean transports of types 1) and 2a) are directly forced by atmospheric freshwater transports while ocean transports of types 2b), 3) and 4) are due to thermal and wind forcings that essentially are independent of the freshwater forcing. In addition to the different types of transports listed above, providing large-scale stirring, diapycnal mixing processes operating on molecular scales are of decisive importance to restore the distribution of salinity in the ocean. Dispersive transports are usually neglected in estimates of freshwater transports but an example from the Baltic showing that it may be a major transport mechanism is given below. In this section, emphasis will be put on transports of types 2a) and 4).

7.1 BAROCLINIC TRANSPORTS DUE TO HORIZONTAL SALINITY (FRESHWATER) GRADIENTS

Horizontal density gradients tend to force baroclinic currents. For the discussion here I will for simplicity consider only two-layer stratification, with a less dense surface layer of reduced salinity S_1 upon seawater of salinity S_0. A two-layer stratification is often a good approximation that captures major features of baroclinic currents. In the sea, also horizontal temperature gradients contribute to the baroclinic currents but this complication is neglected here where the main aim is to study baroclinic transports in response to atmospheric freshwater transports. In cases where rotation can be neglected, e.g. in an estuary with narrow mouth, the baroclinic flow in the mouth obeys the following condition (Stommel and Farmer [14])

$$\frac{u_{1m}^2}{g'H_{1m}} + \frac{u_{0m}^2}{g'H_{0m}} = 1 \tag{5}$$

Here $u_{1m(0m)}$ is the speed and $H_{1m(0m)}$ the thickness of the upper (lower) layer in the mouth, g the acceleration of gravity, $g' = g(\Delta\rho/\rho)$ and $\Delta\rho$ is the density difference between the layers.

The equation of state for brackish water may be written

$$\rho = \rho_f(1 + \beta S) \tag{6}$$

where β is the salt contraction coefficient. Using this equation one may write $g' = g\beta(S_0 - S_1)$.

The vertical section where Eq. (5) applies should be found in the mouth where there is a contraction in width and/or in depth (a sill). This section is called the control section. Equation (5) expresses a dynamical condition on the flow at the open boundary of the estuary (open boundary condition).

For a stationary state conservation of volume and salt is expressed by the so-called Knudsen's relationships

$$Q_1 = Q_a + Q_0 \tag{7}$$

$$Q_1 S_1 = Q_0 S_0 \tag{8}$$

Here $Q_{1(0)} = u_{1m(0m)}H_{1m(0m)}B_m$ is the volume flow of the upper (lower) layer and B_m is the width of the mouth which for simplicity is considered to be rectangular, i.e independent of depth.

For the discussion below, we introduce the height H_{1f} of freshwater contained in the upper layer of the estuary of thickness H_1. Note that H_1 is different from H_{1m} as discussed below.

$$H_{1f} = H_1 \frac{S_0 - S_1}{S_0} \qquad (9)$$

The rate of wind-driven entrainment of seawater into the surface layer is

$$Q_0 = w_e A \qquad (10)$$

where w_e is the so-called entrainment velocity and A is the horizontal surface area of the pycnocline in the estuary. The entrainment velocity can be computed using a formula of the type introduced by Kato and Phillips [15]

$$w_e = \frac{2m_0 u_*^3}{g' H_1} \qquad (11)$$

The friction velocity u_* can be obtained from the wind stress by standard methods. The empirical non-dimensional constant m_0 (~ 0.6) depends on the efficiency of turbulence with respect to diapycnic mixing.

A simple analytical model for the salinity S_1 and thickness H_1 of the upper layer can be obtained for a wide fjord with a deep ($u_{0m} \approx 0$) and narrow mouth under steady state conditions. Using Eqs. (5,6,7,8,9,10,11) and the following relationship between H_{1m} and H_1

$$H_{1m} = \phi H_1 \qquad (12)$$

the present author [16] derived the following analytical expressions for H_1 and S_1

$$H_1 = H_{1f} + \frac{N}{2Q_a g \beta S_0} \qquad (13)$$

$$S_1 = \frac{S_0 N}{N + 2\phi[Q_a^5 (\frac{g \beta S_0}{B_m})^2]^{1/3}} \qquad (14)$$

Here $N = CW^3 A$ and $C \approx 2.510^{-9}$ is an empirical constant, which contains the drag coefficient for airflow over water, the ratio between air and water densities, and W is the wind speed.

Equation (13) shows that the thickness of the upper layer may be expressed as the sum of the freshwater thickness H_{1f}, which is influenced by the hydraulic control in the mouth, and a "mixing" thickness proportional to the Monin-Obukhov length that is proportional to the ratio between the energy flux from the wind and the buoyancy flux due to supply of freshwater; see Stigebrandt [16]. The model predicts that the mixing thickness is independent of the hydraulic control. The salinity of the upper layer, Eq. (14), depends on W^3 and $Q_a^{-5/3}$ and is modified by the topographic parameters B_m and A. It may be pointed out that this model is a model for the estuarine conveyor belt circulation in a fjord.

To investigate the existence of hydraulic control in the mouth of a Norwegian fjord Stigebrandt and Molvær [17] used the property that H_{1f} is independent of the rate of mixing, thus

$$H_{1f} = \phi \left(\frac{1}{g \beta S_0 B_m^2} \right)^{1/3} Q_a^{2/3} \tag{15}$$

Eq. (15) clearly shows that the amount of freshwater stored in the fjord is regulated by the width of the mouth and the freshwater supply. This equation shows that H_{1f} is independent of the mixing and that the flux of freshwater out of the estuary is proportional to $H_{1f}^{3/2}$.

If Q_a ($=P-E+R$) is negative in the basin inside the mouth, the basin will produce a water saltier than seawater and the circulation will be reversed compared to that considered above. This occurs for instance in the Mediterranean which produces its own deepwater that also is exported to the Atlantic Ocean. Together with Knudsen's equations, Eq. (5) can be applied to the steady-state flow through the Strait of Gibraltar where the flow thus is assumed to be topographically controlled. If it is assumed that $H_{1m} \approx H_{0m}$ one obtains a realistic salinity of the out-flowing salty Mediterranean water, e.g. Assaf and Hecht [18].

When rotation cannot be neglected, i.e. in open and coastal oceans and in estuaries with wide mouths, the volume transport Q_1 of the upper layer upon a lower layer at rest is given by

$$Q_1 = \frac{g' H_1^2}{2f} \tag{16}$$

where f the Coriolis parameter. The freshwater transport by the baroclinic current is $Q_f = Q_1(\Delta S/S_{ref}) = g\beta(\Delta S)^2 H_1^2/(2fS_{ref})$. The magnitudes of $\Delta S = S_0 - S_1$ and H_1 are strongly dependent on the mixing intensity in the system. Defining $H_{1f} = H_1 \Delta S/S_{ref}$, which is the freshwater content of the layer, one finds that the freshwater transport Q_f is proportional to H_{1f} squared. For a steady-state situation $Q_f = Q_a$ and for this case $Q_a \sim H_{1f}^2$. Thus, both under rotating and non-rotating conditions baroclinic currents due to freshwater supply provide a strong mechanism to restore horizontal salinity gradients in the sea due to atmospheric freshwater transports.

Baroclinic transports due to freshwater runoff from continents are often evident in coastal areas. Gustafsson and Stigebrandt [19] analysed the wind-driven circulation of the freshwater influenced surface layers in the Skagerrak, a through-flow area for freshwater supplied to the Baltic Sea and the North Sea. They used hydrographic data to first compute the freshwater content and the potential energy of the freshwater influenced surface layers. From these quantities, they could describe freshwater transports and volume transports (gradients in potential energy are related to volume transports). Among other things they found that the cyclonic circulation in the surface layers of Skagerrak is forced by wind-driven Ekman transports (independent of the freshwater content) converging at the coasts (downwelling) causing strong baroclinic coastal currents driven by the accumulated freshwater buoyancy. The estimated long-term along-shore gradients in steric sea level, due to gradients in the freshwater content, are in agreement with geodetically determined coastal sea level gradients.

Eq. (16) has been applied to the inflow of relatively warm and salty Atlantic water into the Nordic Seas putting H_1 equal to the sill depth of the Greenland-Iceland-Scotland ridge, implying that the flow is assumed to be topographically controlled by the height of the ridge. One may then investigate how the salinity of the dense deepwater exported from the Nordic Seas varies with the mixing of freshwater into the in-flowing water under the assumption that the mixture is cooled to a certain temperature. Realistic results for the rate of deepwater production may be obtained (Stigebrandt [20]). About two thirds of the cooling takes place in the Norwegian and Barents Seas (Simonsen and Haugan [21]) why it seems that the deepwater formation is particularly sensitive to increased supply of freshwater to these seas. If too much of freshwater is mixed into the in-flowing Atlantic water, as further touched upon in section 8 below, deepwater production should stop and the Nordic Seas would go into a state of estuarine circulation.

The examples given in this section show that baroclinic currents due to horizontal freshwater (salt) gradients tend to restore the freshwater and salt balances of the oceans. The response is complicated by diapycnal mixing, rotation and topography. The baroclinic response seems to dominate in semi-enclosed basins but in the open ocean also other transport mechanisms, acting independently of salinity gradients like wind-driven Ekman transports and baroclinic currents forced by thermal gradients, might be important for the salinity response to atmospheric freshwater forcing.

The super-fjord heat pump concept was introduced by Berger and Jansen [2] to explain temperature variations in the Nordic Seas during the last deglaciation. The authors assume that, due to the large freshwater supply during the deglaciation, the Nordic Seas and the Arctic Ocean are in a state of estuarine circulation with outflow of low-saline water at the sea surface and inflow of sea water below. They also assumed that the inflow of (warm) seawater increases with the supply of freshwater. Thus, the greater the supply of freshwater the greater the inflow of sea water and heat into the Nordic Seas, i.e. $Q_0 \sim Q_a^b$ where the exponent $b > 0$. This means that there should be a positive feedback between freshwater supply and the heat supply to the Nordic Seas. Referring to decreased supply of freshwater during the Younger Dryas episode the super-fjord heat pump model may explain why this episode was anomalously cold. Now when we have the formalism of a quantitative model for freshwater-driven fjord-type circulation at hand it is easy to check if the super-fjord heat pump works as supposed by Berger and Jansen [2].

We are thus interested to see how freshwater supply influences the inflow of seawater to fjords. Making use of Equations (9), (10), (11) and (15) and $g' = g\beta(S_0 - S_1)$ one finds that the inflow of seawater to non-rotating fjords is $Q_0 \sim Q_a^{-2/3}$. To obtain the inflow to a rotating fjord one has to replace Eq. (15) with the corresponding result for rotating systems which is written down a few lines below Eq. (16). It is then readily seen that $Q_0 \sim Q_a^{-1/2}$ in the rotating case. Since the exponent is negative for both rotating and non-rotating fjords ($b < 0$) one may conclude that in fjord-type circulation the inflow of seawater evidently decreases with increasing freshwater supply and there is thus negative feed-back between freshwater supply and heat flow. The assumption of positive feed-back, i.e. $b > 0$, by Berger and Jansen [2] is apparently wrong and fjord-type circulation does not work as a heat pump with positive feed-back.

7.2 DIFFUSIVE TRANSPORTS - SPREADING OF JUVENILE FRESHWATER IN THE BALTIC PROPER

The concept of juvenile freshwater, introduced by Eilola and Stigebrandt [22], is defined as freshwater trapped by a seasonal stratification when entering the sea. The content of juvenile freshwater in a layer of thickness h is defined by the freshwater height F

$$F = \int_0^h \frac{S_{ref} - S(z)}{S_{ref}} dz \qquad (17)$$

In the surface waters of the Baltic proper, of salinity ≈ 7, the buoyancy of freshwater is only about 20% of that in oceanic water (salinity ≈ 35) and juvenile freshwater may be regarded as an almost non-buoyant tracer. Eilola and Stigebrandt [22] computed monthly means of $S(z)$ for $1° \times 1°$ squares in the Baltic proper (data from the period 1959-1995). For each square, S_{ref} was chosen as the salinity in March (i.e. before the seasonal stratification becomes established). F was computed for the subsequent months with h equal to the depth of the seasonal pycnocline 25 m. Fig 9 shows the mean distribution of juvenile freshwater in the Baltic proper in April and in July. The authors showed that the increase of juvenile freshwater is in agreement with the supply from rivers and adjacent seas. It can be seen (Fig. 9b) that much of the juvenile freshwater spreads southwards along the central parts of the Baltic proper from the dominating sources in the north.

It was concluded that the spreading of juvenile freshwater might be regarded as a large-scale "natural diffusion experiment" repeated each year. The empirical relationship $\sigma^2 = 0.0108t^{2.34}$ between the spatial variance σ^2 of a tracer cloud and the time-scale t for diffusion (Okubo [23]) gives a number for the length-scale L of diffusion ($L = 3\sigma$) which seems to apply also to the spreading of juvenile freshwater in the Baltic proper. The corresponding spreading velocity is about 5 cm s^{-1}. Eilola and Stigebrandt [22] concluded that juvenile freshwater is spread mainly by fluctuating winds. Mean circulation and long-term mean Ekman transports contribute only minor parts to the spreading. It was also concluded that the average spreading pattern of juvenile freshwater in the Baltic proper should provide a challenging test case for state-of-the-art 3-D ocean circulation models. In previous salt and freshwater transport estimates for vertical sections across open oceans diffusive transports of "Okubo-type" seem to have been neglected which possibly may imply serious errors in the estimates.

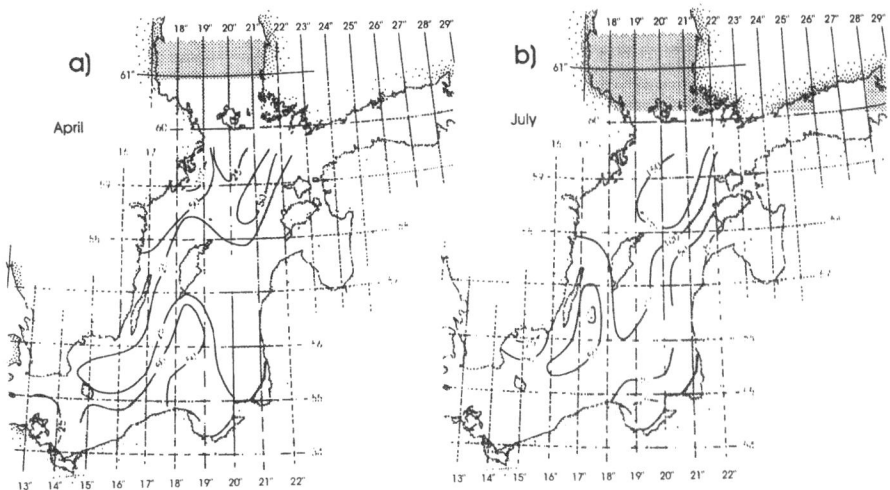

Figure 9. Height (cm) of juvenile freshwater in the Baltic proper in a) April and b) July. Shading indicates squares lacking data. (From Eilola and Stigebrandt [22]).

8. Consequences of Atmospheric Freshwater Transports for the Vertical Circulation of the Ocean

In a steady-state vertical circulation of the ocean, deep water should be produced and 'consumed' at equal rates. Deepwater production takes place at the sea surface by the action of thermohaline processes driven by energy and water exchange at the air sea (ice) interface. Deepwater consumption, however, is by diapycnal mixing in the abyss that tends to make the deepwater less dense and by that paves the way for later formed deepwater. Steady-state vertical circulation of the ocean may thus be expected to require a rate of deepwater production dictated by the rate of loss of deepwater due to diapycnal mixing. The consumption of deepwater in the ocean should be rather constant in time and independent of the state of the climate system because the energy for abyssal diapycnal mixing seems to come essentially from astronomical energy (tides) while the climate system (winds) probably contributes only a smaller fraction, Sjöberg & Stigebrandt [24].

Deepwater sources should be located where sufficiently dense water may be produced at rates required by the abyssal diapycnal mixing. In the present day oceans, only the Atlantic is salty enough to produce deepwater by cooling at high latitudes. Thus, inter-oceanic salinity differences caused by inter-oceanic atmospheric moisture transports set the rules of the game in the competition between oceans to produce deepwater.

As discussed in section 7.1 the deepwater formation in the Nordic Seas might be sensitive to increased freshwater supply to especially the Norwegian and Barents Seas. Large amounts of water of low salinity from the Polar Ocean pass the Nordic Seas along the East Coast of Greenland. If less than half of this was forced eastward to the Norwegian Sea the deepwater production might stop. Such an eastward transport of surface water would occur if southerly winds became more frequent at the expense of northerly in the western part of the Nordic Seas as suggested in Stigebrandt [20]. It is known that relatively large salinity variations may occur both in the in-flowing Atlantic water and in the Nordic Seas, see e.g. Dickson et al. [25] for a description of the "Great Salt Anomaly".

Deepwater may be produced not only by cooling of sufficiently saline water. It may also be produced by processes increasing the salinity, either due to evaporation in semi-enclosed seas like the Mediterranean or ice production in shallow high-latitude shelf areas like those found in the Arctic Ocean and the Weddell Sea. Deepwater production by ice production also means production of freshwater (almost fresh ice) which tends to decrease the surface salinity in areas where the ice melts.

9. Future Research Priorities

Atmospheric freshwater transports tend to change the distribution of salinity in the ocean. This tendency is counteracted by a complex system of oceanic transports. Some of the transport modes are forced by the resulting salinity gradients. However, there are also transport modes driven by heating/cooling and winds and tides working independently of the salinity gradients.

In this paper, I have tried to identify major processes for the large-scale circulation of freshwater in the sea. Some aspects of the freshwater influenced circulation were illustrated by examples from estuaries and coastal seas. The estuary examples clearly demonstrate that diapycnal mixing drives the baroclinic circulation (of conveyor-belt-type) but the magnitude of this is modified by the rate of freshwater supply and by topographic parameters. It was also shown that there is a rather strong negative feedback between freshwater supply and inflow of seawater to estuaries. The negative feedback is the reason why the super-fjord heat pump does not work in the way suggested by Berger and Jansen [2]. Studies of dispersion of juvenile freshwater in the Baltic proper suggest that Okubo-type of spreading might be important also in the open oceans.

The very large-scale picture of the oceanic response to atmospheric freshwater transports is as follows. The Pacific has lower surface salinity than the other oceans due to convergence of atmospheric moisture transports. A consequence of this is that the sea level stands higher in the North Pacific than in the North Atlantic, which drives the barotropic flow of low-saline water through the Bering Strait. The convergence-divergence of barotropic flow in the oceans may be estimated by integration of the net loss of water to the atmosphere in latitudinal bands starting from Bering Strait. Furthermore, having the generally highest salinity the Atlantic Ocean will provide the deepwater for the deep conveyor belt circulation in the oceans. The magnitude of this circulation is determined by diapycnal mixing essentially driven by tidal dissipation in the abyss.

How sensitive are present day's atmospheric transports of freshwater to perturbations in climate? The atmospheric moisture transports are integrated parts of the response of the climate system to the radiation imbalance at the top of the atmosphere. It is hard to believe that quite different large-scale transport patterns of moisture should evolve as a result of small perturbations of the radiation balance. The circumstance that tropic rain forests and subtropic deserts seem to be very old features supports this view. There are, however, many uncertain features of the circulation of freshwater in both past and present day's atmosphere and oceans and a short list of important questions to answer by future research is given below.

-   Are there important feedback loops between the atmospheric freshwater transports and the oceanic thermohaline circulation? How sensitive is present day's thermohaline circulation to perturbations of the systematic atmospheric transports of freshwater? Can small changes of the atmospheric transports of freshwater destroy the conditions for deepwater production in present day's locations? If this happens, which areas will then become likely locations for deepwater production?
-   Was the atmospheric freshwater forcing of the oceans at high latitudes during the last glaciation much different from present day's forcing? If so, which were the differences and how was the oceanic response? In periods when Bering Strait was closed large differences in ocean salinity may have resulted, particularly in the N. Atlantic and the Arctic Oceans, see e.g. Stigebrandt [12, 20] and Shaffer and Bendtsen [26]. What is the intrinsic oceanic time-scale for relaxation of the salinity difference between the N. Pacific and the N. Atlantic?
-   Are dispersive transports by unsteady motions of importance for transports of salt and freshwater in the open ocean?
-   Is El Nino triggered by an oceanic response to zonal atmospheric transports of freshwater in the tropical parts of the oceans? What is the typical oceanic volume transport of freshwater from west to east during one event?

Acknowledgements

This work was supported by the Swedish Natural Science Research Council (NFR).

10. References

1. Wijffels, S.E., Schmitt, R.W., Bryden, H.L. and Stigebrandt, A., 1992: Transport of freshwater by the oceans. J. Phys. Oceanogr., 22, 155-162.
2. Berger, W. H. and Jansen, E., 1995: Younger Dryas episode: Ice collapse and superfjord heat pump. In *The Younger Dryas* (Troelstra, S.R., van Hinte, J.E. and Ganssen, G.M., editors). 61-105.
3. Trenberth, K.E. and Solomon, A., 1994: The global heat balance: heat transports in the atmosphere and ocean. Climate Dynamics, 10, 107-134.
4. Keith, D.W., 1995: Meridional energy transport:; uncertainty in zonal means. Tellus, 47A, 30-44.
5. Baumgartner, A. and Reichel, E., 1975: The world water balance. Elsevier, 179 pp.
6. Schmitt, R.W., Bogden, P.S. and Dorman, C.E., 1989: Evaporation minus precipitation and density fluxes for the North Atlantic. J. Phys. Oceanogr., 19, 1208-1221.
7. Roach, A., Aagaard, K., Pease, C., Salo, S., Weingartner, T., Pavlov, V. and Kulakov, M., 1995: Direct measurements of transport of water and properties through Bering Strait. J. Geophys. Res., 100, 18443-18457.
8. Chen, T.-C., Pfaendtner, J. and Weng, S.-P., 1994: Aspects of the hydrological cycle of the ocean-atmosphere system. J. Phys. Oceanogr., 24, 1827-1833.
9. Susskind, J. and Pfaendtner, J., 1989: Impact of interactive physical retrievals on NWP. Report on the *Joint ECMWF/EUMETSAT Workshop on the Use of Satellite Data in Operational Weather Prediction: 1989-1993*. Vol. I, Shinfield Park, Reading, UK, ECMWF, 245-270.
10. Jourdan, D., Peterson, P. and Gautier, C., 1997: Oceanic freshwater budget and transport as derived from satellite radiometric data. J. Phys. Oceanogr., 27, 457-467.
11. Tomczak, M. And Godfrey, J.S., 1994: Regional Oceanography: an Introduction. Pergamon Press. 422 pp.
12. Stigebrandt, A., 1984: The North Pacific: A global-scale Estuary. J. Phys. Oceanogr., 14, 464-470.

13. Stigebrandt, A., 1981a: Is the magnitude of the salinity difference between the North Atlantic and the North Pacific controlled by the topography of the Bering Strait? Univ. of Gothenburg, Dept. of Oceanogr., Rep. No. 39, 9 pp.

14. Stommel, H. and Farmer, H.G., 1953: Control of salinity in an estuary by a transition. J. Mar. Res., 12, 13-20.

15. Kato, H. and Phillips, O.M., 1969: On the penetration of a turbulent layer into a stratified fluid. J. Fluid Mech., 37, 643-655.

16. Stigebrandt, A., 1981b: A mechanism governing the estuarine circulation in deep, strongly stratified fjords.Estuarine, Coastal & Shelf Sci., 13, 197-211.

17. Stigebrandt, A., and Molvær, J., 1996: Evidence for hydraulically controlled outflow of brackish water from Holandsfjord, Norway. J. Phys. Oceanogr., 26, 257-266.

18. Assaf, G. and Hecht, A., 1974: Sea straits: A dynamical model. Deep-Sea Res., 21, 947- 958.

19. Gustafsson, B. and Stigebrandt, A,. 1996: Dynamics of the freshwater-influenced surface layers in the Skagerrak. J. Sea Res., 35, 39-53.

20. Stigebrandt, A., 1985: On the hydrographic and ice conditions in the northern North Atlantic during different phases of a glaciation cycle. Paleogeography, Palaeoclimatology, Palaeoecology, 50, 303-321.

21. Simonsen, K. and Haugan, P., 1996: J. Geophys. Res., 101, 6553-6576.

22. Eilola, K. and Stigebrandt, A., 1998: Spreading of juvenile freshwater in the Baltic proper. J. Geophys. Res., 103, 27795-27807.

23. Okubo, A., 1974: Some speculations on oceanic diffusion diagrams. Rapp. P.-v. Réun. Cons. Int. Explor. Mer, 167, 77-85.

24. Sjöberg, B., and Stigebrandt, A., 1992: Computations of the geographical distribution of the energy flux to mixing processes via internal tides: its horizontal distribution and the associated vertical circulation in the ocean. Deep-Sea Res., 39, 269-291.

25. Dickson, R.R., Meincke, J., Malmberg, S.A. and Lee, A.J., Progr. Oceanogr., 20, 103-151.

26. Shaffer, G. and Bendtsen, J., 1994: Role of the Bering Strait in controlling North Atlantic ocean circulation and climate. Nature, 367, 354-357.

GLOBAL ATMOSPHERIC CIRCULATION PATTERNS AND RELATIONSHIPS TO ARCTIC FRESHWATER FLUXES

J.E. WALSH
Department of Atmospheric Sciences
University of Illinois
Urbana, IL 61801 U.S.A.

1. Introduction

The large-scale atmospheric circulation affects the Arctic Ocean's freshwater fluxes in at least three ways. First, the circulation controls the direct fluxes of precipitation (P) and evaporation (E) at the Arctic Ocean's surface through the pattern of atmospheric moisture inflow and its convergence over the Arctic Ocean. Second, the circulation determines P and E over the Arctic terrestrial watersheds; the net P-E eventually reaches the Arctic Ocean as river runoff. Third, the large-scale circulation provides the wind-forcing that drives the advective fluxes of sea ice and ocean freshwater anomalies into and out of the Arctic (e.g., through Bering and Fram Straits) as well as from one portion of the Arctic Ocean to another. The atmospheric thermal anomalies resulting from the large-scale circulation also modify the freshwater fluxes in the Arctic Ocean through the freezing and melting of sea ice. This thermodynamic forcing is often correlated with the wind-driven forcing on interannual timescales.

Given these various influences of the atmospheric circulation on Arctic freshwater fluxes, it follows that the large-scale atmospheric circulation is a primary driver of Arctic hydrologic variability over a variety of timescales, including synoptic (several days), seasonal, interannual, decadal and longer. In this review, the focus will be on the interannual to decadal timescales. In surveying the variability over these timescales, we will attempt to place the forcing of Arctic hydrologic variations into a broader perspective of large-scale (i.e., regional to hemispheric) atmospheric variability. This perspective will include reviews of large-scale patterns of variability, spatial teleconnections involving the Arctic, and temporal changes that may be manifestations of secular trends or low-frequency variability in the atmosphere-ocean-ice system. In the course of the review, we will also mention some of the patterns of change in the greenhouse scenarios projected by global climate models. While attribution of trends in the observational record is beyond the scope of this paper, the magnitudes and patterns of the projected greenhouse changes provide a perspective for the large-scale atmospheric variations of the past century or so.

Of the three categories of processes by which the large-scale atmospheric circulation affects the Arctic Ocean freshwater fluxes, the first two noted above involve moisture fluxes in the atmosphere. Discussion of these two will comprise the major portion of this paper. Before addressing the associated mechanisms and effects, however, we present two examples of the third category of processes by which the atmosphere drives

E.L. Lewis et al. (eds.), The Freshwater Budget of the Arctic Ocean, 21-43.

22

oceanic freshwater fluxes through forcing at the ocean surface. Figure 1 shows the patterns of annually-averaged Arctic sea level pressure for two different years, 1968 and 1984, together with corresponding trajectories from sea ice model simulations forced by the wind fields of the same years. Sea ice outflow and the implied freshwater export from the Arctic Ocean to the Atlantic Ocean through Fram Strait are strikingly greater in the 1968 simulation, while much of the ice remains in the western Arctic gyre during 1984. The implied freshwater export to the North Atlantic is small in 1984. If either pattern in Figure 1a or 1b were to dominate the Arctic atmospheric circulation for multiyear periods, it is likely that significant anomalies of the freshwater content of the upper water column would develop in the Arctic Ocean and in the North Atlantic subpolar seas in response to the altered freshwater flux through Fram Strait. With regard to the later discussion of Arctic pressure anomalies, we note that the 1968 pattern in Fig. 1a is characterized by a Barents/Norwegian Sea low pressure system and cyclonic

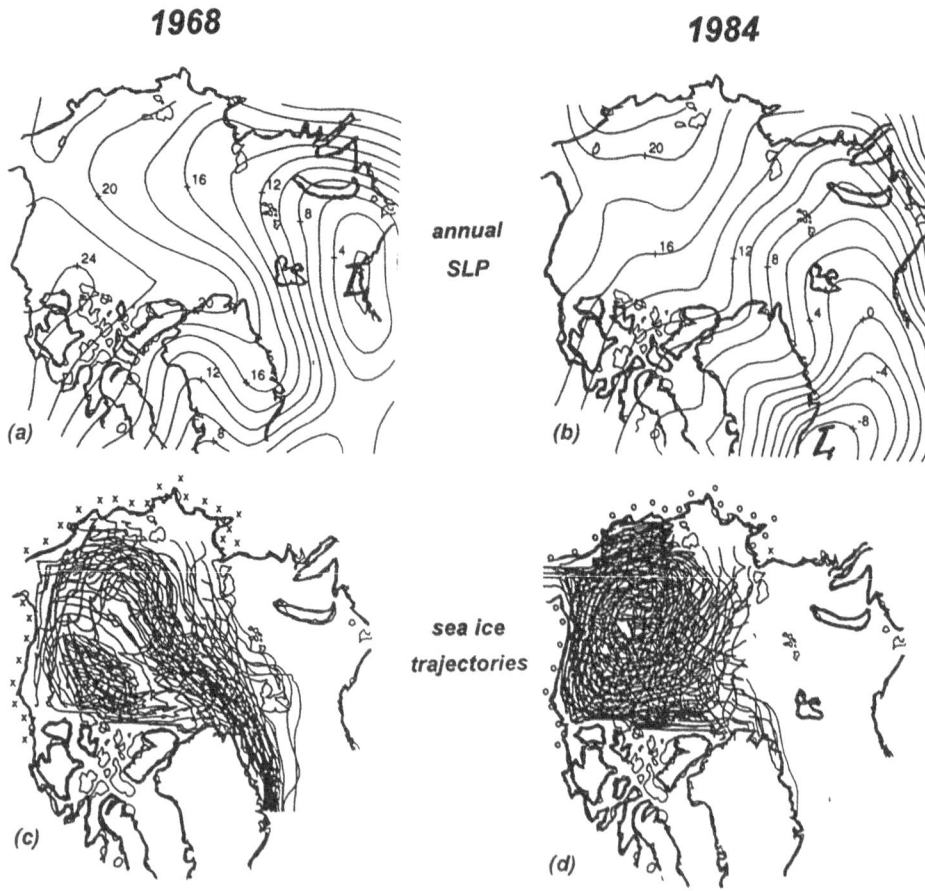

Figure 1. Annual mean Arctic sea level pressures for 1968 and 1984 (upper panels) and corresponding simulated trajectories of sea ice originating at coastal locations denoted by "x" and "o" (lower panels). From Tremblay et al. [26].

curvature of the SLP isobars over most of the Arctic Ocean; the 1984 pattern is characterized by a strengthening (by ~10 mb) and a retreat of the Atlantic subpolar low to the Greenland-Iceland corridor, and by a generally anticyclonic curvature of the SLP isobars over much of the Arctic Ocean. During the recent decades (1979–1995), there has been a shift toward a predominance of the pattern of Fig. 1a in the central Arctic (Walsh et al. [27]). This pattern implies a larger wind-driven export of sea ice from the central Arctic to the Greenland Sea (Dickson et al. [4]; Kwok and Rothrock [16]).

The impact of a prevailing mode of atmospheric circulation on Arctic freshwater transport is apparent in the model results of Hakkinen [9], whose ice-ocean model simulations were forced by several decades of atmospheric observational data. Figure 2 shows that the simulated annual outflows of freshwater from the Arctic through Fram Strait were unusually large from the early 1960s through the late 1960s, including 1968. The earliest signs of the Great Salinity Anomaly (Dickson et al. [5]) appeared near the end of this period of enhanced freshwater outflow from the Arctic. There are suggestive indications (e.g., Fig. 1 above; Mysak et al. [19]) that the Arctic Ocean freshwater export was a key factor in the development of the Great Salinity Anomaly, although other, more localized factors may also have contributed.

In the remainder of this paper, we review the variability of the atmospheric moisture fluxes that ultimately affect the Arctic ice/ocean through the direct exchanges, precipitation and evaporation, included in the first of the three categories of circulation effects listed earlier. Specifically, we address the interannual variability of Arctic P (or

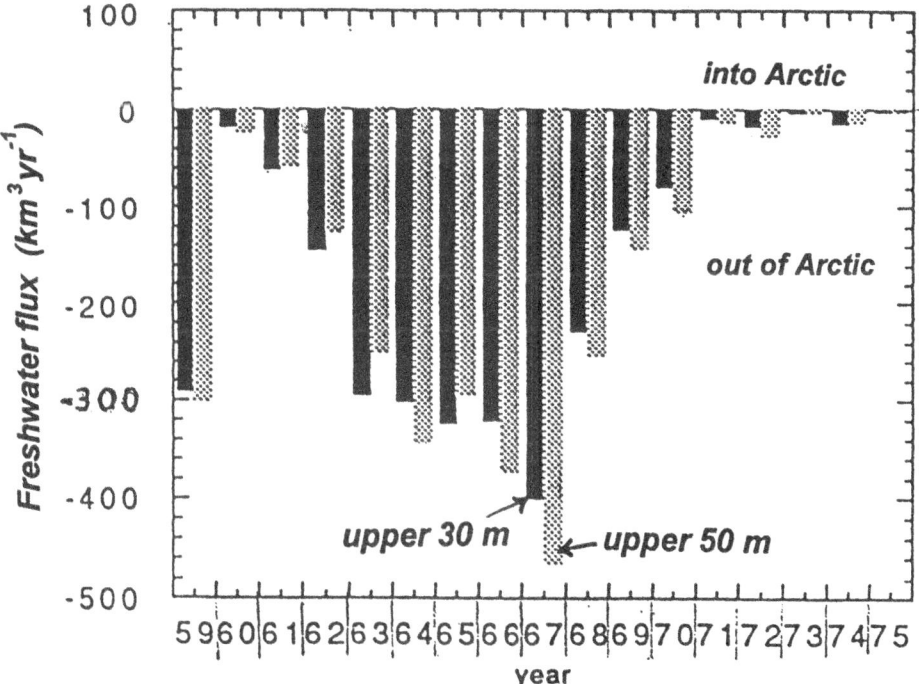

Figure 2. Model-derived annual values of net freshwater flux (km^3 yr^{-1}) through Fram Strait in upper 30 m (solid bars) and upper 50 m (dotted bars) of ocean. From Hakkinen [9].

P-E) in the context of the high-latitude atmospheric circulation and its relationships to the large-scale circulation patterns of middle and lower latitudes. The large-scale circulation patterns will be examined in terms of their "teleconnection" patterns, their seasonality, and their low-frequency variations, which can appear as trends in short records. The emphasis will be on large-scale atmospheric associations with P rather than with E because estimates of E have much greater uncertainties, whether they are deduced as residuals in moisture budget studies (Walsh et al. [28]) or evaluated from model parameterizations that are generally untested in high latitudes.

2. Atmospheric Moisture Fluxes, Precipitation, and Evaporation

In order to provide a one-dimensional spatial perspective on the linkage between the circulation variability and the Arctic hydrologic cycle, we show in Fig. 3 the longitudinal dependence of the annual averages of the net atmospheric moisture flux across 70°N. The longitudinal profiles for individual years (1973–1990) as well as for the 18-year mean were evaluated from rawinsonde data (Walsh et al. [28], Serreze et al. [24]). When averaged over longitude, the net flux is positive (northward), representing an influx of moisture equivalent to an areally-averaged surface flux of 16.9 cm yr^{-1} for the polar cap north of 70°N. This moisture flux convergence from the atmospheric component of the hydrologic cycle represents a freshwater source for the Arctic Ocean. The gain of freshwater from this source (and from river discharge) must be offset by a net export of freshwater by the ocean currents and the associated sea ice outflow. However, the longitudinal profiles contain pronounced maxima, which appear as peaks in Fig. 3. The strongest peaks appear in the Atlantic sector between 70°W and 40°E; the double-maximum occurs because Greenland effectively blocks the northward flux between 40°W and 20°W. While this double-maximum in the Atlantic sector represents the largest atmospheric moisture influx to the Arctic, another pronounced maximum occurs in the Pacific longitudes, 160°E–220°E, centered on the Bering Strait. Figure 3 also shows that there is a net outflow of moisture from the Arctic across 70°N in the western Canadian sector, 80°W–120°W. This outflow occurs, even though the air is

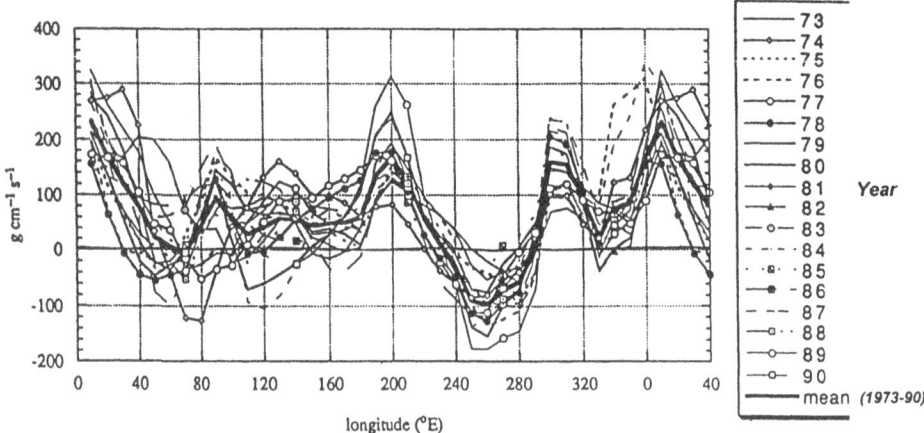

Figure 3. Longitudinal dependence of the net moisture flux across 70°N in individual years, 1973–1990. Heavy solid line is 18-year mean. From Walsh et al. [28].

cold and dry, because southward airflow predominates in this region. Figure 3 also shows that there are large interannual variations of the fluxes at all longitudes; the magnitudes of the yearly anomalies often exceed 50% of the mean values. With regard to the discussions in the following subsections, the salient feature of Fig. 3 is that the primary contribution to the variability of atmospheric moisture influx to the Arctic occurs in the Atlantic longitudes (70°W–40°E) and a narrow Pacific sector (160°E–140°W) encompassing the Bering Strait. Thus the atmospheric teleconnections involving these regions are of particular importance to Arctic hydrology.

Elsewhere in this volume, Bromwich et al. [2] describe similar computations based on the atmospheric reanalyses of the National Centers for Environmental Prediction (NCEP) and the European Centers for Medium-Range Weather Forecasts (ECMWF). The respective reanalysis-derived estimates of moisture flux convergence poleward of 70°N are equivalent to P-E of 19.4 cm/yr^{-1} (NCEP) and 18.2 cm/yr^{-1} (ECMWF). Subsequent work by Bromwich et al. has shown that the rawinsonde-derived estimates are adversely impacted by the sparseness of the rawinsonde network, particularly in the Atlantic sector. Moreover, changes in the rawinsonde network over time degrade the depiction of interannual variations in the rawinsonde-derived results. For these reasons, the reanalysis-derived values of 18-20 cm/yr^{-1} are likely to be the more accurate estimates of P-E for the Arctic polar cap, implying that the errors in the rawinsonde estimates are approximately 10% of the long-term mean and somewhat larger percentages of the values for individual years and months.

The temporal variability of regional-scale Arctic precipitation is illustrated in Fig. 4, which is a time series for 1949–1991 of precipitation for October-March averaged over the land area poleward of 55°N between 90°W and 55°E (northern Canada and Europe as well as the North Atlantic). Even over this broad area, the annual mean P can vary by 20–30% from one year to the next. Figure 4 shows a substantial increase of precipitation over this area between the beginning and end of the period. The 1995 IPCC assessment [12] also shows an increase of P during this period for the broader area, 55°–85°N (Fig. 5). The recent increase in Fig. 5 is part of a broader positive trend dating back to the earlier decades of the 20th century. Elsewhere in this volume, Kattsov et al. [15] show that the trend in Fig. 5 appears to be at least partly attributable to the changing nature of the network of station gauges in high latitudes. If there has been an increase of high-latitude P in recent decades, the role of any corresponding changes in the large-scale atmospheric circulation must be addressed in the quest for attribution. Moreover, Fig. 4 shows the variations of precipitation averaged over northeastern North America and northern Europe. A characteristic of the North Atlantic Oscillation is an out-of-phase relationship between wintertime precipitation in these two areas (e.g., Hurrell [10]), implying that the variance of the time series in Fig. 4 may well be reduced by the most prominent large-scale mode of atmospheric variability in the North Atlantic sector.

Figure 6 shows that trends in the large-scale atmospheric circulation are indeed apparent over the past several decades. The pattern of wintertime sea level pressure change (Fig. 6a) is dominated by a decrease in the subpolar North Atlantic, where the maximum change is approximately 0.75 mb decade^{-1} or about 3 mb over four decades. There is an increase of pressure to the south over the mid-latitude North Atlantic and Europe. The pattern of change implies an enhancement of the flux of Atlantic moisture to the Eurasian Arctic and its terrestrial watershed, especially if the pressure decreases from Iceland to the eastern Siberian coastal region are indicative of increased cyclone frequencies. In the Pacific sector, the east-west couplet of positive and negative pressure

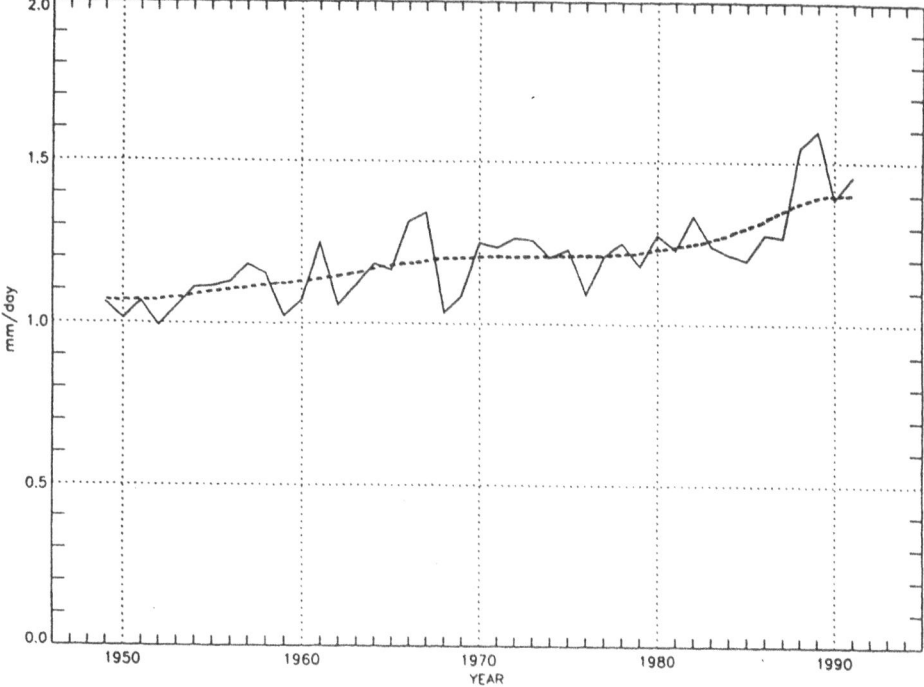

Figure 4. Wintertime (October–March) precipitation over high-latitude land areas poleward of 55°N and between 90°W and 55°E. Values are based on land-station measurements. Dashed line is decadal running mean. [Figure from M. Hulme, Univ. of East Anglia].

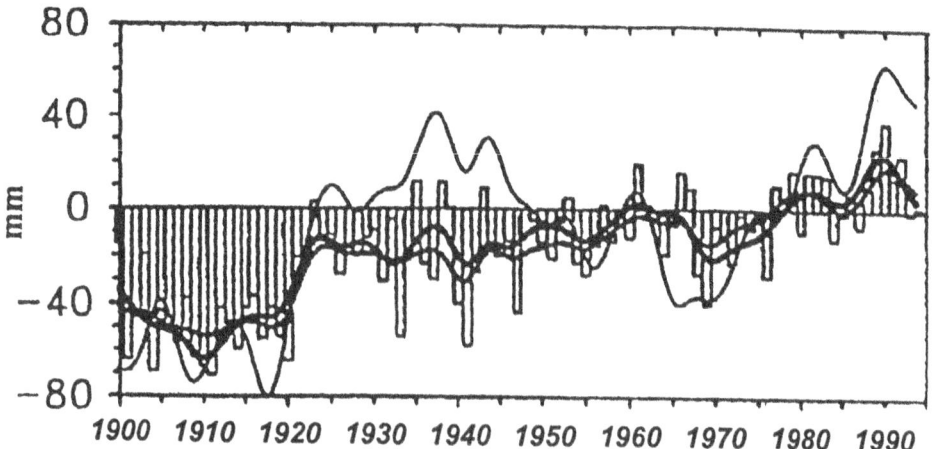

Figure 5. Annual precipitation anomalies (relative to 1960–1990 mean) for zonal belt between 55°N and 85°N. Heavy solid lines are running mean precipitation anomalies; light solid line is running mean surface air temperature anomaly. From IPCC [12].

changes south of Alaska implies an enhanced inflow of atmospheric moisture into the Arctic in the Bering longitudes. However, these Pacific features are considerably weaker than those in the North Atlantic. We also note that the pattern of pressure change in Fig. 6a is consistent with the temperature trends of the past 3–4 decades (IPCC [12]), at least partly because of an atmospheric-circulation-driven warming of northern Eurasia and the Alaska/Yukon region. The recent cooling of southeastern Canada, the Labrador Sea, and parts of Greenland also appears to be circulation-driven. The corresponding summer panel (Fig. 6b) shows a diffuse pattern of change over the subpolar oceans, but a substantial decrease of pressure (and, implicitly, stronger cyclone activity) over the Arctic Ocean. The changes of gradient wind corresponding to Fig. 6b imply a strengthening of the flux of Atlantic moisture into the Arctic via the Norwegian/Barents Sea corridor. Whether the pressure patterns in Fig. 6 reflect secular changes or low-frequency variations is unknown, although a relationship to the recent strengthening of the centers of action of the North Atlantic Oscillation (NAO) will be apparent later in the discussion.

The study of low-frequency variations of the large-scale circulation has a long history in the Russian literature (e.g., [6], [8]). The Russian classification scheme distinguishes zonal flow regimes and several meridional regimes; the meridional regimes are subclassified according to the locations of the troughs and ridges in the quasi-stationary longwaves. For the Arctic freshwater budget, the significance of such schemes is that the poleward moisture fluxes will be greater in the meridional flow regimes (e.g., the Z and C types of Girs [8]) than in the zonal regimes (e.g., Girs' E and Z types). Figure 7 shows that there are pronounced decadal-scale variations in the frequencies of the two major circulation categories, zonal and meridional. Table 1 shows that the two zonal types, E and Z, occurred more frequently during the 1970s and 1980s than during any other decades of the century. These changing frequencies are consistent with the strengthening of the Icelandic and Aleutian lows discussed earlier in conjunction with Fig. 6. This consistency, together with the fact that the Russian classification scheme has been applied to a century of atmospheric fields, calls for an integration of the Russian scheme into a study of decade-to-century-scale moisture flux

TABLE 1. Average number of days per year with the Girs-Vangengeim [6] circulation types. (Table adapted from Radionov et al. [23].)

Period	Type E	Type W	Type C	Type Z	Type M1	Type M2
1904–14	127.7	151.7	85.6	149.3	98.8	116.9
1915–25	133.7	154.1	77.2	118.1	109.6	137.3
1926–36	154.1	130.1	80.8	130.1	94.0	140.9
1937–47	139.7	109.6	115.7	110.9	83.2	178.9
1848–58	151.7	110.9	102.4	122.9	112.0	130.1
1959–69	175.7	88.0	101.3	128.9	95.2	140.9
1970–80	212.9	76.1	76.0	167.3	76.0	121.7
1981–91	185.3	104.9	74.8	193.7	58.0	113.3

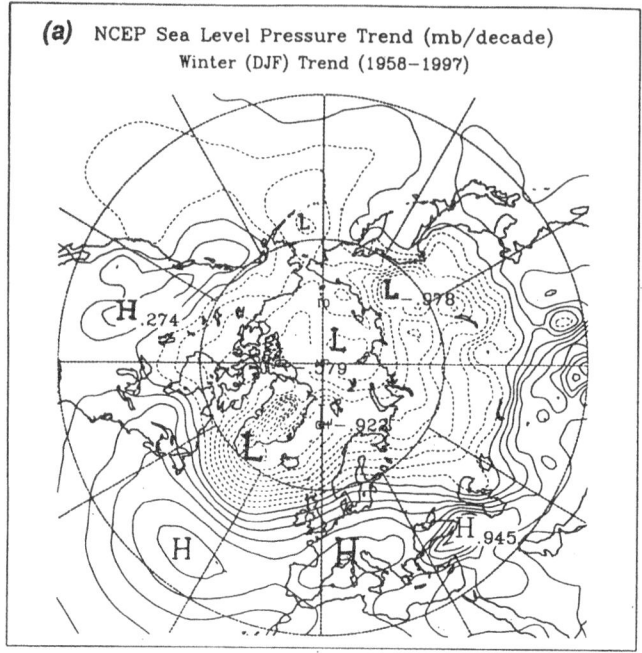

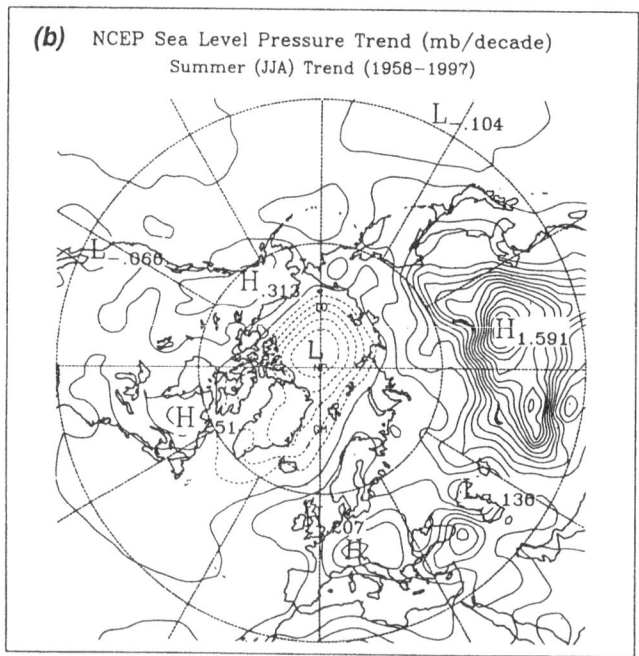

Figure 6.   Linear trends (mb per decade) of sea level pressure for (a) winter, Dec.–Feb. and (b) summer, Jun.–Aug. Contour interval is 0.1 mb per decade.

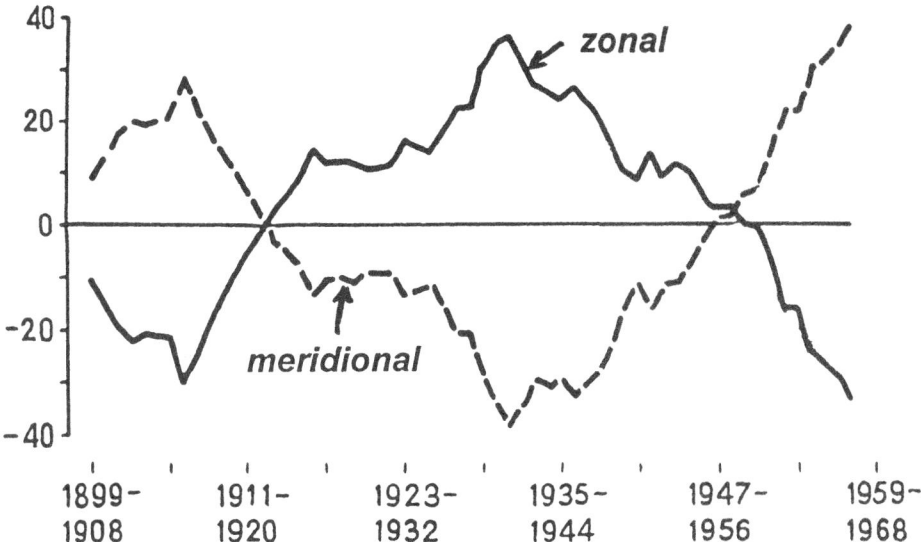

Figure 7. Variations of frequency of zonal (solid line) and meridional (broken line) circulation patterns over the Northern Hemisphere. From Lamb [17] after Dzerdzeevskii [6].

variations. However, the scheme needs to be objectively quantified for digital processing applications. One must also keep in mind that the Russian circulation types noted above are upper-air (500 mb) patterns; the associated lower-tropospheric circulation patterns have the largest effects on moisture transport. Thus it is not surprising that previous studies of the primary moisture flux corridors into the Arctic (i.e., the North Atlantic) have tended to emphasize variations in surface or sea-level pressure (SLP).

The most widely investigated mode of Atlantic SLP variability is the so-called North Atlantic Oscillation (NAO), which is characterized by mutual changes of intensity of the subpolar low and subtropical high in the North Atlantic. A commonly used index of the NAO is the difference between the sea level pressure anomalies in the Azores/Portugal region and the Iceland region. Figure 8 shows the composite difference field of sea-level pressure for the days during 1958–1997 having NAO index values in the top 5% and bottom 5% of all daily values. (The signs in Fig. 8 correspond to the NAO's high-index days.) It is apparent from Fig. 8 that the north-south pressure difference between the Icelandic low and the subtropical high strengthens by about 60 mb during a change from an NAO in the bottom 5% to an NAO in the top 5%. A positive NAO is characterized by strong surface westerlies across the North Atlantic from the Labrador Sea to the U.K., and by stronger easterlies north of the subpolar low and south of the subtropical high. More importantly for the moisture transport into the Arctic, Fig. 8 shows that, during the positive phase of the NAO, there is an enhanced atmospheric moisture flux from the North Atlantic to the Arctic in the region from Iceland eastward to, and including, the Barents Sea. The positive NAO is also associated with an equatorward flux of climatologically drier air in the Baffin Bay/Labrador Sea region. By contrast, a negative NAO implies the opposite pattern: weaker moisture influx to the Arctic east of Iceland and weaker southward flow to the

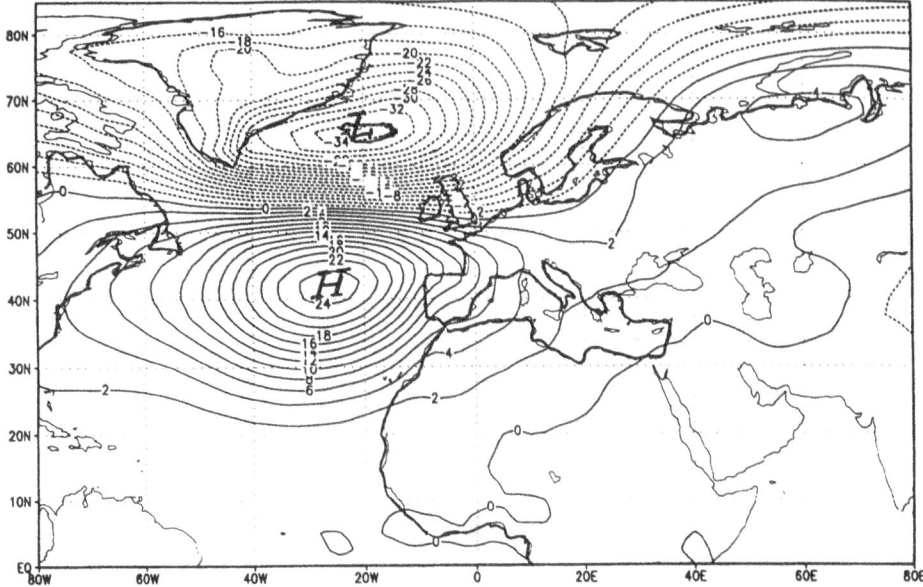

Figure 8. Differences of sea level pressure between days with NAO index in upper 5% and
lower 5% of distribution of daily values of NAO index.

west of Greenland. The positive NAO pattern clearly has the greater exchange of
moisture between the middle latitudes and the Arctic. The NAO-induced changes in the
intensity of the Atlantic moisture flux to the Arctic become relevant to decadal-scale
hydrologic variability in the Arctic when one examines the time series of the annual
winter-averaged NAO index (Fig. 9), which has increased from a period-of-record
minimum in the 1960s to a period-of-record maximum in the 1990s. This variability has
been documented by Dickson et al. [4], Hurrell [10] and others in association with
decadal-scale ocean and atmosphere variations in the Atlantic sector. In particular,
Hurrell [10] has examined the recent northward shift of positive precipitation anomalies
to the Scandinavian region as part of this decadal-scale strengthening of the NAO.
Figure 8 implies that the Arctic is also likely to have been impacted, although the
paucity of direct measurements of precipitation in the central Arctic limits the direct
evidence.

In an ongoing investigation, Thompson and Wallace [25] have identified linkages
between the North Atlantic Oscillation, Arctic sea level pressure and the stratospheric
polar vortex (Fig. 10). The corresponding structure, which emerges as the leading
empirical orthogonal function of wintertime SLP in the northern hemisphere, implies
that the NAO is part of a broader mode than the subpolar-subtropical couplet (e.g., Fig.
8) traditionally regarded as the NAO. The subpolar North Atlantic SLP lobe in Figs.
10a–10c encompasses the Arctic Ocean as well as the North Atlantic; the compensating
mass anomalies are found in the North Pacific and the subtropical North Atlantic.
Because of the Arctic's central location in this pattern, the mode has been termed the
"Arctic Oscillation." Thompson and Wallace [25] interpret this mode as "an equivalent
barotropic signature of modulations in the strength of the polar vortex aloft" [25,
abstract]. They suggest that the recent wintertime warming trend over the northern land

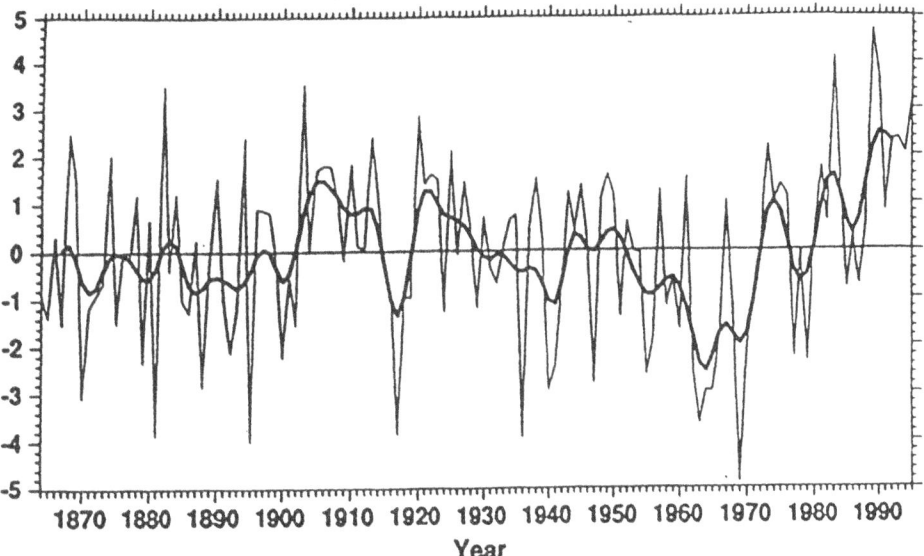

NAO Index (Dec-Mar) 1864-1995

Figure 9. Time series of winter (Dec.–Mar.) NAO index from 1864–1995. Heavy line is the index smoothed with a filter that removes fluctuations with periods less than 4 years. From Hurrell and van Loon [11].

areas is a manifestation of a "systematic bias" of this mode in recent decades. The role of the mode in controlling variations of Arctic freshwater fluxes, including horizontal moisture transport as well as precipitation, remains to be investigated. Relationships to previous circulation "typing" schemes (e.g., the Russian classifications discussed earlier) also need to be explored.

3. Diagnostic Studies of Precipitation-Circulation Relationships

Although the Arctic Oscillation has only recently been identified explicitly, its signature has appeared previously in the literature. For example, Fig. 11 shows Bjornsson et al.'s [1] composite difference field of wintertime 700 mb height corresponding to heavy (minus light) precipitation in the Mackenzie Basin. The similarity between Figs. 10 and 11 is clearly evident, although the Pacific features are more prominent in the Mackenzie-based pattern (Fig. 11), particularly over the Gulf of Alaska. The pattern in Fig. 11 also shows some resemblance to the NAO which, as noted in Section 2, has features in common with the Arctic Oscillation. However, even in the Atlantic sector, there are some important differences between Fig. 11 and the signature of the NAO/Arctic Oscillation. First, the Atlantic features of Fig. 11 are considerably farther north than the corresponding features of the NAO and the Arctic Oscillation. Second, the positive phase of the NAO is characterized by a negative pressure anomaly near Iceland, while Fig. 11 shows negative anomaly centers over Hudson Bay and the Arctic Ocean. Given the disparity in the dominant timescales of the NAO and Deser and Blackmon's [3] NAO-like dipole, it is highly likely that the pattern in Fig. 11 is fundamentally different from the NAO in its temporal characteristics. Nevertheless, all

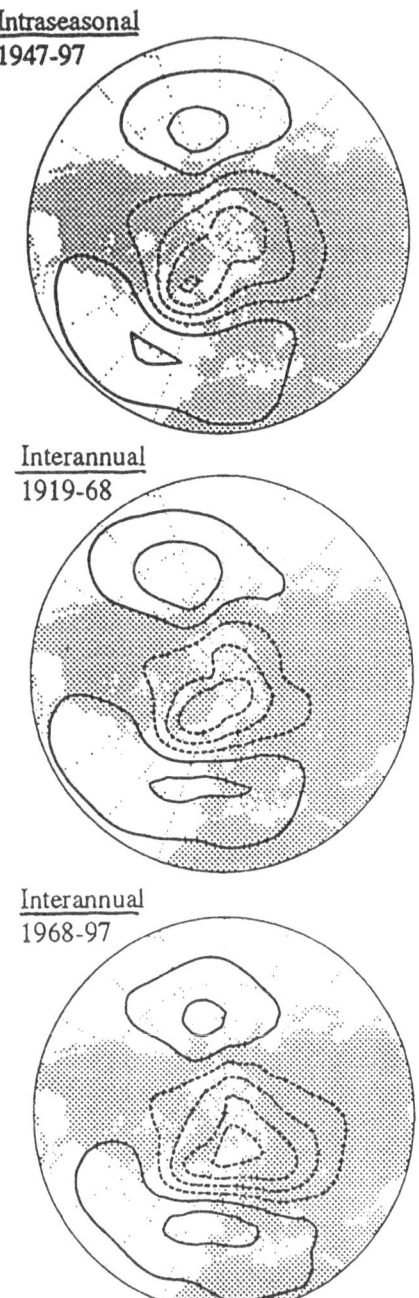

Figure 10. First EOF (Empirical Orthogonal Function) of northern hemisphere sea level pressure based on intraseasonal fluctuations, 1947–1997 (upper panel), interannual fluctuations, 1919–1968 (center panel), and interannual fluctuations, 1968–1997 (lower panel). From Thompson and Wallace [25].

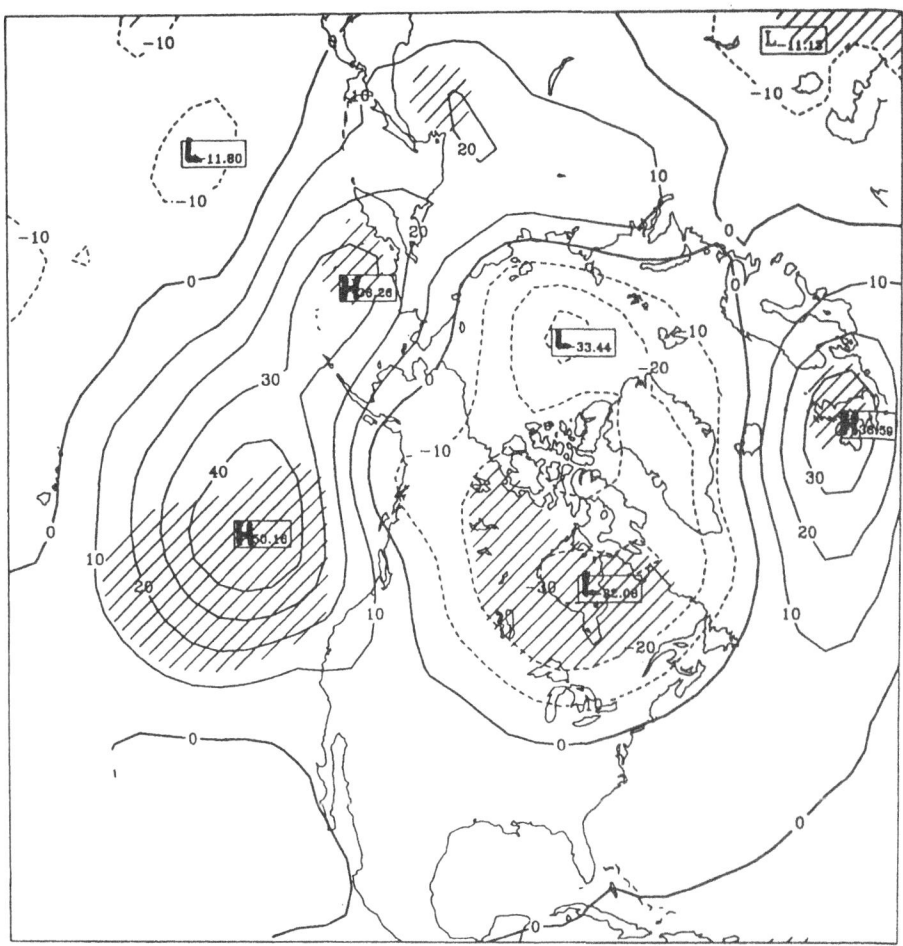

Figure 11. Difference field of 700 mb height obtained by subtracting means for winters with "light" precipitation in Mackenzie Basin from means for winters with "heavy" precipitation in Mackenzie Basin. Shading indicates significance at 95% level. From Bjornsson et al. [1].

these anomaly patterns do imply enhanced influxes of moisture and heat into the Arctic from the North Atlantic.

Bjornsson et al. [1] also showed that above-normal Mackenzie precipitation is associated with a strengthened storm track (greater variance of upper-air pressure) in the corridor from the northwestern Pacific to the Bering Sea, Gulf of Alaska, and British Columbia. In an examination of the seasonal dependence of the precipitation-discharge correlations, Bjornsson et al. found that the strongest correlation was between spring discharge and the autumn-winter-spring mean precipitation. The storage of cold-season precipitation in continental snow cover implies that this seasonality should be typical of rivers that discharge into the Arctic Ocean.

In a study of the atmospheric forcing of precipitation over Russian river drainage basins, Peng and Mysak [22] explored linkages to the North Atlantic. Multiyear periods of below-normal summer runoff (1951–1955) and above-normal summer runoff (1969–1974) in the Ob Basin were found to be associated with persistently above- and below-normal sea surface temperatures in the North Atlantic south of Greenland. The atmospheric linkage is a strengthened North Atlantic storm track (low SLP) from southern Greenland to the Barents/Kara Seas and into northern Asia in years of above-normal Ob discharge. The enhancement of the moisture influx is strongest in the northeastern Atlantic sector (20°W–40°E), corresponding to the longitudes of peak moisture influx to the Arctic (Fig. 3). Thus the importance of this corridor appears to extend to freshwater influxes to the Arctic Ocean from Siberian rivers. However, Peng and Fyfe [21] have shown that the association between North Atlantic sea surface temperatures and the atmospheric circulation undergoes a change in character during the progression from early winter to late winter. Specifically, the early winter is dominated by a dipole structure in which the atmosphere drives the ocean, while in late winter the ocean forces the atmosphere through a monopole mode. Thus the mechanisms underlying the interannual correlations are complex and call for further investigation using observational data as well as controlled model experiments that focus on linkages to precipitation.

Other studies have explored the atmospheric circulation's forcing of precipitation anomalies over Greenland and Scandinavia. Because these regions discharge primarily into water bodies other than the Arctic Ocean, we will not survey those studies here.

4. Atmospheric Reanalysis as a Diagnostic Tool

Diagnostic studies of the atmosphere's role in the Arctic hydrologic cycle have been hampered by the lack of consistent datasets spanning multi-decadal periods. The recent release of products from atmospheric reanalysis projects is a step towards the creation of consistent long-term datasets. The reanalyses of the National Centers for Environmental Prediction (NCEP) and the European Centre for Medium-range Weather Forecasts (ECMWF) span the periods 1958–1997 and 1979–1993, respectively, and are described by Kalnay et al. [13] and Gibson et al. [7]. The output products from both reanalyses are gridded (2.5° latitude × 2.5° longitude) at six-hourly intervals; the ECMWF reanalysis, which was produced by a higher-resolution model, is also available on 1° latitude × 1° longitude grids. Elsewhere in this volume, Serreze et al. and Bromwich et al. show that there are significant deficiencies in these first attempts at atmospheric reanalysis, particularly in the NCEP reanalysis. However, in order to explore uses of an internally consistent database containing Arctic hydrologic variables, we will present several examples of uses of the NCEP reanalysis to document circulation-hydrology linkages, thereby complementing the data-limited studies discussed earlier.

Figure 12a shows a time series of normalized monthly precipitation anomalies averaged over all NCEP reanalysis grid points in the Arctic Ocean. The monthly anomalies show considerable month-to-month and year-to-year scatter, although a weak positive trend (1–2% per decade) is present in the time series. Figure 12b shows the corresponding NCEP time series for the Mackenzie River drainage basin; this time series was obtained by averaging the precipitation at all 2.5-degree grid points in the subdomain shown in Bjornsson et al.'s [1] Fig. 2. In this case, low-frequency (multiyear) variability is apparent, e.g., the relatively dry 1964–69 period vs. the generally wet 1971–77 period. In addition, a strong positive trend is present in the time series, as the mean for the final ten years exceeds the mean for the first ten years by 10–15%. This positive trend is consistent with the results shown earlier (Figs. 4–5) and with

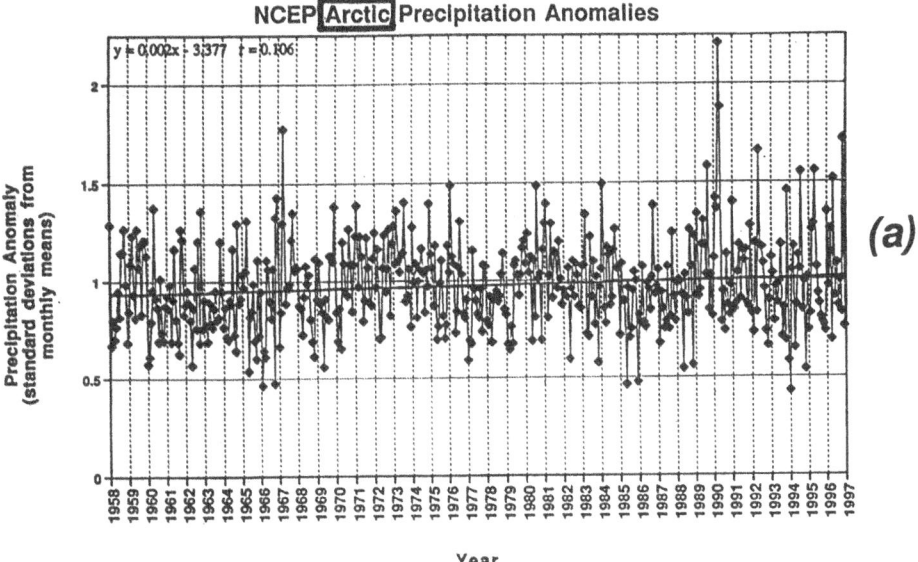

(a)

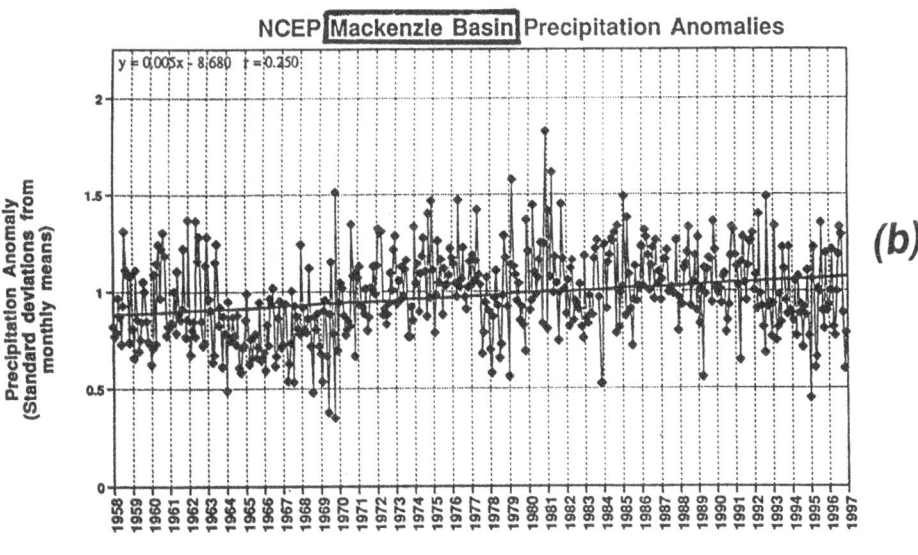

(b)

Figure 12. Standardized monthly precipitation from NCEP reanalyses. Time series are shown for (a) Arctic Ocean and (b) Mackenzie drainage basin.

other assessments of high-latitude precipitation trends (e.g., Karl et al. [14]). The precipitation trend in the NCEP reanalysis for the 1958–1997 period is positive over the polar cap, 55°–90°N. However, the positive trend in precipitation from the NCEP reanalysis is almost entirely attributable to the increases in the western hemisphere. This model-derived finding is somewhat surprising, because the positive trend of the NAO during the reanalysis period would imply an increase of precipitation over northern Eurasia and a decrease over eastern Canada. A possible explanation is that the trends are strongly dependent on precipitation during the summer, when the NAO plays a smaller role in interannual variability. Nevertheless, the spatial distribution of recent trends of precipitation appears to merit further investigation.

The internal consistency of the atmospheric circulation data and the hydrologic variables in the reanalysis provides the potential for a systematic examination of the circulation-rainfall linkages. Figure 13, for example, shows the composite SLP field for the months of "heavy" (highest 10%) Arctic precipitation minus the corresponding composite fields for months of "light" (lowest 10%) Arctic precipitation over the winter half of the year (November–April, inclusive). The signs of the pressure differences in Fig. 13 correspond to the "heavy precipitation" months. The Arctic-North Atlantic low-pressure phase of the Arctic Oscillation is dramatically apparent, as the double lobe of low pressure implies a strong enhancement of the low-level moisture influx to the Arctic in the Greenland-Iceland-Norwegian Seas corridor. The Atlantic inflow dominates the Arctic Ocean, although it is augmented in the Chukchi/Beaufort region by enhanced inflow of Pacific moisture. The positive feature in the Bering Sea in Fig. 13 is accompanied by a weak negative feature in the subtropcial Pacific. The latter feature, which is marginally significant at the 95% level, is one of the few hints of an Arctic connection to the low-latitude Pacific and the ENSO (El Niño/Southern Oscillation) phenomenon. Niebauer [20] has shown that sea ice and the atmospheric circulation in the Bering Sea region are correlated with ENSO, but the nature of the correlations (i.e., their spatial patterns) has changed substantially in recent decades.

The use of a composite difference field (e.g., Fig. 13) as a diagnostic tool raises the issue of possible nonlinearities in the precipitation-circulation association. The composite field (not shown) for the "light precipitation" months shows a close resemblance to the negative of Fig. 13: all five anomaly centers in the Pacific – North America – Europe sectors are present, although the isobars in the important Greenland-Norway corridor have a more zonal orientation than in Fig. 13. In the corresponding "heavy precipitation" composite, both negative anomaly centers of Fig. 13 are also present, but the isobar packing is tighter than in the "light" composite and the orientation implies northward airflow from the Greenland-Norwegian Seas towards the central Arctic. The composite difference fields discussed in the following paragraphs show similar symmetries, which can be regarded as qualitative but not quantitative.

Figure 14 shows that the atmospheric circulation features corresponding to *summer* precipitation anomalies in the Arctic are quite similar to those of winter. In the summer case, the most significant anomalies are in the subtropical Atlantic/Mediterranean region. The corresponding fields for spring and, to a lesser extent, autumn, are similar.

Figures 15a and 15b show the composite difference fields of SLP based on precipitation ("heavy" minus "light") in the Mackenzie River drainage basin. Again, the pattern is seasonally consistent, with a prominent signature of the Arctic Oscillation in both summer and winter. In agreement with Bjornsson et al. [1], a belt of high pressure extends across the Bering Sea, Gulf of Alaska, and into British Columbia. The onshore

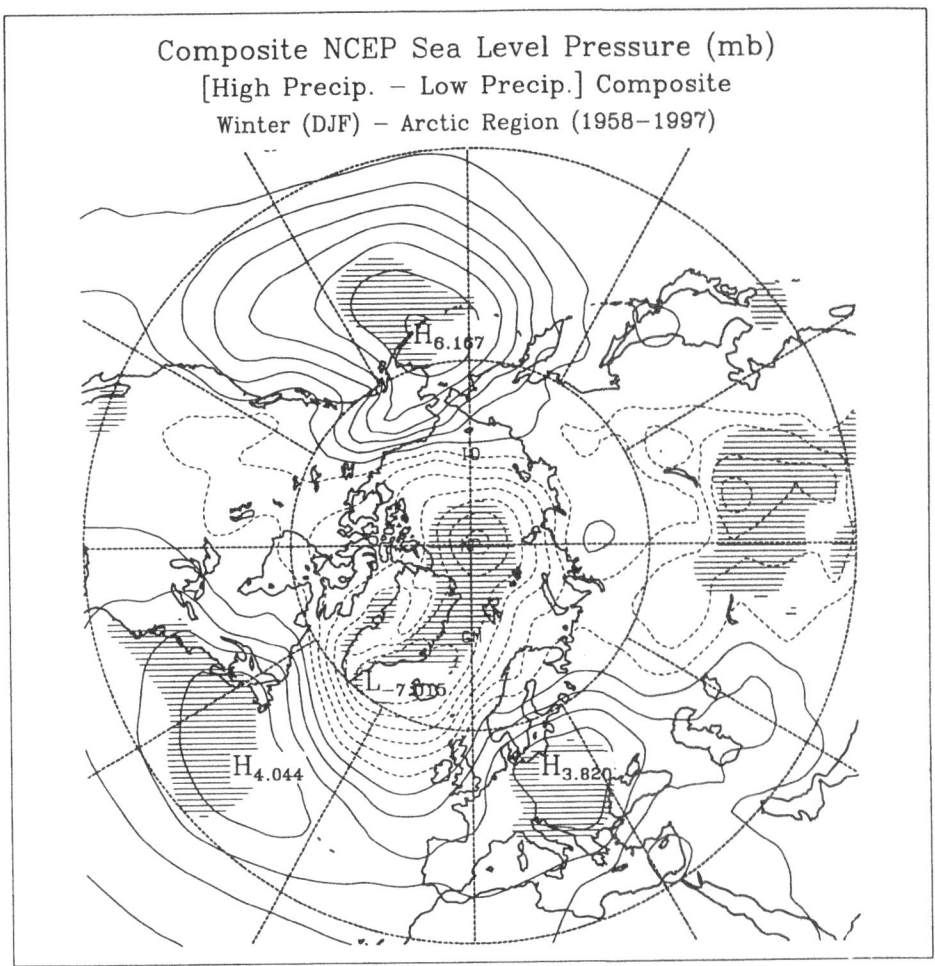

Figure 13. Difference field of sea level pressure obtained by subtracting NCEP reanalysis-derived means for winters (Dec.-Feb.) with light precipitation over Arctic Ocean from means for winters with heavy precipitation over Arctic Ocean. Contour interval is 1 mb. Shading indicates significance at 95% level.

moisture flow on the northern side of this feature is a meteorologically plausible contributor to enhanced precipitation in the Mackenzie domain. The Mackenzie-derived fields in Fig. 15 both contain a Pacific Ocean belt of negative differences centered near 35°N, 180°W. This feature is statistically significant in both seasons, as are the smaller differences in the equatorial Pacific, suggesting that possible linkages to a forcing by ENSO merit further investigation. The fact that the positive differences in the mid-latitude North Atlantic are also statistically significant in both seasons suggests that the forcing may be hemispheric in scale or that the Mackenzie precipitation is part of a broader hemispheric response of the atmosphere to more regional forcing.

38

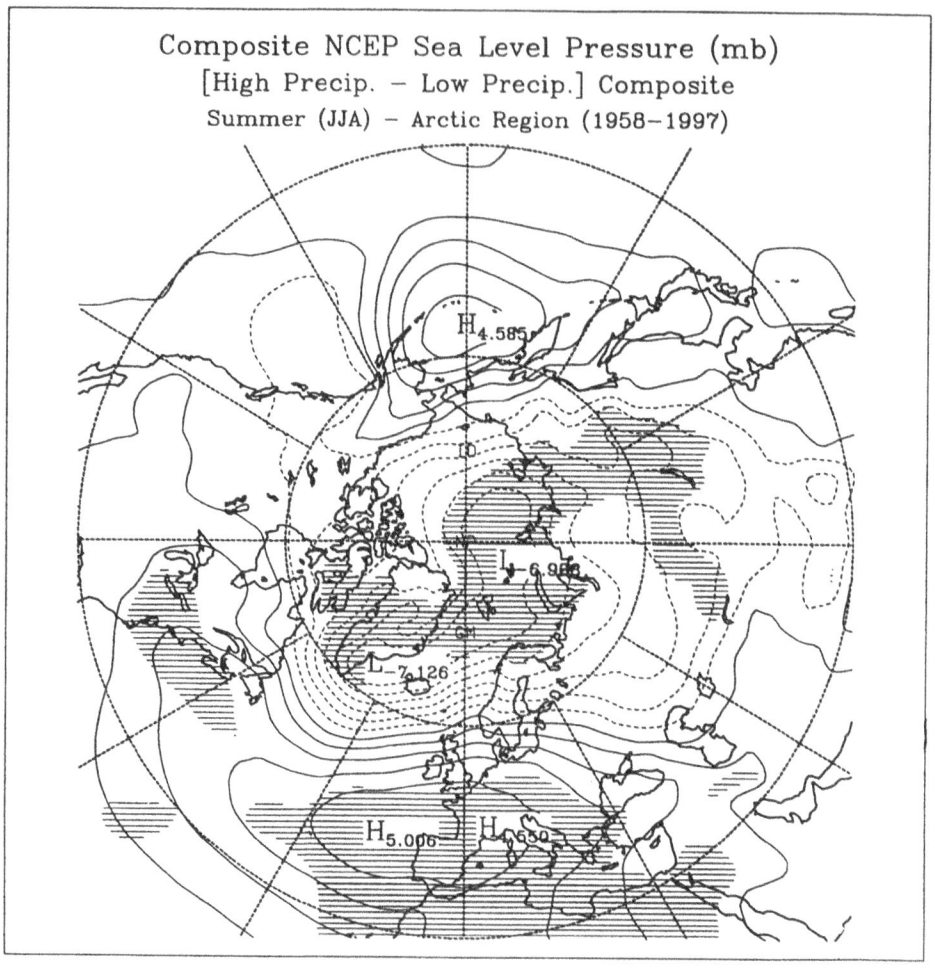

Figure 14. As in Fig. 13, but for summer (Jun.–Aug.). Contour interval is 1 mb. Shading indicates significance at 95% level.

5. Future Research Priorities

This review has focussed on the large-scale atmospheric circulation and its role in Arctic freshwater flux variability. The emphasis has been on the atmospheric fluxes, including horizontal atmospheric transport and the surface/atmosphere exchanges (precipitation, evaporation). We have not addressed in a substantive way the atmospheric forcing of horizontal fluxes of freshwater in the ocean or sea ice.

Our survey has shown that Arctic freshwater fluxes are associated with atmospheric moisture transport in several regions, particularly the North Atlantic. The circulation fields associated with Arctic freshwater flux anomalies, including P and E, are broad

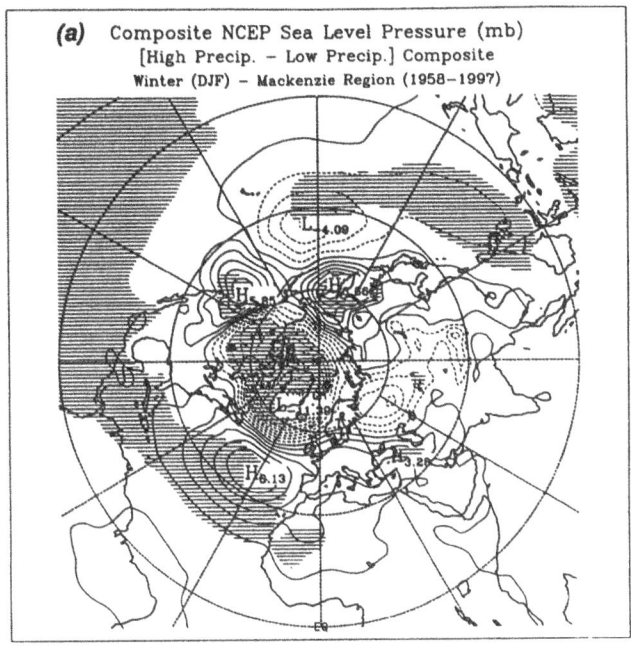

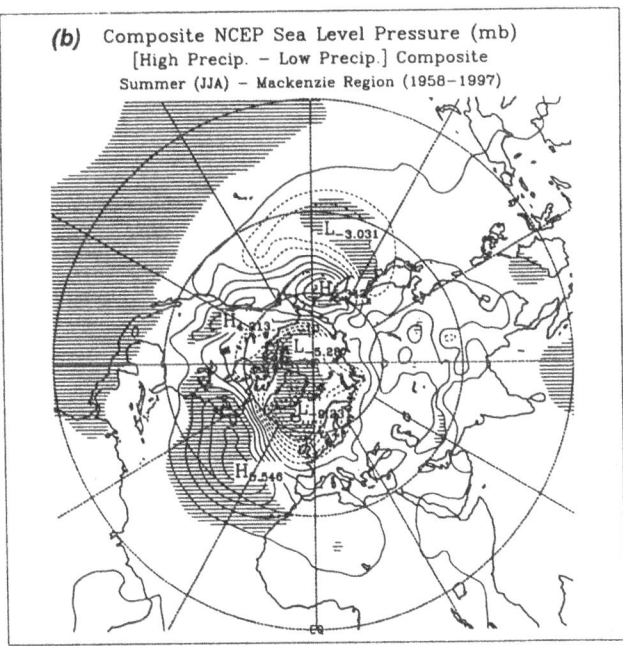

Figure 15. As in Fig. 13, but for the Mackenzie Basin during (a) Dec.-Feb. and (b) Jun.-Aug. Contour interval is 1 mb. Shading indicates significance at 95% level.

and spatially coherent. The atmospheric circulation over the North Atlantic, as part of the so-called "Arctic Oscillation," is a pervasive factor of importance to the Arctic freshwater flux variations. Despite recent advances in the identification of key modes of atmospheric variability in high- and middle-latitudes, several avenues of needed research emerge from the discussion in the preceding sections:

- *Identification of key physical mechanisms underlying the Arctic Oscillation.* While this recently recognized mode is a key contributor to Arctic freshwater flux variations, the origin of its fluctuations is not known. A role of the stratosphere has been proposed by Thompson and Wallace [25]; forcing by Atlantic sea surface temperature anomalies is a distinct possibility; and a link to geographically remote forcing cannot be discounted. This mode is sufficiently new to diagnostic investigation that even its seasonality is not well documented

- *Clarification of the role (if any) of El Niño/Southern Oscillation (ENSO) in the variability of high-latitude freshwater fluxes.* Of the results surveyed here, only the association between Mackenzie Basin precipitation and the large-scale circulation showed a possible ENSO linkage. It would be surprising if ENSO were not a stronger player in the variability of the Arctic hydrologic cycle. Indeed, Livezey at al. [18] have recently shown that sea surface temperature anomalies in the equatorial Pacific are associated with substantial teleconnections to large extratropical areas, including not only the Pacific subarctic and the central Arctic, but also (in some months) the high-latitude North Atlantic. These atmospheric signals vary by calendar month and are not oppositely phased for positive and negative equatorial SST anomalies. Such studies have yet to examine possible linkages to high-latitude precipitation and wind-forcing of the polar ocean.

- *Discrimination of systematic trends and natural variability of atmospheric forcing of the Arctic hydrologic cycle.* The results in the preceding sections point to positive trends in high-latitude precipitation over the past several decades. An accelerated hydrologic cycle is a greenhouse projection of global climate models. However, the primary atmospheric modes that force the moisture fluxes vary naturally over a wide variety of timescales. There needs to be a reconciliation of the apparent hydrologic trends and the circulation variations. In addition, the characteristics and predominance of the atmospheric circulation modes are likely to change under a greenhouse-warming scenario. These changes must be identified and understood in order to permit the anticipation of changes in Arctic hydrologic fluxes over the coming decades.

Several other research needs pertain to the infrastructure that is required in order to facilitate progress in the evaluation and understanding of variations in Arctic hydrologic fluxes:

- *The development of credible validation datasets for Arctic precipitation and evaporation.* The paucity of accurate station measurements of Arctic P and E, together with difficulties in obtaining useful satellite-derived estimates of Arctic P and E, make this undertaking a challenging one. Such undertakings are nevertheless necessary if we are to obtain a meaningful record of Arctic surface flux variations as required for closure of the Arctic hydrologic budget and for a diagnosis of causal mechanisms. Atmospheric reanalysis represents a promising vehicle, but there are significant problems in the P and E fields of the first iteration of the reanalyses, as shown elsewhere in this volume.

- *More effective meshing and utilization of objective strategies for typing the large-scale atmospheric circulation.* The rich history of Russian investigations of Arctic circulation variability has not been tied to more recent studies of large-scale circulation variability. It is unclear whether there are differences in perspective or whether there is sufficient common ground to achieve major advances in circulation diagnostics that are of particular relevance to the Arctic hydrologic cycle.

Finally, as noted earlier, the atmospheric circulation contributes to Arctic hydrologic variations in several ways, e.g., by driving ocean-ice transport variations and by providing variable direct inputs of moisture (P-E). The relative importance of these contributions, and possible synergies or feedbacks between them, have yet to be addressed. A holistic approach to the freshwater budget of the Arctic must include consideration of the interactions between the various hydrologic influences of the large-scale atmospheric circulation.

Acknowledgments

The preparation of this paper was supported by the National Science Foundation's Climate Dynamics Program through Grant ATM-9319952 and Division of Ocean Sciences through Grant OCE-9529763. Thanks are due Norene McGhiey for the word-processing of the manuscript and William Chapman for the preparation of several of the figures. The comments of an anonymous reviewer were especially helpful in improving the manuscript.

42

6. References

1. Bjornsson, H., Mysak, L.A., and Brown, R.D. (1995) On the interannual variability of precipitation and runoff in the Mackenzie drainage basin, *Climate Dynamics* **12**, 67-76.

2. Bromwich, D.H., Cullather, R.I., and Serreze, M.C. (1999) Reanalyses depiction of the Arctic atmospheric moisture budget, *This Volume*.

3. Deser, C., and Blackmon, M.L. (1993) Surface climate variations over the North Atlantic Ocean during winter: 1900-1989, *J. Climate* **6**, 1743-1753.

4. Dickson, R.R., Lazier, J., Meincke, J., Rhines, P., and Swift, J. (1996) Long-term coordinated changes in the convective activity of the North Atlantic, *Progress in Oceanography* **38**, 241–295.

5. Dickson, R.R., Meincke, J., Malmberg, S.-A., and Lee, A.J. (1988) The "Great Salinity Anomaly" in the northern North Atlantic, 1968-1982, *Progress in Oceanography* **20**, 103–151.

6. Dzerdzeevskii, B.L. (1963) Fluctuations of general circulation of the atmosphere and climate in the twentieth century, *Changes of Climate*, UNESCO, 91–141.

7. Gibson, J.K., Kallberg, P., Uppala, S., Hernandez, A., Nomura, A., and Serrano, E. (1997) ECMWF Re-Analysis, Project Report Series: 1. ERA Description. European Centre for Medium-Range Weather Forecasting, Reading, U.K., 72 pp.

8. Girs, A.A. (1956) Long period changes of the atmospheric circulation types and variations of solar activity, *Meteorologiya i Gidrologiya* **10**, 38–44.

9. Hakkinen, S. (1993) An Arctic source for the Great Salinity Anomaly: A simulation of the Arctic ice-ocean system for 1955-1975, *J. Geophysical Research* **98**, 16,397–16,410.

10. Hurrell, J.W. (1995) Decadal trends in the North Atlantic Oscillation: Regional temperatures and precipitation, *Science* **269**, 676–679.

11. Hurrell, J.W., and van Loon, H. (1997) Decadal variations in climate associated with the North Atlantic Oscillation, *Climatic Change* **36**, 301–326.

12. Intergovernmental Panel on Climate Change (1996) *Climate Change 1995: The Science of Climate Change* (J.T. Houghton, L.G. Meira Filho, B.A. Callander, N. Harris, A. Kattenberg, and K. Maskell, Eds.), Cambridge University Press, 572 pp.

13. Kalnay, E.M., Kanamitsu, M., Kistler, R., and 19 others (1996) The NCEP/NCAR 40-year reanalysis project, *Bull. American Meteorological Society* **77**, 437–471.

14. Karl, T.R. (1998) Regional trends and variations of temperature and precipitation. In *The Regional Impacts of Climate Change: An Assessment of Vulnerability* (R.T. Watson, M.C. Zinyowera, R.H. Moss, and D.J. Dokken, Eds.), Cambridge University Press, 412–425.

15. Kattsov, V.M., Walsh, J.E., Dethloff, K., and Rinke, A. (1998) Atmospheric circulation models: Simulation of the Arctic Ocean fresh water budget components, *This Volume*.

16. Kwok, R., and Rothrock, D.A. (1999) Variability of Fram Strait ice flux and North Atlantic Oscillation, *J. Geophysical Research* **104**, 5177-5189.

17. Lamb, H.H. (1972) *Climate: Past, Present and Future, Volume 1: Fundamentals and Climate Now*, Methuen & Co. Ltd., London, 613 pp.

18. Livezey, R.E., Masutani, M., Leetmaa, A., Rui, H., Ji, M., and Kumar, A. (1997) Teleconnective response of the Pacific-North American region atmosphere to large central equatorial Pacific SST anomalies, *J. Climate* **10**, 1787-1820.

19. Mysak, L.A., Manak, D.K., and Marsden, R.F. (1990) Sea ice anomalies observed in the Greenland and Labrador Seas during 1901–1984 and their relation to an interdecadal Arctic climate cycle, *Climate Dynamics* **5**, 111–133.

20. Niebauer, H.J. (1998) Variability in Bering Sea ice cover as affected by a regime shift in the period 1947–96, *J. Geophysical Research* **103**, 27,717-27,737.

21. Peng, S., and Fyfe, J. (1996) The coupled patterns between sea level pressure and sea surface temperature in the midlatitude North Atlantic, *J. Climate* **9**, 1824–1839.

22. Peng, S., and Mysak, L.A. (1993) A teleconnection study of interannual sea surface temperature fluctuations in the northern North Atlantic and precipitation and runoff over western Siberia, *J. Climate* **6**, 876–885.

23. Radionov, V.F., Bryazgin, N.N., and Alexandrov, E.I. (1997) The Snow Cover of the Arctic Basin, Tech. Report APL-UW TR 9701, Applied Physics Laboratory, University of Washington, Seattle, WA, 95 pp.

24. Serreze, M.C., Barry, R.G., and Walsh, J.E. (1995) Atmospheric water vapor characteristics at 70°N, *J. Climate* **8**, 719-731.

25. Thompson, D.W.J., and Wallace, J.M. (1998) Observed linkages between Eurasian surface air temperature, the North Atlantic Oscillation, Arctic sea level pressure and the stratospheric polar vortex, *Geophysical Research Letters* **25**, 1297-1300.

26. Tremblay, L.-B., Mysak, L.A., and Dyke, A.S. (1997) Evidence from driftwood records for century-to-millennial scale variations of the high latitude atmospheric circulation during the Holocene, *Geophysical Research Letters* **24**, 2027–2030.

27. Walsh, J.E., Chapman, W.L., and Shy, T.L. (1996) Recent decrease of sea level pressure in the central Arctic, *J. Climate* **9**, 480–486.

28. Walsh, J.E., Zhou, X., Portis, D., and Serreze, M.C. (1994) Atmospheric contribution to hydrologic variations in the Arctic, *Atmosphere-Ocean* **32**, 733–755.

ATMOSPHERIC COMPONENTS OF THE ARCTIC OCEAN FRESHWATER BALANCE AND THEIR INTERANNUAL VARIABILITY

R.G. BARRY AND M.C. SERREZE
Cryospheric and Polar Processes Division, CIRES, University of Colorado, Boulder, 80309-0449, USA.

1. Abstract

The atmospheric components of the hydrological cycle comprise the following: total precipitation (rainfall, snowfall and other hydrometeors deposited on the Earth's surface), evapo-sublimation, the horizontal advection of water vapour and of water in the liquid or solid phase. The information available on each of these variables is reviewed and assessed. For the Arctic Ocean, the conventional approach to moisture-balance assessment, using data on precipitation and evaporation, is severely constrained by the paucity of direct observations and their severe biases. There are even fewer direct evapo-sublimation measurements, and most rely on bulk aerodynamic formulations. The preferable alternative for the Arctic basin as a whole is to use indirect aerological computations of the net moisture balance. This approach can be supplemented by observations and model-derived data in order to obtain the spatial patterns of moisture balance variations. The data required to calculate atmospheric moisture flux divergence are acquired from atmospheric balloon soundings (rawinsonde ascents) of upper air winds and moisture content and from analyzed atmospheric fields. Calculations of the poleward flux of moisture from rawinsonde data indicate an annual flux convergence of 161 mm across 70°N with a 23-year range from 212 to 106 mm. The net (P-E) over the Arctic Ocean is about half of the calculated runoff into the Arctic and 20 percent of the total input. The potential variability in the atmospheric component, including its spatial characteristics, is assessed relative to that associated with the other components of the freshwater balance.

2. Introduction

The freshwater balance of the Arctic Ocean is determined by three major inputs and the ocean export of ice and water. The inputs are as follows: river discharge, net precipitation minus evaporation (P-E) and the transfer of low-salinity ocean water, primarily through the Bering Strait. These three contributions are in the approximate proportions, of 2:1:1, respectively. This chapter focuses on the atmospheric components net atmospheric moisture transport into the Arctic, total precipitation (liquid and solid precipitation deposited on the surface) and evaporation/sublimation over the Arctic

E.L. Lewis et al. (eds.), The Freshwater Budget of the Arctic Ocean, 45-56.
© 2000 *Kluwer Academic Publishers.*

Ocean itself — and their spatial and temporal variability. The information available on each of these components is reviewed and assessed. The nature of their variability is also compared with that of the other main components of the freshwater balance. First, the magnitudes of the primary components, as identified in current literature, are briefly examined and compared. The Russian Atlas of the Arctic [1] estimates the annual runoff volume into the Arctic seas as 3887 km³. The widely cited work of Aagard and Carmack [2], used by Barry et al. [3], estimates runoff as 3300 km³, corresponding to 347 mm/yr for an ocean area of 9.5 x 10⁶ km². The World Ocean Atlas [4] also estimates an annual freshwater layer of 340 mm, of which 98.5 percent is from surface drainage, including 9 percent from icebergs. More recently, Ivanov and Yankina [5] indicate a volumetric contribution of 3590 km³, excluding discharge into the Norwegian Sea. They state that the component from within the Arctic Circle is 1900 km³, comprising a 1430 km³ liquid contribution and 470 km³ from icebergs, while the large rivers, most of which have drainage basins extending well into middle latitudes, account for the remainder. The estimated land area from which drainage into the Arctic Ocean is derived exceeds 22 million km² [6]. Because permafrost is continuous beneath much of the land area surrounding the Arctic Ocean, and the depth of the active layer is generally 0.5 in or less, most runoff occurs from surface rivers and streams. There is a paucity of data in the North American Arctic where stream-gauging programs are only quite recent. For example such programs began on the coastal plain of Alaska in 1970 [7]. In some areas data are almost entirely absent, as in the Canadian Arctic Islands [8]. For Eurasia, there are long records for the major rivers draining into the Arctic Ocean. For the Arctic Ocean as defined here (comprising the Arctic Basin, Siberian Shelf seas, the Beaufort Sea, the Barents Sea and the White Sea) covering 9.5 million km² , lvanov's [9] data indicate a runoff of 3120 km³, corresponding to a 328 mm layer of freshwater (Table 1).

3. Precipitation

Precipitation data are provided by gauge records at hydrometeorological stations on the mainland and Arctic islands, and by the former North Pole drifting stations operated by the USSR continuously from 1954-1991, with at least one station each year [10, 11, 12]. There are further station measurements of daily snow depth [13] and 10-day snow-course transect data. A Russian program of airborne landings, named "Sever" (which means north), collected additional snow-depth data from multiyear and seasonal ice in the Arctic Basin [14]. These data were collected annually in April during 1959-88. The number of station precipitation records tabulated by Vose et al. [11] for the USSR comprise only 15 percent of the Russian national network [15] although a larger number of stations is represented in the climatologies for the Arctic Ocean published in map form [16, 17]. Burova [18] credits Bryazgin with the map in Gorshkov [4]. The monthly fields of gauge-corrected precipitation north of 70°N prepared by Bryazgin have been digitized onto 2.5° x 2.5° latitude-longitude grids and a derived annual map is presented by Walsh et al. [19]. Legates and Willmott [20] provide a global monthly climatology that incorporates gauge correction, but results for the central Arctic Ocean rely largely on extrapolation from coastal stations. Serreze and Hurst [21] have blended

TABLE 1: Mean monthly water balance components (mm) for the Arctic Ocean

VARIABLES	A	S	O	N	D	J	F	M	A	M	J	J	YR
Precip.(1)	35	33	29	24	20	21	17	17	15	16	21	30	279
Precip.**													252
(P-E)*	19	24	17	14	10	9	10	11	11	11	13	15	161
(P-E)**													119
Evap.***	16	10	11	10	10	12	7	7	5	6	8	15	118
Evap.**													136
Cumul. SWE^		27	52	66	73	80	92	103	106	110	104	24	110
Runoff**	7												328

(1) Serreze and Hurst [21] for the area north of 70°N.
(2) For the Arctic Ocean area (9.5 million km²) [21]
* Calculated from atmospheric moisture flux data north of 70°N (1973-95) [24]
** For the Arctic Ocean area (9.5 million km²) [21]
^ Cumulative SWE of snow cover on the Arctic Ocean ice
*** Calculated as the difference between observed precipitation [21] and P-E from aerological analyses north of 70°N [24]

several hundred additional station records for Russia north of 60°N provided by Groisman et al. [22], 34 stations for the Northwest Territories, Canada (provided by P. Louie, Atmospheric Environment Service, Canada) and North Pole drifting station measurements with the Legates and Willmott analysis. A Cressman interpolation procedure was used to provide improved mean monthly fields on a 100x100 km Lambert equal area projection [23]. The Groisman and Louie data sets incorporate corrections for undercatch caused by wind speed, gauge wetting and the effect of unrecorded trace-precipitation amounts. The Serreze and Hurst [21] annual field is shown in Figure 1. The Bryazgin map of mean annual precipitation (not shown) depicts maximum values over the Norwegian Sea and a minimum over the central Arctic-Beaufort Sea including Greenland. In contrast, the Hurst and Serreze analysis shows a maximum of 1400 mm off southeast Greenland and lowest amounts (<200 mm) over central Arctic Ocean and Canadian Arctic Archipelago that are slightly greater than in Bryazgin's results. Monthly precipitation amounts for the area north of 70°N (Serreze and Barry [24]and for the Arctic Ocean (based on the area tabulation of Ivanov [9] and totaling 9.5 million km^2) are shown in Table 1. The annual totals for these two areas are 279 mm and 252 mm, respectively. Monthly maps of Arctic precipitation are provided by the global reanalyses of NCEP/NCAR and ERA, but at the present these are biased, especially from the NCEP/NCAR model over land in summer [25, 21].

4. Evaporation

Evaporation in the Arctic has been estimated mainly by empirical methods, and there are only limited measurements [26]. Estimates of monthly evaporation are reported by Zubenok [27] based on a relationship between soil moisture and evaporation rate, combining energy and water balance concepts, based on 110 stations located in basins draining into the Arctic Ocean. Adjustments for altitude are incorporated based on vertical gradients of the climatic elements. In winter, evaporation from land area is essentially zero. In early summer, most of the available energy goes into snow melt. Evaporation typically peaks in July, which may account for 60 percent of the annual total. Mean annual evaporation from the more southerly land areas is typically 300-400 mm, with values decreasing poleward. The mean for land areas in the zone 70°N - 83°N, accounting for approximately 50 percent of the surface in this latitudinal belt, is about 90 mm ([27], Figure 2A).

Khrol [28] has calculated evaporation for the Arctic Ocean by a combination of methods using the bulk aerodynamic approach for the ice-covered areas. In general the necessary data on surface and air temperature, moisture content and wind speed were not available, so long-term means were used. For the majority of months, an energy balance formulation is used, based on the data of Marshunova and Chernigovskiy [29]. Allowance is also made for melt ponds and openings using ice-concentration data.

Figure 1. Annual mean precipitation from Serreze and Hurst [21].

Annual amounts in the central Arctic and Beaufort-Chukchi seas are 70 mm or less, but increase rapidly in the Barents Sea and Norwegian Sea to 600 mm and higher (Figure 2B). For ocean surfaces in the zone 70°N-80°N, a zonal mean annual value of 200 mm is determined; the corresponding figure for the entire Arctic Ocean is 227 mm. vaporation / sublimation from the Arctic Ocean has a double maximum, according to Khrol, peaking in May and September and having minima in January to March and July.

5. Precipitation Minus Evaporation

Precipitation minus evaporation (P-E) represents the net freshening or salination of the ocean through surface exchanges with the atmosphere, and the net gain or loss of water by the land surface through exchanges with the atmosphere. Considering the problems

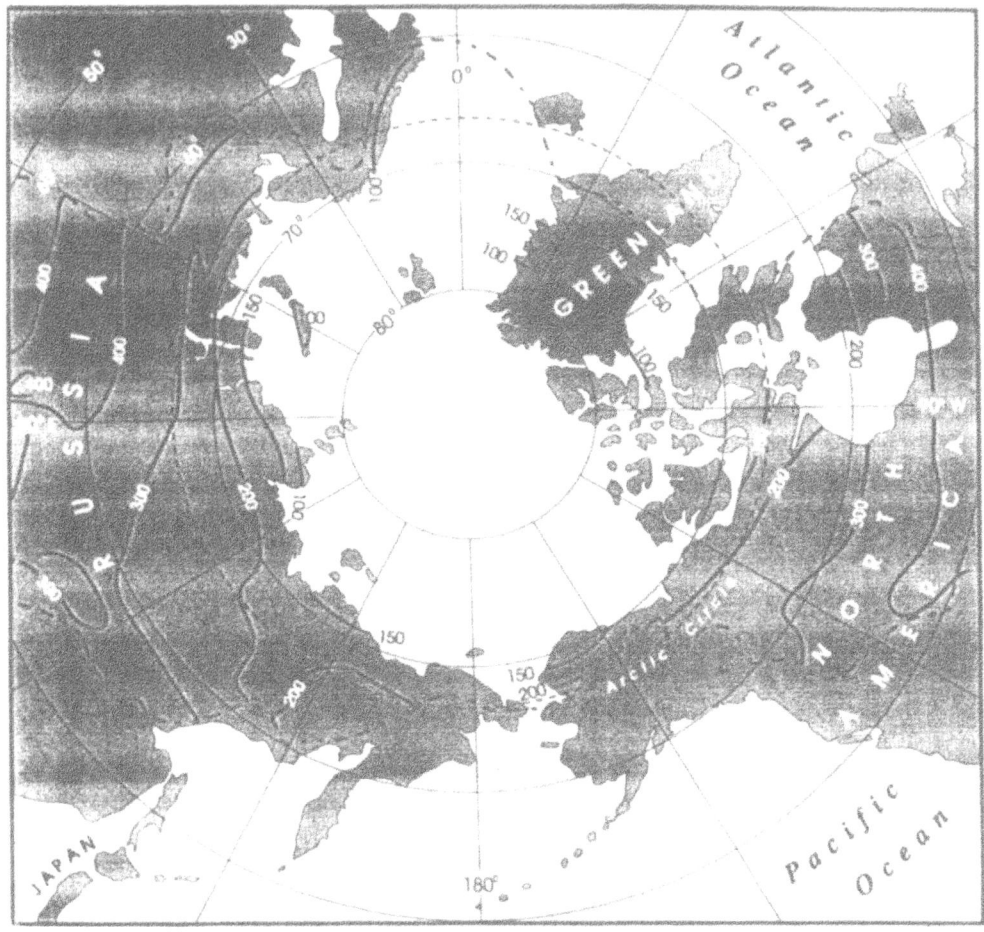

Figure 2 (a). Annual evaporation (mm) estimates for the Arctic land areas (after Zubenok [27]. The dash-dot lines in delineate land areas draining into the Arctic Ocean.

in accurately estimating P and E over the Arctic, P-E has the advantage in that its areal average can be estimated by the "aerological approach." Briefly, vertically-integrated water vapor fluxes from station rawinsonde data can be used to compute the horizontal flux convergence as (e.g., [30]):

$$- \nabla \bullet Q = 1/A \; \nu \; \mathbf{Q} \bullet \mathbf{n} \; dC \qquad (1)$$

where Q is the flux, dC is the segment of a perimeter defined by rawinsonde stations (or from selected locations obtained from interpolation of the station fluxes), **n** is a unit vector normal to the boundary and A is the area of the domain. Over annual time scales, the flux convergence can be considered equivalent to P-E. Over shorter (e.g., monthly) time scales, the flux convergence must be adjusted by the time-change in preciptable water (W) in the domain (also calculated from rawinsonde data) such that:

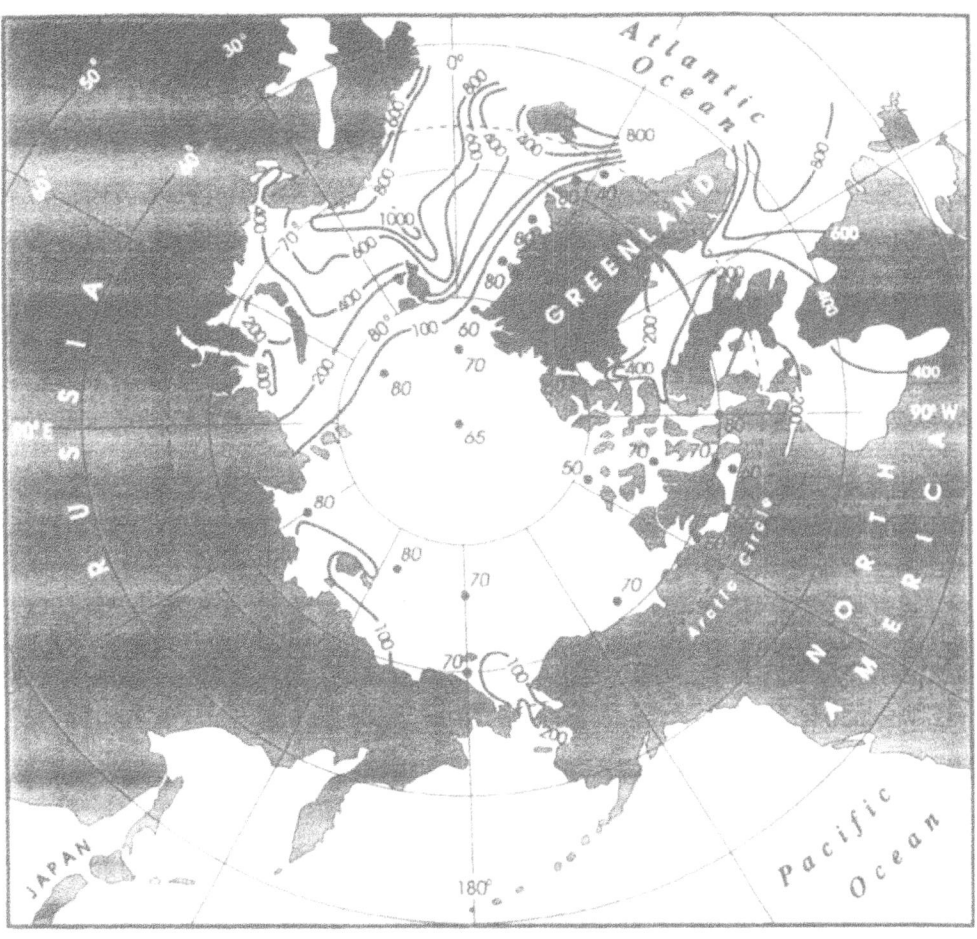

Figure 2 (b). Annual evaporation (mm) estimates for the Arctic Ocean (after Khrol, [28]).

$$P\text{-}E = -\nabla\bullet\mathbf{Q} - \partial W/\partial t \qquad (2)$$

Similar approaches can be applied to analyzed atmosphere fields to provide gridded fields of P-E (Bromwich et al., this volume). The aerological approach ignores the atmospheric flux of water in the liquid and solid phases because the liquid/ice flux is generally significant only in localized regions or for short periods, for example, over warm ocean currents and in cumulus clouds in the tropics. The calculated P-E from the atmospheric moisture flux convergence based on rawinsonde data amounts to 161 mm for the 15.4 million km^2 area north of latitude 70°N. Annual evaporation determined as a residual from the atmospheric moisture budget and the measured precipitation is 118 mm, compared with 136 mm by Ivanov [9]. The aerological P-E estimate for 1973-95 of 161 mm is almost identical to that obtained by Serreze et al. [31] and exceeds by a third the corresponding values of 120 mm obtained by Peixoto and Oort [32] for 1963-73, and

119 mm from Ivanov's data. Bromwich et al. [33] obtain a higher P-E of 184 mm using analyzed fields. Hence, the runoff contribution to the ocean is about three times that from net precipitation.

6. Snow Depth

Three sets of snow depth records available from the Russian North Pole stations have been examined by Colony et al. [34] and Warren et al. [12]. The snow survey lines (500 or 1000 m in length), surveyed at 10-30 day intervals, provide the most comprehensive sampling of snow depth by including drifts, sastrugi and areas of ridged and level ice. The fixed stake measurements are subject to bias due to the influence of the buildings and station activities. However, the incomplete spatial and temporal coverage of the snow survey data necessitates some judicious incorporation of other information. The data coverage also allows only mean geographical and seasonal patterns to be determined. Information on interannual variability can be derived only for the Arctic Ocean mean snow depth. The mean monthly Arctic Ocean snow water equivalent (SWE) values are shown in Table I. The annual maximum is 110 mm water equivalent in May. Combining this with the estimated evaporation, gives 246 mm or considerably more than the aerological P-E.

7. Ocean Exchanges

The oceanic contributions to the fresh water balance comprise the Bering Strait and other inflows of low salinity water and the Fram Strait and Canadian Arctic Archipelago outflows of ice and saline water [35]. The import of low salinity water through Bering Strait and via the Norwegian coastal current accounts for about 180 and 30 mm/year, respectively [2]. The mean northward volume transport through the 85-km wide Bering Strait during 1990-94 was 0.83 Sv, similar to the earlier estimate of Coachman and Aagaard [36], with a freshwater contribution of about 0.11 Sv (Melling, this volume). The ice-ocean-model-derived value obtained by Steele et al. [37] is 290 mm of freshwater, which is considerably larger than Aagaard and Coachman's estimate of 180 mm. The primary export pathway for ice and water is via the Fram Strait. A recalculation of an earlier estimate by Vinje et al. [38], based on satellite and buoy ice velocities and upward-looking sonar ice drafts, indicates an ice contribution of 2850 km^3yr^{-1} across 79°N; thus about 12 percent of the Arctic annual pack ice is exported annually. The mean cross-strait ice thickness, excluding open water, is about 3.26 m. Harder et al. [39], using a sea-ice model forced by daily observed winds, calculate a simulated ice flux of 0.86 Sv (2700 km^3). They suggest that the annual flux varies between 0.06 and 0.11 Sv, with a mean seasonal cycle ranging from almost zero in August to 0.14 Sv in November-December and March. The flux is particularly sensitive to wind speed and ocean current strength, with linear relationships in each case. It should also be noted that estimates of ice flux need to be specified for a particular latitude in view of the progressive southward ice melt.

The outflow of low salinity water in the East Greenland Current is not well determined. Osterhus and Vinje [40] estimate about 1000 km^3 of freshwater annually, accounting for a further 105 mm. The outflow of water and ice through the channels of the Canadian Arctic Archipelago is poorly known [41]. A model estimate by Steele et al. [37] suggests an annual net freshwater outflow through the Archipelago of 280 mm, double the earlier estimate of Aagard and Carmack [2]. The calculations are complicated by the recirculating flows in most of the large channels and sounds. Melling [42] suggests a freshwater flux as ice equivalent to 30 percent of the Fram Strait value, i.e., 90 mm.

8. Annual Mean Freshwater Balance; Uncertanties and Spatio-Temporal Variability

Based on the values presented above, a summary of the components of the Arctic Ocean freshwater balance is given in Table 2. The inflow and outflow components are in close agreement for the annual long-term mean state.

TABLE 2. Summary of the annual mean freshwater balance (mm) and error estimates (in parentheses).

	Inflow		Outflow
P-E	161 (27)	Fram Strait Ice	300 (30)
Bering Strait	290 (60)	East Greenland Current	120 (10)
Spitzbergen+Barents	-70 (10)	Canadian Arctic Arch. - ice	90 (130)
Norwegian coastal current	30 (5)	Canadian Arctic Arch. - Water	280
River runoff	350 (60)		
Total	771		790
RMSE	(90)		(134)

The mean values presented for each of the freshwater budget components are obviously subject to a variety of uncertainties associated with the accuracy of the original measurements, the network coverage and representativeness, and temporal variability as a result of climate trends. Table 3 represents a rather subjective attempt to indicate the sources of uncertainty, the percent range about the mean that may be involved and, in actual units, the interannual variability (IAV). Using the percent range in association with the estimated mean annual values, a root-mean square-error of +/- 90 mm (12 percent) is determined for the total inflow and +/- 134 mm (17 percent) for the total outflow values. These error estimates are included in Table 2. Taking these error estimates into account, the inflow and outflow components are effectively balanced.

TABLE 3. Observational uncertainties for hydrological components over the arctic ocean

Variable	Source	Range in Estimated Mean Value	IAV
P	Undercatch	-20 to -100%	
	Interpolation	+/- 10%	+/- 50 mm
Aerol. (P-E)	Longitudinal station	+/- 15%	
	Coverage of troposph. rivers		+/- 80 mm
E	Few data; residual term	+/- 15%	+/- 20%
Ru	Gauge density (1 station/25,000-70,000 km²)	+/- 10-25%	+/- 15%
	Ungauged areas	+/- 10%	

Freshwater Export:

Fram Str. Ice	Velocity, thickness, concentration, melt	+/- 10%	+/- 1000 to 1500 km³
East Greenland Current		+/- 10%	+/- 150 km³
Bering Strait inflow	Velocity, salinity	+/- 20%	+/- 50%
Can. Arctic Arch.		+/- 50%	unknown

9. Concluding Remarks

This survey of our knowledge of the atmospheric components of the Arctic Ocean freshwater balance in the context of the total net inflow and outflow shows that the mean annual conditions are reasonably well bounded. Uncertainties remain particularly in the runoff and ocean-transport components, but it seems unlikely that substantial errors are fortuitously cancelling one another out. The magnitude of the interannual variability is known approximately, but balance calculations for individual years are not yet possible. The seasonal and spatial distribution of the atmospheric contributions is also reasonably well defined for mean conditions, although this is not the case for runoff and ocean fluxes. Seasonal analyses must also take into account the freezing of ocean water and summer melt of the snow covered sea ice.

9.1 FUTURE RESEARCH NEEDS

Future prospects for improvements in these assessments will depend largely on the use of data assimilation techniques in numerical model studies. Such work is well underway for sea-ice concentration [43], but additional efforts are needed to treat the high spatio-temporal variability of precipitation and atmospheric moisture flux. Unfortunately the termination of the Russian North Pole Station program, and the closure of many high-latitude stations, or their conversion to automatic weather stations, means that observing networks for precipitation, snow-water equivalent and evaporation determinations will be considerably less complete than for the 1951-90 period.

References

1. Treshnikov, A.F., E.S. Korotkevich, Yu. A. Kruchinin and V.F. Markov. eds. (1985) *Atlas Arktiki*, Glaviioi Upravenie Geodezii i Kartografii, USSR, Moscow. Plate 97.
2. Aagaard, K. and Carmack, E.C. (1989) The role of sea ice and other freshwater in the Arctic circulation. J. *Geophys. Res.*, 94(C10) 14,485-14,498.
3. Barry, R.G., Serreze, M.C.and Walsh, J.E. (1996) Atmospheric components of the hydrological cycle in the Arctic. In P. Lemke et al., eds., Proceedings of the ACSYS Conference on the Dynamics of the Arctic Climate System, WCRP-94, WMO/TD No. 760, WMO, Geneva, pp. 24-31.
4. Gorshkov,S.G.,ed. (1983) *World Ocean Atlas. Volume 3: Arctic Ocean.* (in Russian). Pergamon Press, Oxford, 189 P.
5. Ivanov, V.V. and Yankina, V.A. (1991) Vodnye resursy Arktiki, ikh izuchennost'i ocherednye zadachi issledovanii. *Probl. Arkt. Antarkt.*, 66:118-12§.
6. Vuglinsky, V. (1998) River water inflow to the Arctic Ocean - conditions of formation, time variabilty and forecasts. *Proceedings of the ACSYS on Conference Polar Processes and Global Climate*, WCRP-106, WMO/TD. No. 98, WMO, Geneva, pp. 277-78.
7. Dingman, S.L., Barry, R.G., Weller, G., Benson, C.S., LeDrew, E.F. and Goodwin, C.W, (1980) Climate, snow cover, microclimate and hydrology. In Brown, J., Miller. P.C., Tieszen, F.L., and Bunnell, F.L.eds., *An Arctic Ecosystem., The Coastal Tundra at Barrow, Alaska.* Dowden, Hutchinson and Ross, Stroudsburg, PA., pp. 30-65.
8. Wedel, J.H. (1990) Regional hydrology, in T.D. Prowse and C.S.L. Ommanney, eds., *Northern Hydrology: Canadian Perspectives*, NHRI Science Report No. 1, Minister of Supply and Services, Canada, pp. 207-26.
9. Ivanov, V.V. (1998) The contribution of the run-off to the freshwater balance of the Arctic Ocean and its marginal seas. This volume.
10. NSIDC (1997*) Arctic Ocean Snow and Meteorological Observations from Drifting Stations: 1937,1950-1991.* CD-ROM. National Snow and Ice Data Center, University of Colorado at Boulder.
11. Vose, R.S., R.L. Schomyer, P.M. Steurer, T.C. Peterson, R. Heim, T.R. Karl and J.K. Eisheid. (1992) The Global Historical Climatology Network: Long-Term Monthly Temperature, Precipitation, Sea Level Pressure, and Station Pressure Data. Oak Ridge National Laboratory, Environmental Sciences Division Publication No. 3912, 99.pp. plus appendices.
12. Warren, S.G., Rigor, I.G., Untersteiner, N., Radionov, V.F., Bryazgin, N.N., Aleksandrov, Y.I., and Colony, R. (1999) Snow depth on Arctic sea ice, *J. Clim.*, in press.
13. NSIDC (1996) *Historical Soviet Daily Snow Depth (HSDSD).* CD-ROM. National Snow and Ice Data Center University of Colorado at Boulder.
14. Romanov, I.P. (1996) *Atlas of Ice and Snow of the Arctic Basin and Siberian Shelf Sea.* 2nd edn., Backbone Publishing Fairlawn, NJ, 250 charts.
15. Razuvaev, V. (1997) The Arctic runoff data base (ARDB). In R.G. Barry et al., eds., Proceedings of the Workshop on the Implementation of the Arctic Precipitation Data Archive (APDA) at the Global Precipitation Climatology Centre (GPCC). WCRP-98, WMO/TD No.804, WMO, Geneva, pp.41-46.
16. Bryazgin, N.N. (1976) Srednegodovoe kolichestvo osadkov v Arktike s uchetom pogreshnosti osadkomerov (Mean annual precipitation in the Arctic with consideration of gauge biases.) *Trudy Arkt. Antarkt. Nauch.- Isssled. Inst.*, 323:40-74.
17. Korzun,V.1. ed. (1976) *Atlas of World Water Balance.* Gidrometeiozdat, Leningrad. 122p.
18. Burova, L.P. (1983) *Vlagoborot v atmosfere Arktike.* (Moisture in the Arctic atmosphere) Gidrometeiozdat, Leningrad. 128pp.
19. Walsh, J.E., Kattsov, V., Portis, D. and Meleshko, V. (1998) Arctic precipitation and evaporation: model results and observational estimates. *J. Clim.*, 11: 72-87.
20. Legates, D. and Willmott, C. (1990) Mean seasonal and spatial variability in gauge-corrected global precipitation. *Int. J. Climatol.*, 10: 110-127.
21. Serreze, M.C. and Hurst, C.M. (1999) Representation of mean Arctic precipitation from NCEP/NCAR and ERA reanalysis. *J. Clim.*, in press.

56

22. Groisman, P. Y., Koknaeva, V.V., Belokrylova, T.A. and Karl, T.R., (1991) Overcoming biases of precipitation: A history of the USSR experience. *Bull. Amer. Met. Soc.*, 72: 1725-33.
23. Armstrong, R. and Brodzik, M-J. (1995) An earth-gridded SSM/I data set for cryospheric studies and global change monitoring, *Advances in Space Research*, 16 (10): 155-161.
24. Serreze, M.C. and Barry, R.G. (1988) Atmospheric components of the Arctic Ocean hydrological budget assessed from rawinsonde data (this volume).
25. Serreze, M.C. and Maslanik, J.A. (1997) Arctic precipitation as represented in the NCEP/NCAR reanalysis. *Ann. Glaciol.*, 25: 429433.
26. Barry, R.G., Courtin, G.M. and Labine, C. (1981) Tundra climates, in L.D. Bliss, O.W. Heal and J.J. Moore, eds., *Tundra Ecosystems: Comparative Analysis*, Cambridge University Press, pp. 81-114.
27. Zubenok, Z. T. (1976) Isparenie s sushi vodosbornogo basseina Severnogo Ledovitogo Okeana. (Evaporation from the basins draining into the Arctic Ocean.) *Trudy Arkt. Antarkt Nauchno.-Issled. Inst.* 323: 87-100.
28. Khrol, V.P. (1976) Isparenie s poverkhnosti Severnogo Ledovitogo Okeana. (Evaporation from the Arctic Ocean surface), *Trudy Arkt. Antarkt. Nauch-Issled. Inst.*, 323: 148-155.
29. Marshunova, M.S. and Chernigovskiy, N.T. (1978) *The Radiation Regime of the Foreign Arctic.* NSF-TT-72-51034 (NTIS, PB 2864692), 189 pp.
30. Alestalo, M. (1983) The atmospheric water vapor budget over Europe, In A. Street-Perrott, M. Beran and R. Ratcliff. Eds., *Variations of the Global Water Budget*, D. Reidel, Dordrecht, Netherlands, pp. 67-79.
31. Serreze, M.C., Barry, R.G., Rehder, M.C. and Walsh, J.E. (1995) Variability in atmospheric circulation and moisture flux over the Arctic. *Phil. Trans.R. Soc.* Lond., A 352: 215-25.
32. Peixoto, J.P. and Oort, A.H. (1983) The atmospheric branch of the hydrological cycle and climate, in A. Street-Perrott, M. Beran and R. Rackcliff, eds., *Variations of the Global Water Budget*, D. Reidel, Dordrecht, Netherlands, pp. 5-65.
33. Bromwich, D.H., Cullather, R.I., and Serreze, M.C. (this volume).
34. Colony, R., Radionov, V. and Tanis, F.J. (1998) Measurements of precipitation and snow pack at the Russian North Pole drifting stations, *Polar Rec.*, 34: 3-14.
35. Rudels, B. (1987) On the mass balance of the Polar Ocean with special emphasis on the Frain Strait. *Norsk Polar Inst. Skrifter* 188: 1-53.
36. Coachman, L.K. and Aagaard, K. (1988)Transports through the Bering Strait: annual and interannual variability. *J. Geophys. Res.,93:* 15.535-539.
37. Steele, M., Thomas, D., Rothrock, D., and Martin, S. (1996) A simple model study of the Arctic Ocean freshwater balance,1979-1985. *J. Geophys. Res.*, 101 (C9): 20,833-848.
38. Vinje, T., Nordlund, N. and Kvarnbekk, A. (1998) Monitoring ice thickness in Frain Strait. *J. Geophys. Res,,103(C5):* 10,437-449.
39. Harder, M., Lemke, P. and Efilmer, L.(1998) Simulation of sea ice transport through Fram Strait: Natural variability and sensitivity to forcing. *J.Geophvs. Res.*, 103(C3): 5595-5605.
40. Osterhus, S. and Vinje, T. (this volume).
41. Rudels, B. (1986) The outflow of polar water through the Arctic Archipelago and oceanographic conditions in Baffin Bay. *Polar Res.*, 4: 161-180.
42. Melling, H. (this volume).
43. Thomas, D., Martin, S., Rothrock, D. and Steele, M. (1995) Assimilating satellite concentration data into an Arctic sea ice mass balance model, 1979-85. *J. Geophys. Res.*, 101 (C9): 20,849-868.

HYDROCLIMATOLOGY OF THE ARCTIC DRAINAGE BASIN

L.C. Bowling, D.P. Lettenmaier, and B.V. Matheussen
University of Washington
Department of Civil Engineering Box 352700, Seattle, WA 98195

1. Introduction

By most estimates, runoff from the land surface represents the single largest input of freshwater to the Arctic Ocean. Alekseev and Buzuev [1] estimated the total freshwater input to the Arctic Ocean to be between 4300 and 6475 km^3/year based on ice formation data. Treshnikov [2] (reported in [3]) estimated the total freshwater flux from gauged land surface areas to be 3300 km^3/yr, or 51 to 77 percent of the total from [1]. Grabs et al. [4] report that the accumulated sum of mean annual flow from the 35 largest gauged Arctic basins is approximately 202 mm/yr (2603 km^3). When extrapolated to the entire land area draining to the Arctic (approximately 19,300,000 km^2), this yields a total annual freshwater flux from the land surface of approximately 4000 km^3, or between 62 and 93% of the total freshwater input as estimated from [1]. Another estimate of total freshwater flux from land areas into the Arctic Ocean is 4269 km^3/year, with 17 and 24% attributed to ungauged discharge from islands and mainland, respectively [5].

Despite the importance of Arctic land-surface freshwater fluxes to global ocean circulation and the global heat balance, the hydrology of the Arctic basin, taken as a whole, has been largely neglected by hydrologists. The dynamics of land-surface hydrology in the Arctic basin are not well understood due to their dependence on unique features such as the control of extreme seasonal runoff by snowmelt and ice break-up, large-scale redistribution of snow and the effects of ephemeral and permanently frozen soils. An important complication is the paucity of observed data and the difficulties associated with data collection. Although extrapolation of observed streamflow to ungauged areas allows bulk estimates of land surface freshwater fluxes such as those cited above, the temporal and spatial variability of runoff response from approximately 30% of the Arctic drainage area is unknown. Hydrologic models offer the most feasible option for estimating runoff in these ungauged areas. In recognition of this fact, the hydrological programme of the Arctic Climate System Study (ACSYS) has as one of its specific objectives "the development of mathematical models of the hydrological cycle under specific Arctic climate conditions..." [6].

Land surface models that can represent unique northern hydrologic features are in their infancy. Coupled land-atmosphere-ocean models are an important tool for studies of the interaction of land, atmosphere and ocean processes as they affect climate. However, land surface representations in most such models treat cold season and cold region processes quite crudely. For instance, most general circulation models (GCMs) do not simulate frozen soils, nor do they represent the sub-grid variability of snow.

E.L. Lewis et al. (eds.), The Freshwater Budget of the Arctic Ocean, 57-90.
© 2000 *Kluwer Academic Publishers.*

Nevertheless, some early efforts have been made to incorporate cold regions processes into land-surface models. Anisimov and Nelson [7] simulated permafrost extent in conjunction with three global GCMs, using a dimensionless "frost index". Foster et al. [8] compared snow output from seven GCMs with passive-microwave observations of snow depth. The U.S. Air Force snow depth climatology and National Oceanic and Atmospheric Administration (NOAA) satellite observations of snow extent were used as the standard of reference. Although seasonal and interannual snow distributions were simulated fairly well, transition season simulations (spring and fall) were susceptible to large errors.

The relatively crude representation of cold-regions land surface processes in GCMs can result in over-estimation of peak runoff, as well as errors in timing. For instance, Arctic river runoff predictions are too early by up to two months in both the Canadian Climate Centre and the National Aeronautic and Space Administration (NASA)/Goddard GCMs [9, 10]. Kite et al. [9] hypothesized that one factor in these discrepancies may be the lack of model representation of storage of snowmelt in surface depressions, which allow greater evapotranspiration [9]. They found that partial coupling of GCM runoff output with a simple land surface hydrological model corrected volume and phase deficiencies of the GCM simulation.

Although some conventional hydrology models may have more detailed cold-regions process representations, they are generally applicable to small regions and suffer from the paucity of observational data over high latitude land areas with which to force these models. Typical data requirements for hydrology models include the downward components of the water and energy budgets at the land surface, namely shortwave and longwave radiation, air temperature, humidity, wind speed and precipitation. Land surface characteristics, including soil, topography and vegetation are also needed. The single most important variable for streamflow prediction is precipitation. The scarcity of precipitation data and the inconsistency of gauge observations due to variation in gauge catch efficiencies in northern regions has led many researchers to perform climatological studies using precipitation predicted by atmospheric models [11, 12, 13] and convergence fields calculated from radiosonde measurements [14, 15]. These studies have generally used so-called analysis or reanalysis fields. The latter are the results of retrospective runs of weather forecast models using a fixed version of the respective forecast model and archived observations for model initialization, while the former are a by-product of real-time predictions. Three major modeling centers have completed multi-decade reanalysis runs; the European Center for Medium Range Weather Forecasting (ECMWF), the (U.S.) National Centers for Environmental Prediction (NCEP), and the NASA Data Assimilation Office (DAO).

The analysis fields arguably represent the best estimate of conditions in the atmosphere and at the land surface resulting from a combination of model skill and observations. Several recent studies have concluded that the ECMWF reanalysis fields of precipitation less evaporation (P-E) at high latitude have a more realistic spatial variability than those of NCEP [11, 12, 14]. Spatial correlations between the NCEP reanalysis totals and the Legates and Willmott [16] gridded precipitation climatology are between 0.8 - 0.9 in winter, but much lower during summer [12]. Oki et al. [17] found that the global distribution of vapor flux convergence calculated from ECWMF reanalysis corresponded well with Global Runoff Data Center (GRDC) runoff in 70 major world rivers, particularly in the mid- to high-latitudes of the Northern Hemisphere. There was some bias towards underprediction in rivers of all regions [17].

Temporal variations of both the ECMWF and NCEP reanalyses were similar to gauge observations over the Mackenzie [11], although the precipitation maximum over the central Arctic occurred one month too early in the NCEP reanalysis [12, 13]. Overall, the NCEP reanalysis tended to underpredict precipitation over the Atlantic Ocean and overpredict precipitation over land and the central Arctic Ocean, particularly during summer [11, 13, 15, 18].

Reanalysis data are ideal for climate studies because they provide a temporally and spatially complete data source. However, the major motivation for reanalysis efforts is the recognition that changes in the GCM or analysis component of the weather forecast models over time have strongly impacted the resulting climatology [19]. The implication is that the mean climates of the reanalysis models are inextricably linked to not only the accuracy of the physical parameterizations, but also to the data assimilation system. In areas of sparse data such as the Arctic, the assimilation system must rely heavily on the model-simulated fields [19], therefore the process of updating the model state to match observations is not constrained to a water or energy balance and can result in physically unrealistic fields.

The opportunities and limitations presented by observational and model-derived data sets for understanding the freshwater balance of the Arctic basin are reviewed in this paper, from a land surface hydrological perspective. The paper is organized into three sections. The first describes the hydroclimatic zones within the Arctic drainage basin and the cold-regions hydrologic processes that dominate the major Arctic river systems. In the second section, the space-time variability of these processes is examined through large-scale surface water and energy budget analyses. Finally, in the third section, the ability of a state-of-art macroscale hydrologic model to reproduce the observed regime is explored for the Mackenzie and Ob River basins.

2. Background

Runoff from high-latitude rivers represents a significant source of freshwater to the Arctic Ocean. The temporal and spatial variability of fresh water fluxes to the Arctic Ocean are controlled not only by space-time variability in precipitation and evapotranspiration, but by the storage of significant portions of annual precipitation as snow, and in lakes and wetlands. The influence of seasonally frozen ground also affects the interaction of surface and subsurface runoff storage with streamflow production. Much of the available data which would allow examination of processes such as snow accumulation and ablation is from small-scale field observations. However, by combining observations from various field studies, a picture of trends at the large-scale begins to emerge.

2.1. REGIONAL CHARACTERISTICS

The Arctic drainage basin, defined here as the land area draining to the Arctic Ocean, spans 37° of latitude, from 46° N in the Yenisei Basin to 83° N at the tip of Greenland (Figure 1), and several distinct vegetation and climate zones (Figure 2). For simplicity in this paper, we will adopt the definition of Woo and Winter [20] in which arctic refers to those areas underlain by continuous permafrost; sub-arctic is defined by the limit of discontinuous permafrost and the north temperate zone includes the zone of seasonally frozen ground (see Figure 3). The entire Arctic drainage basin lies within one of these three zones.

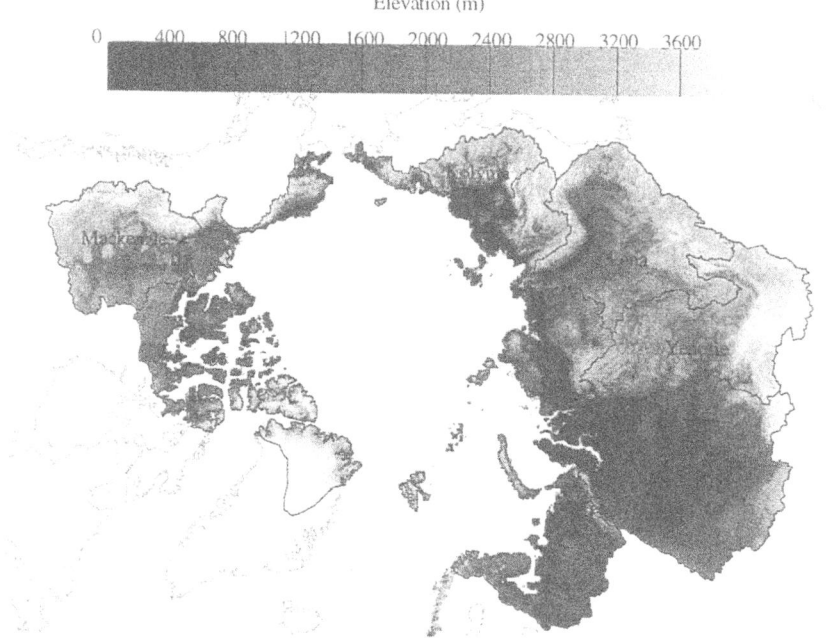

Figure 1. Digital elevation model of the Arctic drainage basin.

All three regions contain multiple landforms and vegetation [21], as shown in Figures 2 and 3. The largest vegetation class in both the Eurasian and North American portions of the Arctic drainage basin is boreal forest, composed of mixed conifer and evergreen forest, and needleleaf taiga. In the south-central Ob and Mackenzie basins, mixed deciduous and coniferous forest transitions into steppe grassland and cropland at the southern extremes. Extensive swamp and marshland within the low altitude taiga forest dominate the central lowlands of the Ob River basin. There is a lesser, but still significant, distribution of wetlands throughout the Lena and Kolyma basins in Eurasia and to a similar extent in the Mackenzie basin in North America. There are negligible wetlands in the Pechora and Yenisei basins.

True arctic tundra represents a relatively small portion of the land area of the Arctic basin, primarily consisting of the Canadian Archipelago and the coastal plains of both

North America and Eurasia. Tundra vegetation tends to transition northward from wooded and shrub tundra to low-lying moss and lichens. Wetlands and thaw ponds dominate the low-topography coastal plain.

Both the Eurasian and the North American portion of the Arctic basin contain alpine zones. In the Mackenzie Basin, this includes a moderate maritime zone of coastal and alpine forest along the Rocky Mountains, as well as high elevation alpine tundra. Pockets of alpine tundra are also located in the Ural Mountains in the western Ob basin and in the Altai Mountains and Aldan Plateau in the headwaters of the Lena and Yenisei Rivers.

Figure 2. General vegetation classes of the Arctic drainage basin
(adapted from the USGS Global Land Cover Characterization [21]).

2.2. PRECIPITATION AND EVAPORATION

The spatial and temporal distribution of runoff source areas can be explored through analysis of the land surface and atmospheric water balances. The distribution of atmospheric convergence (which, in the long term, must be balanced by P-E) can be calculated from radiosonde observations (e.g., as a line integral) or as the spatial integral of the convergence field produced from atmospheric model analysis (e.g., resulting from data assimilation). By using atmospheric soundings, Serreze et al. [22] found that specific humidity generally decreases poleward for all seasons and precipitable water decreases poleward for winter, spring, and autumn on the Atlantic side of the Arctic. Precipitable water ranged from winter lows of 1.8 and 3.8 mm (for 80-90° N and 65-70° N, respectively) to summer maxima of 12.5 and 18.2 mm [20]. However, there was large spatial variability independent of latitude. Land surface water balance calculations from experimental sites also show a decrease in P-E poleward, as summarized in Table 1. Due to the greater availability of data, this section is biased towards North American trends. Although moisture availability decreases with latitude, there is still a net gain of atmospheric fresh water to the polar region.

TABLE 1. Experimental measurements of P-E

Location	Latitude	Longitude	P-E (mm)	Reference
Imnavait Creek	68°30' N	149°15' W	170	[23]
Siksik Creek	68°45' N	133°30' W	90	[24]
Western Siberia	-	-	311	[25]
Marmot Creek	50°57' N	115°10' W	460	[26]

High-latitude vapor convergence reaches a maximum over the pole in late summer, corresponding to a minimum over the Mackenzie basin which sometimes becomes negative due to proportionately larger evapotranspiration over the Mackenzie River than in the central Arctic [27]. An 18-year mean of convergence calculated from radiosonde data implies a mean evaporative fraction (evapotranspiration/precipitation) over the Mackenzie River of 0.26 [27]. This contrasts with water balance estimates, which suggest a much larger evaporative fraction (see Section 2.3).

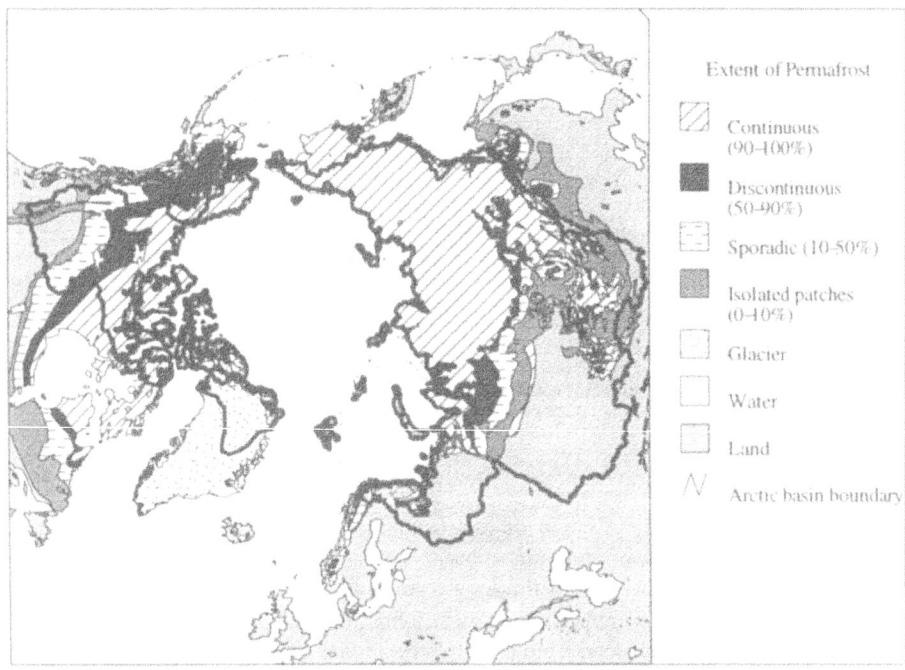

Figure 3. Distribution of continuous and discontinuous permafrost in the Northern Hemisphere (adapted from Circum-Arctic map of permafrost and ground-ice conditions [28])

The variability of vapor convergence correlates well with both precipitation and discharge, but only explains about 50 % of the variability in discharge [27]. Bjornsson [29] found that spring runoff in the Mackenzie River is highly correlated with total precipitation from fall to spring. An analysis of interannual variability of runoff from the 15 largest rivers draining to the Arctic from 1950-1990 indicates a maximum of mean annual total river runoff occurred in the 1970s. However, the maximum runoff rates for rivers of the Kara, East-Siberian and Beaufort Seas (Ob, Yenesei, Kolyma and Mackenzie Rivers, among others) were accompanied by minimum values in land surface runoff to the Barents and Laptev Seas (Pechora and Lena Rivers, among others) [28]. Based on ocean salinity data, discharge into the western Arctic Ocean is more variable than discharge into the eastern Arctic Ocean [31]. Coefficients of variation (C_v, standard deviation divided by mean) for the large river basins are on the order of 0.08 to 0.28 [2, 32]. In particular, the Yenisei, Lena, Mackenzie and Ob Rivers have C_v's of 0.08, 0.12, 0.10 and 0.15 respectively [4]. A more detailed discussion of the variability of these rivers is provided in [4].

2.3. RUNOFF RESPONSE

The seasonal cycle of runoff from northern rivers is strongly controlled by the storage of precipitation as snow during all but the summer season. Large northern rivers tend to have a relatively short period (2 to 4 months) of sustained high flow in late spring/early summer followed by an extended seasonal dry down [32]. Flow may stop completely during the winter. In catchments of up to 100-200 km^2, the annual maximum runoff rate is often associated with rainfall events, because over such small areas the rainfall rate may exceed the maximum rate of snowmelt [33]. With increasing catchment area, this effect is reduced due to the relatively small scale of high-intensity rainfall. Additionally, the likelihood of rain floods exceeding snow floods decreases with latitude, due to the progressive reduction in the length of the rain season [33]. Interannual differences in the runoff response of northern rivers are in large part due to changes in snow accumulation, but spatial differences are often controlled by the distribution of permafrost, frozen ground, lakes and wetlands. The effects of these mechanisms are discussed below.

2.3.1. Permafrost and frozen ground

As shown in Figure 3, the entire Arctic drainage basin is underlain by seasonally frozen ground and permafrost (soils in which the water content is perennially frozen). This results in similar early spring surface water responses for the entire basin. Although meltwater percolation may raise soil surface temperatures to 0° C, soils usually remain frozen until the snow cover is depleted [20, 34]. The presence of frozen water limits infiltration into the soil, and decreases the melt water storage capacity of the soils. Several researchers have found that the upward flux of water vapor from the soil throughout the winter increases the ice-free void space at the soil surface by winter's end, allowing some infiltration of meltwater [34, 35, 36]. However, the initial meltwater refreezes in the soil, completely filling void spaces and preventing any further infiltration [34]. Therefore, the dominance of surface flow is typical of both permafrost and seasonally frozen regimes during snowmelt.

In contrast to seasonally frozen soil, the presence of permafrost continues to inhibit ground water recharge and movement, and restrict plant growth throughout the growing season [37]. All surface and subsurface groundwater interaction will occur in the active layer, the top 1 to 3 m of soil that thaws annually [37]. The uneven thawing of the active layer in response to radiation exposure, meltwater infiltration, and other heat transfer processes results in changes to basin storage capacity which may not reflect the overlying topography [34]. As the depth of thaw increases, non-saturated areas may develop, however the presence of the frost table continues to severely restrict the storage capacity of permafrost soils, resulting in a higher proportion of annual surface runoff [34, 35].

Untersteiner [37] suggested that differences in the amplitude of the seasonal hydrograph of major north flowing rivers may be due in part to the presence of permafrost. Runoff ratios (annual discharge/precipitation) reported for some of the largest Arctic rivers are summarized in Table 2 [10, 17 32]. Presumably, the different ratios result from using different sources of precipitation data. In general, the Yenisei, Lena and Kolyma Rivers tend to have the highest runoff ratios. These rivers are underlain by continuous or discontinuous permafrost for 88% (Yenisei River) to 100% (Kolyma and Lena Rivers) of their length. In contrast, the Ob and Mackenzie rivers traverse permafrost for 27 and 83% of their length, [37]. The Pechora River is an anomaly as it traverses discontinuous permafrost for only 50% of its length but it is also characterized by fewer wetlands than the other major north flowing rivers.

TABLE 2. Runoff Ratios of Major Arctic Rivers

River	Oki et al. [17]	Kuhl and Miller [10]	Vuglinsky [32]
Yenesei	0.52	0.54	0.47
Kolyma	0.53	0.39	0.49
Lena	0.53	0.61	0.46
Ob	0.29	0.33	0.25
Mackenzie	0.46	0.41	0.30
Pechora	0.73	ND	0.58
Severnaya Dvina	ND	0.60	0.43

Notes: Kuhl and Miller [10] used precipitation from Shea [38]. Oki et al. [17] used precipitation from [16]. The source of precipitation for Vuglinsky [32] is unknown. ND indicates no data was provided for that river.

2.3.2. Lakes and wetlands

Seasonal variability is dramatically different between large Eurasian and North American Rivers [32, 39]. Eurasian rivers tend to have a short period (2-3 months) of sustained high flow in late spring and early summer that exceeds the mean low flow discharge by 7 to 10 times [32], although exceedances as high as 40 times have been recorded in the Yenisei and Lena Rivers [39]. North American rivers tend to have an extended seasonal cycle, with one to two more months of sustained high flow. Mean high flow discharge exceeds mean low flow discharge by factors of 4 to 5 in North American rivers [32, 39]. This attenuation is generally thought to be due to storage in lakes and wetlands, which cover a much larger portion of North American as opposed to Eurasian river basins.

Evaporation from small lakes during the ice-free period often exceeds summer and even annual precipitation [40, 41, 42]. This is also the case on the Alaskan coastal plane where Rovansek et al. [41] observed that snowmelt runoff is delayed until the entire storage of ponds and wetlands is refilled.

Wetlands are areas having a water table near or above the mineral soil for most of the thaw season [43]. According to Church [44], northern lakes and wetlands create a muskeg regime, in which high flows are attenuated by the absorption capacity of the muskeg vegetation and resistance to flow by surface conditions. However, the storage capacity of wetlands is often limited by frozen ground [36]. Therefore, flow attenuation by wetlands is usually highest for summer rainfall after the peat is thawed [36, 45].

High ice content wetland soils are slow to thaw since ice requires more energy for melt. Meltwater infiltration is delayed longer, thus the presence of frozen wetlands can actually accentuate the surface water response of frozen ground [20, 36, 46]. These wetlands are often poorly drained, remaining saturated through the summer and refreezing with a high ice content the next winter [46]. In the north temperate zone, the feedback between wetlands and frozen ground can actually be reversed as winter flooding of wetlands can prevent or restrict frost formation [20].

The presence of thousands of small thaw lakes on the coastal plain and river delta regions can also have a profound impact on heat and energy exchange. Conductive heat flow from a lake system can serve as a significant source of heat flow to the atmosphere, maintaining locally moderate temperatures [47, 48].

2.4. SNOW HYDROLOGY

Snow cover accumulation, redistribution and ablation dominate the land surface hydrology of the Arctic. Although rain may account for a fairly large fraction of the annual precipitation, even at high latitude, snow is proportionately more important to runoff due to low rates of winter evapotranspiration and the timing of meltwater release. Snow accumulation and ablation processes are somewhat different for each of the four major Arctic land classifications (Section 2.1) and are discussed separately.

The snow-covered period extends from 6 to 10 months in the tundra zone. Spring melt releases between 35-60% of annual precipitation [49]. Negligible mid-winter melt occurs and winter sublimation in the tundra zone is generally energy limited [23, 50, 51]. However, sublimation during blowing snow events can represent a significant loss of snow over the winter [23, 50]. The initiation and movement of blowing snow depends on wind velocity and duration, vegetation and snow structure. Therefore, the likelihood of transport decreases with increasing air temperature and snow age [52, 53]. The absence of tall vegetation in moss-lichen tundra regions and a dry snowpack facilitates significant wind redistribution of snow, which creates a fairly consistent snow spatial structure with respect to topography [23, 54, 55, 56, 57]. Despite wind redistribution, snow cover is nearly continuous at the start of melt and melt energy is dominated by net radiation and sensible and latent heat fluxes to a lesser degree [49]. As melt progresses, faster melt will occur in thin areas, in part due to lower albedo, longwave emittance from protruding plants and increased absorption of shortwave radiation penetrating to the underlying surface [23, 49]. Longwave radiation and sensible heat from the snow-free ground warms the overlying air and accelerates the melting of surrounding snow, resulting in large spatial variation in snowmelt rates [23, 49].

During snowmelt, evapotranspiration depends on the dryness of air relative to the snow surface, and is therefore bounded by the vapor pressure gradient rather than net radiation [50, 51]. During melt in the tundra region of Spitsbergen (Norway), Takeuchi et al. [58] found that an increase in vapor pressure of the air actually favors condensation. Kane et al. [50] found evapotranspiration during snowmelt of 20 to 47 mm (20 to 34% of average snowpack) in a small Alaskan Arctic catchment.

Although multiple freeze-thaw events are possible in the interior plains [59], annual volumetric transport during blowing snow events in the Canadian Prairies and the Eurasian steppe zone, which represent between 15 to 40% of annual snowfall [60, 61], are typically a much larger term in the winter water budget. Significant sublimation can occur during blowing snow events [62, 63, 64]. Air temperature, humidity and net radiation are the most important factors in determining the rate of phase change from blowing snow [63]. Pomeroy et al. [57] found that 28% of annual snowfall sublimated from tundra surfaces in a low Arctic catchment in northwestern Canada. The areas of maximum water loss for snow during the cold season were greatest in the north central coastal regions of Russia, the area with the greatest number of blowing snow events [53, 65]. The annual average sublimation across Russia from blowing snow events was between 2.5 to 10 mm with a maximum of 60 mm on the Kamchatka Peninsula [65]. Partial sheltering from the wind by open forest stands in the forest steppe/steppe transition zone in European Russia results in greater average accumulation [66].

Snowmelt in interior Canada and Russia generally begins in mid-April and proceeds rapidly often lasting only 1 to 2 weeks. Melt is radiation dominated and is less often affected by advection [67]. Evaporation during melt in the forest-steppe to steppe region of northwestern European Russia was estimated to be 15 to 60 mm/yr, increasing to 35 to 90 mm/yr in the forest region [68]. Are and Petropavlovskaya [67] found that the highest evaporation rates preceded the initiation of melt in the steppe region of central Yakutia, Russia.

The boreal zone contains a mosaic of vegetation varying from closed coniferous canopies, through sparsely vegetated clearings and numerous open lakes [69]. The snow season lasts for approximately 6 months, with melt occurring in late April [69, 70]. Mid-season melt is rare. Solar fluxes dominate melt in the northern extent of the boreal forest, but sensible heat and longwave radiation increase in importance to the south [70].

Interception of snow by boreal forest canopies can store up to 60% of the cumulative winter snowfall [61]. Increased exposure and turbulent transfer can result in sublimation losses from intercepted snow of between 3 to 40% of annual snowfall [68, 71]. The canopy interception capacity is dependent on species, tree area and snow load, with decreasing interception efficiency for increasing snow load [61, 72].

Vegetation of the alpine zone typically contains dense conifers extending to the treeline, where shrubs and grasses give way to talus slope and bare rock [26]. Therefore, the alpine zone contains elements of all of the preceding regions. However, it is further characterized by an increase in snow accumulation with elevation due to orographic enhancement of precipitation and decreasing temperature gradients. The gradient in snow depth increases throughout the snow season, due to continued snowfall at the highest elevations and differential melting in the valleys. The high elevation snowpack generally melts by late July [26]. Along the eastern slope of the Rocky

Mountains, snow melt is enhanced by the occurrence of Chinooks, strong winds from the south accompanied by elevated temperatures. Golding [73] found potential evaporation rates from the snow of 1.2 to 2 mm/day during chinooks, with a maximum rate of 10.4 mm/day. Snow melt during the chinooks ranged from 0.4 to 1.6 mm/day [73].

It is unclear how melt from a spatially varying snow pack, which primarily has been studied at small scales, affects runoff from medium to large Arctic rivers. Variability in snow cover extent is important because of the effect on albedo, the area contributing meltwater during spring melt, and its effect on the length of the melt season. However, the initiation of melt in large basins can vary by a month or more from southern headwaters to the mouth due to the difference in latitude alone. Given this large temporal range in meltwater production, differences in runoff rates on the order of several days to weeks due to small-scale snow distribution may be unimportant. At present, a good understanding of the relative influence of within-basin and between-basin variability in controlling runoff in the major Arctic basins is lacking.

3. Observations of Northern Freshwater Components

The literature reviewed in Section 2. illustrates the unique hydrologic features of the Arctic drainage basin. In Section 3., we explore how such features interact and influence large-scale hydroclimatological trends and the extent to which they are reflected in global data sets essential to hydrologic model forcing and validation. Section 3. utilizes several data sets, as follows:

- **5-minute digital elevation model (DEM):** ETOPO5 DEM (5 minute) from the Earth Resources Observation Systems (EROS) Distributed Active Archive Center (DAAC).
- **2° gridded precipitation:** daily station data from Schnur and Lettenmaier [74] adjusted for systematic error using Global Precipitation Climatology Project (GPCP) version 1A gauge precipitation data set [75].
- **Analysis increments and convergence fields:** output from the Goddard Earth Observing System (GEOS-1) data assimilation system of the NASA DAO reanalysis project.
- **Small basin discharge:** mean annual flow computed from the Arctic Runoff Data Base (ARDB) maintained by the GRDC for 100 basins smaller than 7,000 km^2.
- **Spatially distributed basins:** derived Geographic Information System (GIS) dataset of 43 basins in the ARDB delineated from a 30-arcsecond DEM.
- **Weekly snow extent:** provided by the Earth Observing System (EOS) DAAC at the National Snow and Ice Data Center (NSIDC), University of Colorado at Boulder.

3.1. SPATIAL VARIABILITY

The Arctic drainage basin (Figure 1) was delineated from a 5-minute DEM of the world (ETOPO5 data set), obtained from the EROS Data Center DAAC. As shown in Figure 4, 60% of the drainage area lies south of 65° N and there is a higher proportion of area at lower latitudes in Eurasia than in North America. The distribution of land area is critical because latitude and topography strongly influence the land surface hydrology through (a) orographic effects on precipitation, (b) the seasonal cycle of solar insolation, and (c) the distribution of vegetation.

Figures 5 (Eurasia) and 6 (North America) show the distribution of precipitation, evapotranspiration and runoff with latitude derived from three independent sources. The gridded station precipitation data for 1979 to 1993 from [74] were not corrected for gauge catch deficiencies and therefore probably underestimate precipitation, especially in the cold season. To compensate for this underadjustment, the precipitation data were scaled to match the monthly means from the GPCP station data set. This dataset was corrected (by GPCP) for systematic error following Legates [76].

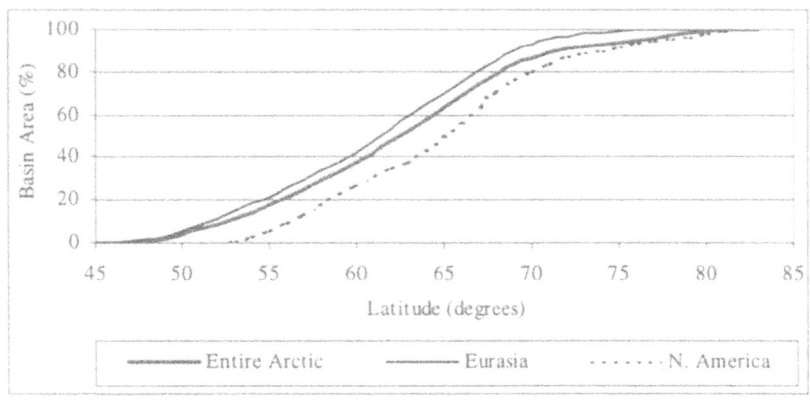

Figure 4. Distribution of Arctic basin area versus latitude.

We computed evapotranspiration from an atmospheric moisture balance using the gridded precipitation and convergence derived from the GEOS-1 data assimilation system of the NASA DAO reanalysis project [77]. Calculated values for convergence were adjusted by the reported GCM analysis increment, which essentially is the amount of water added to, or extracted from, the atmosphere due to the updating process effected by the data assimilation procedure. The influence of the analysis increment is discussed further later in this section. Runoff from small basins was derived using the mean annual flow from 100 basins in the ARDB data set with drainage areas smaller than 7,000 km^2. The emphasis on smaller catchments facilitated assignment of discharges to the latitude of the gauge locations (which were known), rather than the basin centroids (which were not). The random nature of gauge locations means that limited or no data are available for some latitude bands. In addition, updating of the ARDB is dependent on the response of member countries. Therefore, it was impossible to generate values of mean annual discharge from a consistent time period. In order to maximize the available record, mean annual flow throughout this paper was calculated from the entire record of each basin, which may vary from 10 to 100 years.

Figure 5 shows that in Eurasia, precipitation tends to increase with latitude up to 60-62° N above which it begins to decrease. In contrast, North American precipitation begins to decrease with latitude north of 54-56° N. In general, precipitation exceeds evapotranspiration in North America, but for latitudes less than about 62° N in Eurasia, zonally average P-E is negative. Surprisingly, there is no clear pattern in discharge distribution with latitude. This is partially due to statistical problems associated with small sample sizes and the difficulty of assigning discharge to single latitude bins. In addition, no discharge data were available north of 73N (as nearly 95% of the gauged area falls south of 65° N as shown in Figure 7a). The distribution of discharge stations used in Figures 5 and 6, in comparison to the Arctic basin distribution is also shown in Figure 7b.

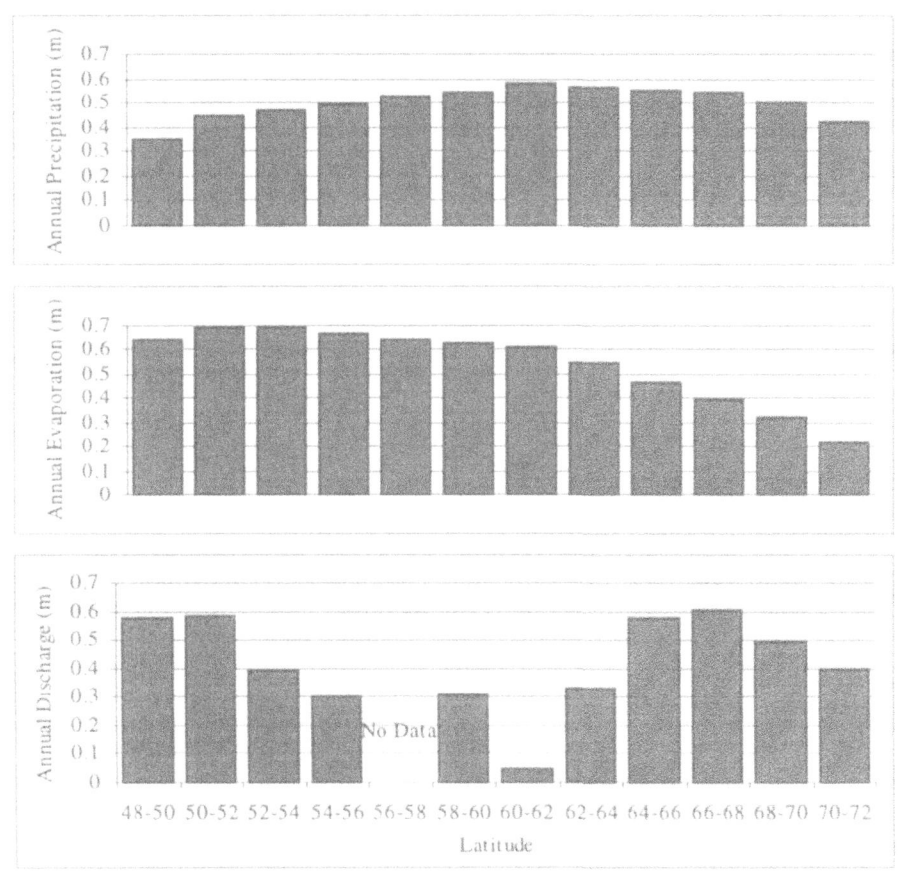

Figure 5. Precipitation, evapotranspiration and discharge versus latitude for Eurasia

Spatial evapotranspiration trends were further analyzed by comparing the evapotranspiration derived from the GEOS-1 data product with the spatial distribution of evapotranspiration derived from basin water balances (Figure 8). Water-balance evapotranspiration was calculated for the 43 spatially distributed ARDB basins delineated from 30-arcsecond DEMs. Mean annual runoff for each of the delineated basins was distributed evenly (as a depth) over the basin area and gridded at a 2° resolution. The runoff depth was subtracted from the precipitation to obtain a gridded average annual evapotranspiration. These same basins are used in Figures 8, 9, 10 and 12 to calculate and display basin average variables.

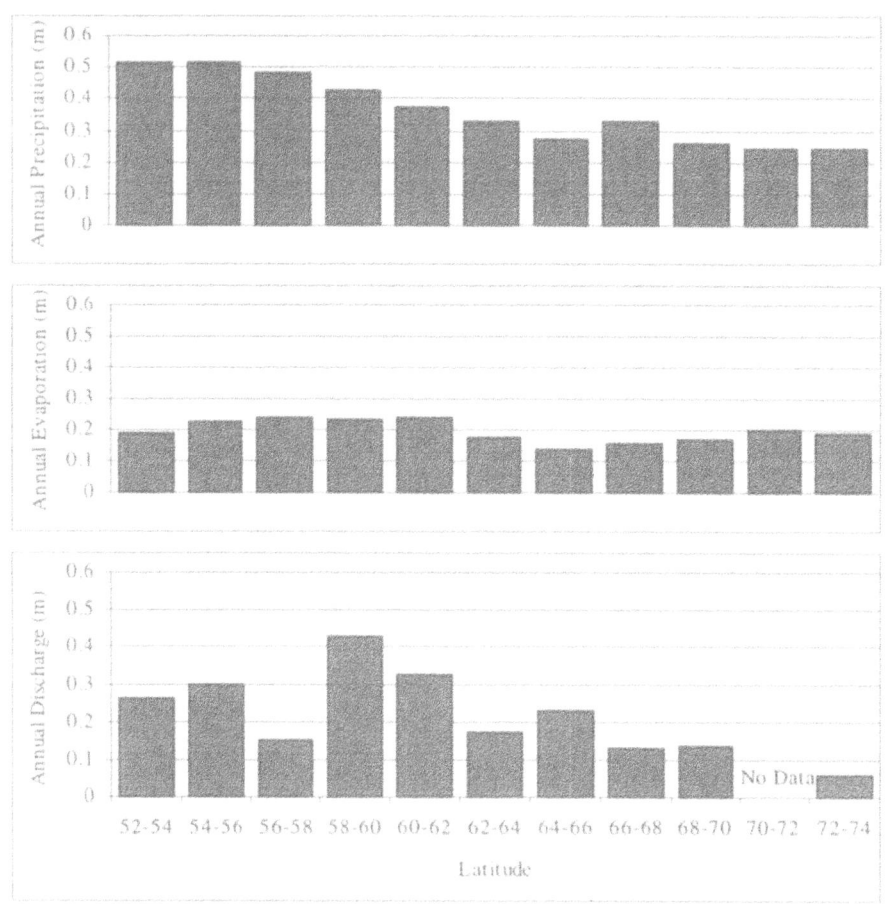

Figure 6. Precipitation, evapotranspiration and discharge versus latitude for North America

The two estimates of evapotranspiration (Figure 8) are roughly equivalent over North America. However, the GEOS-1 derived estimates of evapotranspiration over Eurasia are much higher than the water balance, particularly over the Ob basin where the estimates differ by several hundred millimeters. These discrepancies have to do in part with the analysis increment of the DAO reanalysis. As described by Molod et al. [19], "The analysis increments (the difference between the model first guess and the analyzed state) are normalized and included as a constant forcing term during model reintegration over the six-hour time span centered on the analysis time". The analysis increment is generally negative for most of the higher latitudes and is often of the same order of magnitude as the other terms. Because it is negative, it acts as an atmospheric sink and extra evapotranspiration is generated by GEOS-1 to fulfill this sink. Calculating evapotranspiration as the residual of precipitation minus divergence plus the analysis increment generates high values for evapotranspiration, by essentially assigning the entire error term to the evapotranspiration.

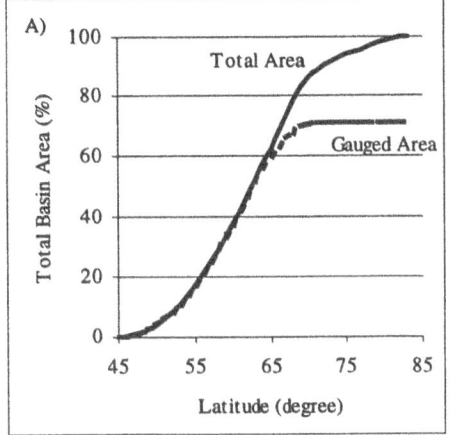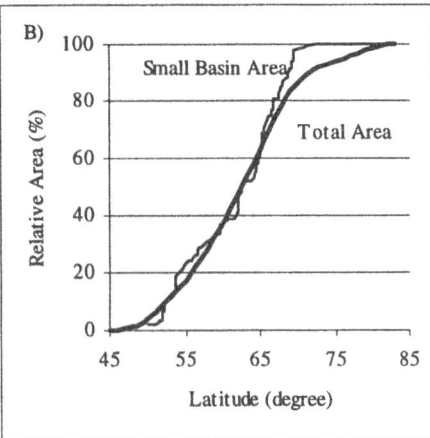

Figure 7. Distribution of A) gauged area versus latitude as a percentage of total basin area and B) area of small basins relative to total small basin area.

The relationship of discharge in the small basins shown in Figures 5 and 6 to geographic factors is further complicated by the effects of topography on the spatial distribution of precipitation. Precipitation maxima for the Arctic drainage basin are over the Rocky Mountains in Canada and the Ural Mountains in western Russia (Figure 9a), creating a strong west-east gradient in precipitation and evaporation in these basins, which somewhat confounds the north-south trend shown in Figures 5 and 6. (See the Hydrological Atlas of Canada [78] for a more complete discussion of the maritime to continental transition in the Mackenzie basin). The distribution of mean annual discharge (Figure 9b) coincides closely with the distribution of precipitation. Relatively high discharge occurs in the Pechora basin, with low discharge in the southern Yenisei basin. Higher discharge occurs in the mountainous areas of the Mackenzie River, although the response relative to precipitation is not as extreme as in the Eurasian rivers, as illustrated by the runoff ratio (annual discharge/precipitation) in Figure 9c. The consistency of the runoff ratio throughout Eurasia confirms the control of precipitation on discharge spatial variability, notwithstanding a possible increasing northward trend due to reduced evapotranspiration at high latitude (Figure 5).

3.2.   TEMPORAL VARIABILITY

As shown in the previous section, the variability in the discharge of the spatially distributed basins is strongly correlated with variations in latitude and elevation. There is also a large component of interseasonal variability which is associated primarily with latitude and catchment response to snow accumulation and melt. As noted in Section

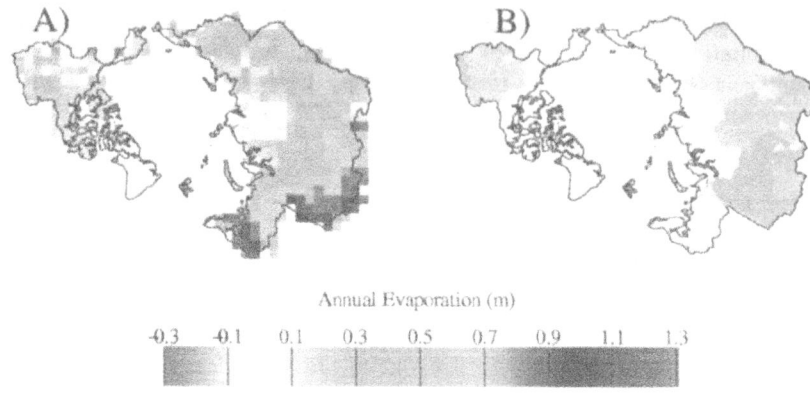

Figure 8. Average annual evaporation computed from
A) atmospheric water balance and B) basin water balance

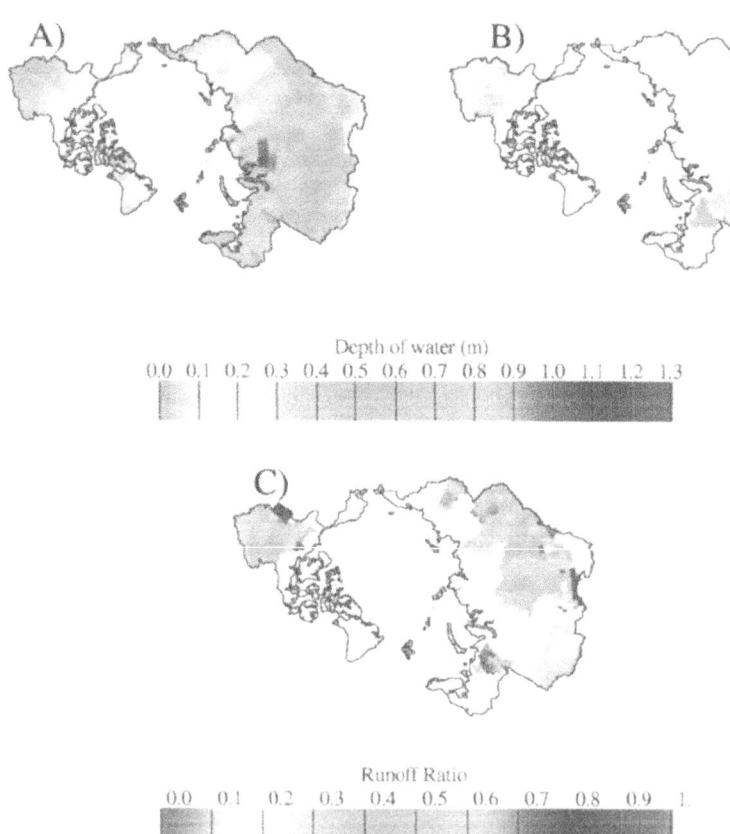

Figure 9. A) Average annual precipitation (1971–1994), B) Average
annual discharge (variable records), C) Average annual runoff ratio

(For a colour version of these figures, see p. 611)

2.1, Eurasian basins generally are characterized by a very rapid snowmelt response followed by 6 to 7 months of low flow. North American rivers tend to have a lengthier period of seasonal high flow, followed by a shorter period of low flow. This is reflected in the average ratio of minimum monthly discharge/maximum monthly discharge (Figure 10a). The data show quite a bit of scatter, but in general, Eurasian basins have a larger range between maximum and minimum annual discharge (smaller values of min/max discharge).

In Figure 10b the ratio of precipitation occurring as snow to annual runoff is shown. Snow was defined as daily precipitation occurring when the average daily air temperature was below $0°$ C. Figure 10b is influenced by both the percentage of precipitation which falls as snow (controlled primarily by latitude) and the percentage of meltwater subject to evapotranspiration (controlled by vapor pressure gradient and net radiation) [50], [51]. Although it is somewhat difficult to distinguish from basinwide averages, there appears to be a slight increasing northward trend in Figure 8b due to the increasing fraction of precipitation occurring as snow. In general, the highest (Ob River) and lowest (Lena and Kolyma Rivers) values in Figure 10b correspond with areas of high and low precipitation (Figure 9a), respectively. This is presumably because the higher vapor pressure in the Ob basin suppresses evapotranspiration during melt, yielding a higher percentage of runoff from the snowpack.

Variability of snow extent also plays a large part in determining interannual variability of runoff response. Figure 11 shows the minimum, maximum and average number of days with snow cover throughout the Arctic basin from 1971-1995 derived from snow extent data provided by the EOS DAAC. The largest differences can be seen in coastal areas and the mountainous regions of Asia and North America. Figure 12 illustrates the relationship between snow and discharge interannual variability by plotting the coefficient of variation of discharge versus the average range (maximum – minimum) of snow covered days from 1971-1994. The number of snow covered days was computed as a spatial average for each of the 43 spatially distributed basins. There is no clear relationship between discharge variability and variability in the annual length of snow cover for basins larger than 100,000 km^2 (Figure 12a). There is a decrease in discharge variability with increasing variation in snow cover, albeit with more scatter, for smaller basins (Figure 12b). This may reflect a greater sensitivity to rainfall variability rather than snow in smaller basins, due to the relatively small-scale of high-intensity rainfall.

4. Hydrological Modeling

The exploratory data analysis of Section 3., taken together with a knowledge of fundamental northern hydrologic regimes, as reviewed in Section 2., suggest that the most important controls on runoff of medium to large Arctic rivers are latitude, orography and surface storage. However, there remains considerable variability between catchments that could be controlled by a number of possible mechanisms. For instance, it is not clear from the data analysis if discharge variability is more a function of rain or snow precipitation and how that difference varies over the region. Likewise, the effect of spatial trends in precipitation and albedo on snowcover and runoff

A)
B)
Min/Max Discharge
0 4 8 12 16 20 24 28
Snow precip./runoff
0 1 2 3 4

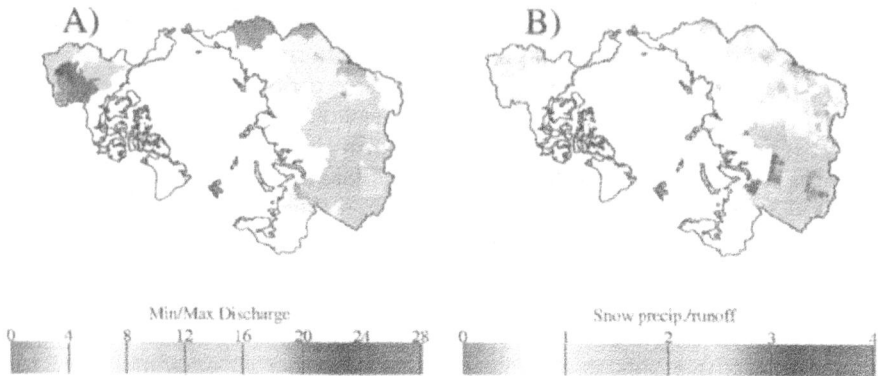

Figure 10: A) Mean ratio of minimum monthly discharge to maximum monthly discharge, and B) Ratio of snow precipitation to runoff

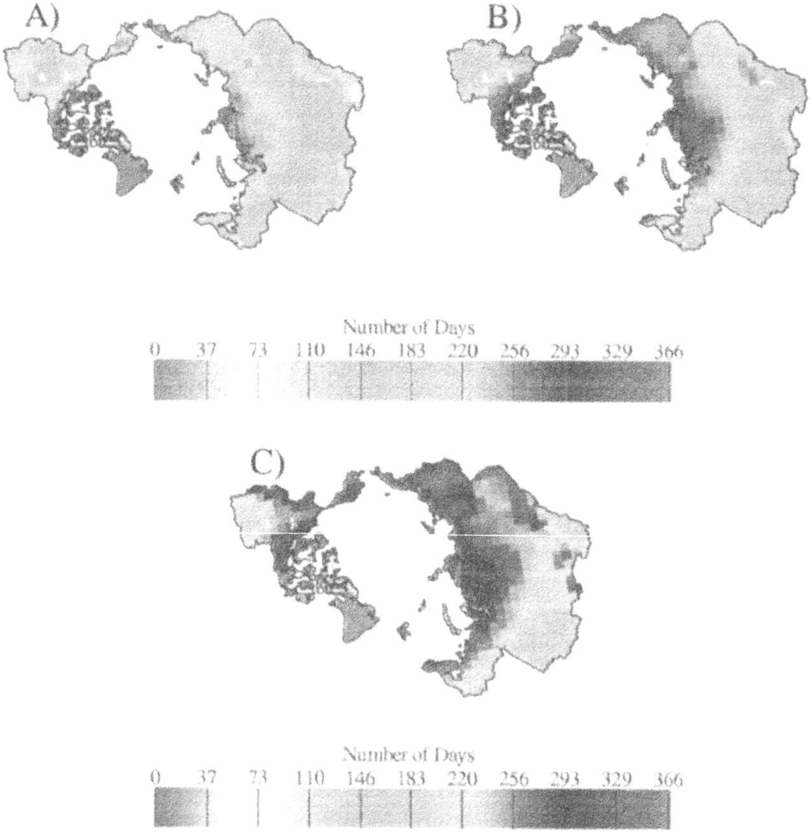

A)
B)
Number of Days
0 37 73 110 146 183 220 256 293 329 366
C)
Number of Days
0 37 73 110 146 183 220 256 293 329 366

Figure 11. Number of days with snowcover (1971 –1994) A) Minimum B) Average and C) Maximum

(For a colour version of these figures, see p. 612)

generation are not well understood. Such questions further complicate the estimation of freshwater fluxes from the ungauged portion of the Arctic basin. To examine the potential of hydrologic models as diagnostic tools in ungauged portions of the Arctic drainage basins, a macroscale hydrological model, the Variable Infiltration Capacity (VIC) model [79], [80] was applied to the Mackenzie and Ob River basins.

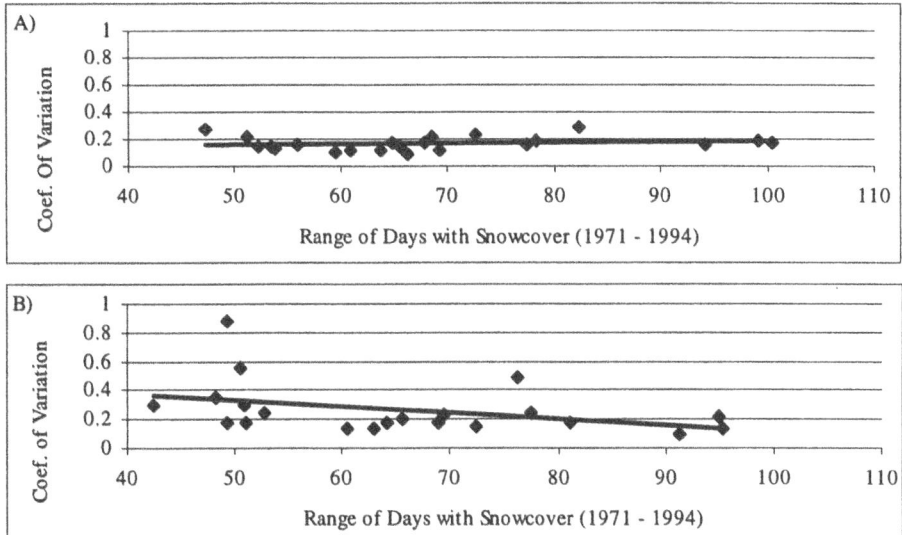

Figure 12. Coefficient of variation of discharge versus the average range of snow covered days: A) basins larger than 100,000 km^2, B) basins smaller than 100,000 km^2

4.1. MODEL DESCRIPTION

4.1.1. VIC Model

The VIC model is a grid-based Soil-Vegetation-Atmosphere Transfer (SVAT) scheme designed both for inclusion in GCMs, and for use as a stand-alone macroscale hydrological model [79, 80]. The model has two modes of operation: a full water and energy balance mode, in which the surface energy budget is closed by iterating for surface temperature, and a water balance mode, in which surface temperature is assumed to equal air temperature. The water balance mode was used for this application. The model partitions the soil column into multiple soil layers, which allows representation of the rapid dynamics of soil moisture movement during storm events in the upper root zones, and the slower deep interstorm response in the bottom layer. Each grid cell is assigned a partial coverage of a user-specified number of vegetation or bare soil land-use types. The parameters associated with each land-use type are time-varying leaf area index (LAI), the relative fraction of roots in each soil layer, canopy resistance, minimum stomatal resistance, roughness and displacement length. A more complete description of model processes can be found in [79] and [80].

A recent addition to the VIC model incorporates a two-layer energy balance snow model [81, 82], originally developed for the Distributed Hydrology-Soil-Vegetation Model. The model employs two snow layers of variable thickness. The thin surface layer is used to solve the surface energy balance, while the bottom or pack layer is used to simulate deeper snowpacks. The model accounts for the energy advected by rain, throughfall or drip, as well as net radiation and sensible and latent heat. Snow interception by the overstory, if present, is calculated as a function of LAI. Intercepted snow can be removed from the canopy through snow melt, sublimation and mass release. The model does not account for redistribution or sublimation due to blowing snow.

The affects of topography on air temperature and precipitation are represented through snow elevation bands. For each model grid cell, snow accumulation and melt is calculated separately for a variable number of bands. Fractional area, average elevation and the fraction of grid cell precipitation falling in the band is specified for each band. The sum of all fractions must be one. A maximum of five bands was used for this application, with fewer bands in areas of low topographic relief. Each grid cell is divided into equal elevation width bins based on the maximum and minimum elevation at 5-minute spatial resolution. The area and average elevation of the 5-minute cells that fall into each bin are equal to the band elevation and area. Bands that were separated by less than 500 m were merged together for computational efficiency. Temperature was lapsed using the adiabatic lapse rate. Precipitation was divided among the bands in proportion to the elevation difference for each band. These elevation bands are the only way in which partial grid cell snow cover is represented. The inherent patchiness of a melting snow cover and resulting effect on area of energy exchange with the snow pack and local advection are not currently represented.

4.1.2. Routing Model

When the VIC model is implemented over a grid mesh (i.e. watershed), evapotranspiration, energy fluxes, runoff and baseflow are predicted independently for each grid cell. Streamflow is predicted at a specified location by routing surface and subsurface grid cell runoff using the method of Lohmann et al. [83, 84]. To account for differences in travel time from different areas within a cell, the single runoff time-series produced for each cell by VIC is routed to the cell outlet using a triangular unit hydrograph [83, 84]. The hydrographs for the individual cells are routed to the basin outlet through a channel network, in which cells are connected in the direction of major flow to one of the 8 adjacent cells (Figure 13). All flow from a cell is routed in a single direction. For cells at the edge of the basin, the relative proportion lying within the basin may be specified, allowing the conservation of the true basin area. The routing network for this application was developed manually using GIS overlays of the main river systems on the delineated basins. Flow directions for each grid cell were determined by visual inspection.

The routing model is operated independently from the VIC hydrologic model. Therefore, although the unit hydrograph can be used to control the response time from each model grid cell, it does not represent the ponding and continued evaporation of surface water in low topography areas. The same is true for the effects of large lakes on the inter-cell channel routing scheme. As such, the current model does not explicitly represent the effects of lakes and wetlands.

4.1.3. Model implementation

The VIC model was applied to the Mackenzie and Ob River basins at 2 degree spatial resolution (see Figure 13) and daily temporal resolution. Two southerly sub-basins of the Ob and Mackenzie Rivers, the Irtish and the Athabasca, respectively, were also modeled (hashed in Figure 13). Simulation of smaller sub-basins facilitates model implementation by decreasing computation time at the calibration stage. Also, it allows assessment of the transferability of the more empirical model parameters. In particular, the Irtish and Athabasca sub-basins were selected for their southerly locations and contrasting land forms, as described below.

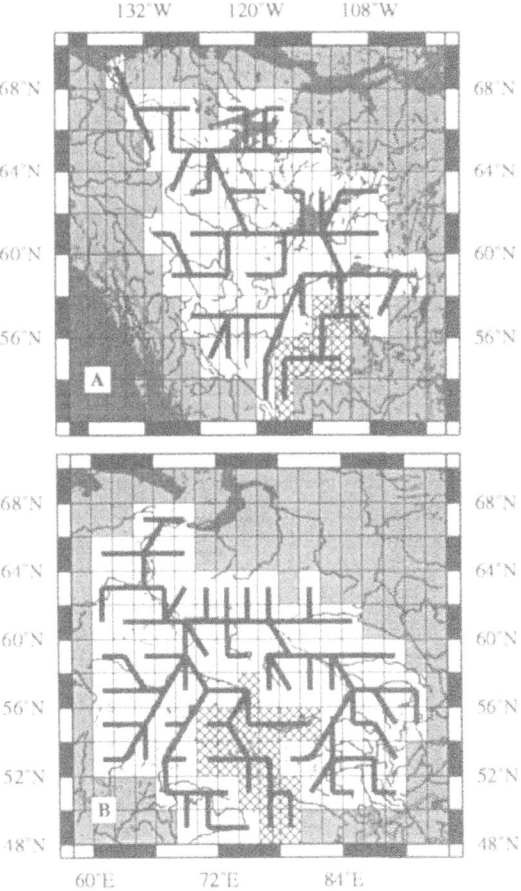

Figure 13. Routing networks for A) the Mackenzie (white) and Athabasca (hashed) Rivers and B) the Ob (white) and Irtish (hashed) Rivers

The Mackenzie basin drains an area of 1,660,000 km^2 in northern Canada, centered at approximately 115° W and 63° N. It includes the Great Bear Lake, Great Slave Lake and Lake Athabasca, which account for about 4% of the drainage area. Water flows north from latitude 49° N to the mouth at the Beaufort Sea at latitude 69° N. Basin elevations range from over 3500 m in the Rocky and Mackenzie Mountains in the west to sea level. The Athabasca River above McMurray flows east from headwaters in the Rocky Mountains and drains an area of 133,000 km^2. The basin contains some isolated permafrost patches in the mountains and vegetation consists of primarily coniferous and mixed coniferous forest

The Ob River (basin centroid 78° E and 59° N), drains an area of 2,950,000 km^2 of the Siberian plains. The Ural Mountains form the western basin boundary, but because of the relatively flat landscape the eastern boundary is somewhat unclear. The Ob River originates in China at latitude 48° N and flows north to 69° N were it enters the Gulf of Ob. The Irtish River above Omsk drains 321,000 km^2 of wooded steppe and steppe highlands in the southernmost Ob basin. The Irtish basin is not significantly affected by permafrost.

The required meteorological inputs for the water balance mode of the VIC model are daily minimum and maximum air temperature (also used to infer solar and longwave radiation) and precipitation. Precipitation for the period 1979 to 1993 was taken from [74] and adjusted for gauge catch deficiences as described in Section 3.0. Gridded minimum and maximum temperature station data were adjusted to match the mean monthly temperature climatology of [85]. A five-year period from January 1979 to December 1983 was used for calibration, however the first year (1979) was used for model initialization and is not shown in subsequent results. The remaining ten years from January 1984 through December 1993 were used for model evaluation. This ten-year period was used in the model interpretation in Section 5.

4.2.   MODEL CALIBRATION

Soil and vegetation parameters were based on a data set produced by the International Satellite Land Surface Climatology Project (ISLSCP), also used by [74] in their global simulations. The vegetation parameters were left unchanged, as follows:
*   Architectural resistance: 40 s/m (overstory), 10 s/m (understory);
*   Minimum stomatal resistance: 80-140 s/m (overstory), 80 s/m (understory);
*   Albedo: 20 %; and
*   LAI: variable.

Saturated hydraulic conductivity and bulk density were assigned categorical values based on soil texture suggested by the ISLSCP. The maximum velocity of baseflow (D_{smax}) was calculated from average surface slope and saturated hydraulic conductivity. These values were not adjusted during calibration. The most sensitive soil parameters in this work were the infiltration capacity curve shape factor, b_{inf}, the baseflow parameter D_s, the fraction of maximum baseflow at which the baseflow formulation changes from a linear to a nonlinear function of soil moisture W_s, and the upper and lower soil depths. These parameters were changed during calibration. Table 3 shows the values of the parameters used. Constant values were used throughout each of the simulated basins since no coherent procedure was available for adjusting them spatially based on available information. The depth of the top soil layers is realistic only if viewed as a surrogate for surface water storage, which is not represented by the model. Baseflow is only removed from the bottom layer, so moisture retained in the top layer is available for evapotranspiration as it slowly percolates to the bottom layer.

TABLE 3. Model Soil Parameters

Soil parameters	Ob	Irtish	Mackenzie	Athabasca
Infiltration parameter, b_i	0.1	0.1	0.25	0.1
Fraction of maximum soil moisture, Ws	0.99	0.98	0.99	0.8
Fraction of maximum baseflow, Ds	0.008	0.01	0.001	0.0275
Top layer depth (m)	2.0	1.0	1.5	0.8
Bottom layer depth (m)	1.0	0.5	0.5	0.1

The VIC model was calibrated for the Mackenzie and Ob basins, as well as the Athabasca and Irtish sub-basins (Figures 14 and 15). The mean monthly flows for the Mackenzie River from 1980 – 1993 (Figure 14a) indicate that the volume of the snow-melt peak is captured fairly well, although the initial rise is too slow, resulting in an extension of the peak flow into July. This may be due to over compensation for surface storage effects in the top soil layer, which are actually due to storage in the large lakes of the lower Mackenzie Basin. The constant values of soil storage that are not correlated with true wetland locations prevents quick runoff response from non-wetland areas. The same problem is not evident in the Athabasca River (Figure 14b) in which the simulated annual peak and recession is too abrupt. It is interesting to note that although the Athabasca sub-basin is at the southern end of the Mackenzie basin and upstream of the three major lakes, the annual peak occurs about one month after that of the Mackenzie. For the most part the Athabasca basin is not underlain by permafrost, while a significant portion of the Mackenzie Basin is. In addition, the difference in simulated and observed hydrographs may reflect errors in representation of snowmelt at the highest latitudes.

Although the timing of the snowmelt peak is well-matched for both the Ob and the Irtish Rivers, Figures 15a and 15b show that in both basins the rising and recession limbs are too steep. Although the Ob basin is not effected by large lake storage to the extent of the Mackenzie basin, it does contain much more extensive wetlands. The overly steep recession may be due to a lack of representation of this storage effect or due to an over-prediction of the rate of melt of high-altitude snow cover in the southern steppe region. The monthly hydrograph for the Irtish River (Figure 15d) indicates that both the simulated and observed series have more interannual variability than do the other rivers. However, the simulated and observed variability are out-of-phase, which suggests that the model is less responsive to changes in the meteorological inputs than the natural system.

The modeled and simulated mean number of days with snow cover are shown in Figure 16. Since the snow model is a physically-based energy balance model, there is little calibration involved. The maximum temperature at which precipitation can fall as snow and the minimum temperature at which precipitation can fall as rain were fixed at 1.5 and $-0.5°$ C, respectively. Snow surface roughness was fixed at 1 mm. On average, the length of the simulated snow season is slightly over-estimated, but maintains the correct spatial distribution. There appears to be greater spatial variability in the simulated snow cover extent than in the observations (Figures 16 a and b, respectively).

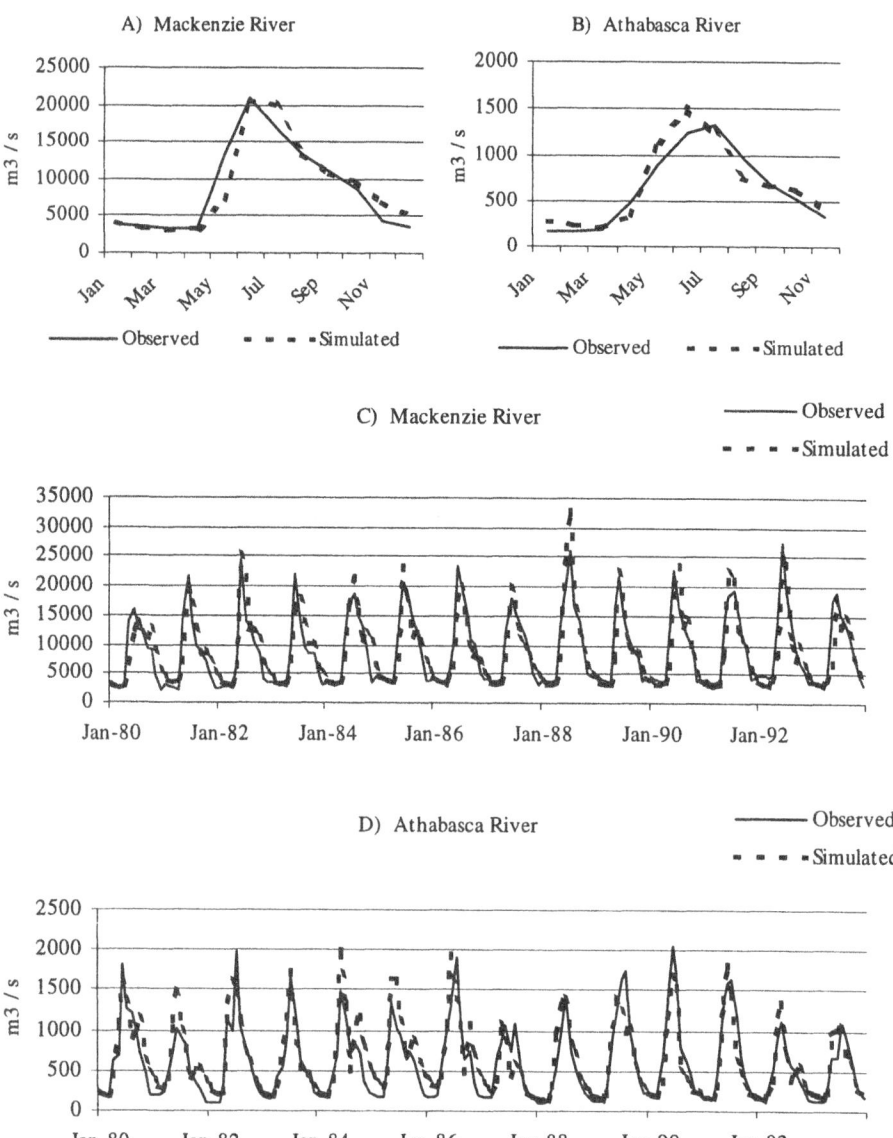

Figure 14. Observed versus simulated hydrographs 1980-1993: A&B) Mackenzie and Athabasca Rivers mean seasonal hydrograph, C&D) Mackenzie and Athabasca monthly hydrographs

Although sublimation directly from the snowpack and from canopy interception are represented by the VIC snow model, the enhanced rate of sublimation during blowing snow events is not. Modeled sublimation ranged spatially from –9 to 64% of snow precipitation (average of 27%) for the Ob basin and between –6 and 56% (average of 37%) for the Mackenzie basin. Areas of high sublimation coincide with areas of dense forest.

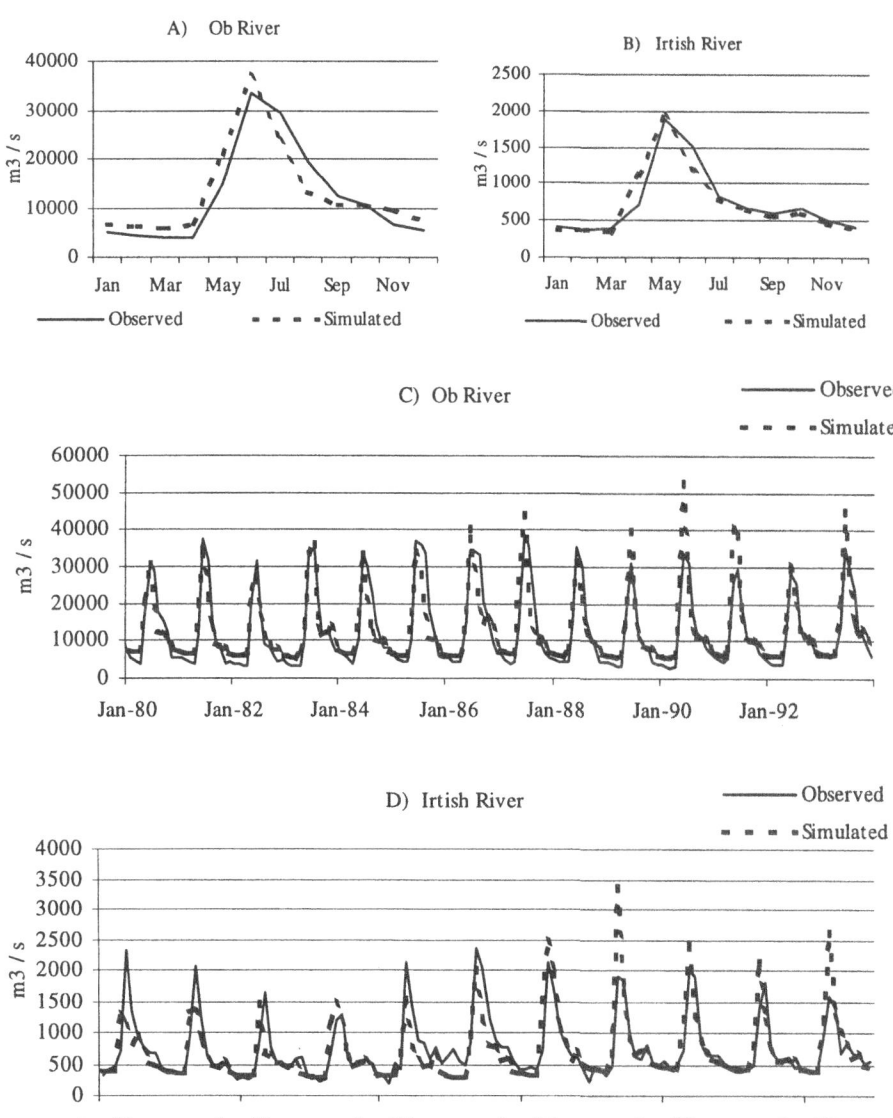

Figure 15. Observed versus simulated hydrographs 1980-1993: A&B) Ob and Irtish mean seasonal hydrographs, C&D) Ob and Irtish monthly hydrographs

82

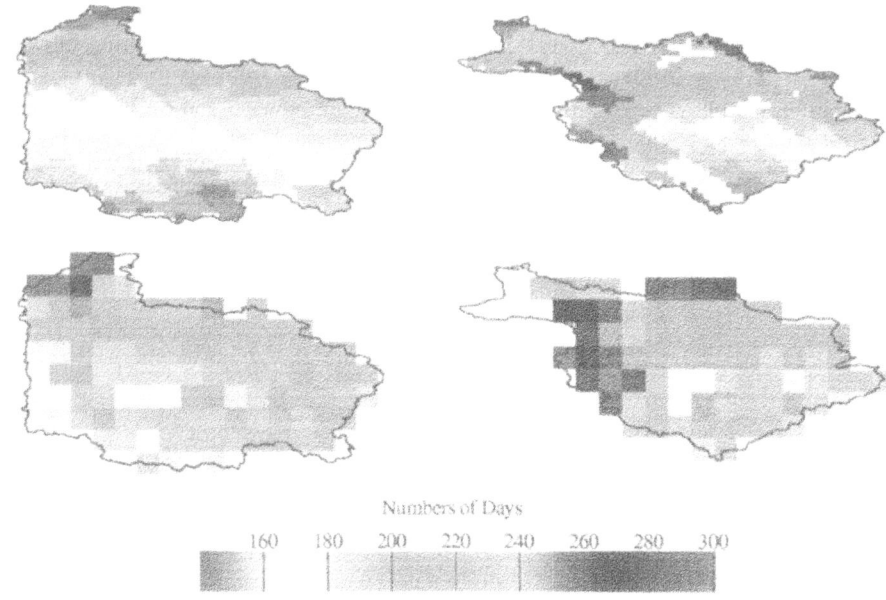

Numbers of Days

160   180   200   220   240   260   280   300

Figure 16. Remotely sensed (top) and simulated (bottom) number of days
with snowcover. A) Ob River and B) Mackenzie River

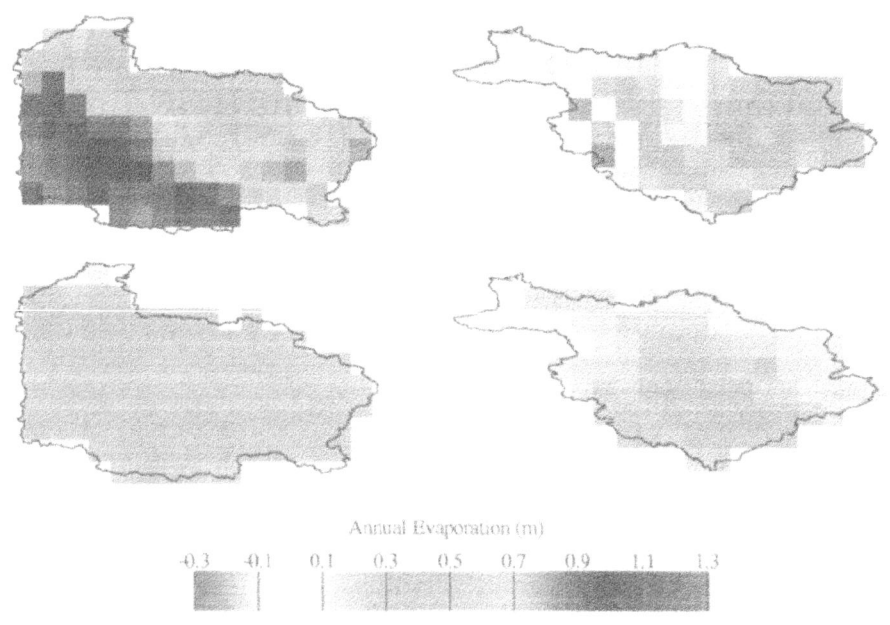

Annual Evaporation (m)

-0.3   -0.1   0.1   0.3   0.5   0.7   0.9   1.1   1.3

Figure 17. Annual evaporation derived from DAO reanalysis fields (top)
and simulated (bottom). A) Ob River and B) Mackenzie River

(For a colour version of these figures, see p. 613)

5. Interpretation of Simulation Results

Interpretation of hydrologic model results facilitate analysis of the macroscale hydrology of the Arctic for variables not directly observed at a useful resolution, and also serves as a useful diagnostic tool for existing data sets and global models. For example, Figure 17 shows model-simulated evapotranspiration in comparison to the DAO derived evapotranspiration. The DAO evapotranspiration is probably unrealistically high in the Ob basin, for reasons discussed in part in Section 3. In general, the spatial patterns and magnitudes of evapotranspiration from the hydrology model are more believable, perhaps in part because the model results are constrained by runoff and precipitation, at least at the catchment scale.

Figure 18a shows the percent runoff ratio for the Ob and Mackenzie River basins over the period of simulation. This figure is similar to Figure 9c, but at a higher spatial resolution. It shows that a higher percent of precipitation results in discharge at higher latitudes in the Ob basin than in the Mackenzie basin. In the Ob basin, evapotranspiration seems to be controlled more by latitude than by elevation. This trend is not clear in the Mackenzie basin, where increased storage creates a lower runoff ratio in the central part of the basin. The maximum rate of baseflow in the model is controlled by the topographic gradient. In the flat, central part of the Mackenzie basin, baseflow is lower, so moisture stored at or near the surface is available for evapotranspiration for a longer time period.

The percentage of runoff derived from snow (mean maximum snow water equivalent (SWE)/annual runoff) is shown in Figure 16b. These results neglect the effect of infiltration and evapotranspiration of meltwater, and therefore it is possible to have SWE greater than the observed discharge. The greatest percentages of SWE over runoff coincide with the areas of smallest percent runoff in Figure 18a (in particular in the southern steppe region of the Ob basin). Higher values of maximum SWE relative to runoff are expected in low runoff areas, where meltwater is subject to greater evaporation. However, snow accumulations approaching 500% of observed runoff may not be explainable in terms of evaporation alone. The high percentages occur in a region of cold temperatures resulting in a higher proportion of snow accumulation (Figure 18c). In addition, since this is a non-forested area, it may be subject to significant sublimation from blowing snow. Since this effect is not represented, the resulting overprediction of SWE could explain the excessively high percentage of snow accumulation relative to runoff. Figure 18c shows that the mean maximum SWE as a percentage of annual precipitation decreases with latitude in the Mackenzie basin, starting with 50% at the northernmost latitudes. This trend is not as clear in the Ob basin.

The spatial and interannual variability of runoff is explored in Figure 19 through the coefficients of variability of runoff ratio, runoff due to precipitation and runoff due to snowmelt. The average C_vs of runoff ratios in both the Ob (mean = 0.27) and Mackenzie (mean = 0.40) basins are higher than those reported by Grabs et al. [4] for total basin runoff. However, many of the individual grid cell values are consistent with estimates from discharge at the basin outlet and a few regions of high variability elevate

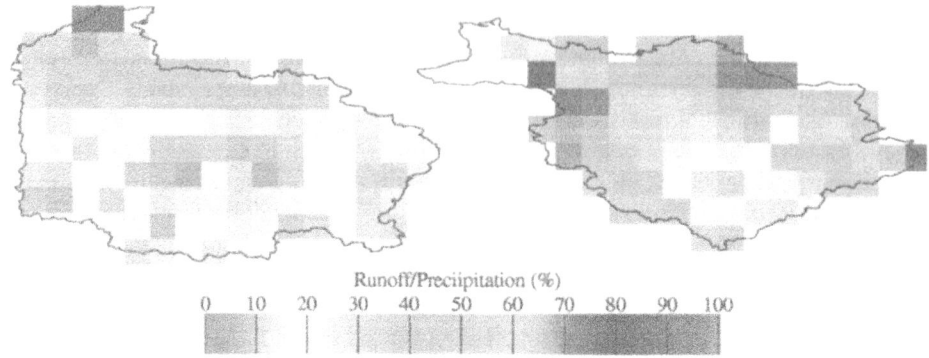

A) Annual Runoff/Annual Precipitation: Ob River (left) and Mackenzie River (right)

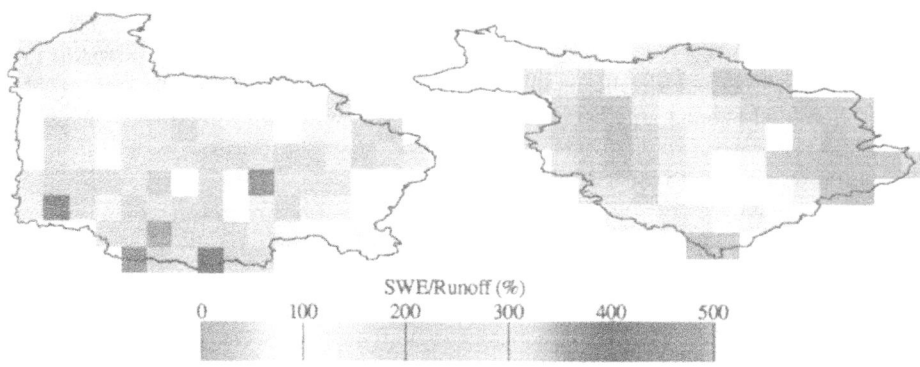

B) Maximum Snow Water Equivalent (SWE)/Annual Runoff: Ob River (left) and Mackenzie River (right)

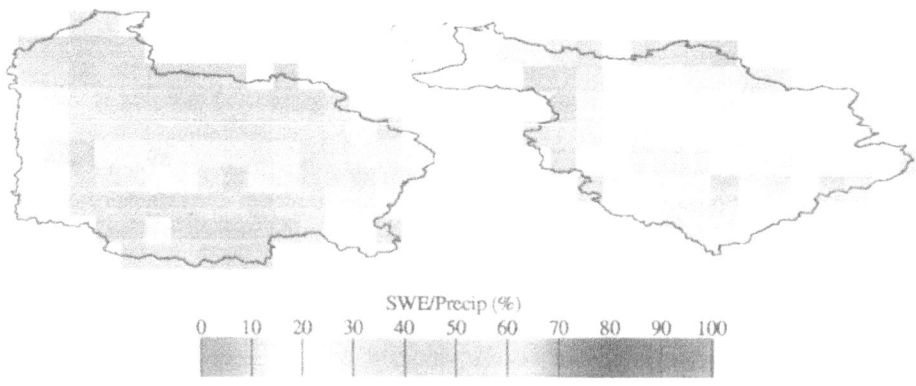

C) Maximum SWE/Annual Precip: Ob River (left) and Mackenzie River (right)

Figure 18: Simulated Water Balance Components

(For a colour version of these figures, see p. 614)

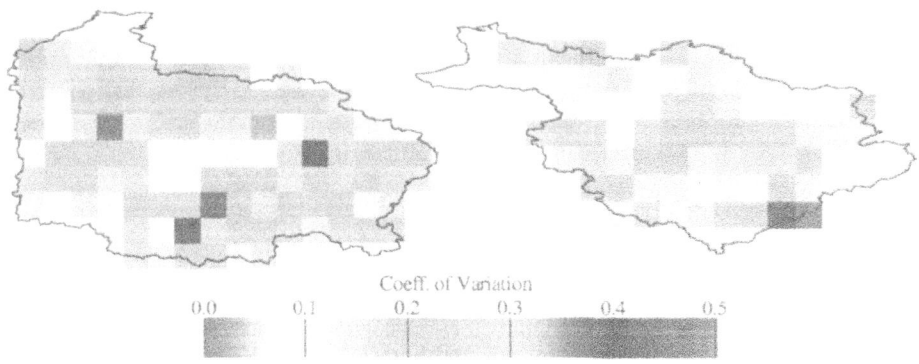

A) Runoff ratio: Ob River (left) and Mackenzie River (right)

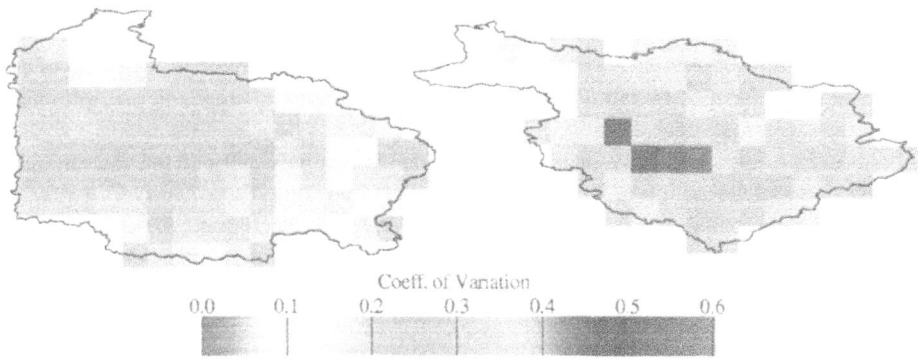

B) Maximum Snow Water Equivalent (SWE): Ob River (left) and Mackenzie River (right)

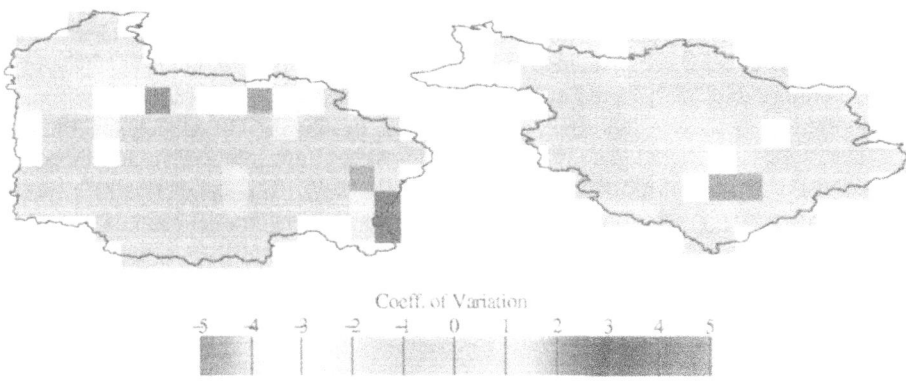

C) Annual Runoff –Maximum SWE: Ob River (left) and Mackenzie River (right)

Figure 19: Coefficient of Variation of Simulated Water Balance Components

(For a colour version of these figures, see p. 615)

the average (Figure 19a). The Mackenzie basin contains an area of very low variability consistent with the low rates of base flow in the approximate location of the Great Bear and Great Slave Lakes. The variability in runoff is predominately due to variation in annual runoff minus the maximum annual SWE (surrogate for runoff due to rain precipitation - Figure 19c). This value is much more variable than the maximum annual SWE (Figure 19b), which corresponds to the runoff due to snowmelt.

6.   Conclusions

The Arctic drainage basin is defined as all land surfaces that drain to the Arctic Ocean. It spans 37° of latitude and 360° of longitude and encompasses several mountain ranges, as well as extensive coastal plains and lowland boreal forests. Vegetation varies from boreal forest in the south to sparse Arctic tundra in the north.

The controls on runoff of medium to large Arctic rivers in this vast area were investigated through review of field experiments, exploratory data analysis and interpretation of hydrologic model results. This analysis suggests the following:

- The spatial distribution of runoff production within the Arctic drainage basin is primarily controlled by latitude and orography, as a result of trends in precipitation and evaporation.
- Surface storage, primarily in lakes and wetlands, suppresses both the interannual and seasonal variability of North American rivers relative to Eurasian rivers.
- Runoff interannual variability is only weakly related to interannual variability in snow extent and only for large rivers. The majority of interannual variability in runoff is explained by variations in precipitation.
- Preliminary application of a macroscale hydrologic model to the Mackenzie and Ob basins demonstrates the potential of hydrologic models to provide a continuous spatial and temporal trace of hydrologic variables, such as evapotranspiration and snow cover, that can be used as diagnostic tools in data scarce regions. As is the case with atmospheric data assimilation, the results are somewhat model dependent. However, in contrast to atmospheric assimilation methods, the macroscale hydrology model is constrained to preserve a moisture balance at the land surface.
- Exploratory data analysis in Sections 2 and 3 of this paper suggests the importance of physical processes that are not represented by the current model generation. These processes include sublimation from blowing snow, surface storage in lakes and wetlands and infiltration limitation by frozen soils. The extent to which more detailed process representations will improve current model simulations is, however, unclear due to the paucity of observational data.

7.   Future Research Priorities

The analysis done to date has demonstrated that the VIC macroscale hydrologic model can replicate the timing and variability of fresh water fluxes to the Arctic Ocean from large northern river systems. However, it has also illuminated potential concerns, which may increase in importance when transferring the model to the smaller, ungauged catchments of the Arctic drainage basin. Resolution of these concerns points to the following research needs:

- Future work in this area is constrained by the availability of surface meteorological data at appropriate space-time resolutions. The primary need is to assemble a retrospective precipitation data set of the highest possible quality including gauge correction, at least one decade in duration.

- Understanding of land surface processes should be developed from both models and data, with a focus on surface energy exchange, and in particular the snow vapor flux, as well as the water cycle.

- Future modeling must focus on the land surface processes of the entire Arctic basin. The spatial variability of surface water impoundment by lakes and wetlands, and frozen soil cannot be captured by parameterizations that represent spatially averaged processes at the resolution of a model grid cell. Computationally efficient process representations of these processes, perhaps in a derived distribution context, are needed.

8.   References

1.   Alekseev, G. and Buzuev, A. (1973) On the evolution of the ice-surface layer of the ocean system in the region of drift of the Severniya Polyus-16 station, *Problems of the Arctic and Antarctic* **42**, 37-43.
2.   Treshnikov, A.F. (1985) Main stages and prospects of study in the polar regions of the Earth, in *Problems of the Arctic and the Antarctic, Collection of Articles* **57**, 1-19.
3.   Aagaard and Carmack (1989) The role of sea ice and other freshwater in the Arctic circulation, *Journal of Geophysical Research*, **94** (c), 14485-14498.
4.   Grabs, W.E., Portmann, F., de Couet, T. (1999) Discharge observation networks in Arctic regions: computation of the river runoff into the Arctic Ocean, its seasonality and variability, this volume.
5.   WCRP (1996) Report of the fourth session of the WCRP ACSYS scientific steering group, Toronto, Canada, October 11-14, 1995, WCRP informal report N10.
6.   WCRP (1994) Arctic Climate System Study (ACSYS) Initial Implementation Plan, WCRP-85, WMO/TD-No. 627.
7.   Anisimov, O.A. and Nelson, F.E. (1997) Permafrost zonation and climate change in the northern hemisphere: results from transient general circulation models *Climate Change* **35**, 241-58.
8.   Foster J., Liston, G., Koster, R. Essery, R., Behr, H., Dumenil, L. Verseghy, D., Thompson, S., Pollard, D. and Cohen, J. (1996) Snow cover and snow mass intercomparisons of general circulation models and remotely sensed datasets *Journal of Climate* **9**, 409-26.
9.   Kite, G.W., Dalton, A. and Dion, K. (1994) Simulation of streamflow in a macroscale watershed using general circulation model data, *Water Resources Research* **30**, 1547-1559.
10.   Kuhl, S.C. and Miller, J.R. (1992) Seasonal river runoff calculated from a global atmospheric model, *Water Resources Research*, **28**, 2029-2039.
11.   Arpe, K., Behr, H. and Dumenil, L. (1996) Validation of the ECHAM4 climate model and re-analyses data in the Arctic region, *Proceedings of the workshop on the implementation of the Arctic Precipitation Data Archive at the Global Precipitation Climatology Centre*, WMO/TD No. 804, 31-40.
12.   Hurst, C. and Serreze, M.C. (1997) The utility of NCEP/NCAR reanalysis for Arctic precipitation studies, *Proceedings, Conference on Polar Processes and Global Climate*, Part I of II, 100-102.
13.   Serreze, M.C. and Maslanik, J.A. (1997) Arctic precipitation as represented in the NCEP/NCAR reanalysis, *Annals of Glaciology*, **25**.
14.   Bromwich, D.H. (1997) The atmospheric moisture budget of the Arctic and Antarctic from atmospheric numerical analyses, *Proceedings, Conference on Polar Processes and Global Climate*, Part I of II, 30-32.

15. Genthon, C. (1997) Convergence of energy and moisture to the Arctic and Antarctic polar caps from ECMWF re-analyses, *Proceedings, Conference on Polar Processes and Global Climate*, Part I of II, 60-62.

16. Legates, D.R. and Willmott, C.J. (1990) Mean seasonal and spatial variability in gauge-corrected, global precipitation, *International Journal of Climatology* **10**, 111-27.

17. Oki, T., Musiake, K., Matsuyama, H. and Masuda, K. (1995) Global atmospheric water balance and runoff from large river basins, *Hydrological Processes* **9**, 655-678.

18. Walsh, J.E. and Kattsov, V. (1996) Data requirements for validation of climate model simulations and reanalyses of arctic precipitation, *Proceedings of the workshop on the implementation of the Arctic Precipitation Data Archive at the Global Precipitation Climatology Centre*, WMO/TD No. 804, 19-25.

19. Molod A., Helfand, H.M. and Takacs, L.L. (1996) The climatology of parameterized physical processes in the GEOS-1 GCM and their impacts on the GEOS-1 Data Assimilation system, *Journal of Climate* **9**, 764-785.

20. Woo, M.K. and Winter, T.C. (1993) The role of permafrost and seasonal frost in the hydrology of northern wetlands in North America, *Journal of Hydrology* **141**, 5-31.

21. Anderson, J.R., Hardy, E.E., Roach, J.T. and Witmer, R.E. (1976) A land use and land cover classification system for use with remote sensor data, U.S. Geological Survey Professional Paper 964.

22. Serreze, M.C., Rehder, M.C., Barry, R.G. and Kahl, J.D. (1995) The distribution and transport of atmospheric water vapour over the Arctic basin, *International Journal of Climatology*, **15**, 709-27.

23. Hinzman, L.D., Kane, D.L., Benson, C.S. and Everett, K.R. (1996) Energy balance and hydrological processes in an Arctic watershed, *Ecological Studies* **120**, 131-154.

24. Marsh, P., Quinton, W. and Pomeroy, J. (1994) Hydrological processes and runoff at the Arctic treeline in northwestern Canada, *Proceedings, 10th Inter. Northern Research Basins Symp. And Workshop,* Trondheim, Norway, Norwegian Institute of Technology, 368-397.

25. Moskvin, Y.P. (1989) Runoff from hummocky marshes of western Siberia, *Soviet Meteorology and Hydrology* **3**, 73-80.

26. Storr, D. and Golding, D.L. (1974) A preliminary water balance evaluation of an intensive snow survey in a mountainous watershed, in *Advanced Concepts and Techniques in the Study of Snow and Ice Resources*, Nat. Acad. Sci., Washington, D.C., pp. 294-303.

27. Walsh, J.E., Zhou, X., Portis, D. and Serreze, M.C. (1994) Atmospheric contribution to hydrologic variations in the Arctic, *Atmosphere-Ocean* **32**, 733-755.

28. Brown, J., Ferrians, O.J. Jr., Heginbottom, J.A. and Melnikov, E.S. (1997) Circum-Arctic map of permafrost and ground-ice conditions, United States Geological Survey, Circum-Pacific Map Series, CP-45, Reston, VA, USA.

29. Bjornsson, H., Mysak, L.A. and Brown, R.D. (1995) On the interannual variability of precipitation and runoff in the Mackenzie drainage basin, *Climate Dynamics* **12**, 67-76.

30. Pavlov, V.K. and Stanovoy, V.V. (1997). Climatic signal in the fluctuations of the sea level and river run-off in the Arctic Ocean, *Proceedings, Conference on Polar Processes and Global Climate*, Part II of II, 184-186.

31. Stele, M. (1996) A simple model of the Arctic Ocean freshwater balance, 1979-1985, *Journal of Geophysical Research* **101c**, 20833-20848.

32. Vuglinsky, V.S. (1997). River water inflow to the Arctic Ocean - conditions of formation, time variability and forecasts, *Proceedings, Conference on Polar Processes and Global Climate*, Part II of II, 275-276.

33. Sokolov, A.A. (1965) On the excess of maximum discharges of summer-fall rain floods over discharges of spring high water, *Soviet Hydrology* **5**, 476-482.

34. Woo, M.K. and Steer, P. (1983) Slope hydrology as influenced by thawing of the active layer, Resolute, N.W.T., *Canadian Journal of Earth Sciences* **20**, 978-86.

35. Kane, D.L. and Hinzman, L.D. (1988) Permafrost hydrology of a small Arctic watershed, In: B.A. Senneset-Kaare (ed), *Permafrost; fifth international conference* **1**, International Conference on Permafrost, Proceedings, pp. 590-595.

36. Roulet, N.T. and Woo, M.K. (1986) Hydrology of a wetland in the continuous permafrost region, *Journal of Hydrology* **89**, 73-91.

37. Untersteiner, N. (1984) The cryosphere, in *Global Climate*, Cambridge, Cambridge, U.K., pp. 121-40.

38. Shea, D. (1986) Climatological atlas: 1950-1979, surface air temperature, precipitation, sea-level pressure and sea-surface temperature (45°S–90°N), *Tech. Note/TN-269+STR*, Natl. Center for Atmospheric Research, Boulder, Colorado.

39. Cattle, H. (1985). Diverting Soviet rivers: some possible repercussions for the Arctic Ocean, *Polar Record* 22, 485-498.

40. Marsh, P. and Bigras, S.C. (1988) Evaporation from Mackenzie delta lakes, N.W.T., Canada, *Arctic and Alpine Research* 20, 220-229.

41. Rovansek, R.J., Hinzman, L.D. and Kane, D.L. (1996) Hydrology of a tundra wetland complex in the Alaskan Arctic coastal plain, U.S.A., *Arctic and Alpine Research*, 28, 311-317.

42. Bigras, S.C. (1990) Hydrological regime of lakes in the Mackenzie delta, Northwest Territories, Canada, *Arctic and Alpine Research* 22, 163-174.

43. Zoltai, S.C. (1979) An outline of the wetland regions of Canada, in *Proc. workshop on Canadian wetlands* 12, Environ. Can., Lands Dir., Ecol. Land Classif. Ser., pp. 1-8.

44. Church, M. (1974) Hydrology and permafrost with referece to northern North America, in *Proceedings Workshop Seminar on Permafrost Hydrology*, Can. Nat. Comm., IHD, Ottawa, pp. 7-20.

45. Woo, M.K. (1988) Wetland runoff regime in northern Canada, in *Permafrost: fifth international conference* 1, Proceedings, International conference on permafrost, pp. 644-649.

46. Woo, M.K. and Xia, Z. (1996) Effects of hydrology on the thermal conditions of the active layer, *Nordic Hydrology* 27, 129-142.

47. Jacobs, J.D. and Grandin, L.D. (1988) The influence of an Arctic large-lakes system on mesoclimate in south-central Baffin Island, N.W.T., Canada, *Arctic and Alpine Reseearch* 20, 212-219.

48. Jeffries, M.O., Zhang, T., Frey, K. and Kozlenko, N. (1999) Estimating late winter heat flow to the atmosphere from the lake-dominated Alaskan north slope, *Journal of Glaciology*, in press.

49. Marsh, P. and Pomeroy, J. (1996) Meltwater fluxes at an Arctic forest-tundra site, *Hydrological Processes* 10, 1383-1400.

50. Kane, D.L., Hinzman, L.D., Benson, C.S. and Liston, G.E. (1991) Snow hydrology of a headwater Arctic basin, I, *Water Resources Research* 27, 1099-1109.

51. Woo, M.K. (1982). Snow hydrology in the high Arctic, presented at the western Snow Conference, Reno, Nevada, April 20-23.

52. Li, L. and Pomeroy, J.W. (1997) Probability of occurrence of blowing snow, *Journal of Geophysical Research* 102d, 21955-64.

53. Mikhel, V.M. and Rudneva, A.V. (1967) Regionalization of the USSR according to the transport of snow, 5 441-449.

54. Shook, K. and Gray, D.M. (1996) Small-scale spatial structure of shallow snowcovers *Hydrological Processes* 10, 1283-1292.

55. Woo, M.K., Heron, R., Marsh, P. and Steer, P. (1983) Comparison of weather station snowfall with winter snow accumulation in High Arctic basins, *Atmosphere-Ocean* 21, 312-325.

56. Woo, M.K. (1976) Hydrology of a small Canadian high Arctic basin during the snowmelt period, *Catena* 3, 155-168.

57. Pomeroy, J.W., Marsh, P. and Gray, D.M. (1997) Application of a distributed blowing snow model to the Arctic, *Hydrological Processes* 11, 1451-1464.

58. Takeuchi, Y., Kodama, Y. and Nakabayashi, H. (1995) Characteristics of evaporation in Spitsbergen in the snowmelt season, 1993, *Proceedings of the NIPR Symposium on Polar Meteor. And Glaciol.*

59. Pomeroy, J.W. and Li, L. (1997) Development of the Prairie Blowing Snow Model for application in climatological and hydrological models, in *Proceedings of the Eastern Snow Conference* 54, pp. 186-197.

60. Pomeroy, J.W., Gray, D.M. and Landine, P.G. (1993) The Prairie Blowing Snow model: characteristics, validation, operation, *Journal of Hydrology* 144, 165-192.

61. Pomeroy, J.W. and Gray, D.M. (1995) Snowcover accumulation, relocation and management, *National Hydrology Research Institute Science Report No. 7*, NHRI Environment Canada, Saskatoon, 144 pp.

62. Dyunin, A.K. (1959) Fundamentals of the theory of snow drifting, *Izvest. Sibirsk, Otdel, Akad. Nauk. USSR* 12, 11-24. [English translation by Belkov, G. 1961 Technical Translation 952, National Research Council of Canada, Ottawa.

63. Schmidt, R.A. (1972) Sublimation of wind-transported snow – a model, Research Paper RM-90, USDA Forest Service, Rocky Mountain Forest and Range Experimental Station, Fort Collins, 24 p.

64. Pomeroy, J.W. (1989) A process-based model of snow drifting, *Ann. Glaciol.* **13**, 237-240.

65. Groisman, P.Y., Golubev, V.S., Genikhovich, E.L. and Bomin, S. (1997) Evaporation from snow cover: an empirical study, in *Proceedings, Conference on Polar Processes and Global Climate*, Part I of II, pp. 72-73.

66. Deryugin, A.A. (1990) Snow cover on small forest and field watersheds in the taiga zone of European USSR, *Soviet Hydrology and Meteorology* **1**, 101-5.

67. Are, A.L. and Petropavlovskaya, M.S. (1982) Spring snow melting and evaporation of snow in central Yakutia, *Soviet Meteorology and Hydrology* **2**, 72-76.

68. Krestovskiy, O.I., Postnikov, A.N. and Sergeyeva, A.G. (1972) Evaporation during the snow melting and flood period in spring, *Soviet Hydrology* **5**, 439-451.

69. Harding, R.J. and Pomeroy, J.W. (1996) The energy balance of the winter boreal landscape, *Journal of Climate* **9**, 2778-87.

70. Davis, R.E., Hardy, J.P., Ni, W., Woodcock, C., McKenzie, J.C., Jordan, R. and Li, X. (1997) Variation of snow cover ablation in the boreal forest: a sensitivity study on the effects of conifer canopy, *Journal of Geophysical Research – Atmospheres* **102**, 29389-95.

71. Pomeroy, J.W. and Schmidt, R.J. (1993) The use of fractal geometry in modelling intercepted snow accumulation and sublimation, in *Proc. Eastern Snow Conference* **50**, pp. 1-10.

72. Hedstrom, N.R. and Pomeroy, J.W. (1998) Measurements and modelling of snow interception in the boreal forest, *Hydrological Processes* **12**, 1611-1625.

73. Golding, D.L. (1978) Calculated snowpack evaporation during chinooks along the eastern slopes of the Rocky Mountains in Alberta, *Journal of Applied Meteorology* **17**, 1647-51.

74. Schnur R. and Lettenmaier, D.P. (1997) A global gridded data set of soil moisture for use in General Circulation Models, Poster presented at the 13th conference on hydrology, 77th AMS annual meeting, Long Beach, CA, February 7.

75. Huffman, G.J., Adler, R.F., Arkin, P.A., Chang, A., Ferraro, R., Gruber, A., Janowiak, J., Joyce, R.J., McNab, A., Rudolf, B., Schnieder, U. and Xie, P. (1997) The Global Precipitation Climatology Project (GPCP) Combined Precipitation data set, *Bulletin of the American Meteorological Society* **78**, 5-20.

76. Legates, D.R. (1987) A climatology of global precipitation, *Climatology* **40**.

77. Schubert, S.D., Rood, R.B. and Pfaendtner, J. (1993) An assimilated data set for earth science applications, *Bulletin of the American Meteorological Society* **74**, 2331-2342.

78. --- (1978) *Hydrological Atlas of Canada*, Secretariat of the Canadian National Committee for the International Hydrological Decade.

79. Liang, X., Lettenmaier, D.P., Wood, E.F. and Burges, S.J. (1994) A simple hydrologically based model of land surface water and energy fluxes for general circulation models, *Journal of Geophysical Research* **99** 415-428.

80. Liang, X., Lettenmaier, D.P. and Wood, E.F. (1996) One-dimensional statistical dynamic representation of subgrid spatial variability of precipitation in the two-layer Variable Infiltration Capacity model, *Journal of Geophysical Research* **101d**,403-21,422.

81. Cherkauer, K.A. and Lettenmaier, D.P. (1999) Hydrologic effects of frozen soils in the Upper Mississippi River basin, *Journal of Geophysical Research* **GCIP Special Issue**, in press.

82. Storck, P. and Lettenmaier, D.P. (1999) Predicting the effect of a forest canopy on ground snow accumulation and ablation in maritime climates, in C. Troendle (ed), *67th Western Snow Conference*, Colorado State University.

83. Lohmann, D., Raschke, E., Nijssen, B. and Lettenmaier, D.P. (1998a) Regional scale hydrology: I. Formulation of the VIC-2L model coupled to a routing model, *Hydrological Sciences Journal* **43**, 131-141.

84. Lohmann, D., Raschke, E., Nijssen, B. and Lettenmaier, D.P. (1998b) Regional scale hydrology: II. Application of the VIC-2L model to the Weser River, Germany, *Hydrological Sciences Journal* **43**, 143-157.

85. Jones, P. D. (1994) Hemispheric surface air temperature variations: A reanalyis and an update to 1993, *Journal of Climate* **7**, 1794-1802.

THE ARCTIC OCEAN'S FRESHWATER BUDGET: SOURCES, STORAGE AND EXPORT

EDDY C. CARMACK
Institute of Ocean Sciences
9860 West Saanich Road
Sidney, British Columbia, V8L 4B2
Canada

1. Introduction

Freshwater components are delivered to the Arctic Ocean by atmospheric transport and by ocean and river inflows. Further net distillation of fresh water may occur within the Arctic during the melt/freeze cycle of sea ice, provided that the ice and rejected brine formed by freezing in winter can be separated and exported before they are reunited by melting and mixing the following summer. Once in the Arctic fresh water is stored within the various layers above and within the halocline, the latter serving as an extremely complex and poorly understood reservoir. The ultimate sink of fresh water is its export southwards into the North Atlantic via Fram Strait and the Canadian Arctic Archipelago, for it must ultimately replace the fresh water evaporating from low latitude oceans to close the global freshwater budget. To close this loop, this return of fresh water to the climate system must re-enter the thermohaline circulation either as deep or intermediate waters, and then return to the surface at low latitudes. In this chapter we address in various degrees of detail a number of questions about the disposition of fresh water in the upper layers of the Arctic Ocean: What are the sources and pathways of moisture that enter arctic drainage basins? How are the various components that comprise the freshwater budget (e.g. precipitation, ocean and river inflow, ice-melt) partitioned? What structures and processes affect the distribution and residence time of fresh water within the arctic halocline? What role (major or minor) do adjacent shelves play in halocline formation? What are the ways and means of freshwater export from the Arctic Ocean? Where does fresh water go when it re-enters the World Ocean? Finally, how stable are present-day distributions and trans-

E.L. Lewis et al. (eds.), The Freshwater Budget of the Arctic Ocean, 91-126.

port patterns to change? Such questions require good answers before we put faith in predictions of global change.

The motivation for understanding the freshwater budget of the Arctic Ocean is clear; it is salt stratified rather than temperature stratified, so that the addition or removal of fresh water will have a major effect on ocean dynamics. Indeed, the surface and halocline layers of the Arctic Ocean (Figure 1) are part of a much larger, salt-stratified system that covers the high-latitude oceans of the northern hemisphere between the subarctic fronts of the Pacific and Atlantic oceans. Within the global climate system the Arctic Ocean receives an excess of precipitation and runoff over evaporation, a requirement if it is to assist the Earth's climate system in transporting heat poleward. Its salt-stratified upper layers provide this ocean with the vertical stability it requires to allow formation of an ice cover. The subsequent export of low-salinity upper waters and sea ice southwards into the convective regions of the North Atlantic couples the Arctic to the World Ocean's thermohaline circulation. The various processes and connections within the high-latitude climate system are shown schematically in Figure 2. Despite these very straightforward concepts, major difficulties exist in understanding the Arctic Ocean's freshwater budget and its relationship to the global climate system.

The task of compiling mass and energy budgets for the Arctic Ocean has long challenged researchers (cf. [1], [2]). Early investigations of the Arctic Ocean's role in climate were largely concerned with volume transports and, especially, the heat budget (cf. [3], [4], [5]). In recent years, however, attention has turned more and more towards defining the freshwater budget. An early motivation for this line of inquiry concerned the relationship between ocean stratification and ice cover in view of the impact of proposed Eurasian river diversion schemes [6], [7], [8]. Even greater interest appeared when investigators began to consider the role of freshwater fluxes in high latitudes on global scale climate (cf. [9], [10], [11], [12]). Owing to both modelling [13], [14], [15], [16] and observational [17] work, this latter issue drives much of current research in arctic oceanography.

It is currently thought that there are two main ways in which the Arctic Ocean might impact on global climate, and both relate to the high-latitude hydrological cycle. The first is through its effect on the surface heat balance, i.e. the role of the ocean, sea ice and clouds in the local radiation balance and albedo (Figure 3a). The second is through its effect on the global thermohaline circulation by regulating deepwater formation rates (Figure 3b). Here, the strength of the overturning circulation, and thus the oceanic heat transport, depend on small density differences, which in turn depend upon a balance between the joint effects of cooling and freshening at high latitudes. To investigate potential large-scale consequences of freshwater export, Aa-

Figure 1. Map of the Arctic Ocean showing geographical provinces, major basins and watershed boundaries.

gaard and Carmack [18] compiled a freshwater budget for the Arctic Ocean and the Greenland, Iceland and Norwegian seas. They concluded that the Arctic Ocean exported sufficient freshwater components that even minor perturbations might force events such as the Great Salinity Anomaly in the North Atlantic (cf. [19]), and impact on deep convection in the Greenland Sea. Subsequently, Häkkinen [20] used an ocean general circulation model to demonstrate the plausibility of the first of these speculations, and Schlosser *et al.* [21] used time-dependent geochemical tracers to show inter-decadal variability in deep ventilation of the Greenland Sea. Aagaard and Carmack [22] later argued that perturbations in the freshwater budget may lead to basin-scale shifts in ocean frontal regions and the domains of deep convection. For example, they speculated that the role currently played by the Greenland gyre as a pre-condition to deep convection might move south

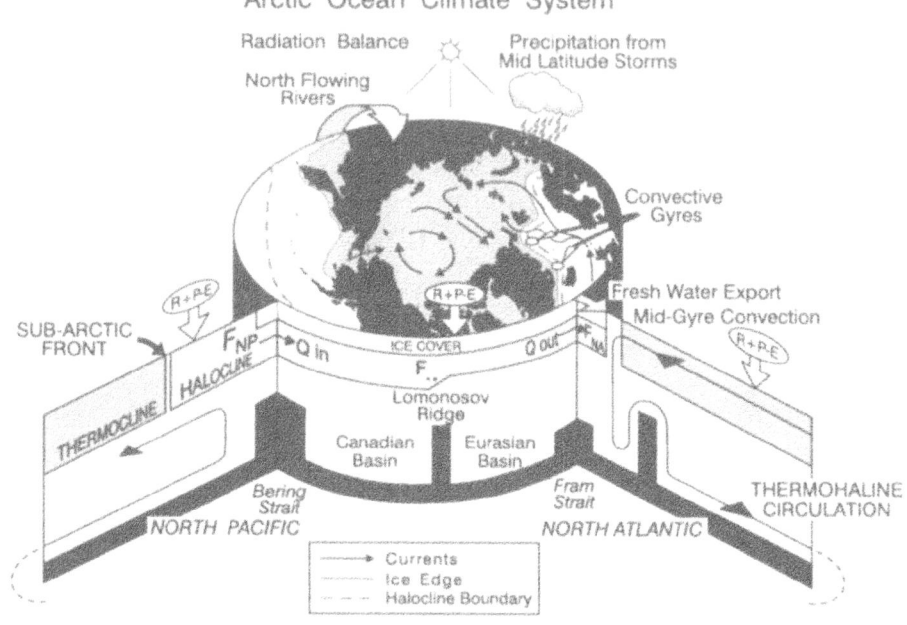

Figure 2. Schematic of the Arctic Ocean climate system. The North Pacific and North Atlantic connect to the Arctic Ocean through Bering and Fram straits, respectively. The components of the freshwater balance include runoff (R), precipitation (P), evaporation (E), storage in the upper North Pacific (FNP), the Arctic Ocean (FAO), and the North Atlantic (FNA), and the horizontal freshwater fluxes by ocean currents (Q_{in} and Q_{out}). In the horizontal plane the extent of sea ice in winter is shown by the shaded region, the mean surface circulation by arrows, and the convective gyres in the Greenland and Iceland seas by the dark shaded ovals. Connections to the global ocean are depicted through intermediate and deepwater formation and circulation. (Adapted from [22].)

into the Norwegian Sea or north into the Eurasian Basin under different regimes of freshwater export.

River runoff is clearly one of the key processes contributing to the Arctic Ocean's freshwater budget. The importance of runoff has prompted consideration of an estuarine-like circulation for the Arctic Ocean, and simple models reflecting this view have been proposed [23], [24]. Investigations using geochemical tracers have advanced our understanding of the riverine circulation within the Arctic Ocean. Anderson *et al.* [25] used alkalinity/salinity relationships to map the spreading pathway of Russian rivers across the Eurasian Basin, and estimated a much shorter residence time for runoff there than in the Canadian Basin. Falkner *et al.* [26] noted that barium could be used to distinguish between runoff of North American and Eurasian origin, owing to their specific watershed characteristics. Schlosser

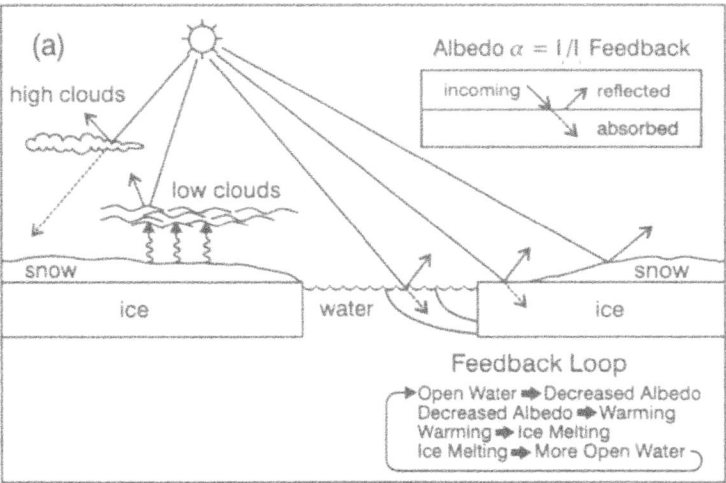

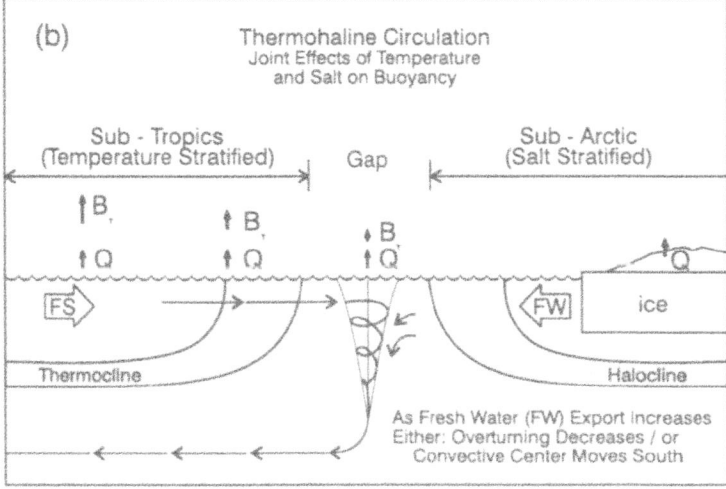

Figure 3. Schematic diagrams illustrating the role of the Arctic Ocean in global climate: (a) effect of the ocean, sea ice and clouds in the local radiation balance and albedo; (b) effect on the global thermohaline circulation by regulating deepwater formation rates.

et al. [27] used isotopes of helium and oxygen to elucidate pathways and time scales for the transit of fresh water through the Arctic Ocean.

Fresh water is stored mainly in the surface and halocline layers of the Arctic. Nansen [28] first noted that since temperatures remain cold within the halocline, its origin cannot be a linear mixture of surface and Atlantic waters. Coachman and Barnes [29] suggested that cold and saline waters draining through submarine canyons adjacent to the Eurasian Basin allowed halocline formation. The idea that off-shelf drainage of brine water derived from ice growth is largely responsible for halocline formation has been suggested by Aagaard et al. [30], Melling and Lewis [31], and Jones and Anderson [32], and modelled by Killworth and Smith [33]. Coachman and Barnes [34] also noted that inflows from the Pacific via Bering Strait contributed to halocline water formation. Melling [35] showed that halocline ventilation varied from year to year, and argued that in addition to sea ice formation, a two-step pre-conditioning of shelf waters involving the wind-driven removal of low-salinity surface waters and upwelling of deep waters in autumn was required to force ventilation. Modelling efforts have further elucidated processes of brine drainage from shelves. For example, Gawarkiewicz and Chapman [36] address the role of flow instabilities, Chapman and Gawarkiewicz [37] consider the influence of submarine canyons in offshore transport, and Chapman and Gawarkiewicz [38] examine the role of shelf polynyas in promoting rapid ice growth and brine production. Untersteiner [39] and Rudels et al. [40] proposed an additional mechanism for halocline formation in the Eurasian Basin which entails the melting of sea ice exiting the Arctic Ocean by incoming Atlantic water and its subsequent subduction below the mixed layer. Ekwurtzel [41] used geochemical tracers and found halocline tracer (^{3}H/^{3}He) ages on isopycnal (isohaline) surfaces to be similar across the entire basin: near $S = 33.1$ the tracer age was about 4 years, and near $S = 34.2$ the tracer age was about 10 years. She also noted that sea-ice meltwater was the only freshwater component in the southern Nansen Basin, consistent with Atlantic water melting of sea ice as an important process of halocline formation.

Prior to this decade, the water mass structure of the Arctic Ocean was generally believed to be stationary. However, evidence of change is growing. Significant changes were first observed as a warming of the Atlantic layer [42], [43], [44], [45], [46], and is also evident now in ice drift patterns [47]. Such changes may reflect the ocean's link to large-scale changes in atmospheric pressure [48]. More importantly, with regard to the volume storage of fresh water within the Arctic Ocean, is the observation by McLaughlin et al. [49] that halocline frontal systems may re-locate across basin-scale topographic features. Steele and Boyd [50] compared recent and climatological data to conclude that the Eurasian Basin cold halocline has retreated significantly in the 1990s. Proshutinsky and Johnson [51] give evidence that variability in arctic circulation is linked to atmospheric forcing. Steele et al. [52] developed an ice–ocean model initialized by satellite and other data and found interannual variability both in mixed-layer salinity and in freshwater export. Harder et al. [47], Hilmer et al. [53] and Kwok and Rothrock [54] cover the variability in sea-ice transport through Fram Strait. It is now clear that any comprehensive view of the Arctic Ocean's freshwater budget must take temporal and spatial variability into account.

For general background information, it is noted that Baumgartner and Reichel [55], Wijffels *et al.* [56] and Jourdan *et al.* [57] discuss global-scale compilations of freshwater budgets and fluxes. Barron *et al.* [58], Schmitt and Wijffels [59] and Webster [60] give discussions on the role of the hydro-logical cycle in ocean processes. Untersteiner [39], Aagaard and Carmack [22] and Barry *et al.* [61] provide perspectives on the Arctic's relationship to global climate. Reviews of physical processes affecting the Arctic Ocean are given by Carmack [62], [63], Muench [64], McPhee [65] and Padman [66].

2. Freshwater Sources

This section gives a compilation of freshwater imports and their pathways to the Arctic Ocean. For the purposes of this budget the Arctic Ocean is defined as lying north of Bering, Davis and Fram straits, and approximately east of the 15°E meridian joining Norway and Spitsbergen. It is admitted, however, that a more comprehensive budget would include the seasonally ice-covered areas of the Sea of Okhotsk, Hudson Bay and the convective regions of the Greenland and Labrador seas. Following Aagaard and Car-mack [18], all freshwater fractions are taken relative to a salinity of 34.80, estimated as the mean salinity of the Arctic Ocean. Focus is on defining present-day values for net precipitation, runoff, and Bering Strait inflow. Again, it is admitted that the Arctic Ocean's freshwater budget is a prod-uct of moisture transport patterns, especially those delivering water to the great drainage basins, and that such fields are subject to climate variability and feedback processes.

2.1. RIVER RUNOFF

The major rivers entering the Arctic Ocean comprise an area of over 22×106 km^2 and drain watersheds extending as far south as 45°N. Approxi-mately 60% of this area is covered by hydrometric observations. Long-term mean runoff tabulations from the major rivers entering the Arctic Ocean have been compiled by UNESCO [67], Milliman and Meade [68], Treshnikov [69], and Vuglinsky [70], and by various authors contributing to this book including Semiletov *et al.* [71].

Individual contributions from the major rivers entering the Arctic Ocean as tabulated by Vuglinsky [70] are shown schematically in Figure 4. Gauged flows include the Yenisei (630 km^3 yr^{-1}), Lena (532 km^3 yr^{-1}), Ob (404 km^3 yr^{-1}), Pechora (130 km^3 yr^{-1}), Kolyma (128 km^3 yr^{-1}), Severnaya Di-vna (109 km^3 yr^{-1}), Khatanga (88 km^3 yr^{-1}), and Indigirka (54 km^3 yr^{-1}) from Eurasia, and the Mackenzie (262 km^3 yr^{-1}) and other gauged rivers

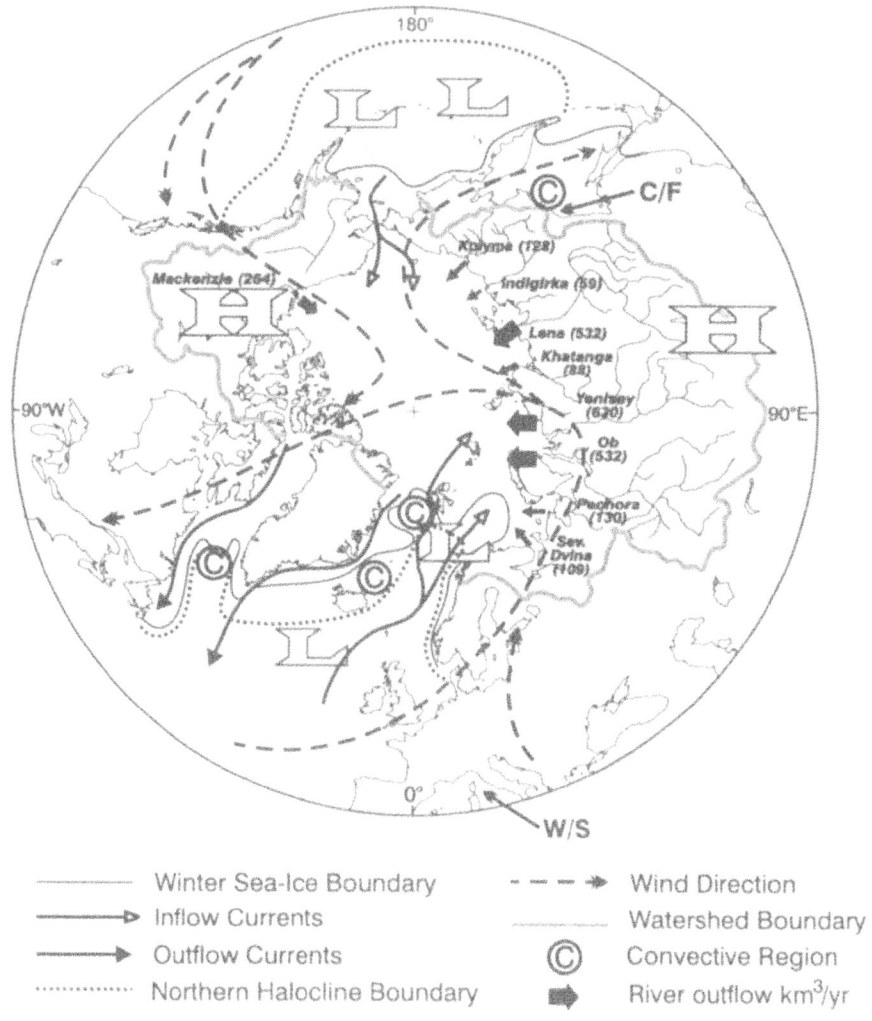

Figure 4. Map summarizing freshwater distributions and sources including rivers, ocean currents and moisture transport by winds. Also shown are approximate boundaries for the wintertime sea ice and the northern halocline, north of which the ocean is salt stratified. Convecting gyres are indicated in the Greenland, Iceland and Labrador seas.

(209 km^3 yr^{-1}) in North America. Ungauged flows surrounding the basin include runoff from both the mainlands (1007 km^3 yr^{-1}) and arctic islands (711 km^3 yr^{-1}). Vuglinsky thus finds the total runoff to the Arctic Ocean to be 3558 km^3 yr^{-1} from the mainland and 711 km^3 yr^{-1} from the arctic islands. A major point brought out by Vuglinsky's analysis is the potential importance of the ungauged flows. For example, he shows the

ungauged runoff from North American sources to be 603 km^3 yr^{-1}, over twice that of the Mackenzie River and greater than any other individual component other than the Yenisei River. The total river inflow is taken to be 4270 km^3 yr^{-1}; however, uncertainty and variability are acknowledged to be high, as discussed below.

Arctic rivers typically exhibit large seasonal variation in discharge (cf. [72], [73], [74]). The variety of the landscape (e.g. vegetation, elevation, and terrain) places enormous variability in the hydrological retention characteristics of these diverse drainage systems. Rivers with major headwater lakes will have different seasonal runoff patterns than those without. The Mackenzie River, for example, draws water from Athabaska, Great Slave and Great Bear lakes to maintain moderate flows during wintertime, while the Lena and Yenisei have no major headwater lakes. As a consequence, the Mackenzie River exhibits about a fivefold variation from summer to winter, while the Lena and Yenisei rivers vary by tenfold or more. Arctic rivers also exhibit interannual and decadal scale variability (cf. [75]). Smaller arctic rivers and those lying at very high latitudes can be expected to experience very large seasonal and interannual variability (cf. [76]). However, Vuglinsky [70] states that the variability of total annual inflow to the Arctic Ocean is small.

An additional point of caution must made concerning the above tabulations, namely, Arctic rivers are difficult to gauge (see [77]). In most cases, a river is flow-metered at two or more times during the year, hopefully representing extreme high and low flows at some hydraulic control point, and the corresponding water level measured. These measurements are then used to construct the so-called stage/discharge curve, which presupposes a relationship between flow and water level. For the remainder of the year, water level only, is recorded and this is used to calculate, indirectly, river flow. This method has severe limits, both practical and theoretical, in arctic rivers. For example, it is far from simple to meter ice-covered northern rivers under minimum flow conditions, as this time usually corresponds to minimum air temperatures. Throughout winter the ice cover varies in roughness (Chezy coefficient) and thus in its hydraulic resistance to flow. Frazil dam formation beneath ice can also alter flow conditions. Measurements at high flow are also difficult and, indeed, dangerous, as this often occurs coincident with ice break-up, a time of wildly fluctuating frictional resistance in river channels, and frequently under conditions of overflow (runoff flowing over the ice cover), all of which confounds runoff estimates at a critical time of the year. Near points of inflow to the ocean measurements of water level may be confounded by tide and wind set-up effects. It is noted that no international standards exist for the gauging of Arctic rivers, nor has a comparative study of methodology been carried out.

2.2. BERING STRAIT INFLOW

The generally northward flow through the shallow (about 50 m) and narrow (about 85 km) Bering Strait connects the Pacific and Arctic oceans. Transport through Bering Strait has been discussed recently by Coachman and Aagaard [78] and Roach et al. [79]. Dynamical considerations are covered by Overland and Roach [80] wherein a northward pressure gradient corresponding to a sea surface slope of order 10^{-6} (cf. [81]) is balanced by southward wind stress and friction. Pacific water inflow occurs via two main branches crossing the Chukchi Shelf [82] and entering the Arctic Basin through Barrow and Herald canyons (cf. [83], [79], [84], [85]).

Coachman and Aagaard [78] calculate a long-term mean flow of 0.8 Sv. The annual cycle is estimated at 0.3 Sv, with a maximum in summer and a minimum in winter. The salinity of the inflow varies from about 31 to 33, and the long term mean is about 32.5. The freshwater import from the Pacific is thus 1670 km^3 yr^{-1} [18]. A more recent analysis by Roach et al. [79] based on direct current measurements between 1990 and 1994 attain a nearly-identical mean volume transport of 0.83 Sv, but suggest larger amplitudes of interannual variability.

Bering Strait inflow is believed to supply the halocline with two distinct water masses, traditionally called Bering Sea Summer Water (BSSW, a relatively warm layer near $S = 32.2$) and Bering Sea Winter Water (BSWW, a cold layer near $S = 33.1$). Alternately, the two water types may represent the two branches of Bering Strait inflow, with BBSW derived from the western flow near Alaska and BSWW derived from the eastern branch near Herald Canyon. Another distinction, yet unexplained, is that while both waters lies within the overall arctic halocline, BBSW lies within a layer of *locally* increased vertical salinity gradients (halocline), while BSWW represents a local mode water (halostad).

2.3. PRECIPITATION LESS EVAPORATION

There is considerable uncertainty regarding the flux of precipitation less evaporation $(P - E)$. For example, Aagaard and Carmack [18] noted the wide differences in the estimates of Vowinckle and Orvig [2] (about 900 km^3 yr^{-1}), Baumgartner and Reichel [55] (about 400 km^3 yr^{-1}) and Burova [86] (about 1400 km^3 yr^{-1}); they accepted an intermediate value of 900 km^3 yr^{-1}. More recent estimates of $P - E$ have been computed from moisture flux convergence based on aerological and reanalysis data given in other chapters in this book.

2.4. OTHER SOURCES

The seasonally variable Norwegian Coastal Current carries fresh water into the Barents Sea. Blindheim [87] used current meter and hydrographic measurements to estimate that the transport into the Barents Sea by the coastal boundary current of low-salinity water off northern Norway is about 0.7 Sv. His data suggest a mean salinity near 34.4. In this manner the contribution of the Norwegian Coastal Current to the Barents Sea and Arctic Ocean is about 330 km^3 yr^{-1}.

Not all fresh waters necessarily enter the Arctic Basin above the halocline. Within the Barents and Kara seas, for example, the incoming Atlantic waters are strongly modified by an unknown fraction of river runoff from the Severnaya Dvina, Pechora, Ob and Yenisei rivers [88], [89], [90], [91], by mixing with the Norwegian Coastal Current, and by freshening due to contact with sea ice (Aagaard, pers. comm.). Subsequent to cooling this water leaves the shelf as the so-called Barents Sea Branch [92], [93]. Blindheim [87] used current measurements across Bear Island Channel and found a net inflow of 1.9 Sv into the Barents Sea. Loeng et al. [89] estimate that the outflow from the Barents Sea varies from 1 to 4 Sv with a maximum in winter. Still, the role of the Barents Sea Branch on the freshwater budget of the Arctic Ocean remains unclear.

Some freshwater sources are believed to be small. One example is the import of sea ice through Bering Strait is neglected; Aagaard and Carmack [18] estimate this source to be only 24 km^3 yr^{-1}. Another unknown but presumed small contribution is the melting of icebergs entering the Arctic Ocean from Greenland and arctic islands. While not strictly with the Arctic Ocean proper, both Hudson Bay and the Sea of Okhotsk are very important components of the high-latitude cryosphere and climate system. Prinsenberg [94], [95] has compiled the freshwater budget for Hudson Bay, and finds the total annual runoff to be 709 km^3 yr^{-1}, far greater than that of the Mackenzie River. The volume ice forming and melting each year is likewise significant, amounting to a 1.45-m equivalent of fresh water assuming an un-deformed ice cover, and an additional 0.4- to 0.6-m equivalent of fresh water, taking into account ridging. A preliminary account of the freshwater budget for the Sea of Okhotsk is given in Rogachev (pers. comm.). Here, the Amur River discharges about 590 km^3 yr^{-1}, net precipitation yields about 1490 km^3 yr^{-1}, and ice melting adds an additional 890 km^3 yr^{-1}. The salt balance is maintained by fluxes through the Kuril and Soya straits.

3. Sea Ice and the Arctic Halocline as Freshwater Storage Reservoirs

The sea-ice cover and the surface and halocline waters of the Arctic Ocean serve as major storage reservoirs of fresh water. Aagaard and Carmack [18] used available hydrographic data to estimate freshwater storage within various regions of the Arctic Basin. They found the mean liquid water storage in the Arctic Ocean to be 80,000 km^3; of this 22,000 km^3 occur on the shelves and 58,000 km^3 in the deep basins. Of the latter, the Canadian Basin contains about 46,000 km^3 and the Eurasian Basin contains 12,000 km^3, owing to the deeper halocline of the Canadian Basin. Hence, the storage varies greatly across the deep basins of the Arctic Ocean, an important consideration in view of the frontal shift observed by McLaughlin et al. [49].

3.1. CIRCULATION AND WATER MASS STRATIFICATION

Two main flow fields characterize the sea-ice drift and surface layer currents in the Arctic Ocean. The first is the Transpolar Drift, in which surface waters of the Eurasian Basin move toward the North Pole and then on toward Greenland and Fram Strait. The second is the anticyclonic flow around the Canadian Basin. Mean speeds are slow in the central ocean, about 0.02 m s^{-1}, but increase as the water exits the Basin as part of the East Greenland Current [96], [97]. Interdecadal variability associated with changes in the atmospheric pressure regime are discussed by Proshutinsky and Johnson [51]; effects of ice transport pathways on the export of sea ice are examined by Steele et al. [52]. Below the surface layer the overall circulation of the Arctic Ocean is believed to be cyclonic and to follow shelf-break and ridge topography [98], [99], [92], [49], [100], [101]. As noted below, this boundary current serves to partition various layers of the arctic halocline according to source and basin geometry.

The basic stratification of the Arctic Ocean consists of an upper mixed layer and halocline complex, and intermediate depth Atlantic layer, and deepwater. The arctic mixed layer consists of the upper 30–50 m where changes involving the annual melt–freeze cycle and river disposition occur. Below this seasonally variable mixed layer lies a complex of cold, salt-stratified layers known collectively as the Arctic halocline. These layers insulate underlying warm saline layers of Atlantic origin from the ice cover and atmosphere. Regionally, the various components of the Arctic halocline can be ascribed to a number of processes that relate circulation patterns to the input of fresh water and to the melting and freezing of sea ice. Water mass boundaries separating waters of different sources are found both in the vertical and the horizontal, and such boundaries are now known to shift on inter-decadal time scales. Below the halocline complex lie the various components of the Atlantic layer, which occupy the water column below the Arctic thermocline and above about 1600 m. Water within this layer enters from the North Atlantic via two prominent branches (Fram Strait and Barents Sea) and moves anti-clockwise as narrow (about 100 km) cores following shelf-break and ridge topography. The 1990s have witnessed a major change within this layer; the spreading of new properties (thermohaline transition) occurs via topographically-steered boundary flows coupled with lateral θ/S intrusions [44], [102].

3.2. THE HALOCLINE COMPLEX

A comparison of representative temperature and salinity profiles (in log-depth scale) is shown in Figure 5 for each of the four major basins of the Arctic Ocean. Note that the salinity decreases from east (Nansen Basin) to west (Canada Basin), and hence freshwater storage varies substantially across the Arctic Ocean. The stratification of the Canada Basin is particularly complex, showing a sequence of modes and clines from the surface to about 300 m. For simplification, McLaughlin et al. [49] distinguished between the halocline structures of the eastern and western domains (Figure 6). The eastern Arctic is characterized by a uniformly cold and moderately fresh mixed layer, a halocline that remains cold as salinity increases to about 34.4, and a thermocline wherein both temperature and salinity increase with depth to the core of the Atlantic layer near 200–240 m depth. In the western Arctic the halocline is influenced by both Pacific and Atlantic inflows, exhibits more vertical structure and extends deeper in the water column before reaching the core of the Atlantic layer. Here, the temperature and salinity structure is characterized by a mixed layer (40–60 m depth, a thin layer at 60–80 m depth of increased temperature and salinity gradients (near $S = 32.4$) called Bering Sea Summer Water, a relatively thick layer at 80–180 m depth of reduced temperature and salinity gradients (near $S = 33.1$) called Bering Sea Winter Water, an upper thermocline of increasing temperatures and salinity gradients in the salinity range 33.1–34.4, and a lower thermocline on increasing temperature and decreasing salinity gradients down to the core of the Atlantic layer near 400–450 m depth. Steele and Boyd [50] make the further distinction between cold and warm haloclines.

At least three main mechanisms exist for the formation of halocline waters: these include (1) salinization of cold surface waters during ice formation [30], [31]; (2) melting of sea ice as the Atlantic waters first enter the ice-covered Arctic basin as a near-surface flow [39], [103], [40]; and (3) inflow of Pacific waters through Bering Strait across the broad Bering and Chukchi shelves [34], [49].

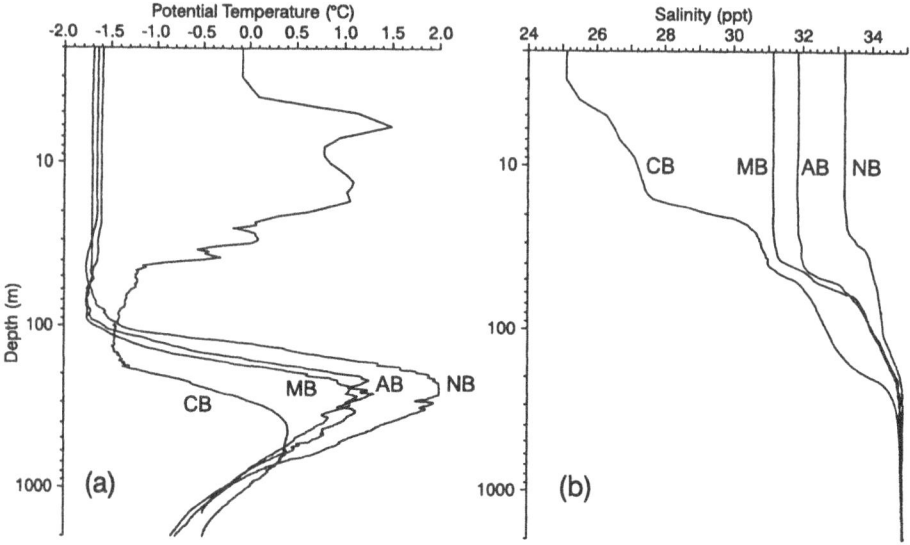

Figure 5. Profiles showing (a) temperature and (b) salinity in each of the four major basins of the Arctic Ocean where NB is Nansen Basin, AB is Amundsen Basin, MB is Makarov Basin and CB is Canada Basin. Note use of a logarithmic depth scale.

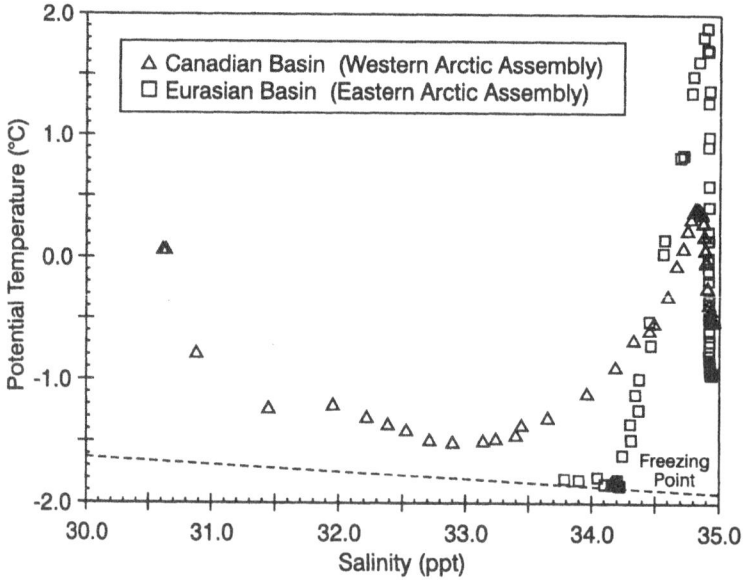

Figure 6. Correlation curves for potential temperature versus salinity illustrating water mass assemblies of the eastern and western domains of the Arctic Ocean. (Adapted from [51].)

3.2.1. *Shelf Processes*

Shelves occupy over 30% of the surface area of the Arctic Ocean, and are the sites where rivers enter and seasonal freezing and melting occur. In summer they are largely ice free and contain freshened waters due to runoff and ice melt. In late fall and winter the open waters freeze rapidly, and re-salinize the shelf waters by brine formation. It is thus impossible to discuss the Arctics freshwater budget without including shelf processes.

A key mechanism for the formation of halocline waters is thought to be the salinization of cold surface waters during ice formation and subsequent flow offshelf [30], [31]. Figure 7 shows a schematic of halocline ventilation by shelf drainage (from [30]). Melling and Lewis [31] and Melling and Moore [104] give observational evidence of offshelf flow into the halocline. Wallace *et al.* [105] discuss associated ventilation rates. Gawarkiewicz and Chapman [36] and Chapman and Gawarkiewicz [37], [38] explore dynamical processes affecting the offshore transport of dense water masses under a variety of forcing situations.

Macdonald and Carmack [106] and Macdonald *et al.* [107], [108] discuss the importance, locally, of under-ice topography in separating the positive and negative estuarine functions of an arctic shelf in winter. They note that the Mackenzie Shelf in the Canadian Beaufort Sea receives large amounts of freshwater runoff in winter and, yet, also produces ventilating water masses by brine rejection from growing sea ice. This is because under-ice topography due to large pressure ridges that form at the boundary between landfast and pack ice separate the shelf into two convective regimes. The inner regime is dominated by the impoundment of river inflow, whereas the outer regime is subject to brine rejection. Seasonal patterns on shelves taking into account river inflow, wind forcing and the annual cycle of ice formation and melting are modelled by Omstedt *et al.* [75].

3.2.2. *Atlantic Inflow and Sea-Ice Melting*

Another possible mechanism for the formation of halocline waters involves the melting of sea-ice as the Atlantic waters first enter the ice-covered Arctic Basin as a near-surface flow (cf. [39], [103]). To account for seasonal processes, Rudels *et al.* [40] present a sequential scenario (Figure 8) for halocline formation which incorporates Atlantic inflow, ice melting and freezing, and river inflow. Atlantic waters first melt ice to form a freshened surface layer, are homogenized by brine-driven convection during winter, and then are capped by incoming river waters.

3.2.3. *Pacific Inflow*

Northward flow through Bering Strait transports waters of Pacific origin into the Arctic Ocean [78], [79]. Inflow of this water occurs via two main

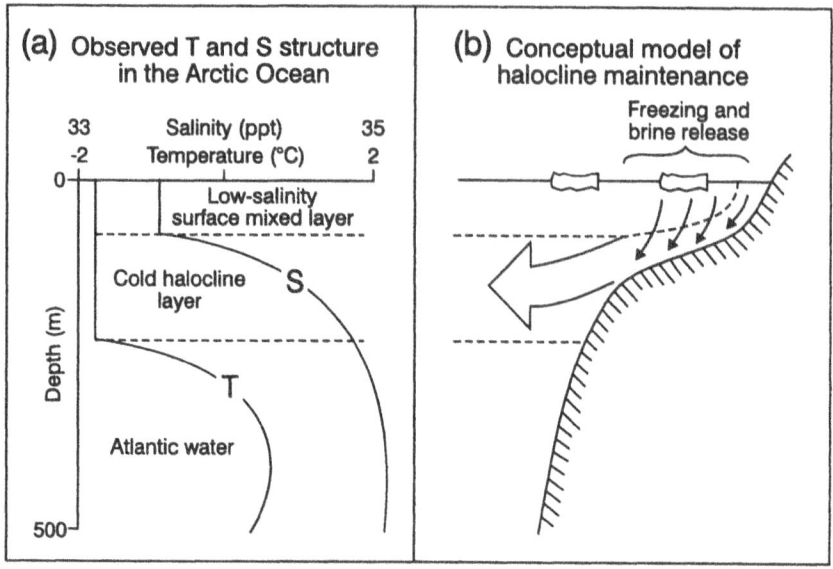

Figure 7. Schematic of halocline ventilation by shelf drainage. (Adapted from [30].)

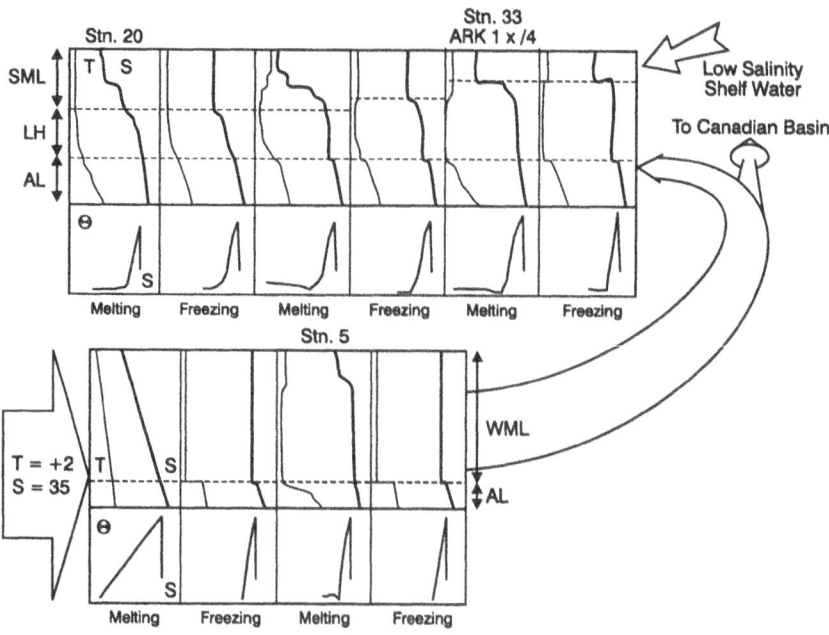

Figure 8. Schematic of halocline formation by sea-ice melting. (Adapted from [42].)

branches crossing the Chukchi Shelf [82] and entering the Arctic Basin through Barrow and Herald canyons. This water has long been recognized by both its T/S signature [34] and high nutrients [109]. Because of the extreme width of the Chukchi Shelf, this water takes on additional, shelf-modified features before entering the basin. One unique feature of this water that it displays an offset N/P ratio from that of other world ocean water masses. Jones et al. [110] argue this is due to de-nitrification; an alternate explanation is that it reflects phosphorus release from organically-rich sediments in the highly productive regions of the Chukchi Shelf. Figure 9 shows the initial spreading of Pacific origin (shelf-modified) waters across the southern Canadian Basin.

3.2.4. *Frontal Variability and Regime Shifts*

McLaughlin et al. [49] noted that arctic halocline structures can be divided into eastern and western assemblies, and that the two assemblies are separated by a frontal system (the Atlantic/Pacific front) that tends to lie parallel to ridge topography. They further noted that this front shifted positions in the late 1980s or 1990s from the Lomonosov to the Mendeleyev Ridge. Figure 10 illustrates the frontal regime shift mechanism and its potential impact on freshwater export [111]. Similarly, Melling [112] observed significant changes in the hydrographic structure of the Canada Basin during the period 1979 to 1996. Zhang et al. [113] argue that the increased inflow of Atlantic water, such as has been observed during the past decade, acts to flush out the overlying cold and fresh Arctic water. This may, in turn, increase the temperature and salinity of the surface layer, decrease stratification, increase oceanic heat flux to the mixed layer, and thus reduce the thickness and extent of ice cover.

3.3. SEA ICE STORAGE

In addition to liquid storage, fresh water is also stored in sea ice. The annual mean ice-covered area of the Arctic Ocean is about 6.5×10^6 km^2 but varies seasonally by 3 to 4×10^6 km^2 [114]. Aagaard and Carmack [18] used a mean sea-ice thickness of 3 m for the Arctic Ocean (taken from [115]) and a mean sea-ice salinity of 4 to compute a mean freshwater volume of 17,300 km^3 stored in sea ice.

The major uncertainty in estimates of sea-ice volume is sea-ice thickness. For example, the above calculation does not account for ice ridging and deformation (cf. [95]). Data from an upward-looking sonar in the Canada Basin, collected between 1990 and 1996 (Figure 11), reveals a thinner ice cover than that shown in most climatologies, and large seasonal variability. Indeed, based on a comparison of ice thickness observations in the Canada

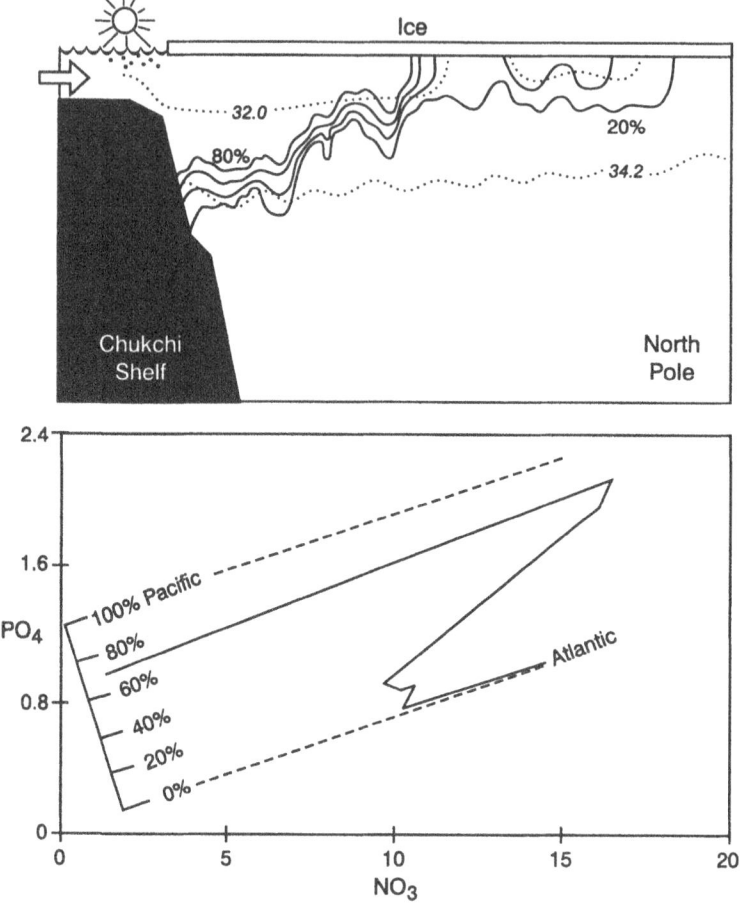

Figure 9. Spreading of Pacific origin (shelf-modified) waters across the Canadian Basin. The section of percent composition of Pacific waters (top) is found using (bottom) a nomogram constructed parallel to the N/P regressions of the two source waters using data from the 1994 Arctic Ocean Section (see [104]).

Basin between the mid-70s and 1997, McPhee *et al.* [116] suggest that the perennial sea-ice cover of the Arctic Ocean now may be vanishing, an hypothesis requiring further examination (but, see [117]).

4. Freshwater Export and Thermohaline Circulation

4.1. VIA FRAM STRAIT TO THE GREENLAND SEA

The 2600-m deep Fram Strait is the main outlet of the Arctic Ocean. Here, Arctic waters combine with the recirculating waters of the West Spitsbergen Current to form the East Greenland Current that flows southward along the Greenland continental margin. Downstream, this current branches along

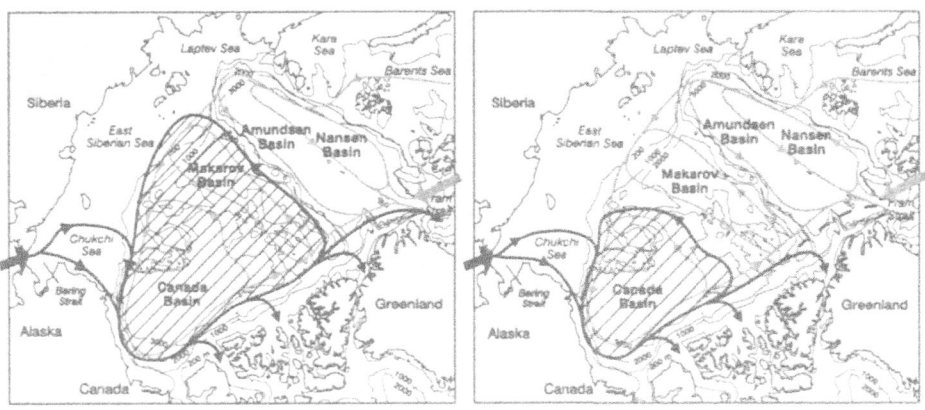

Figure 10. Schematic of the frontal regime shift mechanism from Mode I (Atlantic/Pacific front parallel to the Lomonosov Ridge) to Mode II (Atlantic/Pacific front parallel to the Mendeleyev Ridge). (Adapted from [111].)

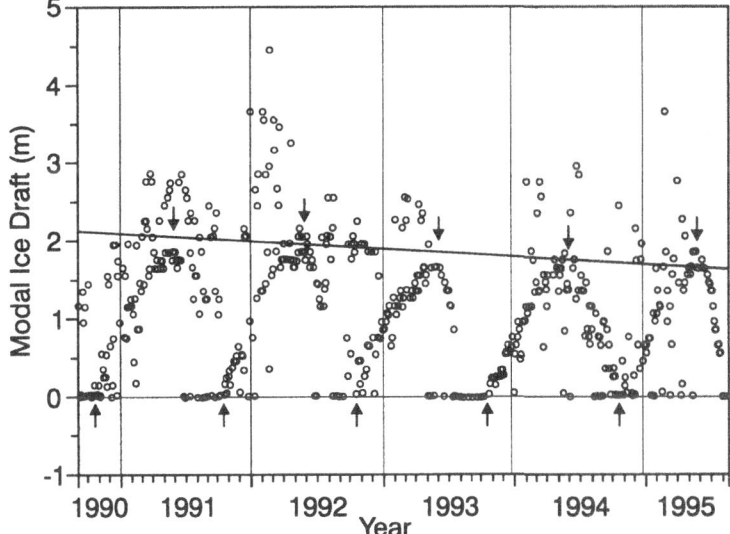

Figure 11. Ice thickness measurements from the Canada Basin, 1990–1996. Arrows denote times of maximum and minimum ice cover (data from R. E. Moritz, pers. comm.).

bathymetric features at the Hovgaard and Jan Mayan fracture zones. The bulk of this current then exits the Arctic Mediterranean through Denmark Strait. Descriptions of the East Greenland Current and water mass transports are given by Aagaard and Coachman [118], [119]; Aagaard *et al.* [120],

Foldvik *et al.* [121], and Aagaard *et al.* [122]. Foldvik *et al.* [121] estimated a transport of Polar Water ($T < 0°C$) from the Arctic Ocean through Fram Strait at 79°N of approximately 1 Sv. Aagaard and Carmack [18] used this value, along with an estimate of transport salinity of 33.7, to arrive at freshwater export to the Greenland Sea (reference salinity, 34.93) of 1110 km^3 yr^{-1}. They argue that the corresponding contribution from deeper waters is small, about 50 km^3 yr^{-1}, yielding an annual liquid freshwater export to the Greenland Sea of 1160 km^3 yr^{-1}.

Ice export appears to contribute the largest export of fresh water. Aagaard and Carmack used the ice export estimates of Vinje and Finnekåsa [96] of 0.16 Sv across 81°N to obtain an export of fresh water to the Greenland Sea of 2790 km^3 yr^{-1}. Hilmer *et al.* [53] examined the variability of the ice volume flux using a dynamic–thermodynamic sea ice model and 40 years (1958–1997) of atmospheric forcing. They found a mean export of 2870 km^3 yr^{-1} and a standard deviation of about 20% of the long-term mean. Recent discussion on export variability and sea-ice trajectories are given by Pfirman *et al.* [97], Vinje *et al.* [123], and Agnew *et al.* [124].

4.2. VIA THE CANADIAN ARCTIC ARCHIPELAGO TO BAFFIN BAY

The Canadian Arctic Archipelago offers a different perspective to shelf-basin interaction processes, for here waters move onto, rather than off, the shelf regions prior to exiting the Arctic Ocean. Collin [125], Sadler [126], Melling *et al.* [127], Fissel *et al.* [128] and Prinsenberg and Bennett [129] have discussed observations of circulation and water mass distributions within the Canadian Arctic Archipelago. Rudels [130] gave an estimate of transport. It is now clear that the Canadian Arctic Archipelago throughflow cannot be treated as simple channel flow through a network of narrow passages. New data shows the Canadian Arctic Archipelago, itself, to be extremely large and complex Arctic shelf, studded with islands, and characterized by interconnected sub-basin circulations (e.g. tidal flows, buoyancy–boundary currents). Figure 12 is a schematic of processes influencing water masses exiting the Arctic Ocean through the Canadian Arctic Archipelago. Three distinct mechanisms affecting throughflow are noted: (1) selective withdrawal, which draws water from the stratified and rotating ocean basin over sills and into the archipelago; (2) formation of buoyancy–boundary currents which act to confine the freshest and fastest moving waters to right-hand flows; (3) and sub-basin circulations, which tend to re-circulate waters within the Canadian Arctic Archipelago. For example, the continuity of water masses moving through the Canadian Arctic Archipelago along a section extending from the Canada Basin to Baffin Bay is shown in Figure 13 (see also [125]). Water within the archipelago largely

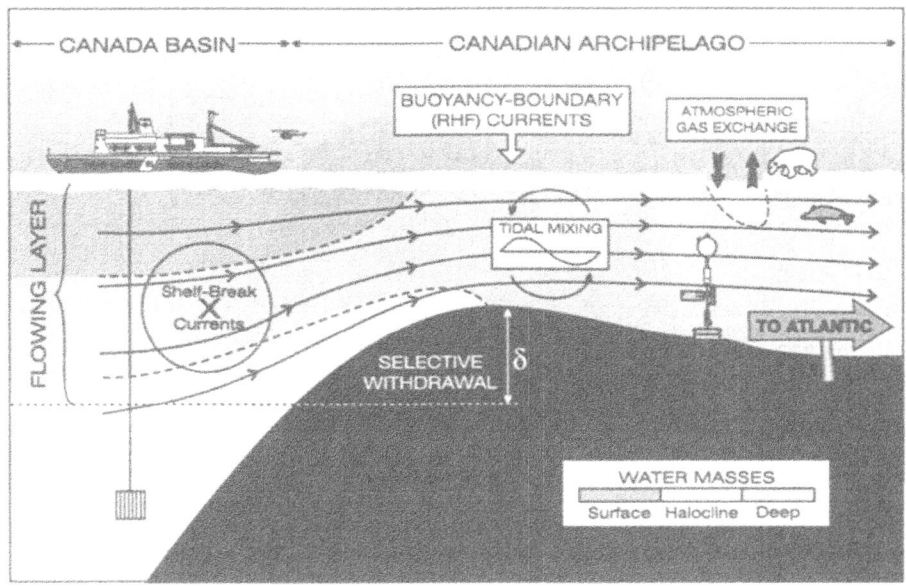

Figure 12. Schematic of processes influencing Canadian Arctic Archipelago through-flow. Water masses move from the Arctic to the North Atlantic forced by barotropic and baroclinic pressure gradients. Flow over sills is controlled by selective withdrawal. Low-salinity flows are further effected by runoff and ice melt to form relatively narrow buoyancy–boundary currents along right-hand shorelines in the direction of flow. Tidal mixing locally modifies the density-driven flow field.

reflects the T/S characteristics of the upper 400 m of the Canada Basin, from which it is drawn, and is quite distinct from the waters of northern Baffin Bay, to which it flows. Evidence of tidal mixing is relatively small, except in the shallow sill regions such as near Barrow Strait. The correlations of anthropogenic CFC-11 versus salinity in Viscount Melville Sound (Figure 14) show the deep waters here to be much older than those of the Canada Basin from which they are derived. Waters with isolation ages of 20 or more years are seen to pool within sub-basins. These sub-basin flows and associated long residence times offer a new potential sink for the high-latitude sequestration of CO_2. Figure 15 shows the structure of buoyancy–boundary currents in Barrow Strait. Here, low salinity waters are seen to exit the archipelago within narrow (order 10 km) and relatively thin (order 10 m) buoyancy-driven flows following their right-hand shoreline in the direction of flow.

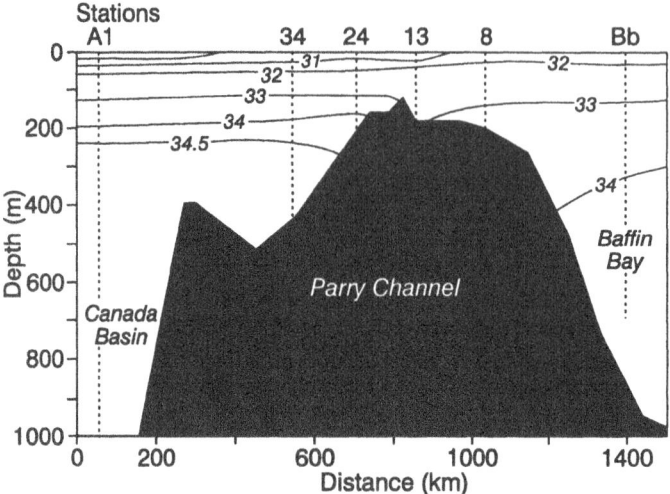

Figure 13. Section of salinity through the Canadian Arctic Archipelago. (Data obtained in summer, 1997 aboard the CCGS *Louis S. St-Laurent*, F.A. McLaughlin, pers. comm.)

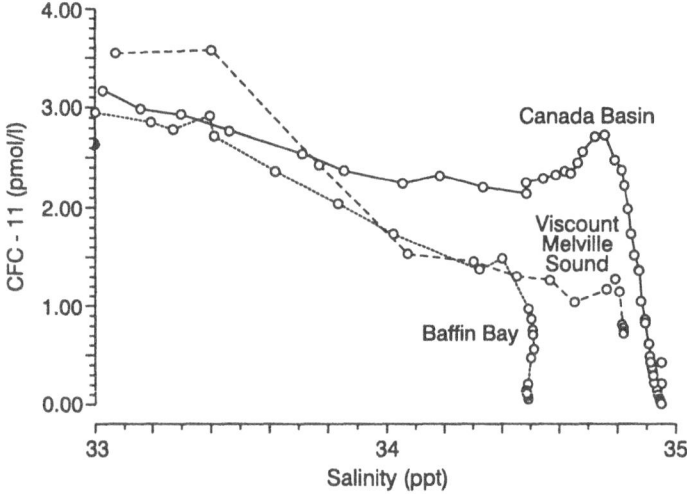

Figure 14. Correlations of CFC-11 versus salinity in the Canada Basin, in Viscount Melville Sound of the Canadian Arctic Archipelago, and in Baffin Bay. (Data obtained in summer, 1997 aboard the CCGS *Louis S. St-Laurent*, F.A. McLaughlin, pers. comm.)

5. Discussion

In this section some of the oceanographic consequences of freshwater export are examined. We will concentrate on three questions. First, what is the influence of freshwater export on convection in the Greenland Sea, and is this system stable to large perturbations? Second, is there an alternate per-

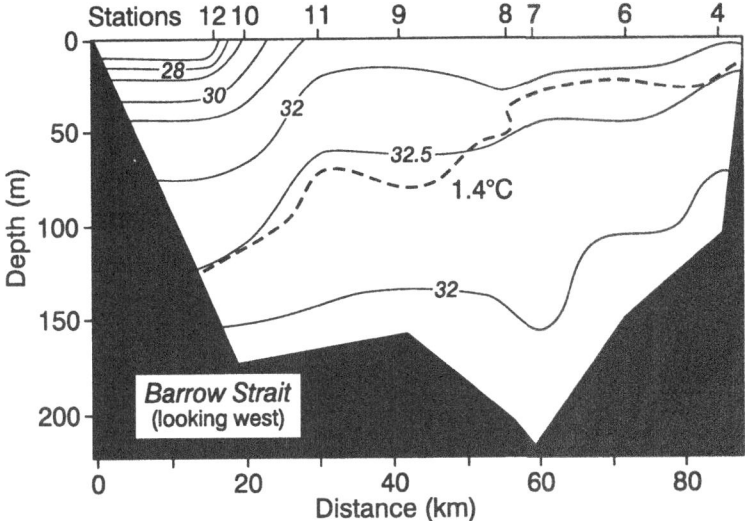

Figure 15. Section of salinity and temperature across Barrow Strait illustrating a buoyancy–boundary current following its right-hand shoreline. (Data obtained in summer, 1997 aboard the CCGS *Louis S. St-Laurent*, F.A. McLaughlin, pers. comm.).

spective to the deep overturning model of thermohaline circulation? Third, how is the freshwater loop closed, and what is the relationship of the Arctic Ocean's halocline to high-latitude stratification? The section concludes with a list of uncertainties.

5.1. OCEAN OVERTURNING

The present evidence is that convection in the high latitudes of the North Atlantic drives much of the global overturning cell in the ocean, but that fluctuations and different stable modes can exist, each associated with a different climate (cf. [14], [15]). In the case of a thermohaline circulation driven vigorously in high northern latitudes, the northern European climate is moderate. On the other hand, if overturning diminishes, the climate cools dramatically. Weakening of the thermohaline circulation has been offered as a cause of major cooling events in the paleoclimate record [131], [10], [12]. Similar mode transitions may be manifested in the Arctic Mediterranean. At present it appears that deep, mid-gyre convection is limited to the Greenland and Iceland seas [132], and that this convection is conditioned by freshwater export from the Arctic Ocean [22]. If export were to increase, convection would likely be diminished, while under conditions of decreased export, convection would likely become more vigorous. Aagaard and Carmack [22] suggested that at some level of increased freshwater export, convection in the Greenland and Iceland seas would cease; the north-

114

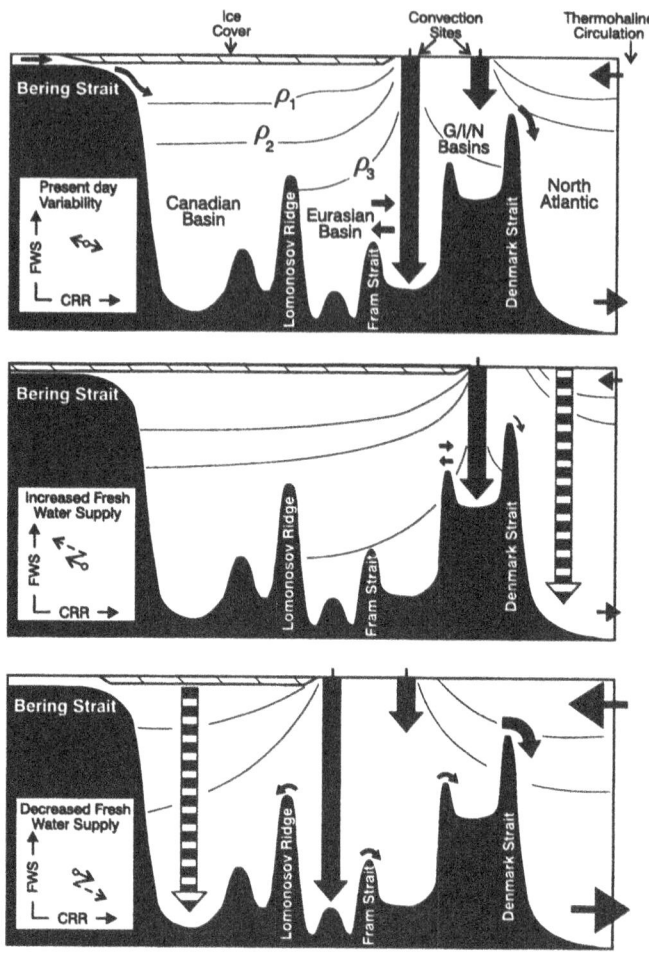

Figure 16. Dependence of convective renewal rate (CRR) on freshwater supply (FWS) from the Arctic Ocean under (top) present conditions, (middle) increased FWS, and (bottom) decreased FWS. The size of the arrows through the right-hand side is representative of the strength of thermohaline circulation forced from far northern seas. The barred arrows represent extreme locations of convection. The solid arrows in the insets indicate the trend of convective renewal with changing freshwater supply, and dashed arrows indicate possible transitions to different circulation modes. (Adapted from [22].)

ward extension of Atlantic water would retreat; and the climatological ice cover would spread southward. In this scenario, the convective region would shift south, and the convective system would assume a new configuration. Conversely, if the supply of fresh water were to decrease, and stratification within the Arctic Ocean to decrease, a shift northwards of the convective regime might be expected (Figure 16).

It is generally taken that freshening of surface layers acts to suppress

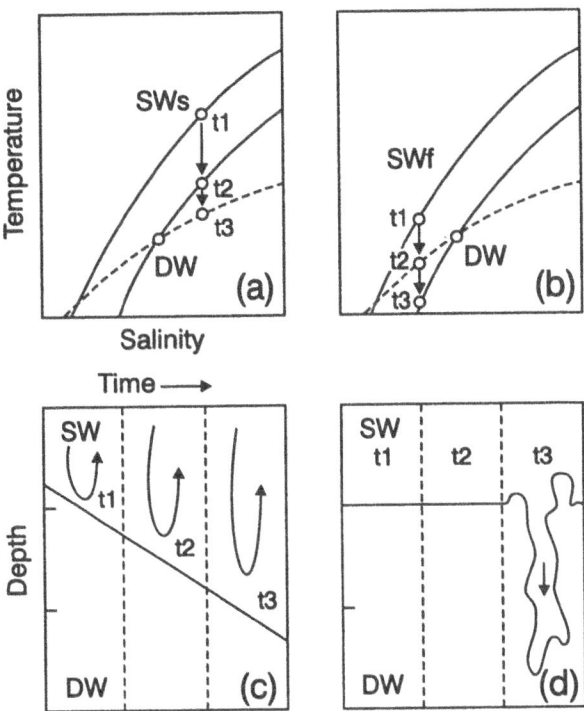

Figure 17. Schema illustrating thermobaric effects and the convective regimes associated with progressive cooling of surface water that is slightly more saline (SW_s, (a)) and slightly less saline (SW_f; (b)) than ambient deep water (DW). Also shown in (a) and (b) are isolines of density relative to near-surface pressures (solid) and near-bottom pressures (dashed). As SW_s cools from an initial temperature, t_1, it eventually reaches a temperature, t_2, at which its density is equal to that of the underlying water, and convection will ensue. However, at near-bottom pressures the SW_s will still be less than the existing DW and further cooling to t_3 is required to drive the convection deeper. Hence a progressive deepening of the upper layer, as shown in (c) will ventilate the water column. When SW_f is cooled to temperature t_3 where its density at near-surface pressures matches that of DW it has already surpassed the density of DW at all greater pressures (d), thus triggering the thermobaric instability. (Adapted from [22]).

deep convection. However, because of the differential compressibility of seawater, with cold water being more compressible (the so-called thermobaric effect) it is likely that within some critical parameter range a small amount of salinity stratification actually serves to promote deep convection (see [18]). This is because the stratification allows the surface layer to be cooled significantly more than the underlying water, so that when convection is initiated, the very cold plume originating above the weak halocline may accelerate downwards as pressure increases during descent (Figure 17). This may, in fact, be a far more efficient ventilating process than the gradual deepening of the mixed layer (see also [133]).

5.2. ALTERNATE SCHEMES OF THERMOHALINE CIRCULATION

That dense waters exit through Denmark Strait into the North Atlantic is clear. Typically, this is viewed as being driven by convention and dense water production in the gyres of the Greenland and Iceland seas (cf. [134]). Mauritzen [135] suggested an alternate mechanism for dense water production. In her model a cyclonic flow, fed by Atlantic water in the Norwegian Atlantic Current, separates into three branches: re-circulating flow in Fram Strait, northward flow through Fram Strait around the Arctic Basin, and flow across the Barents Sea and through the Arctic Ocean. As water moves along each of these paths it becomes progressively denser. The three branches then return southwards as distinct layers within the East Greenland Current and supply the overflow waters to the deep North Atlantic. Rudels *et al.* [101] present similar arguments in which the main water mass transformations in the Arctic Ocean take place in the topographically-steered boundary currents of Atlantic water. Such alternate flow regimes have important implications for the budgets of freshwater and other material properties (cf. [136]).

5.3. THE FRESHWATER LOOP AND HIGH-LATITUDE STRATIFICATION

We now return to the premise that the Arctic halocline is a sub-component (albeit an important sub-component!) of a much larger oceanographic domain; namely, the northern hemisphere halocline. It is this latter feature that ultimately defines the presence or absence of both the ice cover and the global thermohaline convection. To do this requires drawing back from the Arctic Ocean itself, and examining the large-scale stratification of northern hemisphere oceans in relation to global freshwater transport. The northern halocline, itself, is a high-latitude oceanographic domain bounded by sharp fronts in both the Pacific and Atlantic oceans and extending through the Arctic Ocean; this continuous feature thus comprises the main dynamical linkage among the Pacific, Arctic and Atlantic oceans.

The low latitudes of the Earth receive by far the largest amount of incoming solar energy to drive the climate engine. This low-latitude heat source is compensated by a high-latitude heat sink. It is generally assumed that the atmosphere and oceans play roughly equal roles in transporting heat poleward (but, see discussions by Covey [137], Trenberth and Solomon [138], Zaucker *et al.* [139]).

Most of the heat carried by the atmosphere is in the form of water vapor, so again, fresh water plays a key role. Such transport is governed by the Clausius–Clapeyron relationship (see Figure 18a) which states that the saturation vapor pressure is only a function of temperature. This relationship is strongly non-linear (Figure 18b), with maximum curvature between

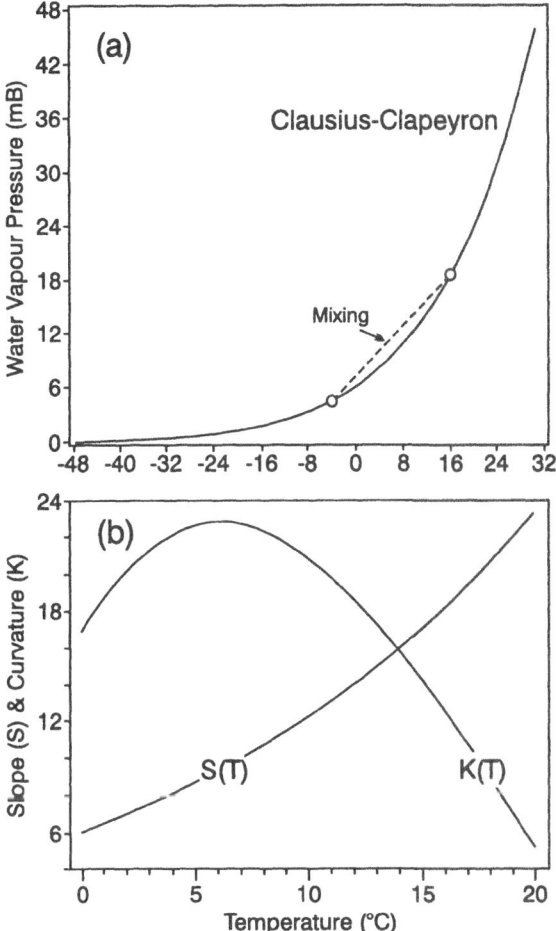

Figure 18. Plots showing (a) the Clausius–Clapeyron saturation vapor pressure curve as a function of temperature and (b) curvature in the Clausius–Clapeyron curve as a function of temperature.

0 and 12°C. Non-linearity in the Clausius–Clapeyron relationship has two major consequences. First, surface vapor pressure in low latitudes is about an order of magnitude greater than in the polar regions. Second, mixing of atmospheric columns near saturation will lead to condensation, a response reminiscent of cabelling in seawater.

A cartoon, which illustrates conservation requirements for the global flux and distribution of heat, air and fresh water, and the distinction between temperature-stratified and salt-stratified oceans, is shown in Figure 19. Excess heat at low latitudes is carried poleward by the atmosphere (air and water vapor) and the ocean (especially in the warm, salty west-

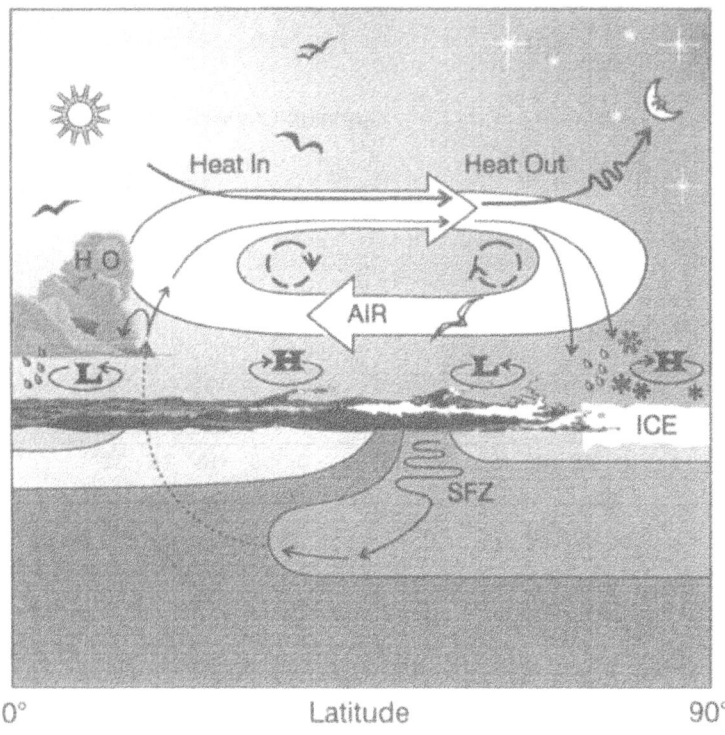

Figure 19. Simplified global climate system showing meridional transports of heat and fresh water by the atmosphere and ocean. High and low, H and L, denote atmospheric pressure centers, respectively. SFZ denotes the subarctic frontal zone where subduction of low-salinity water occurs.

ern boundary currents). The returning air may simply re-circulates equatorward, so there is no continuity problem. However, fresh water, which evaporates in the warm, temperature-stratified oceans and precipitates in the cooler, salt-stratified oceans, must return south. The freshwater loop is closed first by surface and halocline flows in salinity-stratified oceans, and then as both intermediate and deepwater flows in temperature-stratified oceans. The time scales for return flow are vastly different, perhaps 20 to 100 years for intermediate waters (the Lesser Conveyor Belt) and 1000 years for deep waters (the Great Conveyor Belt).

However, what processes maintain the structure of high-latitude haloclines? Because water tends to evaporate in warm regions and condense in cold, because low-latitude oceans gain heat while high-latitude oceans loose heat, and because the thermal expansion coefficient for sea water ($\alpha = \delta\rho/\delta T$) decreases fourfold from subtropical to polar waters while the saline contraction coefficient ($\beta = \delta\rho/\delta S$) varies little, the upper layers of subtropical seas are mainly stratified by temperature ($N_T^2 = \alpha\,dT/dp > 0$) while the upper layers of high-latitude seas are mainly stratified by salt ($N_S^2 = \beta\,dS/dp > 0$). For convenience, these two domains are here referred to as alpha ($\alpha\,dT/dp > 0$)and beta ($\beta\,dS/dp > 0$) oceans. We note that alpha and beta oceans in both hemispheres are typically separated by frontal regions at surface (wintertime) temperatures near 10°C, wherein the surface density reaches a local maximum (e.g. the subarctic front in the North Pacific); however, no dynamical explanation for this observed consistency is available.

5.4. WILD CARDS IN THE FRESHWATER BUDGET

The high-latitude hydrological cycle clearly holds the key to understanding the Arctic's link to climate. Research during the past decade has made it apparent where quantities of fresh water enter, are stored and leave the Arctic Ocean. What remains less apparent are the processes which control the variability of freshwater sources, storage and sinks. The easy questions have been asked and answered. The easy work has now been done. But major challenges remain with respect to the budget, notably:

- to measure ice cover thickness, its temporal and spatial variability, and the volume of ice export from the Arctic (e.g. [97]),
- to understand variability in moisture transport to, and climate within, Arctic Ocean watersheds, and the link to high-latitude teleconnections (e.g. [140]),
- to study the interconnection of the Pacific/Arctic/Atlantic oceans in a systems approach to freshwater transport which incorporates oceanic inflows, the storage of fresh water within layers of the Canadian and Eurasian basins, and export to lower latitudes (e.g. [49] and [113]).

Finally, we must address the questions: Where does all the fresh water that exits the Arctic go? How is fresh water transported from a salt-stratified to a temperature-stratified ocean? And, where would fresh water go if global transport pathways changed? The study of arctic climate is truly about the sources, disposition and export of its freshwater components. As such, understanding the Arctic Ocean's freshwater budget transcends the collection of "numbers" and their manipulation. It is much more about obtaining a fundamental appreciation of the hydrological cycle that explains how a large and complex ocean system works. "Numbers" simply serve as a set of checks for internal consistency within the larger system.

120

References

1. Mosby, H. (1962) Water, salt, and heat balance of the North Polar Sea and of the Norwegian Sea, *Geofysiske Publikasjoner*, **24**, 289–313.
2. Vowinckle, E. and Orvig, S. (1961) Water balance and heat flow in the Arctic Ocean, *Publication in Meteorology No. 44*, McGill University, 35 pp.
3. Fletcher, J.O. (1965) The heat budget of the Arctic Basin and its relation to climate, *R-444-PR*, Rand Corp., Santa Monica, California, 108 pp.
4. Aagaard, K. and Greisman, P. (1975) Toward new mass and heat budgets for the Arctic Ocean, *Journal of Geophysical Research*, **80**, 3821–3827.
5. SCOR Working Group 58, (1979) The Arctic Ocean Heat Budget, *Report No. 52*, Geophysics Institute, University of Bergen, Bergen, Norway.
6. Aagaard, K. and Coachman, L.K. (1975) Toward an ice-free Arctic Ocean, *Eos*, **56**, 484–486.
7. Semtner, A.J. (1984) The climatic response on the Arctic Ocean to Soviet river diversions, *Climate Change*, **6**, 109–130.
8. Cattle, H. (1985) Diverting Soviet rivers: Some possible repercussions for the Arctic Ocean, *Polar Record*, **22**, 485–498.
9. Rooth, C. (1982) Hydrology and ocean circulation, *Progress in Oceanography*, **11**, 131–149.
10. Broecker, W.S., Peteet, D.M., and Rind, D. (1985) Does the ocean–atmosphere system have more than one stable mode of operation? *Nature*, **315**, 21–26.
11. Bryan, F. (1986) High-latitude salinity effects and interhemispheric thermohaline circulations, *Nature*, **323**, 301–304.
12. Broecker, W.S. (1997) Thermohaline circulation the Achilles Heal of our climate system: Will man-made CO_2 upset the current balance? *Science*, **278**, 1582–1588.
13. Manabe, S. and Stouffer, R.J. (1994) Multiple century response of a coupled ocean-atmosphere model to an increase of atmospheric carbon dioxide, *Journal of Climate*, **7**, 5–23.
14. Weaver, A.J. and Sarachik, E.S. (1994) Rapid interglacial climate fluctuations driven by North Atlantic ocean circulation, *Nature*, **367**, 447–450.
15. Rahmstorf, S. (1996) On the freshwater forcing and transport of the Atlantic thermohaline circulation, *Climate Dynamics*, **12**, 799–811.
16. Mauritzen, C. and Häkkinen, S. (1997) Influence of sea ice on the thermohaline circulation in the Arctic-North Atlantic Ocean, *Geophysical Research Letters*, **24**, 3257–3260.
17. Cuffey, K.M., Clow, G.D., Alley, R.B., Stuiver, M., Waddington, E.D., and Saltus, R.W. (1995) Large Arctic temperature changes at the Wisconsin-Holocene glacial transition, *Science*, **270**, 455–458.
18. Aagaard, K. and Carmack, E.C. (1989) On the role of sea-ice and other freshwater, *Journal of Geophyscal Research*, **94**, 14,485–14,498.
19. Dickson, R.R., Meincke, J., Malmberg, S.-A., and Lee, A.J. (1988) The great salinity anomaly in the northern North Atlantic 1968-1982, *Progress in Oceanography*, **20**, 103–151.
20. Häkkinen, S. (1993) An Arctic source for the great salinity anomaly: A simulation of the Arctic ice–ocean system for 1955-1975, *Journal of Geophysical Research*, **98**, 16,397–16,410.
21. Schlosser, P., Bonisch, G., Rhein, M., and Bayer, R. (1991) Reduction of deepwater formation in the Greenland Sea during the 1980s: Evidence from tracer data, *Science*, **251**, 1054–1056.
22. Aagaard, K. and Carmack, E.C. (1994) The Arctic and climate: A perspective, in *The Polar Oceans and Their Role in Shaping the Global Environment*, in O.M. Johannessen, R.D. Muench and J.E. Overland (eds.), *Geophysical Monograph Series*, **85**, AGU, Washington, D.C., pp. 4–20.
23. Stigebrandt, A. (1981) A model for the thickness and salinity of the upper layer in

the Arctic Ocean and the relationship between the ice thickness and some external parameters, *Journal of Physical Oceanography*, **11**, 1407–1422.

24. Björk, G. (1989) A one-dimensional time-dependent model for the vertical stratification of the upper Arctic Ocean, *Journal of Physical Oceanography*, **19**, 52–67.

25. Anderson, L.G., Jones, E.P., Koltermann, K.P., Schlosser, P., Swift, J.H., and Wallace, D.W.R. (1989) The first oceanographic section across the Nansen basin the Arctic Ocean, *Deep-Sea Research*, **36**, 475–482.

26. Falkner, K.K., Macdonald, R.W., Carmack, E.C., and Weingartner, T. (1994) The potential of barium as a tracer of arctic water masses, in O.M. Johannessen, R.D. Muench and J.E. Overland (eds.), *The Polar Oceans and Their Role in Shaping the Global Environment*, *Geophysical Monograph Series*, **85**, AGU, Washington, D.C., pp. 63–76.

27. Schlosser, P., Bauch, D., Fairbanks, R., and Bonisch, G. (1994) Arctic river-runoff: Mean residence time on the shelves and in the halocline, *Deep-Sea Research*, **41**, 1053–1068.

28. Nansen, F. (1906) Northern Waters. *Skrifter Norvegica Videnskabs-Selskabet i Christiania, Kl: Mathematisk-Naturvidenskabelig, K3*, 145 pp.

29. Coachman, L.K. and Barnes, C.A. (1962) Surface water in the Eurasian basin of the Arctic Ocean, *Arctic*, **15**, 251–277.

30. Aagaard, K., Coachman, L.K., and Carmack, E.C. (1981) On the halocline of the Arctic Ocean, *Deep-Sea Research*, **28**, 529–545.

31. Melling, H. and Lewis, E.L. (1982) Shelf drainage flows in the Beaufort Sea and their effect on the Arctic Ocean pycnocline, *Deep-Sea Research*, **29**, 967–985.

32. Jones, E.P. and Anderson, L.G. (1986) On the origin of the chemical properties of the Arctic Ocean halocline, *Journal of Geophysical Research*, **91**, 10,759–10,767.

33. Killworth, P.D. and Smith, J.M. (1984) A one-and-a-half dimensional model for the Arctic halocline, *Deep-Sea Research*, **31**, 271–293.

34. Coachman, L.K. and Barnes, C.A. (1961) The contribution of Bering Sea water to the Arctic Ocean, *Arctic*, **14**, 147–161.

35. Melling, H. (1993) The formation of a haline shelf front in wintertime in an ice-covered arctic sea, *Continental Shelf Research*, **13**, 1123–1147.

36. Gawarkiewicz, G. and Chapman, D.C. (1995) A numerical study of dense water formation and transport on a shallow, sloping, continental shelf, *Journal of Geophysical Research*, **100**, 4489–4507.

37. Chapman, D.C. and Gawarkiewicz, G. (1995) Offshore transport of dense shelf water in the presence of a submarine canyon, *Journal of Geophysical Research*, **100**, 13,373-13,387.

38. Chapman, D.C. and Gawarkiewicz, G. (1997) Shallow convection and buoyancy equilibrium in an idealized coastal polynya, *Journal of Physical Oceanography*, **27**, 555–566.

39. Untersteiner, N. (1988) On the ice and heat balance in Fram Strait, *Journal of Geophysical Research*, **93**, 527–531.

40. Rudels, B., Anderson, L.G., and Jones, E.P. (1996) Formation and evolution of the subsurface mixed layer and the halocline of the Arctic Ocean. *Journal of Geophysical Research*, **101**, 8807–8821.

41. Ekwurtzel, B. (1998) Circulation and mean residence times in the Arctic Ocean derived from tritium, helium, and oxygen-18 tracers, Ph.D. Thesis, Columbia University, New York.

42. Quadfasel, D., Sy, A., Wells, D., and Tunik, A. (1991) Warming in the Arctic, *Nature*, **350**, 385.

43. Carmack, E.C., Macdonald, R.W., Perkin, R.G., and McLaughlin, F.A. (1995) Evidence for warming of Atlantic Water in the southern Canadian Basin, *Geophysical Research Letters*, **22**, 1961–1964.

44. Carmack, E.C., Aagaard, K., Swift, J.H., Perkin, R.G., McLaughlin, F.A., Macdonald, R.W., Jones, E.P., Smith, J., Ellis, K., and Kilius, L. (1997) Changes in

temperature and tracer distributions within the Arctic Ocean: Results from the 1994 Arctic ocean section, *Deep-Sea Research*, **44**, 1487–1502.

45. Morison, J., Steele, M., and Anderson, R. (1998) Hydrography of the upper Arctic Ocean, measured from the nuclear submarine U.S.S. *Pargo*, *Deep-Sea Research*, **45**, 15–38.

46. Grotefendt, K., Logemann, K., Quadfasel, D., and Ronski, S. (1998) Is the Arctic Ocean warming? *Journal of Geophysical Research*, **103**, 27,679–27,687.

47. Harder, M., Lemke, P., and Hilmer, M. (1998) Simulation of Sea ice transport through Fram Strait: Natural variability and sensitivity to forcing, *Journal of Geophysical Research*, **103**, 5595–5606.

48. Walsh, J.E., Chapman, W.L., and Shy, T.L. (1996) Recent decrease of sea level pressure in the central Arctic, *Journal of Climate*, **9**, 480–486.

49. McLaughlin, F.A., Carmack, E.C., Macdonald, R.W., and Bishop, J. (1996) The Atlantic/Pacific water mass boundary in the southern Canadian Basin, *Journal of Geophysical Research*, **101**, 1183–1197.

50. Steele, M. and Boyd, T. (1998) Retreat of the cold halocline layer of the Arctic Ocean, *Journal of Geophysical Research*, **103**, 10,419–10,435.

51. Proshutinsky, A.Y. and Johnson, M.A. (1997) Two circulation regimes of the wind-driven Arctic Ocean, *Journal of Geophysical Research*, **102**, 12,493–12,514.

52. Steele, M., Thomas, D., and Rothrock, D. (1996) A simple model study of the Arctic Ocean freshwater balance, 1979–1985, *Journal of Geophysical Research*, **101**, 20,833-20,848.

53. Hilmer, M., Harder, M., and Lemke, P. (1998) Sea ice transport: A highly variable link between Arctic and North Atlantic, *Geophysical Research Letters*, **25**, 3359–3362.

54. Kwok, R. and Rothrock, D.A. (1999) Variability of Fram Strait ice flux and North Atlantic Oscillation, *Journal of Geophysical Research*, **104**, 5177–5189.

55. Baumgartner, A. and Reichel, E. (1975) *The World Water Balance: Mean Annual Global, Continental and Maritime Precipitation, Evaporation and Run-Off*, Elsevier, New York, 179 pp.

56. Wijffels, S.E., Schmitt, R.W., Bryden, H.L., and Stigebrandt, A. (1992) Transport of freshwater by the oceans, *Journal of Physical Oceanography*, **22**, 155–162.

57. Jourdan, D., Peterson, P., and Gautier, C. (1997) Oceanic freshwater budget and transport as derived from satellite radiometric data, *Journal of Physical Oceanography*, **27**, 457–467.

58. Barron, E.J., Hay, W.W., and Thompson, S. (1989) The hydrologic cycle: A major variable during earth history, *Palaeogeography Palaeoclimatology Palaeoecology*, **75**, 157–174.

59. Schmitt, R.W. and Wijffels, S.E. (1993) The role of the oceans in the global water cycle, *Geophysical Monograph 75*, IUGG Vol. **15**, 77–84.

60. Webster, P.J. (1994) The role of hydrological processes in ocean–atmosphere interactions, *Reviews of Geophysics*, **32**, 427–476.

61. Barry, R.G., Serreze, M.C., and Maslank, J.A. (1993) The Arctic sea ice-climate system: Observations and modeling, *Reviews of Geophysics*, **31**, 397–422.

62. Carmack, E.C. (1986) Circulation and mixing in ice-covered waters, in N. Untersteiner (ed.), *The Geophysics of Sea Ice*, Plenum, New York, pp. 641-712.

63. Carmack, E.C. (1990) Large-scale oceanography, in Smith, W.O. (ed.), *Polar Oceanography, Part A*, Academic, San Diego, Calif., pp. 171–222.

64. Muench, R.D. (1990) Mesoscale phenomena in the polar oceans, in W.O. Smith, Jr. (ed.), *Polar Oceanography, Part A*, Academic, San Diego, Calif., pp. 223–286.

65. McPhee, M.G. (1990) Small-scale processes, in W.O. Smith, Jr. (ed.), *Polar Oceanography, Part A*, Academic, San Diego, Calif., pp. 287–334.

66. Padman, L. (1995) Small-scale physical processes in the Arctic Ocean, in W.O. Smith and J.M Grebmeier (eds.), *Arctic Oceanography: Marginal Ice Zones and Continental Shelves, Coastal and Estuarine Studies*, **49**, AGU, pp. 97–129.

67. UNESCO (United Nations Educational, Scientific, and Cultural Organization). (1978) World water balance and water resources of the Earth, *Study and Report in Hydrology*, **25**, Paris.

68. Milliman, J.D. and Meade, R.H. (1983) World-wide delivery of river sediment to the oceans, *Journal of Geology*, **91**, 1–21, 1983.

69. Treshnikov, A.F. (1985) *Arctic Atlas*, Arkt.-Antarkt. Nauchno-Issled. Inst., Moscow, 204 pp. (in Russian).

70. Vuglinsky, V.S. (1997) River inflow to the Arctic Ocean: Conditions of formation, time variability and forecasts, in *Polar Processes and Global Climate, ACSYS*, Orcas Island, Washington, pp. 275-276.

71. Semiletov, I.P., Savelieva, N.I., Weller, G.E., Pipko, I.I., Pugach, S.P., Gukov, A.Yu., and Vasilevskaya, L.N. (1999) The dispersion of Siberian river flows into coastal waters: Meteorological, hydrographical and hydrochemical aspects, this volume.

72. Serreze, M.C., Rehder, M.C., Barry, R.G., Kahl, J.D., and Zaitseva, N.A. (1995a) The distribution and transport of atmospheric water vapour over the Arctic Basin, *International Journal of Climatology*, **15**, 709–727.

73. Serreze, M.C., Rehder, M.C., Barry, R.G., Walsh, J.E., and Robinson, D.A. (1995b) Variations in aerologically derived Arctic precipitation and snowfall, *Annals of Glaciology*, **21**, 77–82.

74. Bjornsson, Mysak, L.A., and Brown, R.D. (1995) On the interannual variability of precipitation and runoff in the Mackenzie drainage basin, *Climate Dynamics*, **12**, 67–76.

75. Omstedt, A., Carmack, E.C., and Macdonald, R.W. (1994) Modeling the seasonal cycle of salinity in the Mackenzie shelf/estuary, *Journal of Geophysical Research*, **99**, 10,011–10,021.

76. Gushue, W., Carmack, E.C., and Macdonald, R.W. (1996) Freshwater inputs to Husky Lakes and Liverpool Bay, *Canadian Technical Report of Hydrography and Ocean Sciences*, **175**, Institute of Ocean Sciences, Sidney, British Columbia, Canada, 23 pp.

77. Alford, M. and Carmack, E. (1987) Observations on ice cover and streamflow in the Yukon River near Whitehorse during 1983/84. *NHRI Paper No. 32, Scientific Series No 152*, Environment Canada, 63 pp.

78. Coachman, L.K. and Aagaard, K. (1988) Transports through Bering Strait, Annual and interannual variability, *Journal of Geophysical Research*, **93**, 15,535-15,539.

79. Roach, A.T., Aagaard, K., Pease, C.H., Salo, S.A., Weingartner, T., Pavlov, V., and Kulakov, M. (1995) *Journal of Geophysical Research*, **100**, 18,443–18,457.

80. Overland, J.E. and Roach, A.T. (1987) Northward flow in the Bering and Chukchi seas, *Journal of Geophysical Research*, **92**, 7097–7105.

81. Coachman, L.K. and Aagaard, K. (1966) On the water exchange through Bering Strait, *Limnology and Oceanography*, **11**, 44–59.

82. Coachman, L. K. and Aagaard, K. (1981) Reevaluation of water transports in the vicinity of Bering Strait, in D.W. Hood and J. A. Calder (eds.), *The Eastern Bering Sea Shelf: Oceanography and Resources, Vol. 1*, University of Washington Press, Seattle, pp. 95–110.

83. Garrison, G.R. and Becker, P. (1976) The Barrow Submarine Canyon: A drain for the Chukchi Sea, *Journal of Geophysical Research*, **81**, 4445–4453.

84. Muenchow, A. and Carmack, E.C. (1997) Synoptic flow and density observations near an Arctic shelf break, *Journal of Physical Oceanography*, **27**, 1402–1419.

85. Weingartner, T.J., Calivalieri, D.J., Aagaard, K., and Sasaki, Y. (1998) Circulation, dense water formation, and outflow on the northeast Chukchi shelf, *Journal of Geophysical Research*, **103**, 7647–7661.

86. Burova, L.P. (1981) Atmospheric water resources of the Arctic Basin, *Trudy Arkticheskogo I Antarkticheskogo Nauchno Issledovatelskogo Instituta*, **370**, 91–110 (in Russian).

87. Blindheim, J. (1989) Cascading of Barents Sea bottom water into the Norwegian Sea, *Rapports et Procès-Verbaux des Réunions/Conseil Permanent International pour l'Exploration de la Mer*, **17**, 161–189.

88. Hanzlick, D. and Aagaard, K. (1980) Freshwater and Atlantic water in the Kara Sea, *Journal of Geophysical Research*, **85**, 4937–4942.

89. Loeng, H., Ozhigin, V., and Adlandsvik, A. (1997) Water fluxes through the Barents Sea, *ICES Journal of Marine Science*, **54**, 310–317.

90. Harms, I.H. (1997a) Freshwater runoff and ice formation in Arctic shelf seas — Results from a high resolution Kara Sea model, *ACSYS*.

91. Harms, I.H. (1997b) Water mass transformation in the Barents Sea — Application of the Hamburg shelf ocean model (HamSOM), *ICES Journal of Marine Science*, **54**, 351–365.

92. Rudels, B., Jones, E.P., Anderson, L.G., and Kattner, G. (1994) On the intermediate depth waters of the Arctic Ocean, in O.M. Johannessen, R.D. Muench, and J.E. Overland (eds.), *The Polar Oceans and Their Role in Shaping the Global Environment, Geophysical Monograph Series*, **85**, AGU, Washington, D.C., pp. 33–46.

93. Schauer, U., Muench, R.D., Rudels, B., and Timokhov, L. (1997) Impact of eastern Arctic shelf waters on the Nansen Basin, *Journal of Geophysical Research*, **102**, 3371–3382.

94. Prinsenberg, S.J. (1984) Freshwater contents and heat budgets of James Bay and Hudson Bay. *Continental Shelf Research*, **3**, 191–200.

95. Prinsenberg, S.J. (1988) Ice-ridge contributions to the freshwater contents of Hudson Bay and Fox Basin, *Arctic*, **41**, 6–11.

96. Vinje, T. and Finnekåsa, Ø. (1986) The ice transport through the Fram Strait, *Skrifter Norges Polarinstitutt*, **186**, 1–39.

97. Pfirman, S.L., Colony, R., Nurnberg, D., Eicken, H., and Rigor, I. (1997) Reconstructing the origin and trajectory of drifting Arctic sea ice. *Journal of Geophysical Research*, **102**, 12,575–12,586.

98. Aagaard, K. (1984) The Beaufort Undercurrent, in P.W. Barnes, D.M. Schell, and E. Reimnitz (eds.), *The Alaskan Beaufort Sea, Ecosystems and Environments*, Academic, Orlando, Florida, pp. 47–71.

99. Aagaard, K. (1989) A synthesis of the Arctic Ocean circulation, *Rapports et Procès-Verbaux des Réunions/Conseil Permanent International pour l'Exploration de la Mer*, **188**, 11–22.

100. Newton, J.L. and Sotirin, B.J. (1997) Boundary undercurrent and water mass changes in the Lincoln Sea, *Journal of Geophysical Research*, **102**, 3393–3403.

101. Rudels, B., Quadfasel, D., and Friedrich, H.J. (1999) The Northern Circumpolar Boundary Current, *Journal of Geophysical Research*, in press.

102. Swift, J.H., Jones, E.P., Aagaard, K., Carmack, E.C., Hingston, M., Macdonald, R.W., McLaughlin, F.A., and Perkin, R.G. (1997) Waters of the Makarov and Canada basins, *Deep-Sea Research*, **44**, 1503–1530.

103. Steele, M., Morison, J.H., and Curtin, T.B. (1995) Halocline formation in the Barents Sea, *Journal of Geophysical Research*, **100**, 881–894.

104. Melling, H. and Moore, R.M. (1995) Modification of halocline source waters during freezing on the Beaufort Sea shelf: Evidence from oxygen isotopes and dissolved nutrients. *Continental Shelf Research*, **15**, 89–113.

105. Wallace, D.W.R., Moore, R.M., and Jones, E.P. (1987) Ventilation of the Arctic Ocean cold halocline: Rates of diapycnal and isopycnal transport, oxygen utilization and primary production inferred using chlorofluoromethane distributions, *Deep-Sea Research*, **34**, 1957–1979.

106. Macdonald, R.W. and Carmack, E.C. (1991) The role of large-scale under-ice topography in separating estuary and ocean on an arctic shelf, *Atmosphere-Ocean*, **29**, 37–53.

107. Macdonald, R.W., Paton, D.W., and Carmack, E.C. (1995) The freshwater bud-

get and under-ice spreading of Mackenzie River water in the Canadian Beaufort Sea based on salinity and $^{18}O/^{16}O$ measurements in water and ice, *Journal of Geophysical Research*, **100**, 895–919.

108. Macdonald, R.W., Paton, D.W., and Carmack, E.C. (1999a) Using the $\delta^{18}O$ composition in landfast ice as a record of arctic estuarine processes, *Marine Chemistry*, **65**, 3–24.

109. Kinney, P., Arhelger, M.E., and Burrell, D.C. (1970) Chemical characteristics of water masses in the Amerasian Basin of the Arctic Ocean, *Journal of Geophysical Research*, **75**, 4097–4104.

110. Jones, E.P., Anderson, L.G., and Swift, J.H. (1998) Distribution of Atlantic and Pacific waters in the upper Arctic Ocean: Implications for circulation, *Geophysical Research Letters*, **25**, 765–768.

111. McLaughlin, F.A. and Carmack, E.C. (1998) Arctic Ocean circulation: Plumbing the Arctic Ocean's halocline, *Proceedings of the Arctic Seas: Currents of Change Symposium*, Mystic, Connecticut.

112. Melling, H. (1998) Hydrographic changes in the Canada Basin of the Arctic Ocean, *Journal of Geophysical Research*, **103**,

113. Zhang, J., Rothrock, D.A., and Steele, M. (1998) Warming of the Arctic Ocean, *Geophysical Research Letters*, **25**, 1745–1748.

114. Parkinson, C.L., Comiso, J.C., Zwally, H.J., Cavalieri, D.J., Gloersen, P., and Campbell, W.J. (1987) *Arctic Sea Ice, 1973–76: Satellite Passive-Microwave Observations*, NASA, Washington, D.C., 296 pp.

115. Hibler, W.D., III (1979) A dynamic thermodynamic sea ice model, *Journal of Physical Oceanography*, **9**, 815–846.

116. McPhee, M.G., Stanton, T.M., Morison, J.H., and Martinson, D.G. (1998) Freshening of the upper ocean in the Arctic: Is perennial sea ice disappearing? *Geophysical Research Letters*, **25**, 1729–1732.

117. Macdonald, R.W., Carmack, E.C., McLaughlin, F.A., Falkner, K.K., and Swift, J.H. (1999b) Connections among ice, runoff and atmospheric forcing in the Beaufort Gyre, *Geophysical Research Letters*, in press.

118. Aagaard, K. and Coachman, L.K. (1968a) The East Greenland Current north of Denmark Strait, I, *Arctic*, **21**, 181–200.

119. Aagaard, K. and Coachman, L.K. (1968b) The East Greenland Current north of Denmark Strait, II, *Arctic*, **21**, 267–290.

120. Aagaard, K., Foldvik, A., and Hillman, S.R. (1987) The West Spitsbergen Current: Disposition and water mass transformation, *Journal of Geophysical Research*, **92**, 3778–3784.

121. Foldvik, A., Aagaard, K., and Torresen, T. (1988) The velocity field of the east Greenland Current, *Deep-Sea Research*, **35**, 1335–1354.

122. Aagaard, K., Fahrbach, E., Meincke, J., and Swift, J.H. (1991) Saline outflow from the Arctic Ocean: Its contribution to the deep waters of the Greenland, Norwegian and Iceland Seas, *Journal of Geophysical Research*, **96**, 20,433–20,441.

123. Vinje, T., Nordlund, N., and Kvambekk, Å. (1997) Monitoring ice thickness in Fram Strait, *Journal of Geophysical Research*, **103**, 10,437–19,450.

124. Agnew, T. A., Le, H., and Hirose, T. (1997) Estimation of large-scale sea-ice motion from SSM/I 85.5 GHz imagery, *Annals of Glaciology*, **25**, 305–311.

125. Collin, A.E. (1966) Waters of the Canadian Arctic Archipelago, in *Proceedings of the Arctic Basin Symposium*, Arctic Institute of North America, Washington, D.C., pp. 128–136.

126. Sadler, H.E., (1976) Water, heat and salt transports through Nares Strait, Ellesmere Island. *Journal of the Fisheries Research Board of Canada*, **33**, 2286–2295.

127. Melling, H., Lake, R.A., Topham, D.R., and Fissel, D.B. (1984) Oceanic thermal structure in the western Canadian Arctic, *Continental Shelf Research*, **3**, 233–258.

128. Fissel, D.B., Lemon, D.D., Melling, H., and Lake, R.A. (1988) Non-tidal flows

in the Northwest Passage, *Canadian Technical Report of Hydrography and Ocean Sciences*, **98**, Institute of Ocean Sciences, Sidney, British Columbia, Canada.

129. Prinsenberg, S.J. and Bennett, E.B. (1988) Mixing and transports in Barrow Strait, the central part of the Northwest Passage, *Continental Shelf Research*, **7**, 913–935.

130. Rudels, B. (1987) On the mass balance of the polar ocean with special emphasis on the Fram Strait, *Skrifter Norges Polarinstitutt*, **188**, 1–53.

131. Weyl, P.K. (1968) The role of the oceans in climatic change: A theory of the ice ages, *Meteorological Monograph*, **8**, 37–62.

132. Swift, J.H. and Aagaard, K. (1981) Seasonal transitions and water mass formation in the Iceland and Greenland seas, *Deep-Sea Research*, **28**, 1107–1129.

133. Akitomo, K. (1999) Open-ocean deep convection due to thermobaricity 1. Scaling argument, *Journal of Geophysical Research*, **104**, 5225–5234.

134. Aagaard, K., Roach, A.T., and Schumacher, J.D. (1985) On the wind-driven variability of the flow through Bering Strait, *Journal of Geophysical Research*, **90**, 7213–7221.

135. Mauritzen, C. (1996) Production of dense overflow waters feeding the North Atlantic across the Greenland–Scotland Ridge. Part 1: Evidence for a revised circulation scheme, *Deep-Sea Research*, **43**, 769–806.

136. Anderson, L.G., Jones E.P., and Rudels, B. (1999) Ventilation of the Arctic Ocean estimated by a plume entrainment model constrained by CFCs, *Journal of Geophysical Research*, in press.

137. Covey, C. (1988) Atmospheric and oceanic heat transport: Simulation versus observations, *Climate Change*, **13**, 149–159.

138. Trenberth, K.E. and Solomon, A. (1994) The global heat balance: Heat transports in the atmosphere and ocean, *Climate Dynamics*, **10**, 107–134.

139. Zaucker, F., Stocker, T.F., and Broecker, W.S. (1994) Atmospheric freshwater fluxes and their effect on the global thermohaline circulation, *Journal of Geophysical Research*, **99**, 12,443–12,457.

140. Thompson, D.W.J. and Wallace, J.M. (1998) The Arctic oscillation signature in winter time geopotential height and temperature fields, *Geophysical Research Letters*, **25**, 1297–1300.

THE ARCTIC OCEAN FRESHWATER BUDGET OF A CLIMATE GENERAL CIRCULATION MODEL

Howard Cattle
Ocean Applications
Meteorological Office
Bracknell, RG12 2SZ
United Kingdom

Douglas Cresswell
Ocean Applications
Hadley Centre for Climate Prediction and Research
Meteorological Office
Bracknell, RG12 2SY
United Kingdom

1. Introduction

Climate models are major tools for the study of climate, its sensitivity to external and internal forcing factors and the mechanisms of climate variability and change. They attempt to take into account the various processes important for climate involving the atmosphere, the oceans, the land surface and the cryosphere and the interactions between them. As yet, climate modellers have given scant attention to assessment of the components of the Arctic Ocean freshwater budget. In this paper, we consider its simulation in the third generation version (HADCM3) of the coupled climate model which is run at the Hadley Centre at the U.K. Meteorological Office. For those who are unfamiliar with them a brief outline of the general nature of coupled models is provided in the next section. More specific details of HADCM3 are given in Section 3. Section 4 details the model's simulation of the components of the Arctic Ocean freshwater budget which is summarised in Section 5. Finally, some priorities for future research are summarised in Section 6.

2. Coupled climate models

Coupled models are made up of a number of component models (of the atmosphere, ocean, cryosphere and land surface) that are interactively coupled via exchange of data across the interfaces between them. Thus for example the ocean component is driven by the atmospheric fluxes of heat, momentum and freshwater (precipitation less evaporation) simulated by the atmospheric component.   These heat and freshwater fluxes are themselves derived from the sea surface temperatures simulated by the ocean model. Other driving fluxes for the ocean are produced by the brine

E.L. Lewis et al. (eds.), The Freshwater Budget of the Arctic Ocean, 127-139.
© 2000 *British Crown.*

rejection which occurs during sea ice formation, freshwater from sea ice melt and from input of freshwater at the continental boundaries by river inflow.

The atmospheric component of these models enables simulation of the evolution with time of the vertical distributions of the basic parameters of the vector wind, **V**, temperature, T, humidity, q, and of the surface pressure, p_o. This is done by discretisation of the basic equations governing the behaviour of the atmosphere (conservation of mass, momentum, heat and water, plus the atmospheric equation of state) and implementation of these discretised equations on an appropriate computer. The equations are time-stepped forward at intervals which may typically vary from a few minutes to tens of minutes, depending on the model formulation and resolution (see e.g Haltiner and Williams [1]) to produce an evolving simulation of the behaviour of the atmospheric flow and associated temperature, humidity and surface pressure fields.

The model dynamics may be represented either spectrally or on a grid of points covering the globe for various levels of the atmosphere (regional atmospheric models may cover a more restricted domain, but will then need the boundary conditions around their perimeter to be specified). Typically the atmospheric components of the current generation of climate models operate on grids with a spacing of two or three hundred kilometers in the 'horizontal' and a few to tens of levels in the vertical. Various schemes are available for the specification of the vertical coordinates (see [1]).

A key issue is the simulation of the basic physical processes which take place in the atmosphere and which determine many of the feedbacks for climate variability and change. Examples are: the representation of clouds and radiation; dry and moist convective processes; the formation of precipitation and its deposition on the surface as rain or snow; the interactions between the atmosphere and the land surface orography (including the drag on the atmosphere caused by breaking gravity waves) and atmospheric boundary layer processes and their interaction with the surface. Because these processes take place on scales much smaller than the model grid, they must be represented in terms of the large-scale variables in the model (V, T, q and p_o).

Similarly, the ocean component allows simulation of the current, temperature and salinity structure of the ocean and its evolution. Important physical processes are associated with the upper ocean mixed layer and diffusive processes in the ocean. In global models such as we are concerned with here, or in high latitude regional ocean models, account must also be taken of the freezing, melting and dynamics of sea ice and with ice-ocean interactions. Until recently, because of limitations of available computing power, climate models typically ran with the same ocean resolution in the horizontal as the atmospheric components to which they were coupled. Such ocean models have a very poor representation of the large-scale ocean current structure. This is not only because of the lack of resolution of the narrow boundary currents such as the Gulf Stream and the Kuroshio, but also because of the high viscosity coefficients necessary for computational stability (see e.g. Bryan et al. [2]). However, as available computing power has increased, it has become possible to increase the resolution at which the ocean component of coupled models is run to of the order of 1 degree in latitude and longitude. Although this does not allow explicit representation of ocean eddies (a resolution of 1/3 degree is considered 'eddy permitting' and around 1/9 degree or better, eddy-resolving), it does allow much improved representation of current structure.

The land surface components of climate models include representation of the thermal and soil moisture storage properties of the land surface though modelling of its upper layers. Key properties include surface roughness and albedo, specified from global datasets, though models with interactive land surface properties are now being developed. The insulating effects and change of surface albedo due to snow cover are of particular importance for high latitudes as is the ability for river runoff to be represented, especially in the Arctic.

Snow cover and sea ice are the two primary elements of the cryosphere represented in interactive form, though some models do now incorporate explicit parameterisations of permafrost processes (as does the model used here - Section 3). The large ice sheets are represented, if non-interactively, in global 3-d general circulation climate models through the land surface topography and the surface albedo, typically fixed at a value of around 0.8. Likewise there is usually no explicit representation of glacier processes.

As already noted, the component models making up the coupled system are interactively coupled via exchange of data across the interfaces between them. This is done at coupling intervals of typically one day. Models have been shown, particularly through the Atmospheric Model Intercomparison Project (AMIP) (see e.g Gates [3]) to have considerable variations in their ability to represent key climatic fields and coupled models to exhibit a considerable degree of climatic drift. Until recently it has been necessary to make use of so-called 'flux adjustments' (or 'flux corrections', Sausen et al. [4]) to prevent drift in the climate of the coupled system which itself arises from inadequacies in the component models and in the simulated fluxes at their interfaces. These adjustments, when used, are derived as fields of seasonally and geographically varying corrections to the heat and freshwater fluxes between atmosphere and ocean components. These are derived during a calibration run of the coupled model in which the sea surface temperatures and surface salinities are strongly relaxed back to observed climatological datasets of these quantities. These corrections to the fluxes are then applied to succeeding runs of the model to provide improved simulation of the coupled system. Sometimes, flux adjustments have also been applied to the momentum fluxes also. Whilst no flux adjustment is applied over land, it has in the past been necessary to flux-adjust the fields of sea ice concentration and thickness (Cattle et al. [5]). A driver in coupled modelling has been to develop models to the stage where they can now run without flux correction, as is now the case for the latest generation of climate models.

3. The Hadley Centre Coupled Model

We turn now to an outline description of the HADCM3 version of the Hadley Centre coupled model used in the present study. Gordon et al. [6] give more details of this model, which runs with little or no drift for century timescale integrations without use of flux adjustments. They also illustrate its ability to simulate the large-scale features of the global climate system. The model runs within the so-called Meteorological Office unified model system (Cullen [7]) on a regular latitude-longitude grid with an atmospheric resolution of 2.5° by 3.75°. The ocean model resolution is 1.25° in both latitude and longitude and is configured such that six ocean grid squares fit into one atmospheric grid square.   Ocean coastlines are determined by the (coarser) atmospheric model grid. The atmospheric model has 19 unequally-spaced levels in the vertical and the ocean model 20. The atmospheric component incorporates the interactive radiation scheme of Edwards and Slingo [8], a prognostic cloud scheme based on that of Smith [9] (which includes explicit representation of cloud liquid water), a penetrative convection parameterisation based on Gregory and Rowntree [10], but with the addition of a representation of convective downdraughts and a parameterisation of the direct effect of convection on momentum (Gregory et al. [11]), a parameterisation of orographic drag (Milton and Wilson [12]) and the gravity wave drag scheme of Gregory et al. [13].

The ocean model is based on that of Cox [14] with the addition of an ocean mixed layer scheme based on that of Kraus and Turner [15] and including the shear-induced mixing parameterisation of Pacanowski and Philander [16].   Vertical resolution is highest in the upper ocean, the smallest layer thickness being 10m for the top four layers near the surface. Horizontal mixing of tracers uses a version of the Gent and McWilliams [17] adiabatic diffusion with thickness diffusion coefficients determined locally using the formulation of Visbeck et al. [18] and the numerical implementation of Griffies et al. [19]. The convective adjustment scheme used in the model is modified in the region of the Denmark Strait and Iceland-Scotland ridge to better represent downslope mixing of overflow water using a scheme following Roether et al. [20].

A key feature of the high latitude climate is its snow and ice cover. The land surface scheme (Cox et al. [21]) includes the representation of the freezing and melting of soil moisture (permafrost). In the model, snow falling on the land surface accumulates there when the surface temperature is below $0°C$. Snow depth is allowed to change interactively over the model's land surface through the processes of accumulation, sublimation/deposition and melting.   Over land a snow albedo formulation is used which depends on snow depth, the snow-free albedo of the surface, a deep snow albedo for the vegetation type and, near the melting point, on temperature. An equation including the effect of surface heat capacity on the overall surface heat balance is used to determine changes in surface temperature. If snow is present then snow melt occurs when the simulated surface temperature rises above $0°C$. In that case, the surface temperature is reset to the melting point and the excess energy is used to melt snow.

Because of the relatively long timescales in which they respond to change, the major land sheets of Greenland and Antarctica are treated non-interactively in the model with the ice sheet mass held fixed and specified through the model's topography. A constant surface albedo of 0.8 is assumed over land ice.

Sea ice forms within the ocean model when the temperature of its uppermost layer (10m thick) falls below the freezing point of sea water ($-1.8°C$). In that case the layer temperature is reset to the freezing point and a sufficient covering of sea ice is allowed to form of initial mean thickness 0.5m to ensure that conservation of heat is satisfied. Alternatively ice may be advected into a grid square from its neighbours when the ocean top layer temperature is again set to $-1.8°C$ and the ice thickness and concentration adjusted to conserve heat. A simple ice thickness advection scheme is used in the model, which follows that of Bryan et al. [2].

Snow accumulation and melting on the surface of sea ice is also represented in the model. The thermodynamics of the ice is treated using the 'zero layer' formulation of Semtner [22] in which the ice-snow layer is treated as a single slab, the thickness of which changes according to heat balance equations applied at the top and bottom of the slab. The model also allows fractional ice coverage. The treatment of ice concentration in the model is based on that of Hibler [23], which assumes the ice-covered area to have a uniform thickness distribution. The atmospheric model calculates surface turbulent and radiative fluxes separately for ice and open water (leads) over grid squares for which ice cover is present. For ice surface temperatures below $-10°C$, the snow/ice surface albedo is taken to have a constant value of 0.8. Above this temperature, ice albedo decreases linearly to a value of 0.5 at the melting point ($0°C$). The albedo of leads is assumed constant at 0.06.

The heat flux entering (or leaving) the leads is partitioned between ice melt (or ice formation) and warming (or cooling) the upper layer of the ocean. The partitioning between ice melt/formation and ocean warming/cooling is chosen to be directly proportional to the ice area. Ocean surface layer temperatures higher than the freezing point of sea water ($-1.8°C$) result in a heat flux from the ocean to the ice which, if greater (less) than the diffusive heat flux through the ice, contributes to ice melt (formation) at the base of the ice-snow layer. Snow/ice melt occurs at the top of the ice in similar fashion to that for snow melt in the land surface scheme. Any melt water formed is passed to, and freshens, the surface layer of the ocean model. Alternatively brine release associated with ice formation increases surface layer salinity. More generally over the model oceans, salinity can change depending on the sign of the precipitation less evaporation difference. Surface freshening can also occur as a result of runoff from the model's land surface. The formulation of the scheme follows that of Taylor and Bunton [24].

Further details of the mathematical formulation of the sea ice model will be found in Cattle and Crossley [25]. We now turn our attention to the model simulation of Arctic parameters, and in particular of the components of the Arctic Ocean freshwater balance.

4. Model Simulation of Arctic Ocean Freshwater Budget Components

The key components of the freshwater balance of the climate system of the Arctic basin as a whole may be listed as follows:

- River runoff
- Precipitation less evaporation over the basin
- Ocean freshwater transports in and out of the basin
- Export of freshwater as sea ice
- Atmospheric transports in and out of the region.

An additional component internal to the system is the precipitation less evaporation flux (P-E) at the surface. In the long term this should balance the lateral atmospheric transports and contribute to the freshwater balance of the ocean itself, including the contribution made by sea ice. The key components of the Arctic Ocean freshwater balance, which is our concern here, are, therefore:

- River runoff
- Precipitation less evaporation over the basin
- Export of freshwater as sea ice
- Ocean transports.

We deal below with each of these in turn.

4.1 RIVER RUNOFF

The model's river runoff scheme, although simple, allows a representation of the highly seasonal characteristics of Arctic Ocean runoff to be represented in the model. In essence, water at the surface as a result of P-E or snowmelt excess to the needs of the model's soil moisture intake is allowed to directly runoff following the model's surface topography over defined river basins. By way of illustration, the variation in the seasonal and annual mean flow of five Arctic rivers is compared in Table 1 to the

Table 1. HADCM3 modelled versus observed (Mackay and Løken [26]) seasonal and annual mean flow in units of 10^3 m^3 s^{-1} for five Arctic rivers. Modelled flow is top left and observed bottom right of each cell. The highest seasonal value in each case is shown for each river in bold type.

Model vs. Observed	December/ January/ February		March/ April/ May		June/ July/ August		September/ October/ November		ANNUAL	
Mackenzie	1.3		**20.6**		10.0		6.6		9.6	
		8.2		6.9		**16.6**		11.6		10.8
Northern Dvina	2.1		1.4		**10.0**		8.1		8.6	
		1.1		**5.6**		4.1		2.6		3.3
Pechora	0.0		**11.2**		2.8		1.8		4.0	
		0.7		3.4		**6.7**		2.5		3.3
Ob	0.4		**30.7**		10.7		5.1		11.7	
		4.3		6.9		**27.7**		10.0		12.2
Yenisey	0.2		**42.5**		20.3		7.9		17.7	
		4.7		12.8		**41.0**		13.0		17.8

observed values of Mackay and Løken [26]. It will be seen that there is inaccuracy in the phasing and magnitude of the seasonal maximum flow in the model compared to observed (in particular the model tends to peak too early). However, the annual mean is quite well represented.

Figure 1 shows the seasonal mean variation of discharge from all the rivers around the Arctic basin as simulated by HADCM3. The overall annual mean discharge amounts to some 86.8 x 10^3 m^3 s^{-1}, equivalent to 2740 km^3 yr^{-1}, lower than the best estimate value of 3300 km^3 yr^{-1} quoted by Aagaard and Carmack [27] after Treshnikov [28].

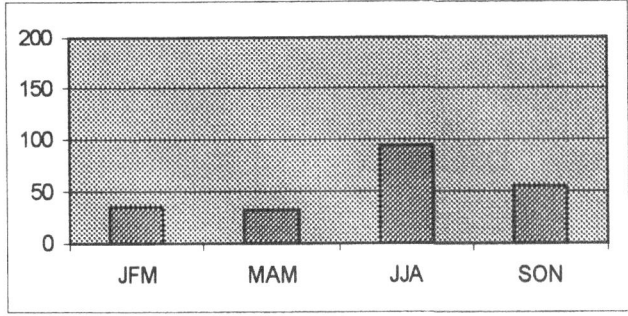

Figure 1. HADCM3 modelled mean seasonal variation of total river input into the Arctic basin.
Units $10^3 \text{km}^3\text{s}^{-1}$.

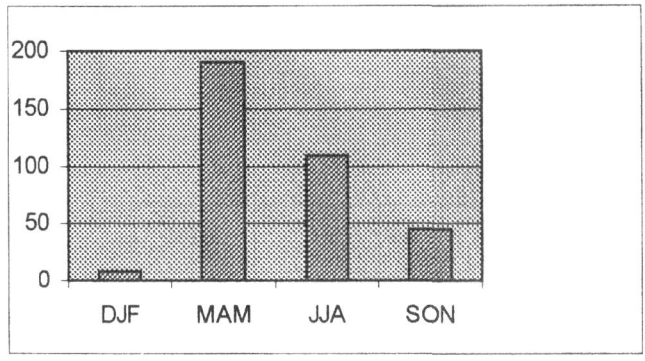

Figure 2. HADCM3 modelled mean seasonal variation of P-E over the Arctic basin.
Units $10^3 \text{km}^3\text{s}^{-1}$.

4.2 PRECIPITATION LESS EVAPORATION OVER THE BASIN

Figure 2 shows the mean seasonal variation of P-E simulated by HADCM3. Highest values occur in the summer months. The annual mean value of P-E amounts to some $53 \times 10^3 \text{ m}^3 \text{ s}^{-1}$, which is equivalent to $1650 \text{ km}^3 \text{ yr}^{-1}$, considerably more than the estimated observed value of $900 \text{ km}^3 \text{ yr}^{-1}$ given by [27] and updated by Serreze (1999, this volume). The model therefore appears to significantly overestimate this quantity whilst underestimating the runoff by about the same amount.

4.3 EXPORT OF FRESHWATER AS SEA ICE

The model's simulation of the mean seasonal variation of Arctic sea ice extent is illustrated in Figure 3. Maximum winter extents are well represented if a little high overall compared to the observed value of 15 x 10^{12} m^2. Summer extents are slightly underestimated compared to the observed minimum ice extent, which is of order 5 x 10^{12} m^2. Figure 3 also illustrates the modelled variability over the multi-decadal timescale. The model exhibits a realistic interannual variability in mean ice area compared to, for example, the data of Bjørgo et al. [29]. It also shows decadal to multidecadal timescale variations.

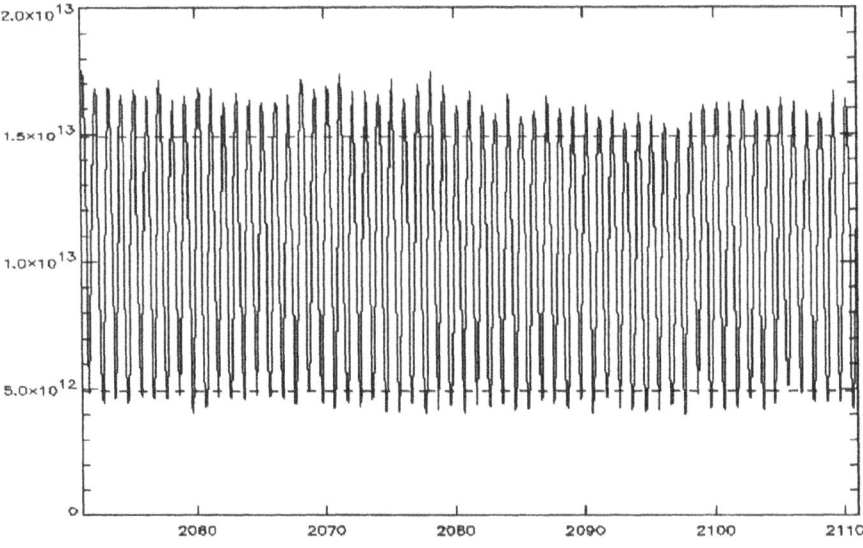

Figure 3: Modelled seasonal variation of sea ice area over 60 years of simulation. Observed mean values are shown as dashed lines

The field of ice thickness is less well represented in the model which has maximum values (typically 3-4 metres) centered over the Beaufort Sea to the north of Alaska. Observations show the maximum values of ice thickness (of over 5 m typically) to be found against the Greenland coast and the Canadian Archipelago. Poorly modelled ice dynamics and wind stresses are both a factor here. Vinje et al., [30] have shown the seasonal variation of monthly mean ice draft in Fram Strait (0° to 15°W) to show a non-systematic seasonal cycle from year to year with an average month to month variation of 0.7 m. Over a four-year period (1990-1994) the average minimum ice draft was observed to be in September (1.8m) and the maximum in July (2.7m). Compared to observed, the modelled ice thickness is too low in this region overall, and the seasonal cycle poorly phased (see Figure 4 showing the model's 50 year average seasonal variation of Fram Strait ice thickness).

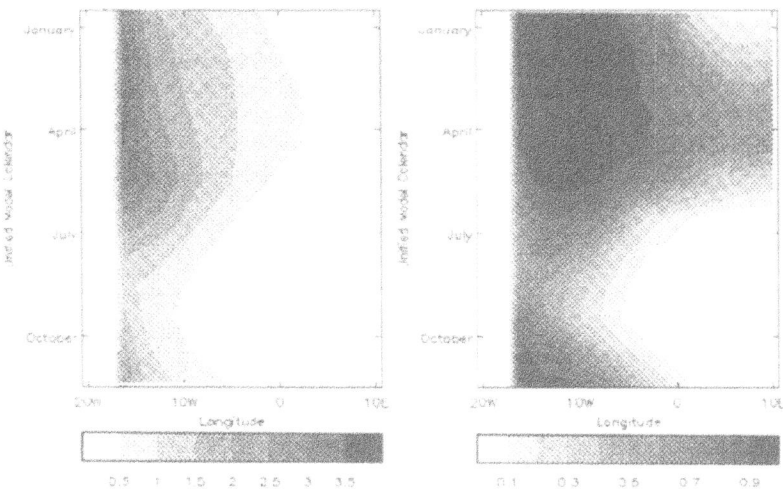

Figure 4. Mean seasonal cycle of modelled ice thickness (left) and concentration (right) across Fram Stait

We have estimated the model's transport of freshwater out of the Arctic basin by sea ice using two approaches. The first simply uses long term annual mean values of ice thickness and velocity across three regions: Fram Strait ($80°$ N), across the Nordic Seas between Spitzbergen and Norway to $15°E$ and through the Bering Strait. Table 2 below shows the derived values. Not surprisingly, it will be seen that the major contribution to the modelled ice flux is through Fram Strait.

Table 2. Estimated freshwater fluxes in the model across the three indicated locations derived from annual mean values of ice thickness and velocity.

Fram Strait	361 km^3 yr^{-1}
Spitzbergen to Norway	4 km^3 yr^{-1}
Bering Strait	7 km^3 yr^{-1}

We have also derived the freshwater flux through Fram Strait from 1 year of daily data on ice thickness and velocity. Based on these daily values, the Fram Strait freshwater flux is 565 km^3 yr^{-1}. This is only a quarter of the value derived by Vinje et al. [30] of 2050 km^3 yr^{-1} for the period 1976-1994 and an even smaller fraction of his value of 3372 km^3 yr^{-1} for the 1990-1994 period. It reflects the too small values in both ice thickness and velocity exhibited by the model in this region.

4.4 OCEAN FRESHWATER TRANSPORTS

The observed Arctic Ocean freshwater budget of Aagaard and Carmack [27] identifies the components associated with the oceanic circulation identified in Table 3:

Table 3. Observed ocean transports of freshwater into the Arctic basin (excluding sea ice transports) from Aagaard and Carmack [27].

Source or sink	Transport km^3 yr^{-1}
Water export through Fram Strait	-820
Water import through Bering Strait	1670
Water export through Canadian Archipelago	-920
Import with Norwegian Coastal Current	250
Saline water import through Barents Sea	-540
Saline water import with West Spitzbergen current	-160

It is evident from Table 4 that freshwater import through the Bering Strait and export through the Canadian Archipelago are major components of the Arctic Ocean freshwater balance. The model has zero barotropic flow through the Bering Straits and as we shall see below little inferred import of freshwater also. The Canadian Archipelago is closed in the model with the result that all its freshwater export from the Arctic basin is driven out through the Atlantic sector. From the model estimates of the salt and mass transports through Fram Strait and the seas from Spitzbergen and Norway (H T Banks, personal communication, Table 4) and given a mean reference

Table 4. HADCM3 mass and salt transports and deduced freshwater flux through Fram Strait and the seas from Spitzbergen to Norway.

Location	Mass transport (M, Sv)	Salt transport (S_t Sv psu)	$S = S_t/M$	Deduced freshwater flux (km^3 yr^{-1})
Fram Strait (80°N)	-3.23	-110	34.056	2748
Spitzbergen to Norway (20°E)	3.22	114	35.404	1171

salinity (S_0) for the model's Arctic basin of 35.0 psu, we deduce the associated freshwater transports in the model to be 2748 km^3 yr^{-1} and 1171 km^3 yr^{-1} respectively by using the formula:

$$\text{Freshwater transport} = M(S - S_0)/S_0 \qquad (1)$$

where M is the net mass transport and S the mean salinity over the section.

The contribution to the total freshwater budget of the Arctic Ocean from both sources is 4030 km^3 yr^{-1}. This is more than three times the value of 1270 km^3 yr^{-1} implied by Aagaard and Carmack for this region (Table 5) and nearly twice the freshwater transports for the Atlantic sector and Canadian Archipelago combined.

5. Summary of the Model Freshwater Budget

Table 5 summarises the components of the model's Arctic Ocean freshwater budget. It will be seen that there is approximate balance of the calculated components. The residual of -90 km^3 yr^{-1} provides an estimate of the implied freshwater input through Bering Strait, which was not explicitly calculated in this study. This value is very much lower than the value of 1670 km^3 yr^{-1} given by Aagaard and Carmack (c.f. Table 3). It also appears to be the case that what the model lacks in terms of export by sea ice and Canadian Archipelago throughflow, it makes up for by exchange with the Nordic Seas through the oceanic circulation.

Table 5. Summary of the components of the HADCM3 Arctic Ocean freshwater budget and implied balance.

Component	Freshwater transport km^3 yr^{-1}
River runoff	2740
P-E	1650
Total: inputs	4390
Sea ice export	- 560
Ocean export (to Nordic Seas)	-3920
Total exports	-4480
Balance	90

6. Future Research Priorities

In this paper we have provided an initial estimate of the components of the Arctic Ocean freshwater budget in the HadCM3 version of the Hadley Centre coupled climate model. It is evident that a more rigorous and definitive examination than was carried out for this NATO ARW needs to be done, including assessment of interannual and interdecadal variability in the terms. However, it is evident from the study carried out here that examination of the freshwater budget components provides a useful diagnostic, which provides pointers for model improvement. This is especially, important if such models are to correctly represent the processes of climate change, and in particular to be used to examine the possible impacts of changes in the freshwater balance for the thermohaline circulation of the North Atlantic. Particular pointers from the present study are the need to:

- Include a representation of Canadian Archipelago throughflow in this model (and in others) to reduce the export of freshwater to the Nordic Seas and provide freshening of the Labrador Sea waters (c.f. Goosse et al. [31])

- Give attention to the representation of the Bering Strait throughflow (c.f. Goosse et al. [32])

- For this model at least, to improve the modelling of sea ice, in particular sea ice dynamics to help obtain an improved geographical distribution of ice thickness and simulation of ice export.

In addition there is a need to carry out an examination of the representation of the Arctic Ocean freshwater budget in a number of other state of the art coupled models, an activity which needs to be further encouraged.

7. References

1. Haltiner, G.J. and Williams, R.T. (1980) *Numerical Prediction and Dynamic Meteorology*, Second Edition. John Wiley and Sons, New York, pp xvii+477
2. Bryan, K., Manabe, S. and Pacanowski, R.C. (1975) A global ocean-atmosphere climate model. II The oceanic circulation, *J.Phys.Oceanogr.*, **5**, 30-46.
3. Gates, W.L., (1995) An overview of AMIP and preliminary results. In Proceedings of the First International AMIP Scientific Conference, Monterey, California, USA, 15-19 May 1995 (W.L. Gates, Ed.), World Meteorological Organisation, WMO/TD-No.732, WCRP-92
4. Sausen, R., Barthel, K. and Hasselmann, K. (1987) A flux correction method for removing the climate drift of coupled models. Max Planck Institute fuer Meteorologie, Hamburg, Report No. 1, pp39.
5. Cattle, H,, Murphy, J.M. and Senior, C.A. (1992) The response of Antarctic climate in general circulation model experiments with transiently increasing carbon dioxide concentrations. *Phil. Trans. R. Soc. Lond. B*, **338**, 209-218.
6. Gordon, C., Cooper, C, Senior, C.A., Banks, H., Gregory, J.M., Johns, T.C., Mitchell, J.F.B., and Wood, R.A. (1999) The simulation of SST, sea ice extents and ocean heat transports in a version of the Hadley Centre coupled model without flux adjustments. *Climate Dynamics* (to appear).
7. Cullen, M.J.P. (1993) The unified forecast/climate model, *Meteorol. Mag.*, **122**, 81-94.
8. Edwards, J.M. and Slingo, A. (1996) Studies with a flexible new radiation code. I: Chosing a configuration for a large scale model. *Q. Jl. R. Meteorol. Soc.*, **122**, 689-719.
9. Smith, R.N.B. (1990) A scheme for predicting layer clouds and their water content in a general circulation model, *Q. Jl. R. Meteorol. Soc.*, **116**, 435-460.
10. Gregory, D. and Rowntree, P.R. (1990) A mass flux convection scheme with representation of cloud ensemble characteristics and stability dependent closure, *Mon. Weather Rev.*, **118**, 1483-1506.
11. Gregory, D., Kershaw, R. and Inness, P.M. (1997) Parameterisation of momentum transport by convection II: Tests in single column and general circulation models. *Q. Jl. R. Meteorol. Soc.*, **123**, 1153-1183.
12. Milton, S.F. and Wilson, C.A. (1996) The impact of parametrised sub-grid scale orographic forcing on systematic errors in a global NWP model. *Mon. Wea. Rev.*, **124**, 2023-2045.
13. Gregory, D., Schutts, G.J. and Mitchell, J.R. (1998) A new gravity wave drag scheme incorporating anisotrophic orography and low level wave breaking: Impact upon the climate of the U.K. Meteorological Office Unified Model. *Q. Jl. R. Meteorol. Soc.*, **124**, 463-493.
14. Cox, M.D. (1984) A primitive equation, 3-dimensional model of the ocean, *Geophysical Fluid Dynamics Laboratory Tech. Rep. No. 1*, Princeton, N.J.

15. Kraus, E.B. and Turner, J.S. (1967) A one dimensional model of the seasonal thermocline. II. The general theory and its consequences, *Tellus*, **19**, 98-106.

16. Pacanowski, R.C. and Philander, S.G.H. (1981) Parameterisation of vertical mixing in numerical models of tropical oceans, *J. Phys. Oceanogr.* **11**, 1143-1451.

17. Gent, P. and McWilliams, J.C. (1990) Isopycnal mixing in ocean circulation models. *J.Phys. Oceanogr.*, **20**, 150-155.

18. Visbeck, M., Marshall, J., Haine, T. and Spall, M. (1997) On the specification of eddy transfer coefficients in coarse resolution ocean circulation models. *J.Phys. Oceanogr.*, **27**, 381-402.

19. Griffies, S.M., Gnanadesikan, A., Pacanowski, R.C., Larichev, V.D., Dukowicz, J.K. and Smith, R.D. (1998) Isoneutral diffusion in a z-coordinate ocean model. *J. Phys. Oceanogr.*, **28**, 805-830.

20. Roether, W., Roussenov, V.M. and Well, R. (1994). A tracer study of the thermohaline circulation of the eastern Mediterranean. In: Ocean Processes in Climate Dynamics: Global and Mediterranean Example (P. Manaotte-Rizzoli and A.R. Robinson Eds.), 371-394, Kluwer Academic Press.

21. Cox, P., Betts, R., Bunton, C., Essery, R, Rowntree, P.R. and Smith, J. (1998) The impact of new land surface physics on the GCM simulation of climate and climate sensitivity. *Climate Dynamics*, (to appear).

22. Semtner, A.J. Jr. (1986) A model for the thermodynamic growth of sea ice in numerical investigations of climate, *J. Phys. Oceanogr.*, **6**, 379-389.

23. Hibler, W.D. III (1979) A dynamic-thermodynamic sea ice model, *J. Phys. Oceanogr.*, **9**, 817-846.

24. Taylor, N.K. and Bunton, C. (1993) River runoff in the new UKMO coupled model, *Research Activities in Atmospheric and Oceanic Modelling (ed. G.J.Boer), Report No. 18.* WMO/TD-No. 553, Geneva: World Meteorological Organisation.

25. Cattle, H. and Crossley, J. (1995) Modelling Arctic climate change, *Phil. Trans. R. Soc. Lond. A*, **352**, 201-213.

26. Mackay D.K. and Løken O.H. (1974) Arctic hydrology. In: *Arctic and Alpine Environments* (Ives, J.D. and Barry, R.G., Ed.), Methuen, London, 111-132.

27. Aagaard, K. and Carmack, E.C. (1989) The role of sea ice and other fresh water in the Arctic Ocean circulation, *J. Geophys. Res.* **94**, **C10**, 14485-14498.

28. Treshnikov, A.F. (1985) Arctic Atlas (in Russian), 204 pp., Arkt.-Antarkt. Nauchuno-Issled. Inst., Moscow.

29. Bjørgo, E., Johannessen, O.M. and Miles, M.W. (1997) Analysis of merged SMMR-SSMI time series of Arctic and Antarctic sea ice parameters 1978-1995. *Geophys. Res. Let.*, **24**, 413-416

30. Vinje, T., Nordlund, N. and Kvambekk, A. (1998) Monitoring ice thickness in Fram Strait. *J. Geophys. Res.*, **103**, **C5**, 10437-10449.

31. Goosse, H, Campin, J.-M, Fichfet, T. and Deleersnijder, E. (1997a) The effects of the water flow through the Canadian Archipelago in a global ice-ocean model. *Geophys. Res. Let.*, **24**, 1507-1510.

32. Goosse, H., Fichefet, T., and Campin, J.-M. (1997b) Sensitivity of a global ice-ocean model to the Bering Strait throughflow. *Climate Dynamics*, **13**, 349-358.

ATMOSPHERIC COMPONENTS OF THE ARCTIC OCEAN HYDROLOGIC BUDGET ASSESSED FROM RAWINSONDE DATA

M.C. SERREZE AND R.G. BARRY

Cooperative Institute for Research in Environmental Sciences, Campus Box 449, University of Colorado, Boulder CO 80309-0449, USA

1. Abstract

The characteristics of precipitation minus evaporation (P-E) for the north Polar Cap (the region north of 70°N), the Arctic Ocean and a Central Arctic Ocean domain are examined by applying the aerological method to rawinsonde data collected between 1973 and 1995. Regional estimates of mean precipitation (P) are obtained using an improved climatology, and evaporation (E) is calculated as a residual. Annual mean P-E, P and E for the Polar Cap are estimated at 161 mm, 279 mm and 118 mm, respectively. Corresponding values for the Arctic Ocean are 153 mm, 252 mm and 99 mm compared to 174 mm, 202 mm and 28 mm for the Central Arctic Ocean. The latter domain excludes the Kara and Barents seas. For all three domains, mean P-E is positive for every month with a winter-half minimum and a late summer and autumn maximum. Precipitation also exhibits a summer maximum, indicating that P is largely a function of the large-scale moisture convergence. This seasonality contrasts sharply with the major terrestrial drainages, which tend to show a summer minimum in P-E and a summer maximum in P, indicating that summer precipitation is strongly related to within-region recycling of water vapor. None of the three domains show trends in P-E. For any season, large values are favored by situations leading to above-average moisture inflows along the Atlantic sector. Our estimate of mean P-E for the Polar Cap is lower than estimates based on gridded wind and moisture fields from "reanalysis" projects.

2. Introduction

The atmospheric contribution to freshwater input into the Arctic Ocean comes from several sources. River runoff, largely dependant on precipitation (P) less evaporation (E) over the terrestrial watersheds (mainly the Ob, Lena, Yenesei and Mackenzie), provides the largest freshwater source to the Arctic Ocean of about 350 mm per year [1]. P-E over the Arctic Ocean itself provides the next largest atmospheric contribution of about 175 mm per year [2]. Changes in atmospheric contributions may influence sea-ice extent and thickness, with consequent impacts on surface albedo and turbulent heat fluxes and potential feedbacks to the atmosphere. Attendant impacts on freshwater outflow through Fram Strait in the form of low-salinity sea ice and water may influence the convective regime in the Greenland-Norwegian-Iceland seas and potentially, the global thermohaline circulation.

141

E.L. Lewis et al. (eds.), The Freshwater Budget of the Arctic Ocean, 141-161.
© 2000 *Kluwer Academic Publishers.*

While the runoff term can be estimated from available gauge records, diagnosing its spatio-temporal variability is made difficult by deficiencies in the terrestrial precipitation station network. Gauge limitations result in severe undercatch of solid precipitation. Although it is possible to apply corrections [3], the network itself is fairly sparse. Precipitation data are particularly meager over the Arctic Ocean itself, largely limited to measurements from the Russian "North Pole" (NP) series of drifting stations or from combining the NP data with measurements from coastal sites, aircraft landings and field programs [4-6]. Estimates of terrestrial and oceanic evaporation are few, and are based on simple energy budget considerations [7-11], the difference between measured runoff and precipitation for individual watersheds [12], or direct measurements from field experiments [13-15].

Modeling approaches show promise in circumventing these problems and have the additional advantage of providing outputs at regular grid points. Interest is growing in the use of Numerical Weather Prediction (NWP) models [6, 16, 17]. NWP models start with a previous forecast as a "first guess" of atmospheric conditions that is then adjusted by observations (e.g., rawinsonde data). Resulting fields (atmospheric analyses) are used to provide short-term forecasts of variables such as P and E. "Reanalysis" efforts by the National Centers for Environmental Prediction/National Center for Atmospheric Research (NCEP/NCAR) and European Reanalysis Agency (ERA) provide time series of analyzed and forecast fields based on fixed models that eliminate pseudo-climate signals introduced by changes in model physics. While the ERA precipitation fields for the Arctic appear promising, large errors with respect to observations are still present [6, 17]. Other studies, such as for Greenland [18] have examined precipitation fields from reanalysis as well as those from enhanced retrievals, which make use of analyzed moisture and wind fields in conjunction with high-resolution topographic data.

Another approach is to estimate P-E via the "aerological" method, using vertical atmospheric profiles of humidity and winds. This has been done in several Arctic studies using the available network of rawinsonde stations (Figure 1), providing assessments for the region north of 70°N, the Arctic Ocean, the Mackenzie basin, NWT of Canada and Greenland [2, 19-24]. The atmospheric moisture budget can be written as:

$$P - E = -\nabla \bullet Q - \partial W / \partial t \tag{1}$$

where the first term on the right represents the horizontal convergence of the vertically-integrated water vapor flux (Q) into the region of interest and the second term is the time change in total column water vapor, W (usually termed precipitable water), averaged over the domain. For a domain, A, the flux divergence term can be written using Gauss' theorem as:

$$\nabla \cdot Q = 1 / A \oint Q \bullet n \, dC \tag{2}$$

where dC is the length of a segment of the domain boundary and n is a unit vector normal to the boundary. Values of Q are taken as averages along each boundary segment. For long-term annual means (and assuming a stationary climate), the last term of Equation 1 may be dropped, such that the flux convergence equals P-E. Similar approaches can be applied to analyzed wind and moisture fields from NWP output to provide gridded fields of P-E [17]. There are several assumptions inherent to the aerological method which are summarized by Walsh et al. [2].

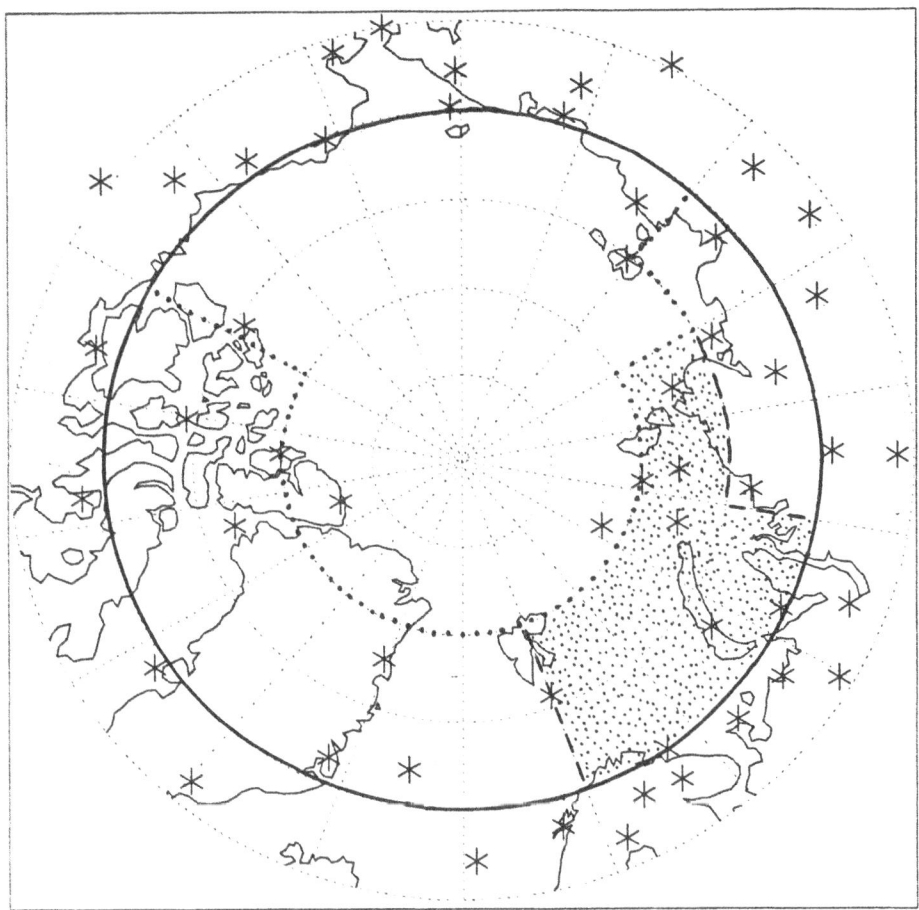

Figure 1: Location map showing the distribution of rawinsonde stations with at least 10 years of data (stars) and the three domains for evaluation of P-E. The bold solid line at 70°N encloses the Polar Cap domain (PC). The Central Arctic Ocean (CAO) domain is enclosed by the solid and dotted lines. The Arctic Ocean domain (AO) includes the CAO and the stippled region.

In practice, the aerological approach as applied to rawinsonde data can only provide estimates for large domains (e.g., exceeding 2.0×10^6 km^2). For smaller areas, errors in the rawinsonde winds, in particular, can cause large errors in the calculated flux convergences. Problems with measuring humidity in cold, dry Arctic conditions, differences between countries in instrument types and reporting practices and changes in instrumentation and reporting through time introduce further uncertainty. Stations variously provide reports from one to four times daily, which because of diurnal variations in winds and moisture can cause sampling problems. The convergence estimates are also influenced by the available station density and temporal changes in the reporting network [2, 24-27]. The quality of analyzed wind and moisture fields from NWP models will be affected by these problems, as well as by biases introduced during the data assimilation process [28].

Here, we use Arctic rawinsonde data (Figure 1) between 1973 and 1995 to examine the mean features and variability in P-E for three domains: (1) the area north of 70°N (defined here as the polar cap, or PC); (2) the Arctic Ocean (AO); (3) the Central Arctic Ocean (CAO). P-E for each domain is compared with available estimates of P with estimates of E computed as a residual. The PC domain is the same as examined by Serreze et al. [23,24] and Walsh et al. [2]. Our AO domain approximates that used in other studies in this volume. The CAO domain is in turn identical to that examined by Walsh et al. [2]. Although our CAO domain is arguably somewhat of a misnomer as it includes the coastal Chukchi and Beaufort seas and part of the Laptev Sea, inclusion of these areas was dictated by the available rawinsonde network. The PC domain is particularly useful - while computation of the meridional fluxes across 70°N provides for a straightforward examination of the primary "pathways" of moisture into and out of the Arctic Basin, the PC results are largely applicable to the irregularly-shaped AO and CAO domains. As such, we focus on the PC , and provide contrasts with the other two domains as appropriate. The present work, in large part, represents an update and extension of earlier studies [2, 23, 24]. Bromwich et al. [17] summarize results for the PC based on reanalysis fields, and make comparisons with rawinsonde-based estimates [see also 29].

3. Primary Data Sets

3.1 HISTORICAL ARCTIC RAWINSONDE ARCHIVE

The Historical Arctic Rawinsonde Archive (HARA) [30] used in our previous work has been updated through June of 1996. The five-volume CD-ROM archive is available from the National Snow and Ice Data Center (NSIDC) in Boulder Colorado, and contains all available soundings for fixed stations north of 65°N (Figure 1). Soundings typically extend to at least 300 hPa. Provided variables include pressure, geopotential height, wind direction and speed, dewpoint depression (the difference between the physical and dewpoint temperature) and associated quality codes. Data are generally reported at both mandatory levels (standard levels such as 700 hPa and 500 hPa) and significant levels (intermediate levels intended to catch "significant" vertical gradients of temperature, humidity or winds). Not all variables are reported at every level. The number of soundings and significant levels reported is greatly reduced prior to 1973. Consequently, we use data from the 1973-1995 period only, at both 0000 and 1200 UTC.

3.2 MONTHLY-MEAN PRECIPITATION

In a related study [6], an improved long-term monthly climatology of Arctic precipitation was assembled by blending the Legates and Willmott [31] gridded product with precipitation measurements from the NP drifting stations and gauge-corrected data for land stations. Thirty NP stations were manned between 1950 and 1991. The average duration of each station was 2.4 years. The records, assembled by Russian scientists at the Arctic and Antarctic Research Institute (AARI), are available on a CD-ROM compiled by University of Washington's Polar Science Center (PSC) and the NSIDC. Colony et al. [5] provide further descriptions of the NP data and their limitations. Here, we use average monthly and annual precipitation over the three domains extracted from the blended data set.

4. Data Processing

Individual soundings were subject to a series of quality control procedures. These largely follow those of Serreze et. al. [23, 24], but with some recent improvements. The procedures consist primarily of a series of limits checks and checks for vertical consistency. Values missing in the original soundings, or flagged as suspect from either the original quality codes or our own error checks were coded as missing and subsequently re-filled through vertical interpolation. Soundings not extending to at least 500 hPa (below which contains over 95% of W) were discarded. For each quality-controlled sounding retained, W and the vertically integrated zonal and meridional vapor fluxes (Q_u and Q_v, respectively) were obtained using the following equations:

$$W = 1/g \int_{P_b}^{P_t} q\, dp \tag{3}$$

$$Q_u = 1/g \int_{P_b}^{P_t} qu\, dp \tag{4}$$

$$Q_v = 1/g \int_{P_b}^{P_t} qv\, dp \tag{5}$$

where g is gravitational acceleration; q, u and v are the specific humidity zonal and meridional components of the wind, respectively, at individual levels; and p_b and p_t are the pressures at the bottom and top level to which the integrations are carried out. Specific humidity, calculated using the dewpoint depression, has units of g kg^{-1}, and represents the ratio of the mass of moist air to the total mass of air. The integrals were computed from the surface to 300 mb by applying the trapezoidal method to both mandatory and significant levels. Each sounding was also interpolated to provide vertical profiles of temperature, winds and vapor fluxes at 100 m increments.

Monthly averages of each variable were then computed at each station. Means were discarded if based on fewer than 20 observations. Mean values were subsequently obtained at the boundaries of the domains shown in Figure 1 at every 2.5° using a Cressman [32] interpolation. Taking the example of precipitable water, the Cressman interpolation has the form:

$$W_p = \sum_{i=1}^{n} W_i\, C_i \bigg/ \sum_{i=1}^{n} C_i \tag{6}$$

$$C_i = (N_2 - d_2) /(N_2 + d_2) \tag{7}$$

where W_p is the precipitable water at a desired point along the domain perimeter; W_i is the precipitable water vapor at a given station location; i, C_i is the Cressman weight given to that observation; d is the distance between the observation and the perimeter point, and N is the search radius at which the weight goes to zero. Typically, an initial search radius is set and all the observations (n) that fall within that radius are used to perform the summations. If no observations fall within the initial radius, the search is extended over larger values of N. For the PC and AO domains, we used an initial search radius of 500 km, extending it to 750 km,

146

1000 km and 1250 km if necessary. For the CAO domain, where the density of rawinsonde stations is lower, it was necessary to use search radii of 750 km, 1000 km, 1250 km and 1500 km. It was rarely necessary to move beyond the second search radius. Nevertheless, results for the CAO contained some missing months for the period 1973-1976, reflecting the less abundant rawinsonde data in the earlier part of the record.

The results obtained are sensitive to the details of the interpolation used, as well as to changes in the station distribution. Experiments were conducted to assess the sensitivity of P-E and the calculated flux convergences to changes in the size of the search radii, the minimum number of stations to be included in each search radius, the use of daily as opposed to monthly mean fluxes, and the filling of temporal gaps in individual station records with climatology and interpolated fluxes from neighboring stations. For individual monthly or annual means, differences can exceed 20%. While it is possible that the use of other interpolations techniques such as kriging or optimal interpolation could offer some improvement, Cullather et al. [29] and Bromwich et al [17], demonstrate that the accuracy of the rawinsonde-based approach is more fundamentally limited by the fairly low density of rawinsonde stations. These limitations may impact assessments of both interannual variability and the mean seasonal cycle.

5. Results

5.1 ANNUAL MEANS

Figure 2 shows the annual mean meridional vapor flux <qv> (the angle brackets denote the time average) interpolated to 70°N (the boundaries of the PC domain) by longitude and height. There are three primary "cores" of poleward transport. The strongest, with transports greater than 6 g kg^{-1} m s^{-1}, is centered near the prime meridian.

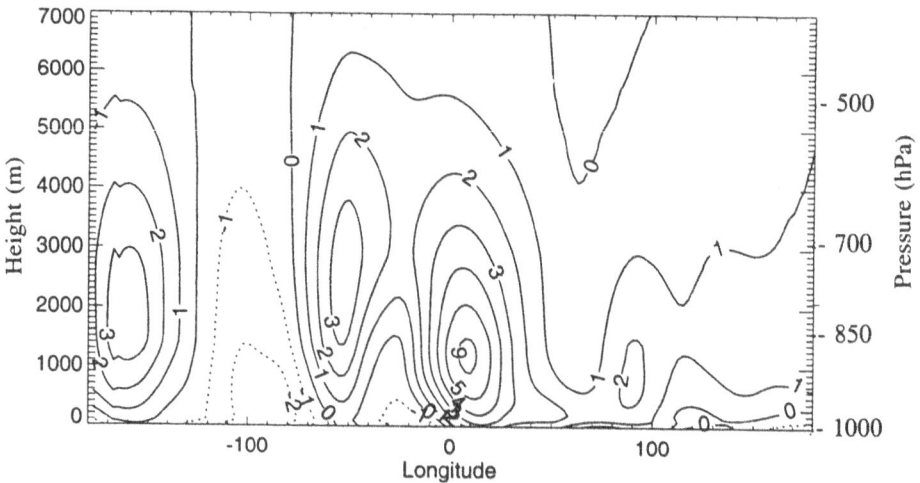

Figure 2: Vertical distribution of annual mean meridional vapor fluxes (g kg^{-1} m s^{-1}) along 70°N. Positive (poleward) fluxes are shown by solid centaurs and negative (equatorward) fluxes are shown by dottted centaurs.

Moving west, the second core, near the longitude of Baffin Bay at about 40-50°W, is characterized by peak transports about half as large. The westernmost core, just east of the date line, has similar flux magnitudes. A subsidiary feature is located near 100°E over Eurasia. The three cores of strongest poleward transport are separated by regions of equatorward transport centered at about 100°W over the Canadian Arctic Archipelago and at about 30°W over Greenland. The latter is clearly a low-level feature, with fluxes turning poleward above 850 hPa.

Peak poleward transports are not found at the surface but in the lower troposphere. This reflects the tendency for the increase in meridional winds (<v>) with height to outweigh the decrease in <q> with height. However, at higher levels, <q> becomes so small that even with increasingly stronger winds, <qv> falls off. Nevertheless, in contrast to the cores of poleward transport, fluxes in the regions of equatorward transport peak at or near the surface, showing that this "tradeoff" between winds and specific humidity does not always hold.

Figure 3 provides the annual mean vertically-integrated meridional vapor flux at 70°N by longitude. The data have been analyzed to separate the contributions to the total flux by the time-mean (or stationary) and transient eddy components. In explanation, the vertically-integrated meridional transport can be expressed as:

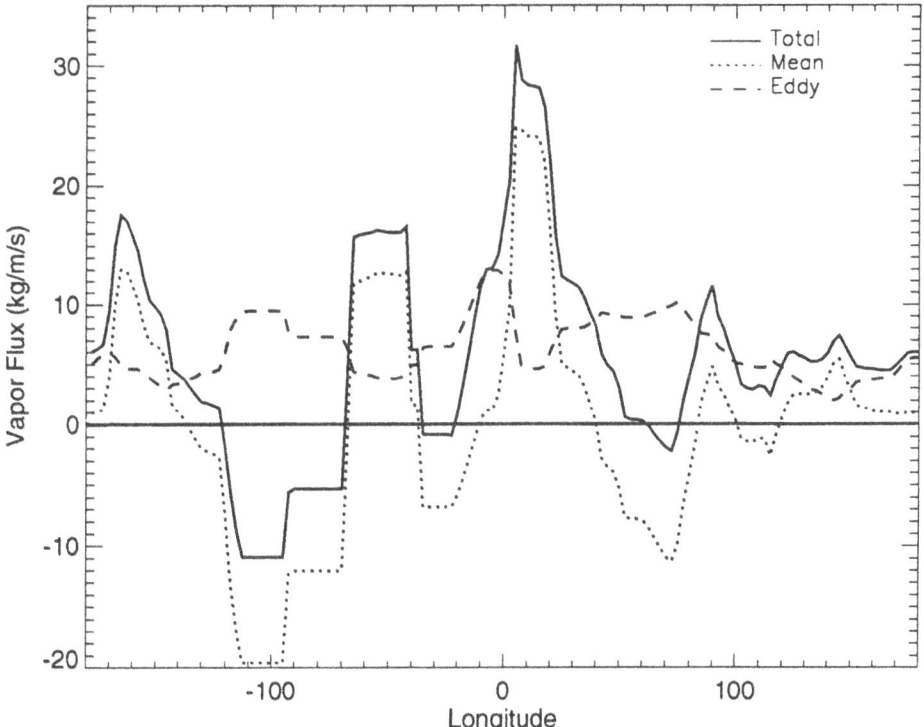

Figure 3. Mean annual vertically-integrated meridional vapor fluxes (kg m⁻¹ s⁻¹) along 70°N broken down into total, stationary and transient eddy components. The transient eddy fluxes capture variability at less than 30 days.

$$\int_{P_b}^{P_t} <qv> \, dp = \int_{P_b}^{P_t} <q> <v> \, dp \; + \; \int_{P_b}^{P_t} <q'v'> \, dp \qquad (8)$$

where primes denote departures from the mean. The term on the left, the total flux, corresponds to Equation 5 and the vertical integral of the results in Figure 2. The first-term on the right, the stationary flux, represents the transport of the mean specific humidity by the mean meridional wind. The second term on the right is the transient eddy flux. The transient eddy component represents the covariance between anomalies of the meridional wind and humidity, e.g., at any level, $<q'v'> = <(q - <q>)(v - <v>)>$.

Positive anomalies of v tend to transport positive anomalies of q while negative anomalies of v tend to transport negative anomalies of q, resulting in net poleward transient eddy transports. For each year, month and longitude, we computed the stationary flux using the average wind and humidity profiles. Using the monthly-mean total fluxes for each year, we then found the transient-eddy terms for each year and month as a residual. We then averaged the monthly values and computed the annual mean. By this technique, the transient eddy term contains the effects of variability over time scales of 12 hours to 30 days.

The longitudinal pattern for all three flux components appears somewhat "blocky", particularly from about 100°W to the prime meridian. This reflects the low density of rawinsonde stations, resulting in the interpolation procedure using values from the same stations for a series of longitudes. The results in Figure 2 and all subsequent vertical cross sections appear less blocky as the data have been further smoothed for clarity (which acts to under-represent the magnitude of the peak transports). The distribution of the vertically integrated total flux clearly captures the results in Figure 2. The largest fluxes, near the prime meridian, exceed 30 kg m^{-1} s^{-1}. The longitudinal pattern of the stationary flux has the same general shape and amplitude as that of the total flux, but is shifted downwards to yield smaller poleward transports and larger equatorward transports. In line with earlier discussion, the transient eddy component is always positive. The magnitudes of the total flux shown in Figure 3 depart somewhat from those reported by Walsh et al. [2] due to differences in data processing and the inclusion of more recent years of data.

The pattern of the stationary flux can be diagnosed from inspection of Figures 4 through 6, which show, respectively, the mean annual 500 hPa height field, the longitudinal distribution of mean annual precipitable water and the vertical cross section of the mean annual meridional wind. In the most fundamental sense, the sign of the stationary transports will reflect the influence of the atmospheric longwaves evident in the mean 500 hPa flow. The mean flow describes the well-known asymmetric circumpolar vortex, with pronounced troughs over Baffin Bay and central Asia. The cores of poleward stationary moisture transport near the prime meridian, Baffin Bay, the date line and over central Eurasia can be associated with the weak poleward geostrophic wind component (as inferred from the height contours) ahead of the mean troughs. In turn, the negative stationary fluxes centered over the Canadian Arctic Archipelago (100°W) and at about 70°E correspond to regions behind the troughs with a northerly wind component. The southward fluxes over Greenland, however, are not readily understood from the mean mid-tropospheric flow.

Precipitable water (Figure 5) is maximized near the prime meridian, in accord with the higher temperatures in this sector [23]. Although both Figure 4 and Figure 6 show that mean poleward winds in this region are light, the abundant water vapor as compared to other

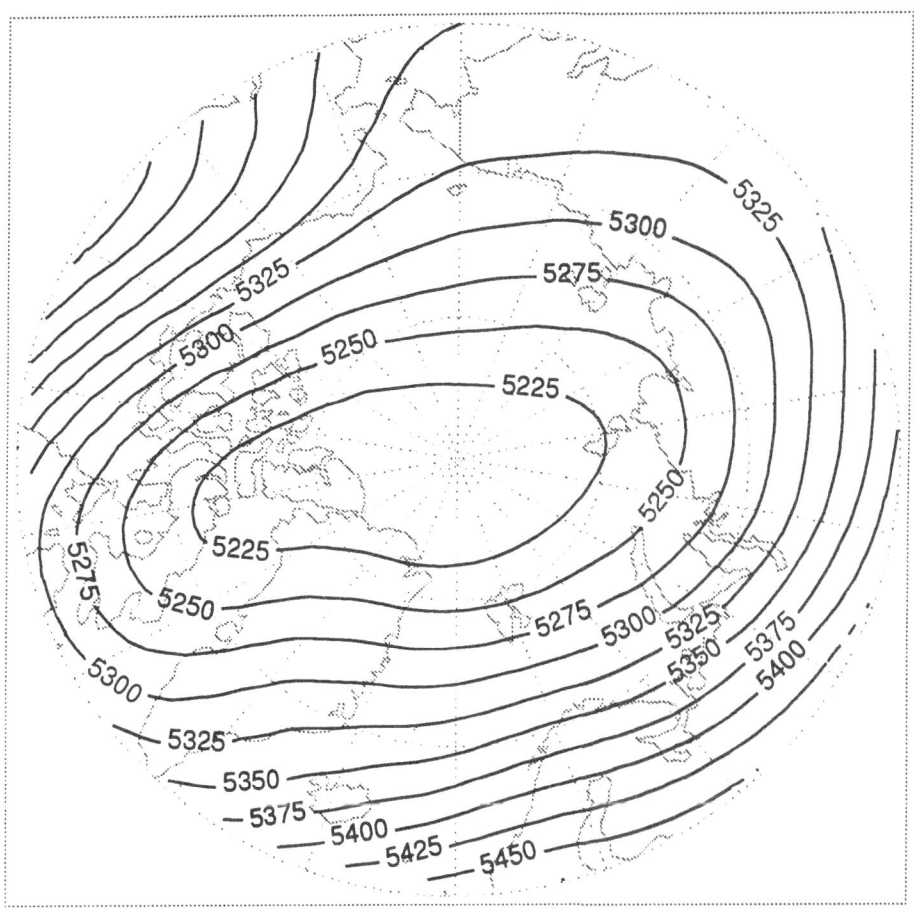

Figure 4: Mean annual 500 hPa height field (m).

sectors allows for strong poleward transports. Precipitable water is much smaller in the Baffin Bay sector [33] but this is compensated for by the stronger mean poleward winds. Similar arguments can be made for the stationary poleward transports near the date line. In turn, while precipitable water is low along the Canadian Arctic, this is offset by the fairly strong mean equatorward winds. Figure 6 shows that the zone of negative stationary transports over Greenland is associated with weak equatorward winds at low levels not evident at 500 hPa.

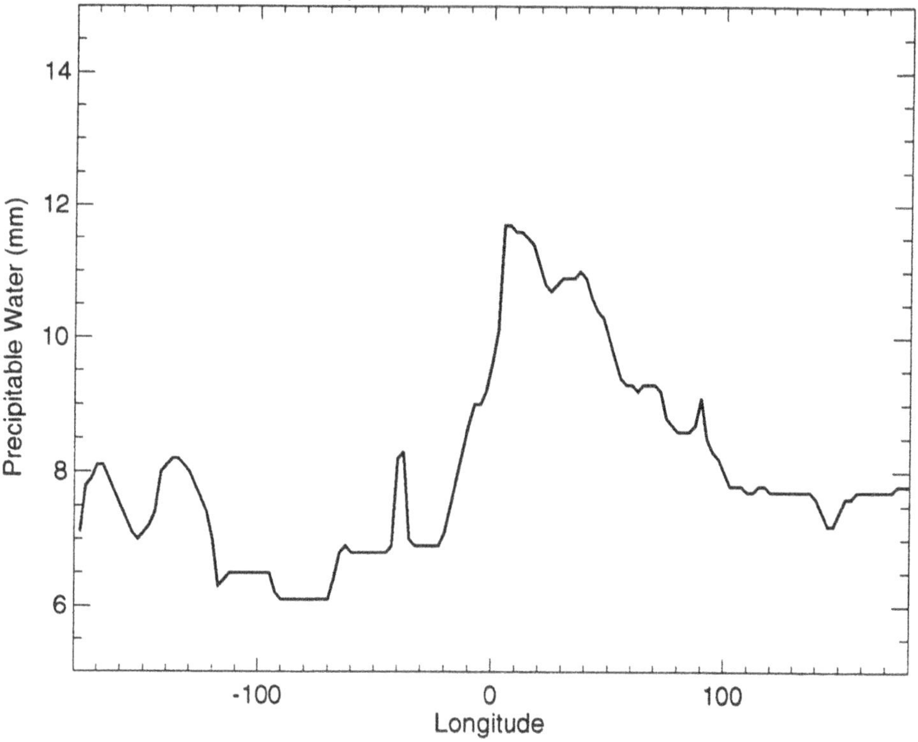

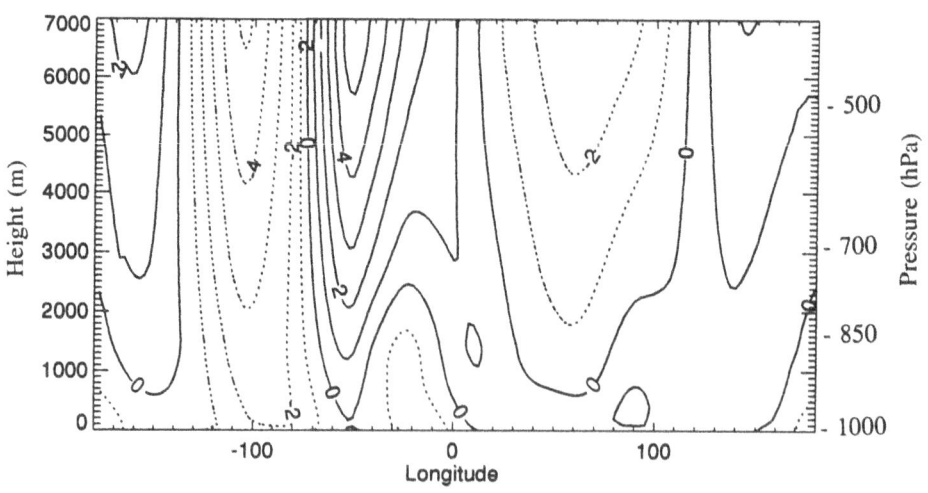

Figure 5: Mean annual precipitable water (mm) along 70°N (g kg^{-1} m s^{-1}).

Figure 6: Vertical distribution of mean annual meridional winds (m s^{-1}) along 70°N.

The transient transports tend to be smaller in magnitude than the time mean transports (Figure 3). A weak peak is found near the prime meridian, seen more clearly in the vertical cross section (Figure 7). The lack of a more pronounced peak in the transient eddy transports in this region is somewhat surprising, given that this area is subject to the frequent passage of extratropical cyclones associated with the North Atlantic cyclone track. Similarly, we see no sharp peaks in transient transports over the Baffin Bay sector or near the date line, which are also known to be areas of strong synoptic activity [34]. This may reflect masking of the eddy flux at synoptic time scales by our use of monthly means — inspection of daily fluxes clearly shows large poleward transports associated with cyclone passages. It is apparent from Figure 3 that in terms of the total flux, the poleward transports outweigh the areas of negative transport, such that the annual mean flux convergence (P-E) must be positive. We obtain an annual value of 161 mm, essentially identical to the value of 163 mm found by Serreze et al. [23] using a shorter record through 1991.

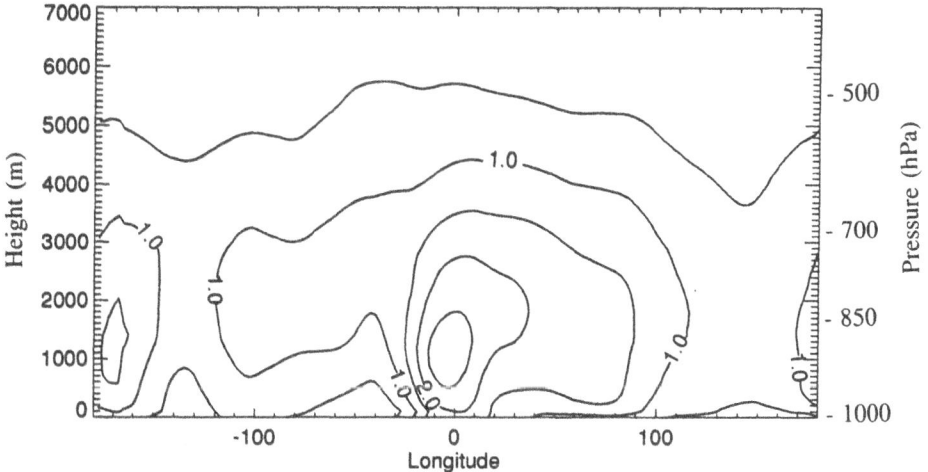

Figure 7: Vertical distribution of mean annual meridional transient eddy vapor fluxes along 70°N

Referring back to Figure 1, it is clear that estimating P-E for the smaller AO and CAO domains requires evaluation of both the meridional and zonal flux components. However, by assessing the fluxes separately for the 2.5° zonal and meridional sectors of each domain, we find that in terms of the flux convergence, the contribution of the zonal flux components is an order of magnitude less important than that of the meridional components. Consequently, P-E for the AO and CAO can be largely understood by only considering the meridional components. To this end, we examined the longitudinal distribution of the vertically-integrated meridional flux along the AO and CAO boundaries, broken down into the total, stationary and transient eddy components (not shown). The fluxes are variously evaluated at 70°N, 75°N and 80oN, depending on the longitude.

Where the three domains share common boundaries, the meridional fluxes are of course identical. Meridional flux patterns along the higher-latitude sectors of the AO and CAO exhibit the same general features seen for the PC, but the magnitudes are smaller. This illustrates that annual P-E for all domains have the same general climatic controls. As for the illustrates that annual P-E for all domains have the same general climatic controls. As for the major differences, the outflow along the Canadian Arctic Archipelago in the AO and CAO domains is very small, and there are smaller inflows near the prime meridian and over Baffin Bay. This reflects the higher latitude at which the fluxes are evaluated, where there is less precipitable water, lighter mean meridional winds and less frequent cyclone activity (not shown). As compared to the PC, the flux averaged across all domain sectors is smaller for the AO, but this is nearly accounted for by the fact that the net inflow is weighted over a smaller area (10.0×10^6 km^2 as opposed to 15.4×10^6 km^2 for the PC), such that P-E is rather similar to the AO at 153 mm. The average flux for the CAO is similar to that for the AO, but weighting by the even smaller area (7.49×10^6 km^2) results in a larger P-E of 174 mm. Of interest are the relative contributions of the stationary and transient eddy transports to annual P-E. As is evident from Figure 3 and from corresponding results for the AO and CAO (not shown), the annual flux convergence from the time-mean transports for all three domains is close to zero. Hence, P-E is essentially determined by the transient eddy component.

The CAO domain differs from the AO in that it excludes the Kara and Barents seas and part of the Laptev Sea. While we have not attempted to evaluate the Kara/Barents sector directly, it is apparent from comparison of the AO and CAO results that P-E must be relatively low in this region. By weighting mean P-E for the AO and CAO by their areas, we obtain a mean for the Kara/Barents sector of 90 mm. This lends support to the Gorshkov [35] climatology, which shows P-E in this ranging from less than -100 mm to 250 mm [see also 17].

By taking the annual precipitation for the three domains from the improved climatology described in Section 3.2 and calculating evaporation as a residual, we can close the water balance. Table 1 summarizes the mean annual P-E, P and E for each domain.

Table 1: Annual Mean P-E, P and E for the Three Domains (mm).

	P-E	P	E
Polar Cap	161	279	118
Arctic Ocean	153	252	99
Central Arctic Ocean	174	202	28

The annual evaporation estimate of 118 mm for the PC compares reasonably well with Korzun's [8] value of 103 mm. Subtracting from Korzun's annual E our annual estimate of P (279 mm) yields a P-E of 176 mm, which is about 10% higher than our value. The P estimate for the Polar Cap taken from the Gorshkov [35] atlas is somewhat higher at 293 mm, which if used with Korzun's E, yields a P-E of 190 mm. Bromwich et al. [this volume], using NCEP/NCAR and ERA analyzed fields, obtain a P-E of 182 mm and 194 mm, respectively. While all of the analyses contain error, our mean of 161 mm is hence on the low end of these different estimates, but considerably larger than Peixoto and Oort's [21] value of 120 mm, which was based on rawinsonde data from 1963-1973. Bromwich et al. [17] provide a more complete summary of available estimates.

The lower precipitation totals for the AO and CAO (Table 1) are understandable because these domains do not include the high precipitation totals over the Norwegian Sea [4, 6]. The lower evaporation totals for these domains reflect the colder conditions, but the annual value of 28 mm for the CAO appears too low. Central Arctic Ocean estimates from the annual map of Khrol [7] range from 50-80 mm while Burova [20] cites an annual total of 70 mm. The NP drifting station data do not contain corrections for gauge undercatch. If P is that P-E is too small would make the residual evaporation term even smaller. In an earlier study [2], the use of Gorshkov's [35] precipitation value yielded an even lower annual E of 22 mm.

5.2 ANNUAL CYCLE

The mean annual cycles of P-E, P and E for the polar cap domain are summarized in Figure 8. On monthly time scales, the flux convergence must be adjusted by the time change in water-vapor column storage to obtain P-E. The adjustment makes P-E smaller than the flux convergence from February through July, when W is increasing, and larger than the flux

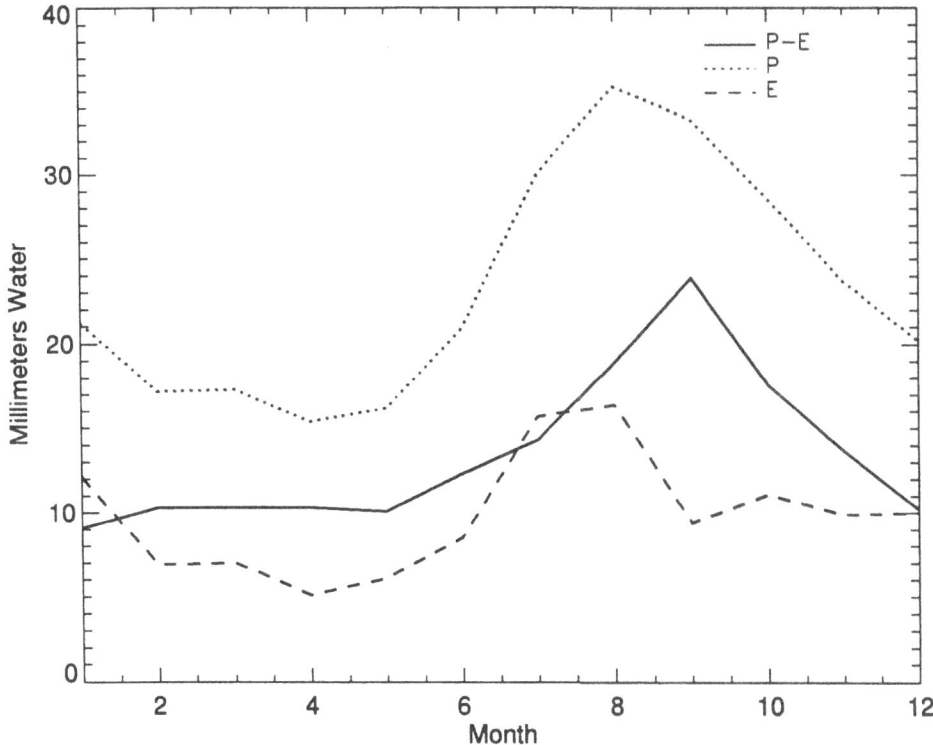

Figure 8: Mean seasonal cycles (mm) of P-E, P and E for the Polar Cap domain.

convergence from August through January, when W is decreasing. The largest adjustments, of approximately 4 mm, occur for June and September.

P-E is at a minimum from December through May and climbs through spring and summer, peaking in September at 24 mm.   P-E follows fairly closely the seasonal precipitation cycle, except that precipitation peaks a month earlier in August. In turn, evaporation peaks in July and August at 16 mm and 17 mm respectively. Our seasonal evaporation cycle differs from Korzun [8] who shows a peak in October of 20 mm.

154

The seasonality in P-E can be understood by examining results for January and September (Figure 9). Precipitable water at 70°N for all longitudes is much higher in September reflecting, of course, the higher temperatures and the ability of the atmosphere to hold more water vapor. Consequently, even if the meridional winds were the same between the two months, the vapor flux convergence would be larger in September.

The pattern of vertically-integrated fluxes for the two months (Figure 10) is also very different. While both months exhibit peak poleward transports near the prime meridian, subsidiary cores of poleward transport near the date line, Baffin Bay and central Eurasia, and equatorward transports over the Canadian Arctic Archipelago, the September transports are much larger in magnitude. This difference is not explained in terms of mean meridional winds, for as shown in Figure 11, they tend to be stronger in January, reflecting the more vigorous winter circulation. Clearly, the larger September transports arise from the greater availability of water vapor. The corresponding January and September meridional flux patterns for the AO and CAO (not shown) echo those for the PC - while exhibiting the same basic pattern, the fluxes are smaller with more subdued longitudinal variations.

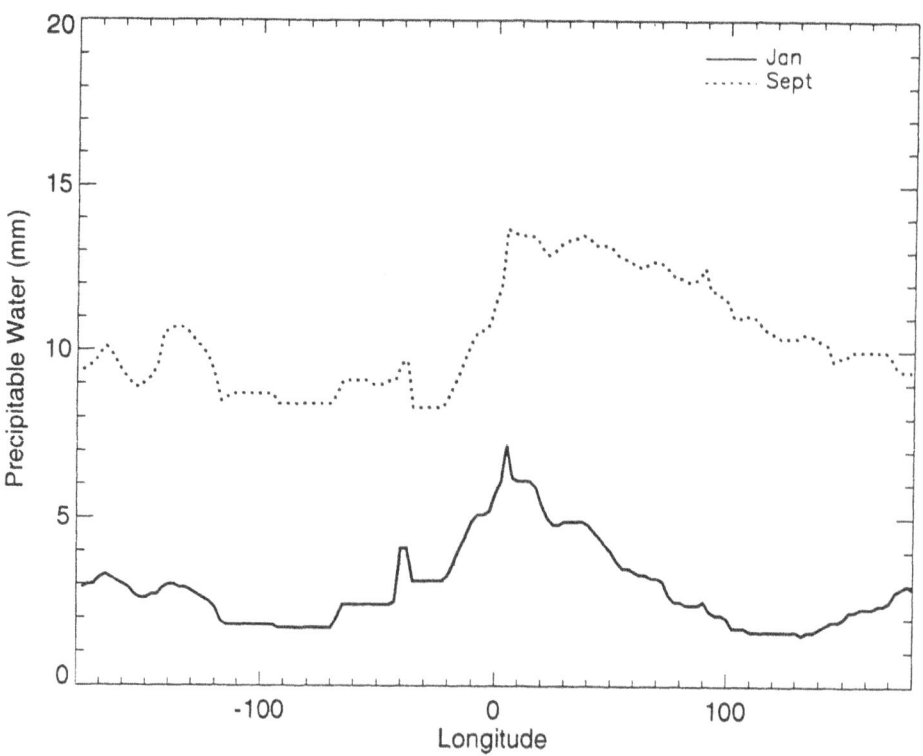

Figure 9: Mean precipitable water (mm) along 70°N for January and September.

As noted earlier, the annual flux convergence (P-E) for all three domains is primarily controlled by the transient eddy transports. This is true of the individual months as well. In fact, the flux convergence by the time mean transports is negative for some months, especially during summer. Inspection of the transient eddy transports for the PC between January and September (Figure 10) confirms that the flux averaged across 70°N is more positive in September, meaning, of course, that the flux convergence is also greater. The peak transports near the prime meridian are somewhat more evident in January than for the annual mean or September, reflecting the stronger North Atlantic storm track for this month. While the larger transient flux at essentially all longitudes in September is consistent with the greater abundance of water vapor, summer and early autumn are known to be characterized by increases in cyclone activity over land areas [34], with these systems acting to transport water vapor poleward.

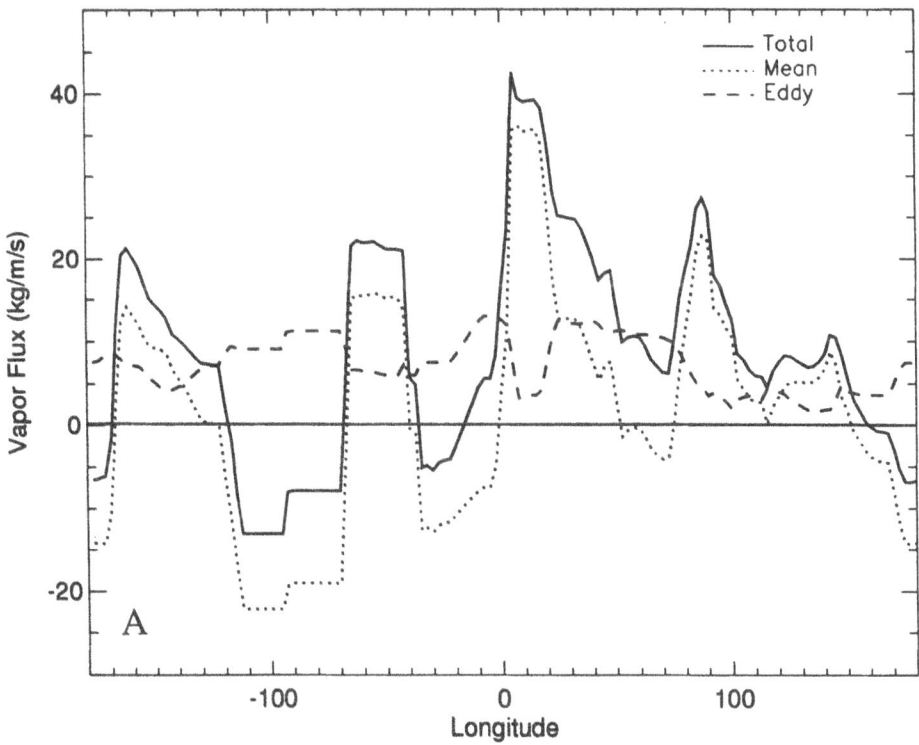

Figure 10. (a) January mean vertically-integrated meridional vapor fluxes (kg m⁻¹ s⁻¹) along 70°N broken down into total, stationary and transient eddy components.

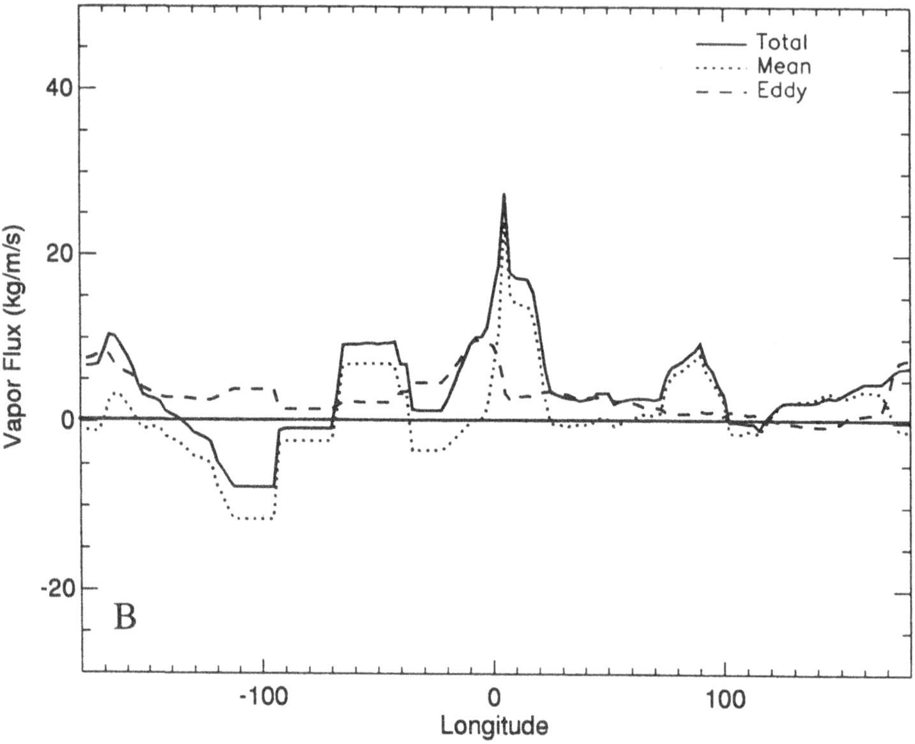

156

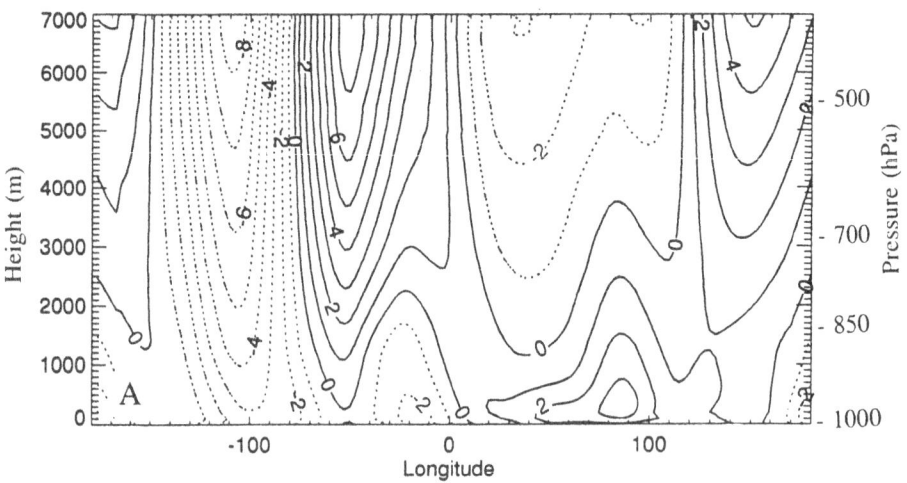

Figure 10. (b) September mean vertically-integrated meridional vapor fluxes (kg m^{-1} s^{-1}) along 70°N broken down into total, stationary and transient eddy components.

Figure 11: January (a) and September (b) vertical distribution of mean meridional winds (m s^{-1}) along 70°N.

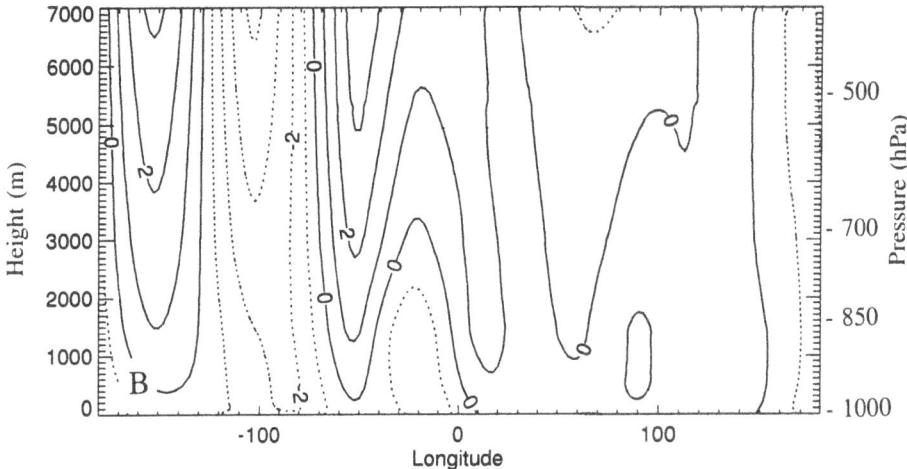

Figure 11: (b) September vertical distribution of mean meridional winds (m s^{-1}) along 70°N

The mean annual cycles of P-E, P and E for the AO and CAO domains (not shown) are again similar to those for the Polar Cap, with minima from December through May and late summer to early autumn peaks. However, the September maximum shown for the PC is not as evident over the CAO. While P for this domain is somewhat lower than for the Polar Cap, the seasonal cycle is the same shape. As a result, the seasonal cycles of P-E and P for the CAO follow each other quite closely. Building on earlier discussion that the annual E of 28 mm seems too low, we find that our residual estimate is slightly negative in February and September. Otherwise, the seasonal cycle is flat.

5.3 INTERANNUAL VARIABILITY

As expected on the basis of results just presented, the time series annual P-E for the three domains (Figure 12) generally follow each other, illustrating that P-E is responding to the same general circulation controls. P-E for the Polar Cap ranges from a low of 106 mm in 1978 to a high of 212 mm in 1989. For the AO, it ranges from low of 73 in 1990 to a high of 243 mm in 1989. For the CAO, the range is 105 mm in 1979 to 234 mm in 1989. There are no obvious trends. Because of missing months, we could not obtain annual results from 1973-1976 for the CAO domain.

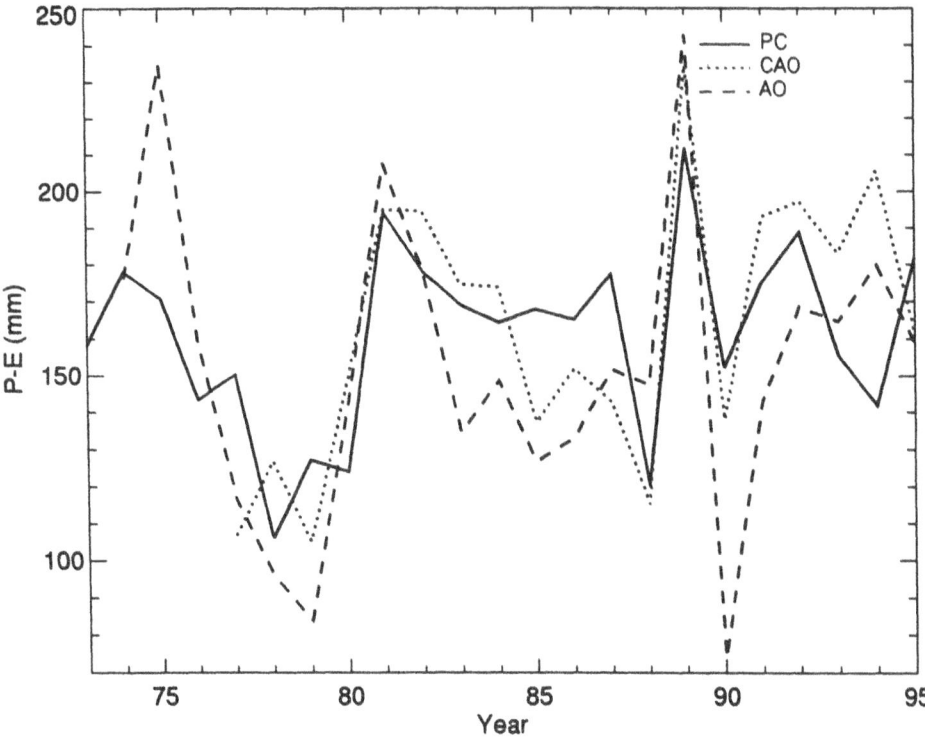

Figure 12: Annual time series of P-E (mm) for the Polar Cap, Arctic Ocean and Central Arctic Ocean domains.

In an earlier effort [24] we examined variability in P-E for the Polar Cap by compositing the longitudinal distribution of the meridional flux by high and low P-E years. It was found that winters with high P-E are associated with greater moisture inflow along the Atlantic sector, partly compensated by greater outflow over the Canadian Arctic Archipelago. Composites for other seasons share these same features, which indicate a control by variability of the atmospheric longwave patterns. Our data indicate that these general relationships hold for the AO and CAO domains. Dickson et al. [36], using our flux data from 1974-1991, show that the winter (DJF) meridional moisture transports at 70°N both along the Atlantic sector and averaged over all longitudes exhibit a positive relationship with the phase of the North Atlantic Oscillation. The NAO has been generally positive since about 1970, with some particularly large positive values since 1980 but turning strongly negative in 1995 [37]. Nevertheless, winter P-E shows no obvious trend, apparently reflecting the large variability in circulation.

6. Summary and Conclusions

We examined precipitation minus evaporation (P-E) for the Polar Cap (defined the region north of 70°N), the Arctic Ocean and a Central Arctic Ocean domain with the aerological approach using Arctic rawinsonde data for the period 1973-1995. We also obtained estimates of monthly and annual P and E. Precipitation was based on an improved data set that includes data from the Russian "North Pole" drifting stations while evaporation was estimated as a residual. The annual values of P-E, P and E for the Polar Cap are 161 mm, 279 mm and 118 mm, respectively. Corresponding values for the Arctic Ocean are 153 mm, 252 mm and 99 mm, compared with those for the Central Arctic Ocean of 174 mm, 202 mm and 28 mm. Our annual E for the Polar Cap compares favorably with a Russian estimate of 103 mm while the Central Arctic Ocean value is likely too low. For each domain, both P-E and P exhibit late summer to early autumn maxima, indicating that P is largely a function of the large-scale moisture convergence.

The pattern of mean annual and monthly moisture fluxes along 70°N, the boundary of the Polar Cap domain, is characterized by large poleward transports in the Atlantic sector from about 0-50°E, near Baffin Bay at about 70°W and near the date line, and a smaller peak in inflow at about 100°E. The only significant area of net moisture outflow is over the Canadian sector. This pattern is primarily controlled by the stationary component of the flux, and reflects the organization of mean winds and precipitable water by the planetary waves. The transient eddy component of the moisture flux is positive at all longitudes, tending to be largest in the Atlantic sector. The vapor flux convergence is primarily controlled by the transient eddy fluxes. The longitudinal pattern of moisture flux at the boundaries of the other domains generally follows that for the Polar Cap. For any season, large flux convergences are favored by situations leading to large moisture inflows along the Atlantic sector, which tend to be partly compensated by increased outflow over the Canadian Arctic Archipelago. Neither annual nor seasonal P-E show obvious trends.

The seasonality of P-E for the three domains contrasts sharply with land areas. Walsh et al. [2], applying the same methods as used here to the Mackenzie Basin, notes that while P-E during summer is near zero, precipitation is at a maximum. This indicates that compared with the marine Arctic, summer precipitation is derived largely from within-region recycling of water vapor. As part of a separate study, we have applied the aerological approach to the Ob, Yenesei and Lena river basins. Results are qualitatively similar to those for the Mackenzie domain.

Our results are sensitive to the details of the interpolation used and to changes in the station distribution. However, results from every sensitivity experiment conducted yielded lower annual and summer-month means of P-E for the Polar Cap and significant differences in interannual variability as compared to estimates using wind and moisture fields from "reanalysis" projects. As discussed by Bromwich and Cullather [17], this may reflect inadequacies in the distribution of rawinsonde stations. This is particularly sobering, as the distribution of rawinsonde stations about 70°N is probably the best on the planet.

7. Future Research Needs

As shown by Bromwich et al. [17] aerological estimates of mean annual P-E for the polar cap from various studies using numerical weather analyses range from 155-217 mm, while those based on rawinsonde data vary from 122-214 mm. Estimates from surface data show an even wider range. While there is clearly a need to reduce the uncertainty in mean

As shown by Bromwich et al. [17] aerological estimates of mean annual P-E for the polar cap from various studies using numerical weather analyses range from 155-217 mm, while those based on rawinsonde data vary from 122-214 mm. Estimates from surface data show an even wider range. While there is clearly a need to reduce the uncertainty in mean values, we must also work towards providing better assessments of spatio-temporal variability in P-E as well as its individual components. Application of gauge corrections to the Russian NP precipitation record may provide a better estimate of mean Arctic Ocean precipitation, but the record, which ends in 1991, is too sparse to provide spatially distributed time series. Accurate spatially -distributed estimates of evaporation are also unlikely to be provided from in situ measurements. It hence does not appear that surface measurements can be used to meet research needs. Regarding aerological retrievals of P-E, there are questions concerning the accuracy of rawinsonde-based estimates, and this approach only provides large areal averages. This leaves the use of analyzed fields from NWP models as perhaps the most viable approach for providing time series of spatially-distributed P-E although further development of satellite techniques for water-vapor retrieval also show promise [38]. NWP models also provide short-term forecasts of P and E. At present, these forecasts have significant shortcomings, but are likely to improve as models are further refined. In this light, we strongly advocate evaluation of output from the upcoming ERA-40 (40-year) reanalysis project. As compared with existing reanalysis effort, the ERA-40 effort will employ a high-resolution model with improvements in the treatment of Arctic boundary conditions. We of course acknowledge that the quality of the output will still be influenced amount and quality of assimilation data.

Acknowledgements

This study was supported by NSF grants OPP-9524740, ATM-9732461 and OPP-9614297.

References

1. Aagaard, K. and Carmack, E.C. (1989) The role of sea ice and other fresh waters in the Arctic circulation, *J. Geophys. Res.* 94(C10), 14,845-14,498.
2. Walsh, J.E., Zhou, X., Portis, D. and Serreze, M.C. (1994) Atmospheric contribution to hydrologic variations in the Arctic, *Atmosphere-Ocean* 34, 733-755.
3. Groisman, P.Y., Koknaeva, V.V., Belokrylova, T.A., and Karl, T.R. (1991) Overcoming biases of precipitation: A history of the USSR experience, *Bull. Amer. Meteorol. Soc.* 72, 1,725-1,733.
4. Bryazgin, N.N. (1976) Srednegodovoe kolichestvo osadkov v Arktike s uchetom pogreshnosti osadkomerov (Mean annual precipitation in the Arctic with consideration of gauge biases), *Trudy Arkt. Antarkt. Nauch.-Issled. Inst.* 323, 40-74.
5. Colony, R., Radionov, V. and Tanis, F.L. (1997) Measurements of precipitation and snow at the Russian North Pole drifting stations, *Polar Geography* 34, 3-14.
6. Serreze, M.C. and Hurst, C.M. (1999) Representation of mean Arctic precipitation from NCEP/NCAR and ERA reanalyses, *J. Climate* (in press).
7. Khrol, V.P. (1976) Isparenie s poverkhnosti Severnogo Ledovitogo Okeana. (Evaporation from the Arctic Ocean surface.), *Trudy Arkt. Antarkt. Nauchno.-issled. Inst.* 323, 148-155.
8. Korzun (1976) *Atlas of World Water Balance* (in Russian), Gidrometeoizdat, Leningrad, 65 pp.
9. Zubenok, Z.T. (1976) Isparenie s sushi vodosbornogo basseina Severnogo Ledovitogo Okeana. (Evaporation from the basins draining into the Arctic), *Trudy Arkt. Antarkt. Nauchno.-issled. Inst.* 323, 87-100.
10. Lydolph, P.E. (1977) *World Survey of Climatology, Vol. 7, Climates of the Soviet Union* (H.E. Landsberg, ed. in Chief), Elsevier, Amsterdam, 443 pp.
11. Fisheries and Environment Canada (1978) *Hydrological Atlas of Canada*, Fisheries and Environment Canada, Minister of Supply and Services, Canada, Ottawa, 31 maps.

12. Kane, D.L., Hinzman, L.D., Woo, M., and Everett, K. (1992) Arctic Hydrology and Climate Change, Chapter 3 in: *Arctic Ecosystems in a Changing Climate*, Academic Press, 35-57.

13. Ohmura, A. (1982) Evaporation from the surface of the Arctic Tundra on Axel Heiburg Island, *Water Resources Research* 18, 291-300.

14. Ohmura, A. (1984) Comparative Energy Balance Study For Arctic Tundra, sea Surface, Glaciers and Boreal Forests, *Geo-Journal* 8 221-228.

15. Weller, G., and Holmgren, B. (1974) The microclimates of the Arctic tundra, *J. Appl. Meteor.* 13, 854-862.

16. Genthon, C., and Braun, A. (1995) ECMWF analyses and predictions of the surface climate of Greenland and Antarctica, *J. Climate* 8, 2324-2332.

17. Bromwich, D.H., Cullather, R.I. and Serreze, M.C. (this volume).

18. Bromwich, D.H., Cullather, R.I., Chen, Q. and Csatho, B. (1998) Evaluation of recent precipitation studies for Greenland Ice Sheet, *J. Geophys. Res.* 103(D20), 26,007-26,024.

19. Barry, R.G. (1972) *The summer Flux Divergence of Atmospheric Water Vapor over the Queen Elizabeth Islands*, Publications in Meteorology No. 104, Arctic Meteorology Research Group, McGill University, Montreal, 24 pp.

20. Burova, L.P. (1983) Rezhim vlazhnosti v troposfere nad Severnym Ledovitym Okean (Moisture regime of the troposphere over the Arctic Ocean), *Trudy Arkt. Antarkt. Nauch.-Issled. Inst.* 381, 60-68.

21. Peixoto, J.P., and Oort, A.H. (1992) *Physics of Climate*, American Inst. of Physics, New York, 520 pp.

22. Robasky, F.M., and Bromwich, D.H. (1994) Greenland precipitation estimates from the atmospheric moisture budget, *Geophysical. Research. Letters.* 21, 2495-2498.

23. Serreze, M.C., Barry, R.G., and Walsh, J.E. (1995a) Atmospheric water vapor characteristics at 70°N, *J. Climate* 8, 719-731.

24. Serreze, M.C., Rehder, M.C., Barry, R.G., Walsh, J.E., and Robinson, D.A. (1995b) Variations in aerologically-derived Arctic precipitation and snowfall, *Annals of Glaciology* 21, 77-82.

25. Gaffen, D.J. (1993) *Historical Changes in Radiosonde Instruments and Practices*, World Meteorological Organization Instruments and Observing Methods Report No. 50, WMO/TD-No. 545, 123 pp. plus microfiche appendices.

26. Elliot, W.P. and Gaffen, D.J. (1991) On the utility of radiosonde humidity archives for climate studies, *Bull. Amer. Meteor. Soc.* 72, 1507-1520.

27. Garand, L., Grassotti, C., Halle, J., and Klein, G. (1992) On differences in radiosonde humidity-reporting practices and their implications for numerical weather prediction and remote sensing, *Bull. Amer. Meteor. Soc.* 73, 1417-1423.

28. Trenberth, K.E., and Guillemot, C.J. (1998) Evaluation of the atmospheric moisuture and hydrological cycle in the NCEP/NCAR reanalyses, *Clim. Dynam.* 14, 213-231.

29. Cullather, R.I., Bromwich, D.H. and Serreze, M.C. (1999) The atmospheric hydrologic cycle over the Arctic basin from reanalyses. Part I. Comparison with observations and previous studies, *J. Climate* (in press).

30. Kahl, J.D., Serreze, M.C., Shiotani, S., Skony, S.M., and Schnell, R.C. (1992) In-situ meteorological sounding archives for Arctic studies, *Bull. Amer. Meteor. Soc.* 73, 1824-1830.

31. Legates, D., and Willmott, C. (1990) Mean seasonal and spatial variability in gauge-corrected global precipitation, *Int. J. Climatol.* 10, 110-127.

32. Cressman, G.P. (1959) An operational objective analysis system, *Mon. Wea. Rev.* 87, 367-374.

33. Barry, R.G., and Fogarosi, S. (1968) *Climatology Studies of Baffin Island, Northwest Territories*, Tech. Bull. No. 13, Inland Water Branch, Department of Energy, Mines and Resources, Ottawa, 106 pp.

34. Serreze, M.C. (1995) Climatological aspects of cyclone development and decay in the Arctic, *Atmosphere-Ocean* 33, 1-23.

35. Gorshkov, S.G., ed. (1983) *World Ocean Atlas, Volume 3: Arctic Ocean* (in Russian), Pergamon Press, Oxford, 189 pp.

36. Dickson, R.R., Osborn, T.J., Hurrell, J.W., Meincke, J. Blindheim, J., Adlandsvik, B., Vinje, T., Alekseev, G., Maslowski, W., and Cattle, H. (1999) The Arctic Ocean response to the North Atlantic Oscillation, *J. Climate* (in press)

37. Hurrell, J.W. (1995) Decadal trends in the North Atlantic Oscillation: regional temperatures and precipitation, *Science* 269, 676-679.

38. Rothrock, D. and Kwok, R. (this volume).

REANALYSES DEPICTIONS OF THE ARCTIC ATMOSPHERIC MOISTURE BUDGET

D.H. BROMWICH and R.I. CULLATHER
Polar Meteorology Group, Byrd Polar Research Center
The Ohio State University
Columbus, OH 43210-1002 USA.

M.C. SERREZE
Cooperative Institute for Research in Environmental Sciences
Division of Cryospheric and Polar Processes
University of Colorado
Boulder, CO 80309-0449 USA.

1. Introduction

The atmospheric hydrologic cycle over the Arctic basin has significant relevance to a variety of research interests. Variability in the sources and sinks of the Arctic freshwater budget has received increased attention as a result of recent studies which suggest the North Atlantic thermohaline circulation as a potential climatic change mechanism. North Atlantic conditions are intimately related to the Arctic via ice and freshwater discharges through the Fram Strait [1]. Modeling studies have also suggested that the Arctic is a location of enhanced sensitivity to possible anthropogenic influences (e.g., Kattenberg et al. [2]). The monitoring of high latitude atmospheric moisture parameters has thus received a high priority from research organizations (e.g., Walsh [3]).

In the Arctic, the presence of a floating ice field creates severe difficulties for *in situ* measurements of atmospheric moisture and precipitation. This has led to increased interest in analysis fields of the operational weather forecasting centers. These gridded depictions of the atmosphere represent a synthesis of available meteorological information, including satellite data. Global atmospheric numerical analyses are routinely produced by operational weather forecasting centers for use as initial conditions in forecast models. The archiving of these fields has produced a valuable data set for use in climate-related investigations [4]. The most recent adaptation of this method is a practice known as "reanalysis" [5, 6], in which the data assimilation system that is utilized is "frozen", meaning that it does not vary with time.

In this paper we examine reanalysis depictions of the atmospheric hydrologic cycle over the Arctic basin using data sets of the European Centre for Medium-Range Weather Forecasts (ECMWF) and the collaborative effort of the U.S. National Centers for Environmental Prediction (NCEP) and the National Center for Atmospheric Research (NCAR). These data are compared with available observations and previous study. As described below in section 2, forecast fields are first compared to available climatic data

163

E.L. Lewis et al. (eds.), The Freshwater Budget of the Arctic Ocean, 163-196.
© 2000 *Kluwer Academic Publishers.*

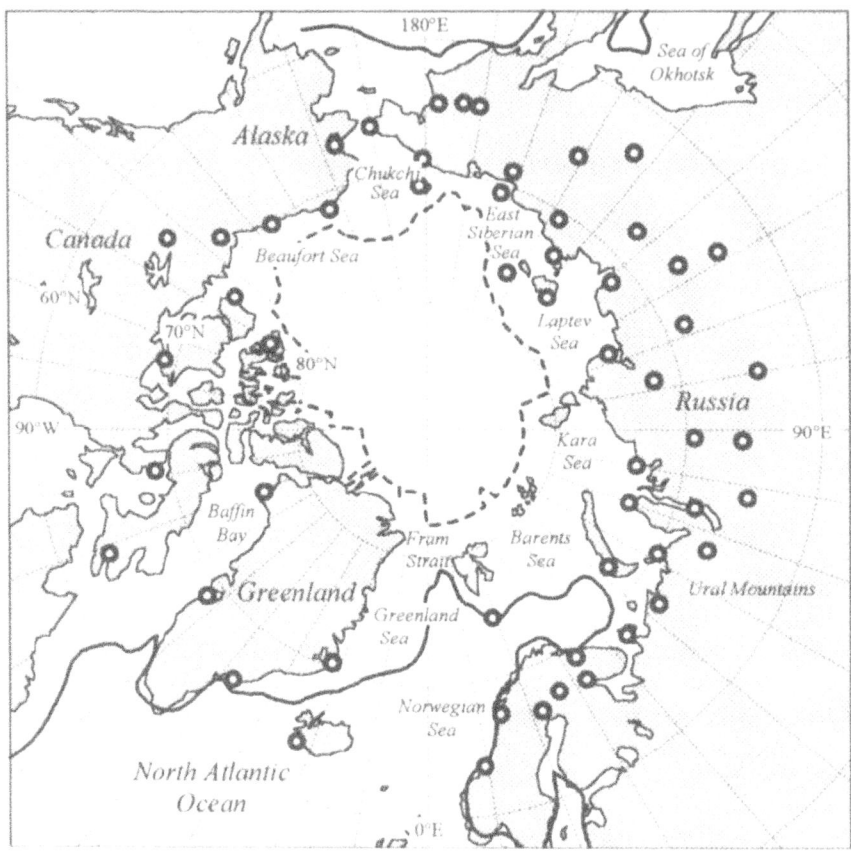

Figure 1. Map of the Arctic basin. Solid and dashed lines indicate the average January and July extent of sea ice, respectively, as defined by the −1.8°C surface temperature contour from the ERA. Rawinsonde stations used for computing the meridional atmospheric moisture transport across 70°N after Serreze et al. [7] are shown as open circles.

for precipitation (P) and evaporation/sublimation (E). An evaluation is then given of the reanalysis precipitable water and computed atmospheric moisture transports. Finally, the hydrologic balance of the two methods is investigated.

A question of fundamental interest relates to the comparison of reanalysis data to atmospheric moisture transports obtained from rawinsonde observations. Serreze et al. [7, 8] used the Historical Arctic Rawinsonde Archive (HARA) of the National Snow and Ice Data Center (NSIDC, [9]) to synthesize the atmospheric moisture transport distribution across 70°N. For reference, Figure 1 shows the distribution of HARA rawinsonde stations used in determining the moisture flux distribution across 70°N for the period 1979-93, as well as the significant geographic features of the Arctic. The HARA data set has been assembled from various sources including the U.S. National Climatic Data Center and NCAR. Preliminary studies on the comparison of the reanalysis data to the rawinsonde synthesis have been presented by Bromwich and Cullather [10] and Cullather and Bromwich [11]; a more comprehensive review has been presented by Cullather et al. [12]. This work is summarized below. In this paper we focus more closely on comparisons of the reanalysis transports at point locations to rawinsonde station data. Aside from various long-term climate atlases and satellite-derived precipitable water estimates, the rawinsonde network is the only source of observational data for the atmospheric hydrologic cycle over the north polar cap (70°N-90°N). In order to attain confidence in the reanalysis data it is therefore particularly important to identify and understand any disagreements with the rawinsonde-based estimates.

2. Data Sets and Methods

2.1 COMPUTATIONAL METHODS

Reanalyses directly address the problem of spurious temporal trends found in archived operational analyses, which are associated with evolving data assimilation techniques. The remaining temporal variability is then either real or the result of changes made to the observational network. Gibson et al. [13] describe the ECMWF Re-Analysis (ERA). The ERA is produced four times daily using a global spectral numerical weather prediction (NWP) model at T106 horizontal resolution with 31 hybrid levels in the vertical. The NCEP/NCAR Reanalysis is also produced four times daily, at T62 horizontal resolution and 28 σ-levels in the vertical [6]. For this study, the ERA basic data set and the NCEP/NCAR Reanalyses were obtained from NCAR at 2.5° × 2.5° resolution at near-surface and standard pressure levels for the concurrent 1979-93. The choice of resolution attempts to strike a balance between the desired full spectral resolution and the practical computational considerations associated with these titanic data sets. The amount of reanalysis data examined here for the period 1979-93 in an uncompressed, binary form is on the order of 330 gigabytes. Additionally, not all of the ERA spectral data had been made available to NCAR at the time of this study.

As part of both reanalysis data sets, a series of short-term forecasts was performed using each analysis center's NWP model, which was initialized using reanalysis fields. This provides supplementary fields, including P and E, that are not directly observed or analyzed. These fields are more dependent on the physics of the NWP model utilized, however.

The computations shown here are similar to those employed in Cullather et al. [14], however the software package of Adams and Swarztrauber [15] is used for moisture convergence computations. This eliminates the need for ad hoc finite difference schemes on the surface of the sphere. The surface moisture balance $P-E$ may be obtained by directly computing the atmospheric moisture budget from the gridded variables. This is done using:

$$P - E = -\frac{\partial W}{\partial t} - \nabla \cdot Q \qquad (1)$$

where W is precipitable water and Q is the atmospheric moisture transport, defined as:

$$Q = \frac{1}{g}\int_{P_{top}}^{P_{sfc}} qV \, dp \qquad (2)$$

where g is the gravity constant, P_{sfc} is surface pressure, q is specific humidity, and V is the horizontal wind vector. The variable P_{top} is the highest level of the atmosphere, which is not zero in the analyses. For annual time scales, the first right-hand-side term, referred to as the storage term, is considered negligible. Using the trapezoidal rule for integrations, equation (1) may be computed at each point of the reanalysis data sets to produce a field of $P-E$. In addition to the residual term obtained from the moisture budget, the flux itself is of interest because it contains information on the atmospheric circulation patterns that result in the distribution of $P-E$. The dominant features of these transport patterns have been referred to as atmospheric "rivers" [16, 17].

As an aside, it is readily seen that equation (1) may be adapted to rawinsonde data by expressing the convergence term as a line integral:

$$\langle P - E \rangle = -\left\langle \frac{\partial W}{\partial t} \right\rangle - \frac{1}{A} \oint Q \cdot n \, dl \qquad (3)$$

where angled brackets indicate an area average, A is the area of interest, and n is the outward-pointing normal to the area periphery. Using rawinsonde stations surrounding the north polar cap, Serreze et al. [7,8] computed an average meridional moisture transport across 70°N. A moisture convergence value for the polar cap is obtained when this value is multiplied by the total length of the 70°N parallel and divided by the area. Serreze et al. [7] assumed that precipitable water estimates from rawinsonde stations near 70°N and a few available stations at higher latitudes were sufficient for determining the area-averaged storage term.

When the forecast P and E fields are commensurate with the right hand side of equation (1) computed from analysis values, the data set is said to be in "hydrologic balance". It is important to recognize the temporal differences between the two methods, however. The forecast fields are accumulated over some period of each forecast, while the atmospheric moisture budget is computed for a specified instant in time. The ERA forecast fields contained in the NCAR archive are obtained from the 12 to 24 hour period of each forecast, while the NCEP/NCAR Reanalysis utilizes the 0 to 6 hour forecast. Thus for ERA forecast fields, an average estimate of $P-E$ for one day is determined from two 12 to 24 hour forecast averages. For NCEP/NCAR Reanalyses, a one-day average estimate of forecast $P-E$ is determined from four 0 to 6 hour forecast averages. For both data sets, $P-E$ computed via the atmospheric moisture budget is discretely sampled four times daily. To mitigate these temporal sampling effects, one typically averages the fields in space and time in order to evaluate the hydrologic balance of a data set. Although it may not be adequate for particular regions associated with unique phenomena, it is generally assumed that four-times daily temporal resolution is sufficient to capture the major hydrologic features on climatic time scales such that equation (1) is expected to be satisfied [18, 19].

As a consequence of the vast quantity and the varied sources of raw data used in producing reanalysis data sets, some production oversights would appear to be inevitable. For the NCEP/NCAR Reanalysis in high northern latitudes, the relevant difficulties that have been identified are that the analyses were not produced using interannually-varying snow cover for the period 1974-94 (E. Kalnay, personal communication, 1997), and that the NWP model's horizontal diffusion parameterization is over-simplified. The latter problem results in the provocative spurious spatial wave pattern at high latitudes (W. Ebisuzaki, personal communication, 1996). This is discussed in more detail below.

Difficulties in the production of the ERA are discussed by Kållberg [20], and include the surface temperature cold bias in northern boreal forests during winter and springtime. In addition, the latent heat of freezing in ERA convection was found to be erroneously set to zero.

2.2 REVIEW OF PREVIOUS STUDIES

Prior to the introduction of the reanalyses, the use of operational numerical analysis data in the study of the global atmospheric moisture budget had received a great deal of attention (e.g., Trenberth and Guillemot [21], Dodd and James [22], Rasmusson and Mo [23]). In the Arctic, Masuda [24] computed transports using ECMWF FGGE data for 1979. An annually averaged moisture convergence value of 15.5 cm yr^{-1} was determined. Larger monthly mean values were also found for the period July to September, producing a significant contrast between winter and summer seasons. Additional studies using operational analyses in the Arctic have been performed for limited regions such as Greenland (e.g., Calanca and Ohmura [25], Bromwich et al. [26]).

A few studies have examined reanalysis depictions of the atmospheric moisture budget from a global perspective (Stendel and Arpe [27], Trenberth and Guillemot [18, 19]). These studies indicate differences with assembled global climatologies as well as demonstrate hydrologic imbalances in reanalysis data. These results highlight the challenges of producing global depictions of atmospheric moisture variables. Several recent studies have investigated the performance of the reanalyses forecast P and E fields in Polar Regions. In particular, Serreze and Hurst [28] have examined the NCEP/NCAR and ERA forecast P in comparison to an improved gauge-based climatology. Both reanalyses were found to underestimate annual values over the Atlantic side of the Arctic. The most significant problem found was a tendency to produce too much convective precipitation over land areas in the NCEP/NCAR forecasts. Serreze and Maslanik [29] have also examined the NCEP/NCAR forecast precipitation fields over the Arctic for the period 1986-93. Serreze and Maslanik [29] compensated for the high latitude spurious wave pattern by applying a Cressman filter to smooth the fields. The resulting fields were found to have a reasonable spatial distribution in comparison to available climatologies, with significant differences in seasonality over the central Arctic basin.

3. Forecast P, E, and $P-E$

3.1 EVALUATION OF MEAN FIELDS

The 1979-93 means fields of forecast $P-E$ from the two reanalyses are plotted in Figure 2. These fields, as well as forecast P and E individually, are qualitatively compared with the climate atlases of Gorshkov [30] and Khrol [31] below. Essentially, this consists of examining the placement and intensity of the prominent large-scale features, which are typically located along the perimeter of the basin. A quantitative evaluation is difficult because of the small number of contour lines shown in the atlases.

Averages of these maps for 70°N-90°N have been estimated by J.E. Walsh (1998, personal communication) however, and these are given in Section 3.3.

In a global investigation of the forecast variables, Stendel and Arpe [27] found the annually-averaged E to be erroneously greater than P over land surfaces in both reanalyses. The spatial distribution of $P–E$ over the whole of the Arctic for the ERA, however, appears to be more plausible. Shown in Figure 2(a), the average spatial distribution of the ERA forecast $P–E$ is dominated by low values of approximately 10 to 20 cm yr^{-1} over the central ice pack. Some of the features of interest include the spatial characteristics over and near the Greenland Ice Sheet, the Scandinavian region, and the Pacific coast of Alaska. The largest values depicted in Figure 2(a) occur along coast of the Gulf of Alaska. For Greenland, the overall ERA forecast $P–E$ distribution is in reasonable agreement with available syntheses of glaciological observations such as that of Ohmura and Reeh [32] and Csathó-PARCA (PARCA: Program in Arctic Regional Climate Assessment; Bromwich et al. [26], Csathó et al. [33]). A concentrated region of maximum $P–E$ occurs along the southeastern coast of Greenland, with values of up to 153 cm yr^{-1}. Large values also extend along the western Greenland coast as far north as Thule (77°N, 69°W), with the smallest values extending over the interior ice sheet plateau. Over the Greenland high elevation plateau regions, the ERA forecast $P–E$ is too low in comparison to glaciology. Annual ERA forecast estimates for Summit (72°N, 38°W) are less than 10 cm yr^{-1}, versus approximately 24 cm yr^{-1} observed [34].

For the ERA data, the Scandinavian region is characterized by a substantial coastal gradient with values of greater than 100 cm yr^{-1} over southern Norway, while a large area of E greater than P extends offshore in the Norwegian and Barents seas between Scandinavia and Svalbard. This is in qualitative agreement with Gorshkov [30]. In general, however, the ERA forecast E is much smaller than P for most of the Arctic. This may be seen in the average annual spatial distribution of forecast E displayed in Figure 3. Over the north polar cap, ERA forecast E depicts the Atlantic sector as the only region where values exceed 30 cm yr^{-1}. ERA values of E exceed 80 cm yr^{-1} in selected locations of the Norwegian Sea. This is actually smaller than values from the Gorshkov atlas [30], which shows contours of 100 cm yr^{-1} for small regions. Over central Greenland, ERA annual values show net sublimation ($E < 0$). The average spatial patterns of the ERA and Gorshkov are very similar, however there are a lack of significant features over the central Arctic Ocean, and thus it is difficult to determine the differences in the average value over large areas.

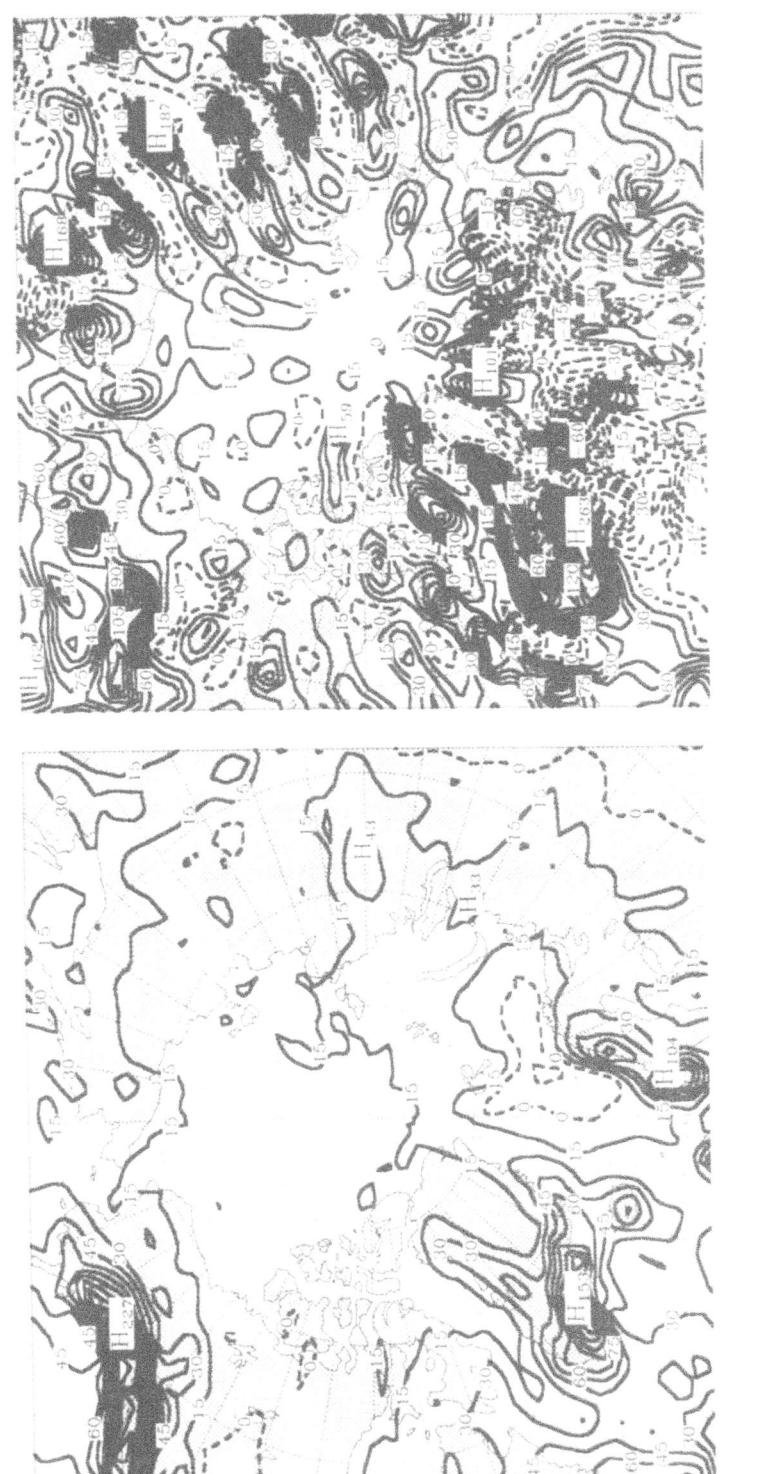

Figure 2. Spatial distribution of forecast *P–E* from (a) ERA and (b) NCEP/NCAR, averaged for 1979-93. The contour interval is 15 cm yr^{-1}.

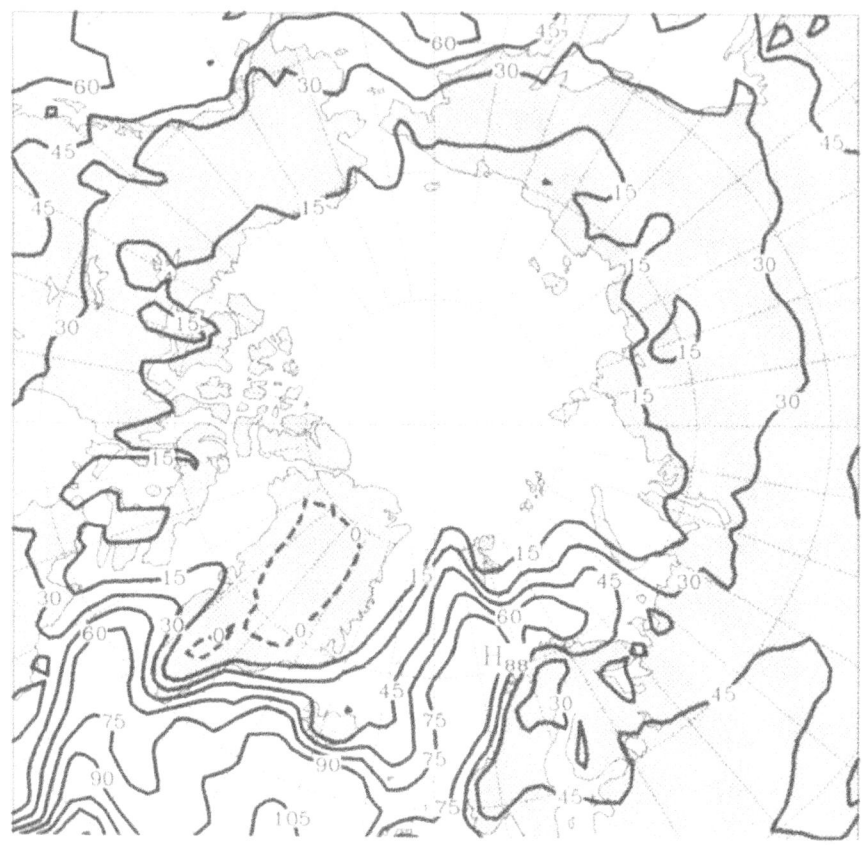

Figure 3. Spatial distribution of annually averaged forecast E from the ERA over the Arctic basin for 1979-93. The contour interval is 15 cm yr^{-1}.

In contrast to the ERA, the average forecast $P-E$ distribution from the NCEP/NCAR Reanalysis shown in Figure 2(b) is a substantially busier field, owing to the spectral wave pattern. In addition to P, the forecast E values also contain the spurious wave pattern, however the wave pattern for the two variables is offset such that large positive $P-E$ values are found in close proximity to locations where E is greater than P. This pattern is particularly evident over Siberia. The high spatial gradients of the NCEP/NCAR forecast $P-E$ in the North Atlantic and over Greenland pose a significant challenge to contouring programs. NCEP has attempted to address the problem via the development of a post-processing algorithm, but for P only (B. Kistler and M. Iredell 1996, personal communication). The correction identifies the bullseyes as spurious moisture sources and removes them in comparison to a diffusion-corrected moisture amount, subject to a temperature threshold. The corrected NCEP P has been examined over Greenland by Bromwich et al. [26], and the 1979-93 average spatial distribution for the Arctic basin is shown in Figure 4. The corrected NCEP forecast P is clearly an improvement over the

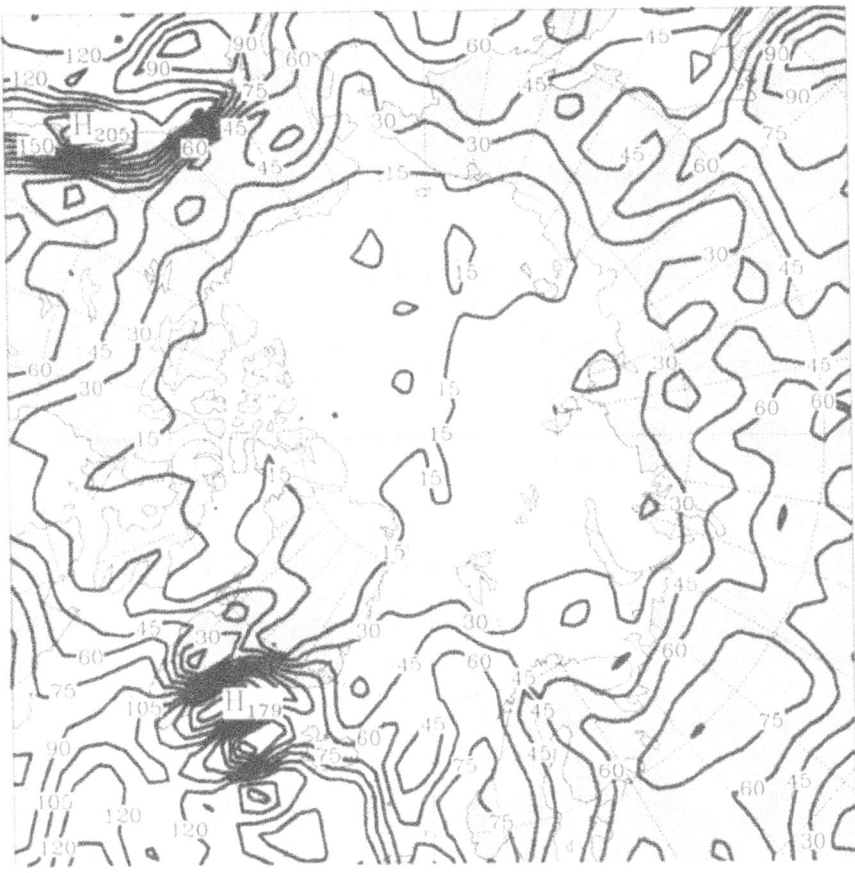

Figure 4. Spatial distribution of corrected P from NCEP post-processing of reanalysis forecasts over the Arctic

original spatial distribution, however it is apparent from further examination that the problem is with the distribution, rather than the total amount of moisture. The corrected precipitation field is found to be overly dry in comparison to the original field and the ERA north of 70°N. This leads to a contrast between northern and southern Greenland that is greater than found in the synthesized climatologies of Ohmura and Reeh [32] and Csathó-PARCA [33]. Additionally, over central Greenland the temperature threshold of the correction does not allow for the complete removal of the spurious maximum that extends over the relatively colder regions of the high plateau. For other locations in the Arctic basin, the corrected spatial distribution is found to be an obvious improvement over the original field, although significant differences exist in comparison to the ERA and the climatological atlases. Values are generally smaller for the corrected NCEP over the central Arctic basin than for the ERA; average values of less than 15 cm yr^{-1} occur over a large area of the central Arctic for the corrected NCEP, while ERA values of P are greater than 20 cm yr^{-1} over most of the Arctic Ocean. Both the Gorshkov and Khrol atlases indicate an intermediate solution, with the 15 cm yr^{-1} contour confined to within the Canadian basin. Significantly, the corrected NCEP field shows smaller precipitation values for Norway, in contrast to the ERA and both atlases, which show values greater than 100 cm yr^{-1} along the western Scandinavian coast. There is also an additional bullseye-minimum for the corrected NCEP data in the Denmark Strait that does not appear to be supported by observations. In general, the spatial distribution of the corrected NCEP P contains discrepancies with the climatological atlases in the North Atlantic.

3.2 COMPARISON WITH DRIFTING ICE STATION GAUGE DATA

Reanalysis values of P have been compared with gauge data for the central Arctic basin. The USSR operated drifting ice stations in the Arctic for the period 1950-1991 [35, 36]. These camps obtained daily gauge estimates of P. Figure 5(a) shows the trajectories of seven stations that were in operation during the 1979-93 time period of the reanalysis data. Six of the seven stations are clustered over the central ice pack north of 80°N, while station NP-31 traveled around the Beaufort gyre. A comparison of ERA and corrected NCEP forecast P from the nearest grid point with the camp data encounters numerous difficulties including location questions, the representativeness of one gauge for a grid box, and the reliability of gauge measurements in polar regions (e.g., Woo et al. [37]). In fact, the day-to-day observations are a poor match with both reanalyses. Typically, the station time series show a series of isolated events, with many reports showing zero precipitation. Both reanalyses show difficulty reproducing the timing, duration, and amount of these events. This is not surprising given the location ambiguities associated with comparing a point measurement with a model box value. When the observations and corresponding reanalysis data are composited into monthly averages however, there is better agreement on the average quantity. Figure 5(b) shows the time series of monthly values for stations NP-26 and NP-31, which operated in the Arctic from July 1983 until December 1984, and January 1989 until March 1991, respectively. Station 26 shows significant discrepancies in the values over a number of months, while station NP-31 shows one summertime maximum that is adequately captured by the ERA. These two stations are representative of the comparisons as a whole. Of the seven stations examined, time series of monthly averages from two (NP-30 and NP-31) show reasonable correlation with the reanalysis data ($r^2 > 0.55$) while the other stations show mediocre to poor agreement ($r^2 < 0.2$ for both data sets). The typical station duration is 25 months. The two stations with reasonable agreement contain the more recent data and were active after November 1987. This suggests that the trend towards better agreement between gauge measurements and forecast P may reflect the increase in the number of meteorological observations available for the central Arctic, particularly from the International Arctic Buoy Program [38]. In Table 1, the average values of the reanalyses forecast P and gauge data are shown for each station. The corrected NCEP values are less than the ERA and typically less than the gauge values by a substantial margin. For example, corrected NCEP values for station NP-26, which operated from July 1983 until December 1984, are 49 percent smaller than the gauge estimate, while the ERA is 25 percent smaller than the average observed value.

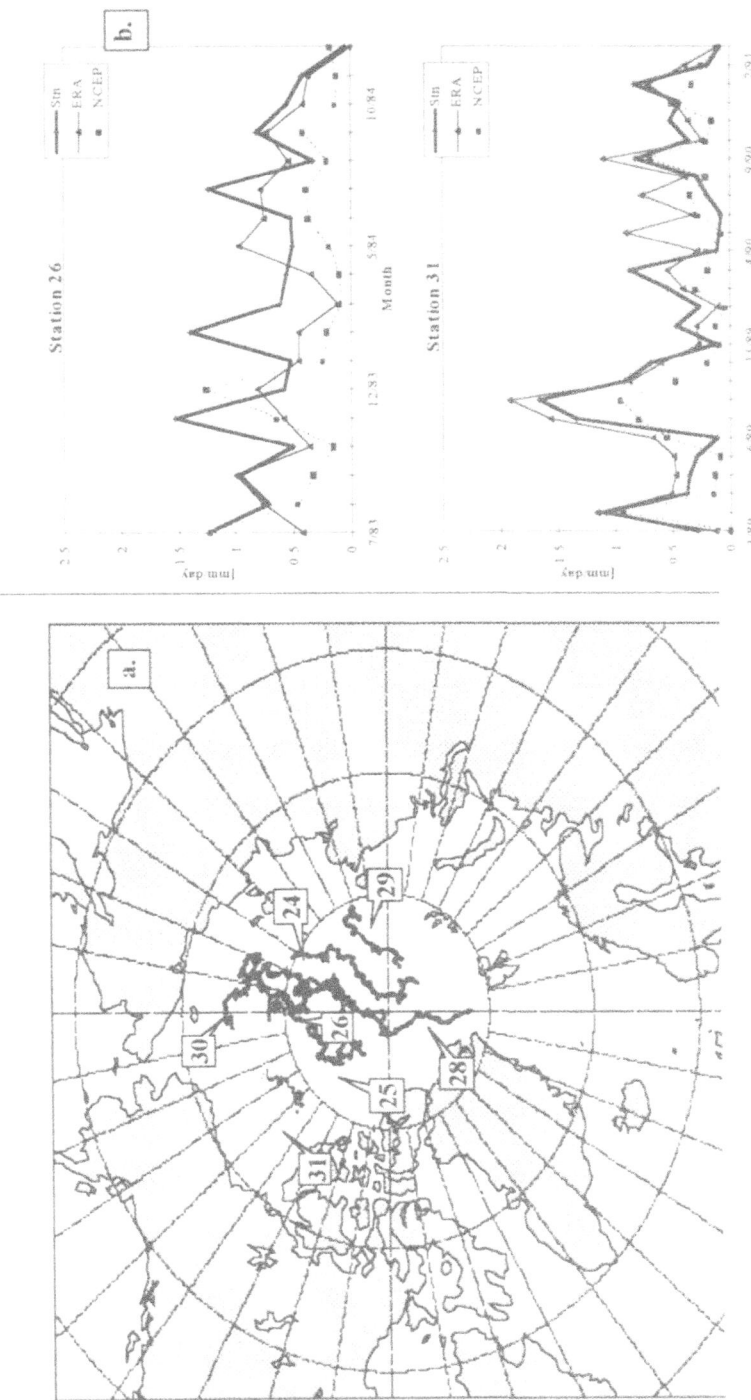

Figure 5. (a) Trajectories of seven USSR drifting ice stations in the central Arctic for the period 1979-1991; (b) Time series of monthly-averaged gauge values for stations NP-26 and NP-31 in comparison to corresponding forecast values of the ERA and NCEP corrected P, in mm day^{-1}.

TABLE 1. A comparison of mean P values for Soviet drift camp gauge measurements and corresponding reanalysis forecast values.

Station	Operating Dates	Mean Precipitation (mm day^{-1})		
		Gauge Value	ERA	Corrected NCEP
NP-24	1/79 - 10/80	0.32	0.44	0.43
NP-25	6/81 - 3/84	0.40	0.46	0.37
NP-26	7/83 - 12/84	0.72	0.54	0.37
NP-28	6/86 - 12/88	0.67	0.61	0.49
NP-29	7/87 - 5/88	0.73	0.62	0.59
NP-30	11/87 - 5/90	0.63	0.60	0.44
NP-31	1/89 - 3/91	0.50	0.59	0.31

3.3 COMPARISON OF AREA-MEAN P AND E FOR THE NORTH POLAR CAP

For the north polar cap region bounded by 70°N, the average forecast P values for 1979-93 are 24.7 cm yr^{-1} for the ERA, 28.6 cm yr^{-1} for the NCEP/NCAR Reanalysis, and 13.1 cm yr^{-1} for the corrected NCEP data. For comparison, an improved climatology using gauge measurements from Soviet drifting stations and gauge-corrected station data for Eurasia and Canada by Serreze and Hurst [28] has an average value of 27.9 cm yr^{-1}, while a value of 29.3 cm yr^{-1} has been computed from the maps of the Khrol atlas (J.E. Walsh, 1998, personal communication). These values, as well as the comparison for central Arctic gauge data, show favorable agreement with the ERA forecast P, while corrected NCEP P values are low. Observational estimates for E are more difficult to obtain, however Khrol [39] has tabulated zonal averages of the surface latent heat flux, which may be directly compared with reanalysis forecast data. Averaged for 70°N-90°N the Khrol estimate is 6.5 W m^{-2}. For the period 1979-93, the average reanalysis forecast values are 9.4 W m^{-2} for the ERA, and 13.8 W m^{-2} for the NCEP/NCAR Reanalysis. It is difficult to assess these comparisons as the atlas estimates are likely to have been determined from observations taken over a long period of time and are not concurrent with the reanalysis data. The relatively close agreement between forecast P and climatology, however, suggests that reanalyses values of E are too high. This is despite comparable spatial patterns of E for the ERA forecasts and the Gorshkov atlas. Differences between reanalyses forecast P and climatology are within 16 percent and 3 percent for the ERA and NCEP/NCAR Reanalyses, respectively, while the differences in the average surface latent heat flux are 45 and 110 percent.

4. Derived Atmospheric Moisture Budget

Previous study of moisture convergence by Serreze et al. [7, 8] employed the HARA data for computing the atmospheric moisture budget over the north polar cap. The HARA data have been compiled primarily from archives at NCAR and the U.S. National

Climatic Data Center. The spatial distribution of rawinsonde stations which were used to compute the meridional flux across 70°N is shown in Figure 1. Figure 1 is slightly misleading because of the large variability in the station network. For a typical month, approximately 30 to 40 of the 56 stations shown are actually available. The HARA data set, however, contains additional high latitude rawinsondes that are not in close proximity to 70°N. A total of about fifty stations poleward of 65°N contain 20 or more soundings per month for a given year. Rawinsonde observations are also available from a few of the Soviet drifting ice camp stations and shipboard platforms in the North Atlantic and the Bering Sea. These in situ measurements form the basis for current knowledge and documentation of Arctic atmospheric circulation. These rawinsonde data, the synthesis performed by Serreze et al. [7, 8], and satellite water vapor retrievals are used below to evaluate the reanalysis moisture budget depictions.

4.1 PRECIPITABLE WATER

In the atmospheric moisture budget (equation 1), the first term of interest is precipitable water (W). Although the storage term is negligible for annual time scales, previous study has suggested that it is significant for individual months in the Arctic [7, 8]. Additionally, examination of the reanalysis precipitable water gives some information on whether potential flux deficiencies are due to the representation of atmospheric moisture content or the winds. Average values of precipitable water for the north polar cap from the reanalyses are first evaluated using a blended data set from the National Aeronautics and Space Administration Water Vapor Project (NVAP) [40]. For the Arctic basin, NVAP is primarily composed of retrievals from the Television and Infrared Operational Satellite (TIROS) Operational Vertical Sounder (TOVS) as well as rawinsonde data using quality control described by Ross and Elliot [41]. The NVAP data set is available for the years 1988-92. A comparison of averaged NVAP data and corresponding values of the reanalyses for the years 1988-92 are shown in Figure 6. Also plotted in Figure 6 are the long-term (1973-91) precipitable water values from Serreze et al. [7]. Serreze et al. [7] used a subset of the HARA stations poleward of 70°N to estimate the precipitable water for the north polar cap. For both reanalyses, precipitable water is an archived variable, and thus the comparisons do not reflect any uncertainty in vertical interpolation. Precipitable water averaged for 70°N-90°N exhibits peak values in July, with a prolonged period of steady but low winter values from November until March. Precipitable water retrievals from TOVS are subject to some uncertainty over sea ice [42]; Figure 6 may then be considered a comparison of methods rather than a validation of the reanalyses. The absolute differences between the two reanalyses and NVAP are at a minimum in June and July. Because of the larger summertime precipitable water values, these differences are very small in terms of percentages, about 1 to 2 percent. In contrast the winter months of November to March show larger differences which are reflected in the discrepancies between the annual values. For the five winter months the NVAP average is 3.1 kg m^{-2} compared with the ERA value of 2.5 kg m^{-2}, NCEP/NCAR, 2.7 kg m^{-2}, and Serreze et al. [7], 2.5 kg m^{-2}. Precipitable water directly synthesized by Serreze et al. [7] compares favorably with both the reanalyses and the NVAP data. The comparison between the ERA and the Serreze et al. data is particularly close for the winter months.

176

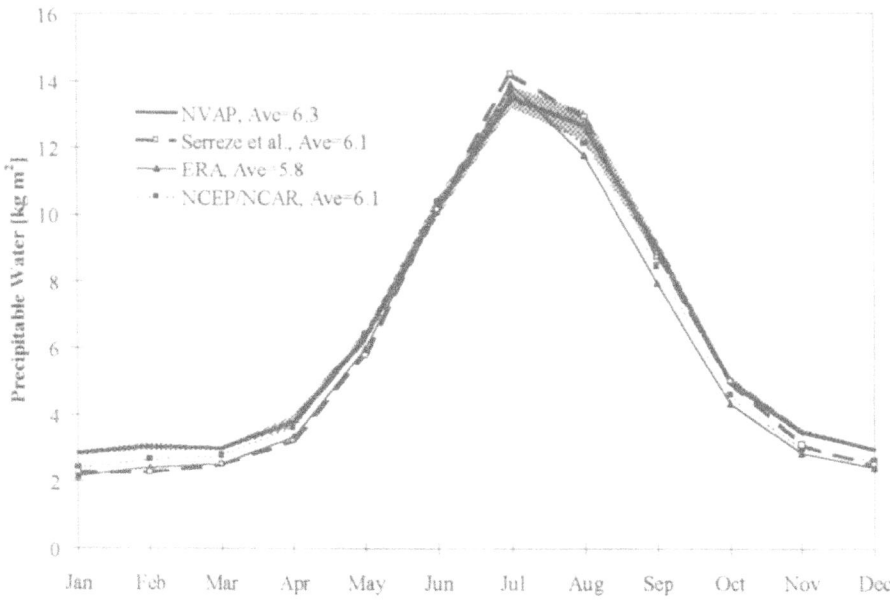

Figure 6. Monthly time series of precipitable water from NVAP data, ERA, and NCEP/NCAR Reanalyses for the region bounded by 70°N-90°N, averaged for 1988-92. Shaded area indicates the standard error of the mean from monthly NVAP data.

Spatially, the annually-averaged precipitable water distribution from NVAP data consists of values less than 6 kg m^{-2} over the central Arctic basin and increasing to greater than 10 kg m^{-2} over the Bering Sea and the North Atlantic. In the North Atlantic, values greater than 10 kg m^{-2} begin south of Svalbard and increase equatorward to greater than 14 kg m^{-2} near the southern Norwegian coast. The ERA annually-averaged precipitable water distribution is lower than NVAP values for most of the region north of 65°N, with the exception of the central Arctic permanent ice pack. These differences are less than 0.5 kg m^{-2} for most regions with the exception of the North Atlantic sector and North Pacific coastal regions. For a small region poleward of 85°N, ERA average precipitable water values exceed NVAP by about 0.5 kg m^{-2}. The NCEP/NCAR average values are also smaller than the NVAP estimates for large regions of the North Atlantic, the Canadian Archipelago, and the Bering Sea regions, with values of up to 1 kg m^{-2} larger than NVAP over the Eurasian land mass, selected areas of northern Canada, and the central Arctic permanent ice pack. For both reanalysis data sets, the largest precipitable water differences with the NVAP occur in the North Atlantic sector. The NVAP data exceeds both reanalyses by greater than 2 kg m^{-2} for the region immediately south of the Fram Strait. This possibly results from uncertainty associated with the positioning of the sea ice front in winter, or the highly persistent cloudiness in the region (e.g., Sheu and Curry [43]). The NVAP data also tends to show substantially larger values over the southern tip of Greenland, while the two reanalyses depictions are more closely related to the topography of South Dome (near 64°N, 45°W). Because of the high elevation of this southern Greenland feature, which is greater than

2500 m, it is speculated that the much lower reanalysis values are more reasonable. The difference between NVAP and reanalysis average values for this area is approximately 5 kg m^{-2} for both the ERA and the NCEP/NCAR Reanalysis. For the region north of 65°N, this is the largest common discrepancy.

Spatial differences between the ERA and NCEP/NCAR Reanalysis precipitable water are more troubling. As shown in Figure 7, a suggestion of the spurious wave pattern in the NCEP/NCAR Reanalysis emerges over Siberia in this difference plot. Discrepancies associated with the NCEP/NCAR horizontal diffusion parameterization are primarily associated with forecast moisture fields such as forecast P, however the use of the NWP model in producing analyzed fields allows for this pattern in areas of sparse data coverage and/or complex topography. An additional example found in the surface temperature climatology is given by Overland et al. ([44], see their appendix section). For the whole of the north polar cap, the ERA precipitable water is smaller than the NCEP/NCAR Reanalysis for the annual average. From Figure 6, it may be seen that the ERA is smaller for most months, with the exception of July. In general, the agreement of the average values of the NVAP and the reanalyses appears to be consistent with the expected uncertainties.

Consistent with the NVAP data around the margins of the Arctic Ocean, precipitable water comparisons at point locations using the HARA data also show the ERA values to be lower than observation. Point comparisons have been made for 1988 with 50 stations from the HARA data that contain at least 20 soundings per month for the year. The rawinsonde values are determined by using all available levels and vertically integrating the data using trapezoidal rule. In addition to the HARA quality control measures applied by Serreze et al. [7], a further restriction is imposed that at least five rawinsonde levels are necessary for a precipitable water value to be determined. The reanalysis data are compared by using only values that are concurrent with the available rawinsonde data. This comparison is summarized in Figure 8 by averaging the data for each month. The ERA values are found to be significantly low for the entire year. In fact, inspection of the individual comparisons found the annually-averaged ERA values are lower than the rawinsonde data for each of 49 stations with the exception of Jan Mayen (71°N, 9°W). The NCEP/NCAR values show a tendency to be lower, but particularly for summer months. Exactly as many rawinsonde stations give annually-averaged precipitable water estimates that are lower than the NCEP/NCAR value as the number of stations with a higher average value. Additionally the locations of the rawinsondes do not suggest a bias pattern for the NCEP/NCAR Reanalysis.

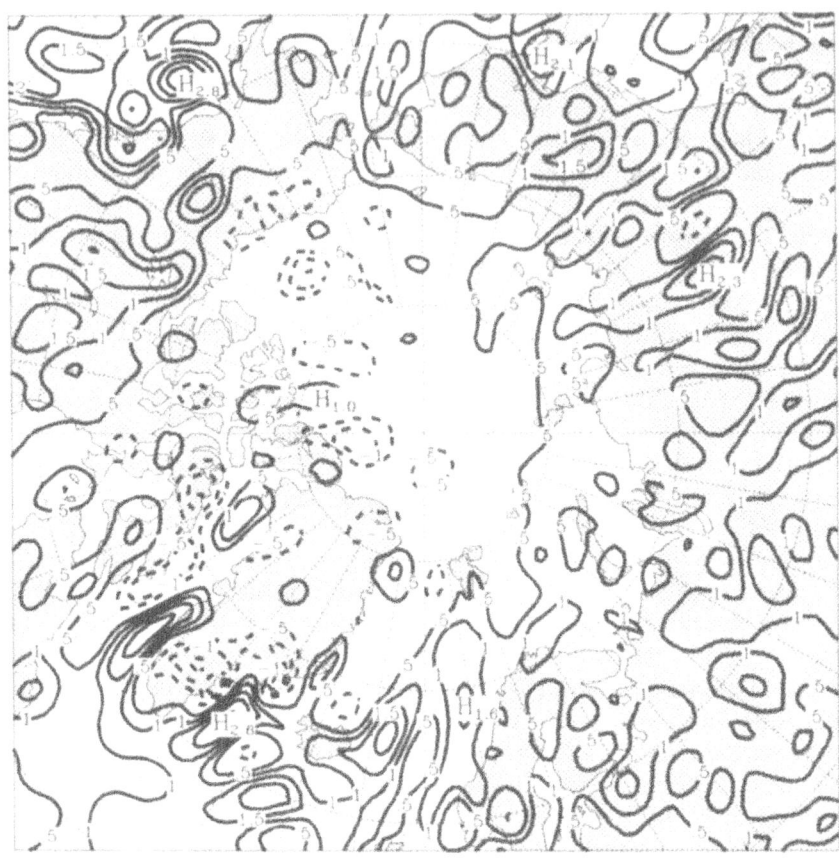

Figure 7. Difference of NCEP/NCAR Reanalysis minus ERA precipitable water, averaged for 1988-92. The contour interval is 0.5 $kg\ m^{-2}$.

Reanalyses precipitable water values for 1988 have also been compared with the NCEP/NCAR Arctic Marine Rawinsonde Archive from NSIDC. These data are more scattered spatially and temporally, however estimates for 570 soundings in the North Atlantic and 25 soundings in the Chukchi Sea region, typically near 70°N, were examined. For the soundings in the North Atlantic, the data from 6 marine platforms compared closely with the ERA. The average observed precipitable water value, 9.6 $kg\ m^{-2}$, is matched by the ERA data, with the NCEP/NCAR value at 10.0 $kg\ m^{-2}$. Observed values for the Chukchi sea region, which were made during July and August, exceed the ERA average by 15 percent and the NCEP/NCAR Reanalysis by 8 percent, however the number of soundings is too small to draw conclusions. For each platform, the average differences are typically less than the standard error of the mean. In general, both the mean values and temporal variability are adequately reproduced by the reanalysis data.

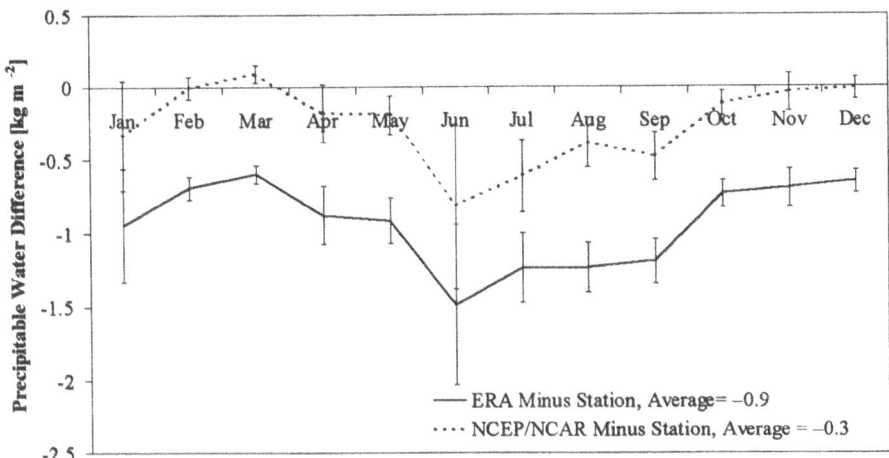

Figure 8. Average monthly difference of reanalyses minus rawinsonde precipitable water values for 1988, in kg m^{-2}. Bars show the standard error of the mean of 50 rawinsonde stations.

Finally, the 1988 reanalysis precipitable water values have also been compared against rawinsondes from the Soviet drifting ice stations. For the full year, station NP-28 was operational north of 82°N (see Figure 5a), and a total of 427 soundings were compared with reanalysis data. It is not clear whether these data were available to the operational forecasting centers for assimilation into the reanalyses data. As shown in Figure 9, there is a substantial discrepancy in summertime values of the NCEP/NCAR Reanalysis. For July 1988, the average precipitable water values for 42 soundings are 14.1 kg m^{-2} for the rawinsonde, 13.7 kg m^{-2} for the ERA, and 10.6 kg m^{-2} for the NCEP/NCAR Reanalysis. In July, the drifting ice station was in close proximity to the pole, at approximately 88°N. It is important to recall that the reanalysis data are

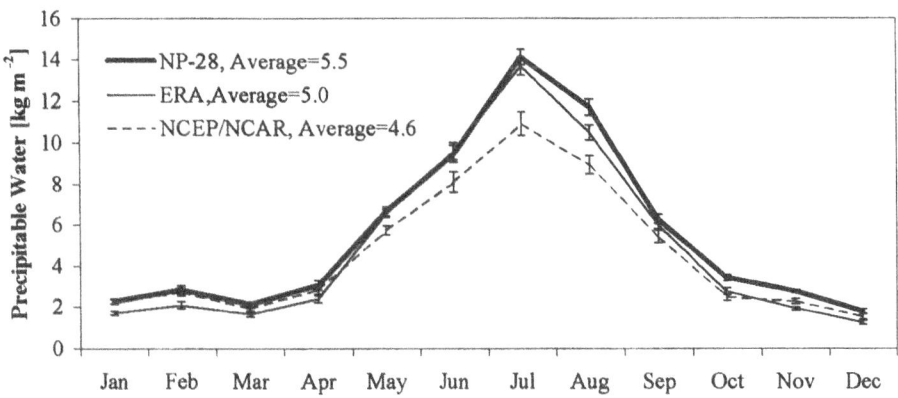

Figure 9. Average monthly precipitable water from Soviet drifting ice station NP-28, and corresponding values of the ERA and NCEP/NCAR Reanalysis for 1988, in kg m^{-2}. Bars indicate the standard error of the mean.

initially produced in a spectral format, which does not contain values for 90°N, and are then interpolated to the 2.5° grid. Averaged precipitable water is a relatively smooth field over the central Arctic basin, however, and the differences shown in Figure 9 are likely to be applicable over the whole of the permanent central ice pack area.

In summary, the available observational evidence indicates that the ERA precipitable water values are too low at the periphery of the north polar cap region, near 70°N. The discrepancy is apparently small but statistically significant over land areas. Figure 8 suggests a deficiency of 0.9 kg m^{-2}. Over the oceanic regions, reanalyses comparisons with the marine rawinsonde data indicate good agreement, while the NVAP data is significantly larger for the North Atlantic. Some difficulties may exist for the NVAP data in the North Atlantic region. Over the central Arctic basin, Soviet drifting ice station rawinsondes and NVAP data both suggest reasonable agreement for the ERA. For the NCEP/NCAR Reanalysis, there is better agreement near 70°N with the land-based rawinsonde data. At higher latitudes the NCEP/NCAR Reanalysis is too low in comparison to the Soviet drifting ice station data.

4.2 ATMOSPHERIC MOISTURE TRANSPORTS

We first present a summary of work presented in Cullather et al. [12], which compares the Serreze et al. synthesized moisture fluxes with the reanalysis data. For computations of the Arctic atmospheric moisture budget using the HARA data, Serreze et al. [7, 8] first determined the distribution of the meridional transport along 70°N for each month, and then averaged these data to produce a multi-year mean annual cycle. The computations were performed in this order for the purpose of retrieving the seasonal to interannual variability. The studies by Serreze et al. [7, 8] used rawinsonde data for the years 1974-91, while the ERA is available for the period 1979-93. To eliminate this time difference, the rawinsonde synthesis has been averaged for the time period concurrent with the reanalysis data. A comparison of the average meridional distribution along 70°N for 1979-93 is shown in Figure 10. The average meridional transport from the rawinsonde synthesis is 6.0 kg m^{-1} s^{-1}, yielding a moisture convergence value for the north polar cap of 17.0 cm yr^{-1}. This compares with the original rawinsonde estimate of 16.3 cm yr^{-1} for the period 1973-91 given by Serreze et al. [7]. For the annual average, it is seen that there are several preferred corridors of poleward (positive) moisture transport near 20°E, 160°W, and 50°W, with average equatorward moisture flux (negative) present only over the western Canadian Archipelago near 110°W. Both of the reanalyses data sets produce a larger average poleward moisture transport than the rawinsonde data. The average ERA meridional transport across 70°N is 6.6 kg m^{-1} s^{-1}; for the NCEP/NCAR Reanalysis, the value is 7.0 kg m^{-1} s^{-1}. This corresponds to convergence values of 18.4 cm yr^{-1} and 19.8 cm yr^{-1} for the ERA and NCEP/NCAR, respectively. It may be immediately seen in Figure 10

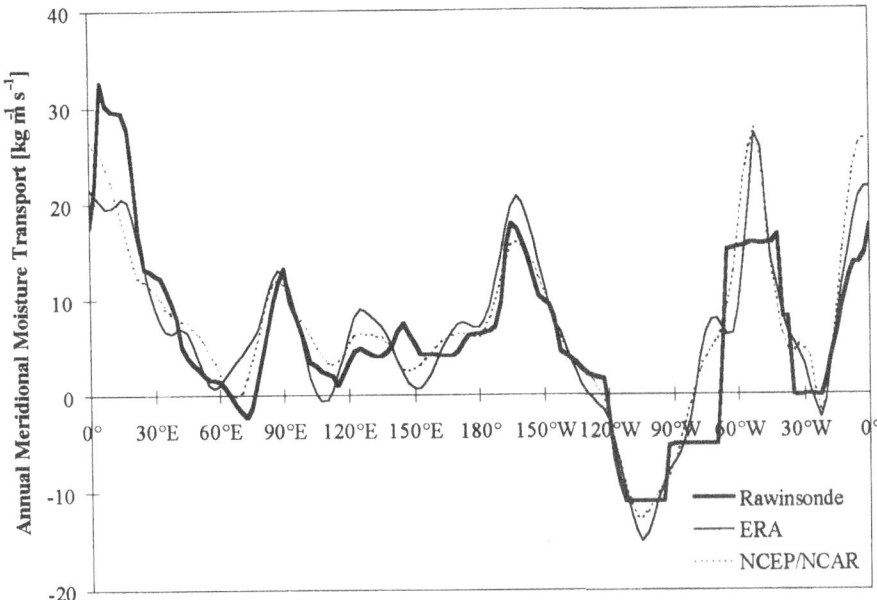

Figure 10. Annually-averaged meridional moisture transport distribution across 70°N from rawinsonde data synthesized after Serreze et al. [7], ERA, and NCEP/NCAR Reanalysis data, for 1979-93 in kg m⁻¹ s⁻¹.

that the rawinsonde data encounter some resolution difficulty over the Baffin Island/Greenland region between 30°W and 120°W.

Figure 11 shows a comparison of the average annual cycle for meridional moisture transport across 70°N. Surprisingly large differences are found between the rawinsonde synthesis and the reanalyses data for summer months, particularly July and August. The reanalyses data then have a fundamentally different depiction of the annual cycle, with a smoothed, unimodal curve and maximum poleward transport in July or August, while the rawinsonde data suggest that maximum values occur during transition months prior to and after the summer. These differences are even more striking when the year-to-year variability in the annual cycle is examined, as shown in Figure 12a,b. In general, the curve shown by the ERA is highly reproducible; the annual cycle for each individual year shows very similar characteristics. Not shown, the NCEP/NCAR Reanalysis annual cycles have very similar characteristics to those shown for the ERA in Figure 12b. In contrast the annual cycles for individual years produced from the rawinsonde data, shown in Figure 12a, have significant year-to-year variability, particularly in the summer months.

For comparison, Figure 11 also shows the annual cycle depicted by Peixoto and Oort ([45], see their Figure 12.21) obtained from rawinsonde data for the period 1963-73. The estimate by Peixoto and Oort [45] of 12.2 cm yr⁻¹ is substantially smaller than that given by Serreze et al. [7]. Serreze et al. [7] attribute the discrepancy to either differences in the stations used, or interannual variability. Two issues that arise from the comparison with Peixoto and Oort [45] are (1) the magnitude and (2) the shape of the annual cycle. While the Peixoto and Oort data contain smaller values in comparison to both the rawinsonde synthesis and the reanalysis data, there is unusual agreement

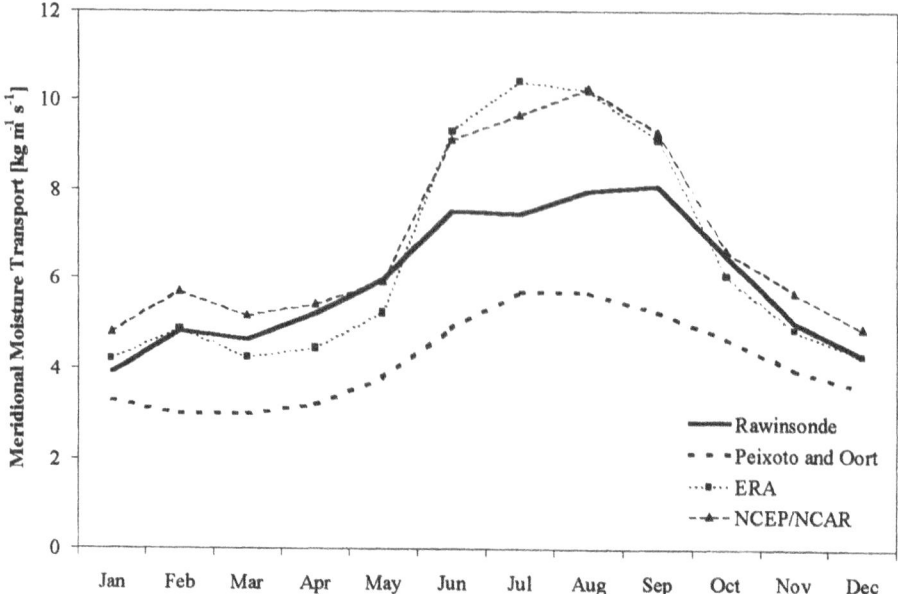

Figure 11. Average annual cycle of vertically-integrated meridional moisture transport across 70°N from reanalyses and synthesized from rawinsonde data, for 1979-93 in kg m^{-1} s^{-1}.

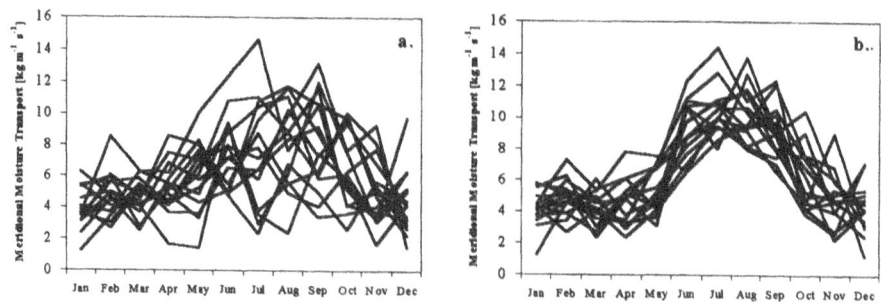

Figure 12. Annual cycles of vertically-integrated meridional moisture transport across 70°N (a) synthesized from rawinsonde data and (b) from the ERA, for individual years from 1979-93 in kg m^{-1} s^{-1}.

between the shape of the annual cycles for Peixoto and Oort [45] and the reanalysis data. The comparison is intriguing, however it is difficult to assess the differences between the Serreze et al. and Peixoto and Oort rawinsonde syntheses without an involved inspection of the original station data. For the present, this investigation is limited to the differences between the Serreze et al. [7] and reanalysis data.

Reanalysis differences with the previous study may first be narrowed in time and then in space. For the fifteen-year period, time series comparisons have been constructed of the year-to-year average transport across 70°N for each month. Both reanalyses and the rawinsonde syntheses agree very closely on the average value and the year-to-year variability from one December to the next December, as well as for the other winter months. Time series for January and July are shown in Figure 13. It is only during the summer months of June to August when the significant discrepancies occur, such as is shown in Figure 13(b) for July. Figures 12 and 13 both show the large interannual variability in the summertime moisture flux depicted by the rawinsonde synthesis as compared to the reanalysis data. The standard deviation of the July values is more than twice as large as for the two reanalyses. The July meridional flux distribution is similar to the annual average shown in Figure 10, but with greater amplitude. If the reanalyses depictions of the meridional transport along 70°N shown in Figure 11 are accurate, then it is not surprising that the differences with the rawinsonde data are largest in the summer months, when the meridional transport gradients along 70°N are substantial and tend to expose the limited data resolution in particular areas.

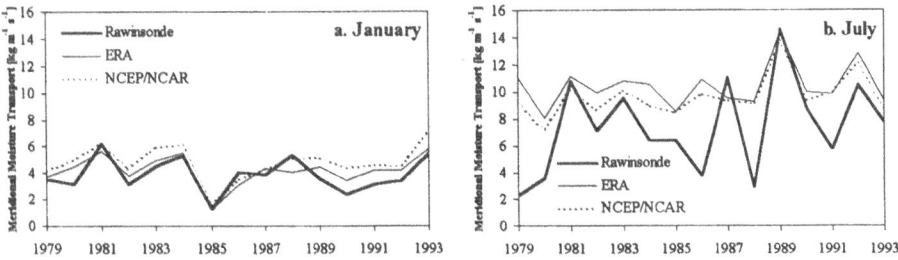

Figure 13. Time series of average vertically-integrated meridional moisture transport across 70°N for (a) January and (b) July, for individual years from 1979-93 in kg m^{-1} s^{-1}.

Because of the stair-stepped pattern of the rawinsonde synthesis, difference plots of the zonal distribution are intricate, but may be simplified by taking differences along 20° longitude sections. These differences for July are shown in Figure 14. Substantial disagreements are found for three regions: western Siberia (80°E-100°E), Chukchi Sea (160°W-180°W), and the Davis Strait (60°W-80°W). For the Chukchi Sea region, the ERA are significantly different from the rawinsonde synthesis, while the NCEP/NCAR Reanalysis is in agreement. In Cullather et al. [12], it was demonstrated that discrepancies for the other two regions resulted from deficiencies in rawinsonde network coverage over these areas. There are no observations which are independent of both methods however, and it is therefore a more difficult task to indicate that the reanalysis depiction is more realistic. It is apparent, however, that the limitations of the rawinsonde

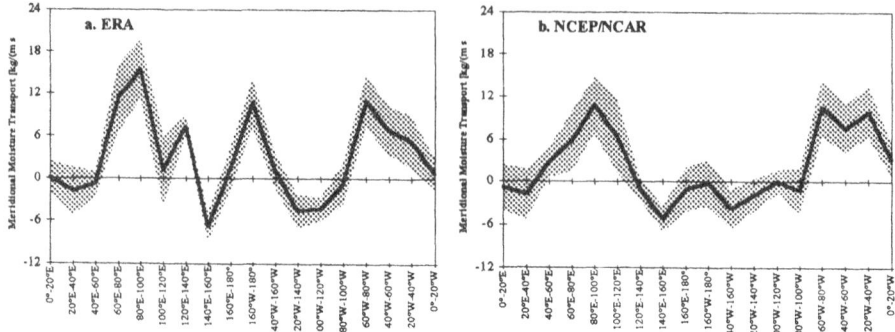

Figure 14. (a) Average differences for 20° longitude intervals of July moisture flux across 70°N for the ERA minus the rawinsonde synthesis, 1979-93 in kg m^{-1} s^{-1}. (b) As in (a) but for the NCEP/NCAR Reanalysis.

network are highlighted during the summer months because of the larger flux magnitudes and spatial gradients. Reanalyses data also incorporate the rawinsonde data and thus may be thought of as a more elaborate synthesis of these data than the method employed by Serreze et al. Because reanalyses also include other factors including additional surface observations, satellite data, topographic forcing, and the downstream advection of observations via the NWP model, it may be readily understood why the reanalysis data may be more able to overcome missing observations and produce a more realistic depiction. The differences shown in Figures 12(a) and 12(b) are compelling.

It is then of interest to examine whether the reanalysis data is reproducing the moisture fluxes at the point locations of the rawinsonde stations in the HARA archive. For the sample year investigated, this generally appears to be the case. Figure 15 shows the differences of the zonal and meridional moisture transports for rawinsonde point locations and the concurrent reanalysis data, averaged for 50 stations from the HARA data set. For the zonal moisture flux, the average from the rawinsonde stations for the year is 23 kg m^{-1} s^{-1}, with a maximum of 54 kg m^{-1} s^{-1} in September and minimum of -2 kg m^{-1} s^{-1} (negative indicating a westward flux) in May. For most of the months in the 1988 sample year, the ERA zonal transport was significantly low by an average of 2 kg m^{-1} s^{-1} in comparison to the rawinsonde station data, while the NCEP/NCAR Reanalysis values are in closer agreement. For the meridional flux, both reanalyses are in close agreement with the rawinsonde data. The average annual meridional moisture transport of the fifty stations examined is 6 kg m^{-1} s^{-1}, with the maximum average value in September at 10 kg m^{-1} s^{-1}, and a minimum of 3 kg m^{-1} s^{-1} in May. It is difficult to encapsulate the HARA evaluations, as the comparison for each station has unique characteristics. For example, some topographic features may be displaced in the spectral format of the reanalyses, or local circulation features may not be adequately resolved at the 2.5° resolution. However, the averaging shown in Figure 15 tends to mitigate these difficulties associated with the point comparisons. In general, Figure 15 is a realistic appraisal of the strengths and weaknesses of the reanalysis transports. For illustrative purposes, Figure 16 shows a comparison of the moisture fluxes for three HARA stations during July and August 1988. Jan Mayen (71°N, 9°W) appears to compare more closely than other stations because of its location away from significant topography. The other two stations have been arbitrarily selected, however. Although the reanalyses show an

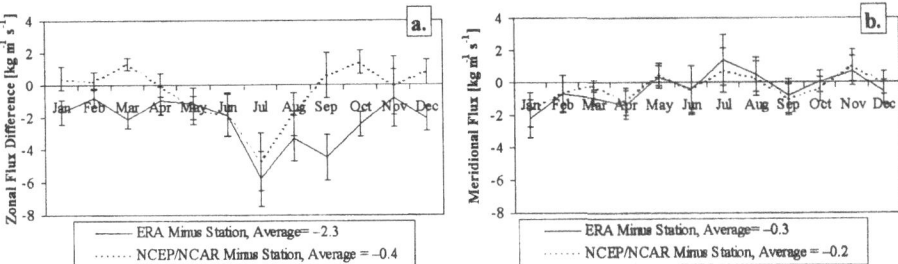

Figure 15. Average difference of reanalysis vertically-integrated moisture transports minus HARA stations for (a) zonal fluxes and (b) meridional, for 1988 in kg m^{-1} s^{-1}.

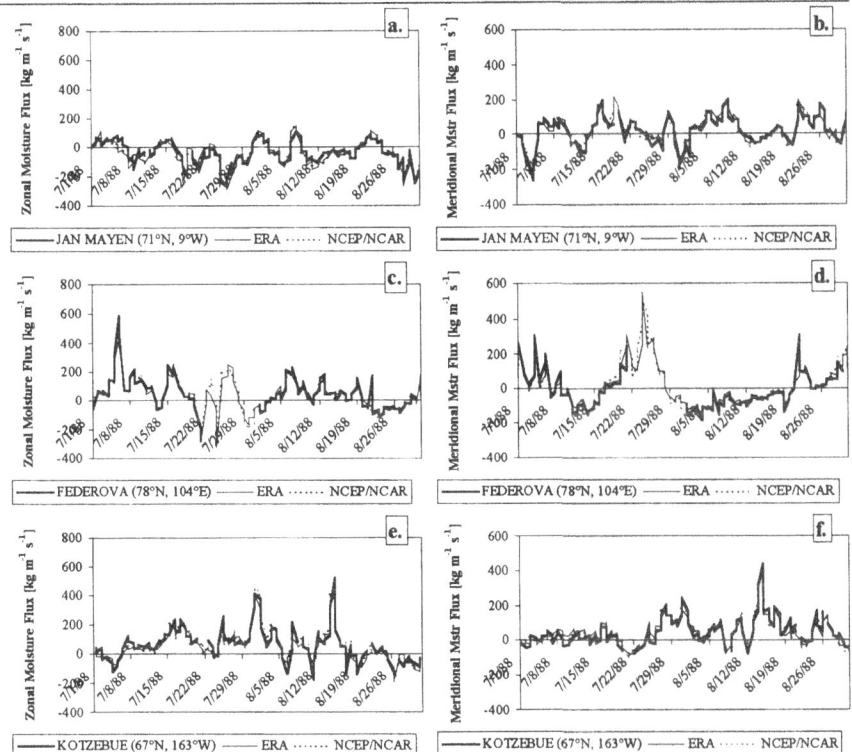

Figure 16. Comparison of 12-hourly zonal and meridional moisture fluxes at 3 stations for July and August 1988, in kg m^{-1} s^{-1}.

apparent tendency to underestimate the large events, it is clear from these examples that the temporal variability of the moisture flux is being reproduced. Comparable to other stations in the area, the two reanalyses at Kotzebue (67°N, 163°W) are similar and do not resolve the differences shown in Figure 14 for the Chukchi Sea region. Although marine rawinsonde data are available for this region, the variability in location and time of the observations does not allow for easy determination. Precipitable water values of the two reanalyses are in reasonable agreement, indicating the winds as the primary cause of the discrepancy.

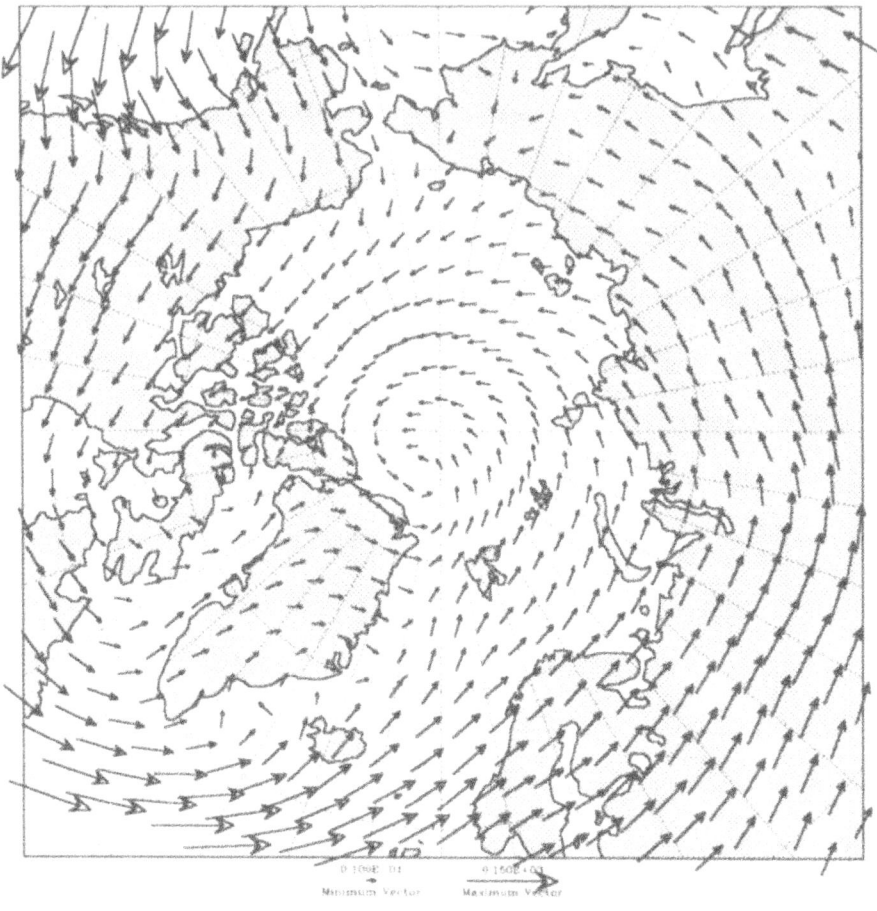

Figure 17. Average vertically integrated moisture fluxes from the NCEP/NCAR Reanalysis for 1979-93. The maximum vector magnitude shown is 150 kg m^{-1} s^{-1}.

Finally, the computed atmospheric moisture flux from the NP-28 Soviet drifting ice station rawinsonde data have been compared with the two reanalyses for the year 1988. Given the station trajectory shown in Figure 5(a), it is apparent that the observed moisture transports experience a directional shifting in close proximity to the pole; differences in the placement of the center of the circulation over the Arctic in the reanalyses will incur a sign differences occur in the transports. Additionally, the monthly mean transport components for the year experience a range of values that are roughly centered on zero. For the period July to September, the station was poleward of 88°N. The largest differences are with the July zonal transport component for the ERA, which is roughly twice as large as the observed value. The NCEP/NCAR Reanalysis is in close agreement, which is surprising given the large difference in the precipitable water values. For the year, the average observed zonal flux for the station is $12 \, kg \, m^{-1} \, s^{-1}$, compared with $14 \, kg \, m^{-1} \, s^{-1}$ for the ERA and $8 \, kg \, m^{-1} \, s^{-1}$ for the NCEP/NCAR Reanalysis. The monthly mean meridional transports for both reanalyses compare more closely with the rawinsonde data. The rawinsonde, ERA, and NCEP/NCAR Reanalyses average meridional transports values for 1988 are all approximately $-3 \, kg \, m^{-1} \, s^{-1}$, and there is close agreement for individual months.

In summary, comparisons of the reanalysis atmospheric moisture fluxes to previous studies indicate large discrepancies that are associated with the station spatial coverage. This is indicated by the relation of these differences to missing station data as shown in Cullather et al. [12], and the close agreement of the reanalysis fluxes at point locations. Over the central Arctic basin, rawinsonde comparisons encounter some directional ambiguity associated with taking flux components in close proximity to the pole. Nevertheless, the comparisons do show significant irregularities for both reanalyses. For the ERA, the zonal flux is too large in July, when the station is in close proximity to the pole. The NCEP/NCAR Reanalysis shows better agreement, but probably for incorrect reasons given the large discrepancy in the precipitable water comparison. These results are tempered by the limited amount of data for the highest latitudes.

4.3 MOISTURE CONVERGENCE AND FLUX PATTERNS

The spatial depictions of the vector-averaged reanalyses moisture transports for 1979-93, shown in Figure 17 for the NCEP/NCAR Reanalysis, reveal a mostly zonal pattern for the interior Arctic basin, with the circulation centered near 90°N. Over most land areas the transport is also approximately zonal, with southerly transport occurring over the Bering Strait and western Alaska, as well as the North Atlantic/Norwegian Sea region. A signature of the Icelandic Low is found southeast of Greenland. The interannual variability in the transport field is very large. The annually averaged central Arctic circulation migrates considerably from one year to the next. A year of interest is 1982, when the center of circulation was located near 85°N and 20°E. Averaged for this year, the atmospheric moisture transport through the Fram Strait was from the north, an unusual climatic anomaly. Differences for the two reanalyses exist for the central Arctic basin, with the NCEP/NCAR Reanalysis depicting moisture transports with a larger meridional component, particularly north of Svalbard. Overall, however, the moisture transport patterns of the two reanalyses are very similar. Although there is general agreement between the reanalyses on the flux patterns, there are significant differences in the convergence fields, shown in Figure 18. At T21 resolution, both reanalyses capture major features along the southeast coast of Greenland and the Gulf of Alaska, however the positioning is not great. For example, in both data sets the southeast Greenland maximum is erroneously centered over the Denmark Strait, while values over the interior plateau are negative. Difficulties also exist in both reanalyses over the Mackenzie River in western Canada, where P–E is less than zero over land for the annual average. These inadequacies are reminiscent of the discussion given by Rasmusson and Mo [23], in which terrain-related biases were noted in the moisture budget of the NCEP operational analyses. The significant difference between the two data sets is found in the North Atlantic, with the $30 \, cm \, yr^{-1}$ contour in the NCEP/NCAR Reanalysis extending from the North Atlantic through the Fram Strait and over the

188

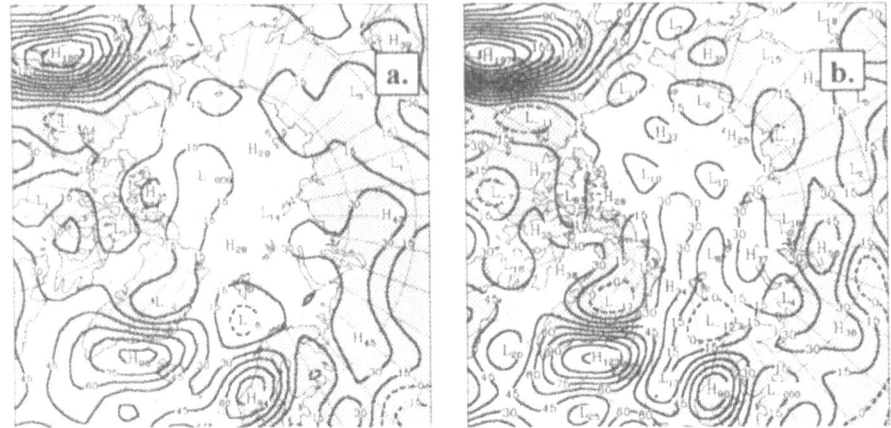

Figure 18 Average annual spatial distribution of moisture convergence computed from (a) ERA and (b) NCEP/NCAR Reanalyses for 1979-93. The contour interval is 15 cm yr⁻¹.

North Pole. ERA values in contrast are generally smaller over the central Arctic basin.

Differences with the rawinsonde synthesis have highlighted the dramatically large variability for the western Siberian region, east of the Urals mountains. Shown in Figure 19, the July meridional moisture transport across 70°N undergoes extensive year-to-year variability over the Eurasian land mass that is associated with the timing and development of the Urals trough. Depending on the year, summertime poleward moisture transport between 60°E and 120°E may be large or non-existent. In Figure 19(a), the average July transport patterns for years of large west Siberian plain poleward flux indicates a prominent river of moisture near 90°E, while most of eastern Europe and Scandinavia experience zonal circulation. This contrasts with other years

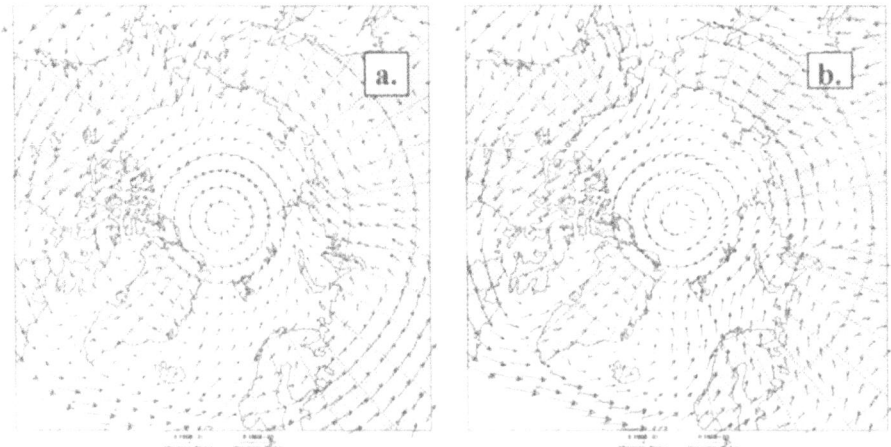

Figure 19. Average vertically integrated moisture fluxes for July from the ERA, averaged for (a) July 1979, 1980, 1984, 1986, 1987, 1991, and 1992, and (b) for 1981, 1982, 1983, 1985, 1988, 1989, 1990, and 1993. The maximum vector shown is 180 kg m⁻¹ s⁻¹.

which are averaged in Figure 19(b). For these years, the poleward flux is more prominent over Scandinavia and regions west of the Urals. This suggests opposing summertime circulation characteristics for Scandinavia and the west Siberian plain, similar to the wintertime effect found by Peng and Mysak [46]. This summer oscillation is intriguing, and will be the subject of future investigation.

5. Comparison of the Two Methods

5.1 COMPARISON WITH PREVIOUS STUDY

Table 2 compares estimates of $P-E$ from the atmospheric moisture budget, as well as the forecast values, to previous studies for the north polar cap bounded by 70°N. The original NCEP/NCAR forecast P is used for consistency with the available forecast E. The reanalysis moisture transport for the near-surface level was used in this study but was not considered by Genthon [53]. Both the ERA and the NCEP/NCAR Reanalysis are not in hydrologic balance. The atmospheric moisture budget results are 38% and 73% larger than the forecast data for the ERA and NCEP/NCAR, respectively. This imbalance is summarized in Figure 20. In Figure 20, the annually-averaged zonal profiles show the moisture budget values are higher than forecast values for both data sets at nearly all latitudes poleward of 50°N. From 75°N to 85°N, the imbalances of the two data sets are similar at about 5 cm yr^{-1}. Equatorward of 75°N, the imbalance in both data sets becomes larger, however this is particularly true of the NCEP/NCAR Reanalysis. The NCEP/NCAR imbalance is consistent with Trenberth and Guillemot [18, 19], who show the discrepancies to be larger in the presence of the higher

TABLE 2. Comparison of $P-E$ estimates for the north polar cap.

		Data set time period	$P-E$, 70°N - 90°N (cm yr^{-1})
Based on surface data			
Sellers [47]		multiyear	5.0
Baumgartner and Reichel [48]		multiyear	6.1
Korzoun et al. [49, 50]		multiyear	20.4*
Based on rawinsonde data			
Peixoto and Oort [51]		1963-73	12.2
Overland and Turet [52]		1965-90	21.4*
Serreze et al. [7, 8]		1974-91	16.3*
Based on numerical analyses			
Masuda [24], ECMWF FGGE analyses		1979	15.5*
Genthon [53], ERA		1979-93	20.7*
This study, ERA		1979-93	18.2*
This study, NCEP/NCAR		1979-93	19.4*
Based on NWP forecasts			
Genthon [53], ERA forecasts		1979-93	12.9
This study, ERA forecasts		1979-93	13.2
This study, NCEP/NCAR forecasts P data)	(original	1979-93	11.2

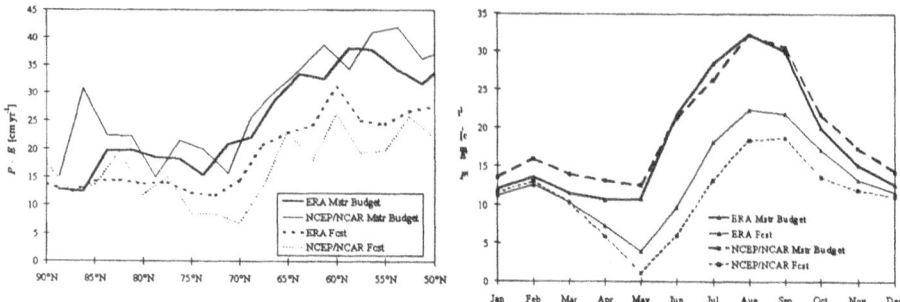

Figure 20. (a) Zonal comparison of average annual $P-E$ computed from ERA and NCEP/NCAR Reanalysis moisture transports and forecast fields, in cm yr^{-1}. (b) Comparison of $P-E$ for the north polar cap from ERA and NCEP/NCAR Reanalysis moisture transports and forecast fields, averaged for 1979-93, in cm yr^{-1}.

magnitude $P-E$ values along the Pacific coast of North America, the southeast coast of Greenland, and the Greenland/Norwegian Seas region. Hydrologic balance does occur in both data sets in close proximity to the pole. In Figure 21, the annual cycles for $P-E$ for both methods and data sets are shown. The hydrologic balance of both reanalysis data sets does not remain constant with the annual cycle. The largest discrepancies occur in the summer months, which is consistent with deficiencies in forecast E suggested earlier in comparison to values given by Khrol [39]. In both data sets, a maximum in forecast E occurs in May, and this is reflected by the very large imbalance for this month. For the ERA, the imbalance ranges from 7 percent of the moisture budget value in January to 64 percent in May, while the NCEP/NCAR imbalance ranges from 15 to 91 percent for the same months.

In Table 2, a general progression may be seen in previous studies from smaller values to contemporary estimates of between 15 and 21 cm yr^{-1}. The surface data studies are based on separate estimates of P and E. For the Korzoun et al. [49, 50] study, which is in better agreement with rawinsonde data, the estimate of P is approximately 73 percent larger than either Sellers [47] or Baumgartner and Reichel [48]. The Korzoun et al. value is also identical to that determined by J.E. Walsh's analysis of the Khrol atlas (J.E. Walsh, 1998, personal communication). Thus preferred values may be selected from Table 2 by eliminating the earlier surface-based studies, the lower values of Peixoto and Oort [51] described earlier, and the NWP forecast values. Using the remaining preferred estimates denoted by the asterisk, an average of 18.9 cm yr^{-1} with a standard deviation of 2.3 cm yr^{-1} is found. This is in reasonable agreement with the model result of the NCAR Community Climate Model version 3 (CCM3) of 18.1 cm yr^{-1} [54]. In comparison, estimates based on NWP forecasts have an average of 12.4±1.0 cm yr^{-1}. This appraisal of Table 2 strongly implies that the estimates based on NWP forecasts are low by at least 3.2 cm yr^{-1} or approximately 17 percent.

In Figure 21, the annual values of the two data sets and methods are shown in comparison to rawinsonde synthesis values of Serreze et al. Despite the large difference in average values between the two methods, there is close agreement on the interannual variability between the four reanalysis data series. The rawinsonde synthesis in general does not capture all of the detail of interannual variability shown, although there is some agreement on larger values for the years 1981 and 1989.

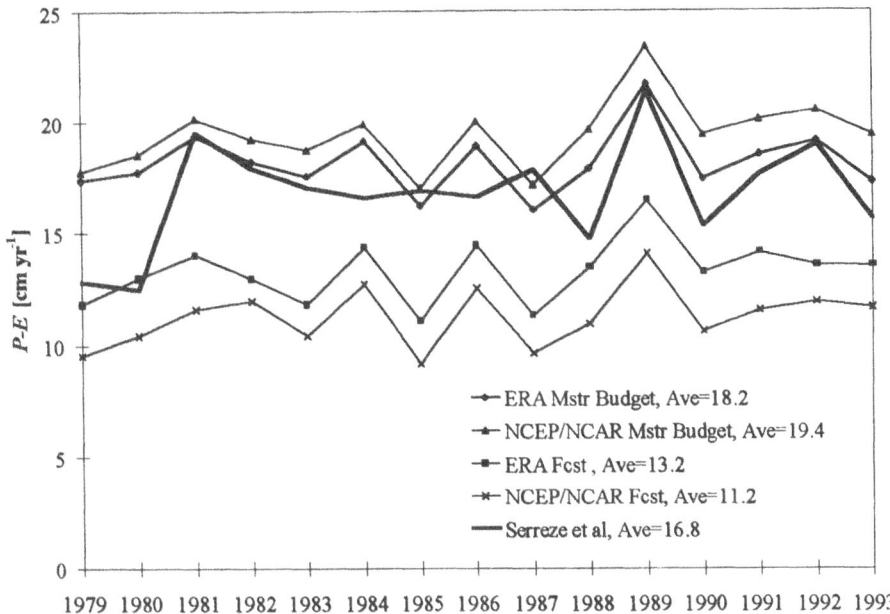

Figure 21. Comparison of annual $P-E$ values computed from reanalysis moisture transports and forecast fields for 1979-93, in cm yr⁻¹.

5.2 COMPARISON WITH OTHER ESTIMATES FROM THIS VOLUME

Of interest to this study is the assessment of moisture budget parameters by Serreze and Barry [55], using available rawinsonde and gauge measurements. The values presented by Serreze and Barry are perhaps the best available long-term estimates that have been directly synthesized from the observational data. Values are presented for three regions as defined by their Figure 1. The interesting feature of their results (see their Table 1) is the strong spatial variability in E, which is about 77% smaller for the central Arctic Ocean domain than for the north polar cap bounded by 70°N. This, combined with the observed smaller values of P over the central Arctic, leaves the observed $P-E$ at about the same value for the two domains. Corresponding values for the reanalysis data have been tabulated in Table 3. Again there is the disagreement for the north polar cap region between the rawinsonde data, the reanalysis moisture budget values, and reanalysis forecast values. Based on the evidence compiled here, it is argued that the slightly larger values of $P-E$ from the reanalyses moisture budget are more realistic, while the area-averaged forecast P values of the two reanalyses are reasonable in comparison to the observed values. Consistent with the previous comparison with data by Khrol [39], the reanalysis values of E are too large. This is particularly true of the central Arctic Ocean domain, where reanalyses forecast E estimates are several times the observed value given by Serreze and Barry.

TABLE 3. Comparison of reanalysis values with Serreze and Barry.

	$P-E$ (cm yr^{-1})	P (cm yr^{-1})	E (cm yr^{-1})
Serreze and Barry			
Polar Cap	16.1	27.9	11.8
Arctic Ocean	15.3	25.2	9.9
Central Arctic Ocean	17.4	20.2	2.8
ERA Moisture Budget			
Polar Cap	18.2	——	——
Arctic Ocean	17.9	——	——
Central Arctic Ocean	16.5	——	——
NCEP/NCAR Moisture Budget			
Polar Cap	19.4	——	——
Arctic Ocean	19.5	——	——
Central Arctic Ocean	19.3	——	——
ERA Forecasts			
Polar Cap	13.2	24.7	11.5
Arctic Ocean	13.3	23.8	10.5
Central Arctic Ocean	12.7	18.9	6.2
NCEP/NCAR Forecasts			
Polar Cap	11.2	28.6	17.4
Arctic Ocean	10.5	26.8	16.3
Central Arctic Ocean	11.9	24.2	12.3

6. Summary

6.1 REVIEW OF RESULTS FROM THIS STUDY

Comparisons of both the forecast and derived moisture budget estimates of $P-E$ from reanalyses show some discrepancies with observational data. For the forecast fields, comparison with surface latent heat flux values from Khrol [39] as well as the comparison in Tables 2 and 3 indicate that forecast E for both reanalyses is too large by a significant amount. Comparisons with climate atlases and in situ gauge measurements indicate reasonable agreement with forecast P from the ERA. There are some discrepancies in temporal comparisons with gauge data, as shown in Figure 5(b). Forecast P from the NCEP/NCAR Reanalysis contains substantial discrepancies associated with the diffusion parameterization in the NWP model. A corrected NCEP forecast P variable is an obvious improvement, however discrepancies remain with climatology, particularly for the North Atlantic.

For the derived atmospheric moisture budget $P-E$, the discrepancy in the shape of the annual cycle with previous rawinsonde analysis is related to inadequacies in the rawinsonde network. As mentioned earlier, Serreze et al. [7, 8] first computed the monthly transport distribution along 70°N for the purpose of retrieving the monthly and interannual variability. When the order of computations is reversed- that is, when long-term flux climatologies are computed for each station, and then a distribution along 70°N is produced- the agreement with the rawinsonde data is not improved. Several other sensitivities have been considered in the rawinsonde synthesis, including the radius of influence for each station, and the spatial resolution of the transport distribution. Sensitivity to these parameters is large but does not alter the overall conclusions. The agreement of the moisture transports, particularly the meridional component, to the rawinsonde data at point locations tends to strengthen the argument that the reanalysis annual cycle depiction is more realistic. Although the reanalyses moisture budgets appear to be superior for large areas, significant differences are present for the convergence field poleward of 80°N, and there are some problems with the positioning of some features. This is perhaps consistent with the general use of the atmospheric moisture budget method. Rasmusson [56] suggested that it is necessary to consider areas larger than 2.0×10^6 km^2. This may be an appropriate limitation for reanalysis computations.

6.2 FUTURE RESEARCH PRIORITIES

Several of the issues raised by this paper have relevance in future study. These issues may be organized by the data sources available. The following recommendations are listed in order of priority.

- *Rawinsonde data.* It remains to be seen whether the rawinsonde estimate can be rectified with a different method, or enhanced using additional observations. It is apparent from the compiled HARA data that several stations contain records that abruptly end and then begin again. Data for missing months and/or additional stations should be sought from alternative sources. If rehabilitation is not possible, then it is concluded that the major features of the poleward moisture flux are not being directly observed, but may be synthetically monitored using reanalysis assimilation methods. This leaves the reanalyses as the only available source for depicting interannual variability. Efforts to shore up the gaps should be considered, such as additional marine soundings.

- *Reanalysis data.* Efforts to identify and work with the major weather forecasting centers in solving analysis deficiencies should continue. Aside from the moisture fields evaluated here, additional work is necessary for the validation of the temperature, wind, pressure, and other variables. It is hoped that the information thus far provided will be useful for the upcoming ERA-40 [57]. Recent meetings between ECMWF personnel and Arctic researchers have been successful in providing input for ERA-40 parameterizations. Although an update of the NCEP/NCAR Reanalysis does not appear to be likely in the near future, dialog with NCEP on polar issues should continue. An area that requires further thought is in how to make these data more "investigator-friendly" for post-processing computations. The availability of vertically-integrated variables, for example, would be a great help. While the reanalysis data sets represent a significant advance towards understanding the atmospheric hydrologic cycle over the Arctic basin, additional efforts are required to understand the strengths and weaknesses of these data.

- *Satellite data.* Continued exploration of satellite techniques for directly retrieving the atmospheric moisture budget is needed, such as the pilot study by Rothrock and Kwok in this volume [58]. This method offers the potential for providing a uniform, high resolution observational depiction of the Arctic atmospheric moisture fluxes. Efforts should be focused on obtaining values from this method over long-term periods for comparison with conventional techniques.

Acknowledgments

The authors are grateful for assistance from J.E. Walsh. ECMWF data and the NCEP/NCAR Reanalyses were obtained from NCAR. NCEP/NCAR Arctic Marine Rawinsonde data and the Soviet drifting ice camp data were obtained by NSIDC. The Soviet rawinsonde data from NSIDC was compiled by J.D. Kahl. This research was sponsored by the National Aeronautics and Space Administration under grant NAGW-3677 to the second author and by the National Science Foundation under grant OPP-9614297 to the third author. Byrd Polar Research Center contribution number 1133.

References

1. Mysak, L.A., Manak, D.K., and Marsden, R.F. (1990) Sea-ice anomalies observed in the Greenland and Labrador seas during 1901-1984 and their relation to an interdecadal Arctic climate cycle, *Climate Dyn.* 5, 111-133.
2. Kattenberg, A., Giorgi, F., Grassl, H., Meehl, G.A., Mitchell, J.F.B., Stouffer, R.J., Tokioka, T., Weaver, A.J., and Wigley, T.M.L. (1995) Climate models- Projections of future climate, in J.T. Houghton, L.G. Meira Filho, B.A. Callander, N. Harris, A. Kattenberg, and K. Maskell (eds.), *The Science of Climate Change: Contribution of Working Group I to the Second Assessment Report of the Intergovernmental Panel on Climate Change*, Cambridge University Press, Cambridge, pp. 285-357.
3. Walsh, J.E. (1995) Recent variations of Arctic climate: The observational evidence, in *Fourth Conference on Polar Meteorology and Oceanography, Dallas Texas*, Amer. Meteorol. Soc., Boston, pp. (J9) 20-24.
4. Trenberth, K.E. (1992) Global analyses from ECMWF and atlas of 1000 to 10 mb circulation statistics, NCAR Tech. Note NCAR/TN-373+STR.
5. Trenberth, K.E. (1995) Atmospheric circulation climate changes, *Climate Change* 31, 427-453.
6. Kalnay, E., Kanamitsu, M., Kistler, R., Collins, W., Deaven, D., Gandin, L., Iredell, M., Saha, S., White, G., Woollen, J., Zhu, Y., Chelliah, M., Ebisuzaki, W., Higgins, W., Janowiak, J., Mo, K.C., Ropelewski, C., Wang, J., Leetmaa, A., Reynolds, R., Jenne, R., and Joseph, D. (1996) The NCEP/NCAR 40-year reanalysis project, *Bull. Amer. Meteorol. Soc.* 77, 437-471.
7. Serreze, M.C., Barry, R.G, and Walsh, J.E. (1995) Atmospheric water vapor characteristics at 70°N, *J. Climate* 8, 719-731.
8. Serreze, M.C., Rehder, M.C., Barry, R.G., Walsh, J.E, and Robinson, D.A. (1995) Variations in aerologically derived Arctic precipitation and snowfall, *Ann. Glaciol.* 21, 77-82.
9. Serreze, M.C., Kahl, J.D., and Shiotani, S. (1992) The Historical Arctic Rawinsonde Archive documentation manual, Special Report No. 2, National Snow and Ice Data Center, University of Colorado.
10. Bromwich, D.H. and Cullather, R.I. (1998) The atmospheric hydrologic cycle over the Arctic basin from ECMWF Re-Analyses, in *First Intl. Conf. on Reanalyses, Silver Spring, Maryland*, WCRP Rep. No. 104 (WMO/TD-No. 876), Geneva, pp. 359-362.
11. Cullather, R.I., and Bromwich, D.H. (1998) ERA and NCEP/NCAR Reanalysis depictions of the Arctic moisture budget, in *First Intl. Conf. on Reanalyses, Silver Spring, Maryland*, WCRP Rep. No. 104 (WMO/TD-No. 876), Geneva, pp. 375-378.
12. Cullather, R.I., Bromwich, D.H., and Serreze, M.C. (2000) The atmospheric hydrologic cycle over the Arctic basin from reanalyses, Part I, Comparison with observations and previous studies, *J. Climate*, in press.
13. Gibson, J.K., Kållberg, P., Uppala, S., Hernandez, A., Nomura, A., and Serrano, E. (1997) ERA description. ECMWF Re-Analysis Project Report Series No. 1.
14. Cullather, R.I., Bromwich, D.H., and Van Woert, M.L. (1998) Spatial and temporal variability of Antarctic precipitation from atmospheric methods, *J. Climate* 11, 334-367.
15. Adams, J.C. and Swarztrauber, P.N. (1997) Spherepack 2.0: A model development facility, NCAR Tech. Note NCAR/TN-436+STR.

16. Newell, R.E., and Zhu, Y. (1994) Tropospheric rivers: A one-year record and a possible application to ice core data, *Geophys. Res. Lett.* **21**, 113-116.
17. Trenberth, K.E. and Guillemot, C.J. (1996) Physical processes involved in the 1988 drought and 1993 floods in North America, *J. Climate* **9**, 1288-1298.
18. Trenberth, K.E. and Guillemot, C.J. (1996) Evaluation of the atmospheric moisture budget and hydrologic cycle in the NCEP reanalysis, NCAR Tech Note NCAR/TN-430+STR.
19. Trenberth, K.E. and Guillemot, C.J. (1998) Evaluation of the atmospheric moisture budget and hydrologic cycle in the NCEP/NCAR reanalysis, *Climate Dyn.* **14**, 213-231.
20. Kållberg, P. (1997) Aspects of the re-analysed climate, ECMWF Re-Analysis Project Report Series No. 2.
21. Trenberth, K.E. and Guillemot, C.J. (1995) Evaluation of the global atmospheric moisture budget as seen from analyses, *J. Climate* **8**, 2255-2272.
22. Dodd, J.P. and James, I.N. (1996) Diagnosing the global hydrological cycle from routine atmospheric analyses, *Quart. J. Roy. Meteor. Soc.* **122**, 1475-1499.
23. Rasmusson, E.M. and Mo, K.C. (1996) Large-scale atmospheric moisture cycling as evaluated from NMC global analysis and forecast products, *J. Climate* **12**, 3276-3297.
24. Masuda, K. (1990) Atmospheric heat and water budgets of polar regions: Analysis of FGGE data, *Proc. NIPR Symp. Polar Meteorol. Glaciol.* **3**, 79-88.
25. Calanca P. and Ohmura, A. (1994) Atmospheric moisture flux convergence and accumulation on the Greenland Ice Sheet, in Jones, H.G., Davies, T.D., Ohmura, A., and Morris, E.M. (eds.), *Snow and Ice Covers. Interactions with the Atmosphere and Ecosystems*, IAHS Publ. No. 223, pp. 77-84.
26. Bromwich, D.H., Cullather, R.I., Chen, Q.-s., and Csathó, B.M. (1998) Evaluation of recent precipitation studies for Greenland Ice Sheet, *J. Geophys. Res.*, **103**, 26007-26024.
27. Stendel, M. and Arpe, K. (1997) Evaluation of the hydrologic cycle in reanalyses and observations. ECMWF Re-Analysis Project Report Series No. 6.
28. Serreze, M.C. and Hurst, C.M. (2000) Representation of mean Arctic precipitation from NCEP/NCAR and ERA reanalyses, *J. Climate.*, **13**, 182-201.
29. Serreze, M.C. and Maslanik, J.A. (1997) Arctic precipitation as represented in the NCEP/NCAR reanalysis, *Ann. Glaciol.* **25**, 429-433.
30. Gorshkov, S.G. (ed.) (1983) *World Ocean Atlas, Vol. 3: Arctic Ocean*, Pergamon Press, New York.
31. Khrol, V.P. (1996) *Atlas of the Water Balance of the Northern Polar Area*, Gidrometeoisdat, St. Petersburg.
32. Ohmura, A. and Reeh, N. (1991) New precipitation and accumulation maps for Greenland *J. Glaciol.* **37**, 140-148.
33. Csathó, B.M., Xu, H., Thomas, R., Bromwich, D., and Chen, Q.-s. (1997) Comparison of accumulation and precipitation maps of the Greenland Ice Sheet, *Eos. Trans. Amer. Geophys. Union* **78**, F9.
34. Bolzan, J.F. and Strobel, M. (1994) Accumulation-rate variations around Summit, Greenland, *J. Glaciol.* **40**, 56-66.
35. Colony, R., Radionov, V., and Tanis, F.J. (1998) Measurements of precipitation and snow pack at Russian north pole drifting stations, *Polar Record* **34**, 3-14.
36. Radionov, V.F., Bryazgin, N.N., and Alexandrov, E.I. (1997) The snow cover of the Arctic basin, Tech. Rpt. APL-UW TR 9701, Applied Physics Laboratory, Univ. Washington.
37. Woo, M.-k., Heron, R., Marsh, P., and Steer, P. (1983) Comparison of weather station snowfall with winter snow accumulation in high Arctic basins, *Atmos.-Ocean* **32**, 733-755.
38. Thorndike, A. S. and Colony, R. (1980) Arctic Ocean Buoy Program, Data Report, 19 January 1979 to 31 December 1979, Polar Science Center, University of Washington.
39. Khrol, V.P. (1996) Energy Balance of the Northern Polar Sea, Gidrometeoisdat, St. Petersburg.
40. Randel, D.J., Vonder Haar, T.H., Ringerud, M.A.,Stephens, G.L., Greenwald, T.J., and Combs, C.L. (1996) A new global water vapor dataset, *Bull. Amer. Meteor. Soc.* **77**, 1233-1246.
41. Ross, R.J. and Elliott, W.P. (1997) Tropospheric water vapor climatology and trends over North America: 1973-93, *J. Climate* **9**, 3561-3574.
42. Francis, J.A. (1994) Improvements to TOVS retrievals over sea ice and applications to estimating Arctic energy fluxes, *J. Geophys. Res.* **99**, 10,395-10,408.
43. Sheu, R.-S., and Curry, J.A. (1992) Interactions between North Atlantic clouds and the large-scale environment, *Mon. Wea. Rev.* **120**, 261-278.
44. Overland, J.E., Miletta Adams, J., and Bond, N.A. (1997) Regional variation of winter temperatures in the Arctic, *J. Climate* **10**, 821-837.
45. Peixoto, J.P. and Oort, A.H. (1992) *Physics of Climate*, Amer. Instit. of Phys., New York.
46. Peng, S. and Mysak, L.A. (1993) A teleconnection study of interannual sea surface temperature fluctuations in the North Atlantic and precipitation and runoff over western Siberia, *J. Climate* **6**, 876-885.
47. Sellers, W.D. (1965) *Physical Climatology*, University of Chicago Press, Chicago.
48. Baumgartner, A. and Reichel, E. (1975) *The World Water Balance*, Elsevier, Amsterdam.

49. Korzoun, V.I., Sokolov, A.A., Voskresensky, K.P., Kalinin, G.P., Konoplyantsev, A.A., Korotkevich, E.S., and Lvovich, M.I. (1977) *World Water Balance and Water Resources of the Earth*, UNESCO Press, Leningrad.

50. Korzoun, V.I., Sokolov, A.A., Voskresensky, K.P., Kalinin, G.P., Konoplyantsev, A.A., Korotkevich, E.S., and Lvovich, M.I. (1977) *Atlas of World Water Balance*, UNESCO Press, Leningrad.

51. Peixoto, J.P. and Oort, A.H. (1983) The atmospheric branch of the hydological cycle and climate, in Street-Perrott, A., Beran, M., and Ratcliffe, R. (eds.), *Variations in the Global Water Budget*, D. Reidel Publ., Dordrecht, pp. 5-65.

52. Overland, J.E. and Turet, P. (1994) Variability of the atmospheric energy flux across 70°N computed from the GFDL data set, in Johannessen, O.M., Muench, R.D., and Overland, J.E. (eds.), *The Polar Oceans and Their Role in Shaping the Global Environment: The Nansen Centennial Volume*, Geophys. Mono., No. 85, Amer. Geophys. Union, Washington, pp. 313-325.

53. Genthon, C. (1998) Energy and moisture flux across 70°N and S from ECMWF Re-Analyses, in *First Intl. Conf. on Reanalyses, Silver Spring, Maryland*, WCRP Rep. No. 104 (WMO/TD-No. 876), Geneva, pp. 371-374.

54. Briegleb, B.P. and Bromwich, D.H. (1998) Polar climate simulation of the NCAR CCM3, *J. Climate* **11**, 1270-1286.

55. Serreze, M.C., and Barry, R.G. (2000) Atmospheric components of the Arctic Ocean hydrologic budget from rawinsonde data, in this volume.

56. Rasmusson, E.M. (1968) Atmospheric water vapor transport and the water balance of North America. II. Large-scale water balance investigations, *Mon. Wea. Rev.* **96**, 720-734.

57. Gibson, J.K. (1998) ECMWF Re-Analysis - Future plans, in *First Intl. Conf. on Reanalyses, Silver Spring, Maryland*, WCRP Rep. No. 104 (WMO/TD-No. 876), Geneva, pp. 406-409.

58. Rothrock, D. and Kwok, R. (2000) Satellite observations bearing on the Arctic Ocean Freshwater balance, in this volume.

MOISTURE TRANSPORT TO ARCTIC DRAINAGE BASINS RELATING TO SIGNIFICANT PRECIPITATION EVENTS AND CYCLOGENESIS

JOHN R. GYAKUM
Department of Atmospheric and Oceanic Sciences
McGill University
805 Sherbrooke Street West
Montreal, QC   H3A 2K6
Canada

1.   Introduction

Moisture transport to high-latitude regions has received substantial recent attention in the refereed literature.   Newell and Zhu [1] have used the term 'tropospheric rivers' to describe the vertically-integrated horizontal moisture transport fields. Such structures can extend from the subtropical latitudes to Arctic regions.   As an example, Newell and Zhu show a tropospheric river that extends from the Caribbean Sea to central Greenland during an especially active cyclogenetic period during the Canadian Atlantic Storms Program II (CASP II; [2]) in the Atlantic provinces [3]. The expression for the vertically-integrated horizontal moisture flux, $\mathbf{Q}$, is:

$$\mathbf{Q} = \frac{1}{g} \int_{0}^{p_0} (q\mathbf{v})dp \tag{1}$$

where g is gravity, p_0 is the surface pressure, q is the specific humidity, and $\mathbf{v}$ is the horizontal wind velocity. This river, seen in Fig. 1a, originates in the subtropical regions north of the Greater Antilles Islands and extends poleward into the Greenland region. The companion figure 1b, showing sea-level pressure for the same time, illustrates the relationship of this river to the lower-tropospheric balanced circulation.   The largest magnitude of $\mathbf{Q}$ appears concentrated in the southwesterly geostrophic flow between the oceanic subtropical surface anticyclone and an intense surface cyclone in the Canadian Atlantic provinces.   The asymmetry in the magnitudes of $\mathbf{Q}$ is related to the larger values of specific humidity associated with the poleward flow, as opposed to the much smaller values of water vapor associated with the equatorward flow westward of the Atlantic Canada cyclone.   The suggestion that surface evaporation occurring in the subtropics is related to arctic precipitation events provides motivation for examining global structures of moisture transports.

E.L. Lewis et al. (eds.), The Freshwater Budget of the Arctic Ocean, 197-208.
© 2000 *Kluwer Academic Publishers.*

2. Earlier Research

Peixoto and Crisi [4] discuss the Northern Hemisphere water vapor flux field for the intensively-studied International Geophysical Year 1958. Figure 2, adapted from their study, shows the year's vertically-integrated mean total meridional flux. Clearly, the most intense meridional transport occurs in the oceanic regions, with the primary northward fluxes across 25° N to be the southwestern Pacific and Atlantic Oceans and the region extending from India into Southeast Asia. The principal regions of poleward transport into the extratropical latitudes include the central and eastern zones of the Atlantic and Pacific basins. Rasmussen [5,6] provided additional detail to the growing library of water vapor flux data for the North American continent.

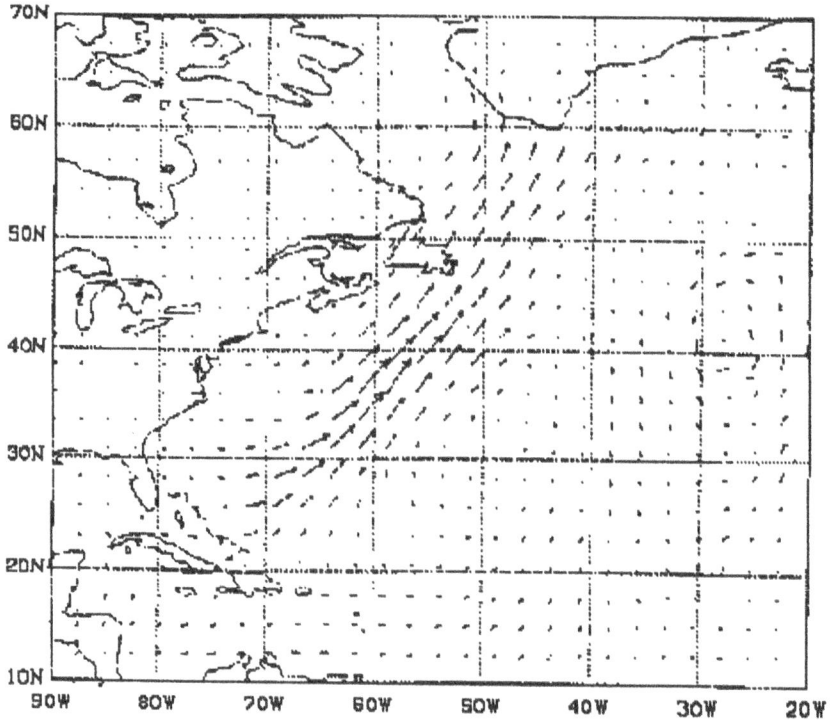

Figure 1a. Water vapor flux for 0000 UTC, 25 January 1992, with the longest arrow approximately corresponding to 2000 kg m⁻¹ sec⁻¹
(after Newell and Zhu [1]).

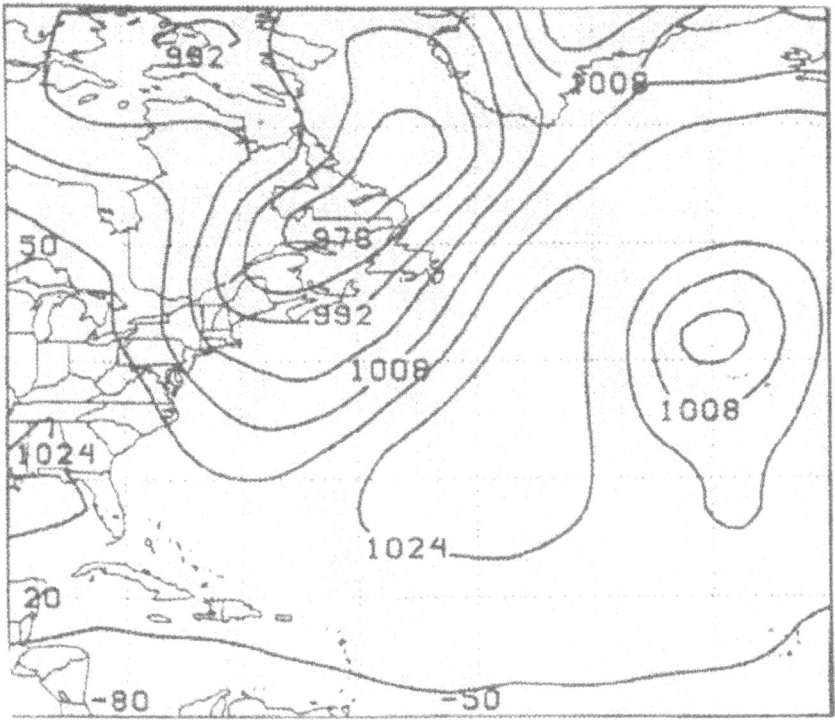

Figure 1b. Sea-level pressure field for 0000 UTC, 25 January 1992
(solid, contour interval of 8 hPa)

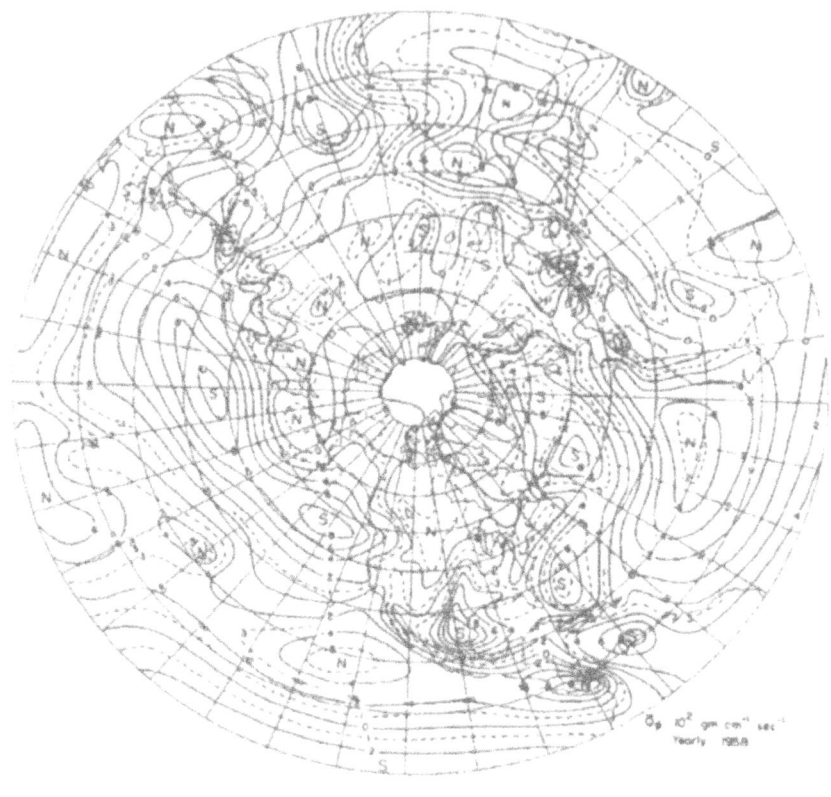

Figure 2. Vertically-integrated mean total meridional water vapor flux for the IGY, 1958. Units shown are 10^2 gm. cm.$^{-1}$ sec.$^{-1}$, with extrema indicated by 'N' for northerly or southward flux, and by 'S' for southerly or northward flux (after Peixoto and Crisi [4]).

Rasmussen [5] confirmed these basic structures for the North American continent in his study of the water vapor fluxes for the period of May 1961 through April 1963. He found a particularly important zone of poleward transport to be along the Pacific coast between 40 and 55° N, and related these to the northeastern extremity of the Pacific cyclone belt.

Partitioning the moisture transports into time mean and eddy components has been shown by Starr and Peixoto [7], and Peixoto and Oort [8], among others, to be an effective means of understanding these transports. The expression for the time-mean vapor transport is:

$$[qv] = [q][v] + [q^*v^*] + [q'v']$$ (2)

where notation [ ] indicates the zonal mean and * is the perturbation from that mean. Similarly, the overbar signifies the time mean and the ' symbol is its perturbation. Therefore, [q][v] represents the transport of water vapor by the mean meridional circulation that dominates in the tropics, $[q^*v^*]$ is that of the mean stationary eddies of the general circulation. Examples of this term include the semipermanent subtropical anticyclones and the semipermanent Aleutian and Icelandic Lows. The final term on the right-hand side of (2), [q'v'], is the meridional transport of water vapor by the transient perturbations that develop along the high-latitude baroclinic zones, and along the intertropical convergence zone.

Therefore, a vertical integration of (2) leads to an elucidation of the total meridional transport as a summation of the three partitioned quantities:

$$[Q_\phi] = \frac{1}{g} \int_0^{p_0} [q][v]dp + \frac{1}{g} \int_0^{p_0} [q^*v^*]dp + \frac{1}{g} \int_0^{p_0} [q'v']dp.$$ (3)

Figure 3, taken from [8], shows the total meridional vapor transport and the contribution by the transient eddies, as represented by the final term on the right-hand side in (3) above. Noteworthy is the maximum of poleward transport that is focused in the northeastern Pacific basin, eastward of the Aleutian Low region. These transports are contributed substantively by the transient eddies, or synoptic-scale cyclones in this region (Fig. 3b).

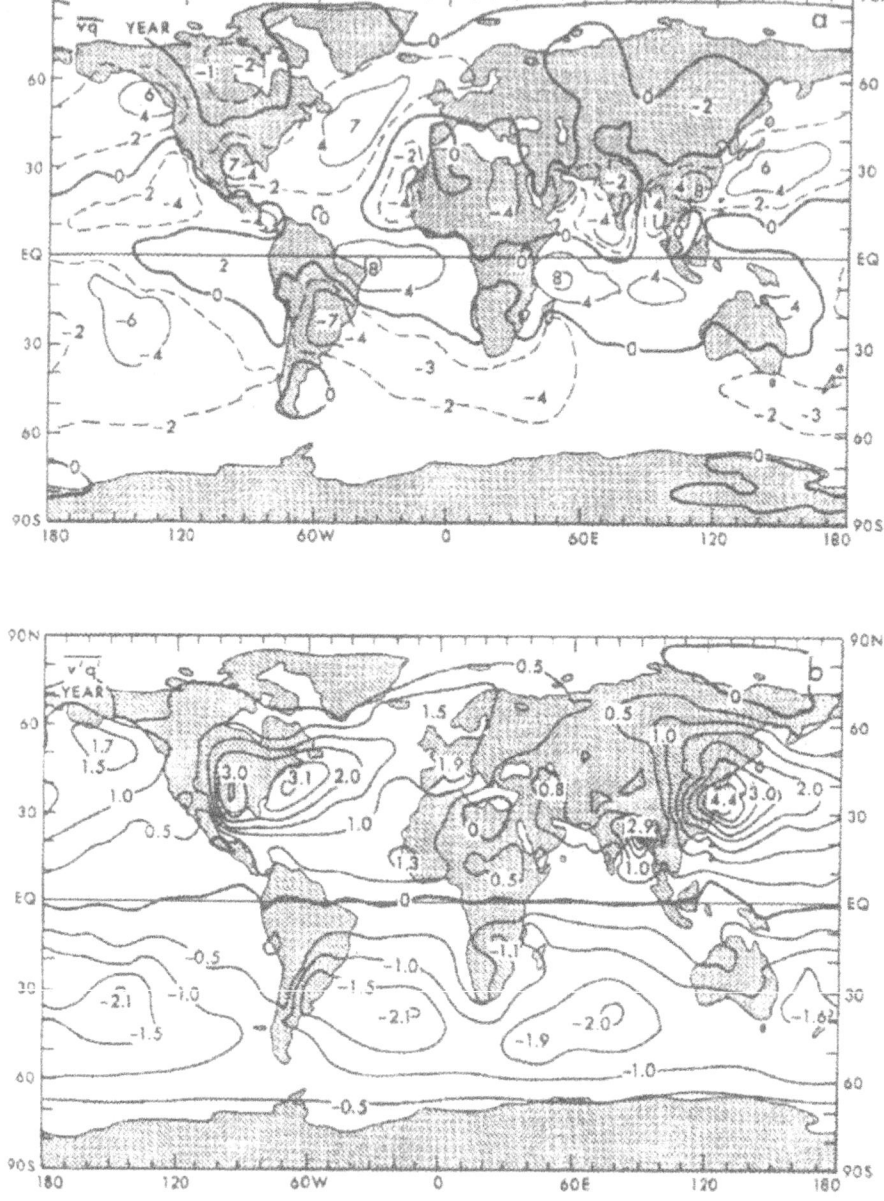

Figure 3. Global distributions of the vertically-averaged meridional transport of water vapor by (a) all motions, and (b) the transient eddies in m s^{-1} g kg^{-1} for annual mean conditions (after [8]).

3.   Relationship to high-latitude precipitation events and to cyclogenesis

The relevance of moisture transport may be related to the hydrological cycle in which, for sufficiently long time scales, the horizontal divergence of the vertically-integrated transport of water vapor (Eq. 1) may be equated to the difference between evaporation and precipitation. This application has been used by Rasmussen [6], among others, for North America, and more recently has been applied by Walsh et al. [9] to the Arctic Ocean and to the Mackenzie River Basin of the Canadian Northwest Territiories. The means by which these authors computed the transports is the so-called 'aerological method' in which rawinsonde data of wind and water vapor are used.

More recently, the availability of long-term gridded datasets, including that of the U. S. National Meteorological Center (NMC), now the National Centers for Environmental Prediction (NCEP), have allowed for the computation of mean, or composite fields for similarly-structured meteorological events. One such study by Lackmann and Gyakum [10] identified a favorable large-scale circulation structure for cold-season (December, January and February) precipitating weather systems in the Mackenzie River Basin (MRB). The 59 cases were chosen with a criterion of at least 2.5 mm of precipitation having fallen in a 24-h period at five or more of the 12 southern and central MRB reporting stations. A further criterion is that a surface low with a central pressure of 996 hPa or lower must exist in the Gulf of Alaska. The technique allows us to identify both the synoptic-scale and planetary-scale signatures associated with these MRB wintertime precipitation events. Figure 4, based upon a 56-case subset from 1963 through 1989, shows the favorable structure evidenced by the presence of large, slowly-varying height anomalies covering much of the Northern Hemisphere. The persistent negative height anomaly in the Gulf of Alaska, coupled with an equivalently-strong positive height anomaly over much of western North America, provides for an extended period of anomalously-strong southwesterly geostrophic flow from the North Pacific Ocean into the MRB.

This anomalously-strong flow that often extends from the Hawaiian Islands northeastwards to the western Washington and British Columbia coasts has been called the 'pineapple express' by weather forecasters. This term is used because of the rapid transport of water vapor from the subtropical latitudes that effects heavy precipitation in these coastal regions [11]. Evidently, an especially persistent 'pineapple express' is responsible for significant precipitation events in the Mackenzie River Basin as well.

204

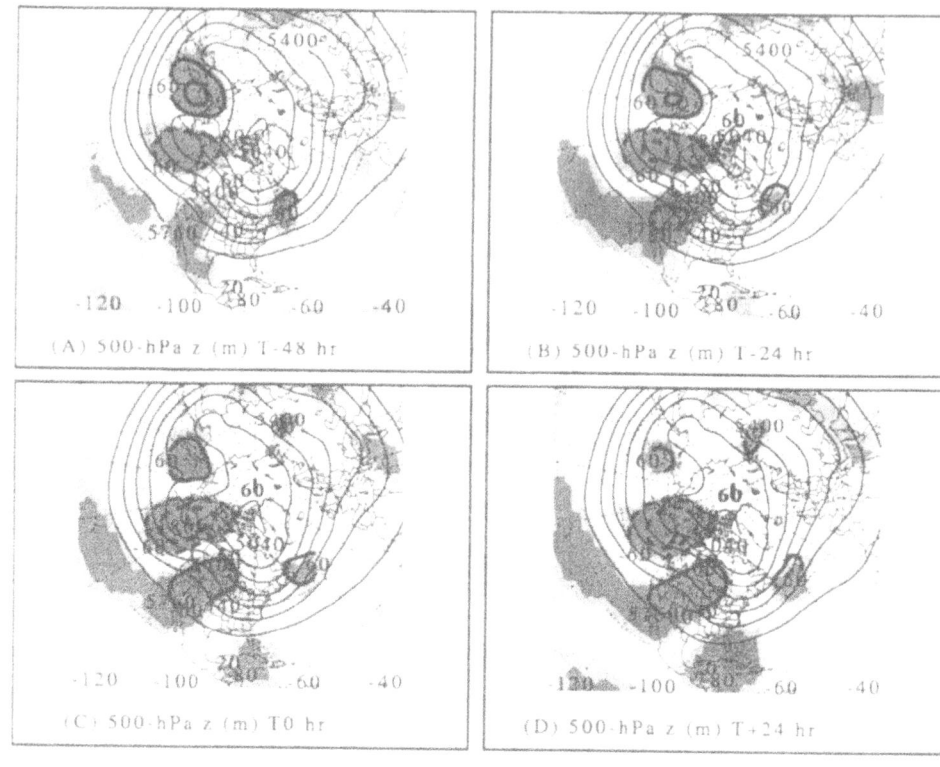

Figure 4. Composite winter MRB significant precipitation 500-hPa geopotential height (thin solid, interval of 120 m) and anomaly (thick, contour interval of 60 m, with positive and negative values respectively solid and dashed and the zero contour omitted) and the statistical significance determined from a two-sided Student's t test (shading corresponding to the 95 and 99% confidence limits shown in the left of each panel). Reference times are shown in each panel (e. g., T-48 corresponds to 48 h prior to the onset of the precipitation event).

A more recent modeling study by Lackmann et al. [12] demonstrates that the climatological-mean southwesterly flow plus the anomalously-strong southwesterly-flow component are responsible for the water vapor transports into the MRB. This technique of quantifying the contributions from the mean and eddy components of the flow is that of potential vorticity inversion [13]. The large-scale persistent southwesterly flow from the Pacific Ocean into the MRB may also be responsible for inducing secondary cyclonic, precipitation-producing disturbances. The authors document an induced lee-side cyclogenesis event that provides its own dynamics in terms of frontogenesis and vertical motions that enhance the MRB precipitation.

Very recently, reanalysis efforts at NCEP [14] have produced 6-hourly global gridded fields of wind and water vapor for the period from 1957 through the present. Similar reanalysis efforts are occurring in the European Centre for Medium-Range Forecasts (ECMWF) and the National Aeronautics and Space Administration (NASA). Such efforts have provided for the possibility of analyzing the long-term interannual variability of water vapor transport for both regional and global studies. However, as pointed out by Trenberth and Guillemot [15], global analyses of water vapor are dependent on the character of the specific model four-dimensional data assimilation (4DDA) and the specifics of the model's moist physical parameterizations. Satellite retrievals of water vapor are available to the global analyses from the TIROS Operational Vertical Sounder (TOVS), but with only limited success [16].

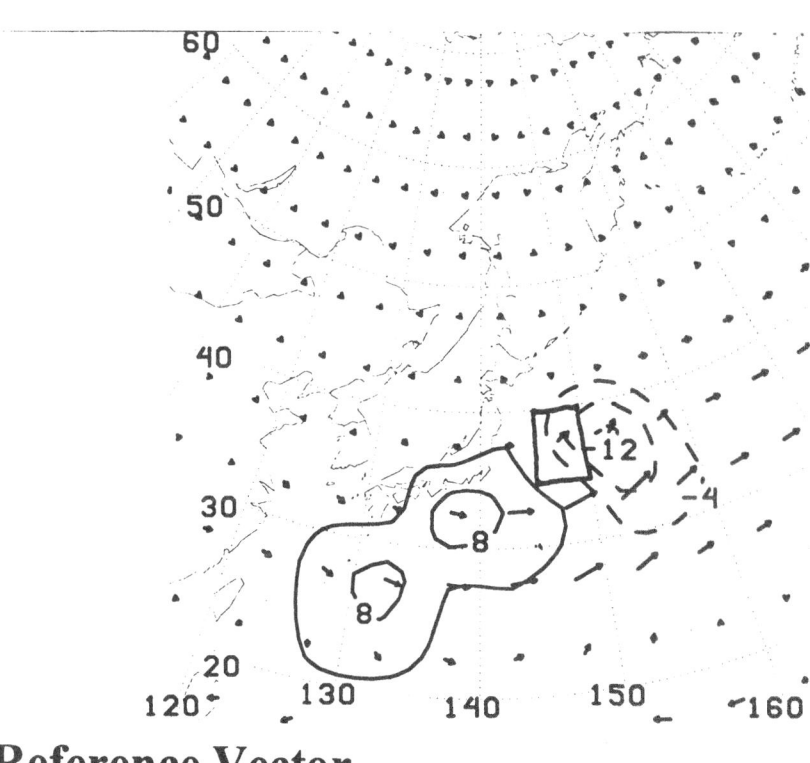

Reference Vector ⟶

Figure 5. Vertical integral of the water vapor transport (reference vector shows 700 kg m^{-1} sec^{-1}), and the horizontal divergence of this quantity (contour interval of 4 mm (12 h)$^{-1}$ with dashed/solid showing negative/positive) for a sample of 17 rapid cyclogenesis cases discussed in [17]. The boxed region shows the area in which the 17 individual cyclones occurred at the onset of their most rapid intensification.

A recent study by Gyakum and Danielson [17] that uses the NCEP reanalysis data, shows the strong subtropical connection of water vapor tranports to rapid surface cyclogenesis (Fig. 5). Since surface cyclogenesis relates very well to latent heat release, it is not surprising that the regions of strongest convergence of the vertically-integrated water vapor tranports are also the regions of cyclogenesis.

Further work concerning the Ice Storm of January 1998, in which more than 75 mm of freezing rain fell across southeastern Canada and New England, reveals the planetary-scale character of the water vapor transport that extends in an anticyclonic gyre from the subtropical regions of the North Atlantic Ocean into the Gulf of Mexico and northward into the northeastern United States and southeastern Canada (Fig. 6). Such cases illustrate the atmospheric 'river' effect discussed earlier in this chapter in reference to Newell and Zhu's work [1].

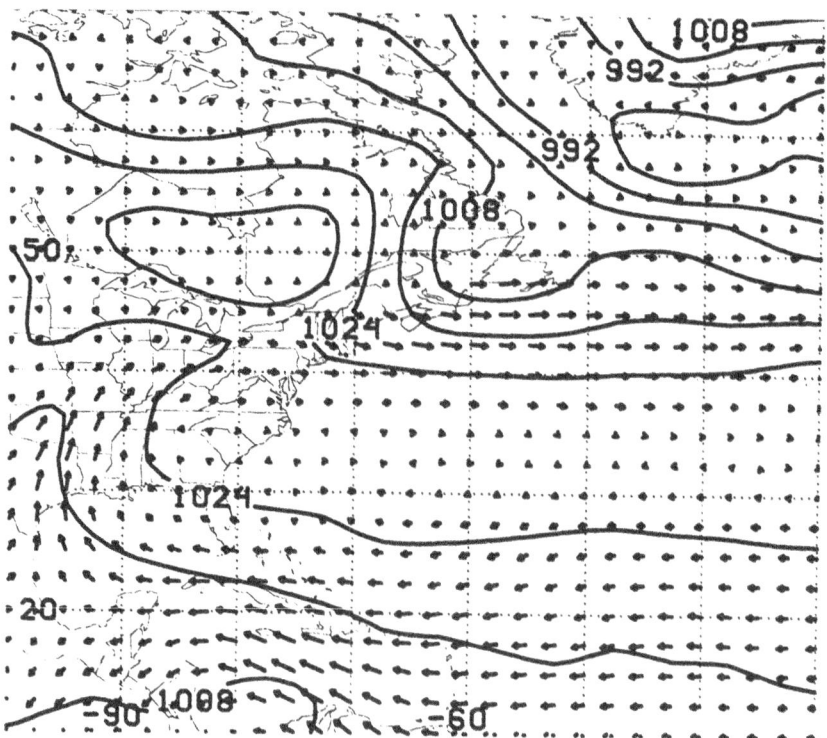

Figure 6. Water vapor tranport vectors (maximum vector length of 950 kg m^{-1} sec^{-1}) with sea-level pressure (solid contours, with interval of 8 hPa) for 1200 UTC, 4 January 1998.

4. Conclusions

The relationship of anomalously-strong transports of water vapor from the subtropics to cyclogenesis and significant precipitation in extratropical and arctic latitudes has been demonstrated. The transports are due to a combination of planetry-scale and synoptic-scale (individual storm scale) flows. The former may be interpreted as a combination of Peixoto and Ort's [8] mean meridional flow and the contribution from the mean stationary eddies, and the latter term is a representation of the transient eddy contribution.

Continuing work concerning the devastating Ice Storm of January 1998 reveals the planetary-scale character of the water vapor □ransports (Figure 6) that extends from the subtropical regions of the North Atlantic Ocean into the Gulf of Mexico and northward into the northeastern United States and southeastern Canada. Cases such as this illustrate the river effect discussed earlier in this chapter in reference to Newell and Zhu's work.

5. Future Research Priorities

Given the relationship of subtropical moisture transports (the Pineapple Express) to significant precipitation events in arctic regions, an important priority in future research should be an assessment of impacts of remote evaporation processes on downstream regions. For example, what impact does a perturbation in the planetary-boundary layer flux of moisture in the Kuroshio Current east of Japan have on precipitation forecasts in downstream regions such as the Mackenzie River Basin?

Additionally, the identification of water vapor source regions for extreme precipitation events and cyclogenesis should be a top research priority. Once these identifications have been made, a focus on processes and better initialization in theses regions is required.

Because of the relatively recent availability of long-term gridded datasets, an improved perspective on precipitation production and cyclogenesis will be provided by studies that focus on water vapor transports.

Large databases that include the gridded reanalyses offer the hope that observational distinctions may be made between ordinary cyclogenesis and prcipitation events and those that are far more extreme and dangerous. The generality of the anomalous moisture transport structure seen in Figure 6 for the Great Ice Storm of January 1998, for example, needs to be established.

6. References

1. Newell, R. E. and Y. Zhu, 1994: Tropospheric rivers: A one-year record and a possible application to ice core data. *Geophys. Res. Lett.*, **21**, 113-116.
2. Stewart, R. E., 1991: Canadian Atlantic Storms Program: Progress and plans for the meteorological component. *Bull. Am. Meteorol. Soc.*, **72**, 364-371.
3. Gyakum, J. R., D.-L. Zhang, J. Witte, K. Thomas and Werner Wintels, 1996: CASP II and the Canadian cyclones during the 1989-92 cold seasons. *Atmos.-Ocean*, **34**, 1-16.
4. Peixoto, J. P. and A. R. Crisi, 1965: Hemispheric humidity conditions during the IGY. *Scintific Report No. 6*, Planetary Circulations Project, Massachusetts Institute of Technology, 166 pp.
5. Rasmussen, E. M, 1968a: Atmospheric water vapor transport and the water balance of North America. I. Characteristics of the water vapor flux field. *Mon. Wea. Rev.*, **96**, 403-426.
6. Rasmussen, E. M., 1968b: Atmospheric water vapor transport and the water balance of North America. II. Large-scale water balance investigations. *Mon. Wea. Rev.*, **96**, 720-734.
7. Starr, V. P., and J. P. Peixoto, 1971: Pole-to-pole eddy transport of water vapor in the atmosphere during the IGY. *Arch. Met. Geophys. Biokl.* **A20**, 85-114.
8. Peixoto, J. P. and A. H. Oort, 1992: *Physics of Climate*, American Institute of Physics, New York, 520 pp.

9. Walsh, J. E., X. Zhou, and D. Portis, 1994: Atmospheric contribution to hydrologic variations in the Arctic. *Atmos.-Ocean*, **21**, 312-322.

10. Lackmann, G. M. and J. R. Gyakum, 1996: The synoptic and planetary-scale signatures of precipitating systems over the Mackenzie River Basin. *Atmos.-Ocean*, **34**, 647-674.

11. Lackmann, G. M. and J. R. Gyakum, 1999: Heavy cold-season precipitation in the northwestern United States: Synoptic climatology and an analysis of the flood of 17-18 January 1986. *Wea. Forecasting*, **14**, 687-700.

12. Lackmann, G. M., J. R. Gyakum and R. Benoit, 1998: Moisture transport diagnosis of a wintertime precipitation event in the Mackenzie River Basin. *Mon. Wea. Rev.*, **126**, 668-691.

13. Davis, C. A. and K. A. Emanuel, 1991: Potential vorticity diagnostics of cyclogenesis. *Mon. Wea. Rev.*, **119**, 1929-1953.

14. Kalnay, E., and co-authors, 1996: The NCEP/NCAR 40-year reanalysis project. *Bull. Amer. Meteor. Soc.*, **77**, 437-471.

15. Trenberth, K. E., and C. J. Guillemot, 1995: Evaluation of the global atmospheric moisture budget as seen from the analyses. *J. Climate*, **8**, 2255-2272.

16. Wittmeyer, I. L., and T. H. Vonder Haar, 1994: Analysis of the global ISCCP TOVS water vapor climatology. *J. Climate*, **7**, 325-333.

17. Gyakum, J. R. and R. E. Danielson, 2000: Analysis of meteorological precursors to ordinary and explosive cyclogenesis in the western North Pacific. *Mon. Wea. Rev.*, in press.

ATMOSPHERIC CLIMATE MODELS:
SIMULATION OF THE ARCTIC OCEAN FRESH WATER BUDGET COMPONENTS

V.M. KATTSOV
Voeikov Main Geophysical Observatory
7 Karbyshev str., St. Petersburg, 194021
Russia

J.E. WALSH
Department of Atmospheric Sciences
University of Illinois at Urbana-Champaign
105 South Gregory Avenue, Urbana, Illinois, 61801
U.S.A.

A. RINKE and K. DETHLOFF
Alfred Wegener Institute for Polar and Marine Research
Telegraphenberg A43, Potsdam, 14401
Germany

1.   Introduction

Atmospheric climate models driven by prescribed sea surface temperature (SST) and sea-ice simulate the two major components of the ocean fresh water budget (FWB): the direct input of fresh water as precipitation less evaporation (P-E) over the ocean surface; and the inflow of river water at the lateral boundaries of the ocean. The latter is ultimately driven by the precipitation and evaporation over the ocean's terrestrial watersheds.   A comprehensive hydrological component-model is required to provide proper migration and retention of the moisture within the land territory in order to simulate a realistic seasonality (and interannual variability) of fresh water inflow into the ocean.

The aim of this study is to draw some general conclusions on capabilities and deficiencies of current atmospheric climate models in their simulation of the primary Arctic Ocean (AO) FWB components -- precipitation

E.L. Lewis et al. (eds.), The Freshwater Budget of the Arctic Ocean, 209-247.
© 2000 *Kluwer Academic Publishers.*

and evaporation -- by intercomparing different model outputs and comparing those against available observation and reanalyses data. We draw upon an array of different types of climate models in order to illustrate the state of the art with regard to simulations of the Arctic FWB.

The rest of this paper is organized as follows. Section 2 summarizes the sources of observational and model data used in the intercomparisons. Section 3 contains selected results from the recent study by Walsh *et al.* [36] in which observational estimates of precipitation and evaporation over the AO and its terrestrial watersheds are compared with corresponding values from a number of global climate model ten-year simulations of the Atmospheric Model Intercomparison Project (AMIP), initiated in 1989 under the auspices of the World Climate Research Programme [13]. Section 4 is devoted to validation of the Arctic hydrological variables simulated by two successive versions of the ECHAM atmospheric general circulation model (GCM) developed at Max-Planck-Institute for Meteorology (MPI), Hamburg, Germany. The difference in physical formulation of the two model versions provides a measure of the sensitivity of the simulated AO FWB atmospheric components to the parameterizations in the model, although only to the aggregates of the parameterizations used in these two versions of the model. The potential use of GCM hydrologic output in simulating the river water discharge into the AO is illustrated by an application of the new global hydrological discharge (HD) model (also developed at MPI) driven by the ECHAM atmospheric model output. Additionally, because the ECHAM model simulation described here spans several decades with prescribed observed sea surface temperature (SST) and atmospheric carbon dioxide concentration, the model results make it possible to assess the interdecadal variation of precipitation and P-E over northern high latitudes. Section 5 deals with early results of HIRHAM high-resolution regional climate model (RCM) of the Arctic atmosphere, developed at Alfred Wegener Institute for Polar and Marine Research (AWI), Potsdam, Germany. The results of the study along with future research priorities are summarized in Section 6.

2. Observational data and model output

2.1 SOURCES OF OBSERVATIONAL DATA

While we do not present here the details of the observational datasets used in this study, we will summarize the primary datasets with an eye toward the limitations they impose on model-data comparisons. By and large, high-latitude observed precipitation climatologies are faced with the following problems: sparseness of surface stations; insufficient representativeness of local measurements; and systematic errors of *in situ* measurements, especially in the case of solid precipitation under windy conditions. With these limitations in mind, we briefly comment on the precipitation data sources.

The primary sources of precipitation data used in this study are the global climatology of Legates and Willmott [25] and the Arctic regional climatology of Bryazgin [5]. Both datasets are gauge-corrected and gridded. For use in this study, Legates-Willmott's original dataset resolution (0.5^{o}x0.5^{o} latitude-longitude grid) has been coarsened to T42 (which is approximately 300 km in the latitude, or 2.8^{o}x2.8^{o}). Bryazgin's climatology has been recently digitized over the polar cap ($70-90^{o}$N) onto a 2.5^{o}x2.5^{o} latitude-longitude grid. It is supplemented by planimetrically-obtained annual means of area-averaged precipitation over individual terrestrial watersheds of the AO of Bryazgin and Shver [6].

A secondary source of precipitation data is the Global Precipitation Climatology Project (GPCP), which is administered by the Global Energy and Water Cycle Experiment (GEWEX). GPCP has produced a monthly mean 2.5 degree gridded precipitation data set for the period July 1987 through December 1994 [18]. This dataset is a result of blending gauge and infrared and microwave satellite estimates of precipitation. A large fraction of the AO is not covered by GPCP data, so the dataset can be used only for the terrestrial watersheds of the AO.

The Global Historical Climate Network (GHCN) monthly dataset [12] with a resolution 5°x5^{o} covers the period from 1901 through 1997 and thus offers the potential for an assessment of interdecadal variability of precipitation.

Additional data on precipitation are the monthly fields of Jaeger [20]; the area-averaged estimates of Vowinckel and Orvig [34] for the AO, including and excluding the Barents and Norwegian Seas; and the World Ocean zonal means from the World Water Balance and Water Resources of the Earth [24], hereafter referred to as WWB. The two latter sources also provide estimates of evaporation for the AO. Additionally, annual means of area-averaged evaporation over the terrestrial watersheds of the AO are obtained from Zubenok [37], whose estimates are based on a parameterized soil moisture that varies temporally in response to observationally specified precipitation [6], radiative fluxes, air temperature, and humidity.

River discharge data for several Arctic river basins are obtained from WWB and Duemenil *et al.* [10]. It should be noted that annual discharge data for major rivers generally have uncertainties of about 10%, owing mainly to errors in the conversion of gauge measurements to volumetric flow rates.

2.2   MODEL-DERIVED CLIMATOLOGIES

An alternative quantification of precipitation in the Arctic region is provided by numerical models of the atmosphere. The models can be run with or without the assimilation of observed atmospheric data in multiyear simulations, in the former case known as reanalyses.

An advantage of the numerical modelling approach is availability of moisture budget components on regular grids from model simulations, thus permitting the closure of the moisture budget over arbitrary temporal and spatial domains. However, simulation of high-latitude atmospheric moisture using global atmospheric models presents certain problems [30]. Finite-difference GCMs require undesirable filtering operations in order to avoid computational instability when a reasonable time-step is used in regions of converging meridians, i.e., the polar regions. While implementing spectral representation of atmospheric fields helps to avoid polar filtering, it in turn results in producing fictitious negative moisture amounts in the dry high-latitude atmosphere, thus calling for correction procedures. Both problems are apparently overcome by application of semi-Lagrangian schemes for moisture advection, which are becoming more widely used in atmospheric GCMs. Meanwhile, in the semi-Lagrangian schemes, the advantages of large time steps and the absence of spurious negative moisture values are partially offset by the lack of exact conservation of the quantity.

2.2.1   *Reanalyses*
The atmospheric reanalyses employed in this study consist of 13 years (1982-94) of output from NCEP/NCAR [21] and 15 years (1979-93) of output from ECMWF [15], hereafter denoted as NRA and ERA, respectively. The resolutions of the models used in the reanalyses are T62 (approximately 210 km) for NRA and T106 (approximately 120 km) for ERA. The resolution of the archived output from both models is 2.5° in latitude and longitude. A detailed discussion of reanalysis-derived depictions of the Arctic moisture budget can be found elsewhere in this volume (e.g., [3]). While several years (1979-1981) of the ERA are not included in the NRA output used here (as well as the NRA 1994 is not included in the ERA output employed), there is no indication of trends during the past two decades that would make the climatologies of the reanalyses fields sensitive to the inclusion or exclusion of a few additional years.

2.2.2 *Climate model simulations*

The 23 AMIP simulations considered here cover one and the same ten-year period (1979-1988) with prescribed observed SST and sea-ice extent, both of which vary temporally. The AMIP GCMs determine their own surface temperatures and snow cover over land areas, where each model has its own specification of surface properties such as albedo, roughness, vegetation, and emissivity. The formulations of physical processes (including cloud formation, precipitation, and evaporation) vary from model to model, as do the horizontal and vertical resolution as well as the choice of spectral or gridpoint representations (Table 1). The details on individual AMIP models are provided in [28]. It should be noted that this study uses the AMIP output that was available in early 1995. Since that time several participating modelling groups have repeated the AMIP simulations with revised versions of their original AMIP models in order to reduce specific systematic errors. Improvement has been achieved in some models and in some variables; however, some models' results have deteriorated [14].

TABLE 1. AMIP simulations [28]

Model	Country	representation/resolution	coordinates/levels
BMRC	Australia	spectral, R31	sigma/9
CCC	Canada	spectral, T32	hybrid/10
CNRM	France	spectral, T42	hybrid/30
COLA	USA	spectral, R40	sigma/18
CSIRO	Australia	spectral, R21	sigma/9
CSU	USA	finite diff., 4 x 5	modified sigma/17
DERF/GFDL	USA	spectral, T42	sigma/18
DNM	Russia	finite diff., 4 x 5	sigma/7
ECMWF	Europe	spectral, T42	hybrid/19
GLA	USA	finite diff., 4 x 5	sigma/17
GSFC	USA	finite diff., 4 x 5	sigma/20
JMA	Japan	spectral, T42	hybrid/21
LMD	France	finite diff., 50slat x 64lon	sigma/11
MGO	Russia	spectral, T30	sigma/14
MPI	Germany	spectral, T42	hybrid/19
MRI	Japan	finite diff., 4 x 5	hybrid/15
NCAR	USA	spectral, T42	hybrid/18
NMC	USA	spectral, T40	sigma/18
NRL	USA	spectral, T47	hybrid/18
UCLA	USA	finite diff., 4 x 5	modified sigma/15
UIUC	USA	finite diff., 4 x 5	sigma/7
UGAMP	UK	spectral, T42	hybrid/19
UKMO	UK	finite diff., 2.5 x 3.75	hybrid/19

Another source of atmospheric climate model data is the output from extended runs with the two versions of ECHAM GCM. The ECHAM-3 version [2] was used in AMIP (identified in Table 1 by its institutional acronym MPI). Relative to that version, a number of substantial changes have been introduced in both the numerics and physics of the more recent model version ECHAM-4 [32]. These two versions of ECHAM GCM were run using the Global sea-Ice and Sea Surface Temperature (GISST) data set [29] with atmospheric carbon dioxide concentration varying temporally in accordance with observations. Both versions have the same spatial resolution, T42L19, as in AMIP. Results of the ECHAM-3 run for 1949-1990 and the ECHAM-4 run for 1903-1994 are considered in this study.

TABLE 2. The summary of the datasets used for the intercomparison in this study

Observational data	
Precipitation	Bryazgin [5]
	Bryazgin and Shver [6]
	Legates and Willmott [25]
	Jaeger [20]
	Vowinckel and Orvig [34]
	GPCP [18]: 1987-1994
	GHCN [12]: 1901-1997
Evaporation	Zubenok [37],
	WWB [24]
River discharge	Duemenil et al. [10]
	WWB [24]
Sea-ice/SST	GISST [29]: 1903-1994
Reanalyses output	
Precipitation and	ECMWF [15]: 1979-1993
evaporation	NCEP/NCAR [21]: 1982-1994
Climate model output	
Precipitation and	23 AMIP-1 AGCMs [28] 1979-1988
evaporation	ECHAM-3 [2]: T42L19 1949-1990
	ECHAM-4 [32]: T42L19 1903-1994
	Arctic HIRHAM-3/4 [9]: 50x50 km 1990; January runs 1985-1995

Finally, our assessment includes output from the HIRHAM regional climate model (RCM) run for the Arctic [9, 31]. The RCM is applied to the polar region north of 65°N with a horizontal resolution 50x50 km and 19 levels in the vertical. The two versions of HIRHAM (hereafter HIRHAM-3 and -4) employ the physical packages of ECHAM-3 and -4, correspondingly. At the lateral boundaries the model is driven by ECMWF analyses updated every 6 hours. At the lower boundary the model is forced by ECMWF-analyzed SST and sea-ice fraction, updated daily. Both HIRHAM versions have been run for the year 1990. Simulations for individual Januarys (1985-1995) have also been performed.

The summary of the datasets used for the intercomparison in this study is given in Table 2.

3.   AMIP GCM simulations of the AO FWB components

In the following discussion of the results, the term "AO" denotes the ocean area (excluding land) poleward of 70°N, unless otherwise noted. The terrestrial watersheds of the AO and of the individual Arctic seas and basins are defined following Ivanov [19] (see also [36]). The three major Arctic watersheds considered here are the European, the Asian (including the American part of the Chukchi Sea watershed) and the American (excluding the Chukchi Sea watershed and Greenland). The Asian watershed also includes the watersheds of the Kara Sea, the Laptev Sea and the East Siberian Sea. The American watershed is subdivided into the watersheds of the Beaufort Sea, the Canada Straits/Foxe Basin, and the Hudson Bay/Straits. Specific intercomparisons are also presented for individual river basins (the Ob/Irtysh, the Yenisey, the Lena and the Mackenzie).

Figure 1 shows the annual cycles of mean monthly precipitation for the AO as simulated by the AMIP models and derived from the observational datasets of Legates-Willmott, Bryazgin, Jaeger and Vowinckel-Orvig. Taken together, the observational estimates span a significant range, which can be taken as a measure of uncertainty of the monthly values. The primary data of Legates-Willmott and Bryazgin are in good agreement with each other and determine the upper limit of the monthly range throughout the year except in winter. It is also in winter that Legates-Willmott and Bryazgin show the largest differences relative to each other (up to 0.15 mm/day in January). Even if the upper limit of the observational range is considered the most plausible observational estimation of precipitation for the AO, the oversimulation of precipitation by most of the AMIP models is apparent in Figure 1. Both the observational data and most of the model simulations show generally similar seasonalities, with maxima in late summer and minima in late spring. This

216

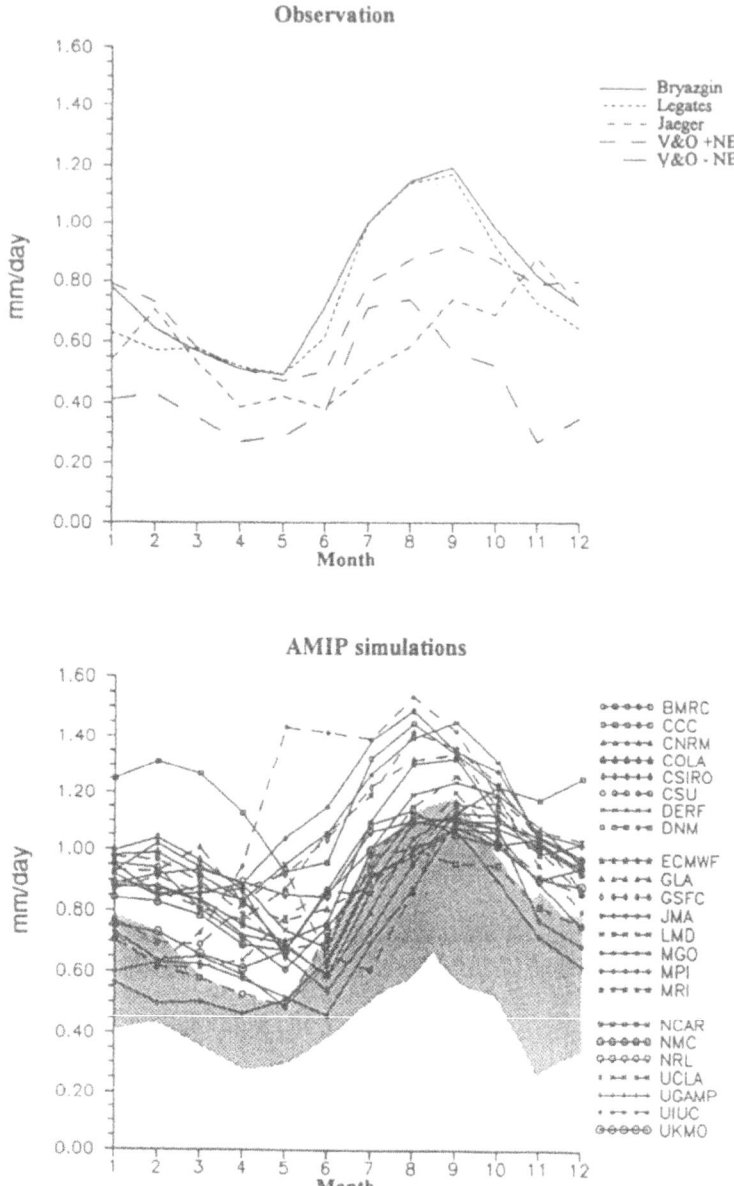

Figure 1. Mean seasonal cycles of area-averaged precipitation (mm/day) for the ocean area poleward of 70°N: (upper panel) observational estimates, and (lower panel) AMIP model results relative to range of observational estimates (shaded area). Shown in (upper panel) are the observational estimates of Vowinckel and Orvig (1970) for the polar cap including and excluding the Norwegian and Barents Seas (from [36]).

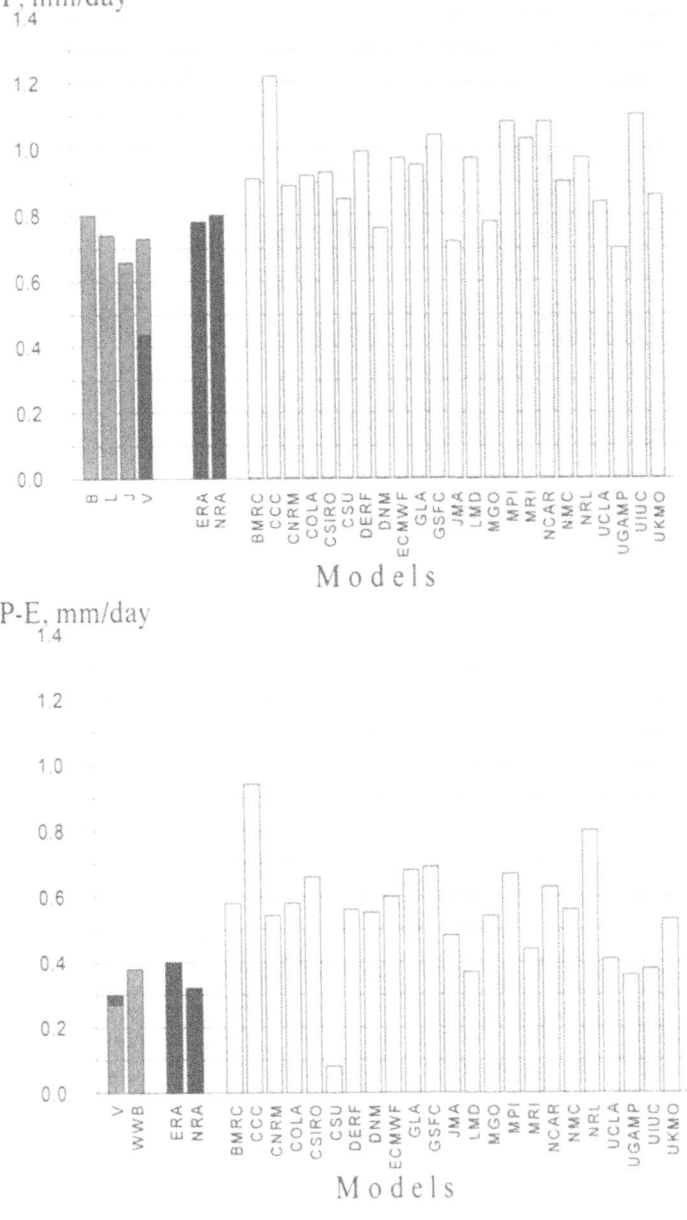

Figure 2. Annual mean rates of (upper panel) precipitation and (lower panel) precipitation minus evaporation for the Arctic Ocean as evaluated from observational estimates, from reanalyses, and from the AMIP models. Observational sources are B=Bryazgin, L=Legates-Willmott, J=Jaeger, and O=Vowinckel-Orvig (darker bar is for domain excluding the Norwegian/Barents Seas). Reanalyses are ERA=ECMWF, and NRA=NCEP/NCAR.

seasonality is consistent with the rawinsonde-derived seasonal cycle of moisture flux convergence across 70°N [33].

Figure 2 shows the decadal (1979-88) annual mean precipitation and P-E from the AMIP models in comparison with the corresponding ERA, NRA, and observational estimates. All but two climate models (JMA and UGAMP) simulate excessive precipitation relative to Legates-Willmott. DNM and MGO models are within the range spanned by Legates-Willmott and Bryazgin's precipitation estimates, as are the two reanalyses. The other 19 AMIP models oversimulate annual mean precipitation; the CCC, MPI, NCAR and UIUC being the wettest. (The oversimulation by the MPI model is noteworthy in the context of the results in the later sections.) The AMIP models' ensemble mean precipitation has a bias of +16% relative to Bryazgin, +25% relative to Leagtes-Willmott, and an even greater percentage relative to other observational estimates. As compared to the WWB-derived estimate and the two estimates of Vowinckel-Orvig, oversimulation of P-E is apparent in the results of most of the AMIP models. The two reanalyses, as well as LMD, UCLA, UGAMP and UIUC models, are rather close to the WWB observational estimate, whereas the CSU model substantially undersimulates the P-E for the AO. The bias of the models' ensemble mean P-E is +44% relative to the WWB estimate.

While the scatter of the AMIP model estimates of P for the Arctic Ocean is large, precipitation for its terrestrial watersheds shows even greater model-to-model variability. It is apparent from the Figure 3 that the decadal means of the AMIP model precipitation are again generally larger than the observational estimates. Although not shown here, the largest seasonal differences between simulation and observation occur in summer, when the precipitation amounts themselves are the largest. The summer oversimulation of precipitation is the primary reason for the excess in the annual means for the watersheds. The differences between Bryazgin-Shver and Legates-Willmott annual means are also substantial for all regions, especially for Europe, where the AMIP model oversimulation of precipitation is open to question. This uncertainty is supported by the GPCP estimate for Europe which, like Bryazgin-Shver's estimate (Figure 3a), is also high. On the other hand, the estimates of area-averaged precipitation over the European AO watershed are relatively sensitive to the resolution of the gridded datasets (mainly due to sharp gradients within the precipitation field over the narrow and mountainous north-west coastal region of Scandinavia; this region is inadequately resolved by AMIP GCMs). It should be noted that the annual means of precipitation from the NRA and ERA reanalyses differ significantly for most of the AO watersheds: the NRA estimates are much higher than those of ERA, and often

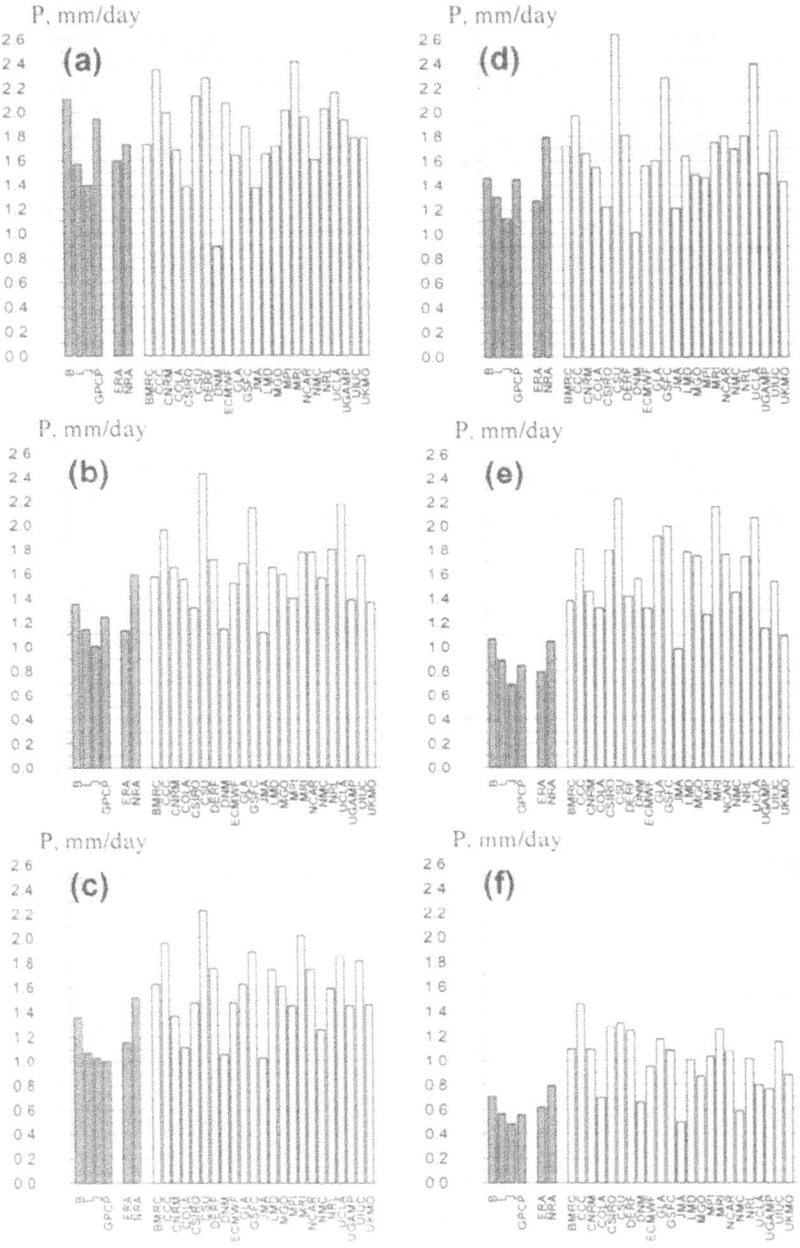

Figure 3. Annual mean precipitation rates for the (a) European, (b) Asian, and (c) American terrestrial watersheds of the Arctic Ocean; (d) the Kara Sea watershed; (e) the East-Siberian Sea watershed; and (f) Canada Straits/Foxe Basin watershed.

220

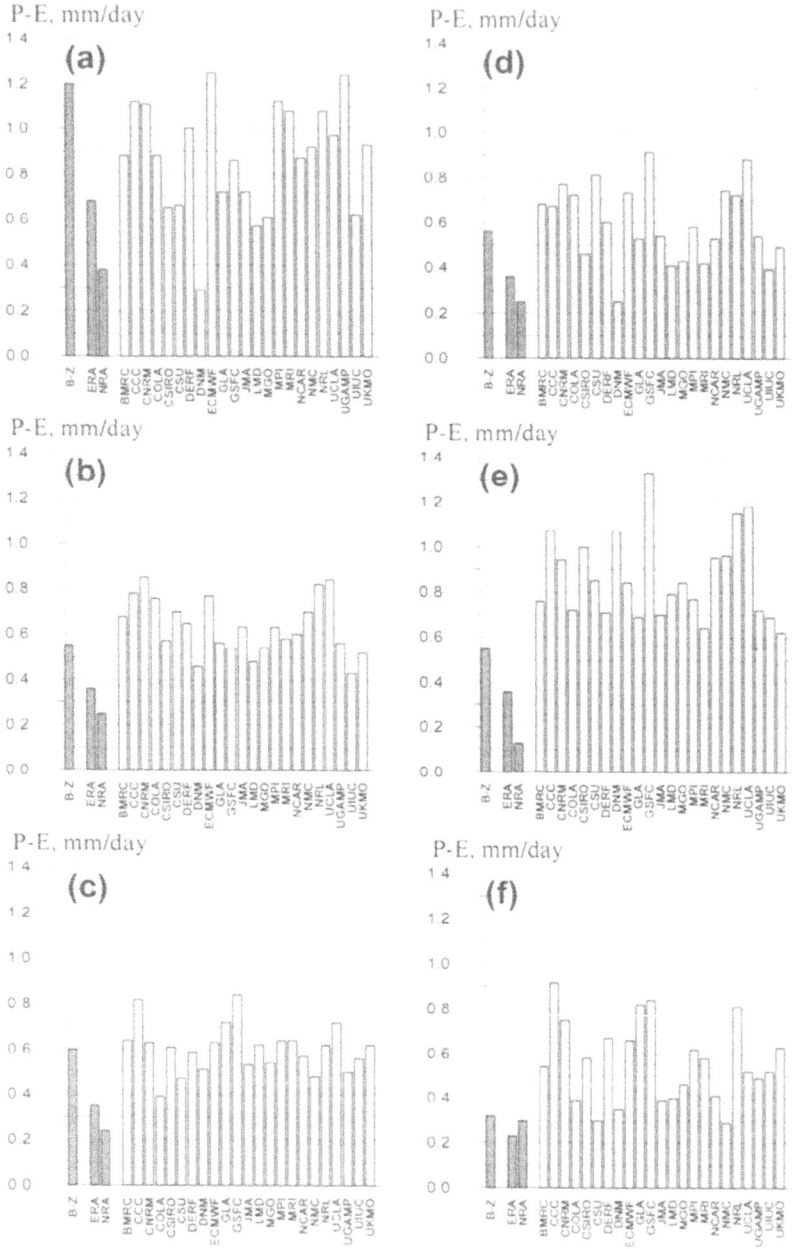

Figure 4. As in Fig. 3, but for precipitation minus evaporation. Observational estimates are areal means of Bryazgin-Shver's precipitation minus Zubenok's evaporation.

higher than the highest (Bryazgin-Shver) observational estimates. The ERA estimates generally remain within the range of uncertainty of the observations, closely matching the Legates-Willmott values.

The difference P-E represents the net input of fresh water to the surface on a local or regional basis. For land areas, P-E is equal to the runoff if there are no long-term changes in soil moisture storage. It should be noted that, due to an insufficient length of the spin-up period, some AMIP models have demonstrated trends in land surface temperature, snow mass, soil moisture and temperature. Nevertheless, the simulated decadal means of P-E for the AO terrestrial watersheds (Figure 4) can be used as "first guesses" in assessments of the AMIP models' fresh water inflow into the ocean.

The annual mean P-E is positive for all models and all regions under consideration. In the AMIP models, as well as in the reanalyses and observations, P-E is generally greater for the comparatively small European watershed than for the Asian and American watersheds. Within the Asian region, all model-derived annual means of P-E for the East-Siberian Sea watershed exceed the observational values and in some cases are more than twice as large as the observational values. Likewise, 21 of 23 AMIP models oversimulate P-E for the Canada Straits/Foxe Basin terrestrial watershed. The biases of approximately 60% in these regions indicate that the AMIP results are characterized by an oversimulation of precipitation and/or undersimulation of evaporation. An oversimulation of evaporation is noteworthy for NRA, since its P-E is usually lower than that of ERA, while its precipitation is much higher.

The individual river basins provide an alternative opportunity for validation of the models' P-E. The area-integrated annual mean P-E can be compared against annual river discharge observations, whose accuracy is superior to that of the above estimates based on areal integration of station-measured and gauge-corrected precipitation (and the even more questionable parameterizations of evaporation). Figure 5 shows a comparison between (1) the AMIP model-derived P-E averaged over the basins of the Ob/Irtysh (2.99 million km^2) and Mackenzie (1.80 million km^2) rivers and (2) the observed mean annual discharges -- 395 km^3 and 350 km^3, respectively, [24]. All AMIP models oversimulate the annual mean P-E in the Ob drainage basin, which is consistent with the general oversimulation of precipitation and P-E for the Asian AO watershed (Figures 3 and 4). The ensemble mean for the AMIP models is 0.60 mm/day, which is 67% higher than the observed discharge-derived 0.36 mm/day. For the Mackenzie basin, the models' ensemble average of 0.61 mm/day is 15% higher than the discharge-derived 0.53 mm/day; however several models underestimate P-E for this region.

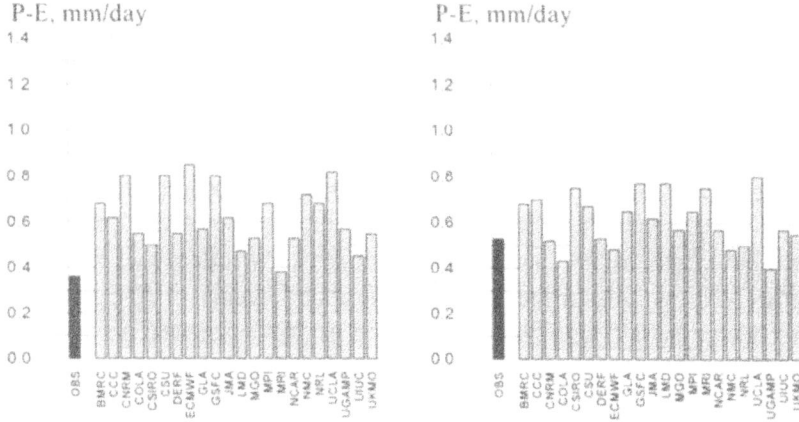

Figure 5. AMIP model-derived precipitation minus evaporation averaged over (left) the Ob Basin and (right) the Mackenzie Basin. Left-most bar of each panel is annual mean river discharge rate (mm^3/day) divided by corresponding area (mm^2) of river basin.

Figure 6 compares the simulated seasonality of P-E for the AO terrestrial watersheds and the seasonality of observed river discharge [10] into the AO. The annual cycles of P-E averaged over the Kara and the Beaufort Seas' land watersheds are nearly opposite in phase to the seasonal discharge of the major Arctic rivers within the watersheds (the Ob/Irtysh and Yenisey total, and the Mackenzie). While the maximum river discharge is observed in May-July, the simulated P-E achieves its minimum (most negative in the AMIP simulations) in June-August. The highest seasonality of discharge is demonstrated by the Lena River, yet the area averaged P-E over the Laptev Sea land watershed shows much less seasonality than the two others.

It seems appropriate here to note that an additional land-surface hydrology model component is required in order to provide a linkage between the two types of seasonality (river discharge into the ocean and P-E simulated by an atmospheric model). This linkage must be included before a model can properly reproduce both the annual cycle and the interannual variability of the river water inflow into the ocean, the latter being an important driver of the thermohaline circulation and sea-ice formation processes within fully coupled climate models. However, the modelling of hydrological land-surface and soil processes remains a major challenge in the northern high latitudes, where the fresh water can be stored in the solid phase (snow and ice on the ground; frozen ground, lakes and rivers). Until recently the above matters have received little attention in atmospheric GCMs [1], particularly in the AMIP models. Section 4 will continue this theme by utilizing recent advances in river-discharge

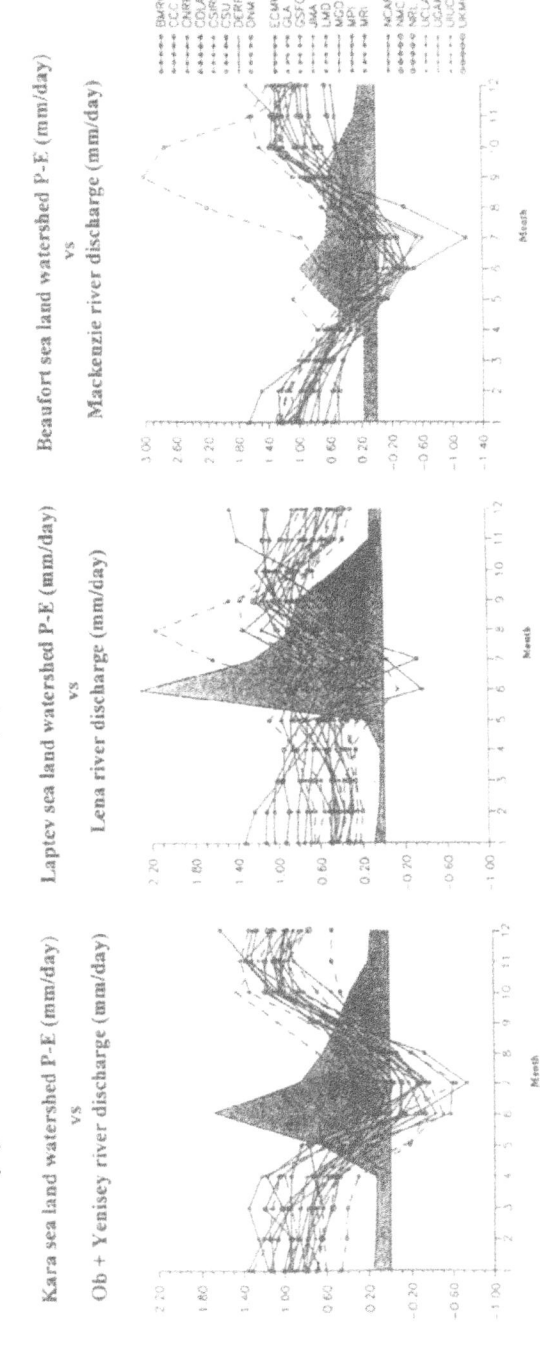

Figure 6. Annual cycles of model-derived precipitation minus evaporation for (a) Kara Sea land watershed, (b) Laptev Sea land watershed and (c) Beaufort Sea land watershed. Shaded are the annual cycles of discharge (equivalent mm/day) of major rivers in each watershed: (a) the Ob and Yenisey (total), (b) the Lena, and (c) the Mackenzie. (from [36]).

224

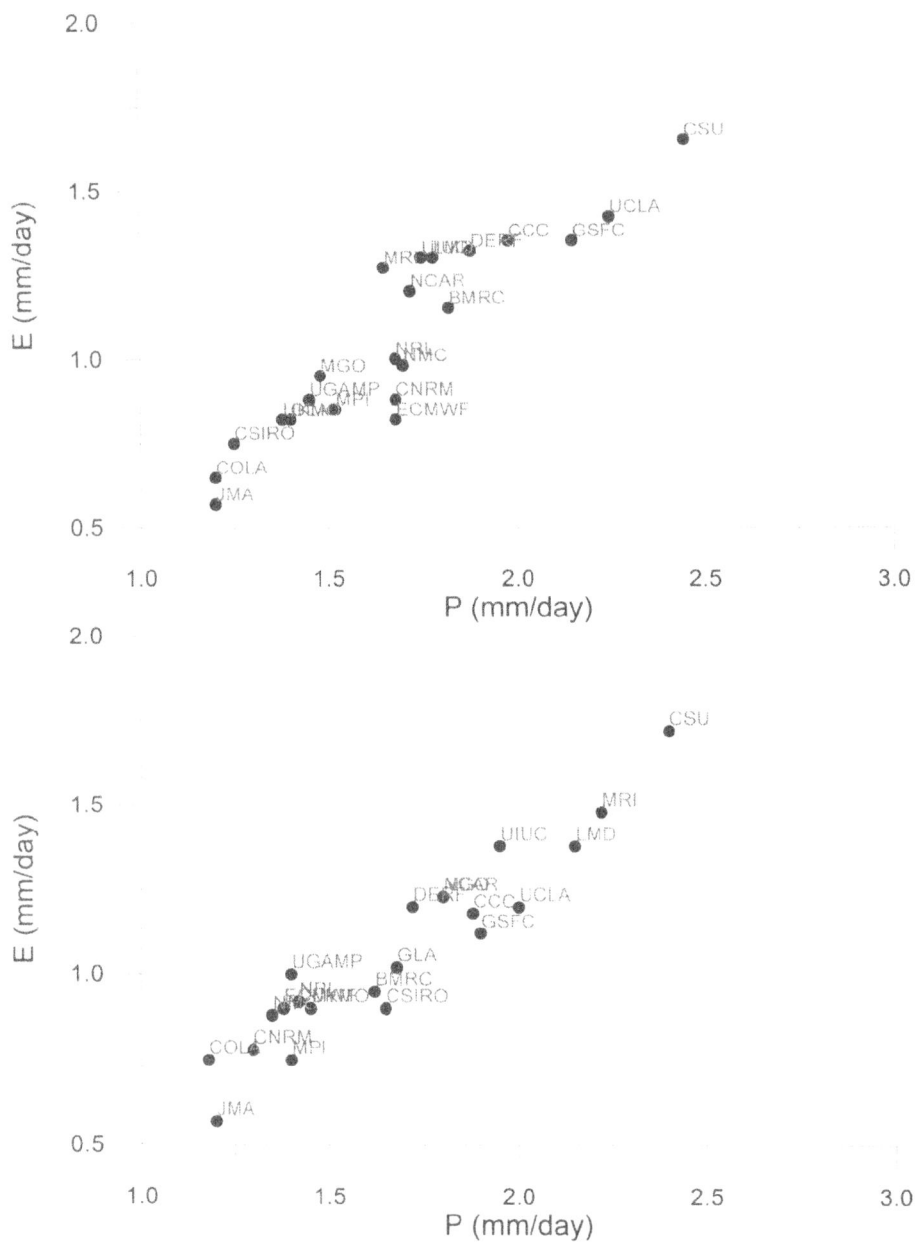

Figure 7. Scatter plots of 10-year (1979-1988) annual mean precipitation vs. annual mean evaporation for the AMIP models over (upper panel) the Ob Basin and (lower panel) the Mackenzie Basin.

modelling.

In an attempt to diagnose the model-to-model differences in precipitation and evaporation for the Arctic drainage basins, the simulated variables were compared across the sample of the models. From the scatterplots of the annual mean precipitation against the annual mean evaporation for the Ob and Mackenzie catchments (Figure 7), it is apparent that the models with the largest (smallest) precipitation are generally those with the largest (smallest) evaporation. The correlation between the annual mean precipitation and evaporation for the AMIP models is about +0.9 for the both watersheds. On a seasonal basis, the correlation is strongest in summer. For example, the correlations between ten-year seasonal mean precipitation and evaporation for the Ob are about 0.5 for winter, 0.6 for spring, 0.9 for summer, and 0.8 for autumn. The positive correlation between the two variables is also found in the interannual variations of individual models. Additionally, Walsh et al. [36] have shown that the monthly precipitation and evaporation are positively correlated in nearly every AMIP model. In many models, the correlation is highest during spring and summer, although some models show the highest correlation during winter. These correlations imply that variations of either precipitation or evaporation contribute to variations of the other (or that both are true).

Although the complex array of differences in the AMIP model formulations makes it difficult to diagnose the model-to-model differences in precipitation and evaporation, it has been found that (1) the models with prescribed soil moisture simulate larger evaporation as well as larger precipitation; and (2) the increase of precipitation is greater [36]. Thus, the specification of soil wetness increases the bias (relative to measured river discharge) of the annual mean P-E in the AMIP models. No other significant dependence of simulated precipitation and evaporation on the AMIP models' numerics or physics has been established, pointing to a need for controlled experiments with individual models.

4. ECHAM GCM simulations of the AO FWB components

During the several years following the AMIP simulation, the ECHAM (MPI) GCM has been subjected to substantial changes [32]: employment of a semi-Lagrangian transport scheme for water vapour, cloud water and trace substances (instead of the previously used spectral transform method); a new radiation scheme (ECMWF) with modifications concerning the water vapour continuum, cloud optical properties and greenhouse gases; a new formulation of the vertical diffusion coefficients as functions of turbulent kinetic energy; a new closure for deep convection based on convective instability instead of moisture convergence. Minor changes concern the parameterizations of horizontal diffusion, stratiform clouds and land surface processes. In addition, a new dataset of land surface parameters has been compiled for the model.

Owing to the above changes, there has been an impressive improvement in the ECHAM simulation of Arctic precipitation. Figure 8 illustrates the seasonal cycle of precipitation over the AO in the ECHAM-4 version (Figure 8, lower panel) in comparison with the ECHAM-3 version used in AMIP (Figure 8, upper panel). Whereas the ECHAM-3 significantly oversimulates the precipitation year-round, the ECHAM-4 does not. The seasonal cycle of ECHAM-4 precipitation is very similar to the observed seasonal cycles of Bryazgin and Legates-Willmott, although the ECHAM-4 cycle appears to lag the observationally-derived cycles by approximately a month. The multiyear (1949-1989) range of precipitation variability in the ECHAM-4 run essentially encompasses the range of the observational climatologies; the latter are entirely within the range of the longer ECHAM-4 run for 1903-1994 (not shown here).

The MPI (ECHAM-3) model, with its decadal mean precipitation of 1.08 mm/day, was one of the "wettest" AMIP GCMs for the AO precipitation. By contrast, the ECHAM-4 multiyear (1949-1989) annual mean (0.78 mm/day) falls exactly within the range of uncertainty of the observationally-derived estimates (0.75 mm/day of Legates-Willmott and 0.80 mm/day of Bryazgin) and is equal to that of ERA (0.78 mm/day). The annual mean evaporation for the AO has decreased from 0.44 mm/day in the AMIP simulation (0.47 mm/day in the ECAHM-3 extended (1949-1989) run) to 0.28 mm/day in the ECHAM-4 run, thus giving a net fresh water input P-E of 0.50 mm/day. This value is substantially closer to the observed 0.38 mm/day (WWB) than is the ECHAM-3 value of 0.66 mm/day.

Figure 9 shows mean seasonal cycles of precipitation for the AO terrestrial watersheds obtained in the ECHAM-3 and -4 runs (1949-1989). Also shown are the seasonal cycles derived from the ERA reanalysis and from observations (Legates-Willmott and GPCP). NRA is not represented here because of its severe oversimulation of precipitation for most of the watersheds, especially in summer. The ECHAM-4 seasonal cycle is generally closer than ECHAM-3 to those of ERA and Legates-Willmott. GPCP seems to give too excessive winter precipitation for the European and Asian watersheds, while its summer values are lower than both ERA and Legates-Willmott.

The multiyear annual means of precipitation and P-E for individual AO terrestrial watersheds are intercompared at Figure 10. Relative to ECHAM-3,

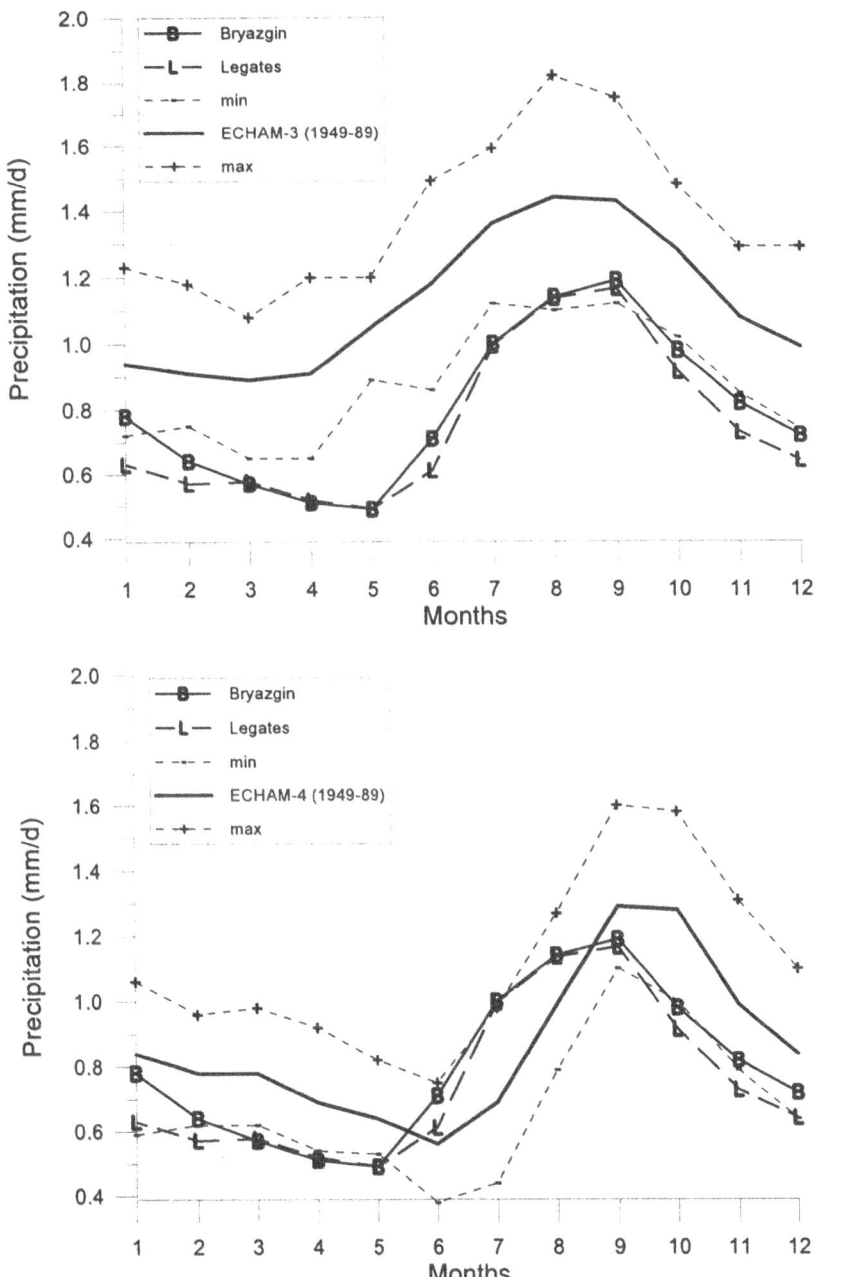

Figure 8. Mean seasonal cycles and ranges of interannual variation during the period 1949-1989 of area-averaged precipitation (mm/day) for the ocean area poleward of 70°N simulated by (upper panel) ECHAM-3, and (lower panel) ECHAM-4 relative to observational estimates of Bryazgin (B) and Legates and Willmott (L).

228

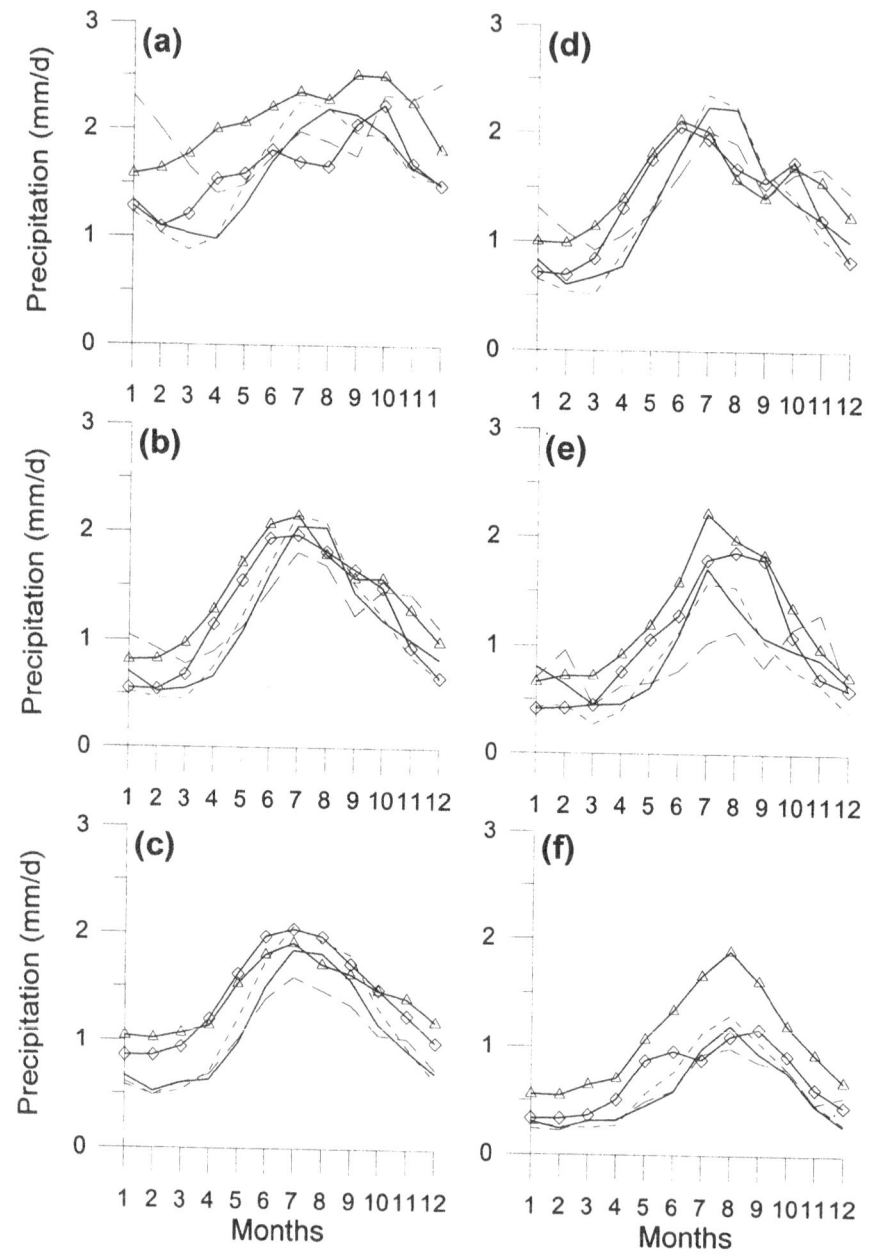

Figure 9. Mean seasonal cycles of area-averaged precipitation (mm/day) simulated by ECHAM-3 (triangles) and ECHAM-4 (squares) for the period 1949-1989 relative to observational estimates of Legates-Willmott (solid), GPCP (long-dashed), and ECMWF reanalysis (short-dashed) for (a) European, (b) Asian, and (c) American terrestrial watersheds of the Arctic Ocean; and (d) the Kara Sea watershed; (e) the East-Siberian Sea watershed; and (f) Canada Straits/Foxe Basin watershed.

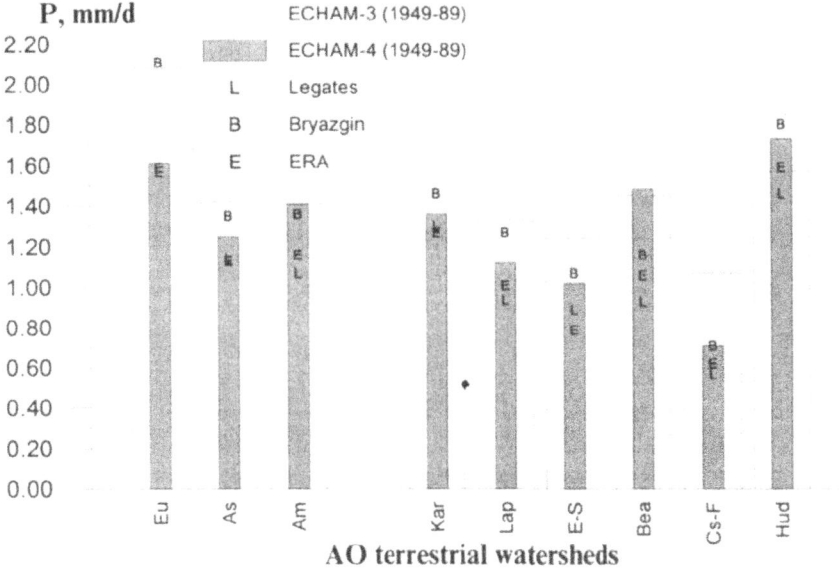

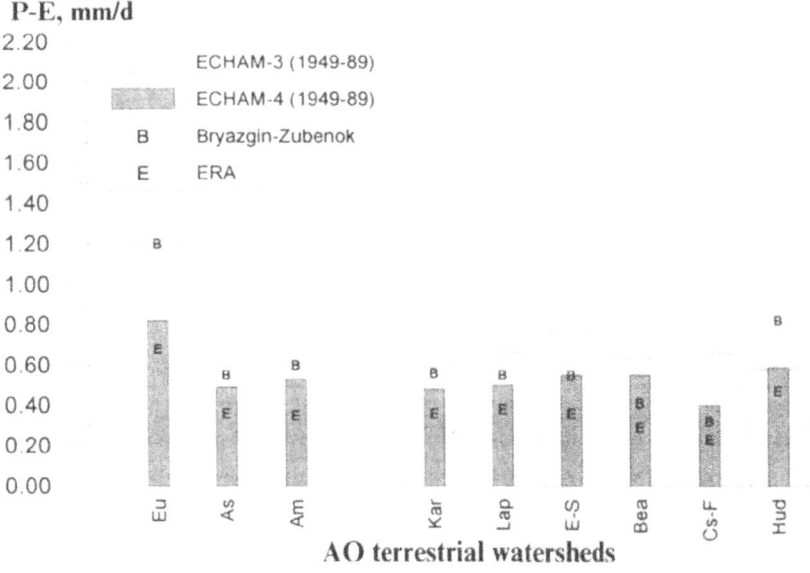

Figure 10. The ECHAM multiyear annual means of precipitation (upper panel) and P-E (lower panel) for individual AO terrestrial watersheds: European (Eu); Asian (As); American (Am); the Kara (Kar), Laptev (Lap), East-Siberian (E-S), and Beaufort (Bea) Seas; Canada Straits/Foxe Basin (Cs-F); and Hudson Bay/Straits (Hud). The centres of the characters (B, L, and E) correspond to observationally- and reanalysis-derived estimates.

precipitation simulated by ECHAM-4 has decreased for Europe and Asia. Precipitation over the American watershed is essentially unchanged due to the concurrent increase over the land watersheds of the Beaufort Sea and Hudson Bay/Straits, and the decrease over the Canada Straits/Foxe Basin. With the exception of the two regions of increase, the ECHAM-4 precipitation has fallen into (or remained within) the range of uncertainty of the observationally derived estimates over all watersheds. The same is true of P-E for all watersheds (Figure 10b), with the minor exceptions of the two watersheds in America (the Beaufort and Canada Straits/Foxe Basin). Even for these watersheds, the P-E from ECHAM-4 fits the observations more closely than does the P-E from ECHAM-3. Thus the improvement is evident throughout the whole region under consideration.

Since ECHAM-4 clearly simulates Arctic precipitation and evaporation more realistically than ECHAM-3, an important question is: What changes of the ECHAM GCM formulation are primarily responsible for the improvement. Actually, all major changes are presumed to be relevant. Controlled experiments with the ECHAM model are a high priority to establish the relative importance of the various model changes. Another promising opportunity is provided by the next phase of AMIP, which was recently initiated. Finally, a noteworthy issue in this regard is the increasing requirement for observational datasets of higher quality. It is apparent from the ECHAM results that the level of uncertainty of the current generation of atmospheric climate models is now comparable to the uncertainties in the observational datasets of Arctic precipitation and evaporation. Further progress in model validation will require a narrowing of the range of the observational estimates.

The positive correlation between precipitation and evaporation, noted earlier for the AMIP simulations, is also found in the extended runs with the ECHAM model. Scatterplots (not shown here) of precipitation and evaporation from the ECHAM-4 run (1903-1994), reveal that the association between the two variables depends on season and region. The highest (and statistically most significant) correlation between precipitation and evaporation from ECHAM-4 is achieved in winter for all regions (the correlation exceeds +0.9 for Asia and Europe); the lowest correlations occur in summer and are even negative for some regions (Europe and the Laptev and East Siberian Sea watersheds). This seasonality is consistent with the seasonality of the correlation between precipitation and evaporation in the AMIP run of ECHAM-3.

Within last few years, considerable progress has been achieved in modelling river water discharge into the ocean on the global scale. Local runoff and drainage from the ECHAM-4 T42L19 integration have been used by Duemenil et al. [11] as input into the hydrological discharge (HD) global model

[17] employing a 0.5°x0.5° resolution. In an application of particular relevance to this review, Duemenil *et al.* [11] used the HD model driven by the ECHAM-4 output to compute river discharge for all basins draining into the AO. The seasonal cycle of the total discharge from the six largest Arctic rivers (the Northern Dvina, Lena, Yenisey, Kolyma, Ob, and Mackenzie) is presented in Figure 11 (from [11]). The summer maximum of the discharge is underestimated by HD. The one-month lag of the HD simulation relative to the observed seasonal cycle can be attributed to the one-month lag of large-scale Eurasian snowmelt in the ECHAM model [17]. Given this limitation, the ability of HD to simulate the seasonality of the river discharge is encouraging. The simulated total discharge by the six rivers into the AO amounts to 2003 km^3, which is quite consistent with the estimate of 1970 km^3 obtained from observations at the gauging stations of the six rivers [10].

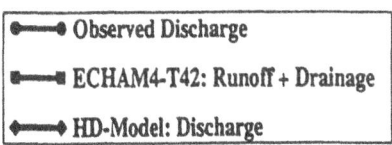

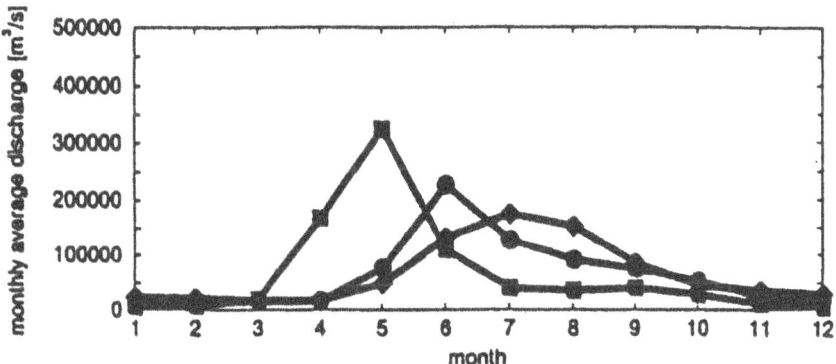

Figure 11. Monthly averaged total discharge from the six largest Arctic rivers. Runoff and drainage from ECHAM-4 is used as input into the lateral transfer scheme by Hagemann and Duemenil [17] (from [11]).

The extended (almost century-long) run of ECHAM-4 driven with observed SST/sea-ice and atmospheric carbon dioxide concentration offers the opportunity of assessing interdecadal variability of precipitation for the Arctic region. Figure 12 compares annual precipitation anomalies relative to the

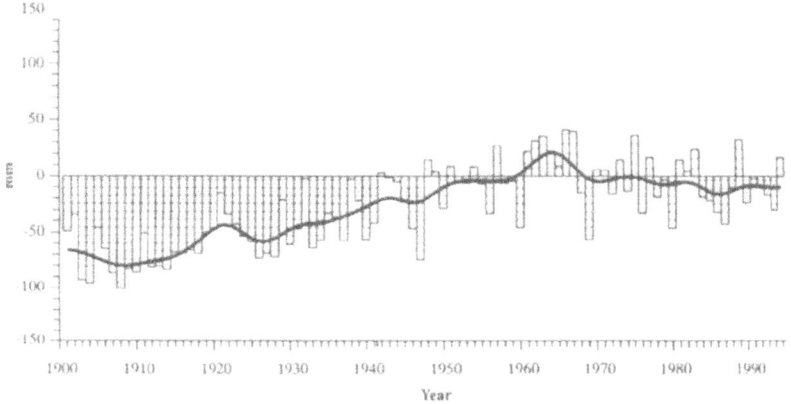

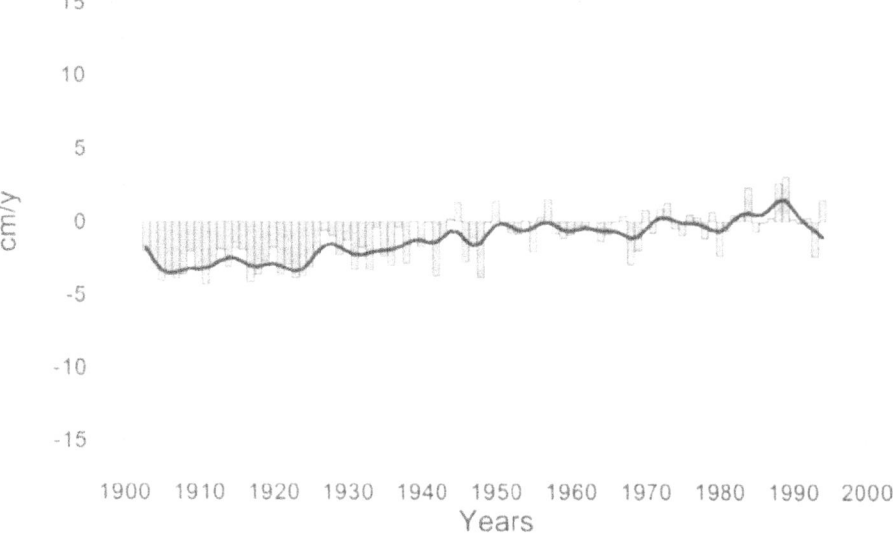

Figure 12. Arctic precipitation annual anomalies from the 1961-1990 means: (upper panel) observed (from [22]), and (lower panel) simulated by ECHAM-4. Smoothed curves were created using a nine-point binomial filter.

1961-1990 mean for the Arctic region (defined by Karl [22]) as derived from observational data and from the ECHAM-4 simulation. Both sets of anomalies show a twentieth-century trend of Arctic precipitation. The consistency is encouraging from the viewpoint of the model's ability to reproduce the observed climatic trends. It is also apparent that the range of the interannual variations of the ECHAM precipitation is smaller than that shown by the observation. However, the underestimation of the variability by the model is open to question because of the insufficient coverage of the observational data in the northern polar region; an analysis of the GHCN dataset shows that the observational coverage of northern high latitudes varies considerably with season, year and decade. The data are virtually nonexistent for the greater part of the AO.

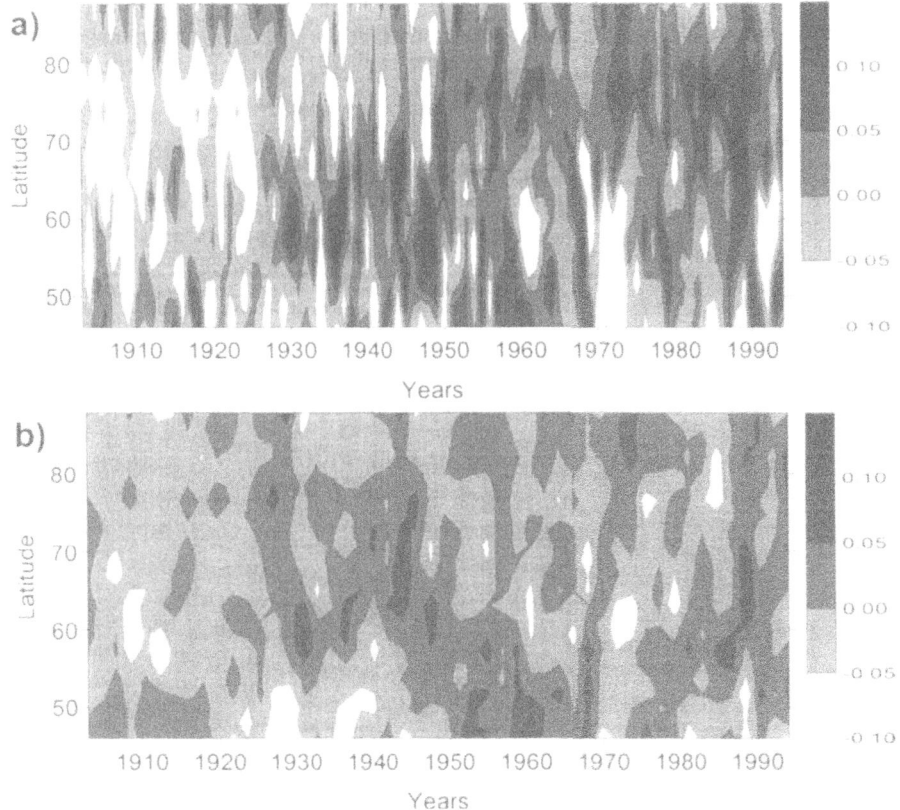

Figure 13. Annual/zonal means (mm/day) of (a) Arctic precipitation and (b) precipitation-minus-evaporation expressed as anomalies from the 1903-1994 zonal means simulated by ECHAM-4.

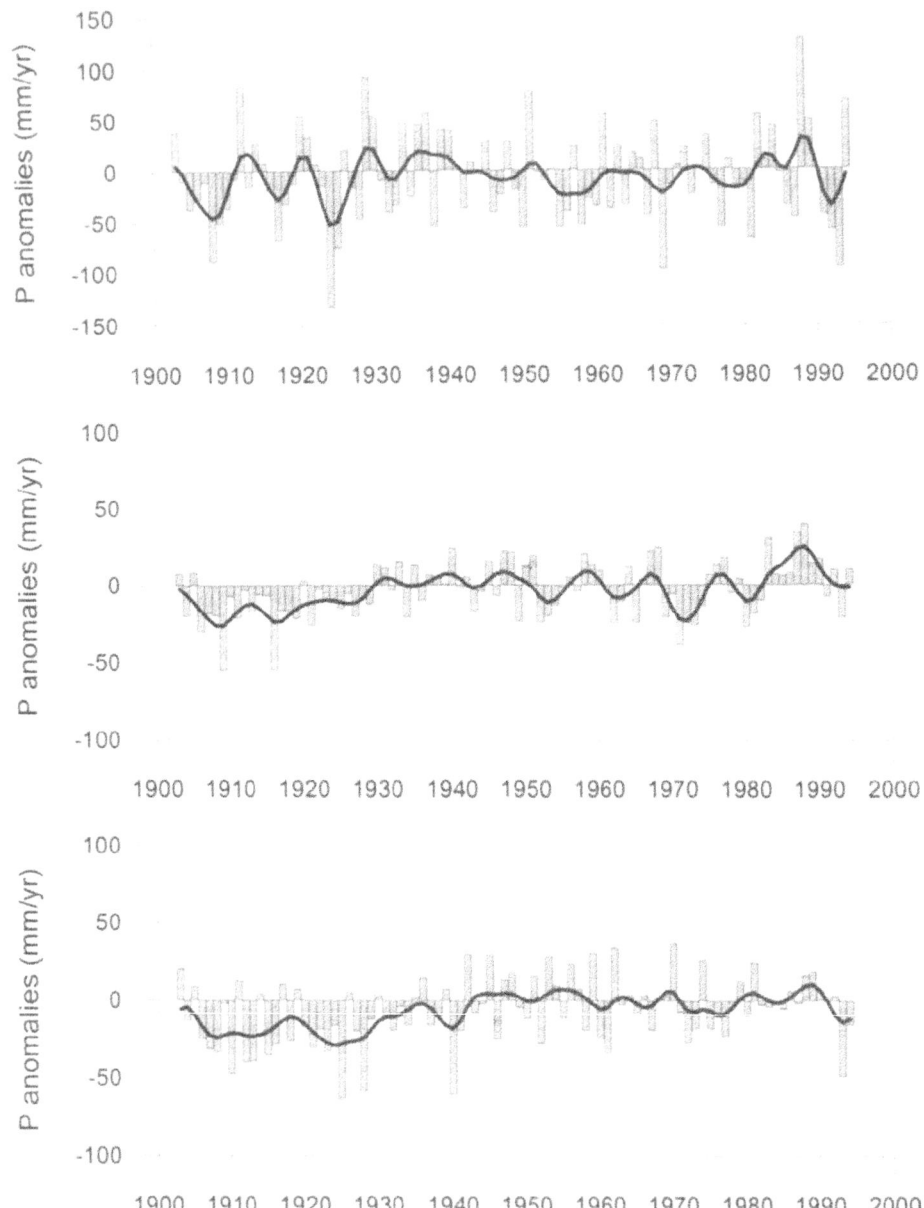

Figure 14. Anomalies of annual precipitation relative to 1961-1990 means over (upper panel) European, (middle panel) Asian and (lower panel) American terrestrial watersheds of the Arctic Ocean. Smoothed curves were created using a nine-point binomial filter.

Figure 13 shows zonally averaged annual anomalies, relative to the 1903-1994 means, of the ECHAM-4 precipitation and P-E for the northern polar region. The gradual increase of precipitation is apparent over the Arctic region (north of 66°N) for the whole period, while in lower latitudes there are two major century maxima, one during 1940-1960 and the other during the decade of the 1980s. The simulated P-E demonstrates less temporal and spatial variability than does the simulated precipitation, although it shows a definite increase over the AO during the second half of the century. Similar precipitation trends are simulated by the ECHAM-4 for the AO terrestrial watersheds. An exception is the Europe watershed, where it is difficult to detect any trend because of the relatively high interannual variability (Figure 14).

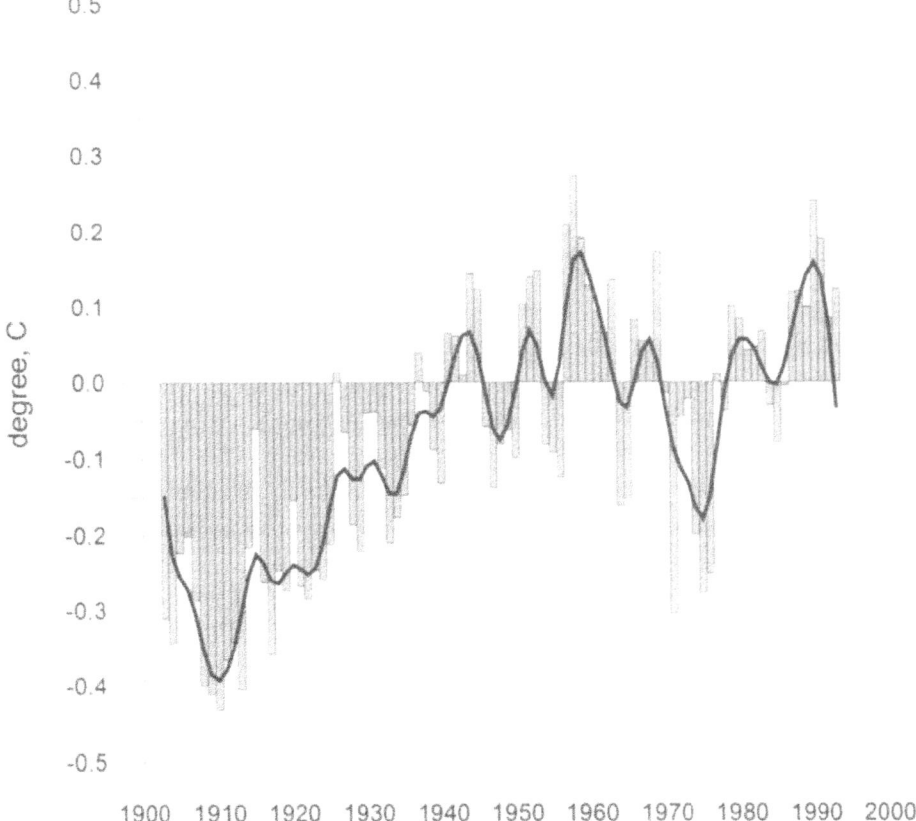

Figure 15. SST (GISST) annual anomalies from 1961-1990 mean averaged over the Northern hemisphere. Smoothed curve was created using a nine-point binomial filter.

The commonality of the trends in the model output and the observational data raises the issue of possible relationships to atmospheric forcing. Two major candidates for a forcing of the precipitation trend of the twentieth century are the SST and greenhouse gas concentrations, both of which have shown positive trends during the past century. Of the two, the former seems to be of primary importance. This hypothesis is supported by the positive twentieth-century trend in the annual anomalies of SST (GISST) area-averaged over the northern hemisphere (Figure 15). In addition, the temporal pattern of the SST variations appears to be the better match to the decadal-scale variations of Arctic precipitation. In any case, sensitivity experiments are required to separate the impacts of the two forcings on the precipitation trend. Specifically, there is a need to perform simulations in which (1) the seasonal cycle of SST is fixed but greenhouse gas concentrations vary, and (2) SST varies interannually as observed but greenhouse gas concentrations are invariant.

5. HIRHAM high-resolution simulations of the AO FWB components

AMIP and other studies have revealed significant biases in the fields of Arctic surface air temperature and surface winds simulated by AGCMs [7,35]. The biases in the simulated polar circulations are important sources of the errors in the precipitation simulated for the Arctic region [4]. There are indications that a more realistic representation of the atmospheric general circulation can be achieved by increasing model resolution [16]. The increased model resolution can improve both the large-scale circulation and local surface forcing (topography, coastlines, sea-ice types and margins, etc.). When accompanied by appropriate tuning of the model parameterizations, the finer resolution should be effective in improving the simulation of precipitation and evaporation in a given region, particularly in areas of significant topography and complex coastal boundaries.

Because refining resolution of modern GCMs by a factor of 5 to 10 is computationally prohibitive, the limited area modelling has received increased use in recent years as a technique to downscale meteorological variables simulated by the GCMs. The performance of a high-resolution RCM, driven at its boundaries by model output, is critically dependent on the quality of the driving fields [23]. In this context, the use of an observationally-derived analysis as the RCM driver provides "perfect" boundary conditions, eliminating the influence of systematic large-scale model errors beyond the domain of interest [8].

Because RCMs can capture the fine-scale structure of climate patterns [23] an important issue in RCM validation is the lack of adequately dense observational data. This problem is compounded in the Arctic region by the absence of essentially any observational data for its central part. A related problem is computational expense of RCM simulations, precluding the simulation of multiyear periods for which climatological data could be used for validation.

An opportunity for simulating the Arctic climate at high spatial resolution has been recently arisen with the development of the RCM of the Arctic atmosphere HIRHAM [9,31] and the Arctic region climate system model ARCSyM [26], as well as of RCMs whose domains include terrestrial watersheds of the AO (e.g., the Canadian RCM adapted for hydrological applications in the Mackenzie Basin by MacKay *et al.* [27]). In this study we will concentrate on the HIRHAM model simulations, since the physical packages of its two versions correspond to the versions of the global ECHAM model discussed in Section 4, thus providing comparability of the RCM and the GCM results. It seems appropriate here to mention a major numerical difference between the two models: while the latter is a spectral model, the former is a finite-difference one.

Figure 16 shows January precipitation, area-averaged over the AO northward of 70°N, for individual years as simulated by the two versions of the ECHAM (3 and 4) GCM and the two corresponding versions of the HIRHAM RCM. Both ECHAM versions generally simulate higher January precipitation amounts than those derived from the Legates-Willmott and Bryazgin observational data sets, and also from the reanalyses. The HIRHAM values are close to the reanalyses and thus are generally lower than ECHAM. It is noteworthy that the difference between the two HIRHAM simulations for each individual January is smaller than the corresponding difference between the two ECHAM simulations. An explanation follows from the fact that the most important differences in the physical packages (radiation, convection) of the ECHAM/HIRHAM versions 3 and 4 are expected to manifest themselves in summer (or at lower latitudes, in case of ECHAM), rather than in the polar region in winter. An important difference between the two ECHAM versions -- the semi-Lagrangian transport scheme for some variables vs. their spectral representation -- is not present in the two HIRHAM versions, which both use finite-difference representations of all variables. Thus, the winter simulations of precipitation with the two HIRHAM versions should be sensitive primarily to the driving fields (which are the same ECMWF analysis) rather than to the model formulation differences. The more pronounced differences between ECHAM-3 and -4 January precipitation simulations are likely attributable to

238

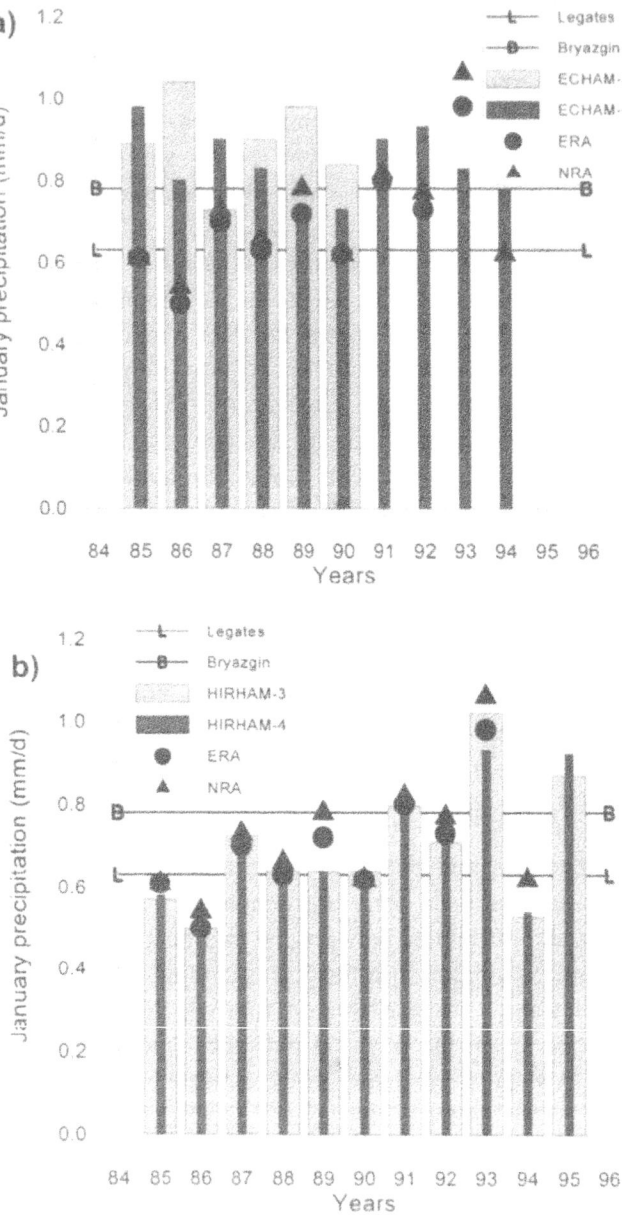

Figure16. Arctic Ocean January precipitation for individual years: (a) ECHAM-4 vs. ECHAM-3, and (b) HIRHAM-4 vs. HIRHAM-3. Shown are observational January means of Bryazgin (B), and Legates and Willmott (L), as well as ECMWF reanalysis (circles) and NCEP/NCAR reanalysis (triangles).

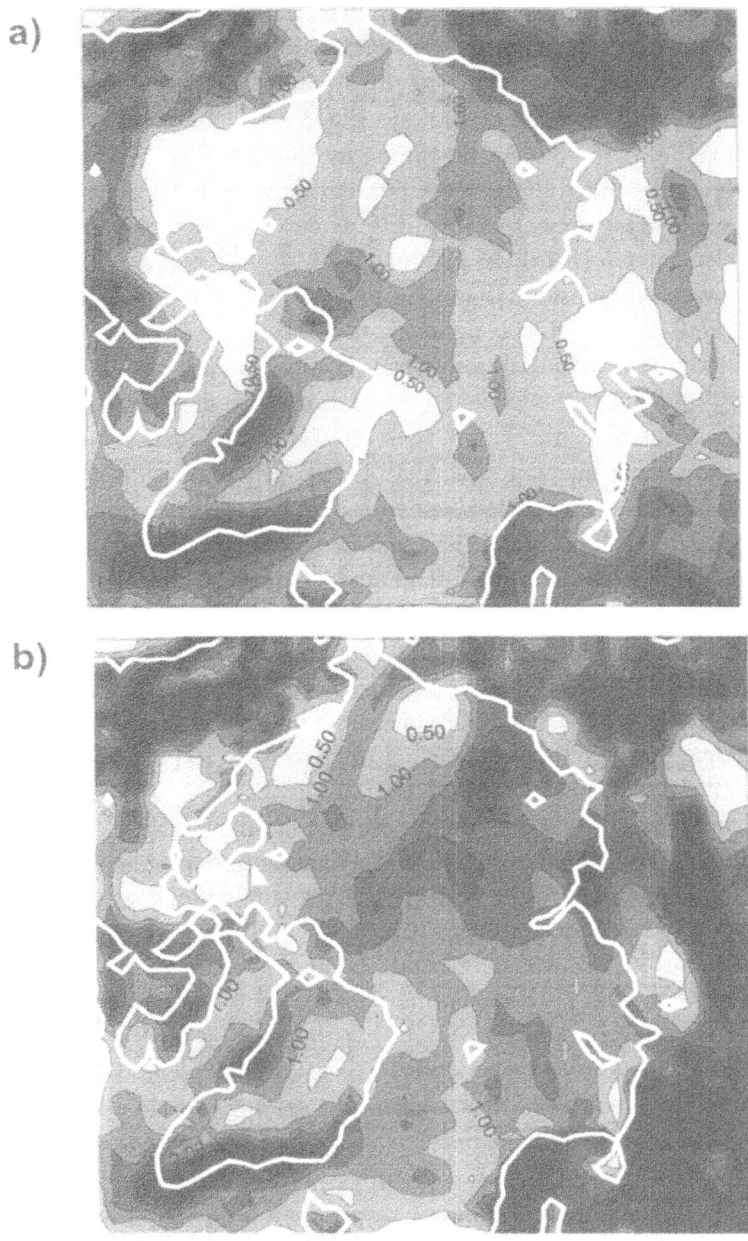

Figure17. Precipitation (mm/day) monthly mean in July 1990 simulated by (a) HIRHAM-4, and (b) HIRHAM-3.

general circulation differences beyond the Arctic region and occurrence of planetary-scale teleconnection patterns in the global model.

A completely different type of situation occurs with the HIRHAM simulations of summer precipitation. Figure 17 compares the July 1990 precipitation fields for the Arctic region as simulated by HIRHAM-3 and -4. The former shows much higher precipitation than the latter almost everywhere in the Arctic. The impact of the differences in physical parameterizations employed by the two HIRHAM versions is illustrated by the Figure 18: in July, the differences are substantial for the greater part of the model domain over both land and ocean, while in January, the differences are substantial only in the vicinity of the North Atlantic boundary of the domain.

In Figure 19, the annual (1990) cycles of the area-averaged Arctic precipitation obtained from the two HIRHAM simulations are compared sequentially against the two ECHAM (1990) simulations, the ERA and NRA reanalyses for the same period, and seasonal cycles of precipitation derived from the Legates-Willmott and Bryazgin climatologies. The monthly means of the HIRHAM winter precipitation almost coincide with each other. These means are somewhat lower than ECHAM and are very close to the reanalyses. With exception of January, the HIRHAM winter precipitation is higher than the observationally-derived climatologies. For the rest of the year 1990, the seasonal cycles of HIRHAM precipitation are highly correlated with each other, the difference between them increasing through May and decreasing after July. The seasonal range of HIRHAM is generally lower than that of ECHAM and is higher than those of the reanalyses as well as the climatologies. An important feature of the 1990 annual cycle is the April maximum of precipitation which is more or less pronounced in both HIRHAM simulations as well as in the ECHAM-3 and ERA simulations. The April 1990 increase of precipitation is entirely absent in the two climatologies and appears to be attributable to interannual variability. For summer 1990, the HIRHAM-3 precipitation exceeds the observationally-derived estimates, while the HIRHAM-4 precipitation is somewhat lower than the observational estimates. For the three autumn months, the HIRHAM-4 areal mean precipitation is virtually identical to the corresponding climatic values.

The 1990 annual means of precipitation simulated by HIRHAM-3 and -4 are 1.02 mm/day and 0.84 mm/day, respectively. The latter is lower than the ECHAM-4 annual mean of 0.88 mm/day for 1990 and closer to observationally-derived climatological annual means of 0.80 mm/day (Bryazgin) and 0.75 mm/day (Legates-Willmott). Meanwhile, the 1990 annual P-E is the same in the both HIRHAM simulations (0.54 mm/day) and is slightly lower than the ECHAM-4 annual precipitation of 0.57 mm/day for 1990.

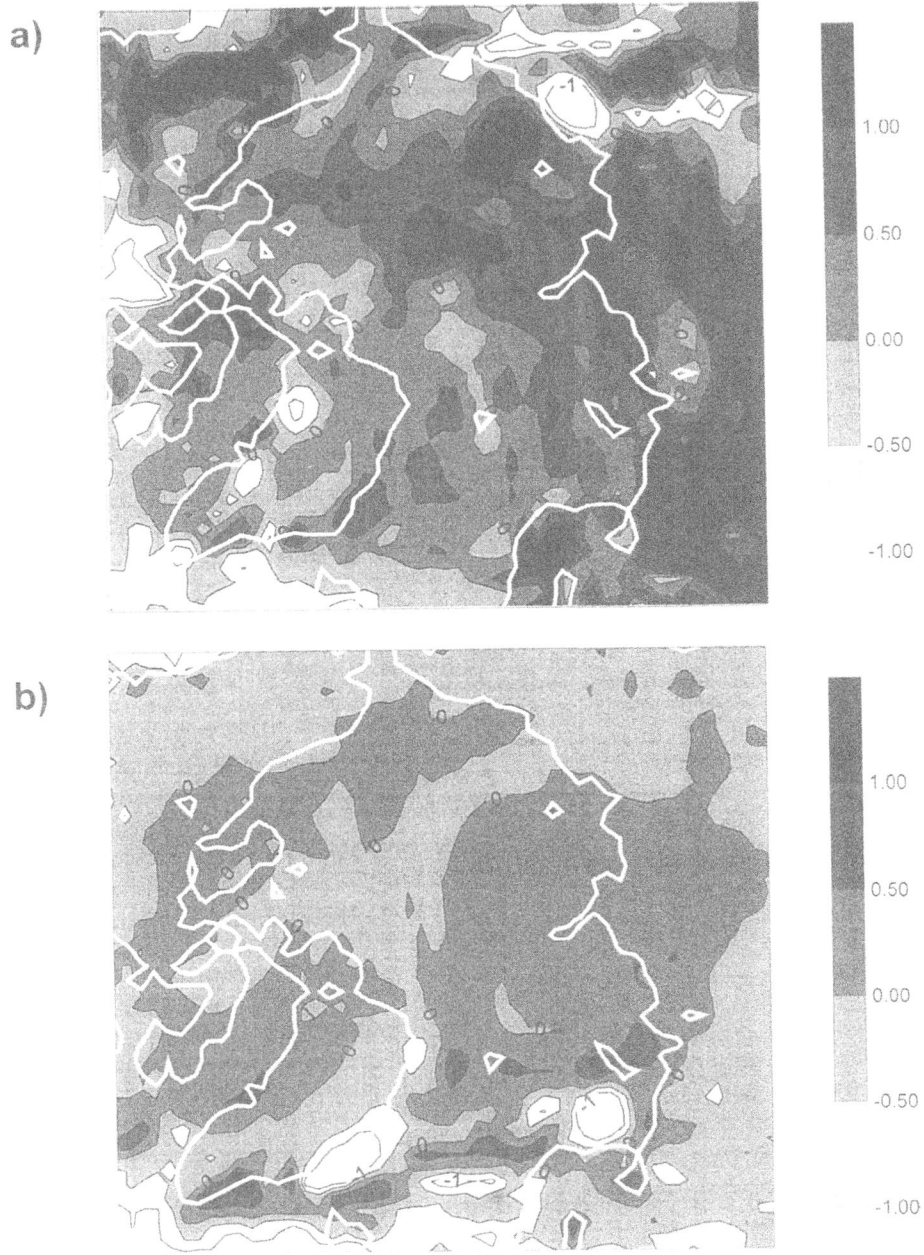

Figure 18. Precipitation (mm/day) fields difference (HIRHAM-3 minus HIRHAM-4): (a) July 1990, and (b) January 1990.

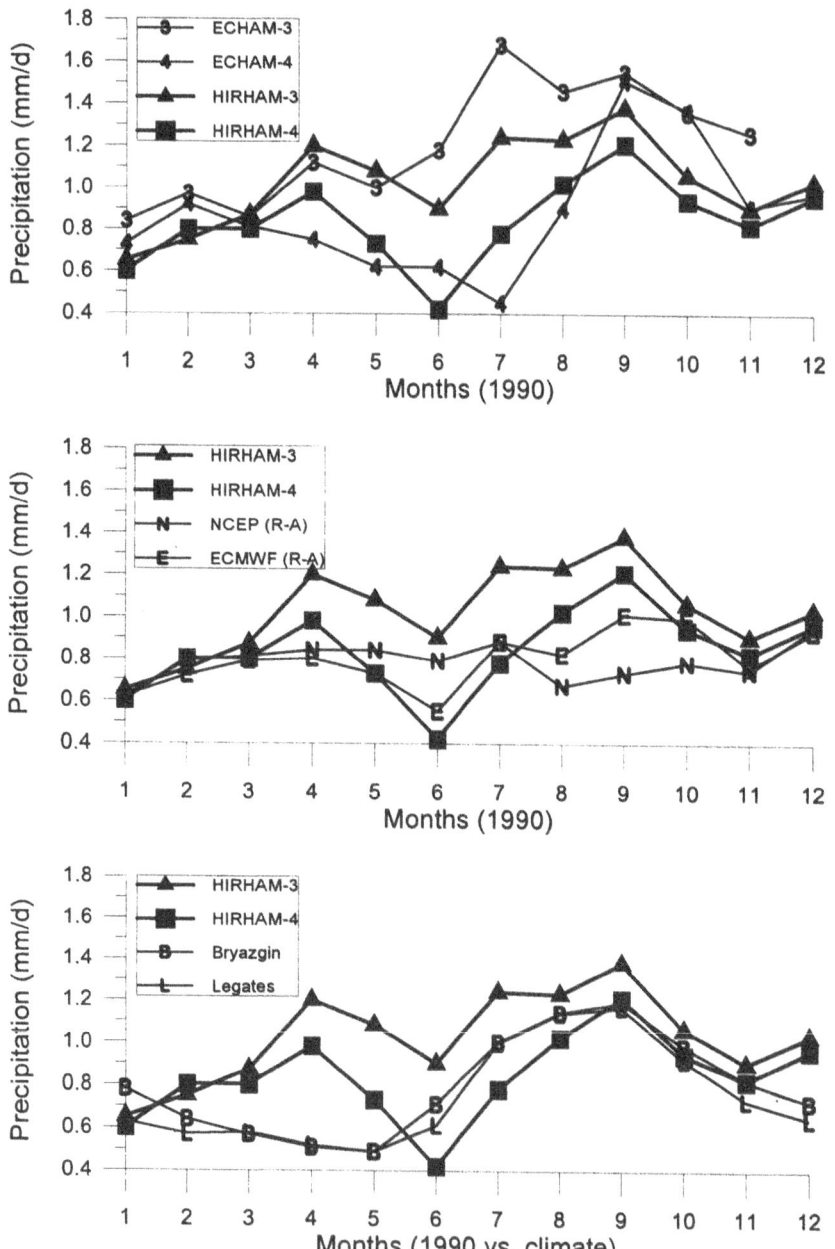

Figure 19. Seasonal cycle (1990) of area-averaged precipitation (mm/day) over the Arctic Ocean poleward of 70°N simulated by HIRHAM-3 and -4 vs.: (upper panel) ECHAM-3 and -4, 1990; (middle panel) ECMWF and NCEP/NCAR reanalyses, 1990; and (lower panel) Bryazgin and Legates-Willmott climatologies.

6.   Future research priorities

AMIP characterizes the recent state of the art of atmospheric climate modelling using GCMs. AMIP simulations of precipitation and evaporation, which are of direct relevance to the AO FWB, reveal significant model-to-model scatter in each region considered here (the ocean and its terrestrial watersheds). In a qualitative sense, the AMIP models generally show seasonal cycles of precipitation similar to those observed. Even after allowing for the divergence of the existing observationally-derived high-latitude climatologies, it is apparent that the AMIP models tend to oversimulate precipitation throughout the year for the most of the regions of this study. The excessive model precipitation is compensated, to some extent, by a general oversimulation of evaporation. Nevertheless, the AMIP models' total freshwater input into the AO is oversimulated. This oversimulation of P-E is a consistent finding of studies that have examined the high-latitude performance of the AMIP models, and an explanation of this oversimulation is a high priority for research.

The simulated precipitation and evaporation are highly correlated across the models, and they are positively correlated over time in most of the individual AMIP simulations. Thus the local hydrologic processes are strongly coupled in the models. Of the various features (both numerics and physics) of the model formulations, the only one common to the high-precipitation (and high-evaporation) AMIP GCMs is the specification of soil moisture. In this context, controlled experiments with individual models would be a useful approach to establishing the dependence of simulated precipitation and evaporation on other features of the model design and formulation. Another promising opportunity is the next phase of AMIP (AMIP-II), which is now underway.

The substantial changes introduced in the ECHAM model's numerics and physics have brought the model's simulation of annual mean precipitation and the mean seasonal cycle of precipitation into much better agreement with observations, for both the AO and its terrestrial watersheds. However, without controlled experiments, one cannot determine which of the concurrent changes to the model formulation were most responsible for this improvement.

The ECHAM-4 extended run (1903-1994) with prescribed observed SST and atmospheric carbon dioxide concentration shows an increase of precipitation in northern high latitudes during the past several decades. This is in agreement with the observational record. The SST is presumed to be a primary driver of the twentieth-century increase in the ECHAM precipitation. However, sensitivity experiments are required to place confidence in this conclusion. In addition, the validity of the observationally-derived trend requires further investigation.

The HD model simulation of the lateral discharge into the AO from its terrestrial watersheds demonstrate that encouraging progress has been achieved in this field within last few years. However, there is still a considerable need for improvements in the GCMs that drive the discharge models. Such improvements are necessary, particularly in the hydrologic components of the GCMs, in order to quantify accurately the river water inflow to the AO on seasonal through interannual timescales. This line of development should receive a high priority in the context of modelling the AO FWB using fully coupled GCMs.

The infancy of regional climate modelling of the Arctic atmosphere is marked with 3 *pro*'s: *pro*gress, *pro*blems and *pro*mise. The HIRHAM model driven with ECMWF analyses produces a credible simulation of Arctic precipitation in recent monthly (January) and yearly (1990) runs. In the HIRHAM model, the implementation of the ECHAM-4 physical package in place of the ECHAM-3 package resulted in an apparent improvement of the simulation of summer precipitation. On the other hand, the HIRHAM winter simulations are far less sensitive to the formulation changes than both the HIRHAM summer and the ECHAM winter simulations. However, a much longer simulation time is required to obtain more robust assessment of the model performance.

An important problem in the validation of the hydrologic output of climate models is the adequacy of the observational data. This problem is especially severe in high latitudes, where the observations are sparse, if available at all. In this context, high-resolution reanalyses can provide potentially important enhancements of the available database.

Finally, an Arctic RCM intercomparison (similar in concept to AMIP) might be extremely fruitful in detecting the models' capabilities and deficiencies, thereby leading to improvements in the models. There is also the possibility that the RCMs will be the most useful vehicles for regional reanalyses designed specifically for the Arctic. Such regional reanalyses could provide high-resolution fields of the hydrologic variables needed for an accurate and detailed assessment of the Arctic FWB.

Acknowledgements

Klaus Arpe of MPI (with assistance of Petr Sporyshev) made available the ECHAM model output used in this study. GHCN precipitation data set was received from Thomas Karl of NCDC (with assistance of Dick Knight). The study benefited from comments, discussions and information provided by: Klaus Arpe of MPI, Nikolay Bryazgin of AARI, Lydia Duemenil of MPI, Vladimir Ivanov of AARI, Peter Lemke of IFM (Kiel), Amanda Lynch of NCAR, Dimitry Mironov of AWI (Bremerhaven), and Yuri Pichugin of MGO. Some computational assistance was provided by Diane Portis of UIUC, as well as by Valentina Margasova, Victoria Mirvis, Igor Shkolnik and Stas Vavulin of MGO. All the listed colleagues are gratefully acknowledged.

Section 3 of this chapter was performed as a part of AMIP Diagnostic Subproject No. 8, which was coordinated by the PCMDI group at the Lawrence Livermore National Laboratory.

The MGO portion of this study was supported by the Russian Foundation for Basic Research through Grant 96-05-65020. The Illinois portion was supported by the National Science Foundation through Grant ATM-9319952. The AWI portion was supported by German Ministry of Education and Research through Contract 07VKV01/1.

References

1. Arpe, K., Behr, H., and Duemenil, L. (1997) Validation of the ECHAM4 climate model and re-analyses data in the Arctic region. *Proc. Workshop on the Implementation of the Arctic Precipitation data Archive (APDA) at the Global Precipitation Climatology Centre (GPCC), Offenbach, Germany,* World Climate Research Programme WCRP-98, WMO/TD No.804, 31-40.

2. Bengtsson, L., Arpe, K., Roeckner, E., Schulzweida, U. (1996) Climate predictability experiments with a general circulation model. *Climate Dyn.*, **12**, 261-278.

3. Bromwich, D.H., Cullather, R.I., and Serreze, M.C. (1999) The Atmospheric Hydrologic Cycle over the Arctic Basin from Reanalysis Data. *This issue.*

4. Bromwich, D.H., Tzeng, R.Y., Parish, T.R. (1994) Simulation of the modern Arctic climate by the NCAR CCM1. *J. Climate*, 7, 1050-1069.

5. Bryazgin, N.N. (1976) Yearly mean precipitation in the Arctic region accounting for measurement errors (in Russian). *Proc. Arctic Ant. Res. Inst.*, **323**, 40-74.

6. Bryazgin, N.N., and Shver, Ts.A. (1976) Atmospheric precipitation over the Arctic land watershed basin (in Russian). *Proc. Arctic Ant. Res. Inst.*, **323**, 75-86.

7. Chen, B., Bromwich, D.H., Hines, K.M., Pan, X. (1995) Simulations of the 1979-1988 polar climates by global climate models. *Ann. Glaciol.*, **21**, 83-90.

8. Christensen, J.H., Machenhauer, B., Jones, R.G., Schaer, C., Ruti, P.M., Castro, M., Visconti, G. (1997) Validation of present-day regional climate simulations over Europe: LAM simulations with observed boundary conditions. *Climate Dyn.*, **13**, 489-506.

9. Dethloff, K., Rinke, A., Lehmann, R., Christensen, J.H., Botzet, M., Machenhauer, B. (1996) Regional climate model of the Arctic atmosphere. *J. Geophys. Res.*, **101**, 23401-23422.

10. Duemenil, L., Isele, K., Liebscher, H.-J., Schroeder, U., Schumacher, M., and Wilke, K. (1993) Discharge data from 50 selected rivers for GCM validation. Max-Planck-Institute for Meteorology/Global Runoff Data Centre Rep. 100, 61 pp.

11. Duemenil, L., Hagemann, S., and Arpe, K. (1997) Validation of the hydrological cycle in the

Arctic using river discharge data. *Proc. Workshop on Polar Processes in Global Climate (13-15 November 1996, Cancun, Mexico),* AMS, Boston, USA.

12. Eischeid, J.K., Baker, C.B., Karl, T.R., and Diaz, H.F. (1995) The quality control of long-term climatological data using objective data analysis. *J. Applied Meteor.,* **34**, 2787-2795.

13. Gates, W.L., 1992: AMIP: The Atmospheric Model Intercomparison Project. *Bull. Amer. Meteor. Soc.,* **73**, 1962-1970.

14. Gates, W.L., and Coauthors (1999) An Overview of the Results of the Atmospheric Model Intercomparison Project (AMIP). *Bull. Amer. Meteor. Soc.,* **80**, 29-55.

15. Gibson, J.K., Kallberg, P., Uppala, S., Hernandez, A., Nomura, A., and Serrano, E. (1997) ECMWF Re-Analysis, Project Report Series: 1. ERA Description. European Centre for Medium-Range Weather Forecasting, Reading UK, 72 pp.

16. Giorgi, F., and Marinucci, M.R. (1996) An investigation of the sensitivity of simulated precipitation to model resolution and its implication for climate studies. *Mon. Weather Rev.,* **124**, 148-166.

17. Hagemann, S., and Duemenil, L. (1998) A parameterization of the lateral waterflow for the global scale. *Climate Dyn.,* **14**, 17-31.

18. Huffman, G.J., Adler, R.F, Arkin, P., Chang, A., Ferraro, R., Gruber, A., Janowiak, J., McNab, A., Rudolf, B., and Schneider, U. (1996) The global precipitation climatology project (GPCP) combined precipitation data set. *Bull. Amer. Meteor. Soc.,* **78**, 5-20.

19. Ivanov, V.V. (1976) Freshwater balance of the Arctic Ocean (in Russian). *Proc. Arctic Ant. Res. Inst.,* **323**, 138-147.

20. Jaeger, L. (1983) Monthly and areal patterns of mean global precipitation, in A. Street-Perrot, M. Beran, and R. Ratcliffe (eds.), *Variations in the Global Water Budget,* D. Reidel, 129-140.

21. Kalnay, E., and Coauthors (1996) The NCEP/NCAR 40-year reanalysis project. *Bull. Amer. Meteor. Soc.,* **77**, 437-471.

22. Karl, T.R. (1998) Regional Trends and Variations of Temperature and Precipitation, in R.T. Watson, M.C. Zinyowera, R.H. Moss, D.J. Dokken (eds.), *The Regional Impacts of Climate Change: An Assessment of Vulnerability,* Cambridge University Press, 412-425.

23. Kattenberg, A., Giorgi, F., Grassl, H., Meehl, G.A., Mitchell, J.F.B., Stouffer, R.J., Tokioka, T., Weaver, A.J., Wigley, T.M.L. (1996) Climate Models -- Projections of Future Climate, in J.T. Houghton, L.G. Meira Filho, B.A. Callander, N. Harris, A. Kattenberg, and K. Maskell (eds.), *Climate Change 1995: The Science of Climate Change,* Camridge Uniresity Press, 285-357.

24. Korzun, V.I. (ed.) (1978) *World Water Balance and Water Resourses of the Earth.* UNESCO Press, 663 pp.

25. Legates, D.R., and Willmott, C.L. (1990) Mean seasonal and spatial variability in gauge-corrected global precipitation. *Int. J. Climatol.,* **10**, 111-133.

26. Lynch, A.H., Chapman, W.L., Walsh, J.E., and Weller, G. (1995) Development of a regional climate model of the Western Arctic. *J. Climate,* **8**, 1555-1570.

27. MacKay, M.D., Stewart, R.E, and Bergeron, G. (1998) Downscaling the Hydrological Cycle in the Mackenzie River Basin with the Canadian Regional Climate Model. *Atmos.-Ocean,* **36**, 179-211.

28. Phillips, T.J. (1994): A summary documentation of the AMIP Models. Program for Climate Model Diagnosis and Intercomparison Rep. 18, Lawrence Livermore National Laboratory, UCRL-ID-116384, 343 pp.

29. Parker, D.E., and Jackson, M. (1995) The standard GISST data sets: Versions 1 and 2, in C.K. Folland and D.P. Rowell (eds.), *Workshop on Simulations of the Climate in the Twentieth Century using GISST,* CRTN-56, Hadley Centre, Bracknell, U.K., 50-51.

30. Randall, D., Curry, J., Battisti, D., Flato, G., Grumbine, R., Hakkinen, S., Martinson, D., Preller, R., Walsh, J., and Weatherly, J. (1998) Status of and Outlook for Large-Scale Modeling of Atmosphere-Ice-Ocean Interactions in the Arctic. *Bull. Amer. Meteor. Soc.*, **79**, 197-219.

31. Rinke, A., Dethloff, K., Christensen, J.H., Botzet, M., Machenhauer, B. (1997) Simulation and validation of Arctic radiation and clouds in a regional climate model. *J. Geophys. Res.*, **102**, 29833-29847.

32. Roeckner, E., Arpe, K., Bengtsson, L., Christoph, M., Claussen, M., Duemenil, L., Esch, M., Giorgetta, M., Schlese, U., Schulzweida, U. (1996) The atmospheric general circulation model ECHAM-4: Model description and simulation of present-day climate. Max-Planck-Institute for meteorology Rep. 218, 90 pp.

33. Serreze, M.C., Barry, R.G., and Walsh, J.E. (1995) Atmospheric water vapor characteristics at 70°N. *J. Climate*, **8**, 719-731.

34. Vowinckel, E., and Orvig, S. (1970) The climate of the North Polar Basin, in S. Orvig (ed.) *Climates of the Polar Regions* (*World Survey of Climatology*, **14**), Elsevier, 129-252.

35. Walsh, J.E., and Crane, R.G. (1992) A comparison of GCM simulations of arctic climate. *Geophys. Res. Lett.*, **19**, 29-33.

36. Walsh, J.E., Kattsov, V., Portis, D., Meleshko, V. (1998) Arctic Precipitation and Evaporation: Model Results and Observational Estimates. *J. Climate*, **11**, 72-87.

37. Zubenok, L.I. (1976) Evaporation over the land watershed basin of the Arctic Ocean. *Proc. Arctic Ant. Res. Inst.*, **323**, 87-100.

DISCHARGE OBSERVATION NETWORKS IN ARCTIC REGIONS: COMPUTATION OF THE RIVER RUNOFF INTO THE ARCTIC OCEAN, ITS SEASONALITY AND VARIABILITY

W.E. GRABS, F. PORTMANN, T. DE COUET
Global Runoff Data Centre (GRDC) in the
Federal Institute of Hydrology
Kaiserin-Augusta-Anlagen 15-17, D-56068 Koblenz, Germany

1. Introduction

Knowledge about freshwater flux is of importance to develop coupled landsurface-atmosphere models which are able to yield satisfactory results when they are calibrated with observed river discharge. The development of hydrological models is also described by other authors in this volume, i.e. Bowling *et al.* Shiklomanov *et al.* and Stewart. Discharge observation networks in the Arctic region provide an observational basis for the assessment of possible long-term trends and variations of the components of the fresh water balance in the Arctic region and the surface water flows into the Arctic Ocean. The knowledge of river discharge in the Arctic is also important in areas such as the navigability of coastal areas and the tracing of pollutants. The change of surface salinity especially in the coastal areas of large rivers has a strong influence on sea-ice formation. The Arctic Ocean is also an important ecological habitat with high economic importance, i.e. for fishery. The Arctic River Database (ARDB) which is being compiled by the Global Runoff Data Centre (GRDC) is a major source of information about river discharges to support research into the Arctic Ocean Freshwater Budget (AOFB).

2. Discharge Observation Network on Arctic Rivers

For the purpose of this paper, the delineation of the Arctic region is defined by the boundaries drawn in Figure 1 taken from WCRP 1998 on the basis of the areal definition by WCRP 1997. Intercomparison between areas cited for the Arctic region is difficult because of different figures cited in the literature for defining the Arctic region. The area covered by the map is 18146855 km^2.

Discharge observation networks on Arctic rivers were established in the past mainly for the assessment of freshwater resources associated with the economic development and resource utilization of these areas. For a number of rivers, discharge observations date back to the 1930s. As environmental conditions are extreme and the logistical support of these stations is very difficult, their maintenance is directly linked

249

E.L. Lewis et al. (eds.), The Freshwater Budget of the Arctic Ocean, 249-267.
© 2000 *Kluwer Academic Publishers.*

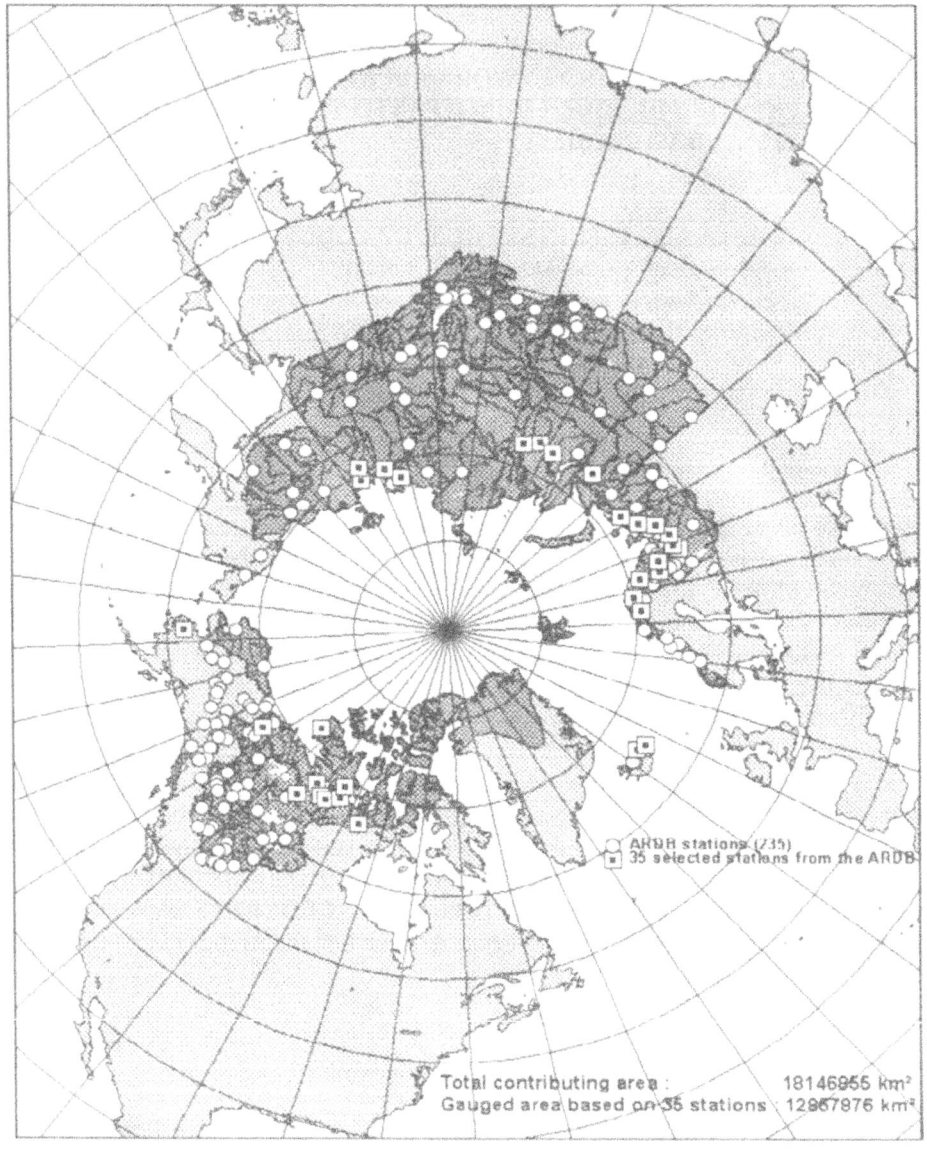

Figure 1. Distribution of stations in the Arctic River Database (ARDB)

to the economic interest to obtain information about water resources. With the beginning of de-population of high latitude areas, especially in recent years in the former Soviet Union, the station network has been decreasing and information from existing stations is difficult to obtain. The GRDC currently has 235 gauging stations associated with the Arctic River Database (ARDB) of the Arctic Climate System Project (ACSYS). However, updating of this database is becoming increasingly difficult because of high operating costs of these stations in a time, when funds to maintain this important network are diminishing. Difficult environmental and infrastructure conditions in high latitudes hamper the establishment, operation and maintenance of hydrometric stations. Amongst these are: supply of materials for the observation of stations, instrument problems, observer availability, measurement difficulties due to ice cover of rivers, lack of observations of discharge over ice sheets during the beginning of the melt-period, impoundment of discharge due to ice conditions and backwater effects as a result of ice-jamming and strong tidal influences. As a general assessment, the probable scale of error is in the order of 15 % for hinterland stations and up to 30 % for near coastal stations. The concept of "error bars" is difficult to apply for hydrological records because the complete station history including validated rating curves are often not available.

2.1. DEFINITION AND SPATIAL DISTRIBUTION OF THE DISCHARGE OBSERVATION NETWORK

Figure 1 shows the location of the 235 gauging stations that constitute the ARDB. The map also shows the location of 35 stations that have been selected for the computation of river discharge into the Arctic Ocean. All stations belong to rivers or tributaries of rivers that ultimately drain into the Arctic Ocean or are near to the mouth of rivers that influence freshwater transport into/out of the Arctic Region. The seasonal and interannual variability of freshwater discharge into the Arctic Ocean is of particular interest to quantify the flux of cold, low-saline Arctic water into the Pacific. Therefore, the Yukon river has been included for the computation of discharge into the Arctic Ocean. Long time series for the Kobuk and Noatak rivers were not available. The map shows that some gauging stations fall outside the regional boundaries defined in the map: They had been selected for inclusion into the ARDB because of the interest of modelers into land-based processes, especially permafrost. For North America, parts of Alaska and the heavily fragmented Northern shoreline of Canada the surface water discharge is not represented. However, a significant freshwater flow into the Arctic Ocean from these mostly borderline, relatively small basins is questionable. For the Northern rim of Canada, it is extremely difficult to estimate the freshwater discharge into the Ocean, as the topography of the fragmented islands and cold climate conditions throughout the year are impeding exact measurements and calculation of flow direction. Furthermore, the data are sparse: only a very limited number of gauges are present there, often not operating continuously throughout the year, limiting the reliability of any analysis to a great extent. It can be seen from the map that many of the furthest downstream stations are still not close to the mouth of the respective rivers into the Arctic Ocean due to a number of causes such as: no economic need to establish stations close to the mouth of rivers, site problems to establish stations in braided river systems or near the river mouths. Based on the map, 12867876 km^2 of the hydrographic area defined in the map, or about 30 % of the area is ungauged and has not been quantitatively considered in the computation of discharge into the Arctic Ocean. Most of the ungauged area belongs to parts of drainage basins downstream gauging stations near - or as close as available - to river mouths into the Arctic Ocean. The location of gauging stations upstream the mouth of rivers constitutes the infrastructural limit to the availability of data.

3. Components of Freshwater Discharge Into the Arctic Ocean

The total freshwater discharge from the land area into the Arctic Ocean is the sum of river discharge into the ocean, subsurface water flows mainly from the freeze-melt cycle in the active layer of permafrost soils and groundwater flows into the ocean. The contribution of freshwater from the freeze-melt cycle cannot currently be determined with a reliable degree of certainty. For hydrological modeling however, a better knowledge of permafrost and its spatial extent and varying depth are of paramount interest. In general, the subsurface and groundwater flow is considered to be magnitudes of orders lower than river discharge due to the immobilization of large parts of subsurface and groundwater in permafrost regions. Direct groundwater flow into the Baltic Sea from the Polish territory has been calculated by [1] at 1.33 % of the total volume of river discharge from Polish territory which has been estimated at 58.6 km^3 / year. This figure is indicative of the order of magnitude of the groundwater contribution to the freshwater budget of the Arctic Ocean.

4. Computation of River Discharge Into the Arctic Ocean

The quality of discharge records as assessed by GRDC [2] is generally satisfactory within the error bars indicated above and cited in WCRP 1998. Record availability ranges from a few years to more than 60 years. The overlap of time series of river discharge is however problematic as the time of available records is not homogenous for all rivers in the ARDB. The computation of the year-to-year variability of cumulated discharge into the Arctic Ocean is therefore not possible for the entire time series of rivers in the ARDB.

On a long-term basis, the computation of river discharge into the Arctic Ocean should be embedded in a monitoring system for the Arctic Ocean. This requires a selection of rivers from the entire database using the following criteria: All major rivers should be incorporated with their gauging stations closest to the mouth into the Arctic Ocean and the cumulated areas of the drainage basins should cover as much as possible the total contributing area of river discharge into the Arctic Ocean. The number of stations needed for the computation of discharge is obtained by plotting the depth of runoff and discharge of river basins ranked in the order of the basin size, starting from the largest drainage basin. Based on available records in the ARDB, Figure 2 shows that only 35 rivers are needed to determine the river discharge into the Arctic Ocean with reasonable accuracy for monitoring purposes: The 35 rivers drain about 70 % of the total contributing area as defined in Figure 1; most of the ungauged area of about 30 % is situated downstream of the stations closest to the mouth of the selected rivers and stations. The stations after rank 24 have basin sizes of less than 7000 km². Figure 2 shows also, that the inclusion of more rivers for computational purposes will not significantly increase depth of runoff and the volume-sum of discharge. Table 1 identifies the 35 rivers selected for the computation. Note, that for reasons explained above, the Yukon river has been included in the computation despite the fact that it lies outside the defined region. The resulting discharge values of the computations are shown in Table 2.

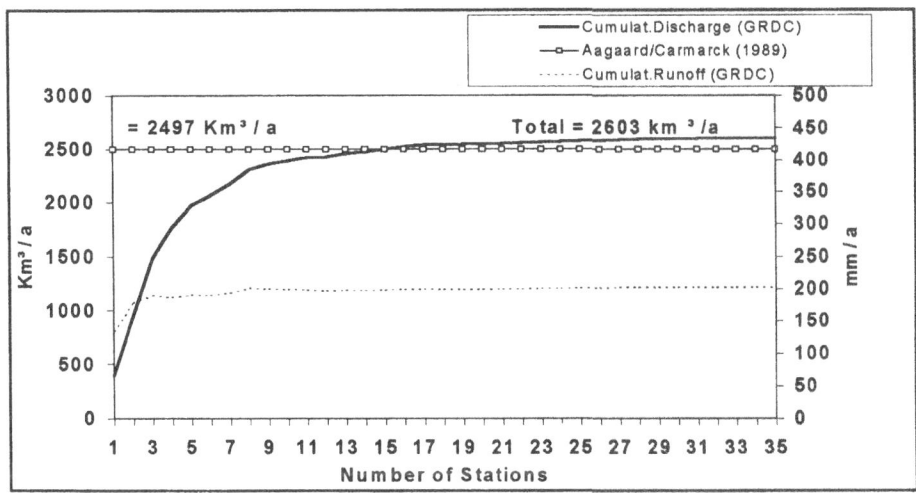

Figure 2. Volume-sum discharge and depth of runoff of selected rivers

TABLE 2. Discharge into the Arctic Ocean based on 35 rivers with gauging stations close to the mouth

Variable	Value based on 35 stations
GRDC-basin area (km²)	12867876
Depth of runoff (mm/year)	202
Volume-Average (km³/year)	74
Volume-Sum (km³/year)	2603

TABLE 1. 35 selected GRDC stations

No.	GRDC-No.	River	Station	Country	Latitude	Longitude	Basin size (km²)	Begin	End
1	2909150	Yenisei	Igarka	Russia	67°48'N	86°50'E	2440000	1/1936	12/1995
2	2903420	Lena	Kusur	Russia	70°70'N	127°65'E	2430000	1/1978	12/1994
3	6971600	Varzuga	Varzuga	Russia	66°40'N	36°63'E	7940	1/1979	12/1988
4	2912600	Ob	Salekhard	Russia	66°57'N	66°53'E	2949998	1/1930	12/1994
5	2998110	Yana	Ubileynaya	Russia	70°75'N	136°08'E	224000	1/1978	12/1994
6	2998150	Omoloy	Namu	Russia	69°38'N	134°62'E	108000	1/1979	12/1993
7	2998400	Indigirka	Vorontsovo	Russia	69°58'N	147°35'E	305000	1/1978	12/1994
8	2998510	Kolyma	Kolymskaya	Russia	68°73'N	158°72'E	526000	1/1978	12/1994
9	2999250	Taz	Sidorovsk	Russia	66°60'N	82°28'E	100000	1/1978	12/1994
10	2999500	Pur	Samburg	Russia	67°08'N	78°15'E	95100	1/1978	12/1990
11	2999910	Olenek	7.5 km downstream of mouth of River Pur	Russia	72°12'N	123°22'E	198000	1/1965	12/1984
12	4103200	Yukon	Pilot Station	United States	61°93'N	162°88'W	831390	10/1975	9/1993
13	4208025	Mackenzie River	Arctic Red River	Canada	67°46'N	133°74'W	1660000	8/1972	12/1992
14	4209400	Coppermine River	Point Lake Outlet	Canada	65°41'N	114°00'W	19300	7/1965	12/1992
15	4209450	Big River	above Egg River	Canada	72°48'N	123°40'W	3640	7/1975	12/1988
16	4209500	Tree River	near the mouth	Canada	67°63'N	111°88'W	5960	12/1968	12/1992
17	4209550	Burnside River	near the mouth	Canada	66°74'N	108°82'W	16800	9/1976	12/1992
18	4209580	Gordon River	near the mouth	Canada	66°81'N	107°10'W	1530	8/1977	12/1992
19	4209600	Ellice River	near the mouth	Canada	67°71'N	104°14'W	16900	1/1971	12/1992
20	4209650	Freshwater Creek	near Cambridge Bay	Canada	69°13'N	104°99'W	1490	7/1970	12/1992
21	4209800	Back	below Deep Rose Lake	Canada	66°08'N	96°50'W	98200	1/1966	12/1984
22	6401090	Oelfusa	Selfoss	Iceland	63°94'N	21°01'W	5760	1/1950	12/1992
23	6401120	Thjorsa	Urridafoss	Iceland	63°93'N	20°60'W	7200	4/1947	12/1993
24	6401700	Joekulsa a Fjollum	Dettifoss	Iceland	66°03'N	16°45'W	7000	9/1939	12/1984
25	6401800	LagarfIjot	Lagarfoss	Iceland	65°50'N	14°37'W	2800	9/1949	12/1993
26	6730500	Tana	Polmak	Norway	70°07'N	28°05'E	14005	1/1912	12/1987
27	6731950	Altaelv	Masi	Norway	69°42'N	23°63'E	5693	1/1978	12/1983
28	6970100	Onega	Porog	Russia	63°80'N	38°27'E	55770	1/1978	10/1993
29	6970200	Solza	Soukhie Porogui	Russia	64°31'N	39°48'E	1190	1/1978	12/1987
30	6970250	Northern Dvina (Severnaya D.)	Ust-Pinega	Russia	64°10'N	42°17'E	348000	1/1883	12/1993
31	6970500	Mezen	Malonisogorskaya	Russia	64°95'N	45°67'E	56400	1/1978	12/1993
32	6970630	Pesha	Volokovaya	Russia	66°50'N	48°25'E	2780	1/1978	12/1993
33	6970710	Pechora	Oksino	Russia	67°63'N	52°18'E	312000	1/1989	12/1993
34	6971100	Kola	Oktiabrsky Railway km 1429	Russia	68°88'N	33°05'E	3780	1/1979	12/1988
35	6971150	Umba	Paialka	Russia	66°64'N	34°08'E	6250	1/1979	12/1988

4.1. VARIABILITY OF TOTAL DISCHARGE INTO THE ARCTIC OCEAN

The variability of total surface water discharge into the Arctic Ocean is illustrated for the rivers Lena, Ob and Yenisei in Table 3 and Figure 3. Unfortunately, data for the Mackenzie River have been available only since 1972. The Mackenzie contributes on average 283 km^3 to the total discharge. When combined with the three Siberian rivers the total discharge of these four rivers amounts to 1772 km^3 or 68 % of the calculated volume-sum discharge. The Lena, Ob, Yenisei alone contribute on average 57 % of the calculated river discharge into the Arctic Ocean.

TABLE 3. Variability of annual river discharge into the Arctic Ocean based on three Siberian rivers

Parameter	Geographical extent				
	Lena	Yenisei	Ob	Sum	Total sum based on 57 % contribution of the three rivers
	(km^3)	(km^3)	(km^3)	(km^3)	(km^3)
Mean	523	567	399	1489	2603
Maximum (year 1978)	602	654	430	1686	2947
Minimum (year 1954)	417	508	339	1264	2211

The data in Table 3 show for the total discharge sum a total span of 27 % between maximum flows (+12 %) and minimum flows (-15 %) as compared to the average total flow of 2603 km^3 per year.

4.2. TERMINOLOGY FOR THE COMPUTATION

To avoid misunderstanding, the used terminology is briefly explained: The total measured discharge volume of all rivers results in the "volume-sum" discharge. The average discharge-volume per river is noted as "volume-average". The depth of runoff is derived as ratio of the volume-sum divided by the basin area.

256

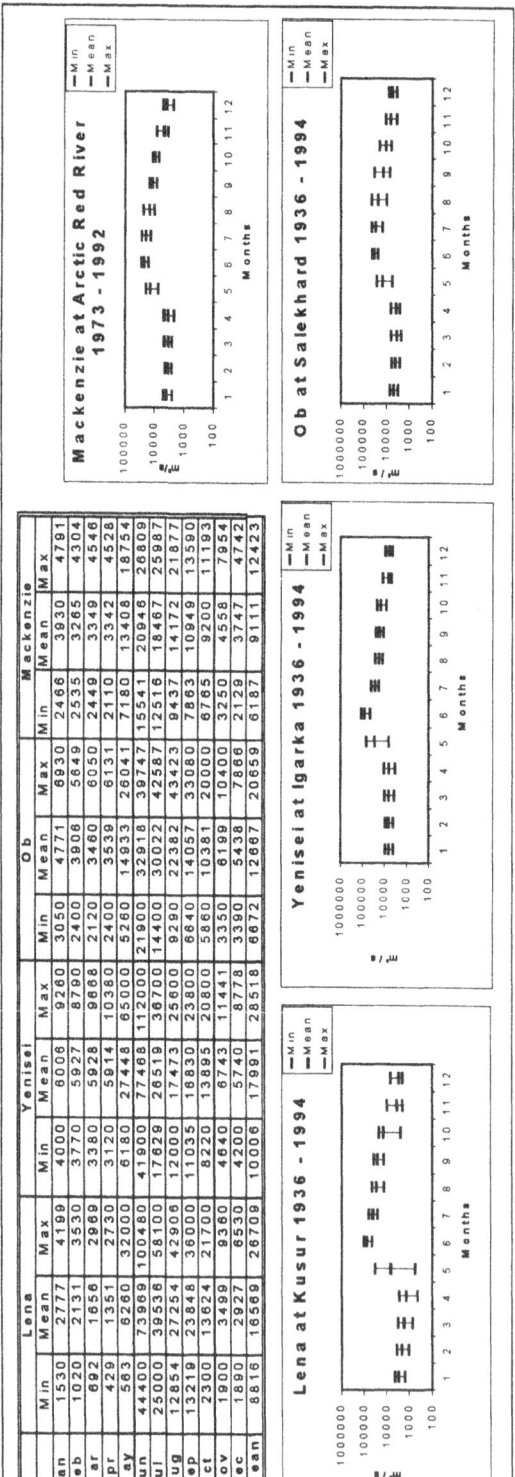

Figure 3. Variability of long-term mean monthly discharge of selected rivers

4.3. COMPARISON BETWEEN OBSERVED DISCHARGE AND PRECIPITATION

Discharge represents the integrated basin response to precipitation, evaporation, groundwater storage and storage of precipitation as snow and ice, including the freeze-melt processes in the Arctic region. In general, the measurement of discharge is more accurate than the derived areal depth of precipitation from point measurements. The data stored in the ARDB therefore allows the validation of other components of the water balance which has great significance for the improved estimation of the Arctic Ocean Freshwater Budget. Discharge data are also used for the development of re-analysis data sets which have become an integral part of hydrological models (see Bowling *et al.*, this volume). Discharge and precipitation can only be compared on an annual basis because there is no direct correlation between observed discharge and precipitation on a monthly basis: The temperature regime of high latitudes effectively de-couples precipitation-discharge relationships due to the storage of precipitation as snow and ice during more than 7 months of the year and subsequent discharge peaks during ice break-up and snowmelt from late May - June and flow recession in the following months.

As a first comparison of the order of magnitude of observed precipitation and discharge, the computed freshwater discharge is compared with precipitation estimates given by Radionov [3]. According to Radionov, the basin area of the Arctic amounts to 14056000 km^2 versus 12867876 km^2 land area of gauged drainage basins. Figure 4 below shows the comparison between discharge and precipitation based on an average annual depth of precipitation of 169 mm after Radionov. The respective volume-sum of precipitation amounts to 2375 km^3 and is in the same order of magnitude as the volume-sum of river discharge (2603 km^3). The months June - September contribute 51 % of the total yearly precipitation while 79 % of the discharge into the Arctic Ocean occurs during the months May - September. Realizing, that in the calculation of surface water discharge the subsurface and groundwater flow has not been included, the precipitation figure given by Radionov seems to underestimate the actual precipitation.

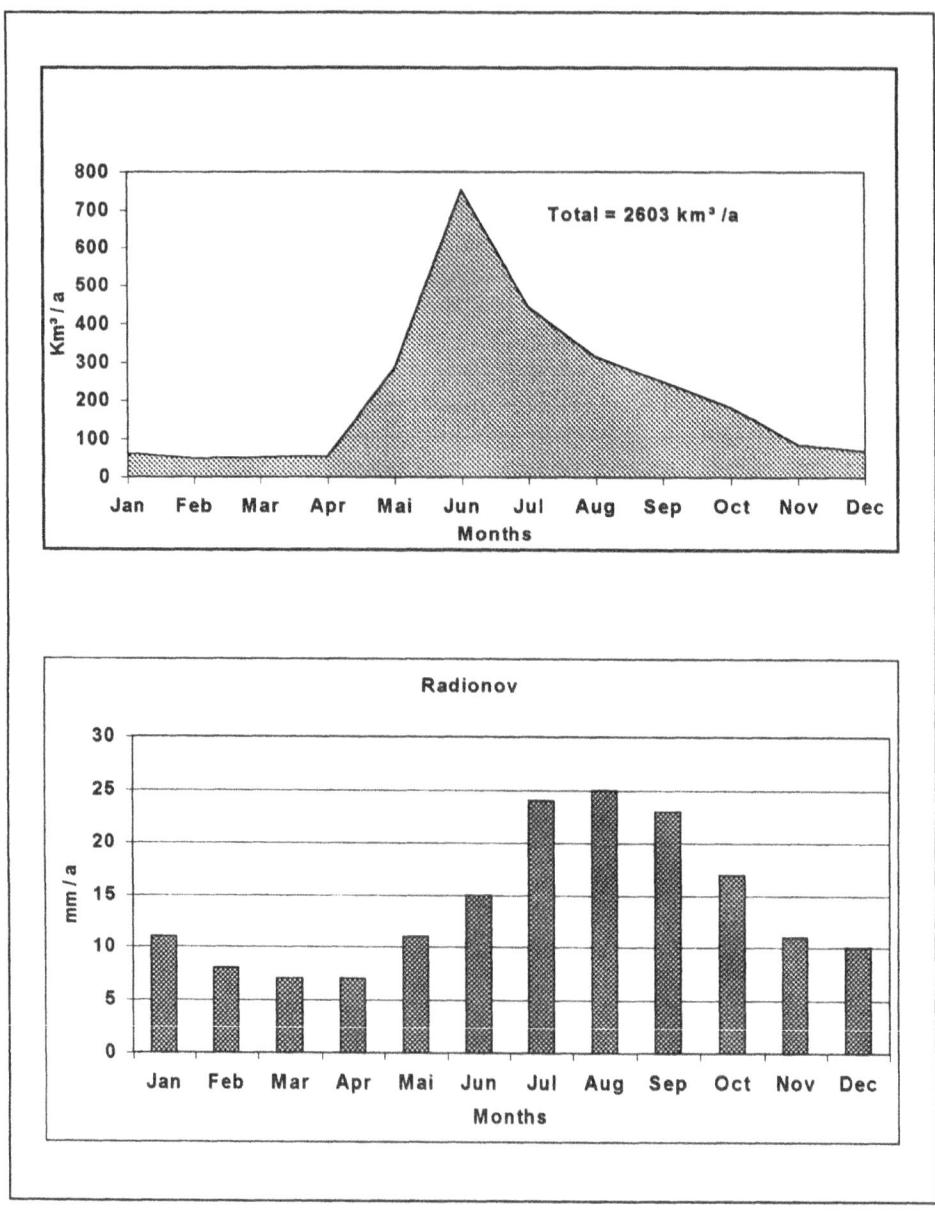

Figure 4. Volume-sum discharge and monthly precipitation

5. Seasonality and Variability of Discharge into the Arctic Ocean

The regimes of Arctic rivers are strongly seasonal and dominated by the spring snow and ice melt. Peak discharge is usually reached in June with a long tail of declining discharge until November and "baseflow" conditions during the months of December through April. As has been shown with available data, nearly 80 % of the discharge into the Arctic Ocean occurs during the months May - September. Due to the incomplete overlap of time series of gauging stations, discharge variability cannot be computed accurately on a monthly basis for the entire data set. Seasonality and variability is therefore illustrated for a selection of rivers. The rivers Lena, Ob, Yenisei, and the Mackenzie River were selected, because they have the best long-term complete records of all the major rivers while on the basis of available time series they contribute 68 % to the total river discharge into the Arctic Ocean. Table 4 gives a description of stations of the selected rivers.

TABLE 4. Station characteristics of selected Arctic rivers

River	Station	Basin size (km²)	Selected period	
			First year	Last year
Lena	Kusur	2430000	1936	1994
Yenisei	Igarka	2440000	1936	1994
Ob	Salekhard	2949998	1936	1994
Mackenzie River	Arctic Red River	1660000	1973	1992

It must be noted that the Siberian basins are of similar size with the largest being the Ob, while the Mackenzie has roughly half that size.

The representativeness of the four selected rivers is fairly good with respect to the area discharging into the Arctic Ocean.

5.1. CONSISTENCY AND HOMOGENEITY

Basic assumptions of consistency and homogeneity within the records have to be met before starting any statistical interpretation of results. Therefore, the series were checked for normal distribution, independence and changes in mean and variance and in special cases also for changes in spectrum. As a measure of normality, the Kolmogorov-Smirnov test for deviations of normal distribution was applied to the annual series, rejecting normality slightly in the case of River Yenisei with an error probability of 5 %. For a detailed description of the statistical tests see [4].

5.2.   ANNUAL VARIABILITY

Considering annual variability, all the annual series of the *Siberian* Arctic (represented by the Lena, Ob and Yenisei) show considerable high mean annual discharge between roughly 10000 m³/s and 20000 m³/s. The minimum and maximum values of the series have dimensions respective to the mean annual value, with a 10 - 15 % coefficient of variation. Of the Siberian rivers, the Ob has least discharge and least minimum discharge, also characterized by the highest discharge variability of all the series. The *North American* Mackenzie River has less discharge, coinciding with the smallest variability.

5.3.   MONTHLY VARIABILITY

For monthly variability, the statistical properties are explained below.

In *Siberia*, two different groups are prominent - the Lena and Yenisei on one side, and the Ob on the other.

In the first group, the *Lena and Yenisei* show similar monthly variability, characterized by a mean discharge within the range of 17000 ± 1000 m³/s and the highest mean monthly discharge in June within the range of 76000 ± 2000 m³/s, which is more than four times the mean annual runoff. Minimum monthly discharge is least in April, the Lena having about 8 times less discharge (about 430 m³/s) than the Yenisei (about 3120 m³/s), each with a maximum to minimum ratio of 230 and 36, respectively. This is a consequence of the increasing continental climate from the Yenisei to the Lena. Following the continental climate pattern, the smallest and highest monthly value (about 430 m³/s and 112000 m³/s at the Lena) are observed just two months apart, in April and June, indicating most variable conditions during the period of these spring and early summer months when snow and ice melt. However, the highest coefficients of variation are observed in May and during the two months prior. The Lena with an extremely high coefficient of variation of 1.1 for May illustrates the highly variable flow regime as a result of climatic variability.

The highest positive skew and kurtosis occur in November, the Lena having much higher coefficients than the Yenisei (skew 2.7 vs. 1.2, kurtosis 16.2 vs. 5.1). A positive skew means, that in contrast to many values around the mean, more high values occur than at a normal distribution with zero coefficient of skew. A kurtosis above 3 means, that mean values occur more often than at a normal distribution, while fewer extreme positive or negative values exist. Therefore, the parameter values indicate that in November and May flows extremely high relative to the monthly mean occur occasionally, in November less often than in May. In short, both months are transient months between summer and winter flow conditions.

By contrast, the *Ob* with a mean discharge of 12500 m³/s, has only about 25 % less discharge, but a more even runoff regime and constitutes the second group. Minimum monthly discharge is observed in March (2100 m³/s) and not in April as the Lena and Yenisei. June has the highest mean monthly discharge (22000 m³/s) like the Lena and Yenisei, together with one of the highest maximum monthly discharges (40000 m³/s). Those maximum values are almost constant during July and August (43000 ± 500 m³/s) and are peaking in August, a fact not observed on the other Siberian rivers where they peak in June. The maximum flows in August also exhibit the highest monthly coefficient of variation of 0.4 together with September. This is equivalent to a inter-monthly maximum to minimum ratio of about 20.

Nevertheless, a negative skew is observed in June and July (-0.7 and -0.4, respectively) pointing at relatively few very small values. The highest kurtosis, combined with an elevated positive skew (+1.2), is observed in October, a relationship also present in March, April and September. This means that in late winter & spring, and early winter & autumn, many monthly discharges close to the mean are combined with extreme, mostly larger discharge values. This happens about one month earlier on the Ob than on the Lena and Yenisei for both seasons.

In *North America*, the conditions for the Mackenzie are different in some aspects: the mean discharge at about 9000 m³/s is even less than that of the Ob, while the coefficients of variation are small, being only 23 % in May and November. Nevertheless, as for the Siberian rivers, the absolute minimum is observed in April, and the mean minimum occurs in the period from February up to April, while the month with the highest mean and maximum discharge is June, a condition to be violated only by the Ob. In contrast to the other rivers, the combination of extreme kurtosis and prominent right-sided, positive skew occurs only in November and in August, the second period replacing May or April in case of the Siberian rivers. For all other months, the skew is only small (< 0.4), and the kurtosis is typical of a normal distribution, showing only slightly peaked periods in September and December. This indicates the occurrence of relatively large flows not in spring like in Siberia, but in summer, early winter being a period common to both regions.

Characteristics of the annual time series, recalculated from the monthly data for the four rivers are shown in Table 5 and characteristics of the monthly series also appear in Figure 3. Flow data unit is m³/s. An explanation of the respective statistical terms is given in Table 6. For specific definitions and notations see [5], [4] and [6].

TABLE 5. Statistics of annual series (explanations are given in Table 6)

River	Begin year	End Year	No. of years	Mean	Min	Max	SE Mean	Std Dev	CF Var	CF Skw	CF Kur
Mackenzie	1973	1992	20	8971	7480	10462	209	936	0.104	-0.138	2.374
Lena	1936	1994	59	16569	12478	22626	259	1986	0.120	0.525	3.426
Ob	1936	1994	59	12667	8791	17812	252	1937	0.153	0.278	2.856
Yenisei	1936	1994	59	17991	15543	20966	179	1377	0.077	0.210	2.294

TABLE 6. Explanations of statistical parameters

Abbre-viation of parameter	Parameter	Type of measure for distribution	Meaning
Mean	Arithmetic mean	Location	Value closest to all sampled values
Min	Minimum	Location	Minimum value of the series
Max	Maximum	Location	Maximum value of the series; Range = Maximum – Minimum
SEMean	Standard error of the mean	Location	Deviation of the mean of the sample from the true mean, based on the assumption of a statistical distribution of the means
StdDev	Standard deviation	Shape	Average amount that individual data deviate from their mean
CFVar	Coefficient of variation	Shape	Standard deviation divided by the mean; often expressed in percentage; a measure of relative deviations of data from the mean (interpretation: < 0.3 small; 0.3 - 0.6 medium; > 0.6 big)
CFSkw	Coefficient of skewness	Shape	Skewness means lack of symmetry (> 0 distribution skewed to the left, the gravitational center is located on the left side while some extreme values on the right side occur; 0 no skew; < 0 skewed to the right, the gravitational center is located on the right side)
CFKur	Coefficient of kurtosis	Shape	Peakedness; a measure serving to differentiate between a flat or sharply peaked distribution

5.4.   FLOW DURATION CURVES

Flow duration curves (FDC) provide a valuable tool for the assessment of water availability over the year. The FDC for the Lena, Ob and Yenisei reveal a very uneven flow regime, as is depicted in Figure 5. The flow regime may be characterized by the median value (50 % of the values over/below that threshold, 50 % percentile, median) and the most extreme 10 % (10 % of the time exceeded). The FDC regimes of the Lena, Yenisei, Ob, and Mackenzie shows a median of approximately 5000 m^3/s, 10000 m^3/s, 7000 m^3/s and 6000 m^3/s, respectively. On the other hand, the most extreme 10 % of the values exceed approximately 40000 m^3/s, 40000 m^3/s, 32000 m^3/s and 18000 m^3/s, respectively. The FDC depicts a sharp decline for the Lena, a more even distribution for the Yenisei and especially the Ob, while the Mackenzie has a sharp decline together with a broad plateau around 50 % between 5000 m^3/s and 9000 m^3/s. Once again, climatic influence on thawing and ice break-up is the cause. The smallest values, given as a sharp rise of the FDC, represent the months at the end of the winter, which are dry in snowmelt. The highest values are associated with spring ice break-up, depleting the storage volume accumulated during winter.

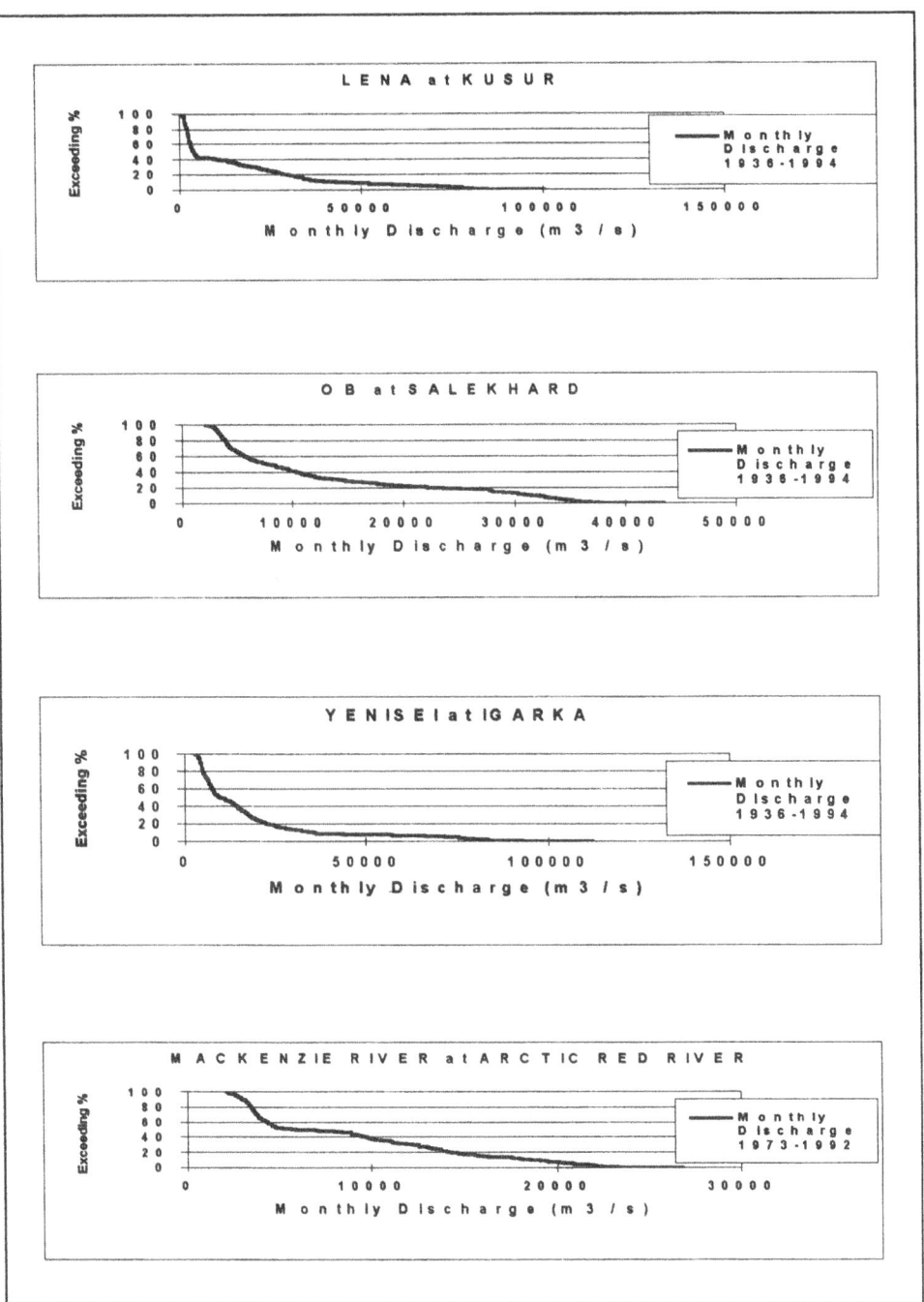

Figure 5. Flow duration curves of mean monthly discharge of selected rivers

5.5.   SUMMARY ON SEASONALITY AND VARIABILITY

Summing previous information on means and extremes in Figure 3, it is evident that the largest flow variability of the Lena occurs during the start of the spring melt in early April and during the time of ice break-up of the river in May. There is much less variability in all other months, with the exception of October. Similarly, for the Yenisei, the largest discharge variability occurs during the break-up in the second half of May. The Ob shows a significant deviation from this pattern as it is characterized by high variability of flows during all months of the year. The seasonal hydrograph of the Mackenzie is similar to that of the Ob, although it displays the overall smallest variability in discharge, even during the melt period in May. Overall, the Lena has the highest overall variability of the three rivers in May. For all four rivers, least variability occurs during the month of peak discharge in June.

6.   Periodicities, Jumps and Trends

Overlapping time series for the Lena, Ob, Yenisei and Mackenzie permit the determination of jumps, trends and periodicities in the time series. The analysis of periodicities, jumps and trends allows the detection of changes in the time series of the hydrological data. These changes may be a result of natural changes in the hydrological regimes and/or anthropogenic changes.

No significant **periodicities** were found in the annual time series of the four main rivers, although, some jumps and trends were observed.

Jumps occur in the monthly series of all four rivers. In Siberia and North America, the conditions are different. The three *Siberian* rivers all show a statistically significant jump in the first four months of the year in the mid-1960s. Another significant jump is also observed on the Yenisei for all months with the exception of May and June, leading to a significant change of the mean in the annual series, too. For the Lena and Ob, additional jumps are observed in 2 to 3 other month periods (December being common to both series), but with different starting years. For the *North American* Mackenzie River, all jumps start only in the 1980s. The winter months (December to March) are suspected to have a jump between the years 1983 and 1984, and specifically in January probably starting a year earlier. In general, the jumps occur in the same months as for the Siberian rivers, but are absent in some of the other. Furthermore, the timing of the first jump (1982/1983) is also observed in the other Siberian river series, but in different months, November for the Ob and September for the Lena. In the annual series, significant jumps are only present on the Yenisei in the beginning of the 1970s.

The overall annual **trend** for the *Siberian* rivers is positive, with a slope most intense for the River Lena. The tendency for each month is also much stronger than for the annual series. In contrast to the overall positive annual trend, negative trends are also evident mostly in the second half of the year, but also in July and May. Positive trends occur from December to April. Statistically significant positive trends for individual months occur from November to April for the Lena and Yenisei, for the Ob only in the sub-periods of November and March. Negative trends (marked with an asterisk in Table 7) occur in September or from July to September, respectively.

The *North American* Mackenzie, however, shows only significant positive trends from February to May. For sub-periods of annual series, a significant trend is observed only on the Yenisei from 1966 up to 1994, pointing to a marked change beginning in the mid-1960s and leading to the significant jump in the mean as early as 1970/71. The start of the significant jumps and of the trends are tabulated in Table 7.

6.1. SUMMARY ON PERIODICITIES, JUMPS AND TRENDS

In summary, statistically significant increases in discharge have occurred in Siberia and North America during the winter months - in Siberia from November up to April and in North America between February and May. A respective significant decrease is only observed in Siberia in late summer (September), or occasionally also earlier in summer (July - September) for the River Ob, the river with least overall trend and significant positive trends only in March and November. The causality of trends and jumps could not be determined.

7. Conclusions

On the basis of observed river discharge, our studies contribute to a better quantification of the Arctic Ocean Freshwater Budget. Based on 235 gauging stations in the Arctic River Database, 35 rivers have been selected on the basis of representativeness for the calculation of the volume-sum of discharge into the Arctic Ocean. The selected rivers drain an area of 12.9 million km^2 with an annual total discharge of 2603 km^3 into the Arctic Ocean. The gauged area of the selected rivers accounts for 70 % of the Arctic region depicted in Figure 1. The error bandwidth resulting from discharge measurements, the use of different time series and the lack of observations from melting processes and subsurface flows add to the uncertainties associated with the computation of total river discharge into the Arctic Ocean. For the analysis of variability and seasonality as well as for tests on trends and jumps in the time series, four rivers have been selected which account for about 68 % of the volume-sum of river discharge into the Arctic Ocean. Climatic variability markedly influences the timing of the melt period around the month of May, and hence river discharge into the Arctic Ocean. Approximately 79 % of the yearly runoff occurs in the months May - September. The more continental the climate, the more extreme the ratio of minimum to maximum discharge. In monthly series, increases in discharge have occurred from November to April in Siberia and from February to May in North America. A decrease in discharge is only observed in Siberia between July and September, the latter month being common to all rivers. The monthly variability of the volume-sum discharge into the Arctic Ocean cannot be computed with sufficient accuracy due to the inadequate overlap of the time series of the available river records. Only for the Yenisei river significant jumps and trends occur in annual series. The observed jumps and trends in time series need to be further examined for their causality.

TABLE 7. Summary of jumps and trends

River	Month	Jumps	Trends
(Period of record)		Year of first jump	Starting year
Mackenzie	JAN	1982/83	-
(1973-1992)	FEB	1983/84	1983
	MAR	1983/84	1983
	APR	-	1983
	MAY	-	1983
	JUN	-	-
	JUL	-	-
	AUG	-	-
	SEP	-	-
	OCT	-	-
	NOV	-	-
	DEC	1983/84	-
	ANNUAL	-	-
Ob	JAN	1965/66	-
(1936-1994)	FEB	1965/66	-
	MAR	1965/66	1967
	APR	1967/68 (1965/66)	-
	MAY		-
	JUN	1969/70	-
	JUL	-	*1969
	AUG	-	*1969
	SEP	-	*1969
	OCT	-	-
	NOV	1982/83	1966
	DEC	1967/68	-
	ANNUAL	-	-
Yenisei	JAN	1965/66	1966 (1975)
(1936-1994)	FEB	1965/66	1966
	MAR	1965/66	(1966) 1982 (1979)
	APR	1965/66	(1966, 1979) 1982
	MAY	-	-
	JUN	-	-
	JUL	1965/66	-
	AUG	1965/66	-
	SEP	1965/66 (1968/69)	*1985
	OCT	1965/66	-
	NOV	1965/66	(1968) 1978
	DEC	1965/66	(1966) 1976 (1978)
	ANNUAL	1970/71	1966
Lena	JAN	1966/67	(1970) 1974
(1936-1994)	FEB	1965/66	(1969) 1973
	MAR	1965/66	(1970) 1973
	APR	1965/66	(1970, 1973) 1986
	MAY	-	-
	JUN	-	-
	JUL	-	-
	AUG	-	-
	SEP	1982/83	*(1970) 1973
	OCT	-	-
	NOV	-	(1970, 1975) 1977
	DEC	1979/80	1969 (1973)
	ANNUAL	-	-

8.   Future Research and Recommendations

The direction of future research requirements in the Arctic region is adequately covered in the implementation plan of the Arctic Climate System Study (ACSYS). In general, Arctic research aims to adequately represent Arctic processes in global climate models. The specific objectives of the ACSYS hydrological program are [7]:
- Determining the elements of the fresh water cycle in the Arctic region and their time and space variability;
- Quantifying the role of atmospheric, hydrological and landsurface processes in the exchanges between different elements of the hydrological cycle;
- Developing mathematical models of the hydrological cycle under specific Arctic climate conditions, suitable for inclusion in coupled climate models;
- Providing an observational basis for the assessment of possible long-term trends of the components of the fresh water balance in the Arctic region under changing climate.

With regard to total freshwater discharge into the Arctic Ocean, we recommend, that the contribution of freshwater from permafrost soils and groundwater seepage needs to be quantified.

With regard to river discharge observations into the Arctic Ocean, the presently available database and observing systems are not adequate to provide the information necessary to achieve the research objectives mentioned above. An adequate monitoring program is a prerequisite for detecting the variability and possible changes in the freshwater flux that might result from i.e. climate change. We therefore recommended that an Arctic Hydrological Monitoring Program should be established, building on existing infrastructure and upgraded technology both, in data collection platforms and data transmission. A long-term commitment for the operation of such a system is required to obtain the necessary results needed for modeling and a better understanding of the influence of Arctic processes on the climate of the world.

9.   References

1. Czeslaw, P. and Müller, H.E. (1985) The direct groundwater flow into the Baltic Sea from quartenary aquifers along the Polish coast, *Beiträge zur Hydrologie* 2, 57-72.
2. GRDC – Global Runoff Data Centre (1996) *Second interim report on the Arctic Climate System Study (ACSYS)*, GRDC – Report No. 12, Koblenz, July 1996.
3. Radionov, V.F. (1997) *The snow cover of the Arctic basin*, Technical report TR 9701, Polar Science Centre, Applied Physics Laboratory, University of Washington, March 1997.
4. Portmann, F. (1998) *Evaluation of statistical properties of discharge data of stations discharging into the oceans - Europe and selected world-wide stations*, GRDC – Report No. 19, Koblenz, June 1998.
5. Maidment, D. R. (1992) *Handbook of hydrology*, McGraw-Hill, New York.
6. WMO - World Meteorological Organization (1988) *Analyzing long time series of hydrological data with respect to climate variability - Project description. WCAP – 3*, WMO/TD-No. 224, Geneva 1988.
7. WCRP – World Climate Research Program (1994) *Arctic Climate System Study (ACSYS). Initial Implementation Plan. WCRP-85*, WMO/TD-627, Geneva 1994.

ARCTIC RIVER FLOW: A REVIEW OF CONTRIBUTING AREAS

T.D. Prowse and P.O. Flegg
National Water Research Institute
11 Innovation Boulevard
Saskatoon, CANADA

1. Introduction

A circumpolar ring of land practically surrounds the Arctic Ocean. The areal extent of the landmass, which transfers flow to this ocean, is substantial and exceeds the ocean area to which it contributes. For example, headwater streams as far south as 46°N in Russia eventually empty into this relatively confined northern basin. As noted by Lewis *et al.* [1], the total volume and temporal variability of freshwater discharge to the Arctic Ocean (AO) is of critical concern to the ocean freshwater budget (FWB) and subsequently, via atmospheric and oceanographic feedbacks, to global climate. In general, considerable research continues to refine the size and variability of the various components that comprise the AOFWB, some of which are difficult to quantify. Despite the relatively large database for the river-flow component, there exists in the literature considerable variation in the estimates of its contribution. The objectives of this paper were to quantify the magnitude of river flow entering the Arctic Ocean according to the ocean definition of [1], to compare this value with other recent estimates, and to identify reasons for differences in the estimates. A detailed analysis of historical trends in Arctic flow are not included since this is provided by Shiklomanov *et al.* [2] and Grabs *et al.* [3], elsewhere in this volume. The potential impact of future climate change on total Arctic river flow, however, is the focus of a final objective. Recommendations for future research conclude the manuscript.

2. Recent Estimates

Although numerous estimates of total Arctic river flow have been made over the past several decades, the most reliable are those produced in the last few years and it is these cited most often in oceanographic analyses and in reviews of the AOFWB. Even then, however, the flow values can vary dramatically and sometimes little information has been provided about how the numbers were derived. To provide clarification, this study completed a detailed review and re-analysis of the data. The results indicate that the major source of variability stems from different geographical definitions of the receiving body of water - the Arctic Ocean, and hence the contributing land area.

E.L. Lewis et al. (eds.), The Freshwater Budget of the Arctic Ocean, 269-280.
© 2000 *Kluwer Academic Publishers.*

270

Figure 1 shows four alternate contributing land areas based on different geographical definitions of the ocean boundaries. Figure 1a refers to the broadest Arctic definition employed by Shiklomanov *et al.* [2] (i.e., "All Arctic Regions" or AAR) in

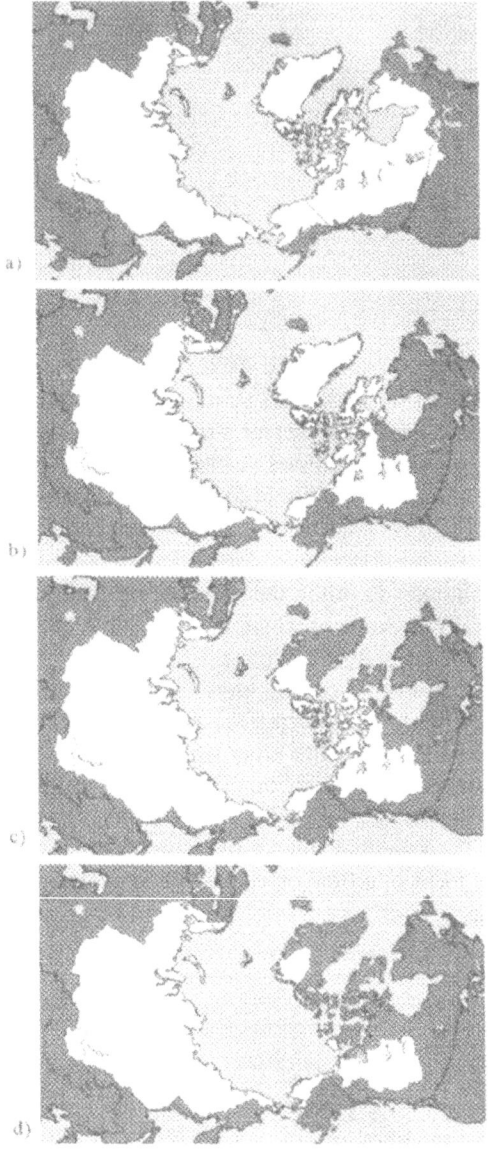

Figure 1. Geographical extent of river basins contributing flow to the Arctic Ocean based on four different definitions: a) "All Arctic Regions" (AAR, [2]); b) "Arctic Ocean Basin" (AOB, [2]); c) "Arctic Climate System" (ACSYS, [5]); and d) "Arctic Ocean River Basins", (AORB, [1]).

which flow is included from river basins in North America that drain into Hudson Bay and from the Yukon River (Alaska) and Anadyrsky River (Russia) that drain into the north Pacific just south of the Bering Strait. Arguably, much of the flow from such rivers influence the freshwater budget of the Arctic Ocean but the effect is indirect resulting from the lowering of the salinity of major ocean inputs, such as through the Bering Strait. Also included in the AAR definition is flow from all of Greenland, the eastern edge of Norway and the Canadian Arctic Archipelago, giving a total contributing area of 23.7 x10^6 km^2 (For reference, Vuglinsky [4] employs a somewhat similar total of approximately 22 x 10^6 km^2 which approximately corresponds to the AAR definition without Greenland, although the exact areas included by [4] are not described). Figure 1b shows the smaller "Arctic Ocean Basin" (AOB) area (18.9 x 10^6 km^2) also defined by Shiklomanov *et al.* [2] which does not include rivers that empty into the North Pacific or Hudson Bay. The contributing area defined by the Arctic Climate System Study (ACSYS) of the World Climate Research Programme (WCRP) [5] further reduces the total catchment area through exclusion of the southern portions of Greenland and the Canadian Arctic Archipelago (Figure 1c). Based on the geographical delineation provided in [5], this study calculated the ACSYS contributing area to be 17.1 x 10^6 km^2 although a larger value of approximately 18 x 10^6 km^2 is noted in [5]. Reasons for the difference are unknown but may be related to the inclusion of the Yukon River. An ever more restrictive geographical definition was adopted at the North Atlantic Treaty Organization (NATO) Research Workshop on the AOFWB [1]. Specifically, it states that the "Arctic Ocean is defined as being bounded by:- the Russian mainland, a line across Being Strait, the north coast of Alaska and the northernmost limit of the islands in the Canadian Arctic Archipelago, then across Kennedy Channel to Peary Land, across from Svalbard, down to Nordkapp of Norway and so back to the Russian coast." The major regions excluded by this definition compared to that for ACSYS include a majority of the Canadian Arctic Archipelago and the relatively short rivers that feed the Canadian Arctic coast east of Cape Bathurst. The oceanographic rationale for such a definition was based on the assumption that the majority of flow carried by channels within the Arctic islands moves southward away from the Arctic Ocean. As such, the only land areas contributing flow would be the coastal basins that drain directly towards the ocean from the northern edges of the Canadian islands, and the northern tip of the Greenland ice cap. Figure 1d shows the contributing river area that results from the Lewis *et al.* [1] definition and is herein referred to as the "Arctic Ocean River Basin" (AORB).

3. AORB Estimate

To calculate the flow entering the Arctic Ocean from the AORB, thirteen regions were first defined based on river-basin boundaries and differences in hydroclimatology. Note that islands located within the Lewis *et al.* [1] definition were not included. It was assumed that their freshwater contributions to the AOFWB would be calculated as part of the central ocean input normally derived from atmospheric precipitation-evaporation (P-E) calculations.

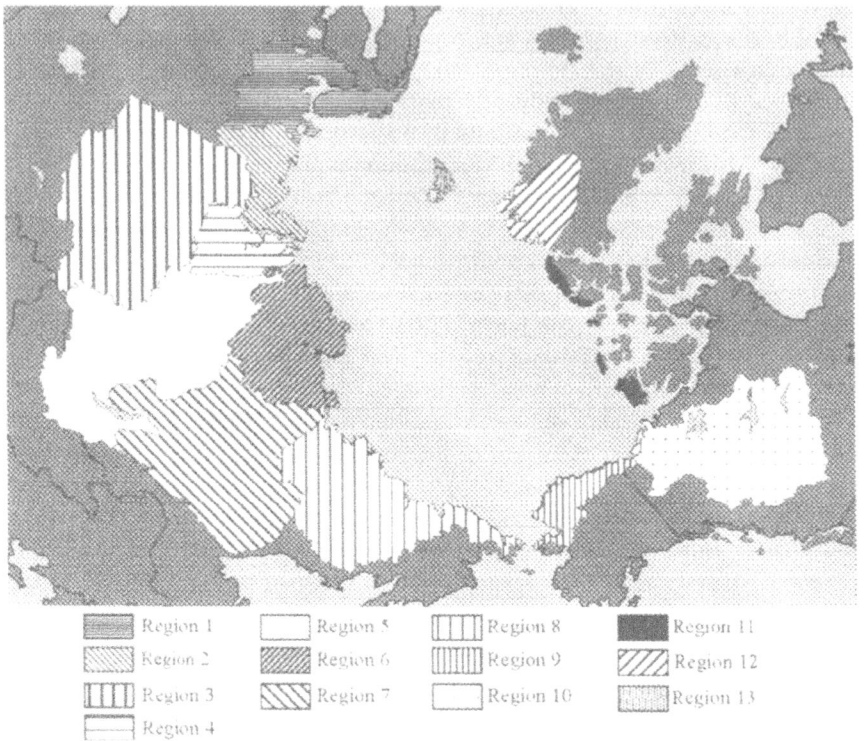

Figure 2. Circumpolar hydrologic regions defined for this analysis. See Table 1 for details

The thirteen regions are shown in Figure 2 and their areal dimensions in Table 1, the total contributing area being 15.5×10^6 km^2. To provide an estimate of the total contributing flow to the Arctic Ocean as defined by [1], a reanalysis was completed of all existing flow data. Given the large area in many regions without hydrometric gauging, an assessment was first made of historical records to define a 10-year period with maximum spatial coverage. Figure 3 shows the published periods of record for all major Arctic rivers. Length of the bars refers to the length of record and the width to its percentage of the total Arctic contributing area (also shown next to the bars). The number next to each river refers to the percentage that the gauged river area represents of the respective region total area. Data were derived from the most recent records available including Environmental Working Group [6], Vörösmarty *et al.* [7], Water Survey of Canada [8], and United States Geological Survey [9]. Data records for most Russian rivers are truncated at 1993, the most recent year available from the comprehensive Russian database [6]. In reviewing the data summarized in Figure 3, it was determined that the 10-year period with the largest total gauged area (71%) was 1975-84. Most of the areas ungauged for at least a portion of this ten-year period lie along the Arctic coast and belong to small Arctic catchments or the mouth portions of large northern rivers (Figure 4).

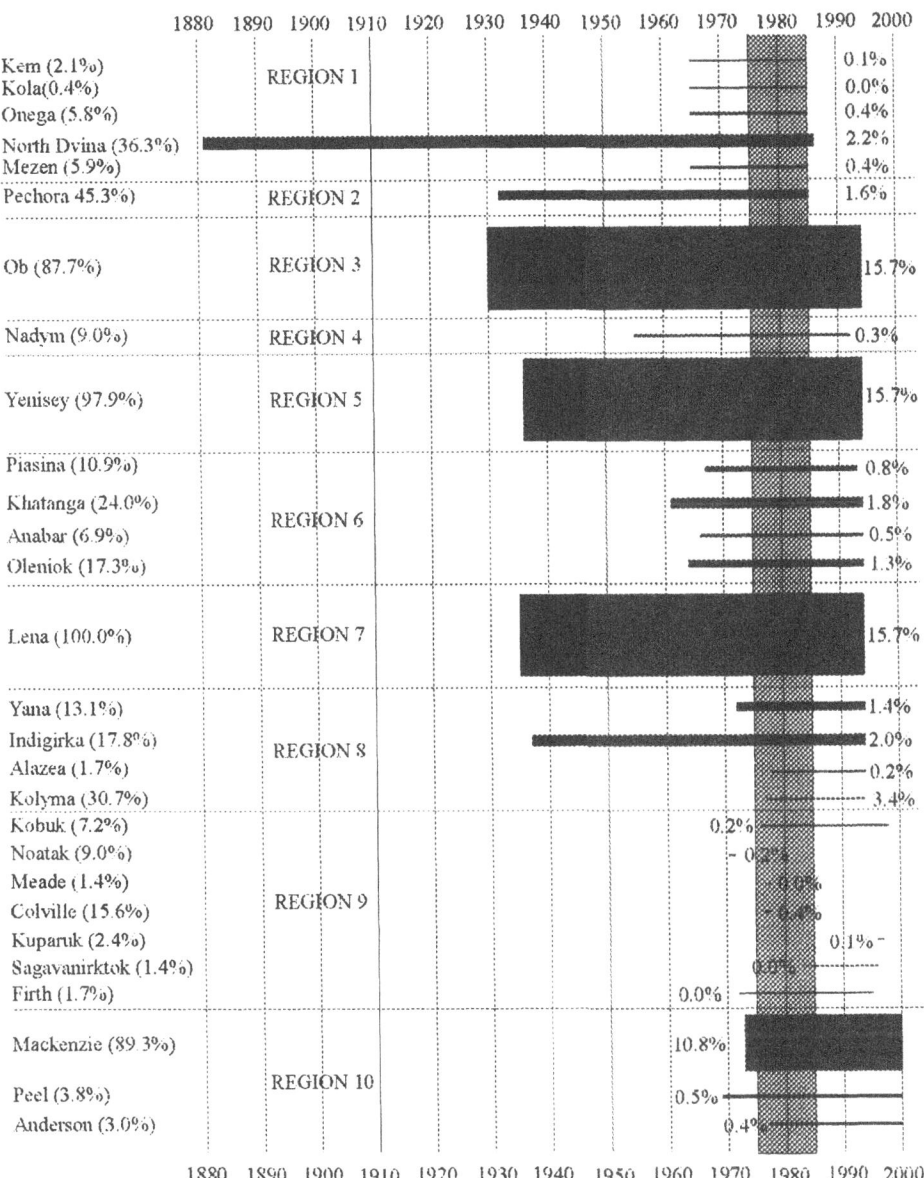

Figure 3. Length of discharge record for major Arctic rivers by hydrologic region. Bar length refers to the length of record and bar width to the percentage of the total AORB contributing area (also shown next to the bars). The number next to each river refers to the percentage that the gauged river area represents of the respective region total area. The shaded vertical bar refers to the 10-year period employed in this analysis.

274

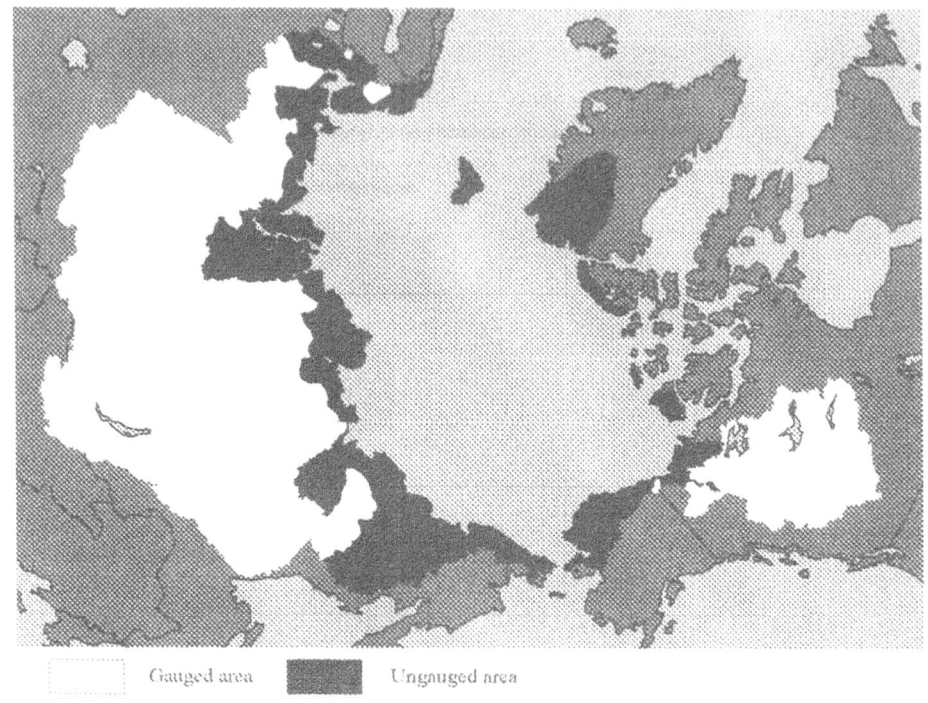

Gauged area   Ungauged area

Figure 4. Gauged and ungauged regions of the AORB for the period 1975-1984.
"Ungauged" refers to catchments without a complete record for this ten year interval.

For the selected 10-year period, it was found that the average gauged inflow to the Arctic Ocean was 2338 km^3/yr ranging from 2169 to 2531 km^3/yr. Assuming that runoff per unit area was equivalent for gauged and ungauged areas within each region, the regional gauged outflow values were increased in direct proportion to the amount of ungauged area (Table 1). In the case of Svalbard, northern Greenland and portions of the Canadian Archipelago, discharges were estimated from annual depth of runoff figures (approximately 100-150 mm/yr) provided in Baumgartner and Reichel [10], Fisheries and Environment Canada [11], Killingtveit *et al.* [12], Thomsen [13] and Woo *et al.* [14]. For glaciated catchments in these regions, such values are intended to reflect conditions where glaciers feeding runoff are characterized by an annual zero net-balance. Notably, however, some glaciers of these areas have been experiencing relatively consistent negative mass balances in the recent past and these runoff figures probably are underestimates of the total regional flow contribution for the sample 10-year period (e.g., [12]). Given that these regions represent only 4.4% of the total contributing area, however, the assumed values will not significantly affect the total flow figure.

TABLE 1. Calculated flow volumes for regions of the AORB for the period 1975-1984.

Region	Gauged Area (km^2)	Total Area (km^2)	Calculated flow volume (km^3/yr)
1	484000	959000	280
2	248000	548000	243
3	2432000	2773000	458
4	48000	534000	162
5	2440000	2492000	596
6	677000	1147000	272
7	2430000	2430000	548
8	529000	1714000	261
9	6000	344000	77
10	1751000	1882000	318
11	---	118000	12
12	---	503000	63
13	---	60000	8
Total	11045000	15504000	3299

Based on the extrapolated data in Table 1, the total 10-year average inflow to the Arctic Ocean basin defined by [1] amounts to 3299 km^3/yr with a coefficient of variation of 0.054 and ranges from a low flow of 3043 km^3 in 1980 to a maximum of 3546 km^3 in 1975. This total figure and its 10-year range are plotted with the average gauged figure on Figure 5 relative to their respective contributing areas. Also shown are the 10-year mean values for the four largest rivers, the Lena, Mackenzie, Ob and Yenisey, and their average combined flow with the 10-year range.

4. Comparisons of Estimates

For comparison, three other total inflow values for different geographical definitions of the Arctic Ocean receiving basin are plotted on Figure 5. These include:

a) AAR, with the largest value of 5250 km^3/yr,

b) AOB, with a value of 4300 km^3/yr lying mid-way between the AORB and the AAR, and

c) GRDC, representing a volume of 2603 km^3/yr provided by [3] that represents the data from 35 gauged northern rivers archived at the Global Runoff Data Centre, although it includes the Yukon River which enters the northern Pacific Ocean.

Also shown are the flow ranges reported for the AAR and AOB areas [2]. Primarily as a result of containing longer term records for some rivers and/or regions, their maximum to minimum range of values are larger than that for the sample 10-year period. Notably, however, they do not represent data for one consistent period from all regions. For details on long term flow variability see [2] and [3].

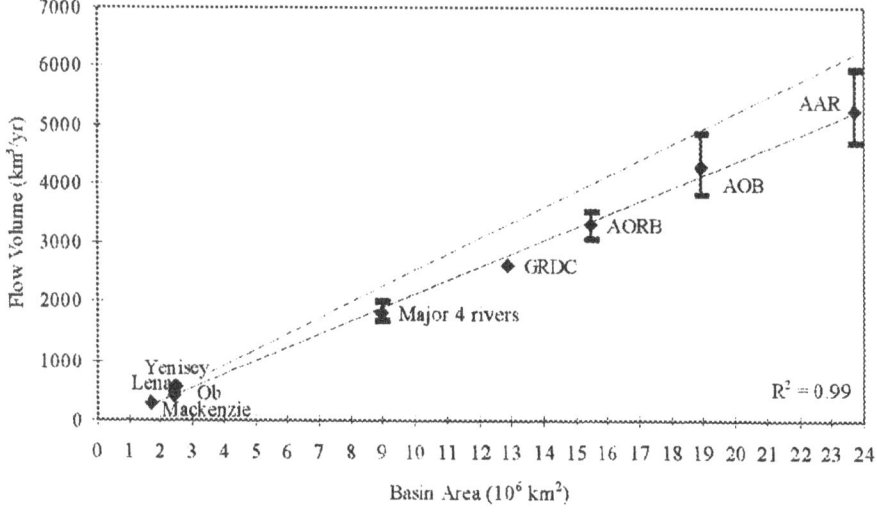

Figure 5. Basin area to annual flow volume relationship for varying geographical regions contributing flow to the Arctic. AAR and AOB refers to the "All Arctic Regions" and "Arctic Ocean Basin" definitions of [2]. GRDC refers to data for 35 gauged Arctic rivers analyzed by [3] and AORB to the "Arctic Ocean River Basins" based on the definition of [1].

Notably, the inflow to area relationship resulting from the various data sources is remarkably linear (i.e., with an $r^2 = 0.99$). In general, this means that the runoff yields of the hydrologic regimes that comprise the spatially conservative definitions are comparable to the additional regions of the broader definitions. For example, much of the additional contributing area of the AAR is provided by river catchments that flow into the North Pacific and Hudson Bay hydrologic systems with yields in a similar range to those of the other large rivers that comprise the AORB.

Figure 5 clarifies why there exists in the literature considerable variability about the magnitude of the river inflow to the ocean. Most of this appears to be simply because of differences in geographical definition. Notably, differences in average total flow values among the three major definitions are large relative to the total range in flows found in the 10-year sample record for the AORB and approximately the same as the range for the longer term record (but of varying length by region) of the AAR and AOB definitions. This points to the necessity of both properly defining the appropriate contributing area for the Arctic Ocean and to employing appropriate average or specific year flow values depending on the time step over which the overall freshwater budget is to be calculated.

5. Effects of Predicted Climate Change

Recent research evaluating the potential effects of climate change suggests that runoff from Arctic catchments could be significantly modified, both in terms of quantity and timing. Various attempts have been made to predict changes in the flow of Arctic rivers employing atmospheric general circulation models (GCM) and simplified hydrologic routines. In general, first-order predictions from such models suggest there will be increased annual flow from Arctic rivers under a 2 x CO_2 climate. Data from Miller and Russell [15], for example, suggest that the combined flow of the four major rivers (Lena, Mackenzie, Ob, Yenisey) will increase above the current measured average by approximately 46%, although the range was from +14% (Yenisey) to +103% (Mackenzie). However, the model also over predicted the observed flow for the present climate by an average 10% (varying from –20% for the Yenisey to +64% for the Mackenzie).

Much of the apparent over prediction by GCM-based models is because the hydrologic routines have been relatively simple and focused primarily on vertical water exchanges at the grid-scale resolution of the atmospheric models and do not adequately consider the complex land-phase lateral transfers of water. Employing the Mackenzie River as an example, Kite et al. [16] showed that such model deficiencies result in severe underestimation of evapotranspiration and, overestimation of precipitation and water available for runoff. In recognition of such problems, recent research has focused on designing more complex hydrologic routines and companion models that can be combined or interfaced at suitable grid-scale resolution with the atmospheric models (e.g., Hagemann and Dümenil, [17]). These remain, however, notably deficient at modelling northern rivers where snowmelt dominates the annual discharge peak (Hagemann and Dümenil, [18]). As proposed by Kite [19], a better solution might be to rely on the wealth of experience of the behaviour of inhomogeneous conventional hydrological models and to scale these physically based macroscale models into continental-scale land-phase components that can be linked to suitable atmospheric and ocean components to form a global model of the complete hydrological cycle.

For northern regions, achieving accurate predictions of river runoff is further complicated by the complexity and dominance of cryospheric components, most of which are predicted to be highly sensitive to the effects of climate change (Rouse et al. [20]). Reviews of the scope of research unknowns facing hydrological science of northern areas are contained in, for example, Bowling et al. [21], Prowse and Ommanney [22], Prowse et al. [23] and Stewart [24]. Major international programs designed to deal with the prediction of northern runoff include the World Climate Research Program (WCRP) Global Energy and Water Exchange Program (GEWEX) studies of eastern Siberia and the Mackenzie River (e.g., [25] and [26]).

Although significant advancements still need to be made before accurate prediction of the effects of climate change on Arctic runoff can be made, it is useful to compare the magnitude of first-order approximations with the historical range of flow. While the hydrologically-simple GCM flow predictions noted above are likely overestimates, some basin-specific predictions have also been made employing climate change scenarios generated by GCMs coupled as input to well-tested basin hydrologic models. The prediction of Shiklomanov et al. [2] (see also [27] and [28]) is herein used as one such example. Based on paleoreconstructions of climate and various GCM predictions linked to water balance models, they found that the annual flow of the Yenisey River would increased by 15-20% under a 2 x CO_2 climate. For illustrative purposes, a 20% increase line has been added to the regression line on Figure 5. Notably, the average flow resulting from such an increase is above the maximum total flow recorded according to any of the geographical definitions. While it must be recognized that much more research is required to accurately quantify the changes in northern hydrologic processes and associated Arctic Ocean river flow that will result from climate change, first-order estimates do suggest that flows under a 2 x CO_2 may on average exceed the maximum values that have occurred in the recorded history of these rivers.

6. Conclusions and Future Research Priorities

Based on this review, it is clear that a precise definition needs to be made about the receiving area of land runoff that flows into the Arctic Ocean. Once defined, the contribution of Arctic river flow contributing to the freshwater budget of the Arctic Ocean can be quantified within a reasonably narrow range from Figure 5. While the effect of rivers such as the Yukon and the Anadyrsky are accounted for in flow and salinity measurements at the Bering Strait, the role of land runoff originating from the Canadian Archipelago and rivers along the Canadian Arctic coast remains in some doubt because of remaining unknowns about the direction and/or seasonality of flow in the marine channels of the Canadian Archipelago. This needs to be resolved given that the difference in average annual flow originating from the AOB and AORB definitions varies by approximately 1000 km^3 and most of this originates from the additional 3.5 x 10^6 km^2 area belonging to major portions of Greenland and the Arctic Archipelago, and northern coastal rivers between Cape Bathurst and the Boothia Peninsula. Hence, two related primary research objectives are:

> 1) For the purpose of defining the contributing area of rivers that feed the AOFWB, it needs to be determined whether runoff from the Canadian Archipelago and rivers along the central Canadian Arctic coast is carried to or away from the Arctic Ocean basin in which the freshwater budget is to be calculated.

> 2) Given that a significant portion of the contributing area to the Arctic Ocean remains ungauged, specifically including unique hydrologic regions of the Canadian Archipelago and along the Arctic coasts, research should be focused on defining the hydrologic processes controlling these regimes and the development of appropriate hydrologic models for runoff prediction.

With the potential of climate change producing significant changes to the runoff regime of northern rivers, this will require an improvement in our ability to model such systems dominated by cryospheric processes and to be able to couple atmospheric and hydrologic models. As such, two additional research objectives are:

> 3) Our basic understanding of the physical processes controlling runoff in cold-regions systems needs to be expanded and incorporated into meso-scale hydrologic models. Such improvements should also be used in the scaling up of models for application to large-scale river basins.

4) Research should continue to improve the linkage between atmospheric and land-phase hydrologic models for the prediction of the impacts of climate change on northern river flow.

Acknowledgments

The authors would like to thank Drs. P. Marsh and W. Quinton for their review of this manuscript and the provision of numerous helpful comments. Assistance in locating some of the data sources provided by L. Bowling is also gratefully acknowledged.

7. References

1. Lewis, E..L. (2000) Introduction, in *The Freshwater Budget of the Arctic Ocean,* Kluwer Academic Publishers, Dordrecht, (this volume).
2. Shiklomanov, I.A., Shiklomanov, A.I., Lammers, R.B., Peterson, B.J. and Vörösmarty, C.J. (2000) The dynamics of river water inflow to the Arctic Ocean, in *The Freshwater Budget of the Arctic Ocean*, Kluwer Academic Publishers, Dordrecht, (this volume).
3. Grabs, W.E., Portmann, F. and De Couet, T. (2000) Discharge observation networks in Arctic Regions:Computation of the river runoff into the Arctic Ocean, its seasonality and variability, in *The Freshwater Budget of the Arctic Ocean*, Kluwer Academic Publishers, Dordrecht, (this volume).
4. Vuglinsky, V.S. (1997) River water inflow to the Arctic Ocean - conditions of formation, time variability and forecasts, in *Proceedings of the Conference on Polar Processes and Global Climate*, Vol. II, pp. 275-276.
5. Arctic Climate System Study (1998) *Report of the ACSYS Workshop on Status and Directions for the Arctic Runoff Data Base*, Arctic Climate System Study, World Climate Research Programme, WCRP Informal Report, June 1998.
6. Environmental Working Group (1998) *Oceanography Atlas for the Summer Period* (CD-ROM Version 1.0), National Snow and Ice Data Center, Boulder, CO, USA.
7. Vörösmarty, C.J., Fekete, B. and Tucker, B.A. (1998) *River Discharge Database* (Version 1.1 (RivDIS v1.0 supplement)). University of New Hampshire, Durham (at http://pyramid.sr.unh.edu/ csrc/hydro/).
8. Water Survey of Canada (1998) *Hydat* (CD-ROM Version 96 - 1.04, Surface water and sediment data), Environment Canada, Downsview, Canada.
9. United States Geological Survey (1999) *Alaska Water Index* (at http://www-water-ak.usgs.gov/ Data/wtrindex.htm).
10. Baumgartner, A. and Reichel, E. (1975) *The World Water Balance*, Elsevier, New York.
11. Fisheries and Environment Canada (1978) *Hydrological Atlas of Canada*, Ministry of Supply and Services Canada.
12. Killingtveit, Å., Petterson, L.A., and Sand, K. (1994) Water balance studies at Spitsbergen, Svalbard, in S. Sand and Å. Killingtveit (eds.), *Proceedings, Tenth International Northern Research Basins Symposium and Workshop*, pp. 77-94.
13. Thomsen, T. (1994) Northern hydrology in Greenland, in T.D. Prowse, C.S.L. Ommanney and L.E. Watson (eds.), *Northern Hydrology: International Perspectives*, Environment Canada, National Hydrology Research Institute, NHRI Science Report No. 3, pp. 21-39.
14. Woo, M.K., Marsh, P. and Steer, P. (1983) Basin water balance in a continuous permafrost environment, in *Permafrost: Fourth International Conference, Proceedings*, National Academy Press, Washington, D.C. pp. 1407-1411.
15. Miller, J.R and Russell, G.L. (1992) The impact of global warming on river runoff, *Journal of Geophysical Research*, 97, D3, 2757-2764.
16. Kite, G., Dalton, W.A. and Dion, K. (1994) Simulations of streamflow in macro-scale watersheds using GCM data, *Water Resources Research*, 30, 1547-1559.
17. Hagemann, S. and Dümenil, L. (1999) Application of a Global Discharge Model to Atmospheric Model Simulations in the BALTEX Region, *Nordic Hydrology*, 30, 209-230.
18. Hagemann, S. and Dümenil, L. (1998) A parametrization of the lateral waterflow for the global scale, *Climate Dynamics*, 14, 17-31.

19. Kite, G. (1998) Land surface parameterizations of GCMs and macroscale hydrological models, *Journal of the American Water Resources Association*, **34**, 1247-1254.
20. Rouse, W.R., Douglas, M.S.V., Hecky, R.E., Hershey, A.E., Kling, G.W., Lesack, L., Marsh, P., McDonald, M., Nicholson, B.J., Roulet, N.T. and Smol, J.P. (1998) Effects of climate change on the fresh waters of Arctic and Subarctic North America, *Journal of Hydrological Processes*, **11**, 873-902.
21. Bowling, L.C., Lettenmaier, D.P., and Matheussen, B.V. (2000) Hydroclimatology of the arctic drainage basins, in *The Freshwater Budget of the Arctic Ocean*, Kluwer Academic Publishers, Dordrecht, (this volume).
22. Prowse, T.D. and Ommanney, C.S.L. (eds) (1991) *Canadian Hydrology, Selected Perspectives*, National Hydrology Research Institute, Environment Canada, NHRI Symposium No. 6, 532 pp.
23. Prowse, T.D., Ommanney, C.S.L. and Watson, L.E. (eds) (1994) *Northern Hydrology: International Perspectives*, National Hydrology Research Institute, Environment Canada, NHRI Science Report No. 3, 215 pp.
24. Stewart, R.E. (2000) The variable climate of the Mackenzie River basin: its water cycle and freshwater discharge, in *Freshwater Budget of the Arctic Ocean*, Kluwer Academic Publishers, Dordrecht, (this volume).
25. GAME International Science Panel (1998) *GEWEX Asian Monsoon Experiment (GAME) implementation plan*, Game International Project Office, Nagoya University, Nagoya, Japan.
26. Stewart, R., Leighton, H.G., Marsh, P., Moore, G.W.K., Ritchie, H., Rouse, W.R., Soulis, E.D., Strong, G.S., Crawford, R.W. and Kotchtubajda, B. (1998) The Mackenzie GEWEX study: the water and energy cycles of a major North American river basin, *Bulletin of the American Meteorological Society*, 79, 2665-2683.
27. Shiklomanov, A.I. (1994) The impact of global climate anthropogenic changes on runoff in the Yenisei River basin, *Russian Meteorology and Hydrology*, **2**, 68-75.
28. Geogievskii, V. Yu., Ezhov, A.V., Shalygin, A.L., Shiklomanov, I.A., and Shiklomanov, A.I. (1996) Assessment of the effect of possible climate changes on hydrological regime and water resources of rivers in the former USSR, *Russian Meteorology and Hydrology*, **11**, 66-74.

THE DYNAMICS OF RIVER WATER INFLOW TO THE ARCTIC OCEAN

SHIKLOMANOV I.A.*, A.I. SHIKLOMANOV**,
R.B. LAMMERS***, B.J. PETERSON****,
C.J. VOROSMARTY***
* State Hydrological Institute, St. Petersburg, Russia 199053
** The Arctic and Antarctic Research Institute, St. Petersburg, Russia
199397
*** Institute for the Study of Earth, Oceans and Space, University of New
Hampshire, Durham, NH, USA 03824
**** The Ecosystems Center, Marine Biological Laboratory, Woods Hole,
MA, USA 02543

1. Introduction

One of the most important components of the water budget in any ocean is the river inflow, which significantly affects the physical, chemical and biological processes of the ocean. River inflow is of special importance in the Arctic Ocean (Table 1) because although the Arctic Ocean contains only 1.0% of the world ocean water, it receives 11% of the world river runoff [1]. It is also the ocean with the largest contributing basin area to water surface area ratio (i.e. 1.5:1.0).

TABLE 1. Major morphometric and hydrological characteristics of the World's Oceans

Ocean	Total area (With islands)	Area of water surface	Area of catchment	Water volume		River runoff inflow	
	10^6 km^2	10^6 km^2	10^6 km^2	10^6 km^3	%	km^3/year	%
Pacific	182.6	178.7	24.9	707.1	53.4	10530	26.6
Atlantic	92.7	91.7	50.7	330.1	24.6	20190	51.1
Indian	77.0	76.2	20.9	284.6	21.0	4530	11.5
Arctic	18.5	14.7	22.5	16.7	1.0	4280	10.8
World Ocean	370.8	361.3	119.0	1338.5	100.0	39530	100.0

The Arctic Ocean with the islands and the northern margins of Eurasia and North America occupies the central part of the northern polar region of the Earth called the Arctic. The southern boundary of the Arctic on the continent is considered to be the line

E.L. Lewis et al. (eds.), The Freshwater Budget of the Arctic Ocean, 281-296.

separating the tundra and forest zones close to the July $+10^{\circ}C$ isotherm. However, the major river water inflow to the Arctic Ocean originates far beyond the limits of the Arctic territory, from vast areas of the northern continents and as far south as $45^{\circ}N$.

2. Hydrometric Data Base

Estimates of the total river-water inflow to the Arctic Ocean differ considerably when averaged over long-term periods [2,3,4,5]. Sometimes the Hudson Bay and Hudson Strait basins are included in the Arctic basin in spite of the fact that their waters flow predominantly to the Atlantic Ocean. Similarly, the Yukon and Anadyr rivers flow into the Anadyr Bay of the Bering Sea, ultimately entering the Arctic Ocean via the Bering Strait. These river waters affect processes in the Arctic Ocean, but are not geographically parts of the direct contributing area.

The areas of the Arctic Ocean basin and its individual parts, whose margins were determined in accordance with the Arctic Atlas [4], are given in Table 2. The total area of the Arctic Ocean contributing basin is more than 19 million km^2, which is considerably greater than the area of the ocean itself (14 million km^2). The majority of the basin (58%) is in the territory of Asia where the largest amount of river water inflow originates.

The hydrometric network in the basin serves as the basis for estimating the dynamics of fresh water inflow to the Arctic Ocean. The data base of monthly and annual river runoff used for the estimates includes approximately 3500 observation stations spread over the various regions as listed in Table 2. This data base has been compiled within the National Science Foundation international project studying water and contaminant runoff from pan-Arctic rivers [6].

TABLE 2. Drainage areas, the number of gauges, unobserved and observed areas of various regions contributing to the Arctic Ocean

Region	Drainage area, $(10^3 km^2)$	The number of gauges	Unobserved area, $(10^3 km^2)$	Observed area, %
North America	4012	552	1997	50.2
North America*	8646	2120	3452	60.1
Asia	11186	1131	1668	85.1
Asia (including Anadyrsky Bay basin)	11409	1139	1735	84.8
Europe	1501	280	440	70.7
Greenland	2176		2176	0.0
Non-Arctic Ocean Basin	4869	1576	1531	68.6
Arctic Ocean Basin	18875	1963	6281	66.7
All Arctic regions**	23732	3539	7803	67.1
All Arctic regions without Greenland	21556	3539	5628	73.9

Note: *   including Hudson Bay, Hudson Strait, Norton Sound and Yukon River basins; ** including Hudson Bay, Hudson Strait, Norton Sound, Yukon River and Anadyrsky Bay basins

The greatest numbers of stations are located in North America, primarily in the basins of the Hudson Bay and Strait that, as previously mentioned, do not directly contribute flow to the Arctic Ocean. Considering the North American territory that does contribute directly, the number of hydrometric sites per unit area is greater than that for the larger Asian territory. The highest density of gauge locations is located in the European part of the basin, while in Greenland and in the Canadian Arctic Archipelago they are practically absent.

The latitudinal distribution of hydrological sites is even more uneven. Approximately two thirds are located in the latitudinal zone between 49 and 58°N; only 6% of all gauges are found beyond the Arctic Circle. The greatest numbers of observed stations (more than 50%) exist for rivers with basin areas of 1000 to 30,000 km^2; only about 5.5% of basins have areas of more than 100,000 km^2 (Fig. 1).

Number of gauges

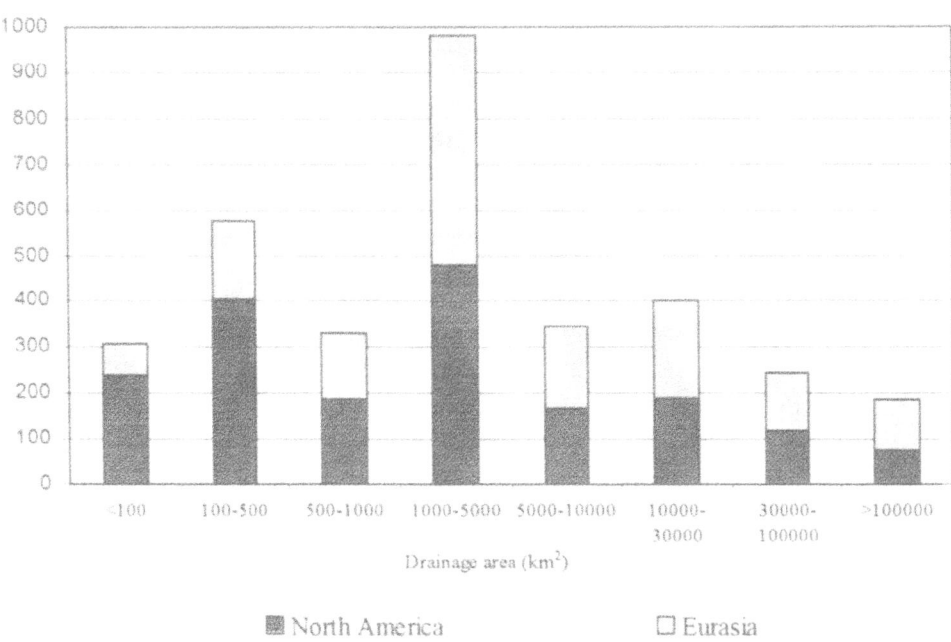

Figure 1. Hydrometric station distribution by drainage area ranges in the Arctic basin

The hydrometric network in the Arctic Ocean basin is comprised of relatively short-term records. This pertains especially to North America where the best records are only 30-40 years long and are limited to only a few years at many others. In the Asian part of the basin, record lengths are of greater duration. With many, there is a large number of hydrometric sites with time series of 50-55 years. In the European territory, there are gauges with time series more than 60 years long.

Another indicator of hydrological data reliability is the absence or presence of gaps in observations. Unfortunately, there are quite a number of gaps in observations, especially in the Canadian winter records. At some sites, winter discharge is not measured at all. Overall, about 20% of the monthly data are unavailable from North American sites; for the European territory, this is within 5-10%.

Despite the large number of hydrometric stations located in pan-Arctic basin, there are vast areas without a measurement network including almost half the area of the Arctic Ocean basin in North America (see Table 2). This is primarily due to the absence of stations on Greenland and their rarity on Canadian Arctic Islands. In the European part of the basin approximately 30% of the area is not monitored, mainly because of the absence of stations on the European Arctic islands (Novaya Zemlya, Spitsbergen, etc.) as well as on the tundra part of the continental coast of the northern European territory of Russia.

It is the Asian part of the Arctic Ocean basin which is best monitored, 85% of the area being gauged. The ungauged portions of the region are near the mouths of large Siberian rivers, as well as on islands and many small and medium tundra-zone catchments with rivers flowing directly to the ocean. Overall, riverflow is not monitored over about one third of the Arctic Ocean. Despite the lack of stations and their record deficiencies, the data over the last two to three decades form a valuable record of flow dynamics to the Arctic Ocean, especially when compared to the situation 50-70 years ago when the greater part of the territory was not monitored at all. The lack of a very long-term record, however, adds uncertainty to achieving reliable estimates of river water inflow to the ocean, especially when attempting to study long-term variations due to natural climate fluctuations and human activities.

3. Time Series Analysis

To study the dynamics of fresh-water inflow to the Arctic Ocean, we attempted to estimate its variations for an extended time period since 1921. This period has been chosen for the following reasons. First, since 1921 the world's renewable water resources and river-water inflow to all the world oceans have been recently assessed [1] and, to compare the results, it is necessary to use the same time period for the Arctic Ocean. Second, the data for the last 75 years are sufficient to assess the inflow dynamics due to global climate fluctuations with long-term periods of increased and decreased air temperatures in the Northern Hemisphere.

Assessing the annual inflow values for this extended long-term period required for all the major river systems and regions of the basin to:
• Update the series of hydrological observations;
• Restore the gaps in observations; and
• Assess river water inflow to the ocean from the continental part of the basin and islands not covered with observations.

To solve the first two problems, wide use has been made of hydrological analogy in combination with linear and multiple correlation. Where suitable discharge time series were available for neighbouring rivers in similar physiographic regions, the data were used. When no analogue site was available, additional information relating the meteorological (monthly air temperature and precipitation) and hydrological time series were used. In most cases, it was possible to obtain relationships between the meteorological data and monthly and annual runoff with correlation coefficients of ≥ 0.8. These were then used to fill gaps in observations or to extend the time series.

Addressing the third problem, calculating runoff from the territories not covered by observation data, required the use of hydrological analogues as well as hydrological mapping and water balance comparisons based on maps of isolines for precipitation and evaporation [2].

4. Time Series Results

By generalising observation data from the hydrological network and applying the above design methods, annual values of river water inflow to the Arctic Ocean were assessed for the period 1921-1996 for all large river systems and particular catchments on the continents and islands. These have been averaged by sea basins and the three major regions: Asia (Siberia), Northern Europe, and North America. For the first twenty years of the study period, when the hydrological network in the basin was very sparse, the results depend heavily on the above methods. The accuracy of the calculated inflow for this period is considered to be much lower than that for the more recent years characterized by much more extensive flow monitoring. The most reliable estimates of Arctic Ocean inflow were obtained for 1960-1990, when information from the hydrometeorological network was most comprehensive.

Statistical data on the inflow to the Arctic Ocean for 1921-1996 from different regions of the basin, as well as runoff from the adjacent territories are given in Table 3.

TABLE 3. Statistics of the annual inflow to the Arctic Ocean for 1921-1996

Basin	long-term inflow km³/year	Coefficient of variation	max inflow		min inflow	
			km³	Year	km³	year
Bering Strait*	305	0.12	333	1984	208	1996
Hudson Bay & Strait	948	0.10	1140	1966	720	1989
North America	1177	0.07	1340	1968	990	1953
North America with Hudson Bay & Strait	2125	0.06	2450	1968	1810	1981
Europe	693	0.08	884	1938	504	1960
Asia	2430	0.06	2890	1974	2100	1953
Arctic Ocean Basin	4300	0.05	4870	1974	3820	1953
Arctic Basin & Hudson Bay and Strait Basins	5250	0.04	5950	1974	4700	1953

*Note: * including, Norton Sound, Yukon River and Anadyrsky Bay basins*

The time series of inflow to the Arctic Ocean for 1921-1996 by individual years and smoothed by a 5-year running mean for individual regions and the entire contributing basin are shown in Fig. 2a,b. Cyclic inflow variations with very small trends are evident for all contributing basins. The lowest total inflow to the ocean took place in the mid-1950s and early 1980s, the highest in the late 1940s - early 1950s, mid-1970s and early-1990s. Relatively synchronous trends are evident in the major regions of the basin (Fig. 2a).

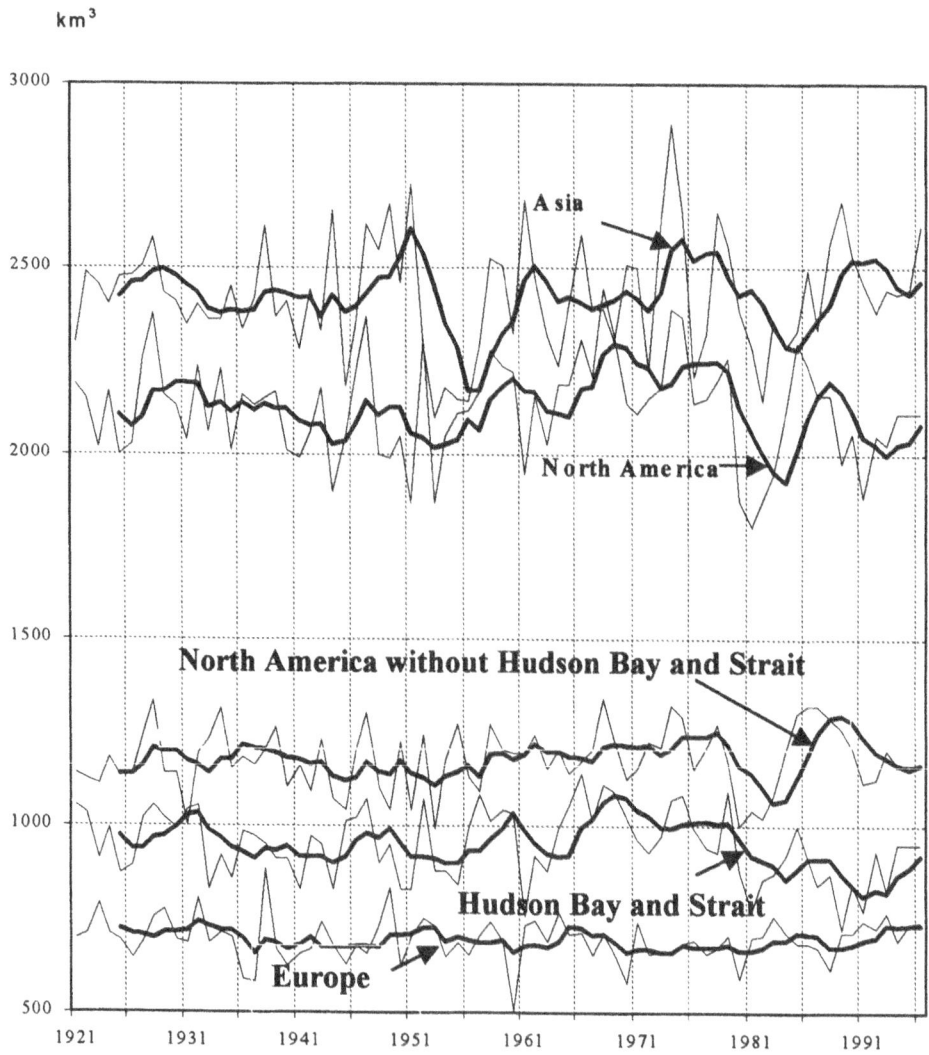

Figure 2a. River inflow into the Arctic Ocean by major regions of the basin

km³

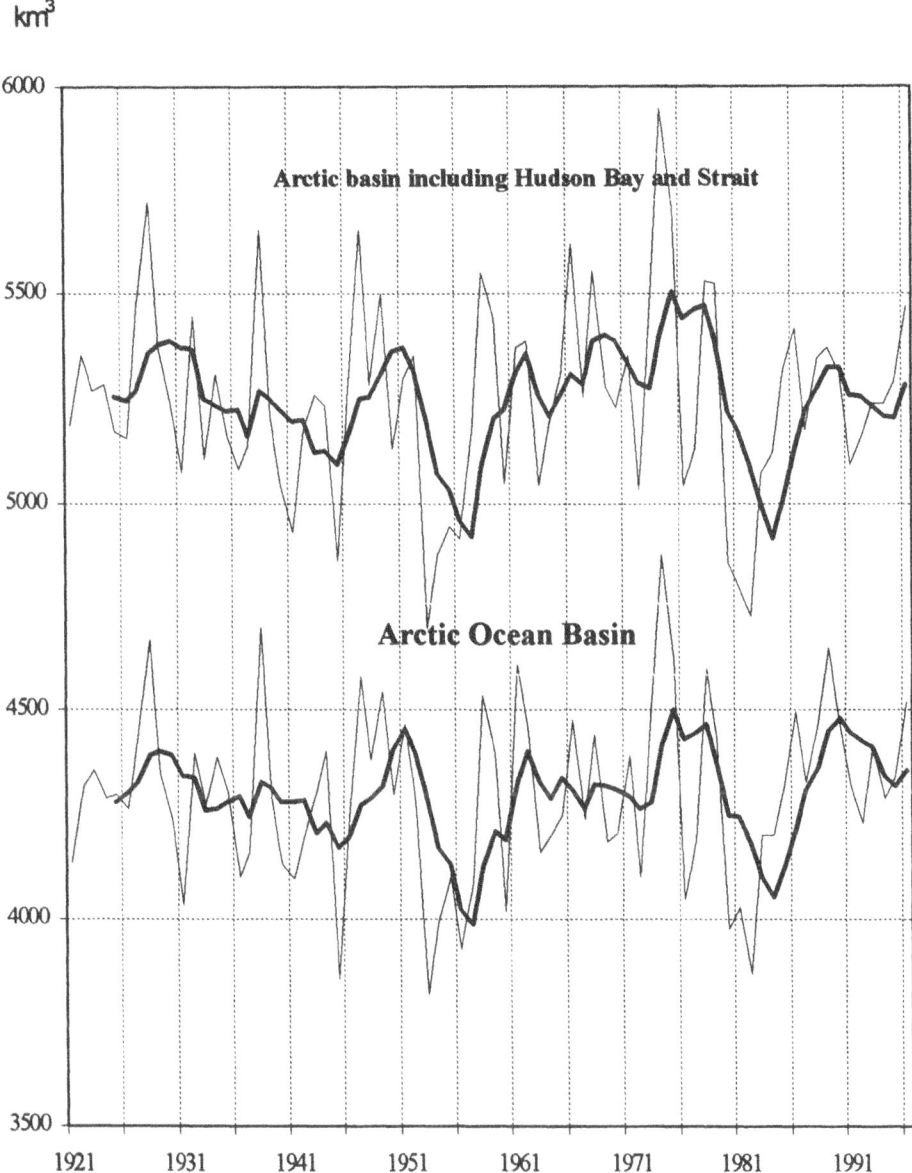

Thick black line is the 5 year running mean

Figure 2b. River inflow into the Arctic Ocean

5. Anthropogenic Influences

5.1   HUMAN IMPACTS ON THE BASIN

Analysing the dynamics of river water inflow to the Arctic Ocean is useful for detecting changes produced by, for example, human impacts on the basin and anthropogenic changes in global climate. Significant human impacts on river flow can be produced by a number of activities including, the construction of large reservoirs, interbasin water diversions, and water withdrawal for municipal, industrial and agricultural needs. However, although all the above economic sectors exist within the basins contributing inflow to the Arctic Ocean, due to low population and minor economic development of the territory, they are relatively minor in scale and unlikely to produce noticeable effects on water inflow to the ocean, even in the near future. This can best be illustrated by considering the current effects experienced on the Yenisey, the largest Arctic Ocean basin, and the one with the world's biggest reservoirs regulating long-term and seasonal runoff.

Among the largest rivers flowing into the Arctic Ocean is the Yenisey. With over 4.5 million people living in the basin, it is also the Arctic Ocean river with the largest potential anthropogenic impact on water resources. The total storage of operating and constructed reservoirs in the basin is 482 km^3, operating - 174 km^3. During the next few decades, there are plans to build four reservoirs more with the total storage of 240 km^3 and operating - 51 km^3 [7].

Detailed studies on the Yenisey basin show that in 1995 the total water withdrawal in the basin for public services, agriculture and industry, together with water loss by evaporation from reservoirs, is 8.7 km^3/yr, water consumption 4.9 km^3/yr. In the future to 2025, these values are expected to grow to 12 and 7 km^3/yr, respectively. Of the annual river runoff measured at the mouth, the current values represent only 0.8-1.4%, and in 2025 1.2-2.0% [8]. Reservoir effects on the Yenisey monthly runoff distribution are shown in Fig. 3. As a result of regulation, winter runoff has increased by 50-60% and spring runoff reduced by about 10%.

In the future, the amount and seasonal effects of runoff regulation will noticeably increase, although in principle the annual river inflow to the ocean will not change. As occurred previously, more than half the annual river runoff will still pass within the period from May to August. Based on this analysis of the Yenisey, it is reasonable to conclude that on all other river systems in the Arctic Ocean basin which have much less reservoir storage, the anthropogenic effect on the inflow regime should be considerably less.

Based on recent State Hydrological Institute studies on the dynamics of global water use by continents, physiographic regions and selected countries [9], we assessed the corresponding values of total water withdrawal and consumption for the Arctic Ocean basin, and its individual parts [9]. The results of this assessment are presented in Table 4. Shown are the total values of water used for public services, industry (including thermal power engineering) and agriculture, as well as water loss by additional evaporation from reservoirs. At the current level, the total water withdrawal in the Arctic Ocean basin and in its large regions is only 0.9-1.4% of the total water inflow to the ocean. For water consumption, these values are even less at 0.2-0.6%. In the future (to 2025), water use is expected to increase by 25-40%, which should not significantly increase the current relatively low effects on river inflow to the ocean, especially since these are within the level of analytical accuracy.

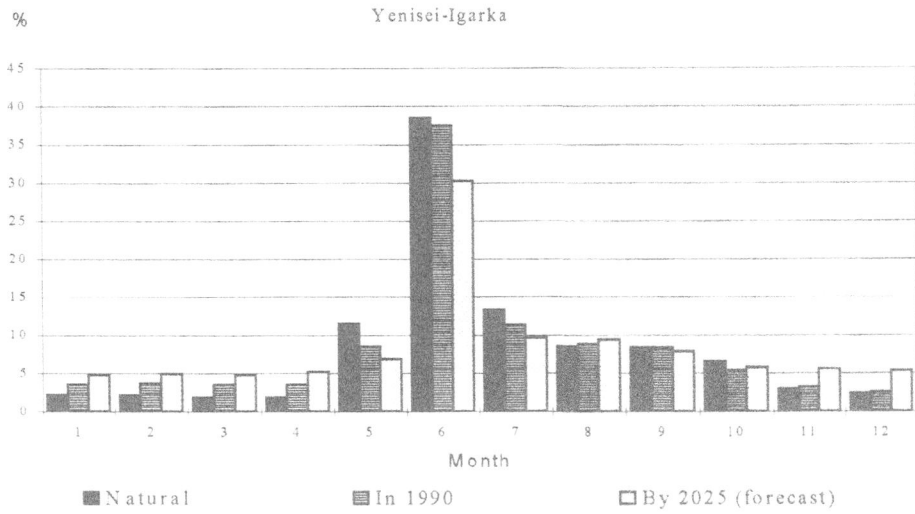

Figure 3. Monthly discharge distribution for different levels of human activity in the basin

TABLE 4. Water use in the Arctic Ocean basin

Region	1900	1950	1960	1970	1980	1990	1995	2000	2010	2025
	Design							Forecast		
North America	0.6	4.0	6.1	8.8	14.0	18.0	19.5	20	22	25
	0.1	0.7	1.0	1.6	2.6	3.6	3.9	4.0	4.5	5.5
North Europe	0.3	0.4	0.9	1.6	7.0	8.2	8.0	8	8.5	10
	0.1	0.1	0.2	0.3	1.0	1.4	1.5	1.6	1.7	2.5
Asia	0.7	5.8	10.9	17.0	27.0	33	33	35	36	40
	0.4	1.4	2.9	8.1	12.0	16	16	17	18	21
Basin as a whole	1.6	10	18	27	48	59	60	63	67	75
(rounded)	0.6	2	4	10	16	21	21	22	24	30

Remarks: in numerator - water withdrawal, in denominator - water consumption

5.2   CLIMATE CHANGE EFFECTS

Considerable emphasis has been recently given worldwide to studies on assessing the anthropogenic climate-change (global warming) effects on hydrological characteristics, water resources and water management in different physiographic and socio-economic conditions. The studies are being conducted in two major directions. First, based on the observed data, the sensitivity of hydrological systems to global warming processes recorded in the world during the past 10-20 years is being assessed. Second, assessments are being made of future hydrological management consequences of global warming for different climate scenarios, and consideration is being given to the most efficient adaptation measures.

Since the 1980s, a period characterized by a noticeable rise in global air temperature, the State Hydrological Institute of Russia has found a considerable increase in annual river runoff in the greater part of the European territory of Russia - approximately 10-12% with a strong seasonal redistribution (increasing winter runoff by up to 30-70%) [10]. This has led to a drastic rise of the Caspian Sea, an internal basin, and caused a whole complex of very acute water-management problems. Similar changes in runoff regime have been recorded in other regions of Eurasia and North America [11,12].

Research has also been undertaken to assess expected changes in river runoff for different regions of the world, including the Arctic Ocean basin and adjacent territories. Generalising the results of these studies together with the analysis of long-term data on river water inflow to the Arctic Ocean allows us to make some preliminary conclusions about climate-induced changes.

Historical time series of air temperature for the Northern Hemisphere, and for specific latitudinal zones of 44^0-64^0N and 64^0-90^0N are shown in Figure 4. The major source of river inflow to the Arctic Ocean originates in the latter latitudinal zone while the former excludes the Arctic Ocean and the northern coastal part of the continents. It is interesting to note that increasing air temperature over the last two decades is more distinct in the 44-64^0N zone, including 9 cases where the air temperature was higher than for the entire observation period since 1866. For the more northern zone (64^0-90^0N), this phenomenon is not observed, but rather air temperature was greater during the 1930-40s than during the last two decades.

It should be mentioned that the above latitudinal differences in air temperature trends do not agree with all the predicted scenarios of future climate based on atmospheric general circulation models (GCM) and palaeoclimatic reconstructions. According to these scenarios the greatest warming will occur at high latitudes.

Proceeding from the diagrams presented in Figure 4, one can isolate the three long-term periods with considerably different air temperature trends in the Arctic Ocean and its basin: 1926-1966 and 1977-1996 are the periods with above normal (relative to the long term 1866 to 1966 average) air temperature, and 1967-1976 is the period with below normal air temperature. The river inflows to the Arctic Ocean for all these periods from all major regions of the basin are shown in Table 5.

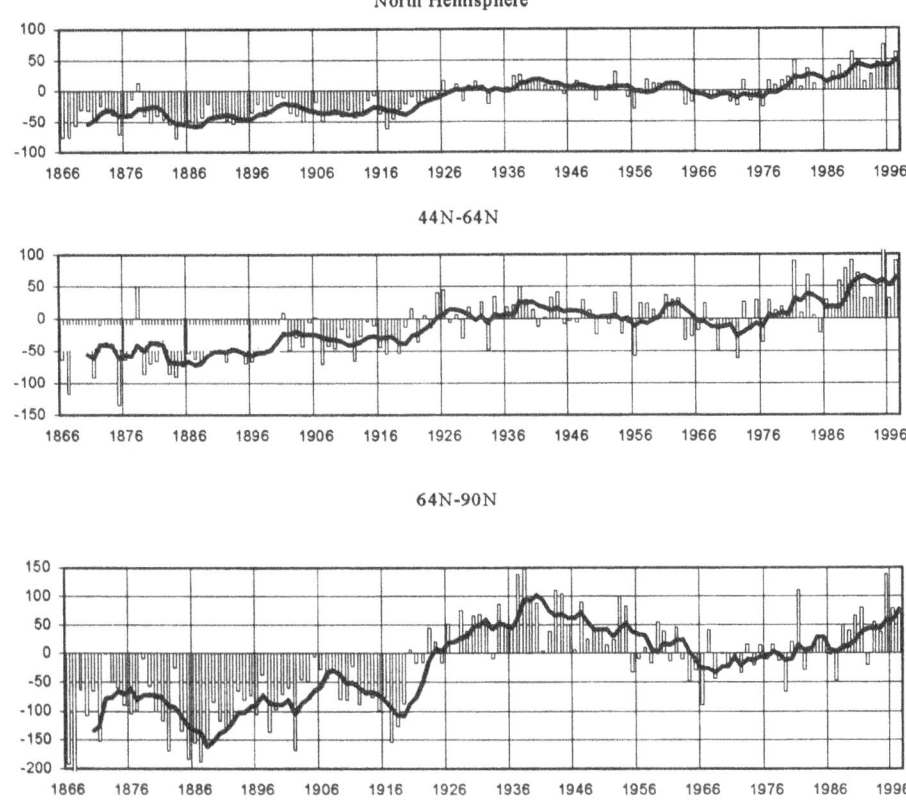

Figure 4. Annual mean temperature anomalies in 0.1 °C selected zonal means: sources: National Oceanic and Atmospheric Administration (NOAA) - 11/1997; base period: 1951-1980.

TABLE 5. Statistics of the annual inflow to the Arctic Ocean for different selected time periods

Basin	Periods	Average inflow, km³/year	Module coefficient*
Hudson Bay and Strait	1926-	950	1.00
	1965	1034	1.09
	1966-	882	0.93
North America Basin	1926-	1168	0.99
	1965	1218	1.03
	1966-	1183	1.01
North America with Hudson Bay	1926-	2118	1.00
	1965	2251	1.06
	1966-	2065	0.97
Europe	1926-	696	1.00
	1965	674	0.97
	1966-	700	1.01
Asia	1926-	2410	0.99
	1965	2482	1.02
	1966-	2436	1.00
Whole Arctic Basin	1926-	4275	0.99
	1965	4370	1.02
	1966-	4318	1.00
Whole Arctic Basin & Hudson Bay	1926-	5224	1.00
	1965	5408	1.03
	1966-	5209	0.99

* - fraction of long-term mean value

The data show that past changes in air temperature, in the range of -0.5°C to +1.0°C, insignificantly affected river inflow to the Arctic Ocean. For all the periods from all regions, the inflow departures from the average value are within 3-5%, and can be explained by natural climatic fluctuations. Furthermore, as Figure 5 shows, no noticeable changes occurred in the seasonal distribution of river inflow. During the three-month period, June to August, the Arctic Ocean receives 63-67% of Siberian river runoff and for the slightly earlier May-July period 55-60% of river runoff from Northern Europe and North America. During the winter to early spring months, December to April, only 8-10% of the annual river flow enters the ocean from the contributing areas of Asia and North America, and 12-14% from Northern Europe. More detailed studies conducted by SHI for the Yenisey, which experienced a general rise in air temperature of 2-3°C over the last 15 years, could also reveal no noticeable changes in the annual or monthly river flow [13].

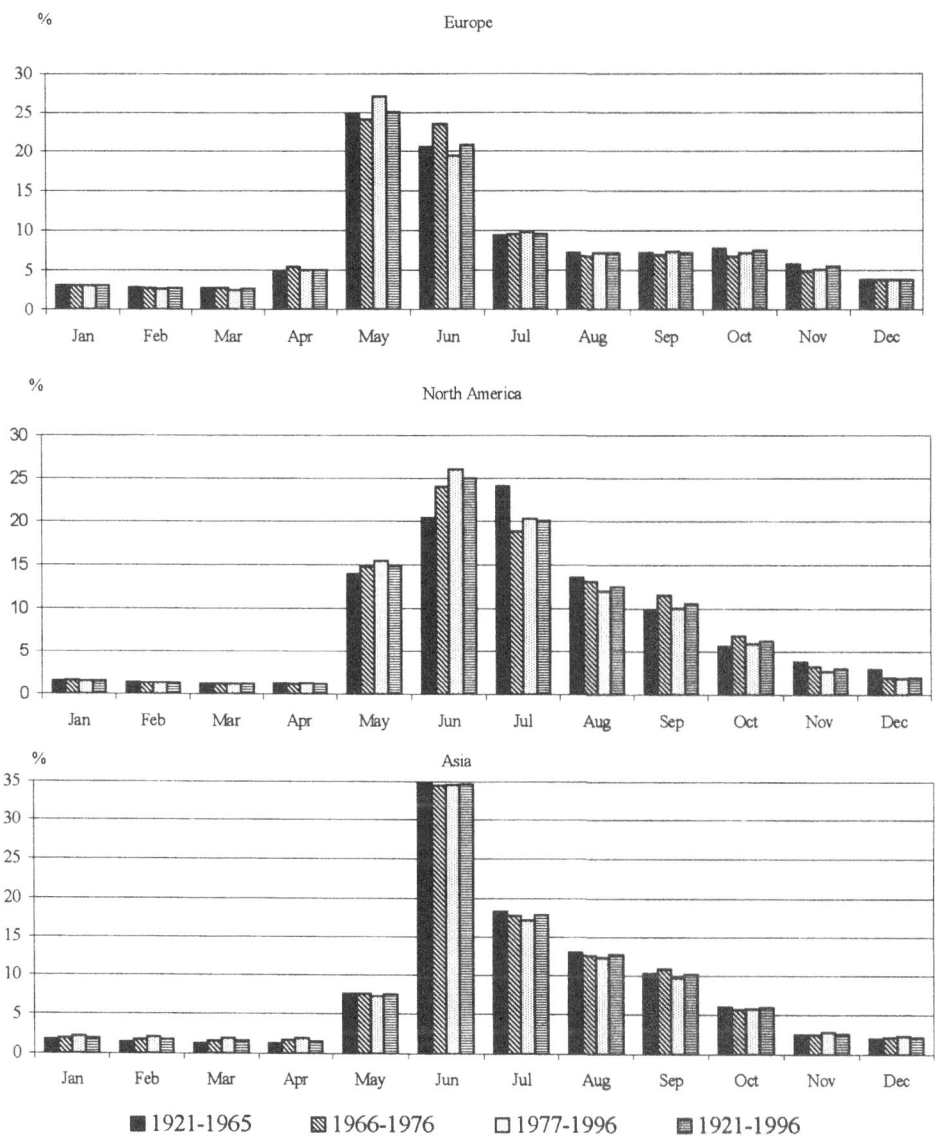

Figure 5. Mean monthly distribution of the river inflow into the Arctic Ocean

The absence of any noticeable change in river inflow to the Arctic Ocean with relatively small variations in air temperature is quite explicable for cold-climate regions characterized by very low winter air temperatures. In such regions, increases in winter temperatures of even 2-3^{0}C exert little effect on runoff generation and subsequent inflow to the ocean, as evidenced by the historical datasets.

To assess further potential changes from global warming on Arctic Ocean river inflow, a range of climatic scenarios has been evaluated based on GCM and palaeoclimatic reconstructions of past warm epochs. In Russia, such detailed studies using different hydrological models have been conducted for the basins of the Yenisey, Ob and the North European rivers [14]. Using 2 x CO_2 scenarios, Miller and Russel [15] have generated hydrological estimates for most of the large Russian and Canadian rivers that feed the Arctic Ocean.

Despite the differences in the climatic scenarios and methodological approaches, the various 2 x CO_2 results are fairly similar for all the studied, large Arctic Ocean rivers. A noticeable increase in annual river runoff with a considerable change in its seasonal distribution was found. This primarily applies to large Siberian rivers, river systems of northern Europe as well as the greater part of the contributing area of Canada. For example, Table 6 shows the results in annual and seasonal runoff for the Yenisey with different global warming scenarios [11]. The most realistic scenarios for this river show increasing annual runoff with 2 x CO_2 of 15-20%.

TABLE 6. Yenisey runoff volume with global climate change, km^3

Climate	Spring (Apr-Jun)	Summer-Autumn (Jul-Oct)	Winter (Nov-Mar)	Annual
Natural*	327	233	70	630
SHI 1^{0}C	350	243	94	687
SHI 2^{0}C	385	188	113	686
SHI 3-4^{0}C	325	171	228	724
GFDL, 2xCO$_2$	403	227	119	749
UKMO, 2xCO$_2$	553	238	126	917

Note: * Average seasonal runoff distribution for the period before the beginning of significant flow regulation (before 1959).

Figure 6 displays annual hydrographs for the Yenisey (at the mouth) obtained from hydrological models using different climatic scenarios. In general, future winter runoff should increase significantly and the major flow period occurs about one month earlier. Similar expected trends have been obtained for other river systems of the Arctic drainage basin. Overall, the results indicate that the annual river inflow to the Arctic Ocean will increase with an atmospheric CO_2 doubling by approximately 10-20% and be accompanied by a 1.5-2.0-fold rise in winter runoff and the majority of runoff (55-60%) will be delivered to the ocean between April and July. However, according to the most recent IPCC estimates, atmospheric CO_2 doubling and corresponding changes in the global air temperature and precipitation will most likely not occur until the end of the next century [16]. Therefore, in the short-term, there should be no significant change to river inflow to the Arctic Ocean due to human activities on the catchment or from anthropogenic global climate change. This is confirmed by the extremely insignificant air temperature rise in the Arctic zone during the last two decades.

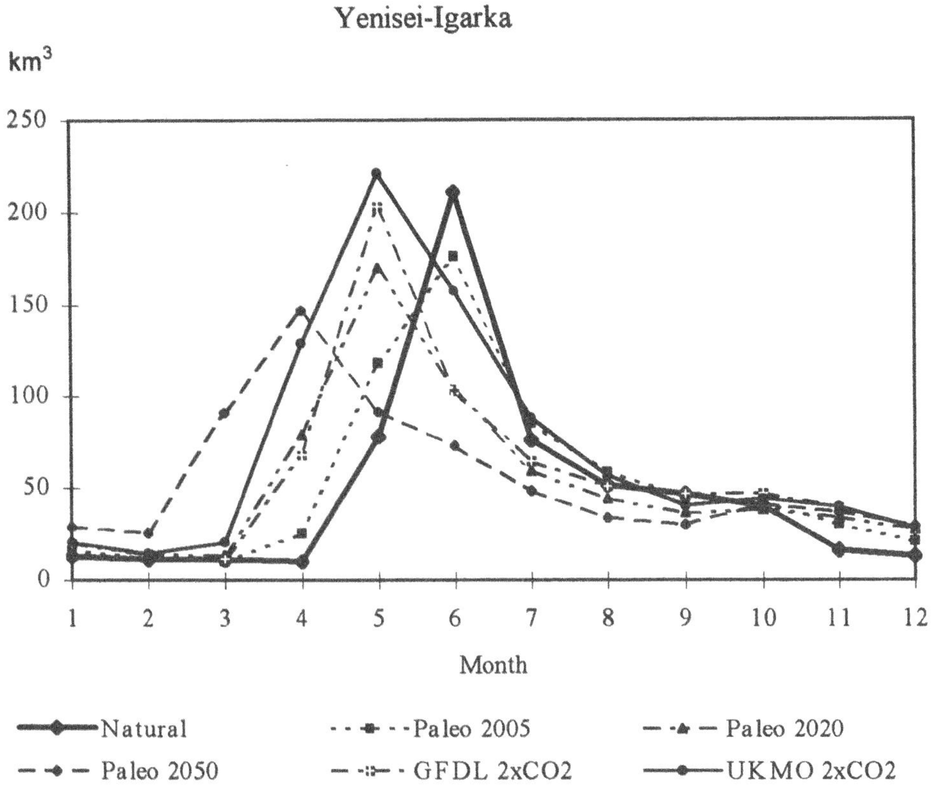

Yenisei-Igarka

km³

Figure 6. *Runoff hydrographs for different climatic scenarios*

6. Conclusions and Future Research Needs

In conclusion, it should be mentioned that the above results on the dynamics of river inflow to the Arctic Ocean and changes in its characteristics due to human impact on the catchment and global warming are very preliminary. To obtain more reliable results, it is necessary
- To develop reliable models of large river systems in the Arctic Ocean basin to assess runoff with different variants of climate change and human impact including land use;
- To develop models to reliably estimate inflow from territories not currently monitored;
- To include remote sensing methods with hydrological observations;
- To develop reliable and detailed forecasts of anthropogenic climate change for the Arctic Ocean basin, as a whole, and for its individual regions.

7. References

1. Shiklomanov, I.A. (1998) *Comprehensive Assessment of the Freshwater Resources of the World: Assessment of Water Resources and Water Availability in the World*. WMO, UNDP, UNED, FAO, UNESCO, World Bank, WHO, UNIDO, SEI. Published by WMO, Geneva, 88 p.
2. *World Water Balance and Water Resources of the Earth* (1974) Leningrad, Gidrometeoizdat, 638 p.
3. Ivanov V.V. (1976) Fresh water balance of the Arctic Ocean. *Proc. AARI*, **Vol.323**, p.138-147.
4. AARI, (1985). *Arctic Atlas*. Arctic and Antarctic Research Institute, Moscow, Russia, 204 p.
5. Grabs W.E. (1998) Discharge Observation Networks in Arctic Regions: Computations of the Total River Runoff into the Arctic Ocean, its seasonality and variability. *NATO Advanced Research Workshop 971307, 25 April to 2 May 1998. Abstracts.* Tallinn, Estonia, p.24-25.
6. Lammers R.B., Shiklomanov A.I., Vorosmarty C.J., Fekete B.M. and Peterson B.J. (in press). Assessment of Contemporary Arctic River Runoff Based on Observational Discharge Records. Submitted to *Geophysical Research*.
7. Malik L.K. (1990) *Geographical forecasts of the consequences of the hydroenergy constructions in the Siberia and Far East*. Moscow, Academy of Science, Institute of Geography. 317 p.
8. Shiklomanov A.I. (1995) Estimate of the change in the Yenisey river inflow to the Kara Sea as affected by industrial activities in its basin. *Abstracts of Scientific seminar "Nature conditions of the Kara and Barents Seas"*, St.Petersburg, Russia,1995.
9. Shiklomanov I.A. (1998) Global renewable water resources. In: *Water a looming crisis? IHP-V. Technical Documents in Hydrology*, **No.18**, p.3-14.
10. Georgievsky V.Yu (1998) On global climate warming effects on water resources. In: *Water: a looming crisis? .Technical Documents in Hydrology*, **No.18**, UNESCO, Paris, p.37-46
11. Shiklomanov A.I. (1994*) Influence of anthropogenic changes in global climate on the Yenisey River Runoff. "Meterology and Hydrology"*. #**2**, p.84-93.
12. Lins, H.F and Michaels, P.J. (1994) Increasing streamflow in the United States. *EOS*, **75**, p.281-286.
13. Shiklomanov A.I. (1997*)*. On the effect on anthropogenic change in the global climate on river runoff in the Yenisei basin. Runoff Computations for Water Projects. *Proc. of the St. Petersbug Symposium 30 oct.-03nov. 1995.* IHP-V UNESCO Technical Docum. in Hydrology **N9** p.113-119.
14. Georgievsky V.Yu., Ezhov A.V., Shalygin A.L., Shiklomanov I.A., and Shiklomanov A.I. (1996) Evaluation of possible climate change impact on hydrological regime and water resources of the former USSR rivers. *"Meteorology and Hydrology"*, 1996, #**11**, p.89-99. ISSN 0130-2906.
15. Miller J.R and Russell G.L. (1992) The impact of global warming on river runoff. *Journal of Geophysical Research*. **Vol. 97**, No.D3, p.2757-2764.
16. *The regional impacts of climate change*. An Assessment of Vulnerability. (1998). IPCC Report. Cambridge University Press, 517 pp.

RIVER INPUT OF WATER, SEDIMENT, MAJOR IONS, NUTRIENTS AND TRACE METALS FROM RUSSIAN TERRITORY TO THE ARCTIC OCEAN

V.V.GORDEEV
P.P.Shirshov Institute of Oceanology
Russian Academy of Sciences,
Nakhimovsky prospect, 36, Moscow 117851,
Russian Federation

1. Introduction

During few last decades the Arctic Ocean has attracted the attention of many scientists in connection with the global climatic changes and high sensitivity of the region to the impact of pollutants.

In the near future it is planned in the Russian sector of the Arctic to exploit a number of large oil and gas deposits on land and in the sea that may lead to serious consequences for the environment. Therefore, it is very important to assess the present river fluxes of water, sediment, mineral and organic substances including pollutants into the Arctic coastal seas.

In 1996 in American Journal of Science a paper [1] was published with an reassessment of the Eurasian river input of water, sediment, major elements and nutrients to the Arctic Ocean for the time up to 1990. Since that time new data on dissolved forms of nitrogen and phosphorus in about twenty Arctic rivers for period from the mid of 70-s to 1995 became accessible (data of Roskomgidromet), and new reliable determinations of nutrients and trace metals for the biggest Arctic rivers Ob, Yenisey and Lena were received in framework of the Russian-French-Netherlands programme SPASIBA (Scientific Programme on Arctic and Siberian Aquatorium, 1989-1995). A significant part of these data was published [1-14].

The main aim of this chapter is to summarise these new data and to give the present day evaluation of multi-year average fluxes of dissolved and particulate substances from the Russian territory to the Arctic Ocean.

2. Short Description of the Study Area

The Russian part of the Arctic Ocean drainage basin covers about $13 \cdot 10^6$ km^2 (Figure 1). The European part of the basin is characterised by the predominance of plains and lowlands, whereas in the eastern Siberian part a mountainous landscape prevails. The major part of the watershed has a strongly pronounced continental climate with mean January temperature from -15 to $-20°C$ along the White Sea coast to -40 to $-45°C$ in the N-E part of Yakutia.

E.L. Lewis et al. (eds.), The Freshwater Budget of the Arctic Ocean, 297-322.

298

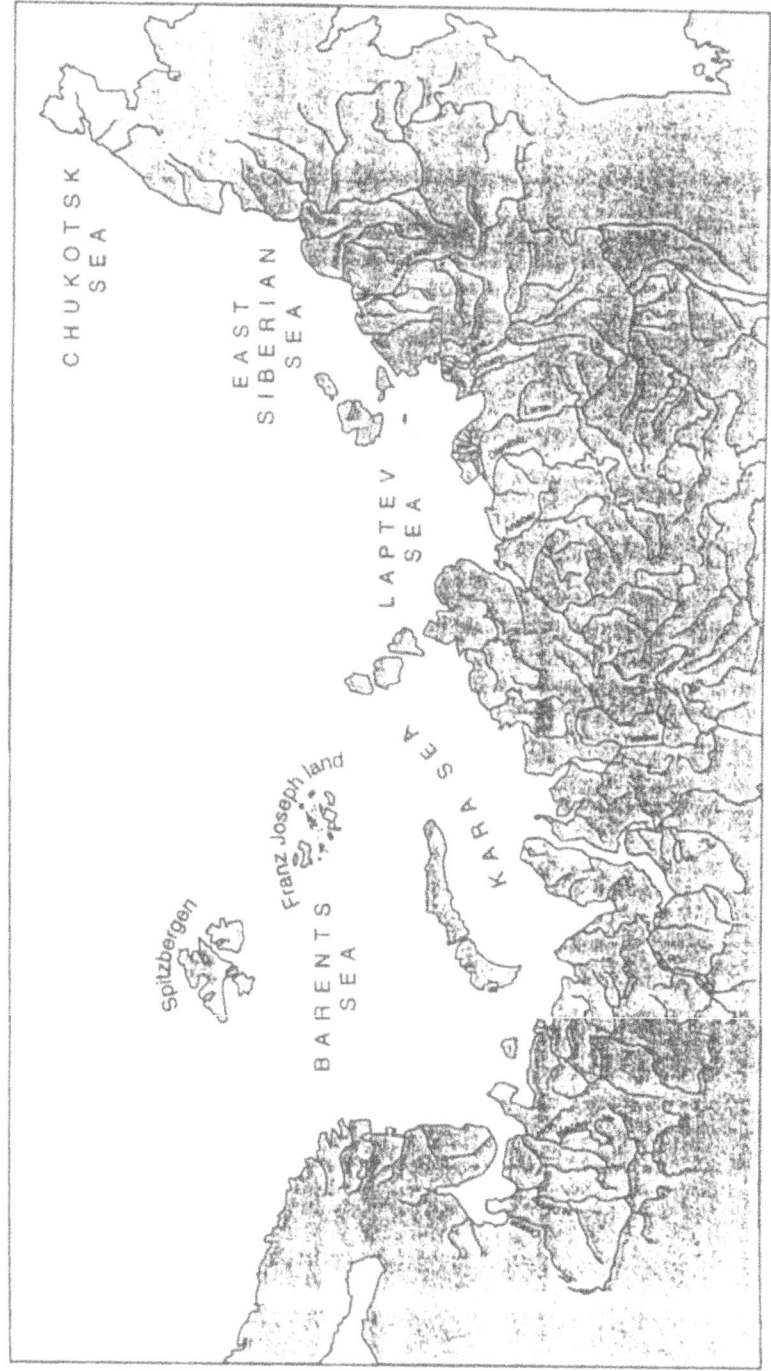

Figure 1. Map of the Russian sector of the Arctic Ocean basin showing its major rivers.

The multi-annual permafrost-permeated rocks are widespread in every the basin and have a substantial impaction the water balance. The main form of water in this area is a solid form, ice, which influences many natural processes. Ice controls the preparation, transportation and deposition of material including pollutants.

The thermal regime of the Arctic Basin rivers depends on climate, areas of permafrost, and on the distribution of swamps, lakes and glaciers. The Siberian rivers, that are meridionally directed, have characteristics of a flood surge wave moving northward, breaking up ice and causing ice dams and high water level. The small Siberian rivers are totally frozen in winter.

3. Materials and Methods

Evaluations of material fluxes in this work are based on the data of Roskomgidromet and Russian-French-Netherlands programme SPASIBA (1989-1995). Water and sediment discharges are taken from [1]. Fluxes of major elements, total organic carbon (TOC), nutrients and trace metals were calculated for the time series listed in Table 1.

TABLE 1. Time series of determinations of water, sediments, major ions, nutrients and trace metals in the Russian Arctic Rivers

Sea basin	Water/ Sediment, up to	Major ions	TOC	Nutrients	Trace metals
White	1988	1976-85	1971-80	1975-95	
Barents	1988	1976-85	1971-80	1979-95	
Kara	1988	1971-80	1971-80	1975-95	1989-93
Laptev	1991	1980-90	1985-90	1984-95	1989-93
E.-Siberian	1991	1980-90	1985-90	1984-95	
Chukchi	1985	1976-85	1991-80	1976-85	

Sampling and analysis of water on major ions, TOC and nutrients were carried out in accordance with standard methods of the State Survey of Observations [15-16]. More information may be found in [1]. Special attention was paid to sampling and analyses of water and sediments for heavy metals. Surface water samples were collected upwind of the bow of a small plastic or rubber boat, sub-surface sampling was done using a Teflon coated Go-Flo water sampler fixed on a plastic hydrowire or with a plastic pump (for determining particulate metals only) [14].

The preliminary treatment of samples (filtration, acidification) was done on board a ship under a laminar air flow clean bench. Analyses were performed in clean laboratories in France, Netherlands, Belgium and Russia.

4. Results and discussion

4.1. WATER AND SEDIMENT FLUXES

The average multi-annual data on the discharge of river water and suspended sediment into the Arctic Seas from the Russian territory are listed in Table 2. The six largest rivers

(Yenisey, Lena, Ob, Kolyma, Pechora and North Dvina) together provide more than 60% of the annual water discharge from the Russian land mass into the Arctic Ocean.

TABLE 2. Total water and suspended matter discharge from the Eurasian land mass into the Arctic Ocean [1].

River	Area $10^3 km^2$	Discharge			Total suspended matter		
		km^3	$m^3 \cdot s^{-1}$	$l \cdot s^{-1} \cdot km^{-2}$	$10^6 t \cdot a^{-1}$	$g \cdot m^{-3}$	$t \cdot km^{-2} \cdot a^{-1}$
Barents and White Seas							
Onega	57	15.9	500	8.8	0.3	18	4.9
N.Dvina	357	110	3470	9.7	3.8	35	10.6
Mezen	78	27.2	860	11.0	0.9	32	11.1
Pechora	324	131	4130	12.7	13.5	80	32.4
Other area	570	179	5690	10.0	3.5	20	6.2
Total	1386	463	14600	10.7	22	47	15.9
Kara Sea							
Ob	2545	429	13500	5.4	16.5	38	6.4
Nadym	64	18	570	8.9	0.4	22	6.2
Pyr	112	34.3	1080	9.8	0.6	18	5.5
Taz	150	44.3	1400	9.5	0.9	21	6.1
Yenisei	2594	620	19600	7.6	5.9	10	2.3
Pyasina	182	86	2730	15	3.4	40	18.8
Other area	867	443	7770	9	5.5	12	6.3
Total	6589	1478	46600	7.2	33.2	22	5
Laptevs Sea							
Khatanga	364	85.3	2700	7.4	1.7	20	4.6
Anabar	100	17.3	550	5.5	0.4	24	4.1
Olenjok	219	35.8	1140	5.2	1.1	31	5.1
Lena	2486	525	16650	6.7	17.6	34	7.1
Omoloy	39	7	220	5.7	0.13	18	3.2
Yana	238	34.3	1090	4.6	3.5	103	14.8
Other area	197	40.3	1280	6.5	0.65	16	3.3
Total	3643	745	23600	6.5	25.1	34	6.9
East-Siberian Sea							
Indigirka	362	61	1930	5.3	12.9	210	35.6
Alazeya	68	8.8	280	4.1	0.7	80	10.2
Kolyma	660	1322	4190	6.3	16.1	120	24.3
Other area	252	48.2	1530	6	3.85	80	15.3
Total	1342	250	7930	5.9	33.6	134	25
Chukchi Sea without Alaska							
Amguema	29.6	9.2	290	9.7	0.05	6	1.8
Other area	64.6	11.2	2050	5.5	0.65	58	10
Total	94.2	20.4	2340	6.8	0.7	34	7.4
For the entire Eurasian Arctic basin							
Total	13054	2960	95770	7.3	115	40	8.8

The specific water discharge decreases from west (9-13 $l \cdot s^{-1} \cdot km^{-2}$) to east (4-7 $l \cdot s^{-1} \cdot km^{-2}$). The maximum specific water discharge is observed for the Pyasina River (15 $l \cdot s^{-1} \cdot km^{-2}$) in relation to the maximum precipitation occurring over its basin. The average specific water discharge for the whole basin is about 7.3 $l \cdot s^{-1} \cdot km^{-2}$. That is lower than the global average (11 $l \cdot s^{-1} \cdot km^{-2}$ [17]).

The seasonal variations of the water discharge by the largest rivers are shown in Figure 2, A. The western rivers (North Dvina, Onega, Mezen) show the maximum discharge in May, for the Siberian rivers, Ob, Yenisey, Lena, Indigirka, et cetera, the highest discharge is observed in June.

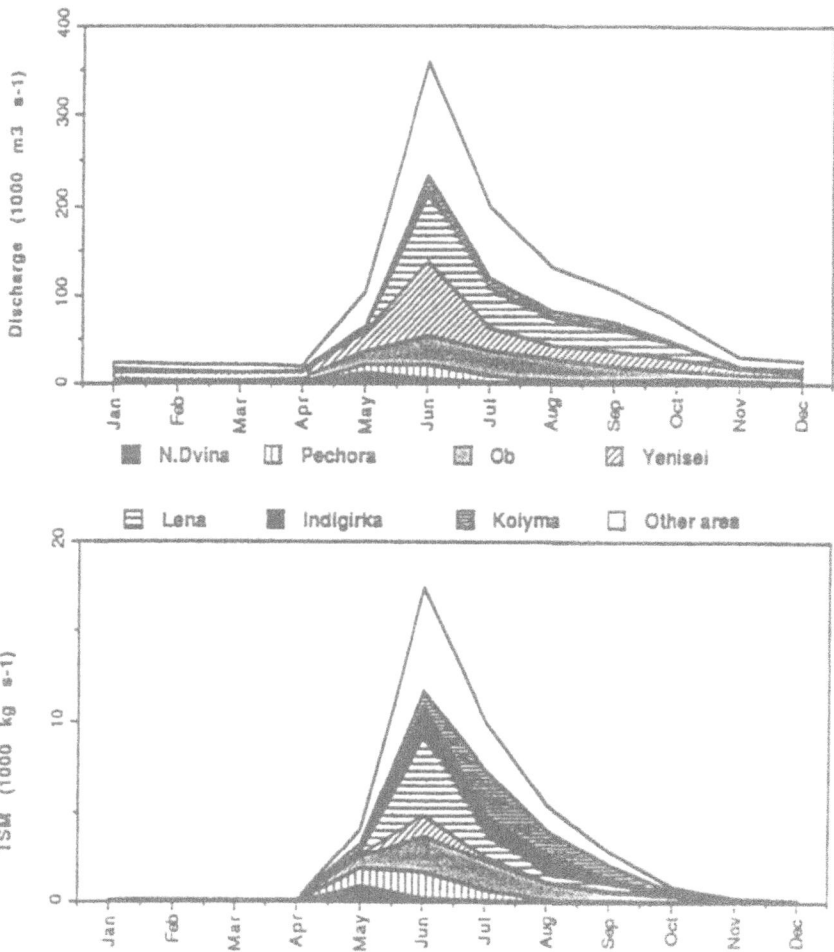

Figure 2. Seasonal variations of the water discharge (A) and total suspended matter (B) by the largest Russian Arctic rivers [1].

During the flood period the Arctic rivers discharge from 45-55% to 60-65% of their annual water discharge, to a maximum of 80% (Olenjok). The total water discharge into the Arctic Ocean is 2960 km^3, that is about 8.5% of global runoff [17].

Average annual concentration of river suspended matter ranges from 6 to 210 mg·l^{-1}, with an average of 38 mg·l^{-1} (Table 2). It is one order of magnitude lower than the average global river turbidity (460 mg·l^{-1}, [18]). The Barents, White and Kara Sea basin rivers are characterised by lower suspended matter concentrations in comparison with the East Siberia rivers (Indigirka, Kolyma - 80-210 mg·l^{-1}).

Seasonal variability of the average total suspended matter (TSM) discharge of the largest Eurasian Arctic rivers is shown in Figure 2, B. As well as the water discharge, the maximum TSM discharges are observed in May for the western rivers and in June for the eastern ones. It is interesting to note a good correlation between the specific TSM discharge and the specific water discharge (Figure 3) Again we see a significant difference between the rivers of west and east of the Russian sector of Arctic. The East Siberian rivers (Yana, Alazeya, Indigirka and Kolyma) show slopes similar to the North American rivers.

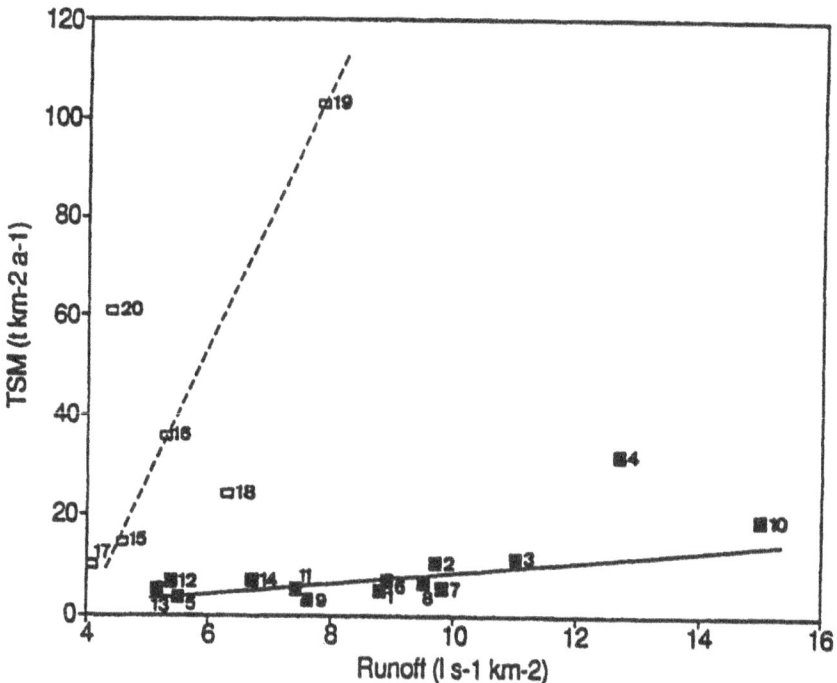

Figure 3. Mean specific annual TSM export by the largest Arctic rivers versus their respective runoff. 1 - Onega, 2 - North Dvina, 3 - Mezen, 4 - Pechora, 5 - Ob, 6 - Nadym, 7 - Pyr, 8 - Taz, 9 - Yenisey, 10 - Pyasina, 11 - Khatanga, 12 - Anabar, 13 - Olenjok, 14 - Lena, 15 - Yana, 16 - Indigirka, 17 - Alazeya, 18 Kolyma, 19 - Mackenzie, 20 - Yukon [1].

The total suspended matter flux from the Russian territory into the Arctic ocean averages $115 \cdot 10^6$t per yr which is only 0.8% of the global TSM discharge ($15 \cdot 10^9$t per yr, [17]).

4.2. DISCHARGE OF MAJOR IONS

In Table 3 the chemical composition of waters of the Russian Arctic Ocean basin rivers is given. The average concentration of dissolved ionic salts (TDS) of the Eurasian Arctic rivers coincides practically with the global average value (90 mg·l^{-1}). The western rivers (North Dvina, Mezen, Onega) have relatively high TDS (140-225 mg·l^{-1}), similar to the TDS of the Mackenzie River. The largest rivers of Siberia - Ob, Yenisey and Lena - are characterised by a little higher TDS (100-120 mg·l^{-1}) than the world average value. Mineralization of the East Siberian rivers is significantly lower (20-50 mg·l^{-1}).

The main type of waters is hydrocarbonate one with prevailing of Ca+Mg above Na+K (Figure 4). However in the small and medium tundra rivers of the Laptev, East Siberian and Chukchi Sea basins the most abundant anion is sulphate. The points, corresponding to these rivers (Omoloy, Ebitem, Alazeya, Amguema), shift in a triangle diagram toward the enhanced Cl+SO$_4$ (Figure 4), corresponding to the chemical composition of the precipitation in their basins.

The chemistry of major ions of the Lena river basin was studied in detail [5]. It is primarily governed by the weathering of carbonate rocks and by ground waters. The seasonal variability of major ions is shown for the Lena River (Figure 5). The average annual mineralization of the Lena River waters is about 114 mg·l^{-1}, the predominant ions are calcium and bicarbonate (carbonate class, calcium group). However, the dissolved solids concentration and the type of waters do not remain constant throughout the year.

At winter time there is a steep drop in water discharge (about 8 km^3·month^{-1}) and an increase in TDS (250 mg·l^{-1}). The Lena River waters pass into the mixed-chloride class, and in April (3.2 km^3·month^{-1}) they are transformed into the chloride class, sodium group. At the height of the flood in June (191 km^3·month^{-1}), i.e. 60 times as much as the minimum discharge in April, the concentration of dissolved solids is the lowest (61 mg·l^{-1}), and again calcium and bicarbonate ions predominate.

In our paper [5] it was shown that the reason for this variability was the influence of groundwater, which frequently has an extremely high mineralization (10-100 g·l^{-1} and higher) and sodium-chloride type.

TABLE 3. Major-ion concentrations for the Arctic rivers [1]

River, area	Cl⁻ mg·l⁻¹	SO₄⁻⁻ mg·l⁻¹	HCO₃⁻ mg·l⁻¹	Mg⁺⁺ mg·l⁻¹	Ca⁺⁺ mg·l⁻¹	Na⁺ mg·l⁻¹	K⁺ mg·l⁻¹	Sum of ions mg·l⁻¹	μeg·l⁻¹
				Barents and White Seas					
Onega	5.1	37.3	102	9.1	32.4	4.6	1.0	191	5180
N.Dvina	8.1	52.0	106	8.8	39.0	9.5	1.3	225	6170
Mezen	6.0	8.9	87.3	5.2	18.8	10.0	1.0	137	3610
Pechora	3.5	7.7	39.5	3.1	10.0	3.2	0.5	67.5	1800
				Kara Sea					
Ob	9.9	4.8	76.3	5.1	19.0	4.2	3.8	123.0	3040
Yenisei	9.9	8.4	55.2	3.3	15.9	6.3	1.0	100.0	2700
Pyr	2.5	1.9	25.6	2.0	4.5	2.9	0.5	39.9	1050
Taz	3.2	5.7	64.9	5.0	12.2	5.3	0.9	97.2	2640
				Laptevs Sea					
Khatanga	17.0	4.7	60.1	4.6	15.1	10.0	4.0	116.0	3140
Anabar	1.5	1.6	34.1	2.2	10.7	0.6	0.1	50.8	1380
Olenjek	5.0	4.0	79.5	4.4	21.8	3.8	0.6	119.0	3230
Lena	17.1	12.3	52.0	4.4	16.0	10.0	1.7	114.0	3230
Yana	2.4	8.4	20.4	1.4	6.1	3.3	0.6	42.6	1160
Omoloy	2.1	17.2	13.1	1.8	7.5	2.7	0.5	44.9	1290
Ebitem	1.4	6.0	18.6	1.4	4.4	2.7	0.5	35.0	930
Tundra	2.0	18.2	13.8	1.9	8.2	2.7	0.4	47.2	1360
				East-Siberian Sea					
Indigirka	1.8	12.3	31.3	2.6	11.9	0.4	0.1	60.4	1650
Alazeya	1.6	5.6	8.9	1.0	3.6	1.2	0.2	22.1	630
Kolyma	2.3	10.2	28.5	1.8	10.5	1.3	0.2	54.8	1480
Tundra	2.3	6.0	12.9	1.2	4.0	2.1	0.3	28.8	780
				Chukchi Sea without Alaska					
Amgyema	1.3	3.9	8.0	0.3	3.0	1.6	0.3	18.4	500
Tundra	1.4	5.2	7.7	0.7	3.0	1.4	0.2	19.6	550
				Eurasian Arctic rivers					
Average	7.6	9.9	48.2	3.6	15.4	4.6	0.8	90	2450
				NORTH AMERICAN ARCTIC BASIN					
				Beaufort Sea					
Mackenzie	10.5	32.6	96.5	9.3	36.5	9.2	1.1	196	5560
				Chukchi Sea, Alaska					
Colvill	1.1	19.4	53.5	4.3	16.4	2.6	0.8	97	2610
Kobuk	0.6	13.7	32.7	3.7	21.6	0.9	0.4	73.6	2260
Kuparuk	1.5	4.8	51.6	1.6	16.4	1.3	0.6	77.7	2010
				WORLD RIVER AVERAGE					
Average	5.8	8.25	52	3.35	13.4	5.15	1.3	89.2	2360

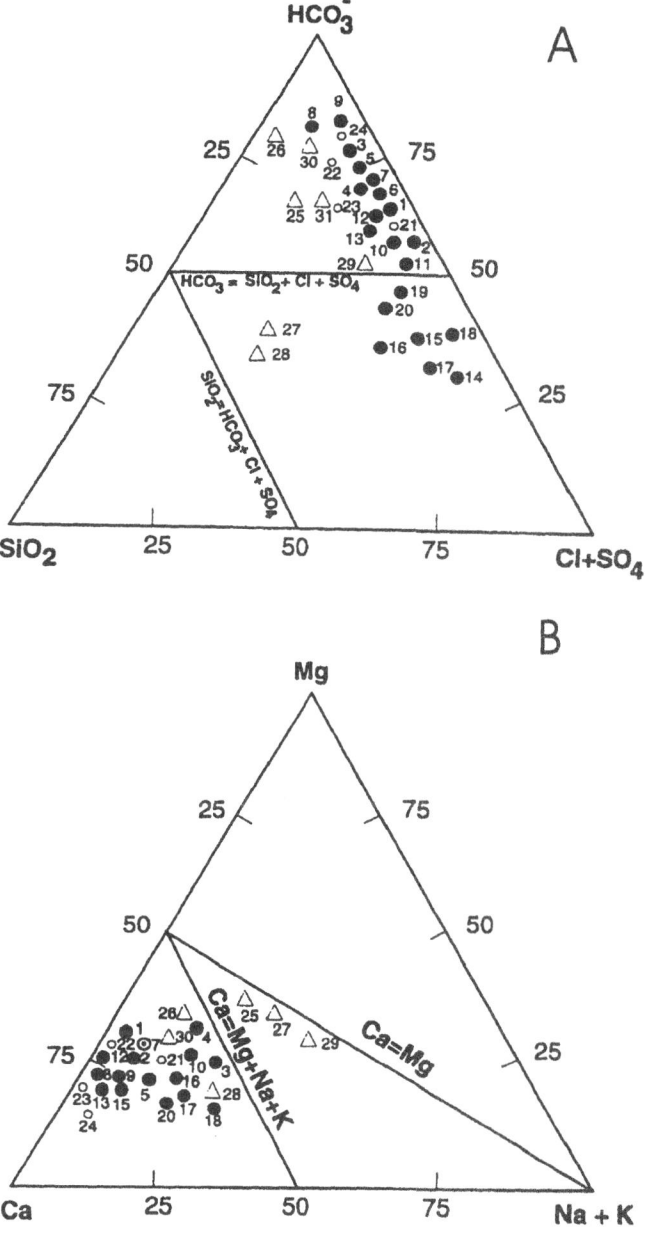

Figure 4. Triangular diagrams of interdependencies HCO_3 - $(Cl+SO_4)$ - SiO_2 (A) and Ca Mg - (K+Na) (B) for river waters of the Arctic basin and the largest rivers of other climatic zones. Russian Arctic rivers: 1 - Onega, 2 - North Dvina, 3 - Mezen, 4 - Pechora, 5 Ob, 6 - Yenisey, 7 - Khatanga, 8 - Anabar, 9 - Olenjok, 10 - Lena, 11 – Yana, 12 - Indigirka, 13 - Kolyma, average and small tundra rivers: 14 - Laptev Sea basin, 15 - East Siberian Sea basin, 16 - Chukchi Sea basin, 17 - Omoloy, 18 Ebitem, 19 - Alazeya, 20 - Amguema; North American Arctic rivers: 21 - Mackenzie, 22 - Yukon: small Alaskian rivers: 23 - Kobuk, 24 - Kuparuk. argest world rivers: 25 - Amazon, 26 - Ganges-Brahmaputra, 27 - Zaire, 28 Orinoko, 29 - Huanghe, 30 - Changjiang; 31 - average for world rivers (all concentrations are expressed in meq/l).

● - Eurasian rivers (1-20); o - North American rivers (21-24); △ - Largest world rivers (25-31) [1].

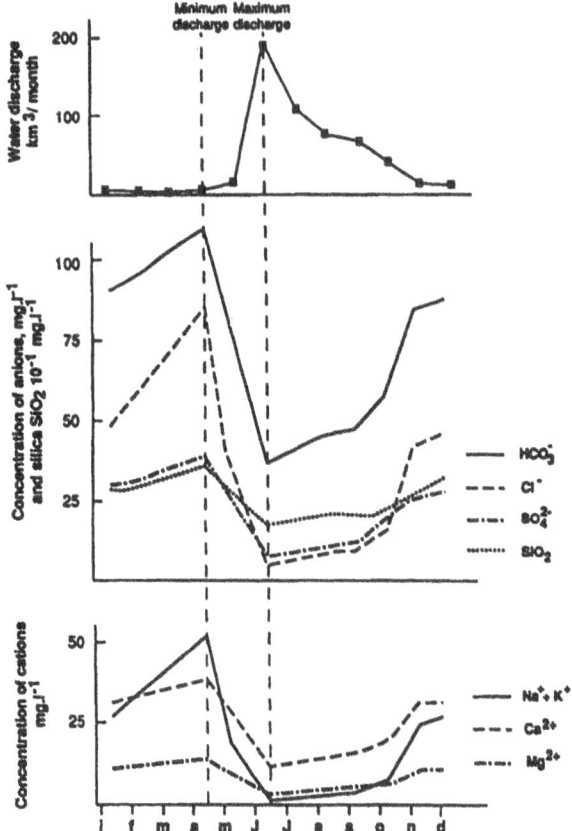

Figure 5. Seasonal variations of water discharge and major ion concentrations in the Lena River water (average for the period 1975-1990) [5].

One can see it in Figure 6, in which the mean annual discharge of cations and anions from the Lena River into the Laptev Sea is shown. Noticeable features are sharp fluctuations of the amounts of Ca and HCO_3 discharged by the river into the sea in correlation with the water runoff. At the same time the variations of the discharged Na+K and Cl are very small. Figure 6 shows that groundwater is discharged during the year with a roughly constant discharge rate. During the winter lean flow the role of groundwater is the most important. During the spring and summer flood the steep increase of the water discharge due to the melting of snow and to atmospheric precipitation suppresses the influence of groundwater input.

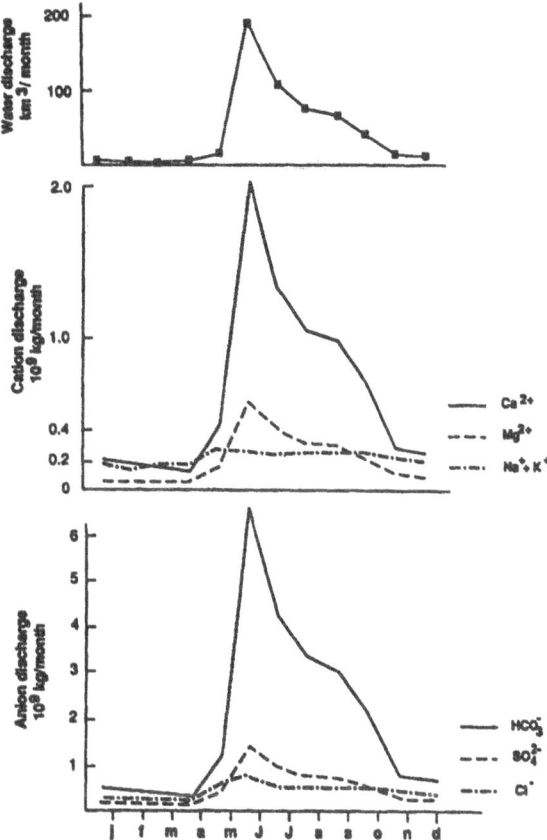

Figure 6. Mean annual discharge of cations and anions (in $10^6 Kg \cdot month^{-1}$) by the Lena River into the Laptevs Sea [5].

The annual chemical denudation rate for the entire Lena basin is about 23.5 $t \cdot km^{-2}$, which is about half of the current annual world average (36 $t \cdot km^{-2}$, [19]).

We can notice again that the specific erosion for the Russian Arctic rivers, as well as specific discharge, decreases from west to east from 10-30 $t \cdot km^{-2}$ per yr for the White Sea basin rivers to 2-5 $t \cdot km^{-2}$ per yr for the Chukchi Sea basin rivers and small tundra rivers of the Laptev, East Siberian and Chukchi Seas (Table 4.).

The total annual flux of TDS by the rivers into the Arctic Ocean is $267 \cdot 10^6$ t, that is about 8% of the global value.

TABLE 4. Multiannual average major-ion export into the Arctic Ocean with the river runoff; TDS: Total dissolved salts; DT: dissolved transport [1]

River, area	Cl⁻	SO₄⁻⁻	HCO³⁻	Mg⁺⁺	Ca⁺⁺	Na⁺	K⁺	TDS	DT
	$10^6 t \cdot a^{-1}$								$t \cdot km^{-2}$
Barents and White Seas									
Onega	0.08	0.6	1.5	0.14	0.51	0.07	0.02	3	52.7
N.Dvina	0.88	5.6	11.5	0.95	4.23	1.01	0.16	24.4	68.2
Mezen	0.16	0.3	2.19	0.13	0.51	0.23	0.03	3.54	45.3
Pechora	0.50	0.4	5.68	0.1	1.52	0.45	0.07	8.71	26.9
Other area	0.98	2.9	10.9	0.74	3.38	0.78	0.13	19.8	34.7
Total	2.60	9.8	31.2.	2.05	10.2	2.54	0.41	59.4	42.9
Kara Sea									
Ob	4.25	2.06	28.9	2.00	7.37	2.29	0.38	47.2	18.5
Yenisei	5.93	5.11	32.8	2.03	9.71	3.77	0.63	59.9	23.1
Pyr	0.09	0.07	0.88	0.07	0.15	0.10	0.02	1.32	11.8
Taz	0.14	0.25	2.88	0.22	0.54	0.23	0.04	4.30	28.7
Other area	1.66	2.01	7.02	0.97	4.17	1.21	0.20	17.2	14.5
Total	12.1	9.5	72.4	5.29	21.9	7.60	1.27	130	19.7
Laptevs Sea									
Khatanga	1.07	0.48	4.09	0.31	1.07	0.66	0.25	7.93	21.8
Anabar	0.03	0.07	0.54	0.04	0.17	0.02	0.003	0.87	8.7
Olenjok	0.17	0.17	2.6	0.15	0.72	0.13	0.02	3.97	18.1
Lena	8.98	6.46	27.3	2.32	8.42	5.26	0.88	59.6	23.4
Yana	0.08	0.31	0.71	0.05	0.21	0.11	0.02	1.49	6.3
Other area	0.09	0.52	1.1	0.09	0.39	0.13	0.02	2.34	9.9
Total	10.4	8.01	36.3	2.96	11	6.31	1.19	76.2	20.9
East-Siberian Sea									
Indigirka	0.11	0.83	1.73	0.14	0.70	0.05	0.01	3.57	9.90
Kolyma	0.30	0.94	3.43	0.25	1.35	0.21	0.03	6.51	9.90
Other area	0.13	0.34	0.79	0.07	0.23	0.12	0.02	1.70	5.30
Total	0.54	2.11	5.95	0.46	2.28	0.38	0.06	11.80	8.90
Chukchi Sea without Alaska									
Amgyema	0.01	0.03	0.07	0.00	0.03	0.02	0.00	0.16	5.40
Other area	0.02	0.06	0.09	0.01	0.04	0.02	0.00	0.24	3.70
Total	0.03	0.09	0.16	0.01	0.07	0.04	0.01	0.40	4.20
For the entire Eurasian Arctic basin									
Total	21.5	29.8	143.5	10.7	45.5	13.8	2.3	267.1	20.5
North American Arctic basin									
Mackenzie	2.08	6.15	22.00	8.2	0.26	2.1	1.88	42.7	23.7

It is interesting to compare the fluxes of dissolved and particulate major cations. Our preliminary estimations show that in general for the Russian Arctic rivers the dissolved input of major cations sharply prevails (for Ca - 93%, Mg - 90%, Na+K - 84% in dissolved form) [7].

There is one more distinction between western and eastern rivers. For the Barents, White, Kara and Laptev Sea basins the TDS exported is greater than the TSM export. The ratios of TSM to TDS are very low (0.1-0.5, except for the Pechora River - 1.5). For rivers of the East Siberian Sea basin the TSM export is greater than over the TDS export, with ratio of 1.6 to 3.6.

4.3. THE DISCHARGE OF ORGANIC CARBON

Data on total organic carbon (TOC), especially on dissolved and particulate organic carbon (DOC and POC), in the Arctic rivers are not enough. The available data are listed in Table 5. According to our estimate, average TOC concentration is about 7.5 mg·l^{-1}. That is lower the global average 9.9 mg·l^{-1} [20].

TABLE 5. Average TOC concentrations and fluxes for the Arctic rivers

Rivers	Discharge km^3	TOC mgl^{-1}	TOC 10^{-6}t·a^{-1}	TOC t·km^{-2}a^{-1}
Barents and White Seas				
Onega	15.9	20.7	0.33	5.8
N.Dvina	110	23.7	2.57	7.2
Mezen	27.2	7.0	0.19	2.4
Pechora	131	13.0	1.7	5.2
Total	430	12.4	5.33	4.3
Kara Sea				
Ob	404	7.1	2.87	0.96
Nadym	18	5.0	0.09	1.4
Pur	44.3	6.7	0.23	2.1
Yenisey	630	7.4	4.66	1.8
Total	1344	6.5	8.74	1.3
Laptevs Sea				
Khatanga	85.3	6.3	0.54	1.5
Anabar	17.3	5.1	0.09	0.9
Olenjok	35.8	7.2	0.26	1.2
Lena	525	7.7	4.04	1.6
Yana	34.3	6.7	0.23	1.0
Total	745	6.9	5.16	1.4
East-Siberion Sea				
Indigirka	61	7.7	0.47	1.3
Kolyma	132	8.1	1.07	1.6
Total	250	7.1	1.78	1.3
Chukotsk Sea (Without Alaska)				
Amgyema	9.2	6.7	0.06	2.1
Total	20	4.7	0.09	1.0
Total Eurasian Arctic	2790	7.5	21.1	1.6

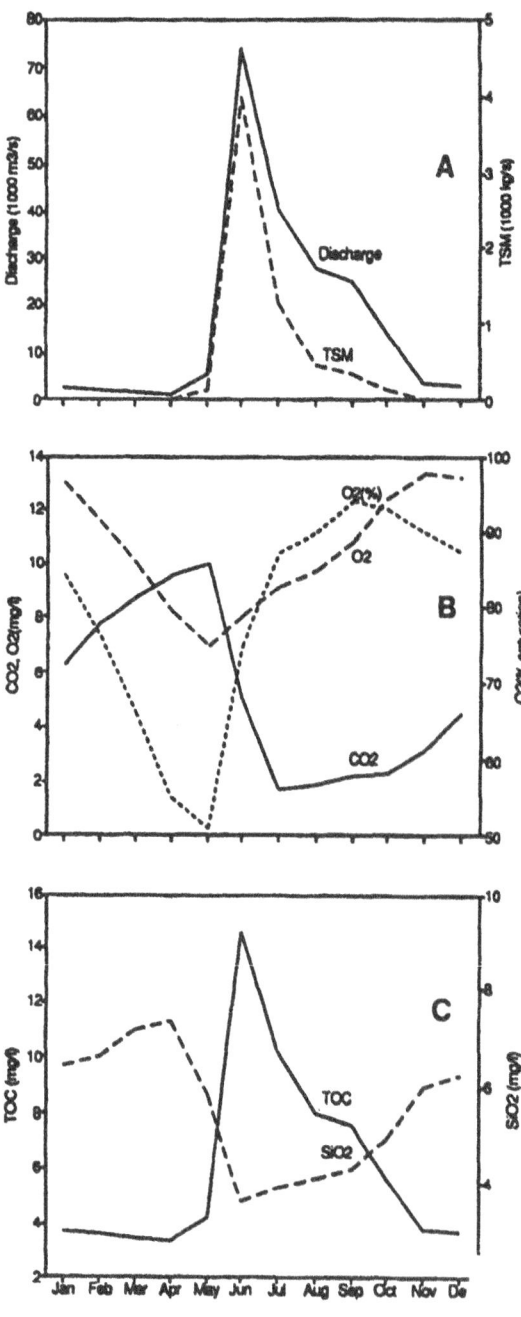

Seasonal variations of TOC for the Lena River are presented on Figure 7,C [2]. The minimum TOC concentration in the lower reaches of the Lena River occurs in winter time (November - May), while the maximum TOC concentration is observed in June. Climatic conditions of winter period prevent almost any biological activity and physical weathering. As a result the concentration and composition of organic matter does not vary. Only 4% of the annual TOC export is discharged during winter. During the flood, TOC concentrations in the Lena delta decrease about 10-15% under the influence of a huge volume of diluted water issued from melting ice. After the high water period, the input of soluble organic matter from soils, rocks and bottom sediments increases TOC concentrations again.

In summer time, more than 50% of annual TOC export of the Lena River enters in its delta. The annual TOC loss per unit area of watershed basin correlates significantly with runoff (Figure 8). The rivers of the Barents and White Sea basin (except the Mezen River) flow through forest and swamps and have the highest TOC yield (5-7 $t \cdot km^{-2}$ per yr). For Siberian rivers, the TOC yield is low and rather uniform (1-2 $t \cdot km^{-2}$ per yr). All Russian Arctic rivers are characterised by relatively high DOC concentrations and low POC concentrations (in $mg \cdot l^{-1}$), while the percentage of POC (dry wt) is

Figure 7. Seasonal variations of water discharge and TSM (A), dissolved oxygen and carbon dioxide (B), TOC and SiO$_2$ concentrations (C) in lower reaches of the Lena River [2].

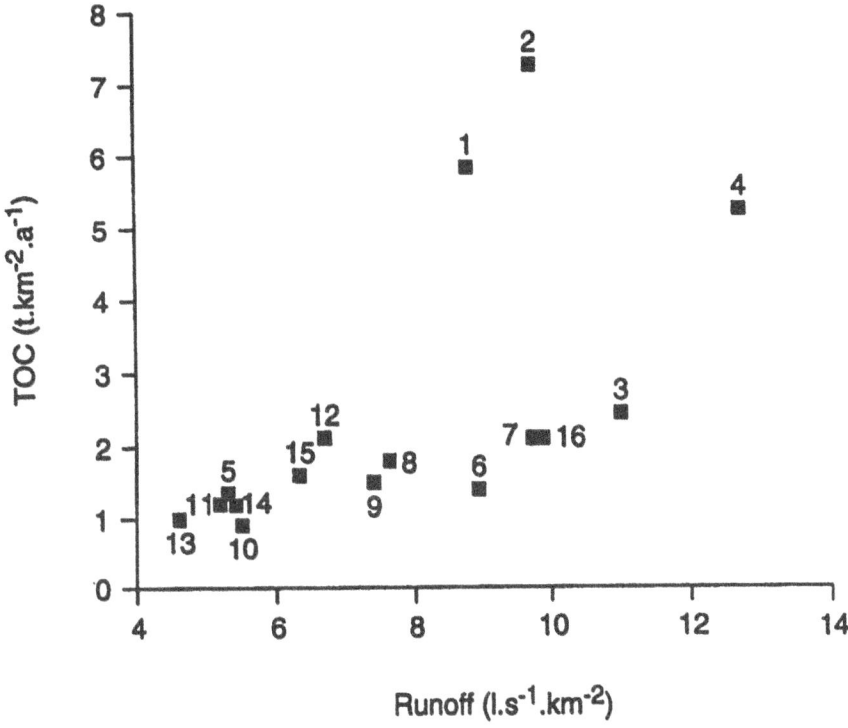

Figure 8. Mean specific annual TOC exported by the Russian Arctic rivers versus their respective runoff: 1 - Onega, 2 - North Dvina, 3 - Mezen, 4 - Pechora, 5 – Ob, 6 - Nadym, 7 - Pur, 8 - Yenisey, 9 - Khatanga, 10 - Anabar, 11 – Olenjok, 12 - Lena, 13 - Yana, 14 - Indigirka, 15 - Kolyma, 16 - Amgyema [1].

relatively high in the European Arctic rivers (16-23%) and much lower in the Siberian rivers (1-4%). In general, DOC contributes 86-91% of TOC export, or twice the Mackenzie River.

According to our new estimate, the total river export of TOC from the Russian Arctic territory into the Arctic Ocean is $21.1 \cdot 10^6$ t per yr, that is about 7.3% from global river TOC discharge [21].

4.4. THE DISCHARGE OF SILICA, NITROGEN AND PHOSPHORUS

4.4.1. Silica

Silica concentration in the Arctic rivers is relatively low: in the White, Barents and Kara Sea basins - 107-164 $\mu g \cdot at \cdot l^{-1}$. In the east lower - 33-98 $\mu g \cdot at \cdot l^{-1}$ (the global river average 170 $\mu g \cdot at \cdot l^{-1}$) (Table 6). Seasonal variations are shown in Figure 7,C for the Lena River. During the flood time, melting ice waters decrease the concentration of dissolved SiO_2 by

TABLE 6. Average concentrations of dissolved nutrients in the Arctic Rivers (μM)

River	SiO$_2$	NH$_4$	NO$_3$	DON	TDN	PO$_4$	DOP	TDP
Barents and White Seas								
N.Dvina	126	14	6			0.7	1.2	1.9
Onega	124	8	7			0.4	0.7	1.1
Mezen	114	8	4			0.8	0.9	1.6
Pechora	109	7	5			1.1	0.6	1.8
Kara Sea								
Ob	164	50	6			2.3		
Nadym	155	25	3			5.6		
Pyr	164	56	2			5.0		
Taz	-	68	2			4.3		
Yenisey	107	24	2			0.4		
Laptevs and East-Siberian Seas								
Khatanga	53	3	2	29	34	0.2	0.2	0.4
Anabar	43	3	2	18	23	0.1	0.1	0.2
Olenjok	45	4	2	29	35	0.1	0.2	0.3
Lena	70	3	3	33	39	0.3	0.7	1.0
Yana	52	3	2	21	26	0.1	0.2	0.3
Indigirka	47	3	2	29	34	0.2	0.3	0.5
Kolyma	67	4	2	29	35	0.3	0.5	0.8
Small rivers of tundra	33	4	2	19	25	0.1	0.1	0.2
Chukchi Sea (without Alaska)								
Amgyema	98		2			0.4		
Eurasian Arctic basin								
Average	95	15	4			0.8		

25-30%. The total discharge of dissolved silica into the Arctic Ocean adds up to $17.3 \cdot 10^6$ t per yr. It is interesting to compare this value with the export of particulate silica, which is the most abundant component of river suspended matter. According to our estimate the gross flux of particulate silica is $31.3 \cdot 10^6$ t per yr, or nearly 81% of total silica export to the Arctic Ocean.

4.4.2. Forms of nitrogen and phosphorus

The average concentrations and export of dissolved forms of nitrogen and phosphorus from the rivers of the Russian sector of the Arctic basin are given in Tables 6 and 7. Average water discharge-weighted concentrations of nitrates, orthophosphates and especially ammonium nitrogen are significantly higher than global average concentrations [20]. According to the data in Tables 6 and 7, the very high volumes of nutrient export are determined primarily by high concentrations of nutrients and huge volumes of water discharge in the Kara Sea basin rivers (they contribute 90% NH$_4$, 60% NO$_3$ and 80% PO$_4$ of the nutrient export into the Arctic Ocean from the Russian territory). One of the main reasons is the higher level of anthropogenic pollution of the western regions of the Russian Arctic in comparison with the East Siberia, clearly seen in Table 6. However there is another, natural reason - abundant swamps in the Kara Sea basin especially and wide distribution of permafrost in the region [22]. At the same time we cannot exclude the existence of uncorrected data for nutrient concentrations, for ammonium especially.

TABLE 7. Nutrient export from the Eurasian territory into the Arctic Ocean (10^3t a^{-1})

River	SiO$_2$	N-NH$_4$	N-NO$_3$	DON	TDN	P-PO$_4$	DOP	TDP
Barents and White Seas								
N.Dvina	820	21.7	9.9			2.3	4.1	6.4
Onega	117	1.7	1.6			0.2	0.35	0.55
Mezen	166	3.0	1.4			0.6	0.7	1.3
Pechora	791	13.3	9.2			4.4	2.5	6.9
Other area	688	2.1	1.2			1.3	1.4	2.7
Total	2580	41.8	23.3			8.8	9.0	17.8
Kara Sea								
Ob	4130	287.4	42			32.0		
Nadym	156	6.5	0.8			3.1		
Pyr	324	22.3	1.5			5.1		
Yenisey	4010	207.8	25.4			8.1	22.3	30.4
Other area	2490	12.7	11.5			11.7		
Total	11100	536.7	81.2			60		
Laptevs Sea								
Khatanga	270	3.4	2.6	34.1	40.1	0.5	0.5	1.0
Anabar	45	0.7	0.5	4.3	5.5	0.1	0.1	0.2
Olenjok	97	1.8	1.1	14.3	17.2	0.1	0.2	0.3
Lena	2200	21	21.7	243	286	4.2	11	15.2
Yana	106	1.4	1.0	10.3	12.7	0.1	0.2	0.3
Other area	94	2.4	1.4	12.8	16.6	0.1	0.2	0.3
Total	2810	30.7	28.3	319	378	5.1	12.3	17.3
East-Siberan Sea								
Indigirka	170	2.4	1.8	24.4	28.6	0.4	0.6	1.0
Kolyma	520	6.6	4.0	52.8	63.4	1.0	2.0	3.0
Other area	110	2.9	1.7	15.4	20.0	0.1	0.2	0.3
Total	800	11.9	7.5	92.6	112.0	1.5	2.8	4.3
Chukchi Sea (without Alaska)								
Amgyema	54		0.9			0.3		
Other area	66		1.9			0.7		
Total	120		2.8			1.0		
Eurasian Arctic basin								
Total	17410	621	143			76		

Seasonal variations of nutrient concentrations are shown for the Pechora River (average annual for period 1972-1995) (Figure 9). During winter time (November-May) NO$_3$ concentrations and in lesser degree NH$_4$ concentrations are relatively high when compared with a period of spring flood and summer - autumn lean flow (accordingly 4-15 μ·M against 0.7-2.0 μ·M for NO$_3$ and 6-10 μ·M against 5-7 μ·M for NH$_4$). In paper [8] it was shown that for the White and Barents Sea basins the groundwater discharge

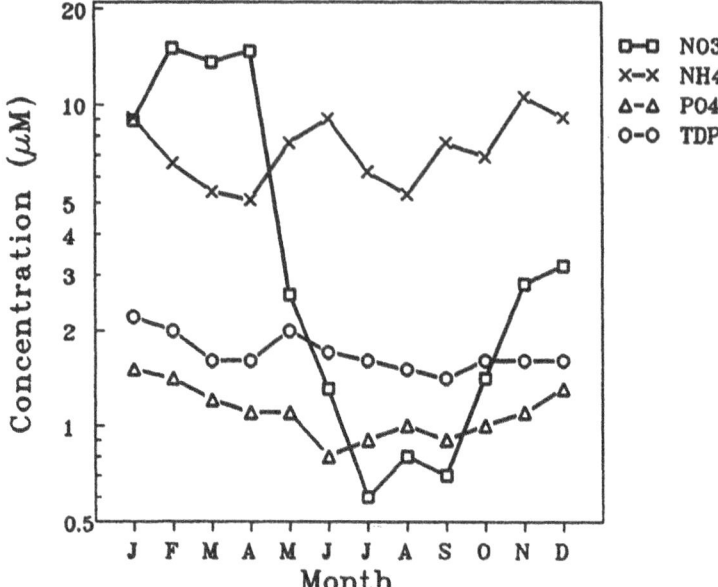

Figure 9. Seasonal variations of dissolved forms of nitrogen and phosphorus in the Pechora River.

contributed about 15% of the river discharge and gave about 20% of dissolved mineral form of nitrogen and 15% of phosphorus from total river export. The important role of the groundwater discharge in wintertime is an explanation of higher concentrations of nutrients at this period.

During the spring flood sharply increased flow of thawed snow and ice waters dilutes the nutrients concentrations and consumption by phytoplankton supports low concentrations of nutrients in summer time.

The data on concentrations of dissolved organic forms of nitrogen and phosphorus are available for the Laptev Sea and East-Siberian Sea basin rivers (Table 6). One can see that the concentrations of the organic forms of nutrients are significantly above mineral ones - for nitrogen 10 times for nitrogen 10 times greater, for phosphorus 2 times greater. No information is available for particulate forms of both elements in the Russian Arctic rivers.

According to our estimate (Table 7) the river export of NO_3, NH_4 and PO_4 from the Russian territory into the Arctic ocean adds up to $143 \cdot 10^3$, $620 \cdot 10^3$ and $76 \cdot 10^3$ t per yr, respectively.

In our previous paper [1], the attempt was made to evaluate the new production of organic matter generated by the river input of nutrient discharge to the Arctic Seas. Assuming the classical Redfield ratios of 42:28:7:1 for C:Si:N:P in phytoplankton (on a mass basis) and an average area of $3.13 \cdot 10^6$ km^2 for the Eurasian arctic seas (depth 0-200 m) and our estimates of the river discharge of nutrients, we computed the new production which can be sustained by this nutrient source. The computations have shown that the river discharge of mineral N and P did not play any significant role on the primary production of the Arctic Seas. The probable estimation is of the order of 10% as compared with other sources, the most important of them is the upwelling of the open sea waters.

4.5. THE DISCHARGE OF TRACE ELEMENTS

Our new evaluations of the river discharge of dissolved and particulate trace elements into the Arctic Ocean are based on the first reliable data for trace metal concentrations, that were obtained in three expeditions to the Laptev and Kara Seas in 1989, 1991 and 1993 (the Programme SPASIBA).

Average concentrations of dissolved and particulate trace elements for three Siberian rivers are given in Table 8. The concentrations of dissolved trace elements in the Arctic rivers are comparable to or even lower than in most major world rivers. Only Cu and Ni in the Ob waters and Zn in the Yenisey waters are 1.5-2.0 times higher than their average world rivers concentration.

The concentrations in the Yenisey (except Zn) and the Lena rivers are overall lower than in the Ob river. It may be not only related to the lower weathering rate but also to the lower biological productivity in the Ob river. We cannot exclude a possibility of anthropogenic pollution due to Norilsk and Murmansk large centres of mining and metallurgic industry, especially through the atmosphere [4].

At the same time, dissolved As in the Lena river (2.0 nM) and Cd in the Ob and Yenisey rivers (5-8 pM and 11-16 pM) were the lowest values reported so far for major world rivers [3].

Also the average content of trace metals in suspended matter of the Siberian rivers is in general below the global average content (Table 8). In a Russian-German expedition to the Lena River basin relatively high content of some heavy metals (Cu, Zn, Cd and other) were found, however the authors [23] explained this fact by natural reasons (correlation with high content of OM and lithology).

TABLE 8. Dissolved (nM) and particulate (ppm) trace metal concentrations in the Ob, Yenisey and Lena river waters [4]

Element		Ob	Yenisey	Lena	World river average
Cu	dis	33	25	14	23.6
	part	50	110	28	100
Pb	dis	0.07	0.03	0.2	0.15
	part	16	30	23	3.5
Zn	dis	6	20	1.2	9.2
	part	104	220	160	250
Cd	dis	0.006	0.014	0.05	0.09
	part	0.26	1.3	0.25	1.2
Ni	dis	22	9	4.4	8.5
	part	38	77	34	90
Hg	dis	0.0028	0.0015	0.005	-
	part	0.05	0.05	0.12	-

We considered above the average values of trace element concentrations. The localities with rather high concentrations of dissolved heavy metals were found in the Yenisey and Ob estuaries. In near bottom waters of one deep in the Yenisey estuary (30 m depth, 3‰ salinity) the measured concentrations were: $Zn > 50$ nM, $Cu > 30$ nM, $Cd > 1$ nM, and $Pb > 2$ nM [10]. This deep is a natural trap for river sediment, enriched by organic matter, and produced in situ by phytoplankton material. A deficit of $O_{2\,diss.}$ (< 4 ml$\cdot$l^{-1}), lower pH, high NH_4, and significant influx of dissolved metals from bottom sediments point to the natural sources of metals in the deep.

The transformation processes of the trace metals forms in the mixing zones of the Ob, Yenisey and Lena rivers were studied [3,9-10,14]. In the Lena River mixing zone relations with salinity show that for Cu, Ni and Cd there is a mobilisation of the dissolved fraction from the suspended matter, with an increase of the dissolved concentration of 1.5, 3 and 6 times, respectively [9]. For Zn and Pb, a simple mixing of the Lena River waters with the Laptev Sea waters was observed. Available information for the particulate form of some metals in the Lena River mixing zone (Scandium-normalisation) indicates an absence of significant variations with salinity (Figure 10). It is necessary to note that the most transformable trace elements are absent in this figure.

In the Ob and Yenisey estuaries the behaviour of truly dissolved, colloidal and so-called "dissolved" forms of Cd, Cu and Fe is presented in Figure 11 ("dissolved" form is a sum of colloidal and truly dissolved one; pre-filtered water using Nucleopore filters - 0.4 μm, 142 mm was further processed through a cross-flow ultrafiltration (CFF) system with a 10^4 Dalton polysulfate membrane-corresponding to a pore size of about 3 nm) [2]. The truly dissolved, colloidal and "dissolved" copper appear to be conservative, while all three forms of Cd show non-conservative behaviour. It is interesting to note that the well-known phenomenon of "dissolved" iron flocculation is explained by flocculation of colloidal iron. The authors [3] stress that "the colloidal material appears to be the most reactive and heterogeneous phase, which determines to a large extent the behaviour of trace metals, either conservative or non-conservative".

Information about trace metals behaviour in the river-sea mixing zones gives a possibility to evaluate not only the gross river flux, but also the so-called net flux with taking in account the processes of sedimentation, absorption-desorption, flocculation and other.

Our evaluations of the gross and net river fluxes of some trace metals and iron in dissolved and particulate form for three Siberian rivers, and also for the whole Russian Arctic are presented in Table 9. In the last case the assumtion was that the Ob, Yenisey and Lena Rivers were representative for the whole Russian Arctic regarding the heavy metals concentrations. We note that for some metals (Cd, Ni, Cu, Fe) the difference between gross and net river flux is quite significant.

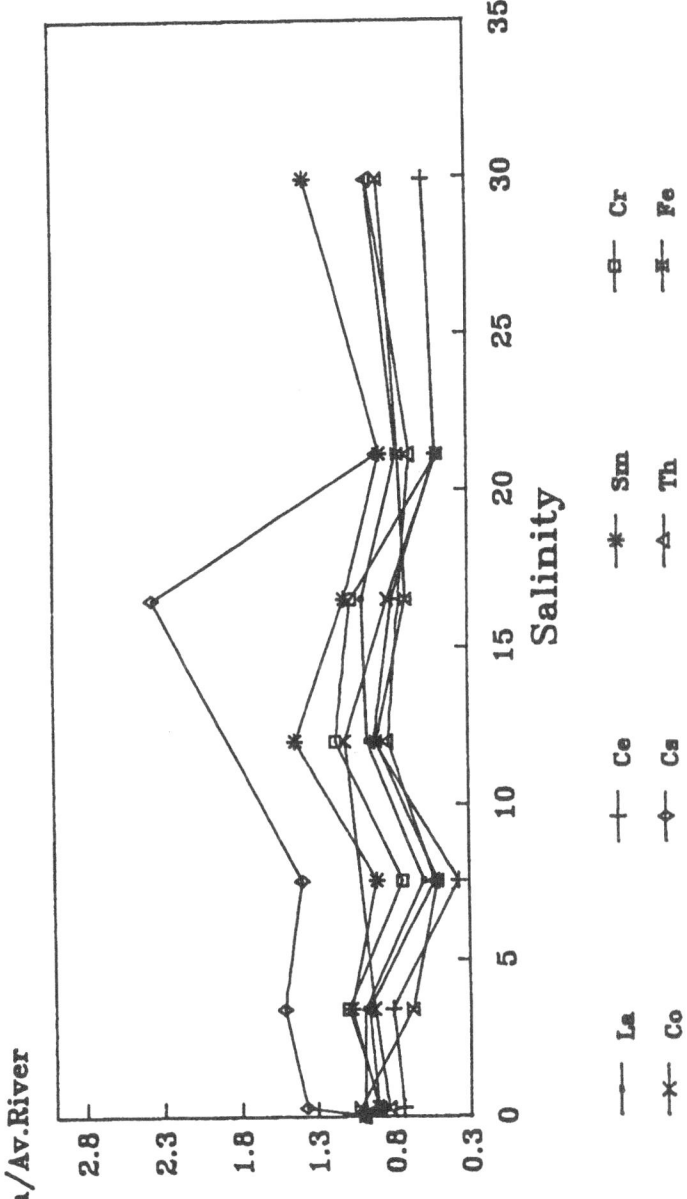

Figure 10. Sc-normalized elements in suspended matter of the Lena River mixing zone related to Sc-normalized elements in the Lena River suspended matter versus salinity [6].

318

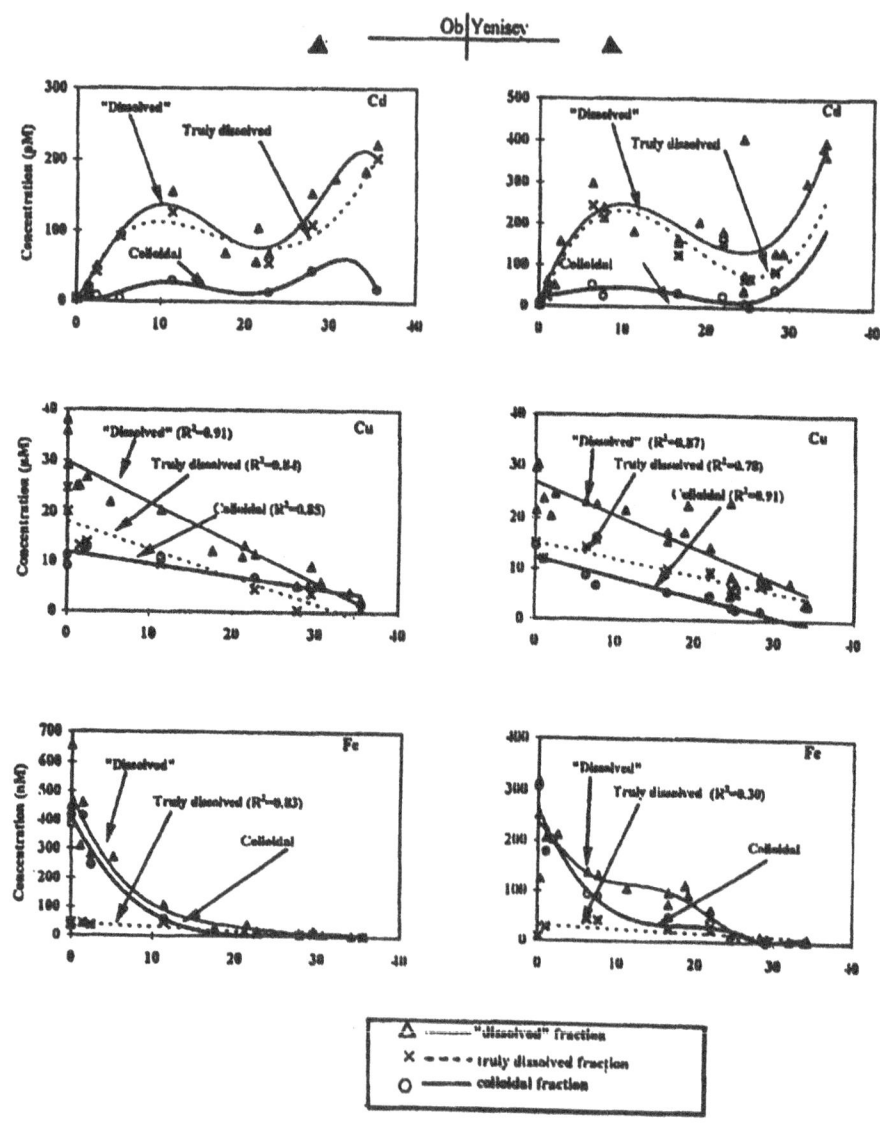

Figure 11. Cu, Cd and Fe in "dissolved" fraction, truly dissolved fraction, and colloidal fraction as a function of salinity in the Ob and Yenisey estuaries [3].

TABLE 9. Riverine gross and net fluxes of dissolved and particulate heavy metals and iron to the Arctic Ocean ($10^3 \text{t} \cdot \text{a}^{-1}$)

River basin	Cu	Pb	Zn	Ni	Cd	Hg	Fe
Ob							
dis	0.85	0.006	0.16	0.53	0.0003	0.0002	12
part	0.84	0.26	1.7	0.63	0.0033	0.0008	940
total gross flux	1.60	0.266	1.86	1.16	0.0036	0.001	952
Yenisey							
dis	1.0	0.004	0.82	0.34	0.001	0.0002	10
part	0.65	0.18	1.3	0.45	0.0136	0.0003	320
total gross flux	1.65	0.184	2.12	0.79	0.0146	0.0005	330
Lena							
dis	0.47	0.027	0.04	0.14	0.0032	0.0004	19
part	0.49	0.40	2.50	0.55	0.0046	0.0040	590
total gross flux	0.96	0.427	2.54	0.69	0.0078	0.0044	610
Whole dis	4.36	0.05	1.92	1.90	0.0085	0.0015	77
Russian part	3.72	1.26	10.34	3.06	0.040	0.0096	3480
Arctic total gross flux	8.08	1.31	12.26	4.96	0.0485	0.011	3560
Whole dis	3.54	0.115-.175	0.69	1.96	0.106	-	7.8
Russian part	0.32-0.64	0.25-0.51	1.6-.18	0.32-0.64	0.003-.006	-	380-760
Arctic total net flux	3.8-4.2	0.37-0.68	2.3-3.9	2.3-2.6	0.11	-	390-770

5. Future Research Priorities

Reliability of the assessments of riverine dissolved and particulate material fluxes into the ocean depends on both the reliability of determinations of water and suspended matter discharge and reliability of the data of river water and suspended matter chemical analyses. At the same time specific distinctions exist between the rivers and estuarine systems in the Arctic and in another climatic zones. There are significant differences between summer and winter regime of river discharge (snow-rain supply of water into a river basin and outflow of the main part of riverine material into the sea in summer, dominant groundwater input and low water and sediment discharge in winter). A specific role play an ice (diffusion of fresh water under ice for a long distance in the sea, a high ability of ice to trap and transport significant amounts of dissolved and suspended material) and seasonal sinking of brines down to the bottom shelf waters in time of ice formation (long distance spreading of a part of riverine sediment material with these heavy saline waters along the bottom). Severe climatic conditions result in practice to that during winter period by duration from 200 to 300 days in year were collected by for lesser analytical data compared with shorter summer period.

The priorities for riverine material fluxes research into the Arctic Ocean are:

1. Water and suspended matter discharge of smaller rivers must be measured more extensively because of significant uncertainty due to their extrapolation (the discharge of these rivers takes about 25% of total discharge into the Russian Arctic).
2. Develop new methods of sampling and analyses of river water and suspended matter in winter time in the Arctic especially. More attention has to be paid to the study of groundwater chemistry.
3. More knowledge of the estuarine processes of riverine material transformation to determine the so-called net input of substances into the ocean after passing of fresh and saline waters mixing zone.

6. Conclusions

If we compare the main characteristics of river basins and water chemistry in western and eastern regions of the Russian Arctic basin (the boundary between two large regions crosses the Laptev Sea basin and coincides with a boundary between the Eurasian and North American tectonic plates) (Table 10) then we see the existence of many differences between two regions.

TABLE 10. Comparative characteristics of river waters and basins of West and East Eurasian regions (to the west and to the east from the Lena River).

Parameter	West Eurasia	East Eurasia	Mackenzie river
Area, $10^6 km^2$	10000	3050	1805
Elevation, m	10-1000	500-3000	1000-3000
Discharge, km^3	2360	600	249
$ls^{-1} \cdot km^{-2}$	5-15	4-9	7.9
TSM, $g \cdot m^{-3}$	10-80, av.29	20-210, av.78	168
$10^6 t \cdot a^{-1}$	68	47	42
$t \cdot km^{-2} a^{-1}$	2-19*, av.6.8	2-36, av.15.4	23
TDS, $mg \cdot l^{-1}$	40-225	18-60	196
$10^6 t \cdot a^{-1}$	234	33	96
$T \cdot km^{-2}$	9-68, av.23.4	4-20, av.11	47.8
TOC, $mg \cdot l^{-1}$	5-24	6-10	12.6
$t \cdot km^{-2} \cdot a^{-1}$	1-7	1-2	1.7
SiO_2, $\mu g \cdot at \cdot l^{-1}$	45-165	30-100	67
NO_3, $\mu g \cdot at \cdot l^{-1}$	2-11	2	3.6
NH_4, $\mu g \cdot at \cdot l^{-1}$	7-70	3-6	3.6
PO_4, $\mu g \cdot at \cdot l^{-1}$	0.3-5.4	0.1-0.4	0.2

* Excluding the Pechora River ($32.4 \ t \cdot km^{-2} \cdot a^{-1}$)

The rivers of the East Siberia in comparison with the rivers of the western part of the Russian Arctic are characterized by lower run-off, higher turbidity, and by significantly lower water mineralization, organic matter and nutrient concentrations. We noted [1] that the East Siberian rivers (Yana, Alazeya, Indigirka, Kolyma) were even more similar to the North American Arctic rivers than to the rivers located westward of the Lena River. It is very interesting to define the place of the rivers of the Russian Arctic basin in the global scale. In Table 11 the main parameters of the Russian Arctic river input and the fluxes of total suspended load and dissolved constituents, related to the unit of watershed area, are given. We see that the fluxes of water, TDS, TOC and TDN for the Russian Arctic rivers are in proportion to area in ratio 0.5-0.6, NO_3 - 0.3, very low proportion for TSS ~ 0.1 and high proportion for PO_4 - 1.46.

TABLE 11. Comparison of the Russian Arctic river input with the global river input

Parameter	Russian Arctic	Whole World	Russian Arctic, % from whole world	Proportion of Russian Arctic to whole world, ratio to area
Area, 10^6 km^2	13	99.9	13	1.0
Water, 10^6km$^3\cdot$a^{-1}	2.96	35.0	8.5	0.65
TSS, 10^9t$\cdot$a^{-1}	0.115	15.0	0.8	0.06
TDS, 10^9t$\cdot$a^{-1}	0.284	3.8	7.5	0.58
TOC, 10^6t$\cdot$a^{-1}	21	370	5.7	0.44
N-NO$_3$, 10^6t$\cdot$a^{-1}	0.15	4.0	3.8	0.3
TDN, 10^6t$\cdot$a^{-1}	1.0	14.5	7.0	0.55
P-PO$_4$, 10^6t$\cdot$a^{-1}	0.076	0.4	19	1.46

The conclusion is from this Table that practically for all parameters (except PO_4) the river fluxes of the Russian sector of the Arctic ocean related to the area unit are significantly lower than the average characteristics of the global river discharge. At the same time, the influence of the Arctic river discharge to the Arctic ocean is much higher than in many other regions of the World Ocean due to very vast shelf and shallow depths of the coastal Arctic Seas.

7. Acknowledgements

This work was supported by the Russian Fund for Basic Research, Grant N 97-05-64576.

8. References

1. Gordeev, V.V., Martin, J.-M., Sidorov, I.S. and Sidorova, M.V. (1996) A reassessment of the Eurasian River input of water, sediment, major elements, and nutrients to the Arctic Ocean, *Amer. J. Sci.* **296**, 664-691.

2. Cauwet, G. and Sidorov, I. (1996) The biogeochemistry of Lena River organic carbon and nutrients distribution, *Marine Chem.* **53**, 211-228.

3. Dai, M. and Martin, J.M. (1995) First data on trace metal level and behaviour in two major Arctic river-estuarine systems (Ob and Yenisey) and in the adjacent Kara Sea, *Earth Planet. Sci. Lett.* **131**, 127-141.

4. Gordeev, V.V. (1997) Heavy metals in water, suspended matter and bottom sediments of the Kara and Laptev Sea, in The AMAP International Symposium on Environmental Pollution of the Arctic. *Extended abstracts, Tromso, Norway,* **1**, 216-217.

5. Gordeev, V.V. and Sidorov, I.S. (1993) Concentrations of major elements and their outflow into the Laptev Sea by the Lena River, *Marine Chem.* **43**, 33-46.

6. Gordeev, V.V. and Schevchenko, V.P. (1995) Chemical composition of suspended sediments in the Lena River and its mixing zone, in H. Kassens, D. Piepenburg, J. Thiede, L. Timokhov, H.-W. Hubberten and S.M. Priamikov (eds.), Russian-German Cooperation: Laptev Sea System, Alfred Wegener Institute for Polar and Marine Research, Bremerhawen, *Report on Polar Research*, **176**, 154-169.

7. Gordeev, V.V. and Tsirkunov, V.V. (1998) River fluxes of dissolved and suspended substances into sea basins, in V.Kimstach, M.Meybeck and E.Baroudy (eds.), *A water quality assessment of the former Soviet Union,* E&FN Spon, London and NewYork, 311-350.

8. Gordeev, V.V., Dzhamalov, R.G., Zektser, I.S. and Zhulidov, A.V. (1999) An assessment of groundwater input of nutrients into the coastal seas of the Russian Arctic, *Water Resources,***38**, 1-6 (in Russian).

9. Guieu, C., Huang, W.W., Martin, J.-M. and Yong, Y.Y. (1996) Outflow of trace metals into the Laptev Sea by the Lena River, *Marine Chem.*, **53**, 255-268.

10. Kravtsov, V.A., Gordeev, V.V. and Pashkina, V.I. (1994) Labile dissolved forms of heavy metals in the Kara Sea waters, *Oceanology,* **34**, 673-680 (in Russian).

11. Letolle, R., Martin, J.-M., Thomas, A.J., Gordeev, V.V., Gusarova, S. and Sidorov, I.S. (1993) [18]O abundance and dissolved silicate in the Lena delta and Laptev Sea (Russia), *Marine Chem.*, **43**, 47-64.

12. Lisitzin, A.P. and Vinogradov, M.E. (1994) The International high-latitude expedition to the Kara Sea (the 49-th cruise of the R/V "Dmitry Mendeleev"), *Oceanology,* **34**, 643-651 (in Russian).

13. Makkaveev, P.N. and Stunzhas, P.A. (1994) Hydrochemical characteristics of the Kara Sea based on the results the 49-th cruise of R/V/ "Dmitry Mendeleev", *Oceanology,* **34**, 662-667 (in Russian).

14. Martin, J.-M., Guan, D.M., Elbaz-Poulichet, F., Thomas, A.J. and Gordeev, V.V. (1993) Preliminary assessment of the distributions of some trace elements (As, Cd, Cu, Fe, Ni, Pb and Zn) in a pristine aquatic environment: the Lena River estuary (Russia), *Marine Chem.*, **43**, 185-200.

15. Lurie, Yu.Yu. (ed.) (1982) Unificated methods of water analyses, Chemistry Publisher, Moscow.

16. Semenov, A.D. (ed.) (1977) A guidance on chemical analyses of surface land waters, Gidrometeoizdat, Leningrad (in Russian).

17. Milliman, J.D. (1991) Flux and fate of fluvial sediment and water in coastal seas, in R.F.C. Mantoura, J.-M.Martin and R.Wollast (eds.) Ocean Margin processes in Global Change, John Wiley and Sons, Chichester-New York - Brisbane - Toronto - Singapore, 69-89.

18. Gordeev, V.V. (1983) River input to the Ocean and its geochemical characteristics, Nauka Publishers, Moscow, 160 pp. (in Russian).

19. Meybeck, M. (1979) Concentrations des eaux fluviales en elements majeurs et apport en solutions aux oceans. *Revue de Geologie Dynamique et Geographie Physique,* **21**, 215-246.

20. Meybeck, M. (1982) Carbon, nitrogen, and phosphorus transport by world rivers, *Amer. J. Sci.,* **282**, 401-450.

21. Meybeck, M. (1988) How to establish and use world budgets of river material, in A.Lerman and M.Meybeck (eds.), *Physical Weathering in Geochemical Cycles,* Kluwer Academic Publisher, Dordrecht -Boston - London, 247-272.

22. Skakalsky, B.G. and Meerovich, L.N. (1991) Modern characteristics of biogenic input from river watersheds of the USSR, European part, in Water quality and scientific principles of its protection. *Proc. V All-Union Congress,* Gidrometeoizdat, Leningrad, **5**, 174-184.(in Russian)

23. Rachold, V., Alabyan, A., Hubberten, H.-W., Korotaev, V.N. and Zaitsev, A.A. (1996) Sediment transport to the Laptev Sea - hydrology and geochemistry of the Lena River, *Polar Research,* **15**, 183-196.

THE DISPERSION OF SIBERIAN RIVER FLOWS INTO COASTAL WATERS: METEOROLOGICAL, HYDROLOGICAL AND HYDROCHEMICAL ASPECTS.

I.P. Semiletov*, N.I. Savelieva*, G.E. Weller**, I.I. Pipko*, S.P. Pugach*, A.Yu Gukov*, L.N. Vasilevskaya*

*Pacific Oceanological Institute, 43 Baltic Street, Vladivostok 690041.
**Cooperative Institute for Arctic Research/Alaska University Fairbanks, P.O. Box 757740, Fairbanks, Alaska 99 775-7740

1. Introduction

Global effects on the Arctic are reflected in regional climate changes and their impacts and consequences. Essentially all cryospheric features of the Arctic will be affected in one way or another. Energy and water fluxes clearly shape the regional temperature regime, which is a primary factor in determining the surface state (frozen vs. thawed), trace gas fluxes, rates of productivity, and the link to regional climate [1,2].

The principal sources of Arctic Ocean surface water are Atlantic water entering through Fram Strait and the Barents Sea, Pacific water entering through Bering Strait, and river runoff [3,4]. This inflow is equilibrated in term of the water balance partly by an outflow of surface water via Transarctic Drift (Current) through the Fram Strait and is strongly tied to the immense freshwater output of the Siberian tributaries, about —2300-2520 km^3 per year [5]. The most recent evaluation of total river discharge from the Eurasia into the Arctic Ocean is 2890 km^3 per year [6], or about —1.6 % of total water circulation budget of the Arctic Ocean [3]. Siberian runoff combined with contribution of the Mackenzie River (340 kin^3) shows that the total riverine inflow entering the Arctic Ocean is about 3230 km^3 per year. A similar value (3300 kin^3) was used early by Treshnikov [7], Aagaard and Carmack [8].

Large amounts of Arctic sea-ice are formed on the Siberian Shelf, underscoring the importance of these processes for the climate system. The Laptevs Sea as source area for the Transarctic Drift is of particular interest. According to Vize [91] and Zakharov [10,11,12], the Laptevs Sea is the most important source of sea ice into the Arctic Basin. Calculation shows that mean winter ice transport out of the Laptevs Sea is about 722 km^3 [10]. This value is the same as mean riverine discharge (1936-1950) entering the Laptevs Sea, 726 km^3 $year^{-1}$ [5]. The latter value is slightly less than the recent evaluation, 745 km^3 $year^{-1}$ 1930-1990 [13], that might be associated with increase in the riverine discharge. Nowadays, the remarkably shallow shelf of the Laptevs Sea is considered as a key region for gaining an understanding of change m the Arctic [11,12,14]. During winter the majority of arctic sea ice for the Transarctic Drift is produced here, thus explaining why the Laptevs Sea is also called "the ice factory of the Arctic Ocean". The ice produced here is carried with the Transarctic Drift across the Arctic Ocean and through the Fram

323

E.L. Lewis et al. (eds.), The Freshwater Budget of the Arctic Ocean, 323-366.
© 2000 Kluwer Academic Publishers.

Strait into the GIN seas for 2 to 3 years 13,12]. Because interannual variations in total riverine input to the Laptevs Sea is up to 37% [5], instability in the riverine input might play a significant role in change of feeding for the Transarctic Drift and thermohaline regime in the GIN seas. Some authors consider that the sub polar sea surface temperature anomalies over the North Atlantic result in the anomalous freshwater/ice transport from the Arctic Ocean via Fram Strait [15,16].

Various attempts have been made to understand better the role of the Arctic Region in the global budget of fresh water, dissolved salts, total carbonate, alkalinity, and silicate [3,17,18,8,19,20,13,4,21]. In general the freshwater balance of the surface mixed layer and halocline of the Arctic Ocean is a product of (1) Rivers that add water at the rims of the basins, (2) low salinity Pacific water, and (3) sea ice melt water that is a secondary product of primary sources (1,2) [4,22]. Clearly, the warm and salty Atlantic Layer is interacted with wintertime high-density shelf waters that is recorded in variability of temperature, salinity, and stratification of the Atlantic Layer and overlaid halocline waters. According to Gudkovich [23] and Nikiforov et al. [24], the variability in regime of the upper and intermediate waters might be related with switch from Az-mode to Zn-mode of atmospheric circulation patterns that might be associated with the Arctic Oscillation in term of Thompson and Wallace [25].

In this paper we examine long-range and interannual variability in the atmospheric circulation-land environment-fresh water input into the Arctic Ocean and its connection to the hydrological and hydrochemical regime on the Siberian Shelf, with a particular eye toward change in land hydrology and atmospheric processes over land and Arctic Basin. In our research, the Lena River is considered as a basic river because changes in the Lena River hydrology play a significant role in feeding for the Transarctic Drift through change in the freshwater inflow to the Laptevs Sea. The Lena River is also a natural boundary between Western and Eastern Siberia: in atmospheric circulation patterns-land hydrology-sea ice condition, tectonic structure, and biological peculiarities onto the Siberian shelf.

2. Area Study

The Siberian shelves are an "estuary" for the Arctic Ocean with 70% of the riverine input into the Arctic derived from the discharge of the Lena (—525 km^3 yeaf'), Ob (—429 km^3 year'), and Yenisey (—620 km^3 year^{-1}) [8,13]. The voluminous fresh water discharge reflects the large catchment area (10.8 x 106 km^2 [6]) for rivers draining into the Kara, Laptevs, and East Siberian Seas, Fig. 1. It is important that watersheds of the East Siberia are situated mainly in the permafrost area that occupies about one half of territory of the Russia. We can see that the Northern European and West Siberian rivers are situated mainly in non-permafrost or/and permafrost islands region (Fig. 1: boundary of the permafrost and permafrost islands regions are taken from Danilov [26]).

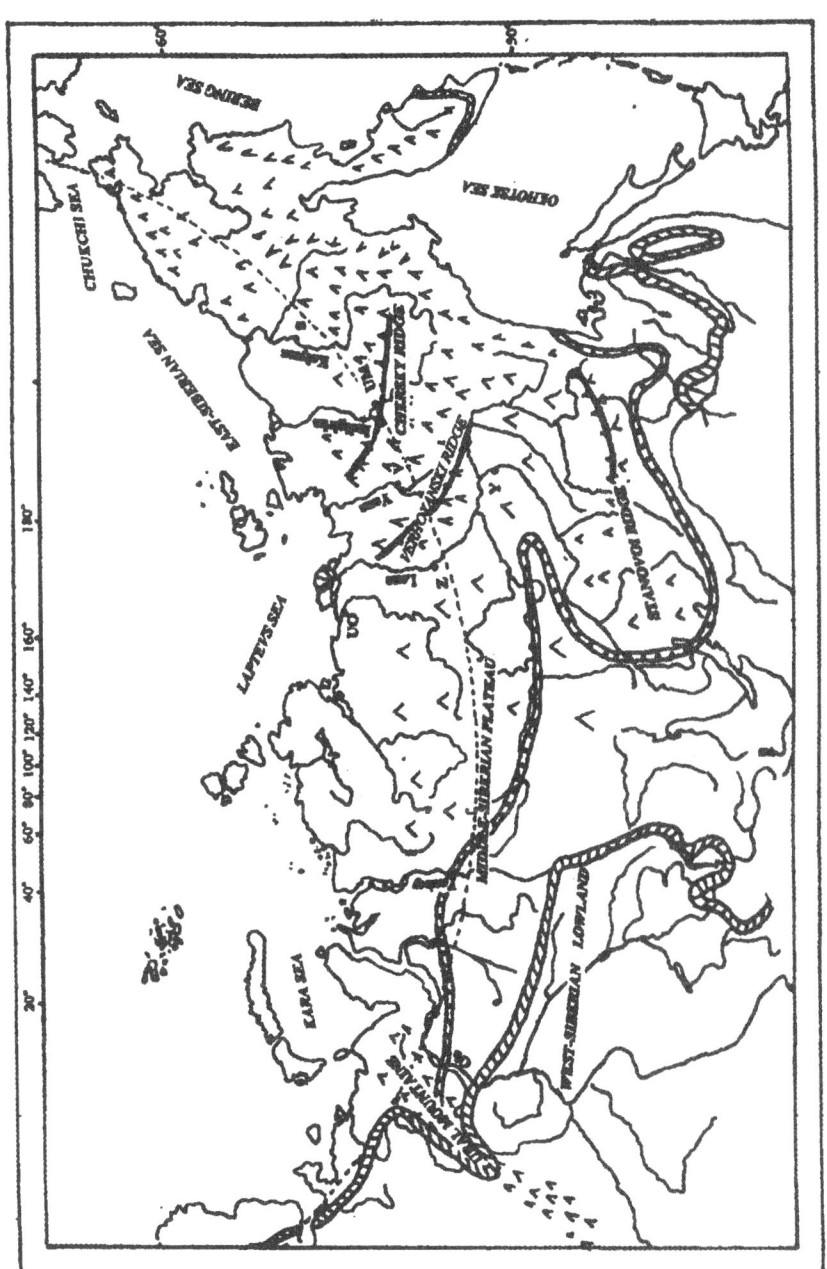

Figure 1. Location map of the study area, where

▨▨▨▨ boundary of discontinuous (island) permafrost

∙∙∙∙∙∙∙ boundary of continuous permafrost.

Permafrost contains a huge amount of ancient organic matter that might be involved in current biogeochemical cycling due to thermokarst, coastal erosion, and increase in summer-thaw layer of permafrost. The recent evaluation shows that the upper 100 m layer of tundra and northern taiga contains at least 9400 Gt of organic carbon that might became available for biotic activity via thermokarst and thaw lake evolution [27]. Two pathways exist for the conversion of buried organic carbon into forms of main greenhouse gases: carbon dioxide is originated mainly due to aerobic oxidation of organic matter; methane is originated only in anaerobic environment [28]. Thus the regime of precipitation and evaporation is very important for regional change in emission of CO_2 (increase in dry environment) and CH_4 (increase in wet environment) that is in positive feedback with the greenhouse warming. Because precipitation over Eurasia is determined mainly by vapor transport from the North Atlantic, the annual precipitation decreases from 750 -800 mm in the southwest to 600 mm in the northeast of the Western Eurasia. The minimum annual precipitation occurs over Siberia, because the vapor burden in the Atlantic air transport is decreased moving from the west to the east. Moreover, mountainous orography is working as a trap for atmospheric vapor transport (Fig. I). The mean annual precipitation (from the 1930s to early 1952s) over the Kara Sea riverine basin was 443 mm, Laptevs Sea 308 mm, East-Siberian Sea 274 mm, Chukchi Sea 424 mm [29]. From Table 1 based on recent available data (see "Materials and Data"), we can see that the minimum annual precipitation in the East Siberia occurs over the Lena — Yana watersheds. Precipitation is increased east of the Lena River, because humid air masses originating in the North Pacific Region influence on water budget of the Eastern Siberia especially of Chukotka Peninsula.

TABLE 1. Mean annual and seasonal of precipitation (P) in Siberian Rivers Basins and their standard deviations (a) (position of meteostations is shown in Fig. 1)

Meteostations, latitude	Annual P	Standard deviations, a	Precipitation (warm period), mm	Standard deviations, a	Precipitation (cold period) mm	Standard deviations, a
Ust-Oleiek (UO), 73°	297	59.1	218	56.8	79	15.8
Zhigansk (Z), 66°	302	62.0	223	57.4	79	21.8
Yakutak (Y), 62°	241	47.6	179	45.8	62	13.5
Verhoynsk (V), 68°	179	43.2	132	32.1	47	16.9
Ust-Monia (UM) 66°	201	44.8	152	42.8	49	18.0
Sredne-kolymsk (S), 68°	230	54.1	144	45.4	86	15.4
Aniguema (A), 68°	426	120.6	221	69.5	205	94.5

Early, Gaigerov [30] considered a land basin of the Laptevs Sea as a boundary between Western and Eastern patterns of atmospheric circulation over the mainland. So, we can consider the Lena River as a transition zone between Atlantic and Pacific atmospheric influence on the water budget over Eurasia. It was confirmed with recent research of Serreze and Maslanik [31].

Gudkovich and Kovalev [32] assumed that the difference in the ice-condition between Eurasian and Amerasian arctic seas was driven by difference between Arctic Ocean inflow of Atlantic waters and inflow of Pacific waters. In order to keep the water balance, increased export of fresh water and ice from the Arctic Ocean through the Fram Strait should be accompanied by decrease in the import of Atlantic waters and an increase in Pacific water inflow [3]. Usually, during years with anomalous high export of fresh water from the Arctic Ocean, the East Greenland Current is increased and enters the Greenland and Iceland Seas, where it plays an important role in the local convection. In this time the Arctic Canadian "opposite" parts in the North-Asia adjacent seas: a) the Kara Sea — the western parcel of the Laptevs Sea; b) the eastern parcel of the Laptevs Sea - Anticyclone is extended in the Amerasian sub-basin and the periphery of the Arctic Anticyclone induces a Trans-Arctic Current that is shifted to the west and ice transport from the Laptevs Sea is increased. Likewise, cyclonic motion is shifted southward and there is increased precipitation over the continental watersheds [33, 32]. According to Zakharov [11], anomalies of the ice export through the Fram Strait are correlated quite well with change in the riverine inflow. Based on all available ice-data sets, Gudkovich and Kovalev [32] identify two East-Siberian-Chukchi Sea. In fact, the Lena River-Laptevs Sea system is considered as a boundary between different circulation patterns, where ice-conditions are developing in opposite manner.

According to Imaev et al. [34] the Laptevs Sea basin is a boundary between the Eurasian and North American tectonic plates. Seismity, seismotectonics, active faults and stress features of the Earth's crust along this boundary are accompanied by anomalously high geothermal flux that should have an influence on formation of taliks (unfrozen ground within permafrost) in the riverine watersheds. Early, Antonov [35] mentioned that high tectonic activity influences significantly the redistribution of riverine feeding between surface (above-permafrost) and underground (beneath permafrost) waters. During wintertime it is more important because in this time underground water feeding through taliks below a riverbed predominates over atmospheric (surface) feeding. The most important water flow from the underground source should be in watersheds of the largest rivers underlying with open taliks. Indeed, this role becomes especially important during the winter lean flows, when the water discharge steeply diminishes (down to 7-10% of annual discharge) and rain and melted snow do not feed the river. Because of winter riverine feeding is determined by a deep (beneath permafrost) ground waters through talik zones, chemical analysis shows that in winter concentration of all chemical species (excepting of oxygen, and pH) are increased many times [36], as a result of chemical weathering being more important in "deep" ground waters in comparison with surface waters. For this reason, winter river input of dissolved material is not so small as seemed from the ratio winter discharge/summer discharge. According to Antonov [35], the Lena River might be considered as a tectonic boundary between the Western Siberia (basin of the Kara Sea: Ob, and Yenisey) and Eastern Siberia (basins situated east of the Lena River). It was suggested that tectonic downwelling in the Western Siberia influence the configuration of mouth of the Ob and Yemsey (avandelta is absent). Likewise, the Lena, Yana, Indigirka, and Kolyma are forming deltas, because of tectonic uplifting there.

A first assessment of the river water and basin characteristics of West Siberia relative to East Siberia has shown very significant chemical differences between these two regions [13]. So, it was found that the East Siberian Rivers (the Yana. Indigirka, Kolyma) are even more similar in chemical regime to the North American arctic rivers than to the rivers located westward of the Lena. Probably it is related to a tectonic factor, because the Yana-Indigirka-Kolyma are situated east of boundary between the Eurasian and North American plates, i.e. these rivers are positioned on the North American plate.

Golilcov and Averintsev [37] emphasized that the Novosibirsky Islands and the Lomonosov Ridge represents the transition area between the Atlantic and Pacific fauna that is reflected also in benthic community of the Laptevs Sea. Thus the Lena-Laptevs Sea system is a "biological boundary" too.

Keeping in mind that all the Eurasian basin belongs to the same climatic zone, but it is bordered by the Lena River in two regions with differences in atmospheric and land-shelf conditions, and tectonics, we consider here the Lena River as a "basic river" in comparison with behavior of other rivers situated in the East, and West Siberian Regions. Some main hydrological features of the Siberian Rivers (mean/maximal/ minimal river discharge, and watershed area) are presented in Table 2.

TABLE 2. Main hydrological features of the Siberian Rivers

Name (gauging station)	Data set period years	Watershed area Km^2	W_m km^3	σ km^3	C_v	W_{max} km^3	Year	W_{min} km^3	Year
Ob (Salehard)	1948-1993	2432000	403	63.26	0.16	577	1979	269	1967
Yenisey (Igarka)	1948-1993	2440000	586	46.07	0.08	698	1974	513	1976
Khatanga(Khatanga)	1961-1995	275000	70	9.86	0.14	94	1992	51	1979
Anabar (Saskyllakh)	1954-1995	78800	14	3.26	0.24	22	1973	5	1979
Olenek(7.5 km upper river mouth Buur)	1952-1995	198000	33	9.48	0.29	58	1991	17	1956
Lena (Kyusyur)	1937-1995	2430000	525	63.09	0.12	729	1989	401	1986
Aniguema (Aniguema)	1943-1986	26700	8	1.64	0.2	13	1962	5	1976
Yana (Yubileynaya)	1971.1995	224000	32	8.80	0.27	56	1985	20	1980
Indigirka (Vorontsovo)	1937-1995	305000	50	10.08	0.22	80	1984	31	1937
Kolyma (Srednekolymak)	1943-1995	526000	74	19.733	0.27	137	1990	41	1973

Wm-mean river discharge, Cv - variation coefficient, σ - standard deviation.

3.   Data and Approach

3.1.   DATA

To understand the mechanism for a long-term hydrological and hydrochemical variability of the shelf water (and Arctic Basin) influenced by the riverine flows we need in a long-term data set for atmospheric regime, river discharge, hydrological and hydrochemical regime in the river-sea system. It is very important to evaluate also the Arctic "domestic" efflux of main greenhouse gases into the atmosphere, because the effect of greenhouse warming may appear not only in the model-predicted temperature increases, but also it may enhance of the natural Arctic circulation that accompanies the warming [38,12,39].

Now we have a long-term data set to investigate variability in atmospheric circulation and riverine discharge. Significantly less data are available for hydrological regime at the Siberian Shelf and Arctic Basixi A little is known about interannual variability of hydrochemical regime of Si (silicates), oxygen and nutrients. There are little or no reliable data for carbonate and dissolved methane [21,27], and no reliable data for variability of the main seawater composition in the Siberian seas influenced strongly by riverine input. Thus, to understand an effect and causes in variability of the hydrological and hydrochemical regime on the shelf we have to base our investigation on a reliable data "Pyramid".

Our primary data source (in table-format) on the regime and resources of surface waters exists as a long-range record of the former USSR Goskomgidromet and later of Roskomgidromet system. Data were published in the State Water Cadastre yearbook, Resources of the Surface Waters of the USSR. and Basic hydrological features (1967, 1978), where the data sets for time spanning for 1935-1961 and 1976 were presented. The data sets for period 1976 and 1987 were taken from "Annual data on the regime and resources of surface terrestrial waters" [40-45,46-53]. Recent data (1986-1995) for the rivers of the Laptevs Sea basin were taken from Regional Tiksi Hydromet Service. Some data of the Ob and Yenisey discharge were obtained in CD-ROM format from Arctic and Antarctic Research Institute (Joint US-Russia Atlas, 1998) [54].

Indexes of atmospheric circulation (see Section 3.2.) in the Northern Hemisphere were obtained from Meteorological Department of Far-Eastern State University, Vladivostok (1935-1945), Far-Eastern Hydrometeorological Research Institute (FERRI), Vladivostok (1945-1986), and Arctic and Antarctic Research Institute (AARI), Sankt-Petersburg (1986-1996). The meteodata (precipitation, air temperature at ground level) were kindly presented by FEI{RI. Atmospheric circulation processes over the Siberia and Arctic Ocean (1954-1989) were investigated with use of Annual Synoptical Bulletin of the former USSR Goskomgitromet and later of Roskomgidromet.

3.2. APPROACH

In 1894 A.I. Voeykov, who was "a pioneer of Russian climotological research", claimed that inland change in climate is driven by transport of air masses that is determined by insolation and interaction between land and ocean [55]. The rivers were considered as a product of climate in their basins. Antonov [56] found that the water balance over basins of the North Asian (Siberian) Rivers (Ob, Yenisey, Lena, Kolyma), and their discharge, is caused by macrosynoptical processes that are determined by interaction in the land-ocean system. To understand better influence of change in atmospheric circulation on water feeding in the Siberia, Antonov [56] used classification of main types of the atmospheric macroprocesses by Vangengeim. Vangengeim [57] classified the atmospheric circulation in the Atlantic-Eurasian sector of the Northern Hemisphere into two meridional patterns and one zonal type (designated E, C, W, respectively). According to this classification at the earth's surface the following conditions will be prevail: below eastern upper-level ridges (western upper-level troughs) there are occur areas of negative temperature anomalies, positive pressure anomalies, and precipitation deficits; associated below western upper-level ridges (eastern upper-level troughs), conversely, are positive temperature anomalies, negative pressure anomalies, and areas of excessive precipitation.

Antonov [56] analysed the main types of the atmospheric circulation that are responsible for vapor transport to the basins of Ob and Yenisey (193 0-1954), Lena, and Kolyma (193 4-1954). It was found that the vapor feeding differed for each watershed.

So, the vapor transport to:

- the Ob watershed was provided mainly by W-circulation in November; and C-circulation in May, and June;
- the Yenisey watershed: by W-circulation in October; E-circulation in April-May; C-circulation in May;
- the Lena watershed: by W-circulation in October and January;
- the Kolyma watershed: by W-circulation in December, and May; C-circulation in June.

Later, X.A. Girs analysed in detail the macrosynoptical processes over Amerasian Sector of the Arctic, and identified three main types (Z, M1, and M2) of atmospheric circulation influenced strongly by the Pacific Ocean. M1, M2, and Z patterns are likewise distinguished in characteristics both of the atmosphere and hydrosphere: an upper-level trough develops at the meridians of Aleutian Islands (the trailing edge of the Aleutian low) when M2 processes are operating; during M1 processes the picture is opposite: an upper-level ridge develops at the meridians of Aleutian Islands, extending the Hawaiian high in this location. During zonal processes (Z) there is usually an eastward shift of Low across the. Aleutian Islands with a subsequent plunge into western America. The study of many-yearly variations in the occurrence of macroprocesses E, C, W, M1, M2, and Z from 1900 to 1971 has shown that the annual frequency of these processes changes distinctly with respect to time. On this basis an epochal classification was made. These epochs caused long-term trends of the sign in the change of a number of characteristics of atmosphere and hydrosphere. Main principles of the Vangengeim classification for the Atlantic-Eurasian sector were combined with classification of Girs for the Amerasian sector (Fig. 2 (a,b,c) [58], and used successfully in AARI for long-range forecasting in the Arctic Region [59].

To link the fate of riverine waters from Siberian watersheds to the shelf seas and pan-arctic transport, we base our approach on analysis of the influence of macrosynoptical processes (by classification of Girs-Vangengeim) for water feeding of the main Siberian Rivers: Ob, Yenisey, Khatanga, Anabar, Olenek, Lena, Yana, Indigirka, Kolyma, and Amguema (listed from west to east), Fig. 1. We consider here the main typical atmospheric macroprocesses that are responsible for water balance m the Siberian watersheds. Orographic barriers (Fig. 1) work as a trap for atmospheric vapor. Water vapor transport to riverine basin of the Laptevs Sea is limited, because mountain ridges on the west, south, and east surround it, although watersheds of the Khatanga, Anabar and Olenek absorb some precipitation originating over the Atlantic Ocean. The Lena River is supplied by precipitation of both Atlantic and Pacific region. Likewise, the Yana, Indigirka, Kolyma, and Amguema receive the precipitation that originated mainly over the Pacific Ocean. Of course, some cyclones come from the North, but the water capacity of this type of cyclone is a small. Therefore, the role of the arctic cyclones in inland water budget is considered as negligible in all seasons [60,61].

Figure 2. Location Scheme of main upper-level (500 mb) ridges and troughs during:

a) W pattern of atmospheric macroprocesses (W_Z, W_{M1}, W_{M2})

b) C pattern of atmospheric macroprocesses (C_Z, C_{M1}, C_{M2})

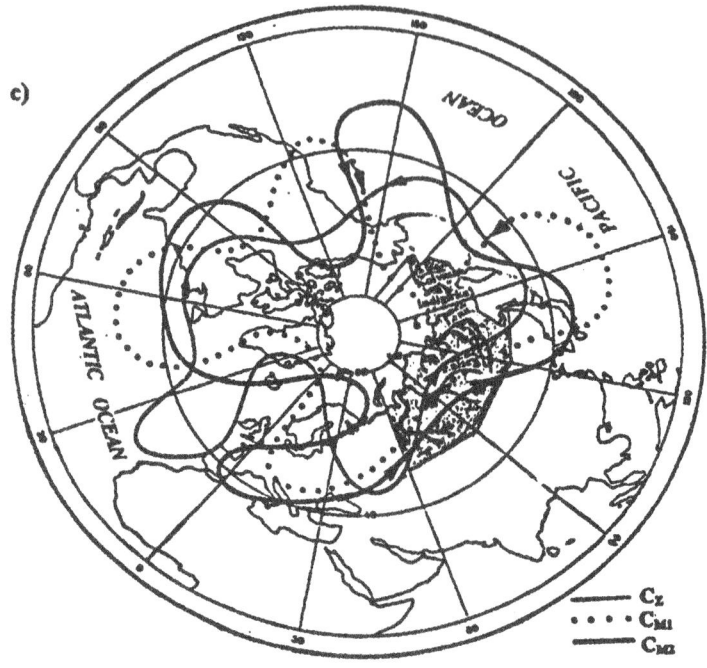

Figure 2. (Continued)
c) E pattern of atmospheric macroprocesses (EZ, EM1, EM2)
(adopted from Girs, A.A. and K. V. Kondratovich, 1978)

Analysis of Fig. 2 (a,b,c) shows schematically the main "routes" of the vapor transport supplied the study area (arrows show routes of cyclones with high water burden that come inland):

- Atlantic cyclones bring water vapor via routes W_z and W_{M2} to the watersheds of the Ob, Yenisey and Lena (increasing of the discharge). At the same time watersheds of Indigirka, Kolyma, Yana, and Amguema are limited by precipitation, because of W_{z-M2} -cyclones are shifted southward. The W_{M1} route provides the vapor transport to the Ob, Yenisey, and Lena (from the Atlantic Ocean) and to Chukotka Peninsula (from the Pacific Ocean);

- Routes "C_z-C_{M1}-C_{M2}" transport provide synchronous change in discharge of the Ob-Yenisey-Lena, "C_{M1}"-transport brings a maximal precipitation in watersheds of all nvers ("Atlantic" feeding for the Ob-Yemsey-Lena, and "Pacific" feeding for the north-eastern Siberia); "C_{M1}-C_{M2}"-transport provides asynchronous water feeding (and riverine discharge) between the Ob-Yemsey-Lena and Lena-Indigirka-Kolyma;

- A switch from "E_z-E_{M2}"-transport on "E_{M1}"-transport may cause shift in regime of maximal discharge from the Ob-Yemsey to the Yenisey only while the Indigirka Kolyma Lowland might be supplied by precipitation exclusively by "E_{M1}"-transport. The watershed of the Lena River might be supplied by precipitation only by "E_{M2}" -transport.

Note that in nature, all types of W, C, E — routes (with Z, M_1, M_2 - modifications) of cyclones might be some shifted northward, and southward, or eastward, and westward, because in Fig. 2 (a,b,c) we consider a scheme of a long-range mean routes of all nine types of atmospheric circulation.

We separate our paper into the next steps of analysis and discussions: 1) the meteorological regimes for watersheds of the Siberian Rivers are analysed; 2) the change in interannual discharge of main Siberian Rivers is considered as an effect of variability in types of the macrosynoptical processes; 3) the influence of change in the riverine input into the Siberian shelf-basin on the ice-condition regime is discussed; 4) an attempt to synthesize change in the atmospheric circulation-land hydrology and shelf environment is done. The Lena River is considered as a "basic" river for our system analysis.

4. Discussion and Results

4.1. METEOROLOGICAL REGIME AND LAND HYDROLOGY IN THE SIBERIAN RIVER'S BASINS

4.1.1. Surface air temperature and riverine discharge: interannual and long-range variability.
A pervasive feature of global climate simulations with enhanced greenhouse forcing is an amplified warming in the polar regions, especially near the surface and in the winter half of the year [38].

To understand recent variations of air temperature over the Eastern Siberia we consider here best-fit linear trends over the period of direct meteorological observations (1947-1995) for all seasons. As shown in Fig. 3 (a,b,c,d) (adopted from [62]):

- **Warming** dominates over continental area (Yakutia) in the <u>winter and spring</u> seasons, as was projected by global climate models; winter **cooling** is indicated over Chukotka Peninsula;
- The area average trend for the <u>summer</u> is nearly zero, except for Chukotka Peninsula, where a positive trend is obtained,
- Little or no warming is indicated over Eastern Siberia and the Far East in <u>autumn.</u>

The recent warming is strongest west of 140 E. The longitudinal dependence of decrease in the warming east of 140-145 E might be caused by thermal inertia of the Pacific Ocean. In general, regional evaluation for the period (1947-1995) agrees with the arctic temperature trends for the period (1961-1990) that are presented by Chapman and Walsh [38]. Based on Fig. 3 (a,b,c,d) and evaluations of Chapman and Walsh [38] we can conclude that watersheds of the West and East Siberia are influenced mainly by the recent warming, though effects decrease as one goes towards the Pacific Rim.

The total regional temperature trends agree with analysis of long-rang mean annual temperature regime at seven stations situated in valleys of the East Sibenan Rivers (Fig. 1) that shows mainly a positive trend (5-years smoothed record); though a negligible trend is observed at Amguema (Fig. 4).

334

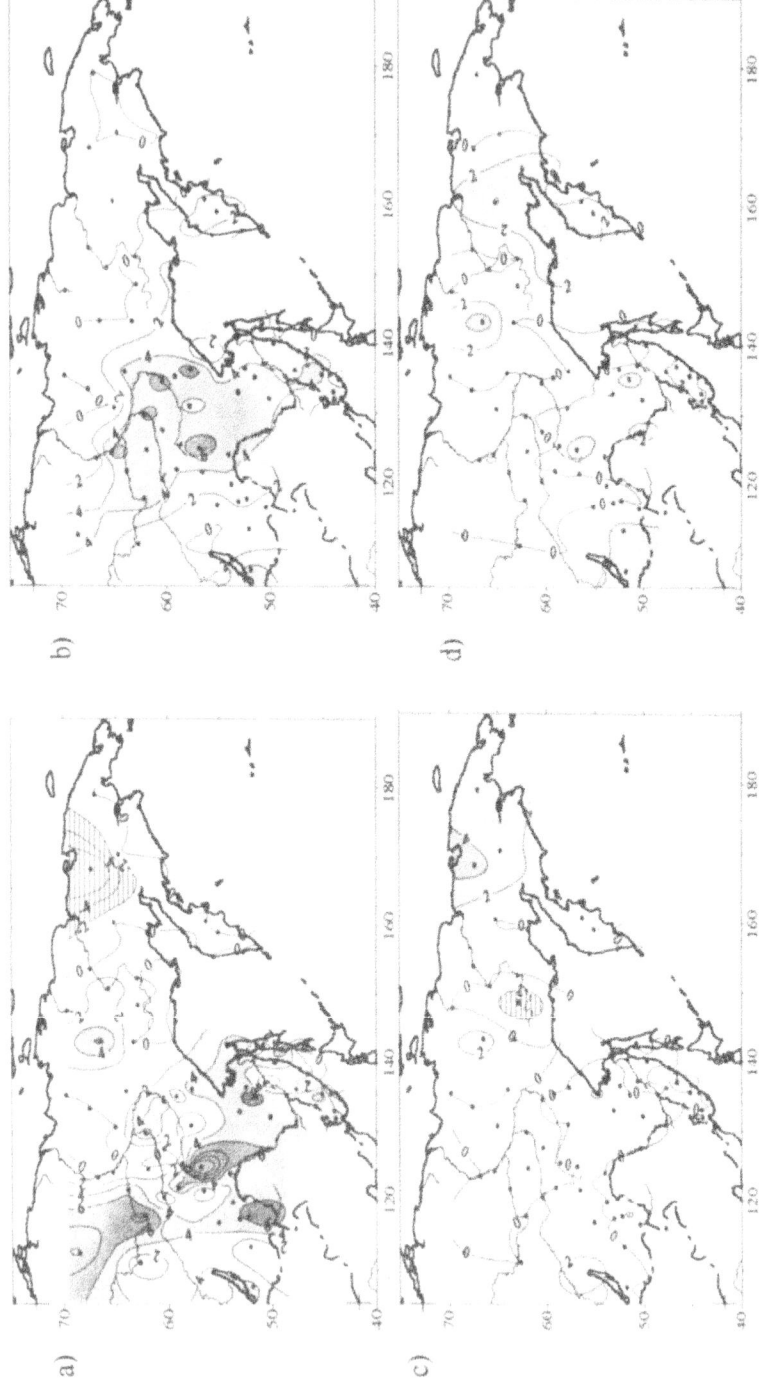

Figure 3. Linear trend estimates for mean seasonal temperatures for 1947-1995 years, 0.01 °C per year (a – winter, b- spring, c- summer, d- autumn). Positive values more than 0.02 °C per year are shaded, area with negative trend is marked by horizontal lines.

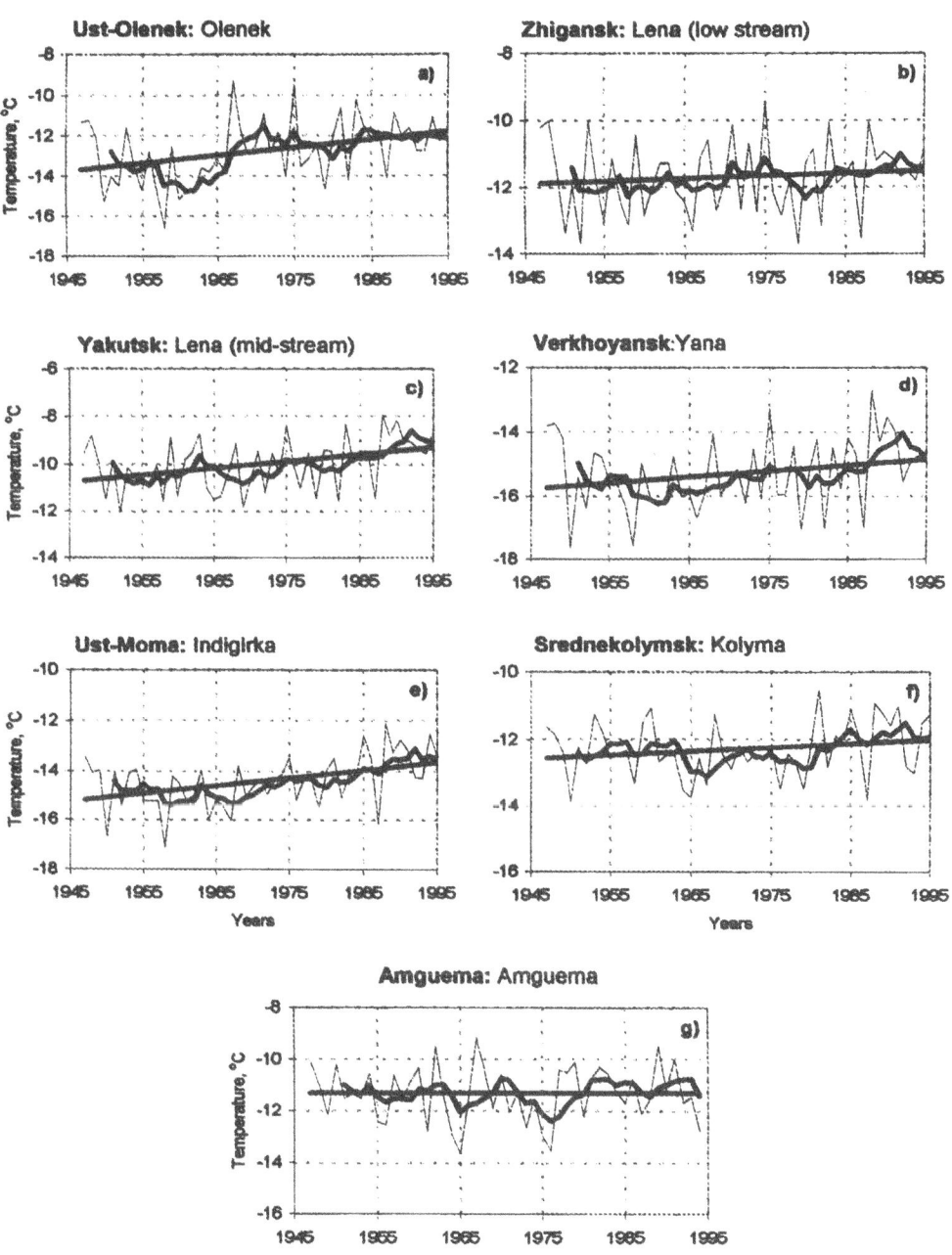

Figure 4. Long-range mean annual temperature (linear and 5-year smooth - in bold line) observed at stations situated in valleys of the East-Siberian Rivers.

        Long-term records (160-286 yr) of freeze up and break up dates for the Eurasian
Rivers (Volga, Dnepr, Neva, Northern (Severnaya) Dvina, Belaya, Yenisey, Ob and Lena)
indicate an early timing of break-up and later timing of freeze up [63]. The timing of
freeze up and break up results from an integration of meteorological conditions (primarily
air temperature) that occur before these events. Therefore change in ice records of great
Eurasian rivers might be considered as an indication of the anticipated greenhouse
warming, not only because of the large response of riverine ice cover to small changes in
air temperature, but also because these records integrate climatic conditions during the
winter-spring seasons when most warming is forecasted to occur. Statistical analysis used
in Ginsburg and Soldatova [63] shows that in 80% of the cases the warning trend's sign
did not change in the 20th century compared to the 19th. This tendency toward warming
is more obvious in the spring than in the fall as projected by global climate models [64].
Therefore we can consider watersheds of the Eurasian rivers as a land influenced mainly
by the recent global warming. Likewise, we should take in our mind, that the North-
Eastern parcel of the study area (Chukotka Peninsula) as a whole is cooling, and the
Kolyma-Indigirka Lowland might be considered as a transitional zone between two
modes "little or no warming" and "a slight or no cooling".

        Recent variations of the long-range mean annual anomaly in the discharge of ten
largest Siberian Rivers is presented in Fig. 5. The linear trends are positive for the
Yemsey, Anabar, Olenek, Lena, Yana, Indigirka, and Kolyma. A negative trend is
obtained only for the Amguema that is situated in Chukotka Peninsula. Little or no trends
were observed in the long-range mean annual anomaly of the Ob and Khatanga discharge.
Both rivers differ significantly in size and geographical position (Fig. 1). The Khatanga is
a small river (basin area is about $2.8x\ 10'$ kin2) in comparison with Ob ($2.4x\ 106\ \sim 2$)
(Table 2), but both rivers flow between a numerous lakes and ponds that are connected
with the riverine stream. Probably, these rivers are regulated (more strongly than other
rivers) by surrounding lakes and ponds that increase "basin's inertia" for global warming.
Note that the amplitude (% of total mean) of riverine discharge (1936-1950) was the
largest in the smaller rivers such as Khatanga (70%), Anabar (72%), Olenek (71%), Yana
(86%), Indigirka (58%), and Kolyma (68%) in comparison with great Siberian Rivers:
Yenisey (19%), Lena (3 7%), and Ob (45%) [51]. Thus an increasing anomaly of the
riverine discharge is observed mainly over an area influenced by the recent warming,
whereas decrease in the anomaly is observed over cooling area. Of course, this conclusion
is very crude, and at present we cannot identify a mechanism for direct influence of small
changes in air winter-spring temperature on response of riverine discharge.

4.1.2. *Precipitation and riverine discharge*

Water balance of the Siberian Rivers situated in the permafrost region differs from the
rivers situated in other regions because the upper surface of permafrost (permafrost table)
beneath the seasonally-thawed layer is working as an water non-permeable sheet that
influences on land hydrology (and hydrochemistry). Indeed, the depth of permafrost table
varies over the range of 10^1-10^2 cm north of the Arctic Circle (and mountainous areas) to
10^3-10^4 cm in the southern area of discontinuous permafrost (permafrost islands).

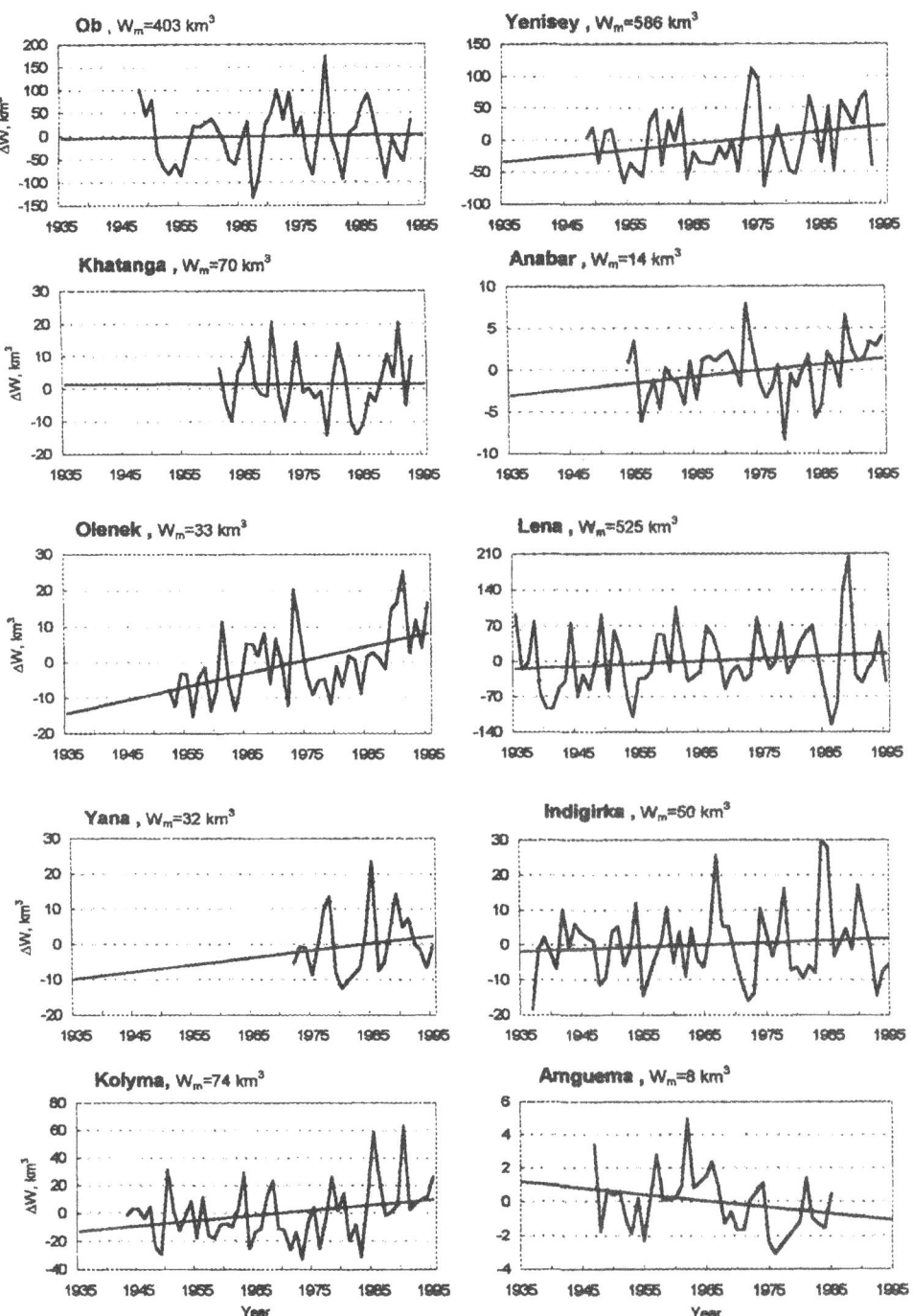

Figure 5. Interannual variability of riverine discharge anomaly (ΔW, km³, $\Delta W = W_l - W_m$, where W_l-discharge value, W_m-multi-annual mean value) and tendency.

Therefore riverine regime of ground feeding is significantly different for rivers situated in different permafrost environments or in non-permafrost environment. Fig. 1 shows that the Khatanga, Anabar, Olenek, Lena, Yana, Indigirka, Kolyma and Amguema are in the permafrost region, whereas the upper and mid-stream of the Yenisey is flowing in region of permafrost islands. The upper and mid-stream of the Ob is flowing in non-permafrost region, and only a small part of the low-stream is situated in the continuous permafrost region. Thus, we can consider "the permafrost rivers" as the rivers with faster "discharge response" (advection dominates here over vertical migration of ground water) initiated by variations in precipitation. Then, discharge (surface runoff)/precipitation (W/P) ratio should be higher in the watersheds underlain by shallow permafrost table (East Siberia) compared with the watersheds with a deeper permafrost table (West Siberia). Indeed, the W/P ratio increases two times from west (European part of Russia) to the east (basin of the Chukchi Sea) from 0.30 to 0.74 [56], whereas the evaporation/precipitation ratio (E/P) decreases from 0.70 to 0.26, respectively. Probably a high value of E/P ratio in the watersheds of the Kara Sea plays an important role in a vapor transport to the eastern watersheds, because about one half (or about 220 mm) of precipitation (mainly Atlantic by origin) is evaporated here and might be transported eastward again. Thus, we can consider this mechanism of water vapor redistribution over the land basin of the Kara Sea as an important factor for water feeding in the basin of the Laptevs Sea, where mean precipitation is low (about 250-300 mm). This crude evaluation indicates that total evaporation from the Kara Sea basin is a value of the same order with total precipitation over the Laptevs Sea basin. Of course, this factor is the most important for the East Siberia in timing when W-types of atmospheric circulation (Fig. 2 (a)) predominate over Siberia. Therefore, we may consider the ratio W/P (and E/P) as an indicator for effect of water removal from the different watersheds to the sea.

Annual records of long-term mean monthly precipitation obtained at seven meteorological stations situated in the East Siberia show that precipitation of warm season (May-October) dominates and contribute about 2/3 of the annual total precipitation (Fig. 6). As shown in Fig. 6, the month of the precipitation maximum is July at all stations (excepting Amguema). The timing of this maximum agrees with the NCEPINCAR reanalysis for the region north of 700N [311. The minimum values were observed from January through April. Note, that in the Lena River watershed only about 9% of total precipitation are of local origin [65], i.e. more then 90% of total precipitation is brought from outside. The twin peak of the maximum precipitation (August and November) was observed for the most north - eastern site of the study area: Amguema basin (Chukchi Sea). In all seasons, this area is influenced strongly by Pacific cyclones entering the Arctic across the Chukotka Peninsula and the Bering Strait [30]. The minimal precipitation was observed in May. Wintertime and summertime precipitation is similar only in the most eastern parcel of the study area (Table 1).

Long-range (time spanning is presented in Table 2) correlation analysis between different stations shows similar mean annual regime of precipitation in the East Siberia (excepting Amguema). The highest coefficient of correlation (r) in winter precipitation was found at the stations situated in the Yana-Indigiika-Kolyma area (r=0.5-0.7), because in cool season this area is influenced by the hollow of Siberian High or the Lena-Kolyma center of high-pressure ("Lensko-Kolymskoe Yadro", in Russian).

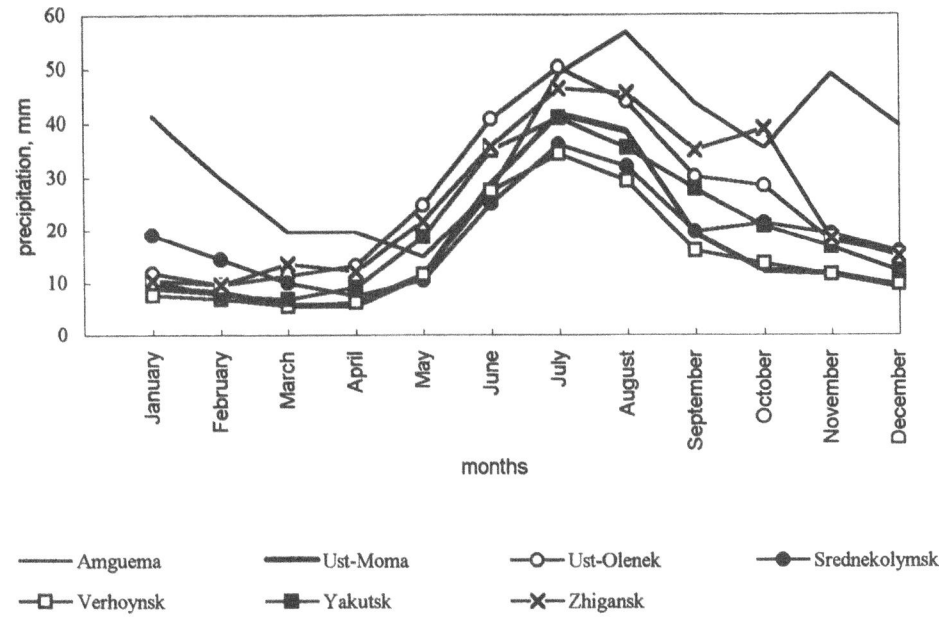

Figure 6. Long-term (1945-1995) mean monthly precipitation in the East Siberian

This statement agrees quite well with analysis of arctic synoptic activity (1952-1989) that was done by Serreze et al. [66], where the mean pressure field shows a closed mean high pressure center over eastern Siberia (1028 mb), and the peak of anticyclone activity occurs near Verkhoyansk station (Yana River valley). In summer (May-October), the mean spatial distribution of precipitation is more inhomogeneous than in winter. A spatially homogeneous distribution was observed only in the Kolyma-Amguema and Olenek-Lena areas. A correlation for summer precipitation between Yana and Indigirka is very low (r=0.2). Different regimes of precipitation (r=0. 1) were found between medium-stream (Yakutsk) and low-stream of the Lena River (Zhigansk) indicating different types for transport of atmospheric water vapour over great watershed of the Lena River, because the Lena River is situated in the different climatological latitude belts. The same long-range difference in precipitation regime was found also for a different part of the Ob River (V. Kuzin, personal communication).

In the West Siberia (the Ob-Yenisey) mean annual precipitation is about 400 mm [56,67]. The water budget in the West Siberia is determined by the Atlantic sector that represents the major water vapour transport in Eurasia and Arctic Basin during all seasons [56,5,58,68,61,69]. About one half of the precipitation is observed here in summer, and one half in winter. Otherwise, in the East Siberia summer precipitation dominates (.~2/3 of total precipitation, Fig. 6) over winter precipitation, with the exception of the Amguema river (Chukotka Peninsula) where the bimodal distribution of precipitation that is obtained is influenced by the Pacific Ocean. The difference in water regime is determined by the change in seasonal cyclonic activity over the West and East

Siberia. According to Gaigerov [30], in summer (July) Atlantic cyclones are entering the West Siberia mainly from the south-west to the north-east. The highest cyclone frequency over Yenisey's watershed is observed in winter (January). A lower cyclone frequency both of "Atlantic" and "Pacific" origin is observed over other riverine basin of the Laptevs Sea. In Yakutia (Lena River), the summer cyclone frequency is higher (15-20%) than in winter season (5-15%).

A long-range mean annual precipitation obtained at gauge stations located in the East-Siberia Rivers indicates a similar variability of precipitation in a smaller river valleys as the Yana-Indigirka-Kolyma, whereas there is no agreement between watersheds of the Great rivers as the Lena with a smaller rivers, Fig. 7.

A comparison of annual precipitation records for the Lena River at Zhigansk (low stream) and Yakutsk (mid-stream) shows a large difference, Fig. 7(b,c). Likewise, the variability of precipitation in low-stream of the Lena has more similarities with the Olenek River that is situated north of the Northern Circle. The record for Amguema River is different from the all others, because the Amguema basin is mainly under the influence of the Pacific atmospheric pattern. A comparison of the annual precipitation records (Fig. 7) with the interannual variability of riverine discharge anomaly (Fig. 5) shows that usually the minimum and maximum riverine discharge is in phase with the annual precipitation. In some dry years, the value of summer precipitation might be similar to the winter precipitation. Such a meteorological situation was observed at the Zhigansk station (Lena) in summer of 1956, 1963, 1972, 1977, and 1984. A comparison of the precipitation and discharge shows high correlation in one time spanning and it is low in others. For instance, during the 1950s - 1970s the correlation between 5-year smoothed records of the summer anomaly in precipitation is correlated well with the annual anomalies of the Lena River discharge ($r=0.78$). Probably such correlation is determined by increasing of anomaly in "meridional" (E+C)-type of atmospheric circulation in this time, Fig.8 (a).

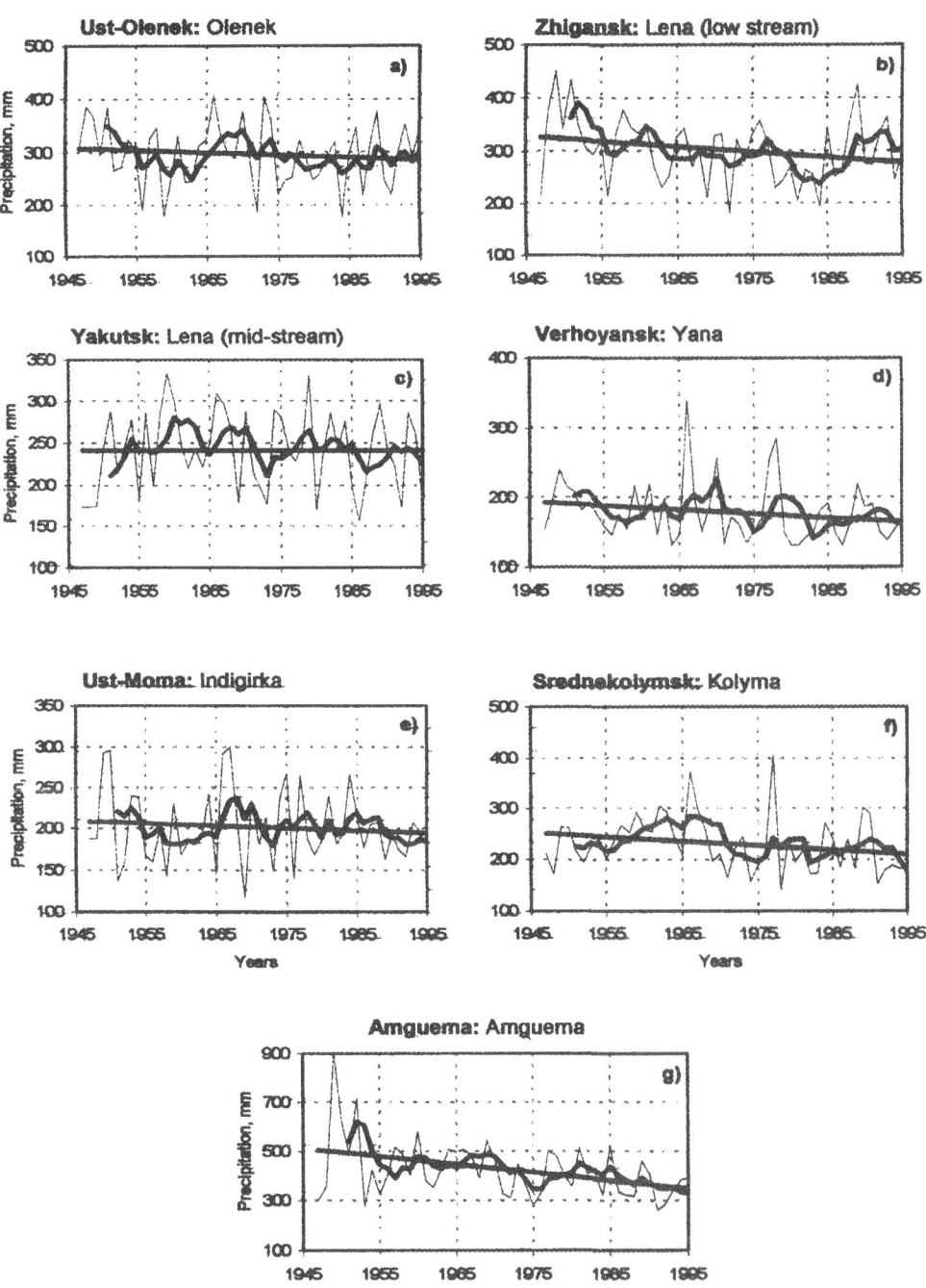

Figure 7. Long-range mean annual precipitation and tendency (linear and 5-year smooth - in bold line) observed at stations situated in valleys of the East-Siberian Rivers.

342

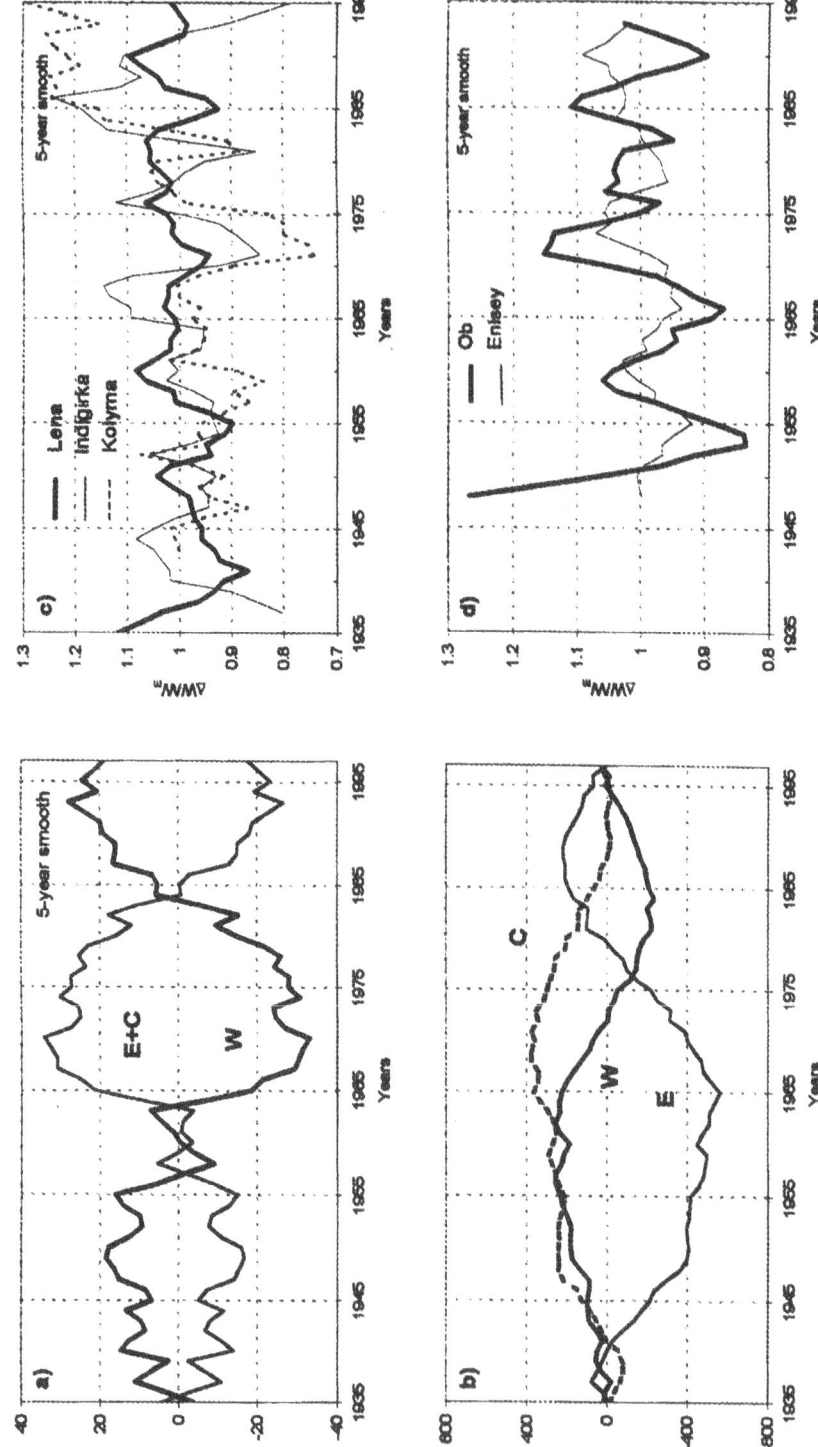

Figure 8. Atmospheric index's anomalies ΔDays (a), their cumulative curves (b) and relative variations of the Siberian River's discharge, ΔW/W$_m$ (c, d).

However, **signs of the long-range tendencies of the riverine discharge (Fig. 9) and their anomalies (Fig. 5), and total annual precipitation (Fig. 7) are differed.** Note, that increase in the riverine discharge coincides with increase in the total sea ice volume and ice growth rates in the Arctic [70,71].

4.1.3. *Riverine discharge and macrosynoptical processes*

It has been known for many tens of years that atmospheric change predetermines change in land hydrology [55], i.e. features of atmospheric circulation are "a cause", and the riverine discharge is "an effect" in this feedback loop.

A spectral analysis of long-range records of annual mean discharge shows maximums with periods of 2-4 yr and 6-8 yr in the all Siberian Rivers, Table 4.

TABLE 4. Spectral maximums of the riverine discharge of the main Siberian Rivers and Vangengeim's indexes of atmospheric circulation

Name	Period		
W	-	4.2	-
E	8	4.2	2.4
C	6.4	-	2.3
Ob	8.7	-	-
Yenisey	10	4.2	2
Hatanga	-	4.2	-
Anabar	6.9	3.2	2.3
Oleniok	7.4	4.0	2.4
Lena	7.1	-	-
Jana	6.9	3.7	-
Indigirka	8.3	3.5	-
Kolyma	5.6	-	2.4
Amguema	8	3.9	2.4

The same periods are found in Vangengeim's indexes of atmospheric circulation showing a similar variability in the atmosphere-land hydrology. Spectral analysis of long-range oceanographic data sets for the Western Bering Sea demonstrates the same periodicity (~2, 3-4, 6-7 yrs) in variability of the sea level, timing of cold conditions (negative temperature of air and sea, existence of sea ice) [72]. Some authors suggest that the two-year period is connected with the two-year variability in the spatial shifts of Atmospheric Action Centers [73,74], and the 6-8 year period may be induced by the Pole Tide due to fluctuation of Earth's axis [75]. In the North Pacific a prominent feature of the low-frequency sea surface temperature field with an oscillatory pattern with a return period of 4-7 years was associated with atmospheric long waves [76]. Thus the periodicity between variability of atmospheric circulation and land hydrology is connected to other natural processes in the Northern Hemisphere.

To understand better the mechanism of atmospheric influence on land hydrology in Siberia, we analyzed long-range variability of the Siberian River's discharge (with 5-year smoothing), Fig. 9. The variability in the Khatanga, Anabar, and Olenek rivers shows synchronous records (r=0.62-0.84). Likewise, a comparison of the records of Lena, Indigirka, and Kolyma shows both synchronous variability and asynchronous variability.

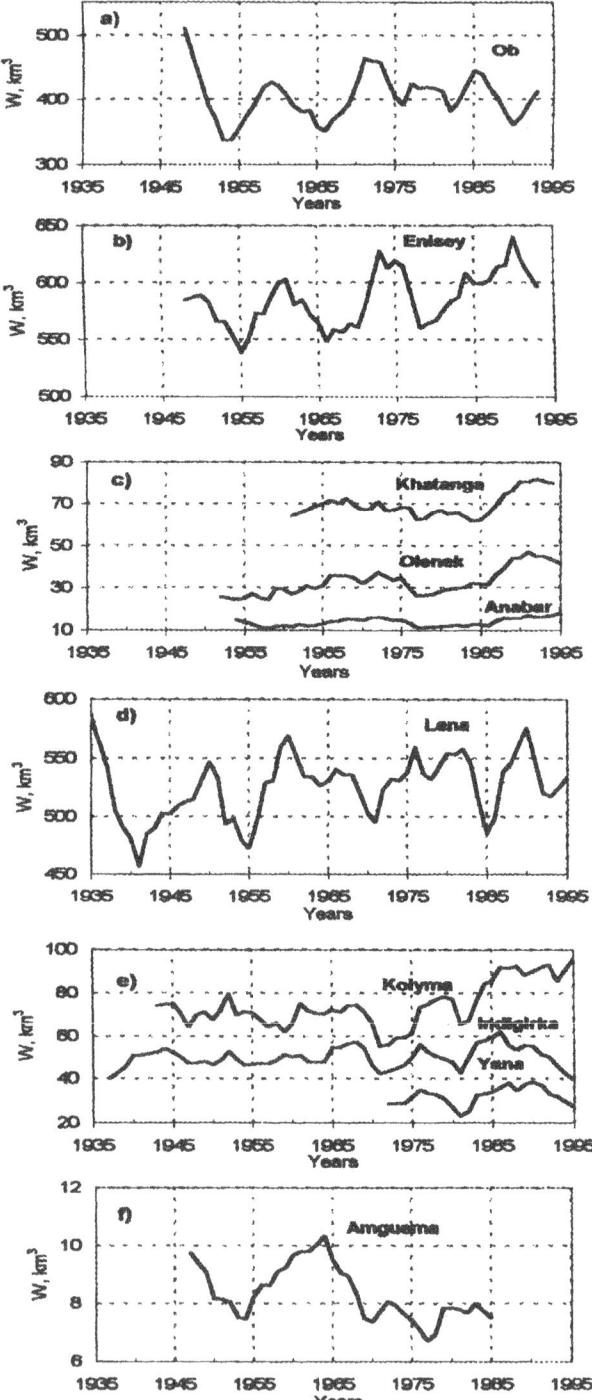

Figure 9. Interannual variability of the Siberian river's discharge (5-year smooth).

For instance, since the mid-1930s until the mid-1950s the variability is opposite in the records of the Lena and Indigirka (r=-O.6 1), and during the next years until mid 1970s, there is synchronous variability (r=0.57). The mid-1970s appear to be the time for the next change in sign of correlation between both rivers. Until the early 1990s the anticorellation was obtained (r-0.58), Fig. 8(c). The change in the synchronous and asynchronous records between the Lena, and Kolyma is similar with a previous river's pair (Lena-Indigirka), though the time when the variability is opposite was started in mid-1930s and continued until early 1960s. If we compare the graphs of the variability of atmospheric index's anomalies (Fig. 8(a)) and the records for the anomaly of the Siberian Rivers discharge (Fig. 8(b)), we see that the opposite regime in the water discharge between the Lena and the Indigirka-Kolyma is observed when anomalies in"zonal" (W-type) of atmospheric circulation anomalies are positive (cumulative W-curves drives up), whereas the synchronous changes coincide with the positive anomalies of (E±C)-tvpe (maximum in C cumulative curve and growth in E cumulative curve) (Fig. 8(c)). We explain this phenomenon by orography in the study area, Fig. 1. Indeed, Verkhoyansky Ridge and Chersky Ridge are working as a barrier to prevent vapor transport from the North Atlantic toward the east by all types of "zonal" W-circulation (Fig. 2(a)), whereas during "meridional" circulation development, when (E+C)-types anomalies dominate, the "mountain barrier" is working not so effectively. Note that the graphs of water discharge anomalies in the Indigirka, and Kolyma show positive correlation only during "meridional"-mode of circulation: in 1948-1971 r=0.53 when (E+C) anomalies are maximal, and in 1972-1992 r0.74 during (E) epoch (growth in cumulative E-curve).

A comparison of the discharge anomaly between the Lena, and Ob and Yenisey (Fig. 8(c,d)) shows synchronous (r=0.63 and 0.89, correspondingly) variability until mid-1960s when anomalies of "zonal" W-circulation were positive (growth of W cumulative curve, Fig. 8(a,b)). As was mentioned above, after this "switch timing" converted "zonal" W-circulation (maximal frequency of anomalies) ("off") to "meridional' (E+C)-circulation ("on"), the vapor feeding of the rivers situated in the East Siberia is quasi-synchronous.

Changes in macrosynoptical processes are very important for the riverine income into the Arctic Ocean because the total riverine inflow is provided mainly by the greatest Siberian Rivers: Yenisey, Lena, and Ob. It is evident that the maximum in their discharge should be obtained during "zonal" W-transport epoch that is bringing air the most enriched by vapor from the North Atlantic to the watersheds of the Ob-YemseyLena. But this scheme is the simplest approach, because meridional processes play an important role at any time, even if the "zonal" circulation is dominating. Indeed, our calculation shows that in period 1935-1996 the mean frequency of W, C, E-types was as follows: W-100 days per year, E-171 days per year, and C-94 days per year; i.e. the sum of "meridional" E- and C-types dominates over W-type, if we consider all available datasets.

In Fig. 8(a,d) we can see that after the "back switch" in the circulation regime in in the mid 1970s, when anomalies of W-type began to increase and anomalies of (E+C)-type decreased, an opposition was obtained between the records of the Ob and Yenisey. After mid 1970s the Ob's graph is in a strong opposition with the Lena's graph. The Yenisey's graph shows some similarities with the graphs of Indigirka-Kolyma (Fig. 8(c,d)).

Probably we can explain this cross-dependence by the increase of E-type of circulation in this time span (Fig. 8(b)). Fig. 2(c) shows that switch from (E_z-E_{M2}) transport to "E_{M1}" transport may cause a shift in the regime of the maximal discharge from the Ob-Yenisey to the Yenisey only. Likewise, the Indigirka-Kolyma Lowland is supplied by precipitation effectively when E_{M1}-type appears. Early, Antonov [56] assumed that E-circulation benefits precipitation income in the Yenisey watershed (mainly in April-May), but there is no benefit for the Ob's feeding. Note, that our preliminary research of the North American Rivers show that hydrological regime in watersheds of the MacKenzie and Yenisey is determined by similar variability in the atmospheric circulation [77].

To understand better the influence of the typical atmospheric circulation processes on land hydrology we consider here the macrosynoptical situation over Siberia in years with maximal (1989) and minimal (1986) annual discharge in the Lena River. Because the summertime precipitation determines the discharge regime in Siberia, we consider here July as "a basic month" with a maximum in precipitation (as was shown in Fig. 6).

An <u>absolute maximum (1935-1995) in annual Lena River discharge was observed in 1989</u>, W=729 km3 (or 136% of mean W=525 km^3). In 1989, the Arctic region north of 70°N was determined by Zn-regime in distribution of the pressure field [59]. So, in 10 months of this year Zn-mode dominated over Arctic Basin ("A-circulation" by terminology used in AARI, [59]). Then in the latitude band of 60°-80°N the zonal atmospheric circulation dominated in 1989. As shown in Fig. 10(a) in July 1989, the fourteen cyclones entered the East Siberia.

In general, cyclonic routes (in the East Siberia) were oriented from west to east. Then in 1989, the discharge in the Kolyma-Indigirka was near or less than "normal" discharge because the mountain barrier trapped the water vapour moving from west to east. Likewise, an anticyclonic pressure field dominated over the West Siberia (mainly over the Ob's watershed) during 21 days of July 1989. For this reason, the extremely low discharge of the Ob was observed (Fig. 5). Because the anomaly of geopotential at level 500 hPa shows the anomalous strong development of E_z-type (with an eastward shift in the cyclonic trajectories) and E_{M2}-type of air vapor transport [78], the positive anomaly in discharge was observed for the Yenisey River (Fig. 5).

An absolute minimum in the Lena River Discharge (1935-1995) was observed in 1986, W=AO1 km3. As shown in Fig. 10(b), in July 1986 no deep cyclones were detected over Yakutia; only one non-deep cyclone went inland from the Arctic Basin and decayed over the Khatanga basin; and some events of cyclogenesis are related with the Okhotsk Sea in the most eastern part of Siberia. In July 1986, E-type (M2, M1, Zmodification) of circulation dominated (Fig. 2(c)). That increased anticyclomc activity over the East Siberia and Yemsey watershed, whereas over Ob basin cyclotuc activity was high: during 21 days in July 1986 cyclones were over the Ob watershed. Therefore a maximum of water discharge was detected for the Ob, and water discharge below the "neutral" was observed for the Yenisey (Fig. 5). Likewise in 1986, water discharge in the Kolyma-Indigirka Lowland was high, because EMI - type of circulation provides income of cyclones from the Okhotsk and Bering Seas (Fig.2 (c) and Fig. 10(b)). Based on detailed

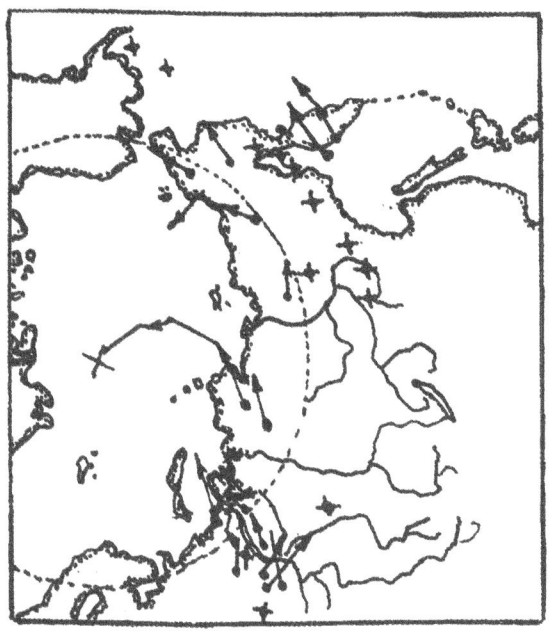

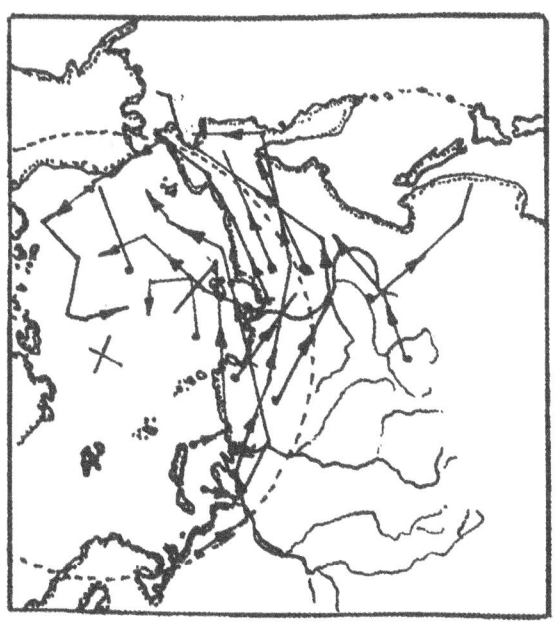

Figure 10. Cyclone trajectory of Eurasia in July 1989 (a) and 1986 (b). (● = cyclone center, × = local cyclone)

analyses of variability of atmospheric processes over Arctic Dniitriev [59] concluded that the period of (1972-1992) might be named as an epoch of E-type of circulation, as illustrated by cumulative curves of the W, C, E-types (Fig. 8(b)). We consider the period from the 1980s to early 1990s as a time with an increase in the anomaly of the "zonal" W-type, and total decrease of anomaly in "meridional" (E+C)types, whereas an anomaly of E-type appear often in this period (total decrease in (E+C)types is determined mainly by decrease in C-type), Fig. 8(a,b). For this reason we consider this period as a "zonal" W-type with an important influence of E-type over Siberia.

As was shown above the different modifications of main E -type (Z, M_1, M_2) might determine an opposite hydrological regime in different rivers of the Siberia, i.e. we can consider E-type of atmospheric circulation as an important factor that might influence the annual discharge through regime of precipitation in July-the month of the maximum precipitation in the study area, especially in the period since the mid-1960s, when E-type (cumulative curve) began to increase, Fig. 8(b).

4.1.4. *Change in the interannual amplitude of river discharge and air temperature.*

Change in amplitudes of a) discharge of the main North Asian Rivers, and b) in air temperature in their watersheds between the time spanning: 1940s - mid 1960s, and mid 1960s - 1990s is shown in Fig. 11.

Analysis of temperature records in the North Asia and America shows that over the last 30 years the warming is as much 2°C/decade in winter [79,80] that agreed well with increase in amplitudes of the riverine discharge as an integrated regional climatic parameter, Fig. 11(a).

It agrees with the NOAA/CMDL air CO_2 data that demonstrates the increase of the CO_2 amplitude of the seasonal cycle at the Northern stations (the rate~0.3 ppm/year) located at Mould Bay (76°N), Barrow (71°N), Cold Bay (55°N), and Ocean Station "M" (66°N) [81]. The difference is significant for both parameters: amplitudes of the riverine discharge and of the air temperature during the last 30 years (except of the Amguema river, Chukotka Peninsula) indicates a general increase in climatic variability in high-latitudes of the Northern Hemisphere. This result demonstrates an accelerated hydrological cycle, and an intensification of the tropospheric circulation, and confirms most model projections of amplified polar warming caused by the increased emissions of greenhouse gases that may have an impact on a variety of climatic variables [38,39] such as temperature, wind, and precipitation with changes not only in the mean quantities but also in the variability. It coincides also with the long-range increase of river-runoff (Fig. 5,9), ice growth rates in the Arctic [71], and strengthening of the Arctic (AO) and North Atlantic (NAO) oscillations after 1970 [70].

A cause for change of the observed amplitudes (Fig. 11), however, is still in question, because in section 4.1.2. (Fig. 5,7,9) it is shown that a negative trend in long-range variability of a total annual precipitation over the riverine watersheds is coincided with a positive trend for a mean annual anomaly in the discharge. To answer on a question: what factor causes this discrepancy, a total water balance should be considered.

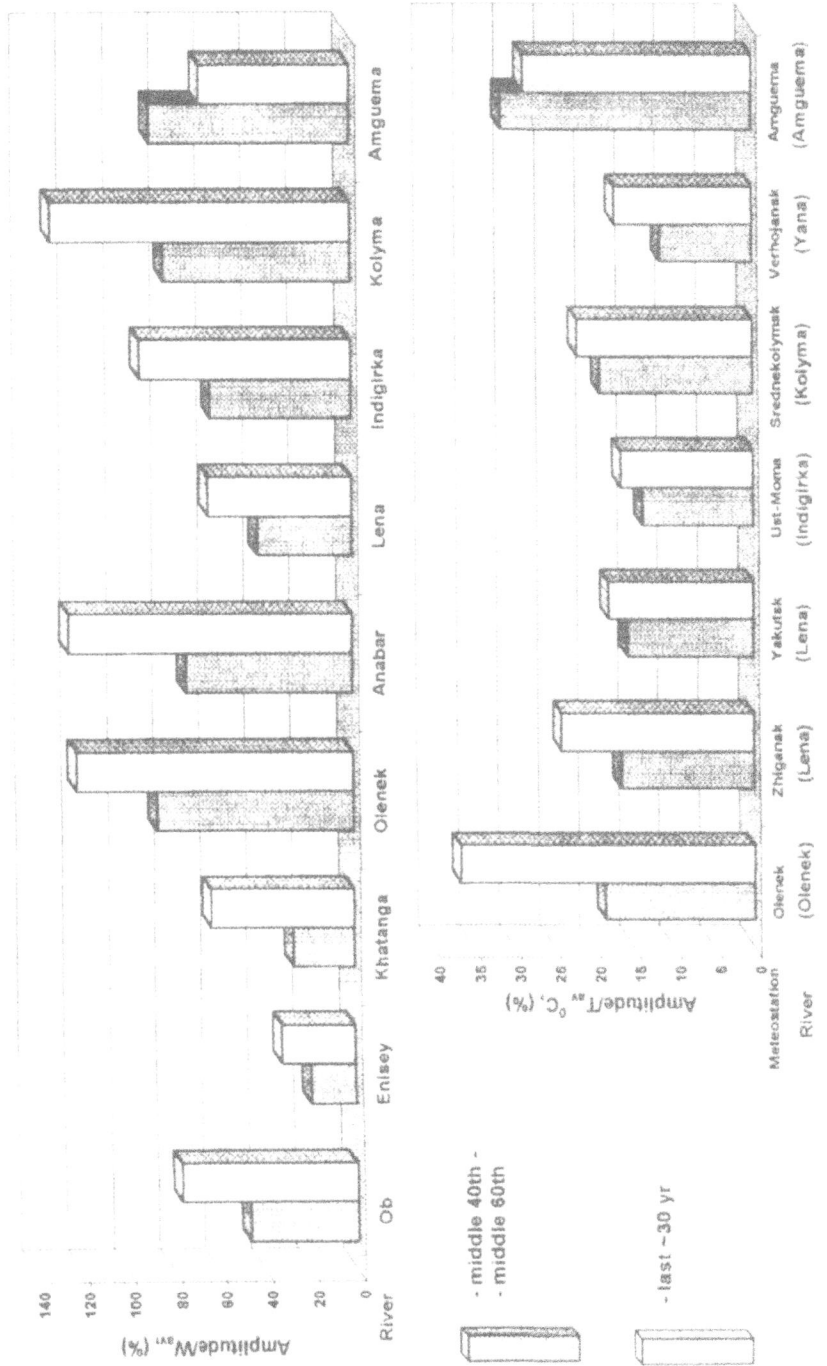

Figure 11. Amplitude change of riverine discharge and air temperature, related with their average value.

A water balance for a watershed isolated from groundwater flow is easy to present as [82]:

$$w \sim P - ET - dS \qquad (1)$$

where w=surface runoff, mm, P=precipitation, mm, ET=evapotranspiration, mm; dS=change in surface and/or soil water storage, mm.

By selecting a time frame where the change in soil storage dS is negligible, it is possible to neglect this term. Thus to keep a balance a positive change in w value accompanied with a negative change in P value should be related with more intensive decrease in ET in comparison with P decrease. Other explanation might be related to increases in the permafrost thawing and thennokarst under regional warming that cause increase in groundwater income, but it is difficult to quantify. Melt of mountainous glaciers (due to the regional warming) might influence the long-range increase in anomaly of river discharge, but we don't see this signal in "the most mountainous river - Indigirka" (Fig.5), that indicates a negligible change in rate of glacier ice-thawing.

4.2. ATMOSPHERE-LAND-SHELF-BASIN SYSTEM: ROLE OF THE RIVER RUN-OFF.

4.2.1. *Some relations between the inflow of the Siberian Rivers and the wind-driven Arctic Circulation*

Many researchers believed that winds are a principal factor forcing the permanent currents in the Arctic Basin, and described different periods of the Arctic Ocean climate variability. So, Nikiforov and Shpaikher [33] suggested that a 5- to 6-year cycle in the Arctic Basin circulation is driven by feedback loops between land hydrological processes and oceanic deep-formation, ice-growth dynamics, and thermohaline circulation, that is similar with main periodicity (6-8 yr) that was found ourselves by spectral analysis of datasets for Vangengeim's indexes of atmospheric circulation and riverine discharge in the main Siberian Rivers (see section 4.1.3).

To answer the question: What drives this periodicity in the atmosphere-land hydrology-Arctic ocean system we should consider the possible mechanism driving this variability in the Arctic. Early Nikiforov and Shpaikher [33] found that circulation and water exchange in the Arctic Ocean are determined by three main factors: anemobaric conditions, freshwater input and thermohaline water characteristics. Winds are one of the principal reasons forcing surface currents while barocinicity and positive freshwater budget stabilize the circulation. It was suggested that a 5- year period in the Arctic Basin circulation is necessary for stabilization of baroclinic circulation, though 6-11 -year periodicity might be observed in the water exchange due to feedback loops between land hydrological processes/oceanic deep-formation and thermohaline circulation. It was found also that the ice drift supported the permanent circulation that is induced in turn by atmospheric circulation. Due to this circulation, ice fields involving into Transarctic current takes 24 years to get to the Fram Strait from the Siberian Seas [29,12,83].

The mechanism proposed by Nikiforov and Sphaikher 133] and Zakharov [12] considers the Arctic as a self regulated closed system. It was speculated that atmospheric circulation, river run-off, freshwater input, ice extension and thickness, ocean-atmosphere heat flux, changes in atmosphere warming/cooling, and changes in the thickness of the warm Atlantic layer along the periphery of the Arctic are linked. Gudkovitch [23],

Nikiforov and Sphaikher [33] discovered two main types of hydrological regime in the Arctic Basin: cyclonic (Zn) when an Icelandic Low prevails in the atmosphere, and anticyclonic (Az), when an Arctic (Canadian) High develops and moves westward. Presently, Zn- and Az-modes of the hydrological regime might be associated with "Arctic Oscillation" as the surface signature of modulations in the strength of the polar vortex [25]. It might be triggered by cold and warm SST (Sea Surface temperature) anomalies in the Northern Atlantic. The results of two-dimensional, wind-forced, barotropic model indicate that wind-driven motion in the Arctic alternates between Az- and Zn-circulation with periodicity of 5-7 years [84]. However, currently the Zn-mode exists during 11 years (since 1988), that is related probably with shift in the general atmospheric circulation in mid and high latitudes of the Northern Hemisphere from E-type to W-type (Fig. 8(a,b)).

We compare here the long-range variability of river inflow in the Kara, Laptevs and East-Siberian seas (Fig. 12(a,b,c,d,e,f)) and two circulation regime of the wind-driven Arctic Ocean simulated by Proshutinsky and Johnson [84], Fig. 12(g).

Some similarities exist between variability of anomalies of the annual riverine inflow into the Siberian Seas and change of anticyclonic and cyclonic regimes. Indeed since the late 1940s variability of river runoff demonstrates an agreement with extremes in intensity of Az-regime or Zn-regime in the Arctic basin: **a positive anomaly in the mean annual inflow corresponds to the Az-regime, and negative anomaly to the Zn-regime**. It means that anticyclonic circulation, dominating over Arctic Ocean for about 10 months from October to July, corresponds mainly to increase in the river discharge in the Siberia. Because summertime precipitation determines river discharge, and the Siberian Rivers discharge about 60% of their total annual water discharge into the Arctic Ocean during the May-July, we can imply that anticyclonic atmospheric circulation over Arctic Ocean is associated with a positive precipitation anomalies over land that results in enhanced river discharge, more fresh water in the near shore surface layer, decreased surface salinity, and increased ice-formation in the shelf seas. In contrast, the cyclonic atmospheric circulation over Arctic Ocean may be associated with negative precipitation over Siberian watersheds that result in reduce river discharge, less fresh water in the shelf surface waters, increased surface salinity, and decreased ice-formation.

Different time spanning of this agreement between the riverine inflow anomaly into the different seas and types of circulation regime of the wind-driven Arctic Ocean was observed. So, during the early 1940s-mid 1980s the record of the extremes of inflow anomaly in the Kara Sea demonstrates a good agreement with intensity of the An-and Zn-circulation, and their periodicity. For the Laptevs sea an agreement with the graph of circulation regimes is observed from the late 1940s to the mid 1960s, i.e. only during a positive mode of "zonal" W-type of atmospheric circulation. Note that in mid-1960s the cumulative curve of W-type decreased below "neutral condition", whereas the E-cumulative curve demonstrated drastic growth and the C-cumulative curve demonstrated reduction in C-type circulation. Riverine anomalies in the East-Siberian Sea demonstrate not so long-time agreement with Az/Zn modes: only during time span since the early 1970s until the early 1980s. Thus, the relations between variability in the riverine input into the Arctic Ocean is weakened from the west toward the east, though agreement between the total riverine input and change in Az/Zn-mode is quite goodl (Fig. 12(f)). This eastward weakening in river runoff signal is determined probably with increase in distance from the North Atlantic that is a main water source for feeding of the Siberian Rivers. Within the one sea basin some differences in graphs of river runoff were found. So, in the Kara Sea basin record of anomaly in the discharge the Ob demonstrates a good

352

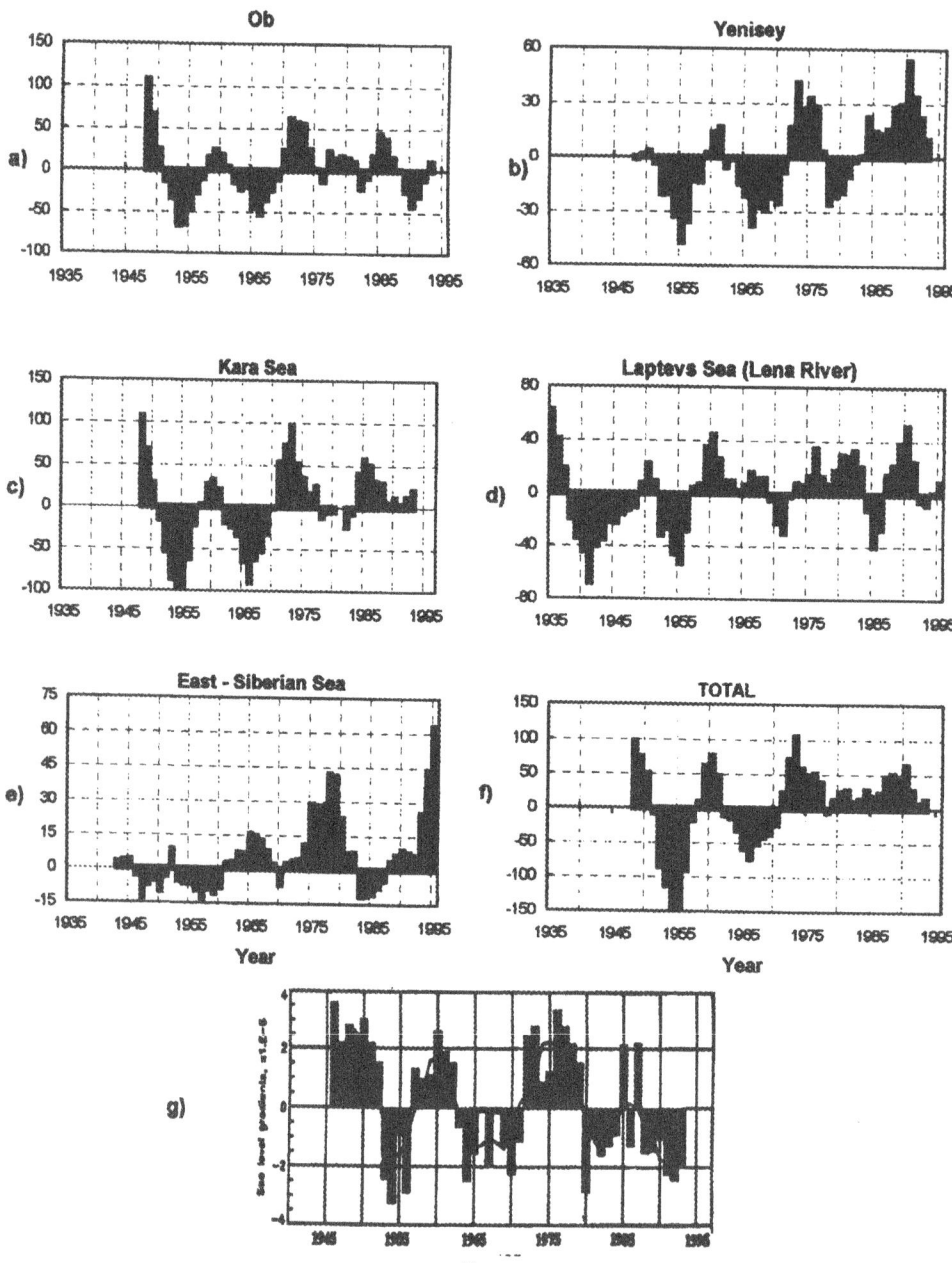

Figure 12. Comparison of anomaly in the river runoff into a) Ob, b) Yenisey, c) Kara Sea, d) Laptevs Sea (Lena river), e) East-Siberian Sea, and f) their total inflow, and g) -two-circulation regimes indicated by gradients of the sea level in the Arctic Basin (positive gradients means anticyclonic circulation and cyclonic circulation - negative values (adopted from Proshutinsky and Jonson,1997)).

agreement with intensity of the An-and Zn-circulation, and their periods during the early 1940s-early 1990s, whereas the Enisey's record demonstrates an agreement with graph Fig. 12(g) only for timing from the late 1940s to the mid 1980s (during a strong mode of "zonar' W-type and increase of E-type of atmospheric circulation). Note that this relationship for a total runoff (Fig. 12(f)) is observing only until the late 1970s, when intensity of "zonal" W transport dropped to the lowest level, and E-type of circulation increased drastically (see cumulative curves of E, W, C - types in Fig. 8).

An important effect in distribution of Atlantic and Pacific waters in the upper Arctic Ocean could be caused by change in river runoff that is associated with a shift from Zn-mode to Az-mode of circulation when the Trans-Arctic Current (northward transport of fresh water from the Asian shelf) is intensified and shifted toward Siberia [33]. To keep the water budget in the Arctic Ocean, inflow of Pacific waters (mainly in winter modification) is enhanced when Az-mode dominates. Likewise, southward inflow of Atlantic Intermediate Water (with concentration more than 75%) into the Kara and Laptevs Seas shelf through the deep canyons (St. Anna, Voronina, Sadko) exists when the Az-circulation regime predominates. In general, the surface salinity in the Siberian Seas should be decreased because an increase in river runoff is obtained (Fig. 12(a,b,c,d,e,f)). However, the surface salinity in the Chukchi and East-Siberian Seas is increased because total inflow of Pacific waters is enhanced. Thickness of the surface freshened layer is increased only in the Kara Sea (no significant northward outflow here), and one is partly decreased in the Laptevs Sea (the strong northward outflow). As a result a depth of convection is decreased in the Kara and Laptevs Seas, and one is increased in the Chukchi and eastern part of the East-Siberian Seas. Then freshening of the deep layers should be decreased when Az-circulation dominates, because the main freshwater storage is keeping in the Laptevs Sea - western part of the East-Siberian Sea, i.e. in the area where ventilation of the Atlantic waters is the most important.

When a **cyclonic regime** (Zn-mode) dominates and river runoff is decreased, total inflow of Atlantic waters is increased, whereas total inflow of Pacific waters and ice transport through Fram Strait are decreased [33]. Due to intensification of Icelandic Low and eastward winds the freshwater transport is shifted to the east and the sea level in the western parts of the Kara and Laptevs Seas is decreased, whereas in the East-Siberian and Chukchi Seas the surface layer is freshened and sea level is raised.

It is attractive to consider here briefly the recent results of Jones et al. [4] concerning the distribution of Atlantic and Pacific waters modified by the river runoff in the upper Arctic Ocean, because this surface circulation pattern was based mainly on the data obtained from 1991 to 1996, i.e. in time when Zn-circulation regime dominates. Jones et al. [4] identified a general cyclonic circulation in the surface layer that is strongest along the Amerasian Rim, with a branch of the mixed Pacific and Atlantic waters that flows across the Lomonosov Ridge through the Fram Strait. Comparison of recent circulation of Pacific and Atlantic waters in the surface layer [4] with results of the hydrodynamic modeling of thermohaline circulation, based on averaging the observation for the period 1955-1990 [85], indicates significant differences. So, the Beaufort Gyre that was produced in the 3D-circulation simulation before 1990 [85] now seems to have weakened [4], that agrees with recent results of the SHEBA project [86] and macrosynoptical research of Walsh et al. [87] that indicates a general decrease of atmospheric pressure in the Arctic, especially over the Canada Basin. Analysis of shift in general atmospheric circulation over the Northern Hemisphere (Fig. 8(a,b)) indicate that the modern change in hydrological regime of the Arctic Ocean might be related to a shift of atmospheric regime from meridional E-type to zonal W-type of circulation. At present, increase in atmospheric circulation is reflected in the amplitudes of riverine discharge (Fig. 11(a)), and eastward cyclone numbers during winter, summer and spring [66], that are obtained not only over the land, but also over the Arctic Ocean [87]. It increases eastward inflow of the subsurface Atlantic water through the Fram that might cause temperature increase in the warm Atlantic layer, and decrease in the surface salinity in the Western Arctic [24], that agrees with recent result of Jones et al. [4], Swift et al. [88], and Carmack et al. [89] indicated by warming in the Atlantic layer and a shift of oceanographic front from the Lomonosov Ridge to the Alpha and Mendeleevs Ridges. The flow pattern in the surface layer suggested by the relative distribution of Pacific and Atlantic source waters [4] agrees with the surface circulation simulated by Proshutinsky and Johnson [84] for a typical Zn-mode of circulation. Historical oceanographic data indicate that during Zn-circulation the depth of convection is decreased in the Chukchi and East-Siberian Seas, whereas one is increased in the Kara and Laptevs Seas [33], that might be associated with freshening in the Canada Basin since 1981 [90]. Of course additional investigation is needed to evaluate intrusion of the Siberian river waters into the deep Arctic Ocean.

Thus, our analysis shows that Az-regime over the Arctic Basin is associated with positive precipitation anomaly over land and increase in riverine discharge, because cyclonic frequency over land is increased at the same time, whereas the Zn-regime over the Arctic Basin is associated with early negative precipitation anomaly over land and decrease in river discharge that agrees with assumption of Nikiforov and Shpaikher [33]. Note, that early Proshutinsky and Johnson [84] proposed that the An-mode is associated with reduced river discharge whereas the Zn-mode - with increased river discharge that differs from our result.

Some typical situations illustrating the feedback relations between change in Az- and Zn-atmospheric regime over the Arctic Ocean (and the wind-driven oceanic circulation) and variability in river discharge and hydrochemistry of the adjacent sea waters are considered in section 4.2.3.

4.2.2. Change in river discharge and ice-condition in the Arctic Ocean

One of the most likely effects of the Arctic Ocean on global climate is the effect on thermohaline circulation through the export of buoyancy (fresh water) from the Arctic Ocean. An increase in fresh water (ice) export through Fram Strait may induce weakening of the thermohaline circulation, and might be responsible for such major disturbances in the paleoclimatic records as the Younger Dryas interruption of the warming trend accompanying the last deglaciation [91]. Zakharov [11] found that a change in river input precedes a change in sea-ice extent, i.e. the export of riverine waters onto the shelf-basin is a cause of variability of sea-ice conditions and the fresh water export to the North Atlantic Seas. Therefore high variability in Siberian River discharge may determine the ice-regime on the shelf. To investigate this relation we calculated coefficient of correlation between cumulative record (1932-1991) of ice-conditions (for July, August, and September) on the Siberian shelf [12] and cumulative records of discharge of the Yenisey, Ob, and Lena. The best correlation was found between cumulative discharge of the Lena (with time lag of two years) and the ice-extent for period of 1932-1992: r=0.82 (July), r=0.77 (August), r=0.74 (September), Fig. 13(a,b).

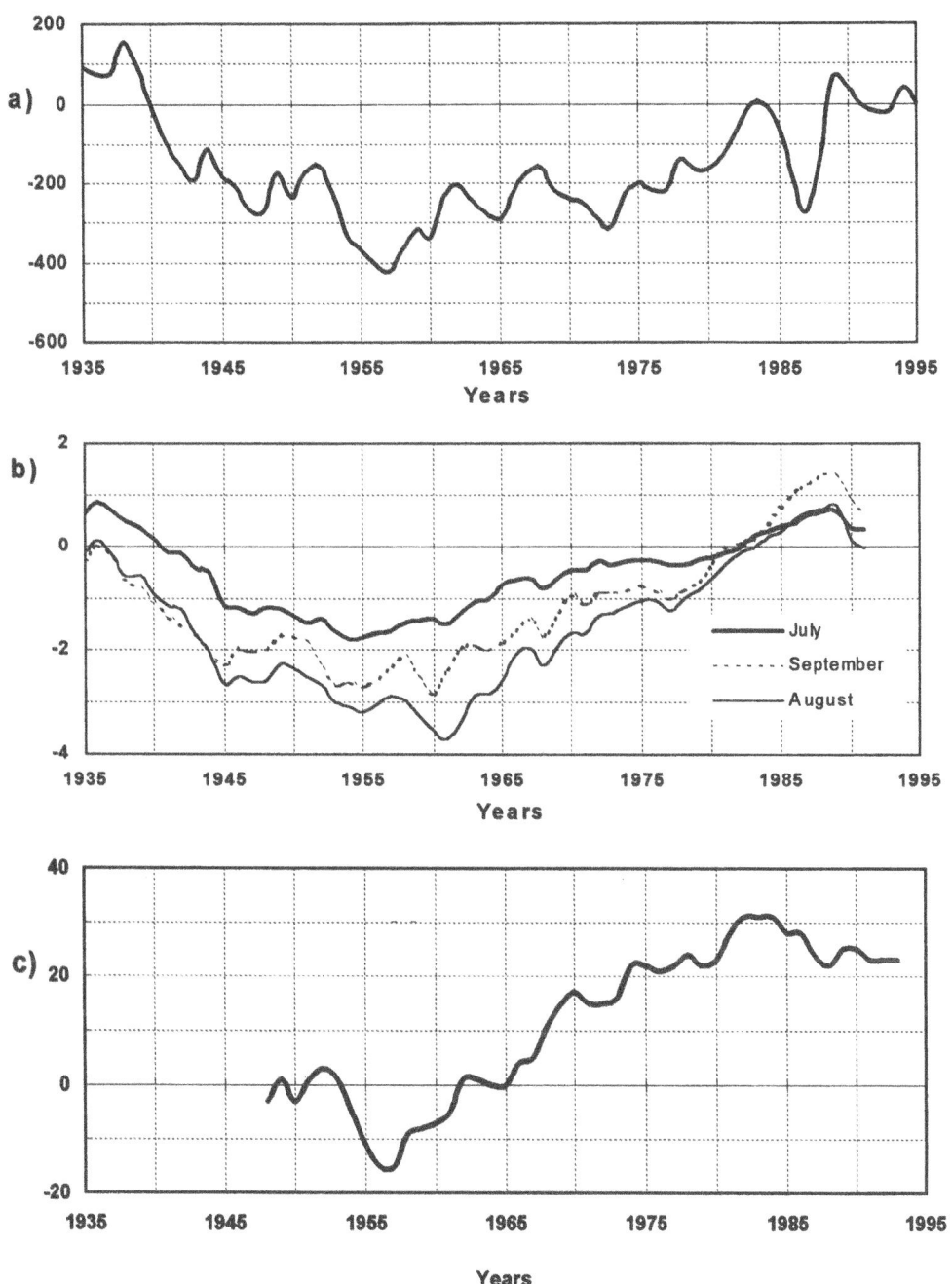

Figure 13. Comparison of cumulative curve of a) the Lena River discharge (km³), b) ice-covered aquatory of the Siberian Seas (km²×10⁶) in July, August, and September and c) index of anomaly of the ice transport (BΔL) through Fram Strait.

The correlation for other rivers (Ob and Yenisey) was significantly lower, because the main haline-forming factor in the Kara Sea is an income of salt waters from the Barents Sea, not inflow of riverine waters from the Yenisey and Ob that is illustrated by negative correlation (r=-0.28) between salinity of the surface water (Kara Sea) and the Riverine income [92]. Note that in the Kara Sea the freshened waters are involved usually in cyclonic circulation west of Trans-Arctic Drift [29,93], i.e. run-off of the Ob and Yenisey do not influence significantly the ice-transport by Transdrift that is confirmed by the recent results of Pfirman et al. [83]. Likewise, the Lena discharge (cumulative curve, Fig. 13(a)) is correlated strongly with ice export ("cumulative index of anomaly of ice export") through Fram Strait, Fig. 13(c) (adopted from [32]). The term "cumulative index of ice integrated anomaly BAL" was proposed by Gudkovich and Kovalev [32] to investigate the long-range variability in ice-conditions of the Arctic Ocean and ice export through the Fram. The cumulative index of ice anomaly is an algebraic sum of the BAL index for each year since 1948. It means that this is a useful approach to investigate a modern history of ice transport through the Fram Strait, because the BA.L index integrates main available long-range ice-data sets. Because ice drifting from the shelf edge of the Laptevs Sea/Eastern part of the East Siberian Sea to the Fram Strait takes some years (about 2-3 years by Antonov [3] or 2-4 years by Pfirman et al. [83]), we calculated the correlation matrix between cumulative curves of ice-condition in the Siberian Seas, the ice export through Fram Strait, and the Lena discharge with time lag in 2,3,4,5,6 years, Table 5.

TABLE 5.   Correlation matrix between cumulative curves of ice-condition (the sum of anomalies of ice extents square in July, August and September) in the Siberian Seas, ice export through the Fram Strait (FRAM), and the Lena discharge (LENA) with time lag in 2,3,4,5, and 6 years (L2. L.3. L4. L5, L6 respectively) **(1948-1992)**

	July	August	September	FRAM	LENA
July	1.00	0.95	0.96	0.93	0.70
August	0.95	1.00	0.98	0.93	0.69
September	0.96	0.98	1.00	0.91	0.70
FRAM	0.93	0/93	0.91	1.00	0.65
LENA	0.70	0.69	0.70	0.65	1.00
L2	0.75	0.72	0.76	0.63	0.56
L3	0.76	0.71	0.74	0.67	0.42
L4	0.75	0.69	0.71	0.70	0.40
L5	0.75	0.71	0.72	0.73	0.46
L6	0.76	0.75	0.75	0.76	0.50

The best correlation between cumulative curves of the Lena discharge and ice export was found for time lag in 4, 5, and 6 years, because it is a sum of the time needed for riverine water extension onto the shelf (1-2 years) and the time for ice transport to the Fram (24 years). Recent measurements support this relation. So, the maximal ice export through the Fram Strait (4700 km^3) was found for a year between August 1994 and August 1995 [94]. That was preceded (with time lag of 5 years) by maximum of total riverine income in the Laptevs Sea, 988 km^3, with absolute maximum in the Lena discharge (1989), 729 km^3 whereas the minimal riverine discharge, 584 km^3, into the Laptevs Sea, and absolute minimum in the Lena River discharge (1986), 401 km^3, preceded the minimal ice transport (August, 1990-August, 1991) through the Fram Strait, 2050 km^3,with time lag of 5 years also.

Thus our approach to investigate connection between historical variability of the Lena River run-off, and sea-ice parameters by [12,32], using cumulative curves, is verified by direct measurement.

Note that the sea-ice data from 1978 through 1995 for the Arctic Ocean and peripheral seas indicate an extreme minimum in the extent of the perennial ice pack in the Laptevs Sea/East-Siberian Sea sector of the Arctic in 1990 [95]. We could relate this phenomenon to an anomalous high discharge of the Lena and other rivers of the Laptevs Sea basin in 1989, though their direct thermal effect is not enough to shift the edge of the perennial ice pack northward (H.Eicken, personal communication, 1998).

Also, some evidence supporting the important role of the Lena River-Laptevs Sea system in the freshwater balance of the Arctic Ocean might be found in the recent literature. So, Closer et al. [96] show that inflow of sea-ice melt water to the total arctic freshwater balance is negligible (less than 2%), but maximal concentration of freshwater in the Arctic Ocean and peripheral seas was determined in the Laptevs Sea (up to 50% of freshwater) that is a main source for Transarctic ice transport [12,83]. The flow pattern of the surface Atlantic and Pacific water modified by riverine runoff follows ice transport along the Lomonosov Ridge toward Fram Strait [4,83]. It is important to note that the surface flow map presented by Jones et al. [4] is based mainly on the data obtained during Zn-circulation regime of the Arctic Ocean (1980, 1983, 1986, 1991, 1993, 1994), i.e. the Transarctic ice/water drift play a significant role in the Arctic ice export not only when Az-regime of circulation dominates and Transarctic current is intensified.

4.2.3.   *What are the implications for Arctic circulation-riverine input, and hydro-chemical parameters?*

To understand better what drives the distribution of riverine waters in the land-shelf-basin system, we should identify at least three possible factors that might be caused by the difference in:
- Distribution of riverine waters on the shelf that is determined by seasonal change in the river discharge;
- Shelf distribution of riverine waters that is induced by interannual variability in the river discharge;
- Shelf-basin distribution of riverine waters that is caused by change in regime of Arctic circulation (An-regime or Zn-regime).

For example, consider the Laptevs Sea system. Because dissolved silicates (and ratio Si/Cl) were used successfully as indicators of riverine waters in the Arctic shelf-basin by AARI, we try to answer these questions based on limited literature data:
- Buinevich et al. [97] found that in general, the distribution of silicates in the Laptevs Sea was similar for summer and fall of 1968. Some differences might be found in the most south-eastern and western part of the sea because variability and inertia of river flow exist, though the "outer" isolines 250, and 500 μg^{-1} between Taymyr and Kotelny I. (Novosibirsky Arc) are quite similar.

- Rusanov and Shpaikher [17] investigated distribution of the surface waters in the Arctic Ocean during 1973, 1974, and 1975 when the <u>An-regime dominated</u>. They identified different types of the surface water masses with core parameters presented in Table 6.

TABLE 6. Parameters of main types of the surface waters in the Arctic Basin

Surface waters	Silicates µM/1	Salinity,‰	Ratio $\frac{Si\,(\mu M/1)\;X\,100}{S(\text{‰})}$
Atlantic	3.2	34.40	9
Pacific	40.0	32.00	125
Riverine Input	24.0	28.70	84
Arctic	3.6	29.50	11

The result of this analysis is shown in Fig. 14(a,b,c).

A belt of the surface arctic waters, that is a result of mixing of riverine waters with saline Atlantic waters and melted ice waters is distributed from the Laptev/EastSiberian Seas toward the Fram Strait although extension of the freshened waters differs from year to year. Comparison of discharge of the Lena River indicates that the minimal discharge was in 1973 (498 km^3) and two previous years (517 km^3 - in 1971, and 505 km^3 - in 1970), i.e. the strong northward flow of the freshened water in 1973 is not related directly to the value of the discharge. It seemed that anomalous intensification of Az-circulation in 1973 [84] determines the extreme northward extension of the shelf water.

- Because the water environment has a significantly higher mass inertia for a change in direction of the main transport we may consider mean distribution (by seasons) of the surface water masses for the years with Zn- and An-regimes as an indicator of mean atmospheric circulation. On the base of T-S and Si-S analysis that was done by Nikiforov and Shpaikher [33] we can see that in summer the distribution of surface water masses are similar in the Laptevs and East Siberian Seas in both years - in the year with mean An-regime and in the year with mean Zn-regime (Fig. 15(b,d)).

Otherwise, in winter An-regime determines a strong northern outflow of the surface waters freshened mainly by the Lena River whereas Zn-regime provides mainly eastward export of freshened waters from the Laptevs Sea (Fig. 15(a,c)). Thus based on Figs. 14 and 15 we can infer that:

- Under influence of two circulation regimes of the wind-driven Arctic Ocean frontal borders between Adantic and Pacific surface water could be shifted significantly.
- Great Salinity Anomaly of the 1970s (1974-1978) [16] was induced by increased transport of fresh water and ice through Transdrift during An-regime of 1970s.

The latter assumption agrees with discussion presented in Sections 4.2.1. and 4.2.2.

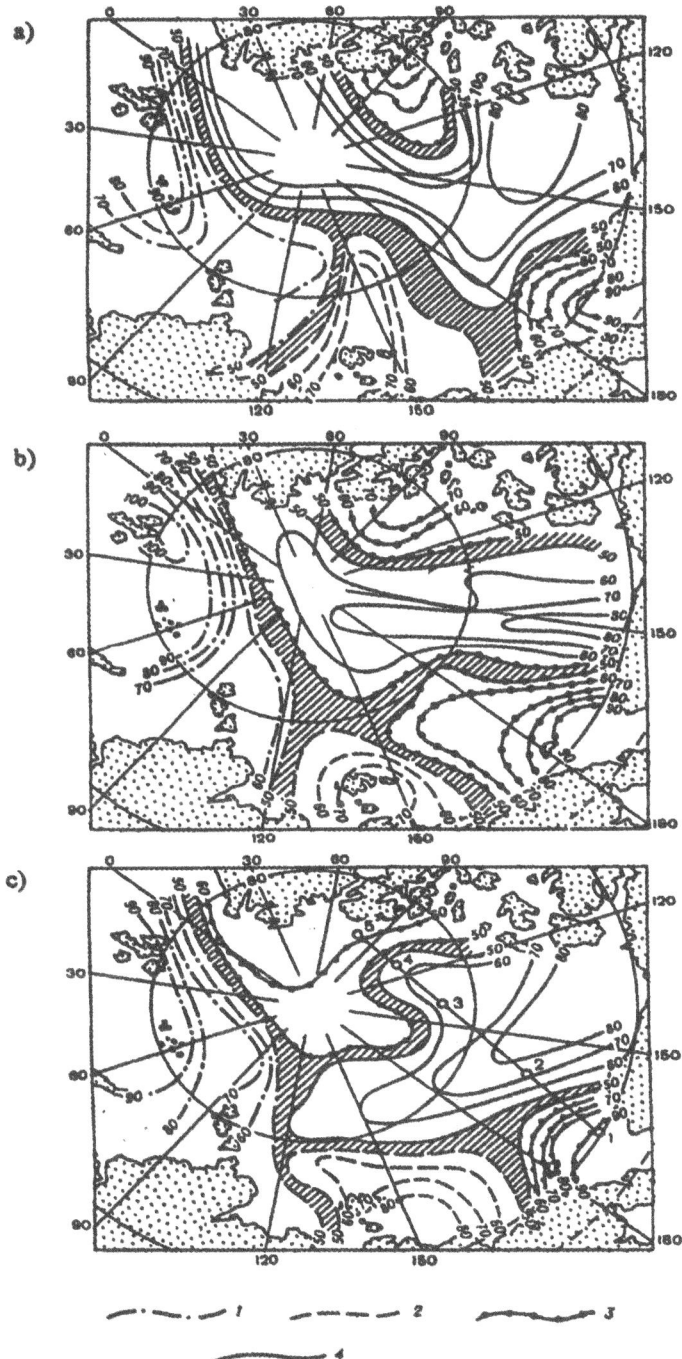

Figure 14. Distribution of water masses (%) in the surface layer of the Arctic Ocean in different years with An-regime: a) 1973; b) 1974; c) 1975

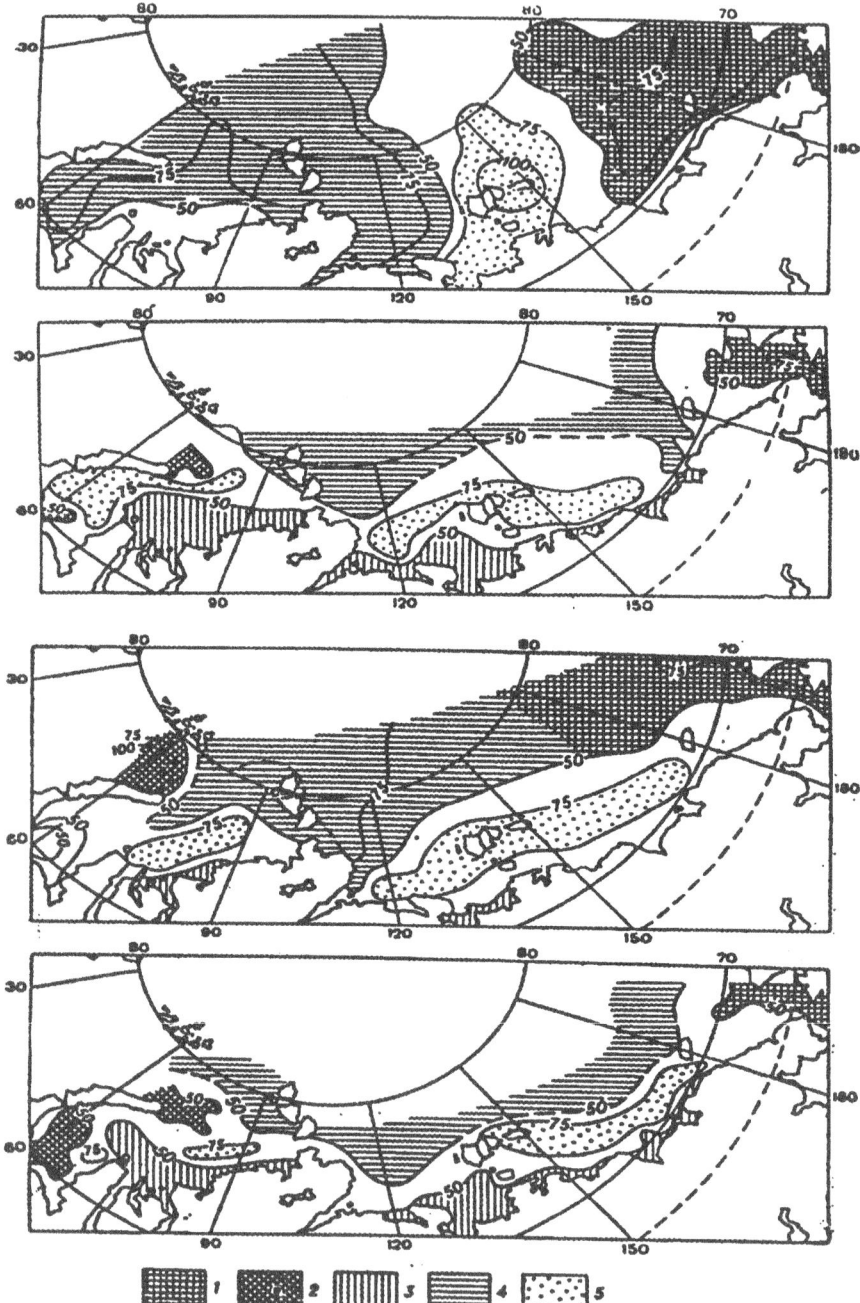

1 - Atlantic water; 2 - riverine water; 3 - Pacific water; 4 - surface arctic water

Figure 15. Mean distribution of the surface water masses (%) of the Siberia:
a) the years when Arctic anticyclonic dominates (winter); c) the years when Iceland Low dominates (winter);
b) the years when Arctic anticyclonic dominates (summer); d) the years when Iceland Low dominates (summer).
1 –Pacific water; 2 – Barents Sea water; 3 – riverine water; 4 – surface arctic waters of the Arctic Basin;
5 – surface arctic waters of the shelf seas.

5. Summary and Conclusions

A macrosynoptical approach has been used to identify the main features in interannual variability of vapor/liquid/ice water cycling in the Arctic atmosphere-land-shelf-basin system.

Using this approach, we show what type of atmospheric circulation causes similarities or differences in precipitation regime of the main Siberian Rivers and their discharge. So, when "zonal" W-type of circulation dominates (late 1930s-1960s) discharge anomalies in the Ob, Yemsey, and Lena are synchronous, while at the same time there is asynchronous discharge behavior between the Lena, and the Indigirka - Kolyma. When an anomaly of "zonal" W-type of circulation is reduced and an anomaly of (C+E)-types enhances, synchronous variability in discharge of the Lena, and the Indigirka-Kolyma is observed. The mountainous barriers play here an important role.

The variability in total (and separate) anomaly of discharge in the main Siberian Rivers agrees with two-circulation regimes of the wind-driven Arctic Ocean: a cyclonic and an anticyclonic circulation. It was found that extremes in intensity of An-regime or Zn-regime and their periodicity agree quite well with positive and negative anomalies in discharge of the great Siberian Rivers (and their total) and their periodicity. A positive anomaly in the mean annual river discharge corresponds to An-regime, and the negative anomaly in river discharge corresponds to Zn-regime. This agreement was found for time period from late 1940s until mid 1970s (until mid 1980s for Yenisey), when the frequency of anomalies in the "zonal" W-type of atmospheric circulation changed from "high" to "low" with a strong negative anomaly in mid 1970s, while "eastern meridional" E-type (integration curve) of circulation enhanced drastically in 1970s-1980s.

During the last 30 years a drastic increase in the interannual amplitude of the river discharge and air temperature agrees with the increase of the CQ amplitude of the seasonal cycle at the Northern Stations, that indicates on a general increase in climatic variability in the Arctic and Subarctic induced by greenhouse warming. It coincides with the strengthening of the Arctic/North Atlantic oscillations after 1970 [70].

High correlation (>0.7) between cumulative records of the Lena River discharge anomalies, ice conditions in the Siberian Seas, and ice export through the Fram Strait was found with a time lag of 4-6 years, because this time is needed for riverine water extension onto the shelf (1-2 years) and ice transport to the Fram Strait (2-3 years). Because the Transdrift ice transport from the freshened Laptevs Sea was enhanced in 1970s as illustrated by distribution of silicates and salinity, the Great Salinity Anomaly of the 70s might be induced this manner.

A positive long-range trend in discharge of the Siberian Rivers and temperature in their basins was found. Increase in river runoff coincides with ice growth rates and increase of the total sea ice volume in the Arctic. The amplified warming over Siberia was identified in the winter half of the year that agreed well with global climate simulations with enhanced greenhouse forcing.

We are led by all these considerations to the conclusion that linkage in the atmosphere-land-shelf-basin is driven by change in atmospheric circulation; and the macrosynoptical approach may be useful instrument to understand the long-range variability in the freshwater budget and environment of the Arctic.

6. Future Research Priorities

- To investigate the relationship between change in the indexes of atmospheric circulation and interannual variability in river runoff of the North American Rivers (Mackenzie, Yukon - Kuskokwim) and their periodicity.
- To investigate in detail water budget in watersheds of the main Arctic Rivers with a special emphasis for orographic effect.
- To evaluate a role of the riverine runoff:
    a) In freshening of the waters in the Canada Basin;
    b) In the process of ice formation / summer melt / refreezing etc. in the Arctic Ocean.

Acknowledgements. We would like to acknowledge support for this work from the International Science Foundation (1SF) under grant No. RJDOOO, and joint grant of 1SF and Russian Government No. RJD3 00. Other financial support came from the Russian Foundation for Basic Research under grants Nos. 93-05-8258, 96-05-66350, 96-05-79143, 97-05-79064, 98-05-79019, 98-05-65673 and Russian Federal Program "Integration between High School and Russian Academy of Sciences " under grant No.726. Partial assistance was available from Cooperative Institute for Arctic Research/Alaska University Fairbanks. We extend our appreciation to our colleagues from the Arctic and Antarctic Research Institute and from Canada, United States and Germany; for very useful information and discussion obtained during NATO ARW: THE FRESHWATER BUDGET OF THE ARCTIC OCEAN that was held in Tallinn, Estonia (25 April -2 May, 1998). We thank E. Peter Jones for helpful comments and English editing of our manuscript.

References

1. Weller. G., Chapin, F.S., Everett, K.R., Hobbie J.E., Kane, D., Oeschel, W.C., Ping, C.L., Reeburgh, W.S., Walker, D., and Walsh, J. (1995) The Arctic flux study: a regional view of trace gas release, *J. Biogeography* **22**, 365-374.
2. Everett, I.T., Fitzharris. R.B., and Maxwell, B. (1998) The Arctic and Antarctic, in R.T. Watson, M.C. Zinyowera and RH. Moss (eds.), *Intergovernmental Panel on Climate Change (IPCC) Special Report on the Regional Impacts of Climate Change*, Cambridge University Press, Cambridge and New York, 85-103.
3. Antonov, V.S. (1968) Nature of water and ice transport in the Arctic Ocean. *Trudi AANII* **285**, 148-177 (in Russian).
4. Jones, E.P., Anderson, L.G., and Swift J.H. (1998) Distribution of Atlantic and Pacific Waters in the upper Arctic Ocean: Implications for circulation. *Geophys. Res. Lett.* **25: 6**, 765-768.
5. Antonov, V.S. and Morozova, V.Ya. (1957) Total riverine runoff in the Arctic Seas, *Trudi AANII* **208: 2**, 13-52, (in Russian).
6. Ivanov, V.V. (1994) River Water Inflow to the Arctic Seas, *Proc. Conf On .4rctic and Nordic Countries*, Goteborg.
7. Treshnikov, A.F. (1985) *Arctic Atlas*, AANII, Moscow, 204 (in Russian).
8. Aagaard, K. and Carmack, E.C. (1989) The role of Sea Ice and other fresh water in the Arctic Circulation, *J. Geoph.Res.* **94**: ClO, 14485-14498.
9. Vize, V.Yu. (1926) Hydrological description for the Laptevs and East-Siberian Seas (in Russian). *Materiali Komissii pa izucheniyu Yakutskoi Avtonomnoi Sovetskoy Socialisticheskoi Respublibliki 5*, Academy of Sciences (USSR) Press, Leningrad.
10. Zakharov, V.F. (1966) Role of Polynias in Hydrological and ice regime of the Laptevs Sea, *Oceanology* **6**, 1014-1022 (English transl.).
11. Zakharov, V.F. (1995) Sea ice in the climatic system, *Arctic and Antarctic problems* **69**, 15-26 (in Russian).

12. Zakharov, V.F. (1996) *Sea ice in the climatic system*, Hydromceteoizdat, Sankt-Petersburgh(in Russian).

13. Gordeev, V.V., Martin, J.M., Sidorov, I.S., and Sidorova, M.V. (1996) A Reassessment of the Eurasian River input of water, sediment, major elements, and nutrients to the Arctic Ocean, *American Journal of Science* **296**, 664-691.

14. Kassens, H. and Dmitrenko, I. (1995) The TRANSDR.IFT II Expedition to the Laptev Sea, *Ber. Polarforch.* **182**, 1-180.

15. Wohlleben. T.M.H. and Weaver, X.I. (1995) Interdecadal climate variability in the subpolar North Atlantic, *Climate Dynamics* **11**, 459-467.

16. Belltin, I.M., Levitus, S., Antonov, J., and Malmberg, S.A. (1998) "Great Salinity Anomalies" in the North Atlantic. *Progress in Oceanography*, **41**:1, 1-68.

17. Rusanov, V.P. and Shpaikher, A.O. (1979) Si-S analysis of surface waters of the Arctic Basin, *Trudi AANII* **361**. 14-23 (in Russian).

18. Jones, E.P. and Anderson, L.G. (1986) On the origin of the chemical properties of the Arctic Ocean halocline, *J. Geophys. Res.* **91**, 10759-10767.

19. Aagaard, K. and Carmack, E.C. (1994) The Arctic Ocean and climate: a Perspective, in *The polar Oceans and Their Role in Shaping the Global Environment, Geophysical Monograph 85*, AGU, 5-20.

20. MacDonald. R.W. and Thomas, D.J. (1991) Chemical interactions and sediments of the Western Canadian Arctic Shelf, *Continental Shelf Research* **11**: 8-10, 843-863.

21. Anderson, L.G., Olsson, K., and Chierici, M. (1998) A carbon budget for the Arctic Ocean. *Global Biogeochemical Cycles* **12**: 3, 455-465.

22. Rudels, B., Anderson, L.G., and Jones, E.P. (1996) Formation and evolution of the surface mixed layer and halocline of the Arctic Ocean,*J. Geophys. Res.* **101**: C4, 8807-8821.

23. Gudkovidi, Z.M. (1961) On basic features of ice-drift in contral Polar basin, in *Trudi Konferensii po probleme vzaimodeistviya atmosferi i gidrosferi v Severnoi chasti Atlanticheskogo okeana*, issue 3-4, 25-42 (in Russian).

24. Nikiforov, E.G., Chaplygin, E.I., and Shpaikher, A.O. (1968) Waters of the shelf slope and atmospheric processes, *Trudi AANII* **285**, 178-188 (in Russian).

25. Thompson, D.W.J. and Wallace, J.M. (1998) The Arctic Oscillation signature in the wintertime geopotontial height and temperature fields. *Geophys.Res.Lett.* **25**: 9, 1297-1300.

26. Danilov, ID. (1990) *Ground Ice*, Nedra Press, Moscow (in Russian).

27. Semilotov, I.P. (1999) On aquatic sources and sinks of CO_2 and CH_4 in the Polar Regions, *J. Atmosph. Sci.*, **56**: 2, 2 86-306.

28. Cicerone, R.J. and Oremland, R.S. (1988) Biogeochemical aspects of atmospheric methane, *Global biogeochemical Cycles* **2**: 4, 299-327.

29. Antonov. V.S. (1957a) Distribution of riverine waters in the Arctic Seas, *Trudi AANII* **208**: 2, 25-52 (in Russian).

30. Gaigerov, S.S. (1962) *Problems of aerological structure, circulation, and climate of the free atmosphere over Central Arctic and Antarctic*, Nauka Press, Moscow (in Russian).

31. Serreze, M.C. and Maslanik, J.A. (1997) Arctic precipitation as represented in the NCEP/NCAR reanalysis, *Annals Glas.* **25**, 429-433.

32. Gudkovich, Z.M. and Kovalev, E.G. (1997) Relations between large-scale processes in atmosphere, ocean, and ice cover in the Northem Polar Region, *Trudi AANII* **437**, 17-29 (in Russian).

33. Nikiforov, E.G. and Shpaikher, A.O. (1980) *Features of the formation of Hydrological Regime Large-scale variations in the Arctic Ocean*, Hydrometeoizdat, Leningrad (in Russian).

34. Imaev, V.S., hnaeva, L.P., Koz'min, B.M., Mackey, K., and Fujita, K. (1998) Seismotectanic processes along the eastern boundary of litospheric plates of the Northeastem Asia and Alaska, *Pacific Geology* (Tikhookeanskaya Geologiya) **17**:2, 3-17 (in Russian).

35. Antonov, V.S. (1964) Anomalies of increased riverine discharge in Arctic and Subarctic zones of the Siberia, *Problemy Arctiki I Anrarctik.* **18**,24-30 (in Russian).

36. Zubakina, A.N. (1979) Features of bydrochemical Regime in the Lena River's mouth area, *Trudi GOIN* **143**,69-76 (in Russian).

37. Golikov A.N. and Averintsev V.G. (1977) In: *Polar Oceans* (M.J. Dunbar, ed), Arctic Institute of North America, Calgary - Montreal. 361-364.

38. Chapman, W.L. and Walsh, JE. (1993) Recent Variations of Sea Ice and Air Temperature in High Latitudes, *Bull. Amer. Meteorol. Soc.* **74**: 1, 33-47.

39. Schubeit, N.I., Perlwitz, J., Blender. R., Fraedrich, K., and Lunkeit, F. (1998) North Atlantic cyclones in CO_2-induced warm climate simulations: frequency, intensity, and tracks. *Climate Dynamics* **14**, 827-837.

40. *Hydrological Year-book (1952-1965)* **8**, Basins of East-Siberian, Chukchi and Bering seas, issue 0-7, Basins of Kolyma River and rivers to the east of it, including rivers of Bering Sea from Dezhnev Cap to south basin of Khatyrka River - 1945-1963, Hydrometeoizdat, Leningrad.

41. *Hydrological Year-book (1967-1980)* **8**, Basins of East-Siberian, Chukchi and Bering seas, issue 0-7, Basins of Kolyma River and rivers to the east of it, including rivers of Bering Sea from Dezhnev Cap to south basin of Khatyrka River -1964-1977, VINITI, Yakutsk.

42. *Hydrological Year-book (1971)* **8**, Basins of East-Siberian, Chukchi and Bering seas, issue 8, Basins of Kolyma River and rivers to the east of it, including rivers of Bering Sea from Dezhnev Cap to south basin of Khatyrka River, Chemysbova M.R. (ed.), Hydrometeoizdat*, Magadan, State Hydromet/Kolymsky Branch, (in Russian).

43. Ibid(1972).

44. Ibid (1973).

45. Ibid (1974).

46. *State Water Cadastre (1979)* Main hydrological characteristica (1971-1975 period and all observation period), **17**, Lena-Indigirka region, Hydrometeoizdat, Leningrad, State Hydromet/Yakutian Branch (in Russian).

47. *State Water Cadastre (1980)* Yearly data of the regime and resources of surface terrestrial waters - 1978, part 1, Rivers and Channels. **17**, issue 0-7, Basins of Laptev, East-Siberian and Chukchi seas. VINITI, Obninsk (in Russian).

48. *State Water Cadastre (1983-1985)* Yearly data of the regime and resources of surface terrestrial waters, **1**, issue 17, Basins of Kolyma river and Magadan's region rivers, Chernyshova M.R.(ed.), Hydrometeoizdat, Magadan (in Russian).

49. *State Water Cadastre (1984-1989)*, Section 1, Surface waters, Series 3, Multi-annual data on the regime and resources of surface terrestrial waters -1984-1987, part 1, Rivers and Channels. **1**, issue 16, Basins of Lena (middle and down stream), Khatanga, Anabar, Olenek, Yana, Indigirka, VINITI, Obninsk.

50. *State Water Cadastre (1985)* Section 1, Surface waters, Series 3, Multi-annual data on the regime and resources of surface terrestrial waters, **1**, issue 17, Basins of Kolyma river and Magadan's region rivers, Hydrometeoizdat, Leningrad, State Hydromet/Kolymsky Branch (in Russian).

51. *State Water Cadastre (1985)* Section 1, Surface waters, Series 3, Multi-annual data on the regime and resources of surface terrestrial waters - 1983, part 1, Rivers and Channels. **1**, issue 16, Basins of Lena (middle and down stream), Khatanga. Anabar, Olenek, Yana, Indigirka, Hydrometeoizdat, Leningrad. VINITI, Yakutsk.

52. *State Water Cadastre (1985)*, Section 1, Surface waters, Series 3, Multi-annual data on the regime and resources of surface terrestrial waters -1983, part 1, Rivers and Channels. **1**, issue 16, Basins of Lena (middle and down stream), Khatanga, Anabar, Olenek, Yana, Indigirka, VINITI, Yakutsk-Obninsk.

53. *State Water Cadastre (1987)*, Section 1, Surface waters, Series 3, Multi-annual data on the regime and resources of surface terrestrial waters, part 1, Rivers and Channels. **1**, issue 16, Basins of Lena (middle and down stream), Khatanga, Anabar. Olenek, Yana, Indigirka, Hydrometeoizdat, Leningrad.

54. *Joint US-Russian Arctic Ocean Atlas* (1998): CD-ROM produced by AARI, The University of Washington.

55. Voeykov, A.I. (1948) Climates of the Earth, and Russia, selected papers/ *Izvestiya Akademii Nauk SSSR*, Moscow-Leningrad (in Russia).

56. Antonov. V.S. (1957b) Climatic reasons for variations of discharge of the main Siberian Rivers, *Trudi AANII* **2O8**: 2, 5-12 (in Russian).

57. Vangengeim, G.Ya. (1940) Long-range forecasting of air temperature and time of riverine breakup. *Trudi Gosudarstv. Gidrologicheskogo Instituta* **10**, Leningrad-Moscow.

58. Girs, A.A. and Kondratovich, K.V. (1978) *Methods of long-range weather forecasting*, Hydrometeoizdat, Leningrad (in Russian).

59. Dmitriev, A.A. (1994) *Variability of atmospheric processes over the Arctic: long-term predictions*, Hydrometeoizdat, Sankt-Petersburgh (in Russian).

60. Bryazgin, N.N. (1976) Mean annual precipitation in the Arctic with consideration of gauge biases, *Trudi AANII* **323**, 40-74 (in Russian).

61. Serreze, M.C., Relider, M.C., Barry, R.G., Kahl, J.D., and Zaitseva, N.A. (1995a) The distribution and transport of atmospheric water vapour over the Arctic Basin, *International J. Climatology* **15**, 709-727.

62. Varlamov, S.M., Kim, E.S., and Khan, E.N. (1998) Modern temperature variability in the Eastern Siberia and Far East. *Meteorological and Hydrological*, **1**, 19-28 (in Russian).

63. Ginzburg, B.M. and Soldatova, I.I.. (1997) Long-term variability of freeze-up and break-up dates as an indicator of climate variations in transitional seasons, *Meteorologiya i Gidrologiya,* (English transl.) **11**, 99-107.

64. IPCC (Intergovernmental Panel on Climate Change) (1996) *Climate Change 1995: The Science of Climate Change*, in J.T. Hougton et al. (eds.), Cambridge Univ. Press, Cambridge.

65. Sorodian, O.G. (1961) Corrected data for water vapor characteristics over the Eastern Siberia and Far East, *Trudi GGO* **3** (in Russian).

66. Serreze, M.C., Box, J.E., Barry, R.O., and Walsh, J.E. (1993) Characteristic of Arctic Synoptic Activity, 1952-1989. *Meteorol. Atmos. Phys.* **51**, 147-164.

67. Aleksandrov, E.II., Bryazgin, N.N., and Lyubarskiy, AN. (1995) Atmospheric precipitation and it's variability in lowstream of the Ob and Yenisey, *Trudi AANII* **434**, 102-110 (in Russian).

68. Serreze, M.C. (1995) Climatological Aspects of Cyclone Development and Decay in the Arctic, *Atmospheric-Ocean* 33:1, 1-23.

69. Serreze, M.C., Rehder, M.C., Barry, R.G., Walsh, J.E., and Robinson, D.A (1995b) Variations in aerologically derived Arctic precipitation and snowfall, *Annals Glac.* **21**, 77-82.

70. Hakkinen, S. and Geiger, C.A. (1998) Low frequency variability of a simulated Arctic ice-ocean system for the period 1951-1993, *EOS Transaction, AGU*, **79**: 45, F415.

71. Pfirman, S.L., Eicken, H., Colony, R., Rigor, I., Schlosser, P.. Mortlock, R., and Bauch, O. (1998) Drifting Sea Ice as an Environment Archive, *EOS Transaction, AGU*, **79**: 45, F436.

72. Ludchin, V.A., Saveliev, A.V., and Radchenko, V. I. (1998) Long-periodical climatic waves in the western Bering Sea and their effect on biological productivity. In: V.F. Kozlov (ed.) Climatic and interannual variability in the atmosphere - land - sea system in the Amerasian Sector of Arctic, *Proc. Arctic Regional Center, Vladivostok*, **1**, 31-40 (in Russian).

73. Sleptsov-Shelevich, B.A (1967) On the study of unstability of heliogeophysical relations, *Trudi AANII* **257**, 93-118 (in Russian).

74. Chaplygin, E.I., and Yanes, A.V. (1968) Cosmic and global factors in the problem of oceanographic predictions, *Trudi AANII* **285**, 223-23 8 (in Russian).

75. Maksimov, I.V. (1970) *Geophysical forcing and ocean waters*, Hydrometeoizdat, Leningrad.

76. Michaelsen J. (1982) A statistical study of large-scale, long-period variability in North Pacific Sea surface temperature anomalies. *J.Phys.Oceanogr.* **12**, 694-703.

77. Semilotov, I.P., Savelieva, N.I., Pipko, I.I., Pugach, S.P., Gukov, A.Yu., and Vasilevskaya, L.N. (1998a) Long-range variations in the atmosphere - land environment - sea system in the North Asian Region. In: V.F. Kozlov (ed.) Climatic and interannual variability in the atmosphere - land - sea system in the Amerasian Sector of Arctic, *Proc. Arctic Regional Center, Vladivostok*, 1, 41-64 (in Russian).

78. Semilotov, I.P., Savelieva, NI., Pipko, I.I., Pugach, S.P., and Weller, G.E. (1998b) On the long-range and interannual variability in the atmosphere - land - shelf system of the Arctic. *EOS Transaction, AGU*, **79**:45, F437.

79. Walsh, J.E., and Chapman, W.L. (1990) Short-term climatic variability of the Arctic, *J.Clim.* **3**, 237-250.

80. Weller, G., Lynch, A., Osterkamp, T., and Wendler, O. (1998) Climate Trends and Scenarios. In: O. Weller and PA. Anderson (eds.) *Implications of Global Change in Alaska and the Bering Sea Region. Proc. Workshop*, June 1997. Center for Global Change and Arctic System Research, AUF, Fairbanks, AK, 15-21.

81. Conway, T.Y., Tans, P.P., Waterman, L.S., Thoning, K.W., Kitzis, D.R., Masarie, K.A, and Zhang, N. (1994) Evidence for Interannual Variability of the Carbon Cycle from the NOAA/CMDL Global Air Sampling Network, *J. Geophysical Res.* **99**: Dl 1,22831-22855.

82. Kane, D.L, Gieck, R.E., and Hinzman, L.D. (1990) Evapotranspiration from a Small Alaskan Arctic Watershed, *Nordic Hydrology* **21**, 253-273.

83. Pfirman, S.L., Colony, R., Nurnberg, D., Eicken, H., and Rigor, I. (1997) Reconstructing the origin and trajectory

84. Proshutinsky, A.Y. and Johnson, M.A. (1997) Two circulation regimes of the wind-driven Arctic Ocean, *J. Geophys. Res.*, **102**:C6, 12493-12514.

85. Polyakov, I.V. and Timokhov, L.A. (1995) Thermohaline circulation of the Arctic Ocean, *Doklady (Transactions) Akademy of Sciences*, **342**: 2, 254-258 (in Russian).

86. Levi, B.G. (1998) A drift on the Ice Pack, Researchers Explore Changes in the Arctic Environment *Physics Today*, American Institute of Physics: November issue, 17-19.

87. Walsh, J.E., Chapman, W.L., and Shy, T.L. (1996) Recent decrease of Sea level pressure in the central Arctic, *J.Clim.* **9**,480-486.

88. Swift, J.H., Jones, E.P., Aagaard, K, Carmack, E.C., Higston, M., MacDonald, R.W., McLaughlin, F.A., and Perkin, R.G. (1997) Waters of the Makarov and Canada basins, *Deep-Sea Res.*, **44**: 8, 1503-1529.

89. Carmack, E.C., Aagaard, K., Swift, J.H., MacDonald, R.W., McLaughlin, F.A., Jones, E.P., Perkin, R.G.. Smith, J.N.. Ellis, K.M., and Killins L R (1997) Changes in temperature and tracer distributions within the Arctic Ocean: results from the 1994 Arctic Ocean Section. *Deep-Sea Res.* **44**: 8, 1487-1502.

90. Melling, H. (1998) Hydrographic changes in the Canada Basin of the Arctic Ocean, 1979-1996, *J. Geophys. Res.* **103**: C4, 7637-7645.

91. Broecker, W.S. (1997) Thermohaline Circulation - the Achilles Heel of our Climate System: will man-made CO_2 Upset the Current Balance? *Science*, **278**, 1582-1588.

92. Appel, I.L. and Gudkovich, Z.M. (1984) Investigation for possible changes of mean salinity of the upper

layer in the Kara Sea that is induced by stable anomaly of river discharge, *Problemi Arktiki i Antarktiki* **58**. 5-14 (in Russian).

93. Nagumy, A.P. and Savchenko, V.G. (1991) Results of computing simulating for variability in riverine inflow into the Arctic Ocean on climate in the Northern Hemisphere, in XF. Treshnikov and G.V. Alekseev (eds.), *Interaction of ocean and atmosphere in the Northern Polar Region*, Hydrometeoizdat, Leningrad, 153-161 (in Russian).

94. Osterhus, S. and Vinje, T. (1998) The export of fresh water and ice from the Arctic Ocean through the Fram Strait and the Barents Sea (plenary report), in *Extended Abstracts of NATO ARW "The Freshwater Budget of the Arctic Ocean"* 25 April - 2 May 1998, Tallinn, Estonia.

95. Maslanik, J.A., Serreze, M.C., and Barry, R.G. (1996) Recent decreases in Arctic summer ice cover and linkages to atmospheric circulation anomalies, *Geophys. Res. Lett.* **23**, 1677-1680.

96. Schlosser, P., Bayer, R., Ekwurzel, B., Falkner, K., Khatiwala, S., and Guay, C. (1998) New insights into the freshwater balance of the Arctic Ocean derived from tracers (plenary report) in *Extended Abstracts of NATO ARW "The Freshwater Budget of the Arctic Ocean"*. 25 April -2 May 1998, Tallinn, Estonia.

97. Buynevich, A.G., Rusanov, V.P., and Smagin, V.M. (1980) Distribution of the riverine waters in the Laptevs Sea by distribution of chemical parameters, *Trudi AANII* **358**, 116-125 (in Russian).

THE VARIABLE CLIMATE OF THE MACKENZIE RIVER BASIN: ITS WATER CYCLE AND FRESH WATER DISCHARGE

R. E. STEWART
Atmospheric Environment Service
Downsview, Ontario Canada

1. Introduction

Fresh water discharge from northern-flowing rivers plays an important role in regulating the thermohaline circulation of the world's oceans [1, 2]. The Mackenzie River of northwestern Canada is one of these rivers. It represents the largest single source of fresh water for the Arctic Ocean from North America, it accounts for approximately 11% of the total gauged river discharge into this ocean [3], and its basin is therefore a critical component of the Arctic climate system.

The examination of the water cycle of basins such as that of the Mackenzie River therefore represents critical contributions to international climate efforts [4]. This includes, for example, the Global Energy and Water Cycle Experiment (GEWEX) which aims to improve our understanding and prediction of the role that the water cycle plays in the climate system. Such improvements are crucial if we are to determine the sensitivity of the climate system to anthropogenic increases in the concentration of greenhouse gases and thereby to reduce the uncertainties in our predictions of climate change [5]. As well, the examination of meteorological and hydrological processes that govern the transport of water into and through the Mackenzie basin represents a contribution to the study of the Arctic climate system as a part of the Arctic Climate System Study (ACSYS).

Over the last few decades, the Mackenzie River basin has also been undergoing considerable change in its climate. The region is experiencing warming, especially during the cold season [6]. Other critical aspects of the basin's climate system (including water vapour, cloud fields, precipitation, and snowcover) are also experiencing some degree of change [7]. This has substantial impacts for the fresh water budget of the Arctic from the atmospheric as well as from the hydrological perspectives.

In this chapter, the water cycle of this region is discussed, including several of its reservoirs, fluxes and controlling processes under a variety of conditions and temporal scales. Information on the magnitudes of these parameters has been obtained from several sources. Water vapour (1956-1994), cloud cover (1953-1997), precipitation types (1950-1995), surface winds (1950-1995), and discharge (1973-1995) were obtained from operational datasets; precipitation and temperature (1950-1994) were adapted from [8];

E.L. Lewis et al. (eds.), The Freshwater Budget of the Arctic Ocean, 367-381.
© 2000 *Kluwer Academic Publishers.*

and snowcover fraction (1977-1995) was calculated from data obtained from the National Snow and Ice Data Center of the United States.

Although one might expect that the generally increasing temperatures of this basin to be closely linked with an accelerated hydrological cycle, this is not necessarily the case. For example, there has been a general change in the large scale atmospheric circulation that has altered advection of warm and cold air into the region and this has also affected the occurrence and strength of basin-scale circulations that lead to adiabatic heating [9]. These secondary circulations develop in part from local topographic influences and are generally capped by a low level temperature inversion. Such variations in large scale and basin-scale circulations are often directly linked with variations in critical parameters such as temperature but they also greatly affect the water budgets and fluxes. The magnitudes of the water cycle components are furthermore affected by other complex processes including, for example, the alteration of the ability of clouds to produce precipitation, the dryness of the boundary layer, changes in the phase of precipitation, the likelihood of summertime convective activity, and the magnitude of evapotranspiration from the variable surface of the basin. Most of these processes affect, in one way or another, the amount and timing of the fresh water discharge from the Mackenzie River into the Arctic Ocean.

It is very difficult to ensure that the Mackenzie River basin's water cycle and its associated discharge are adequately represented within global and regional climate models. The complexity of the coupled atmospheric-surface-hydrological system represents a daunting challenge for modeling such a system and its role in the fresh water budget of the Arctic.

The effort being made in addressing the Mackenzie basin's water budgets is often applicable to other high latitude regions. Many of the same factors are operating in other areas, even though the precise conditions can be different. There should be many similarities for example with the Siberian river basins.

In summary, this chapter is concerned with the variable water budgets and fluxes of the Mackenzie basin, some of the factors influencing these, capabilities for handling them within regional climate models, and implications for the entire Arctic.

2. Mackenzie Basin Climate Characteristics

The Mackenzie River is one of the great rivers of the world, ranking tenth by drainage area. This large drainage basin stretches over 15° of latitude and covers about 1.8 million km^2 or about 20% of the total Canadian land mass (Figure 1).

The Mackenzie basin is composed of 6 main sub-basins, 3 great lakes, and 3 major deltas including one of the world's largest inland freshwater ones (the Peace-Athabasca River Delta). Of the sub-basins, only the Peace River is regulated to a significant degree. The hydrological regime of the basins is influenced by 4 major physiographic regions (Western Cordillera, Interior Plain, Precambrian Shield and Arctic Coastal Plain), by permafrost that underlies a significant portion of the basin, and by vegetation that varies from boreal forest to alpine and arctic tundra.

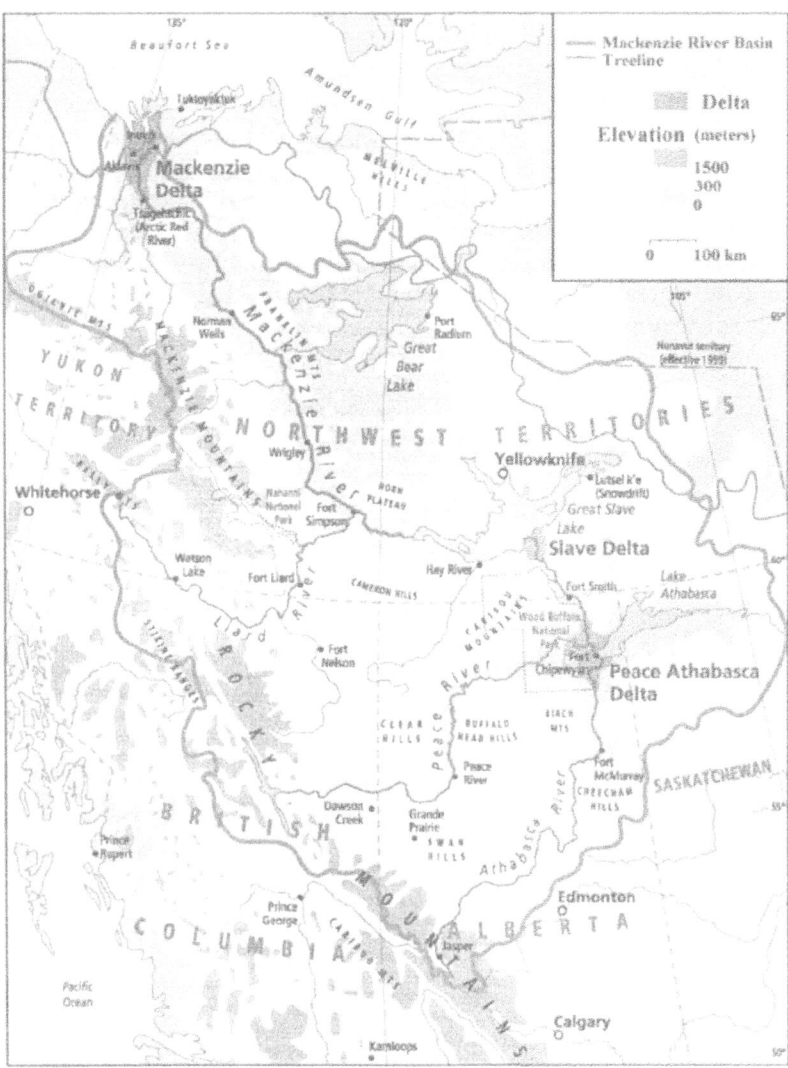

Figure 1: The Mackenzie basin (outlined by the solid line) and some of its features and population centres. This map was adapted from one produced by E. Leinberger for the Mackenzie Basin Impacts Study [10].

There is a very dramatic annual cycle in the surface temperature of this basin. Cold season monthly basin-average temperatures typically reach -25 or -30°C, whereas these average values during the summer reach 15°C. Individual days at specific stations reach much more extreme values with low temperatures approaching -50°C and high temperatures well above 30°C. Basin-average monthly temperatures of at least 0°C take place in the April-June period with northern areas being later, and average temperatures again fall below the freezing point during September-October. Generally, colder temperatures persist longer over northern regions and over higher terrain.

It is estimated that the mean annual precipitation over the basin is about 410 mm/year (690 km^3/year) but this is subject to considerable inter annual variation [8]. In addition, there is considerable uncertainty due to inadequate sampling and errors with gauge measurement in winter conditions [11]. Peak precipitation values generally occur in the summer in association with convective or organized systems and lowest values occur in the February-April period. Precipitation amounts range from less than 300 mm/year in the northwestern sections of the Mackenzie basin, to between 300 to 400 mm/year in the extreme southern sections of the basin, to as high as 1600 mm/year in the mountains on the west side of the Basin.

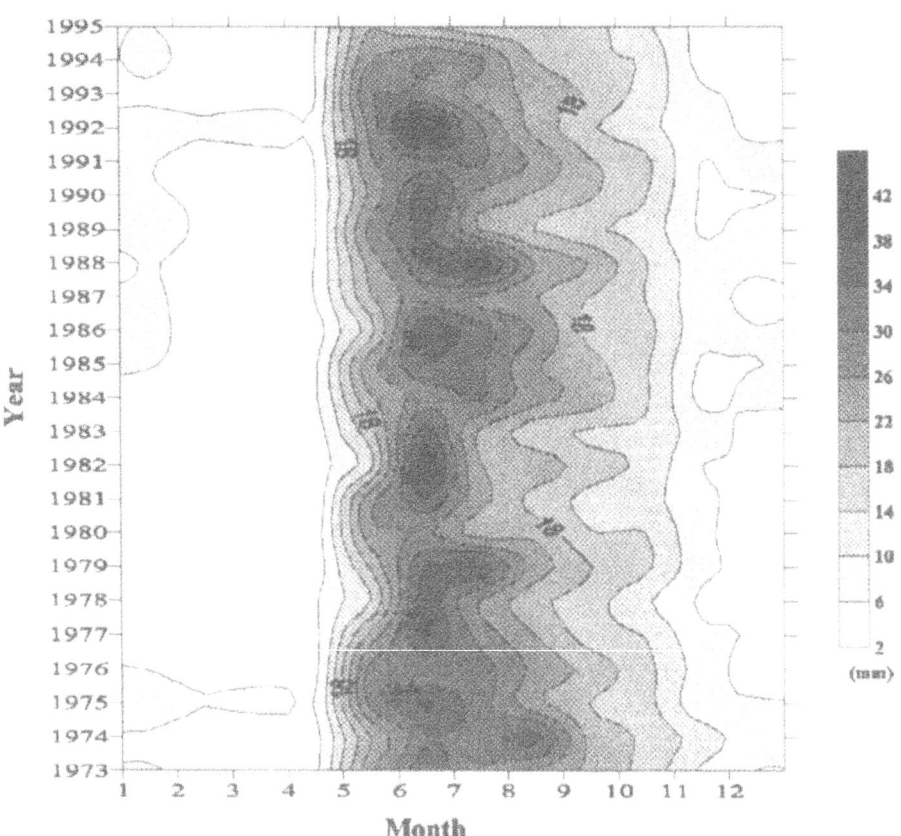

Figure 2: Monthly average discharge for the period from 1973-1995 from the Mackenzie River as inferred from observations at Arctic Red River, but supplemented with those from the Peel River downstream of this location.

Over the Mackenzie basin, a significant portion of the annual precipitation falls as snow, varying from about 30% in southern regions of the basin to about 60% in northern areas. Snow storage of a large portion of the annual precipitation for between 5 and 8 months of the year is therefore an important component of the water cycle of the basin.

The discharge from the basin is one of the most critical parameters associated with the climate of this region and it represents the direct link with the Arctic Ocean (Figure 2). Using the 1973-1995 period as a base, it is estimated that the annual discharge from the Mackenzie is about 170 mm/year (280 km³/year), but difficulties in measurement during the spring melt and winter periods undoubtedly lead to large uncertainties.

Discharge generally displays the characteristic pattern of the seasonal cycle for northern flowing rivers in the Northern Hemisphere. Maximum discharge occurs in association with the spring breakup over selected portions of the basin or after breakup with later snowmelt plus summer rainstorms. The northward advance of the spring break-up occurs at a rate of about 0.3 degrees of latitude per day and the resulting flood wave progresses downstream more rapidly that the melt conditions and this results in large ice jams [12, 13]. This is followed by summer-time flow fed by melting snow in alpine areas and rain that over southern areas can be linked with severe thunderstorms. The remainder of the year is characterized by under-ice flow. However, discharge illustrates a considerable inter-annual variation [3].

There are four distinct periods of the year in regards to the climate system of the Mackenzie basin. Meteorologically, much of the winter snowpack falls in the autumn (September-November) when cyclonic storms affect the area, in the winter (December-March) relatively light snow is more common, in the spring (April-June) warming and rain storms enhance runoff, and in the summer (July-August) convective activity dominates. From the hydrological point of view, summer (July-October) is when rainfall is the main forcing, autumn (October-November) is linked with a recession in flows leading to freeze-up, winter (November-March or April or even may) is when the river and lake system is covered by ice, and spring (April-June) is when melt and ice jams occur. There is of course considerable variation in the exact periods of these events, given the large inter-annual variability and the great expanse of the basin itself.

3. Components of the Water Cycle

The water cycle associated with the Mackenzie basin involves many atmospheric, surface and hydrological components (Figure 3). Water vapour is advected into the basin from large scales, this often forms a variety of cloud fields which act to produce precipitation, although some of that precipitation does not reach the surface due to sublimation or evaporation in the often dry sub cloud region. Since about 50% of the precipitation is in the form of snow, a seasonal snowcover is produced which is nevertheless subject to loss due to factors such as straight sublimation from an intact snowcover or enhanced sublimation during blowing snow episodes [14]. During the summer period, rain falls over the basin and there is substantial evapotranspiration from the variable surface vegetation and from the numerous lakes, wetlands and ponded water within the basin. Lateral water flow occurs through the variable terrain and must overcome spring ice jams to produce the substantial fresh water flow into the Arctic Ocean.

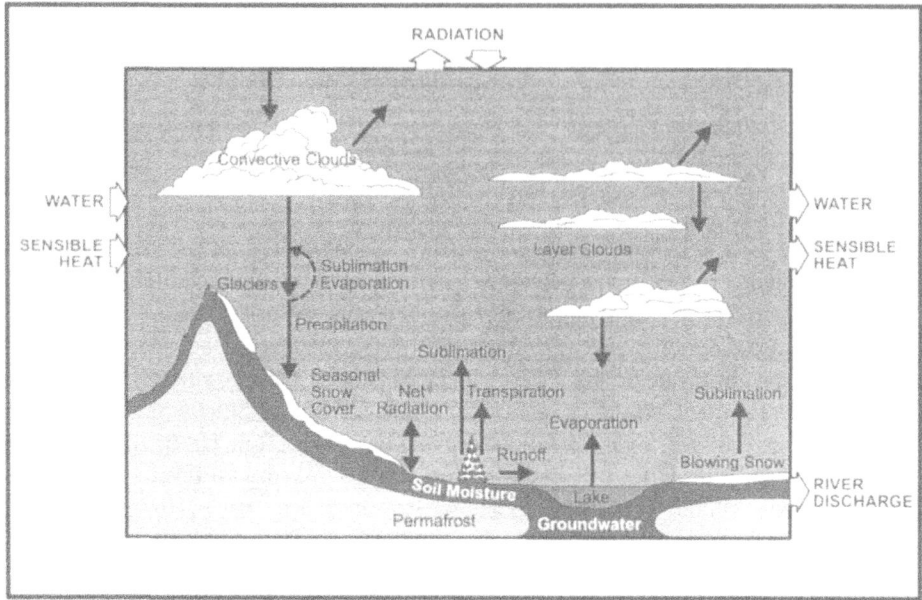

Water and Energy Flows of the Mackenzie Basin

Figure 3: Schematic diagram showing some of the processes affecting the water cycle of the Mackenzie basin.

Figure 4 shows a typical annual cycle in some of the atmospheric water cycle parameters. For most of the year, there is strong convergence over the basin. The basin is generally a net sink for water vapour, but in the summer it is possible that evaporation is larger than precipitation [15].

The water cycle is affected in the atmosphere by, for example, large scale atmospheric circulations, cloud systems, sub-cloud processes, precipitation, and orographic effects. The most common moisture source for the basin is from the north Pacific Ocean, although there are significant contributions from the Beaufort Sea and from the interior of the continent. Much of this moisture is transported to the basin within large and organized frontal systems but these typically undergo major evolution as they pass into and through the basin; this evolution is typically linked with direct and indirect consequences of orography. This evolution includes the common development of dry sub-cloud regions over the basin which limits the amount of precipitation reaching the surface. Such atmospheric factors therefore either define the large scale moisture environment, dictate its regional character, and/or define the rates and limits under which the water cycle can operate.

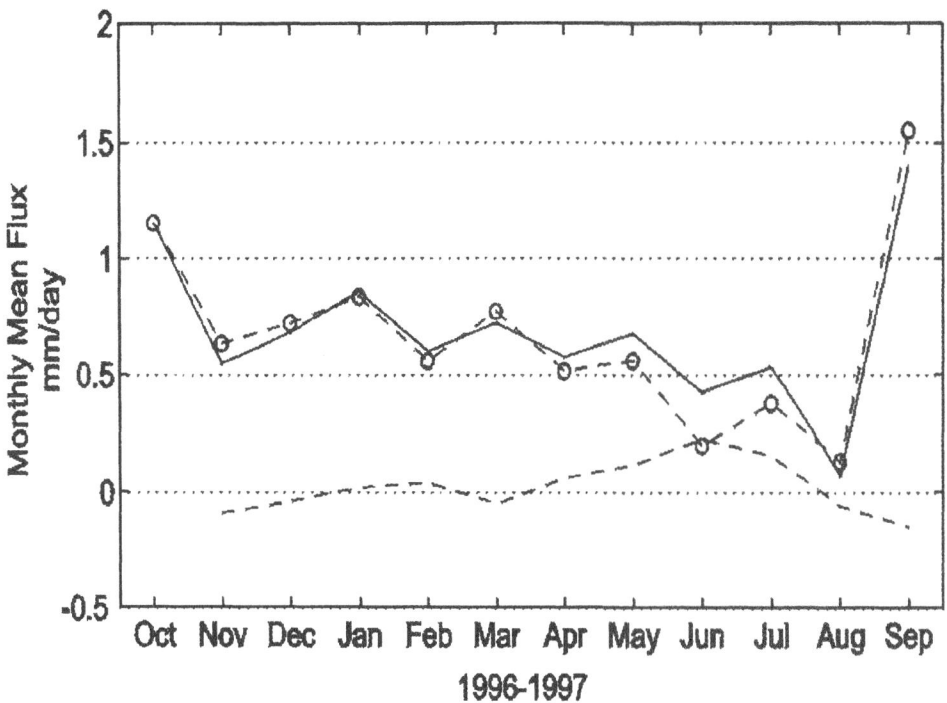

Figure 4: The annual cycle of several of the critical atmospheric water cycle parameters that characterize the Mackenzie River basin for 1995/96. Shown are the monthly means of atmospheric flux convergence (solid line), local rate of change of vertically integrated water vapour (dashed line), and estimated 'precipitation minus evaporation' (dashed line with open circle). Figure courtesy of Geoff Strong.

On the surface, water cycle issues are linked with many factors including the forms of vegetation, the presence or absence of permafrost, and the occurrence of lakes, wetlands and ponded water. The range of vegetation spans from evaporation-efficient tundra [16] to evaporation-inefficient boreal forest [17]. Permafrost underlies about half the basin [18] and this limits the penetration depth of water and so makes a major impact on the basin's water cycle. There are also many lakes and wetlands of varying sizes within the basin and ponded water often occurs; these can represent major sources of moisture through evaporation. Periods of strong evaporation can be especially strong in the late summer or autumn when air temperatures may be much lower than those of the surface water.

It should be noted that the Mackenzie basin is in general a rather inefficient region for converting the available water into discharge. As shown by the schematic diagram of Figure 5a, there are many factors contributing to its low efficiency. Key amongst these is the difficulty for moisture to pass from the ready moisture supply of the Pacific Ocean into the basin. Much of the remaining moisture in fact continues over the basin without falling out as precipitation [19]. The fact that the basin is "open" to the north means that cold conditions readily occur and this is associated with the preferential production of snow. Snow represents a further impediment to the water cycle since it is more able to sublimate before reaching the surface than is the case for rain [20]. As well, much of the moisture evaporated from the surface can be advected out of the basin before it is able to form precipitation [21]. This arises in part because of the topography which places no substantial barrier on the eastern edge of the basin and it is also linked in part with the spatial distribution of the vegetation over the basin. Much of the efficiently evaporating vegetation (typically associated with tundra) occurs on the eastern or north-eastern flank of the basin; it is just a couple of hundred kilometers or so before the moisture from this region is advected completely out of the basin within the prevailing westerly winds.

In contrast, one can imagine a basin with its topography on the easterly side (Figure 5b). Such a basin would be much more likely to capture and to maintain moisture than is the case for the Mackenzie basin. Upslope conditions for the initial advected moisture occur within the basin, and evaporated moisture should be largely captured. In addition, the fact that the basin is "open" on the west to the ocean means that temperatures in general will be warmer and snow will not be as prevalent.

4.   Changes in the Climate System

4.1 TEMPERATURE AND WATER CYCLE FIELDS

Over the last few decades, the Mackenzie basin has been experiencing some of the greatest warming anywhere in the world [6]. This is occurring in particular during the cold season and it is characterized by decadal warming rates of almost 1.5°C.

One though can also look at the temperature trends over an extended period at finer time scales. Figure 6 shows the average temperature anomaly over the Mackenzie Basin for each month over the period 1950-1994. It is apparent that the warming is concentrated within the cold season and this began in the mid-70s; much of the summer is characterized by no significant change. As well, it is evident that there has been an increase in the magnitude of the temperature fluctuations during the cold season. Cold season monthly anomalies have exceeded 10°C over the last couple of decades, but some of the coldest winter periods have also occurred recently. In fact, these cold season fluctuations are believed to be among the largest observed anywhere in the Northern Hemisphere.

If one considers the individual sub-basins of the Mackenzie, there are additional features to the warming. For northern regions of the Mackenzie basin, there has in fact been some warming during the summer period as well. The magnitude of this trend is a maximum of about 0.5°C per decade and the region experiencing this change is very close to the Beaufort Sea.

Some general characteristics of the associated water cycle of the Mackenzie basin have been determined. The integrated amount of water vapour over the basin is typically about 9.8 mm, and there has been a slight increasing trend since about 1965 [7]. Annual mean cloud fraction is about 60%, and this has changed little over the last few decades [7]. Annual precipitation varies by about 25% between years and there is little apparent trend, although this may be masked by the considerable variability [8]. There is a slight trend towards lower values of snowcover over the last couple of decades [22]. Discharge also shows a slight decreasing trend over the period of record. Taken collectively, these features of the water cycle exhibit a relatively stable climate system, although there have been some rather recent systematic trends.

A)

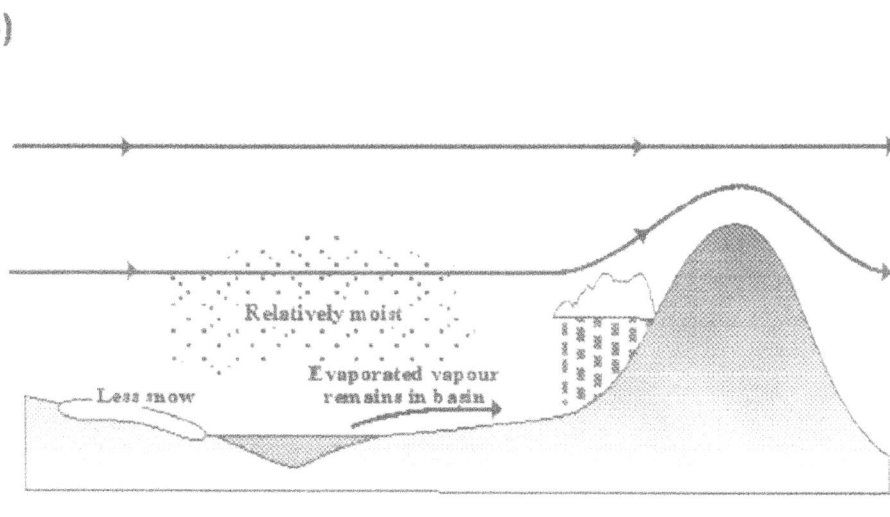

B)

Figure 5: *Simple schematic diagram illustrating some of the features expected to differ between climate systems having differences in terrain arrangement but situated at similar northern latitudes.*

4.2 WATER CYCLE DIFFERENCES

The large fluctuations that occur in annual surface air temperature of the basin can be used as the basis for carrying out simple analyses of associated differences in some characteristics of the water cycle. Here, averages from the 10 warmest and 10 coolest years since 1950 with appropriate data have been compared, and some tentative explanations have been advanced, in order to carry out one such analysis.

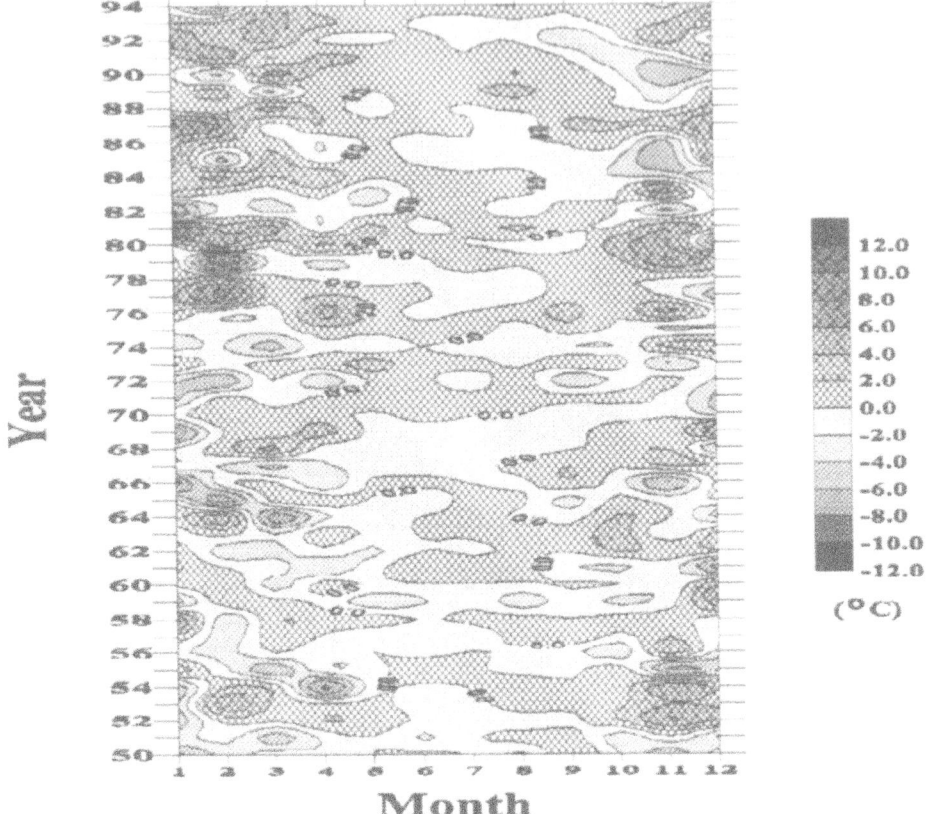

Figure 6: The basin averaged temperature anomaly for each month for the period 1950-1994 over the Mackenzie basin. Anomalies were calculated by subtracting the long-term monthly average from the average for each month.

The differences in several components of the water cycle between the warmest and coolest years are shown schematically in Figure 7. The amount of integrated water vapour increases with surface temperature; this must be largely linked with enhanced advection into the region. In response to the greater amount of water vapour, cloud fraction is increased. However, the amount of precipitation appears to be almost constant, although there is considerable uncertainty in the magnitude of this parameter. The explanations for the little change in precipitation are not clear but one contributing factor may be related to the enhanced vertical descent of air that occurs during warm winters; this acts to maintain dry sub-cloud conditions to allow for more loss of precipitation before it reaches the surface [9, 23]. Snowcover fraction is also reduced with annual surface temperature [22], as expected. The number of blowing snow periods are reduced; this may in part be due to calmer conditions and more frequent freezing rain episodes. Discharge is in general lower in association with warmer temperatures; this may be linked with increased evaporation.

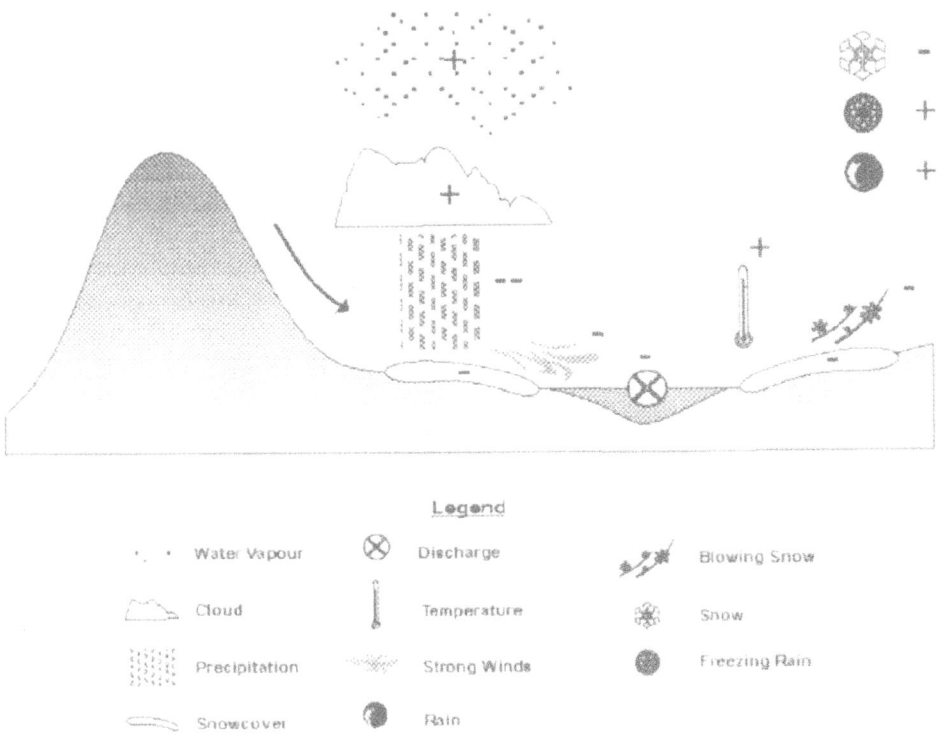

Figure 7: Simple schematic diagram of annual water cycle differences associated with a comparison between cool and warm years. Water cycle variables include temperature, water vapour, cloud fraction, precipitation, snowcover fraction, discharge, and occurrences of blowing snow, precipitation type (snow, freezing precipitation and rain), and strong near-surface winds (> 14 m/s). A plus sign indicates an increase; a minus sign, a decrease; and a dashed line, little change. The black 'atmospheric' arrow indicates the relative difference of cold season vertical motion between cooler and warmer periods during the winter [9].

Some of these results can also be interpreted in terms of the basin's atmospheric recycling rate. This parameter can simply be considered as the amount of precipitation divided by the columnar total amount of water vapour in the atmosphere. On an annual basis, there appears to be a reduction in atmospheric recycling rate. This limited analysis therefore implies that the basin in general becomes less efficient at converting water vapour to precipitation and discharge as surface temperature increases.

Increasing temperatures are expected to be linked with an increasing water cycle [24]. The results of the simple analysis described above, on a regional scale, do not completely support this contention. Although water vapour increases with surface air temperature, the atmospheric recycling rate does not. Other factors therefore come into play on at least the regional scale that may mitigate the full enhancement of the water cycle.

It is certainly recognized, however, that much more work is required to fully explain even these limited results. For example, differences in the annual attributes of the water cycle for similar surface temperatures need to be understood. Future analyses must also more fully address the considerable uncertainty in the values of some of the parameters.

5. Relations to Large Scale Anomalies

Although the main attention of this chapter is the actual water cycle and climate of the Mackenzie basin, it must be realized that the input of moisture to this region is closely connected with large scale atmospheric circulation changes.

One illustration of this is associated with the intensity of the winter-time Gulf of Alaska low pressure system. The intensity of this circulation became stronger and more variable in the 1970s and this situation has persisted in general [25]. Over the Mackenzie basin, this intensification has been linked with warmer and more variable cold season temperatures [9]. When more intense, there is more advection of moisture and heat into the basin. This is directly linked with warmer conditions but it also triggers several other processes that tend to amplify this forcing. One cannot therefore study or predict the climate and water cycle of the Mackenzie basin without properly addressing factors occurring upstream over the Gulf of Alaska.

One of course has to realize that factors occurring over even larger scales affect the Mackenzie basin and the entire Arctic. For example, variations in discharge from the Mackenzie basin are strongly affected by the north Pacific Ocean but they are also linked to Atlantic features [26]. In addition, the basin is affected by El Nino and La Nina events with the mean monthly discharge often being higher during La Nina years [7], and the southern portion of the basin being especially subject to increases in its temperature during El Nino events [27].

6. Global and Regional Climate Model Representations

Climate models have problems simulating the climate of such a basin. In the simulation of present climate, these models are able to account for some of the cold season warming, although they are incapable of accounting for the detailed nature of the warming. The climate models also have a tendency to overpredict the amount of precipitation, and they often produce too little runoff which implies that they also have too much surface evaporation [28]. It is critical that this situation be improved. Unless the critical atmospheric and hydrological processes are being handled properly, one cannot have confidence in climate model predictions of future conditions.

The present Canadian climate model is an example of one that has not been able to adequately replicate some of the critical water cycle fields [21]. Even if the Canadian regional climate model is used over north-western Canada but forced by climate model fields, the over-prediction remains. It has subsequently been shown however that the regional climate model when forced by operational model analyses produced more realistic precipitation over the Mackenzie basin. The main difference in large scale forcing arose from a misrepresentation of the Gulf of Alaska low pressure system; this was too strong in the climate model simulations and this led to the advection of too much moisture into the Mackenzie basin. Climate models must produce the proper large scale forcing fields for successfully modelling the climate of the Mackenzie basin.

Even if the large scale forcing is adequate, there are still many modelling difficulties to overcome in association with processes occurring within the basin itself, as well as over other regions of the Arctic. These include, for example, problems linked with producing accurate seasonal hydrographs. This is due in part to problems linked with precipitation but they are also linked with difficulties associated with the proper routing of surface water and in correctly accounting for the fate of surface water excess.

7. Concluding Remarks

Some aspects of the water cycle of a major river basin draining into the Arctic Ocean have been described and discussed. The processes affecting this cycle are complex but are slowly being appreciated. Nevertheless, there is a long way to go before there is an acceptable ability of global and regional climate models to account for such factors to the appropriate degree of capability.

It is quite possible that many of the issues raised in this chapter can apply to other areas as well. For the atmospheric side in particular, it is quite probable that many of the issues relate to other northern basis although there will be some degree of variation to account for local circumstances. One of the main defining distinctions between basins will be their topographic features; this leads directly or indirectly to many processes in the atmosphere and at the surface that control much of a basin's water cycle.

It should be noted of course that the datasets used in the study have their limitations. Although the ensuing results are believed to be correct to first order, it may be that some systematic differences are affecting results significantly. It is evident that more validation of the data is required in the future so that one has more confidence in the values of the water cycle components.

Nevertheless, this study has illustrated many features of the Mackenzie basin's water cycle that are fundamental to understanding the region's climate and its variations. The fresh water outflow from this basin is equivalent to about 170 mm/year (280 km^3/year) over the basin and it can vary by about 25%. High discharge years are in general linked with a more efficient basin climate system at converting available water vapour into discharge; such years are also typically linked with variations in the atmospheric circulation over the north Pacific Ocean.

Some of the processes and feedbacks operating over the Mackenzie basin may be occurring over other regions of the Arctic and so may also need to be considered before there is a full appreciation of the water cycle of the entire domain. For example, in terms of topography with a major mountain barrier on its upwind side, the Mackenzie basin is somewhat similar to that of the Ob River of Siberia; whereas it is different in this regard from the Lena River basin that has high terrain on its eastern flank. Snow is the common form of precipitation for all areas of the Arctic and so there may well be many implications of the Mackenzie studies on the production of snow and the evolution of the ensuing snowpack. Such similarities and differences in processes and conditions will very likely determine the responses of other areas to increasing temperatures that may occur in the future; these in turn will alter the fresh water inflow to the Arctic Ocean.

In summary, this chapter has examined some aspects of the overall varying climate of the Mackenzie basin and some of the processes that affect this system and its fresh water discharge into the Arctic Ocean. The examination of this basin's water cycle has led to important insights into the fresh water inflow into the Arctic Ocean.

8. Future Research Priorities

On the basis of this work, a number of recommendations can be made to more fully address the overall issue of the fresh water budget of the Arctic Ocean. These recommendations include:

1. The quantification and modeling of the basin's water cycle need to be further pursued through the addition of more information that includes snow water content and sub-surface water storage as well as other parameters over the wide diversity of conditions that has been occurring over the last several decades.
2. The inter-connections between the water cycles of the various basins emptying into the Arctic Ocean need to be well documented so that their collective fresh water impact over a range of time periods can be better appreciated.
3. Much of the insight from this study should be adapted in particular to other river basins emptying into the Arctic Ocean, but the overall approach should be applied in general to the entire Arctic domain to assess its water cycle.

Acknowledgments

The author would like to acknowledge that many of the results presented in this chapter were obtained through the Mackenzie GEWEX Study (MAGS), and he would like to thank his many colleagues within MAGS for their assistance with special thanks to Jason Burford, Bob Crawford, Paul Louie, Murray Mackay, Geoff Strong and Anne Walker. Comments by Terry Prowse are greatly appreciated.

9. References

1. Aagard, K. and E.C. Carmack (1989) The role of sea ice and other fresh water in the Arctic circulation. *J. Geoph. Res.* **94**, 14,485-14,498.
2. Woods, J.D. (1984) The upper ocean and air-sea interaction in global climate, in J.T. Hougton (ed.), *The Global Climate*, Cambridge University Press, Cambridge, 141-187.
3. Grabs, W. (this volume) Discharge observation networks in Arctic Regions: computations of the river runoff into the Arctic Ocean, its seasonality and variability.
4. Stewart, R.E., H.G. Leighton, P. Marsh, G.W.K. Moore, H. Ritchie, W.R. Rouse, E.D. Soulis, G.S. Strong, R.W. Crawford, and B. Kochtubajda (1998) The Mackenzie GEWEX Study: the water and energy cycles of a major North American river basin. *Bull. Amer. Meteor. Soc.* **79**, 2665-2683.
5. Chahine, M.T. (1992) The hydrological cycle and its influence on climate. *Nature* **359**, 373-380.
6. Shabbar, A., K. Higuchi, W. Skinner and J.L. Knox (1997) The association between the BWA index and winter surface temperature variability over eastern Canada and west Greenland. *Int. J. Climate* **17**, 1195-1210.
7. Stewart, R.E., J.E. Burford and R.W. Crawford (1999) On the characteristics of the water cycle of the Mackenzie River basin. *Cont. Atmos. Phys.* (Accepted)
8. Mekis, E. and W.D. Hogg (1999) Rehabilitation and analysis of Canadian daily precipitation time series. Atmos.-Ocean, **37**, 53-85.
9. Cao, Z., R.E. Stewart and W. Hogg (1999) Extreme winter warming events over the Mackenzie basin: dynamic and thermodynamic contributions. *J. Meteor. Soc. Japan* (Accepted)

10. Cohen, S.J. (1997) Mackenzie Basin Impact Study: final report and summary of results. Atmospheric Environment Service, 372 pp
    Goodison, B.E. (1978) Accuracy of Canadian snow gauge measurements. *J. Appl. Meteor.* **17**, 1542-48.
11. Andres, D. and P. F. Doyle (1984) Analysis of breakup and ice jams on the Athabasca River at Fort McMurray, Alberta. *Can. J. Civil Engin.* **11**, 444-458.
12. Prowse, T. D. (1986) Ice jam characteristics, Liard-Mackenzie rivers confluence. *Can. J. Civil Engin.* **13**, 653-665.
13. Pomeroy, J. W., P. Marsh, and D.M. Gray (1997) Application of a distributed blowing snow model to the Arctic. *Hydro. Proc.* **11**, 1451-1464.
14. Walsh, J.E., X. Zhou and M.C. Serreze (1994) Atmospheric contribution to hydrologic variations in the Arctic. *Atmos.-Ocean* **32**, 733-755.
15. Rouse, W.R. (1998) A water balance model for a subarctic sedge fen and its application to climatic change. *Clim. Change* **38**, 207-234.
16. Baldochi, D.D., C.A. Vogel and B. Hall (1997) Seasonal variation of energy and water vapor exchange rates above and below a boreal jack pine forest. *J. Geoph. Res.* **102**, 28939-28952.
17. Church, M. (1974) Hydrology and permafrost with reference to northern North America. Proceedings, Workshop Seminar on Permafrost Hydrology. Canada National Committee for the International Hydrological Decade, 7-20.
18. Szeto, K.K., R.E. Stewart and J.M. Hanesiak (1997) High latitude cold season frontal cloud systems and their precipitation efficiency. *Tellus* **49**, 439-454.
19. Stewart, R.E. (1996) Extratropical cyclones: their mesoscale structure, precipitation and role in the transport of water, in E. Raschke (ed.) *Remote Sensing of Processes Governing Energy and Water Cycles in the Climate System*, NATO-ASI, I45, Springer Verlag, pp. 129-148.
20. Mackay, M. D., R.E. Stewart and G. Bergeron (1998) Downscaling the hydrological cycle in the Mackenzie Basin with the Canadian Regional Climate Model. *Atmos.-Ocean* **36**, 179-211.
21. Brown, R.D. and P. Cote (1992) Interannual variability of landfast ice thickness in the Canadian high Arctic. *Arctic*, **45**, 273-284.
22. Burford, J.E. and R.E. Stewart (1998) The sublimation of falling snow over the Mackenzie River Basin. *Atmos. Res.* **49**, 289-314.
23. Intergovernmental Panel of Climate Change (1996) Climate Change: The Science of Climate Change. Eds. J.T. Houghton, F.G. Meira Filho, B.A. Callander, N. Harris, A. Kattenberg and K. Maskell, Cambridge Univ. Press, Cambridge, U.K., 572 pp.
24. Intergovernmental Panel of Climate Change (1990) Climate Change: The IPCC Scientific Assessment. Eds. J.T. Houghton, G.J. Jenkins and J.J. Ephraums, Cambridge Univ. Press, Cambridge, U.K., 364 pp.
25. Bjornsson, H., L.A. Mysak and R.D. Brown (1995) On the interannual variability of precipitation and runoff in the Mackenzie drainage basin. *Clim. Dyn.* **12**, 67-76.
26. Shabbar, A., B. Bonsal and M. Khandekar (1997) Canadian precipitation patterns associated with the southern oscillation. *J. Clim.* **10**, 3016-3027.
27. Walsh, J.E., V. Kattsov, D. Portis and V. Meleshko (1998) Arctic precipitation and evaporation: model results and observational estimates. *J. Clim.* **11**, 72-87.

ARCTIC ESTUARIES AND ICE:
A POSITIVE–NEGATIVE ESTUARINE COUPLE

R.W. MACDONALD
Institute of Ocean Sciences
P.O. Box 6000
Sidney, British Columbia, V8L 4B2
Canada

1. Introduction

Estuaries have been defined in a variety of ways [1]. From the perspective of fresh water and salt balance, *"An estuary is a semi-enclosed coastal body of water which has a free connexion with the open sea and within which sea water is measurably diluted with fresh water derived from land drainage"* [2]. Within the broadest interpretation of this definition, the whole Arctic Ocean can be considered an estuary with outflows at Fram Strait and through the Canadian Archipelago. The total freshwater inflow of over 3300 km^3 yr^{-1} [3, 4] plus an additional 900 km^3 yr^{-1} of net precipitation produces a grand Arctic Ocean estuary exceeded only by the Amazon estuary (6300 km^3 yr^{-1}). A more traditional view of an estuary is that it comprises a near-shore, low-salinity zone surrounding the mouth of a river. However, a definition that includes only river runoff as it meets the sea is wholly inadequate to describe arctic estuaries in the context of global change and freshwater budget. As will be shown here, arctic estuaries must be viewed at the scale of the shelf to understand their role in the Arctic Ocean's freshwater balance and circulation and how estuaries will respond to change through feedback. Similarly, but on a larger scale, the Arctic Ocean estuary at Fram Strait influences freshwater balance and thermohaline circulation of the global ocean [3].

Rivers impinge on the Arctic Ocean either directly in deltas exposed to the sea (e.g., the Lena and Mackenzie Rivers) or in more sheltered embayments or channels (e.g., the Ob' and Yenisey Rivers). Clearly, the distinguishing feature of arctic estuaries is that they are covered by ice for much of the year. An ice cover that grows in winter and melts in summer has a

E.L. Lewis et al. (eds.), The Freshwater Budget of the Arctic Ocean, 383-407.

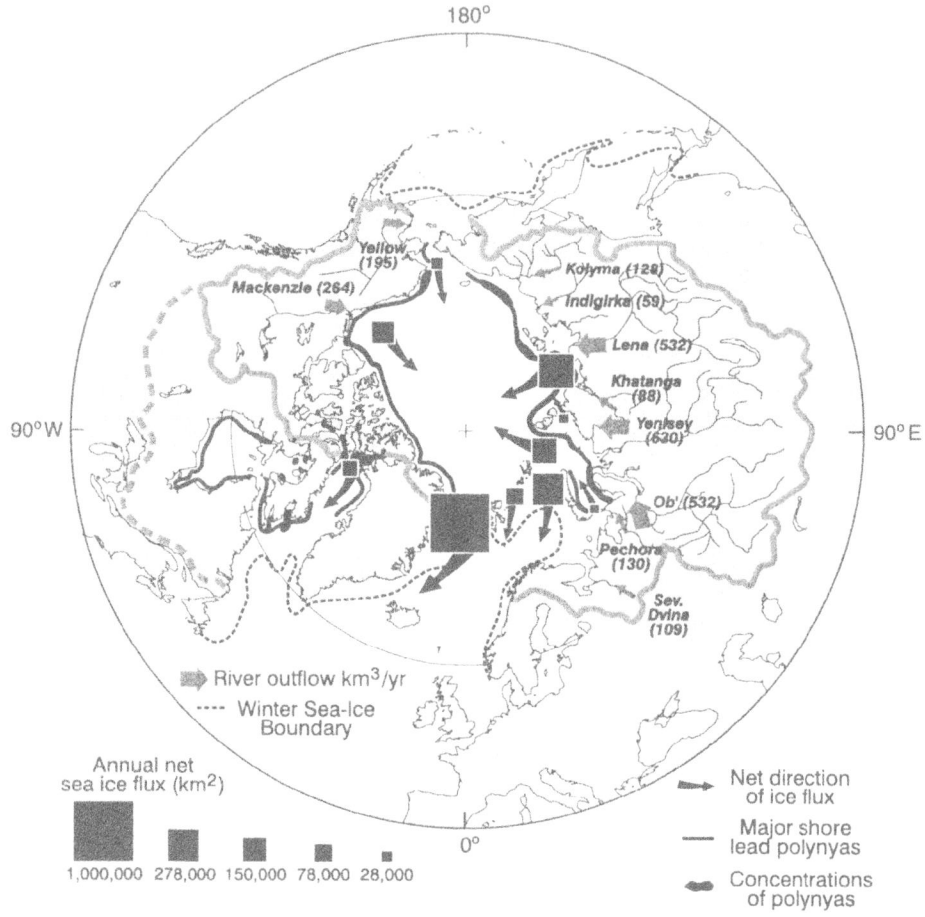

Figure 1. Classifications of estuaries (modified from [13]).

direct, physical effect on the spreading of freshwater runoff into the ocean [5–7] but more importantly ice provides an independent source and sink for the shelf's freshwater budget. A crucial geographical feature of arctic estuaries is that during winter a large flaw lead over the mid shelf lies between the point of entry of runoff and the interior ocean (Figure 1 and see [8–12]). This flaw lead is a region of recurring open water where ice diverges and is exported from the shelf (Figure 1) and where, as a consequence, ice can grow rapidly to produce enhanced amounts of brine.

The word "estuary" is almost always assumed to mean "positive estuary" where freshwater runoff exceeds local evaporation. Although negative freshwater balance (evaporation exceeds runoff) affects circulation in regions as large as the Mediterranean and Red Seas, "negative estuaries" are usually dismissed as unimportant, local curiosities such as coastal lagoons [1]. In the Arctic, regions of open water in winter form a class of negative estuary because exported sea ice leaves behind most of the salt in the seawater from which it was formed thereby acting in a fashion similar to evaporation. However, there is a difference between evaporation and freezing in that the solid phase (sea ice) may remain in place to melt during summer and re-introduce the fresh water withdrawn the previous winter. It is the close juxtaposition of positive and negative estuaries on arctic shelves and the coupling of their distinct seasonal cycles that creates a unique potential for sensitivity to global change, particularly in the context of the salt and freshwater cycle. In this regard, both the Lena and Mackenzie River estuaries offer important study locations because of the proximity of freshwater discharge to a recurring flaw lead where significant amounts of ice are exported (Figure 1).

Our view of positive estuaries derives almost entirely from temperate and tropical locations where they are categorized as salt wedges, partly mixed and well mixed according to the energy of mixing (Figure 2a–c, [13]). Although this scheme could be applied to arctic estuaries during the ice-free season, it does not at all describe how frozen estuaries work [14]. First, arctic rivers tend to have an extremely seasonal discharge (Figure 3) with winter flows contributing only about 5–10% of the total annual flow [15]. In the case of even the largest rivers, over 30% of the annual inflow may come in a single month (e.g., the Lena and Yenisey). The many small rivers whose drainage basins lie wholly within the Arctic and whose inflows collectively contribute a large amount of fresh water to the Arctic Ocean [3, 4] can be expected to freeze almost completely in winter (e.g., see Macdonald *et al.* [14]). Second, air temperatures are below freezing for over half the year (Figure 4) with the result that ice grows on rivers, estuaries and the sea. Fresh water flowing out under an ice cover interacts differently with the sea than does fresh water entering an open, exposed sea surface [7, 16]. Third, ice provides a separate, broadly distributed source of fresh water when it melts in summer and a "sink" for fresh water when it forms in winter. Undoubtedly, the arctic estuary during winter does not fit any standard scheme proposed in Figure 2a, b or c, and a separate diagram (Figure 2d), the subject of this paper, is required.

Arctic estuaries cannot be considered important long-term storage locations for fresh water — they quickly transmit runoff from the river to the sea and respond at a sub-seasonal time scale to freshwater input. There are perhaps a few locations where restricted exchange allows freshwater storage for up to a decade [14] but these include none of the major river estuaries.

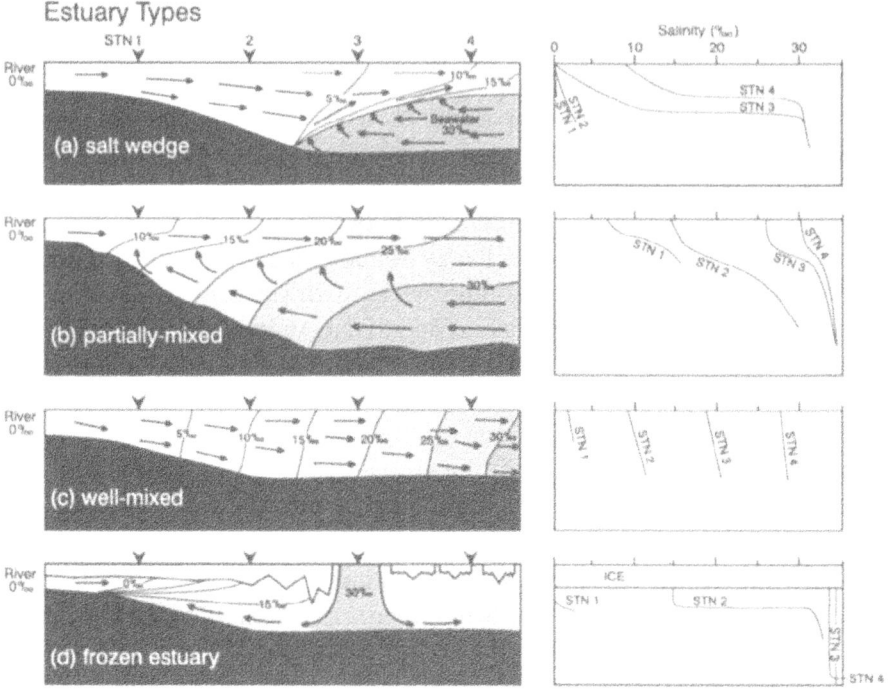

Figure 2. The three standard estuary types (a) salt wedge, (b) partly mixed, (c) well mixed. These estuarine types may be applied to the Arctic during open-water season. However, they do not describe the arctic winter (d) during which river plumes spread under a growing sheet of land-fast ice (positive estuary) while, at the same time, brine is added to the water in the shear zone beyond the land-fast ice from enhanced sea-ice formation (negative estuary). For the Mackenzie shelf, the vertical scale is about 20–40 m.

Characteristic freshwater residence times for the shelves are estimated to be from less than 1 year to as much as 3 to 4 years (Table 1) and for the surface layers of the interior basins perhaps 10–14 years [17, 18]. Estuaries have residence times at or below the short end of these scales depending on how large the estuary is considered to be.

In this paper I will examine the processing of fresh water on arctic shelves, particularly in winter, and the effect of runoff (the positive estuary) on the potential of shelves to separate salt and fresh water through ice formation (the negative estuary). This latter process is central to understanding how the Arctic Ocean might respond in a non-trivial way to global change — i.e., other than a simple change in river inflow. To understand how runoff can impact the negative estuary requires data that (1) extend over a complete year, (2) include at least two "conservative" tracers to distinguish between runoff and ice formation/melting (tradition-

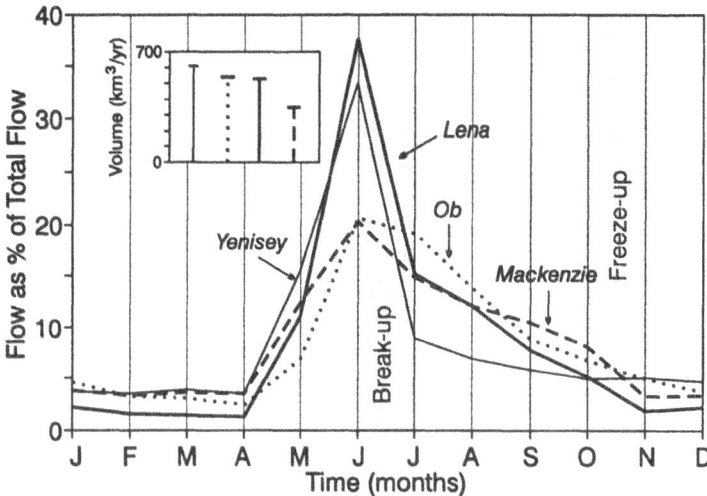

Figure 3. The average monthly flow of the four largest rivers in the Arctic normalized to annual flow (modified from Lawford [30]). The inset shows a histogram of the total volumetric flow for each of the rivers and the approximate dates of freeze-up and ice melt are given as vertical bands.

ally salinity and $\delta^{18}O$ have been used), and (3) include measurements in both water and ice media. That I know of, such oceanographic data have been collected only from the Mackenzie shelf [5, 8, 14]. I will, therefore, rely heavily on data from the Mackenzie shelf to illustrate the interaction between the runoff and ice cycles in an arctic coastal zone but will refer to literature from other arctic shelves where parallels can be drawn. Clearly, there are important differences between the Mackenzie shelf (Figure 5) and other arctic shelves (cf. Table 1); in particular, the volume of freshwater inflow varies from shelf to shelf as do seasonal hydrographs (Figure 3) and freshwater residence times. Nevertheless, the seasons in the western Kara and eastern Barents Seas, as illustrated by satellite imagery [19], appear much the same as those in the Mackenzie shelf, at least in terms of the sea ice and river cycles. Indeed, the Arctic Ocean is unique among world oceans in that the seasonal cycle of runoff is synchronous around the margin, as is the seasonal ice cycle. The Mackenzie shelf is the most estuarine of all arctic shelves with an annual yield (freshwater runoff divided by shelf area) of over 5 m (Table 1) but the Mackenzie shelf also has one of the shortest residence times, and it is a net exporter of ice.

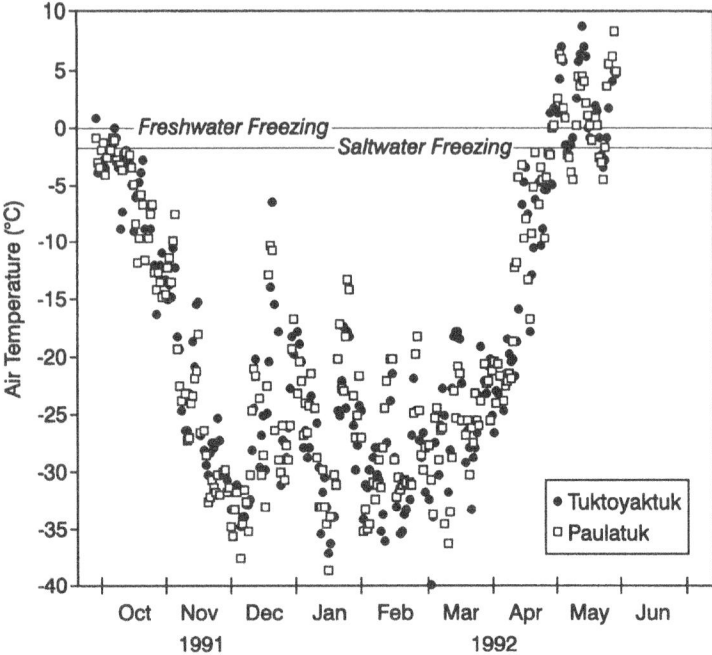

Figure 4. Air temperature records for Tuktoyaktuk and Paulatuk for the period from September 19, 1991 to May 31, 1992 (data from Environment Canada).

TABLE 1. Characteristics of arctic shelves

Shelf region	Major rivers	Area (km^2 ×10^3)	Annual inflow (km^3)	Yield (m)	Residence time (yr)	Total ice export (km^3 yr^{-1})
Kara	Ob', Yenesey	883	1133	1.3	2.5	340
Laptev	Lena, Kotuy	663	767	1.2	3.5 ± 2	480–660
East Siberian	Idigirgka, Kolyma	889	213	0.24	3.5 ± 2	Import
Chukchi	–	587	78	0.13	0.2 – 1.2	Import
Beaufort	Mackenzie	64	330	5.2	0.5 – 1	450
Upper Arctic Ocean	–	9500	3300	0.35	10	2600

Data collated from [3, 9, 12, 14, 18, 20, 21, 44, 45, 46, 47, 48].

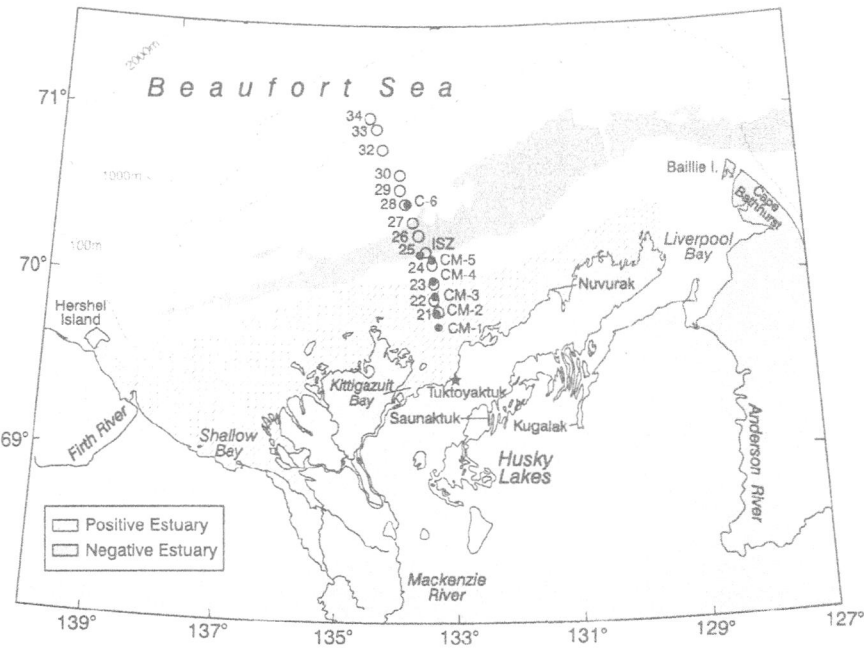

Figure 5. The Mackenzie shelf showing the disposition of freshwater runoff under the ice at the end of winter (the positive estuary) and the flaw lead where divergence during winter enhances ice and brine production (the negative estuary). Open circles (21–34) and closed circles (CM-1 to C-6) refer to stations shown in cross-shelf sections.

2. The Use of Salinity and $\delta^{18}O$ Measurements to Determine Freshwater Sources

The method of using paired salinity and $\delta^{18}O$ measurements to determine the origins of fresh water has been well documented in Arctic Ocean literature [17, 18] and, in particular, for the Mackenzie shelf [5, 8, 20]. Therefore, only a brief outline is provided here to explain how the amounts of runoff and melt water have been derived.

Salinity–$\delta^{18}O$ data from the Mackenzie shelf invariably exhibit the pattern shown in Figure 6. Seawater of some nominal salinity — $\delta^{18}O$ composition provides the saline end of a mixing line (right hand end of the bottom line in Figure 6). Runoff, which is isotopically much lighter than seawater ($-20‰$) and of low salinity (0), provides the other end member. Into this "line" dominated by mixing, the influence of ice is inserted as deviations either to the right or left of the line. When ice forms from sea water, it injects salt into the sea pushing points to the right of the mixing

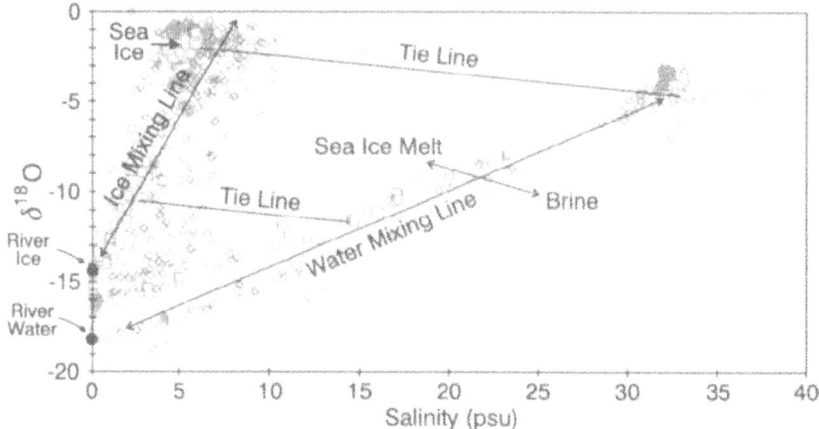

Figure 6. The distribution of $\delta^{18}O$ versus salinity in water and in ice from the Mackenzie estuary and shelf for 1991 (closed diamonds) and 1992 (open squares). The mixing
of isotopically light Mackenzie River water with seawater is seen in the bottom line.
Displacements to the right of the line are produced by the addition of brine when ice
freezes in winter or to the left when sea ice melts in summer. The steeper line to the left
shows that ice also contains a variable mixture of river water and seawater. The tie line
between ice and water has a small negative slope due to isotopic fractionation during ice
formation (ice is heavier than the water it grows from by about 2.6‰). The composition
of any water sample can be expressed algebraically as a mixture of seawater, river water
and sea ice melt.

line — when sea ice melts it injects fresh water (salinity of about 4) with
a relatively heavy $\delta^{18}O$ composition pushing points to the left of the line.
The upper line in Figure 6 shows the relation between salinity and $\delta^{18}O$ in
ice. Most of the noise in this ice line is generated by salinity, which is not
very conservative in ice. In contrast, the $\delta^{18}O$ composition of ice faithfully
records the composition of the water from which it has grown but with a
small fractionation (ice is about 2.6‰ heavier than the water). The wide
range in $\delta^{18}O$ in the "ice" line reflects that estuarine and shelf ice is actually a frozen mixture of runoff and seawater, but the added offset due to
isotopic fractionation causes tie lines between the water and the ice to have
a small negative slope (Figure 6). River ice does not have the same ability
as sea ice to induce convection by excluding brine, as river ice rejects no
salt. Here, we define "sea ice" as ice grown from the pure saline water end
member and, therefore, all ice in an estuarine system will be viewed as a
two-component mixture of river ice and sea ice (cf. Macdonald *et al.* [5] for
a full discussion).

The simplest, and most common, method of determining the respective
contributions of ice and runoff to the freshwater composition of a given

sample is to model the system as a mixture of three end members of known composition — seawater (SW), runoff (R), and sea-ice melt (SIM) using three simultaneous equations and two tracers ($\delta^{18}O$, salinity). Unlike runoff, which has only positive algebraic solutions, the sea-ice melt component can exhibit positive (melt water) and negative (brine) solutions, both of which are valid and meaningful.

3. The Seasonal Cycle of Arctic Estuaries

To understand what is unique about an estuary that grows an ice cover during a substantial portion of the year requires sub-dividing the year into at least four periods (Figure 7). These periods follow naturally from the river's hydrograph (Figure 3), which forces the positive estuary, and the seasonal air temperature cycle (Figure 4) which controls ice formation and, therefore, forces the negative estuary. I start arbitrarily at freeze-up and use the Mackenzie estuary (Figure 5) as a model. Nevertheless, the progress of the seasons listed here will apply equally to other arctic estuaries with some variation in timing and intensity between locations and between years at a given location (for example, see [9, 10, 12, 19, 21]).

3.1. FREEZE-UP

In mid September, a few weeks before freeze-up, ice melt from summer is broadly distributed across the Mackenzie shelf, contributing between 0 to 2 m standing stock of fresh water (Figure 8a). The amount of ice melt is somewhat less than the amount of ice usually present at the end of winter (2 m) because estuarine circulation has carried some ice melt away and winds have blown both ice and ice melt off the shelf [22]. It should be noted, however, that the Mackenzie shelf tends to be a net exporter of ice [23, 24] which is by no means true of all shelves (e.g., see Colony and Thorndike [25]). In addition to ice melt, fresh water from runoff also remains in shelf waters (Figure 8c) contributing an amount depending on the yield (annual volume flow divided by shelf area) and on how much of the season's inflow has escaped the shelf during summer and autumn. For the Mackenzie shelf this works out to perhaps 1–3 m of fresh water which together with the ice melt may account for a total standing stock of 2–4 m of fresh water (Figure 9; September). Storms throughout summer and early autumn mix the ice melt and runoff into the top 10–20 m of the water column and reduce some of the structure although fronts and coastal plumes are still evident [26]. Below 20 m there often remains a remnant of the previous year's polar mixed layer evident as cold saline water containing negative values of ice melt (Figure 8a). Generally for the surface layer, ice melt tends to be a more important freshwater component in the offshore

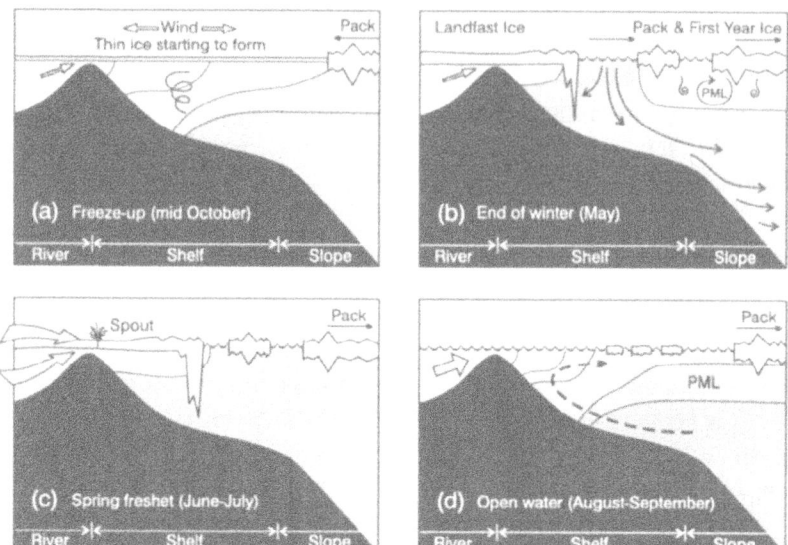

Figure 7. The seasonal cycle of fresh water in the Mackenzie estuary as characterized by inflow and ice cover: (a) freeze-up (low inflow, ice just starting to form over the shelf); (b) the end of winter (low inflow, ice is at its thickest and has essentially stopped growing); (c) spring freshet (high inflow, ice cover remaining intact in the near-shore); and (d) summer (high inflow, most of the ice has melted or been exported from the shelf).

whereas runoff dominates in the near-shore. Most importantly, depending on how ice free the shelf has been during the summer, and the strength, duration and direction of winds, a portion of the fresh water has been forced off the shelf to be replaced by saltier water from the interior ocean. Compare, for example, September, 1986 when the mid shelf contained 3–4 m of freshwater standing stock and September, 1987 when it was less than 2 m (Figure 9). This "pre-conditioning" of the shelf [5, 24] sets the stage for the following winter when the formation of two or more metres of ice withdraws an almost equivalent amount of fresh water. At freeze-up, only a thin layer of ice is present and ice therefore can grow quickly everywhere as temperatures rapidly drop [27, 28]. Later, as the ice thickens, the rate of ice growth slows down except in regions like the flaw lead where divergence presents new open water.

Freeze-up (Figure 7a) commences in early to mid October after air temperatures have dropped below the freezing point of water, the water has cooled to its freezing temperature, and freezing-degree-days start to accumulate (Figure 4). River inflow has by now decreased almost to its low,

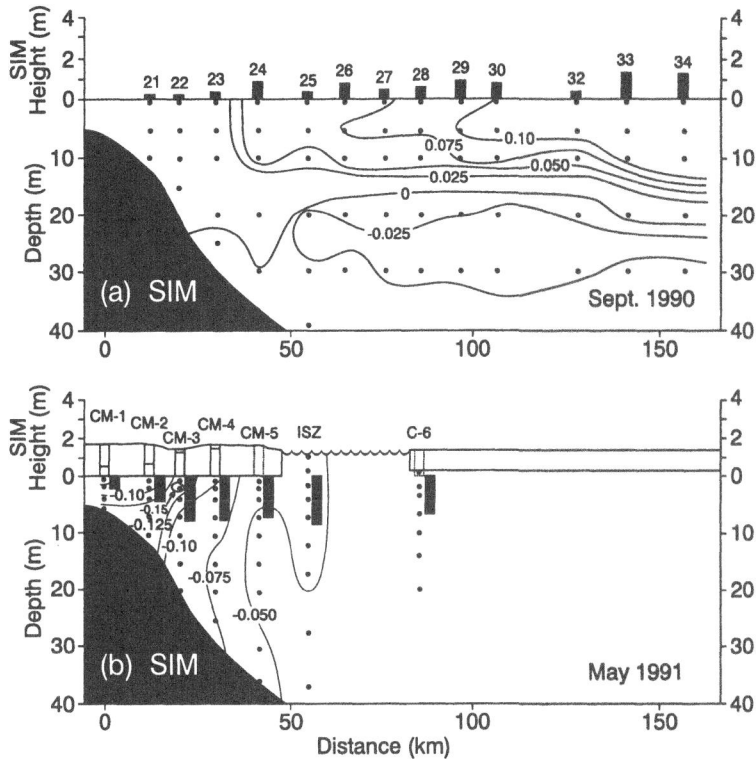

Figure 8. A section across the centre of the Mackenzie shelf (Figure 5) showing disposition of (a) sea ice melt before winter (September, 1990), (b) sea ice melt at the end of winter (May, 1991), (c) runoff (MW - meteoric water) before winter (September, 1990); and (d) runoff at the end of winter (late May, 1991). Isopleths in the water show the fractional composition of fresh water for sea ice melt (a and b) and runoff (c and d) whereas bars show the standing stock of fresh water by source in either water (black bars) or ice (open bars) (adapted from Macdonald *et al.* [5]).

winter values (Figure 3), and goes from entering an open estuary where it can be mixed by winds, to spreading under the near-shore ice in a relatively quiescent environment. Although river inflow in winter is low relative to summer, substantial quantities of runoff may continue to enter the near-shore — over 5000 m^3 s^{-1} in the case of the Lena and Mackenzie Rivers. It is not well known how river and sea meet under the ice, especially in late winter when 2 m of ice covers the estuary. Late in the season, the ice can rest on the bottom in the shallows with the result that tides and storm surges provide important controls on shape and distribution of the flow conduit for water passing between the ice and sediment. Certainly, where more than one river channel leads to the ocean, the ice can modu-

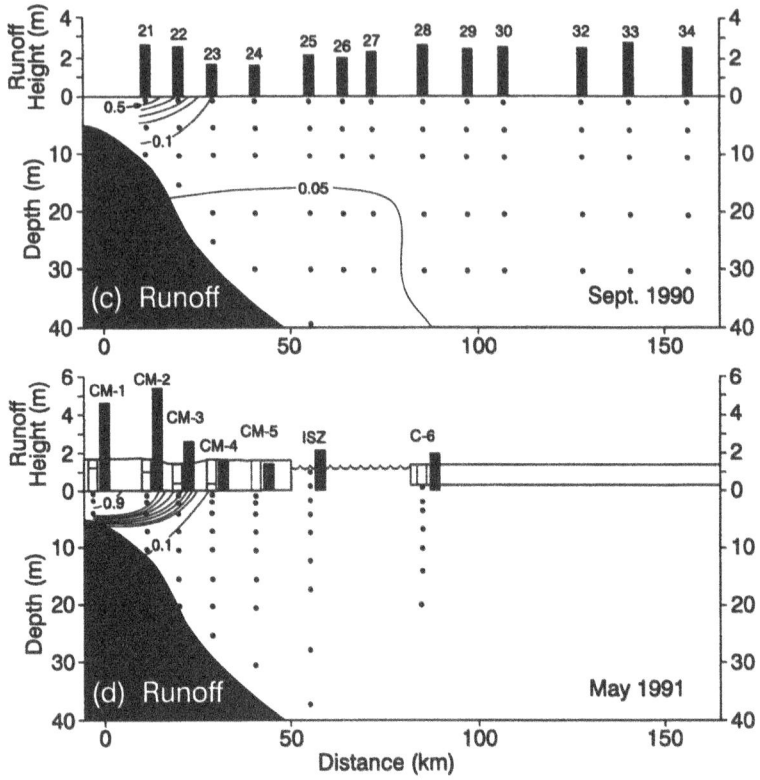

Figure 8. Continued.

late flow between channels based on how channel geometry interacts with a progressively thickening ice cover.

Once the fresh water enters the near-shore, it spreads out under the land-fast ice as an expanding plume (cf. [5, 7, 8]). Using the distribution of $\delta^{18}O$ in land-fast ice at the end of winter, Macdonald *et al.* [5] showed that the Mackenzie plume spreads more rapidly along the coastline (1.3 cm s^{-1}) than away from it (0.2 cm s^{-1}); but in neither case does the plume spread very quickly. At these slow velocities, the plume takes most of the winter to fill the land-fast ice zone out to the rough ice (stamukhi) found at the outer edge at about 20 m depth of water. The disposition of river inflow, land-fast ice, stamukhi at the end of the land-fast ice and flaw lead beyond the stamukhi appear common to all arctic shelves [8, 10, 12]. It may, therefore, be inferred that the scenario described above is duplicated in many places.

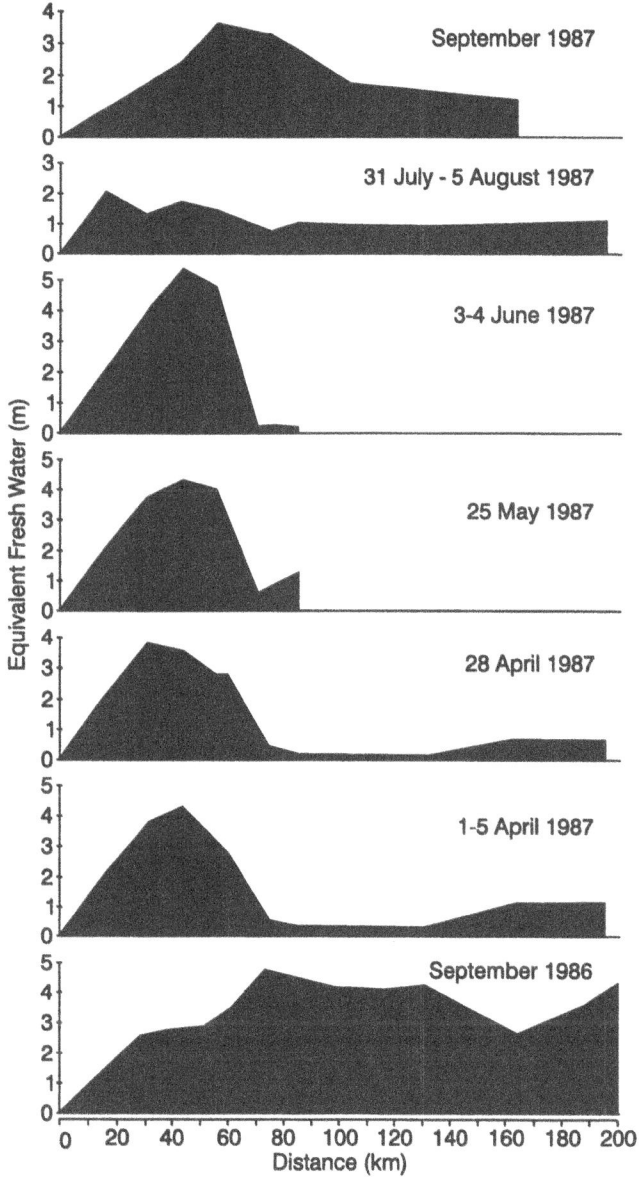

Figure 9. The seasonal cycle of freshwater inventory along a mid shelf section (Figure 5) from September 1986 to September 1987. Standing stocks were calculated using only salinity data (relative to 31.6) and, therefore, combine runoff and sea ice melt. In September 1986, about 3–4 m of fresh water remains over the centre shelf whereas smaller amounts are found in the near-shore due partly to reduced river inflow and partly to limitation by depth. Toward the end of winter (April–May 1987), fresh water has built up under the land-fast ice while at the same time fresh water has been drawn down in the mid shelf. The cycle repeats itself, but in September 1987, mid shelf standing stock was far less than observed in September, 1986.

3.2. END OF WINTER

I define the end of winter as the point at which land-fast ice reaches its maximum thickness and essentially stops growing (Figure 7b). This corresponds approximately to the time when freezing-degree-days stop accumulating, which occurs in about early May (Figure 4), although some lag may occur simply due to the heat capacity of ice cooled well below freezing. At this time, land-fast ice, which extends approximately to the 20 m isobath, is about 2 m thick and more than 2 m of ice has formed in the flaw lead beyond the stamukhi at the end of the land-fast ice although not all of it has remained there. Some of the offshore ice has been piled into ridges and some has been exported [5, 23]. By May, the river's winter inflow has spread out under the land-fast ice now forming a large pool of fresh or brackish water, some of which has been frozen into the ice (Figures 5 and 8). In the case of the Mackenzie estuary, this "floating freshwater lake" contains about 70 km^3 of winter inflow spread over an area of 12,000 km^2. These statistics would rank this "seasonal lake" in the top 20 to 30 lakes of the world by area or volume, respectively.

In the shear zone beyond the stamukhi, ice divergence through winter has enhanced ice production (3 m or more). The salt rejected as brine has helped to mix the surface layer down to 40–50 m, possibly also creating brine plumes that have drained along the bottom to enter the Arctic Ocean halocline or deeper [5, 8, 24, 29]. The mid shelf, therefore, forms a negative estuary through most of the period when air temperatures have been below freezing. This is reflected in the change from positive ice melt in September to negative ice melt in April/May (Figure 8) and in the draw-down in freshwater standing stock (Figure 9).

The rise of air temperature above freezing closes the negative estuary. The ability of ice formation to drive convection and/or produce brine plumes depends upon how much fresh water (buoyancy) has been removed during the autumn pre-conditioning period (Figures 7a and 9). It also critically depends on preventing winter inflow from entering the divergent zone and thereby shutting down convection [8].

3.3. SPRING FRESHET

Breakup commences in the headwaters of the Mackenzie River in late April and progresses northward down the river to the delta where it occurs in late May [30]. Between mid May to June the estuary, therefore, goes rapidly into flood (Figure 3). Because the estuary and the near-shore remain covered with land-fast ice, the river water is forced, often violently, to find away into the near-shore either under the ice or on top of it by over-flooding (Figure 7c). A similar scenario occurs in the estuaries of the Ob' and Yenisey Rivers where land-fast ice remains intact until July [19], well after the onset of freshet (Figure 3). The pulse of turbid water announces the end of winter in the positive estuary and augments the large pool of winter inflow under the land-fast ice. Although the sea ice has not yet begun to melt seriously from solar insolation, latent heat in the river water advances the melting of ice by perhaps as much as two months in the delta and to a lesser degree away from the river mouth [6, 15]. We surmise what happens at this time from the river's hydrology (Figure 3) and remote sensing [6, 19]. However, field data are almost impossible to obtain because both the ice and water are in a such a dynamic state that sampling is hazardous to personnel and moored instruments.

3.4. OPEN-WATER SEASON

As breakup progresses, the land-fast ice disappears and enough heat enters the shelf surface waters, either from the river or the sun's radiation, so that much of the ice melts, leaving its trace as melt water widely distributed across the shelf (Figures 7d and 8a). The river continues to supply a large amount of fresh water (Figure 3) which is evident in plumes [20], fronts [26] and a strongly stratified surface layer (5–10 m). At this time, the Mackenzie estuary looks much like any estuary of a large river impinging on an open shelf, the only difference being an additional, broadly distributed freshwater source of ice melt. The plume invades the near-shore but its distribution and manner of entering the mid shelf or leaving the shelf altogether is very much affected by winds [31]. Plume structure in late summer can often be seen in satellite imagery (and in water properties) extending up to 400 km away from the shore (cf. [19, 20], Engle and Weingartner, pers. comm.). The surface stratification is supported by both runoff and ice melt. At the beginning of this season, in late July or early August, the shelf may still be mostly ice covered whereas by the end of September the shelf is usually completely clear of ice. Considerable variation from year to year occurs in stratification and freshwater distribution forced primarily by winds. In years when ice remains close to shore, the Mackenzie plume likewise is constrained to the near-shore whereas when complete clearing of ice from the shelf waters occurs, the plume often spreads to the edge of the shelf and beyond [32].

4. The Role of Ice in the Freshwater Cycle

Ice formation plays a dual role in the transport of fresh water from the estuary. On one hand, freshwater ice provides a storage location for some of the river's winter inflow perhaps to be transported elsewhere during and after breakup. On the other hand, sea ice creates a brackish (0–5) solid phase from sea water which can then be exported from (or imported to) the shelf. In terms of its potential impact on freshwater and salt inventories, in the first case, ice acts as an almost passive participant in the positive estuary, whereas in the second case freezing and brine rejection makes sea ice an active participant in the negative estuary. In addition to these "phase change" effects, ice also provides a physical barrier between atmosphere and ocean, thereby reducing or delaying mixing of fresh water and altering the pattern and timing of plume advection.

From $\delta^{18}O$ measurements in ice cores, it has been estimated that about 12 km^3 of runoff can be accounted for in the land-fast ice of the Mackenzie estuary at the end of winter — or about 16% of the total winter inflow (73 km^3) from the Mackenzie River [5]. The winter inflow captured by the ice in this manner resides predominantly in the near-shore (Figure 10) due to the interaction of the Mackenzie River plume, as it spreads outward, with the rate of ice growth which is initially rapid but then slows to a halt at the end of winter. Early in winter when ice is growing quickly, the freshwater plume is constrained in the near-shore (Figure 7a). Toward the end of winter when the plume has expanded to its maximum dimensions (Figures 5, 7a and 8b) land-fast ice, which is nearing its maximum thickness, is no longer incorporating much water.

What happens to the winter inflow frozen into the land-fast ice? Satellite imagery suggests that much of this water simply gets melted in situ between late May and early to mid July by heat supplied from the Mackenzie River during the spring freshet [6]. The temperature of the Mackenzie River starts to rise above freezing about mid May, reaching its maximum by the end of June (Figure 11a). Using the average daily flow of the Mackenzie River (Figure 11b) together with the average daily temperature, shows that the river's ice-melting potential rises rapidly from mid May to the end of June, achieving a maximum of about 0.35 km^3 day^{-1}. For a 2-m thick ice cover, this would correspond to melt out of an area 175 km^2 day^{-1} (i.e., a square 13 km on a side). Similarly, the cumulative melt potential reaches a value of over 5000 km^2 by early July (Figure 11d) which corresponds to about a third of the land-fast ice area. Clearly, the conclusion of Dean *et al.* [6] that much of the land-fast ice gets melted by heat supplied by the Mackenzie River is supported by the capacity of the river to deliver heat. Because most of the winter inflow captured by ice resides in the near-shore (Figure 10) it can be concluded that little of the ice formed from winter inflow survives breakup and almost all of it gets converted back to liquid. Therefore, the land-fast ice provides, at the longest, a 7–8 month storage reservoir for runoff, releasing most of it back into the estuary during spring freshet. In addition to heat, the Mackenzie also delivers turbidity and stratification both of which assist solar insolation to accelerate the melt process [6].

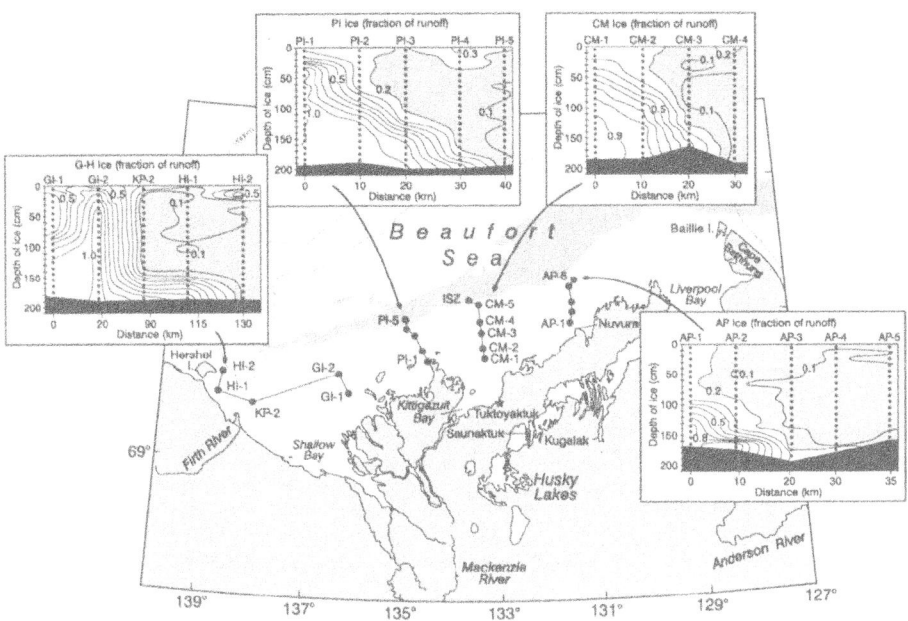

Figure 10. The disposition of runoff incorporated into ice (i.e., river ice) at the end of winter shown as sectional contours of the fraction of river water in land-fast ice. Estimates are based on the $\delta^{18}O$ composition of the ice in ice cores collected at the end of winter (late April–early May).

The second role of ice — capturing fresh water from the sea and releasing the brine to the negative estuary — is more complicated, has a greater significance to freshwater budgets, and may serve under certain circumstances as a cumulative source of fresh water to the surface ocean. However, to understand how sea ice might impact freshwater budgets, care must be taken to distinguish between (1) a separation of fresh water and salt that is only seasonal and (2) a separation of fresh water and salt that is more permanent. In the first instance, ice melt is mixed in summer by wind to form a brackish surface layer (10–20 m) which contains also some portion of runoff. During the following winter, ice withdraws fresh water (as negative ice melt) adds salt to the surface layer and produces an ever deepening winter mixed layer which, eventually, reconnects with the remnant of the previous year's polar mixed layer (e.g., the region containing negative sea ice melt in Figure 8a). By the end of winter, this layer might typically be 40–50 m deep and contain negative quantities of ice melt indicating the complete withdrawal of the summer's ice melt back into the ice. During the following summer the cycle repeats when melting occurs. Over the long term one simply sees an oscillation in the water column be-

400

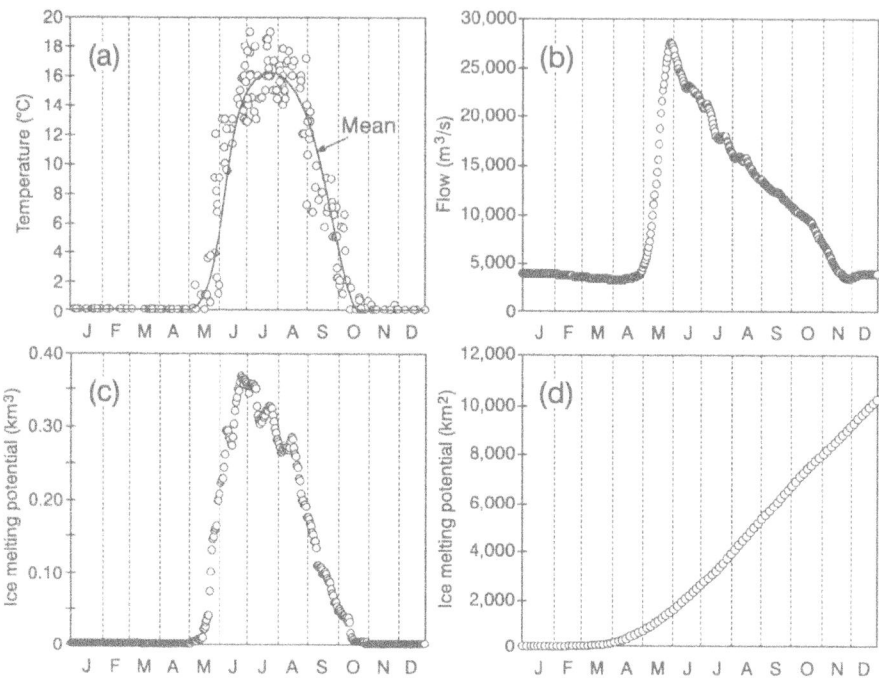

Figure 11. Heat supplied by the Mackenzie River that could contribute to melting ice. (a) The temperature record assimilated from spot readings at Arctic Red River taken from 1960 to 1993 with a smoothed line produced from these data; (b) the average flow of the Mackenzie River from 1973 to 1993; (c) the heat supplied on a daily basis, calculated from the temperature and flow records, presented in terms of the volume of ice that could be melted per day; and (d) the cumulative area of a 2-m ice cover that could be melted as a function of time. All data were obtained from Environment Canada (Water Survey).

tween positive sea-ice melt in summer to negative sea-ice melt (i.e., brine) in winter matched by an opposing cycle in sea ice. Fresh water from runoff can be effectively mixed downward by the ice/melt cycle but it follows its own seasonal cycle dominated by inflow. On a long-term basis, the salt and fresh water cycled by sea ice are being separated by the ice only seasonally, and there is actually no net production of either salt or freshwater in the top 50 m of the water column. One would expect such behaviour, for example, under conditions of thermodynamic ice thickness within the permanent pack.

In contrast to haline convection leading only to mixing just beneath the ice, as discussed above, ice growth coupled to *penetrative* convection introduces asymmetry into the annual freeze–thaw cycle [33]. In this case, a more permanent separation of fresh water and salt occurs [29, 34] because salt rejected from sea ice sinks as brine below the polar mixed layer where it can no longer mix with ice melt the following summer. The salt lost to deeper water during this process is mirrored by a net freshwater gain in the surface layer, either as ice or, when the ice eventually melts, as fresh water. In effect, penetrative convection produces stratification. It is well known that polynyas, or regions of diverging ice cover in winter, are particularly effective in releasing salt and, hence, convecting water masses [24, 35–37]. For the Mackenzie shelf, and arctic shelves in general, the most favourable location to produce convecting water is over the mid shelf just beyond the stamukhi where ice growth is enhanced, while, at the same time, winter runoff is excluded (cf. Figure 2d). Although the outer shelf and interior ocean are even more isolated from winter runoff, they are not as effective locations for growing ice. The land-fast zone is neither a good ice producer nor is it protected from the stratifying influence of winter runoff [5].

Melling [24] concluded both from data and models that cold, dense convecting water produced by sea ice formation carries a salinity anomaly of perhaps 1 to 2.5. Assuming that sea ice grows from water of a salinity of about 30, an anomaly of 1 to 2.5 implies that for each km^3 of water supplied to the halocline through penetrative convection, an equivalent 0.03–0.08 km^3 of brackish water is produced as ice or ice melt. The Mackenzie shelf, which does not produce convecting water masses in every year, is estimated to contribute perhaps 0.04 Sv (1300 km^3 yr^{-1}) of cold saline water to the halocline on average [38]. This, therefore, is mirrored by the net production of fresh water (sea ice) in the order of 40–100 km^3 $year^{-1}$ (cf. [5, 22, 23, 36, 39]). Although the fresh water produced in this way is smaller than the Mackenzie River annual inflow (330 km^3 yr^{-1}), the flow carrying the brine below the mixed layer (1300 km^3 yr^{-1}) is three times as large.

The Mackenzie and its bordering shelf are small by arctic standards making other shelves, especially those off Russia, more important in their potential to influence freshwater budgets through positive–negative coupling. The geographical juxtaposition of inflow and flaw leads imply such coupling, and the Laptev Sea is a leading candidate location simply because it exports enormous quantities of ice (Figure 1). However, as noted in the introduction, we lack entirely the sort of data required to evaluate if, where, when or how much these shelves produce penetrative convection. The freshwater input to Russian shelves would not, by itself, prevent penetrative convection because the yield is considerably less than that of the Mackenzie shelf (Table 1) which still manages to produce deep convection [29]. However, the Russian shelves have longer residence times and greater length scales (Table 1), both of which imply that it will be more difficult for these shelves to eliminate fresh water. Nevertheless, there is intriguing evidence that penetrative convection occurs somewhere on this side of the Arctic Ocean. Swift et al. [40] found a bolus of cold, recently ventilated water between 800–1300 m on the margin of the Makarov basin. The origin of this water was unclear but shelves were clearly implicated and the Barents, Kara and Laptev Seas were suggested to play a major role in ventilating the Arctic Ocean interior.

An important question is what quantities of ice, salt and fresh water are involved in penetrative convection on shelves for the entire Arctic Ocean. Presently, due to the lack of data, we can estimate these figures indirectly and with a great deal of uncertainty. Melling [38] suggests that the overall rate of renewal of waters in the halocline from convecting brine is about 1 Sv $(32{,}000 \text{ km}^3 \text{ yr}^{-1})$ which, following the same reasoning used above for the Mackenzie shelf, implies an equivalent net freshwater (sea ice) production of 950–2500 km^3 yr^{-1}. If the fluxes supporting the halocline are as high as these estimates suggest, they are a significant component of the Arctic Ocean's freshwater budget, and are sensitive to change.

5. The Leverage for Change in Arctic Estuaries

Projections of existing trends on a global scale suggest that two important changes face arctic estuaries, one being the quantity of runoff and the other being the temperature and its effect on sea ice. With doubled atmospheric CO_2, annual runoff is predicted to increase by an average of 25% for rivers in high northern latitudes due to increased precipitation brought on by global warming [41, 42]. Such an increase would deliver an extra 80 km^3 of runoff to the Mackenzie shelf and 800 km^3 to the entire Arctic Ocean. Given that a 25% increase for the Mackenzie River flow is equivalent to about a 1-m yield for the shelf, which would then require more than an extra metre of ice growth just to maintain status quo, this type of change appears significant for this shelf. If, in the extreme case, a 25% increase in freshwater runoff completely stalled convection, it would turn off the 1300 km^3 yr^{-1} this shelf is estimated to contribute to the halocline. Turning off penetrative convection would, as discussed above, also turn off the net production of 40–100 km^3 of fresh water in the form of ice separated from salt. Countering the projected increase in runoff, is the potential loss of inflow due to activities within drainage basins to divert or use the water otherwise (cf. [43]), estimated at perhaps 2% of the total flow [4]. This magnitude of change is well below the inter-annual variance of northern rivers as represented by the Mackenzie over the past 25 years and would be difficult to detect without more than a decade of data.

The timing of entry of freshwater inflow is likely to be a more important consideration than total inflow. Increased runoff, particularly if it occurs just before or during freeze-up, would provide buoyancy and therefore tend to reduce or prevent water convection from ice growth the following winter.

Similarly, the alteration of the seasonal hydrograph to enhance winter inflow at the expense of summer inflow, a by-product of damming for power, could also stall convection on the shelf. The degree to which these kinds of changes would affect convection has been little considered and would perhaps best be investigated at present using models (e.g., [22]). Intermittent penetrative convection controlled by freshwater balance over the mid shelf in the Arctic clearly falls into the category of a "flickering light" and is, therefore, a process where change can most easily be effected.

The second projected change (warming) would lead to a shorter ice-growth season due either to later freeze-up or earlier breakup. These two possible outcomes of warming may well have opposite effects, at least initially. Delayed freeze-up (Figure 7a) will allow additional time for the shelf to clear itself of the fresh water that entered during spring and summer, thereby pre-conditioning shelf waters toward a saltier state. With a more saline starting point, ice production during the ensuing winter then has a better chance of producing enough brine to overcome the remaining fresh-water standing stock and so produce penetrative convection. In contrast, earlier breakup would allow fresh water to enter the mid shelf in late winter. This is the critical time when ice formation has done the work of removing excess fresh water and the system is now ready to produce deeper convection. Adding fresh water at this time could therefore stall penetrative convection just as the system is poised to do it.

The negative estuary is influenced both by the positive estuary, which supplies buoyancy, and other factors that remove buoyancy by the loss of fresh water from the shelf (flushing). Indeed, at the scale of the shelf, the negative estuary can be seen as embedded within the positive estuary (Figure 2d). The response of the shelf to lower runoff inventories in winter is an enhanced capacity to produce penetrative convection during sea ice formation and, with it, enhanced production of fresh water at the surface (Figure 12). Therefore, the coupling between the positive estuary and the negative estuary produces a form of negative feedback with regard to stratification and freshwater balance of the Arctic Ocean surface. While it seems clear that both warming and changes in the seasonal cycle of river inflow could produce large changes in this couple, the crucial element — timing — is not sufficiently well understood to allow us to predict these changes with any certainty. We can envisage two diametrically opposed states for the Arctic Ocean — an ice-free state and one in which ice cover is perennial and locked rigidly in place. Neither of these two extremes would generate thermohaline circulation, the first because ice would not be formed and the second because ice growth and melt would become small and in thermo-dynamic balance (somewhat like present conditions under the permanent pack). The optimum conditions for brine production lie somewhere between

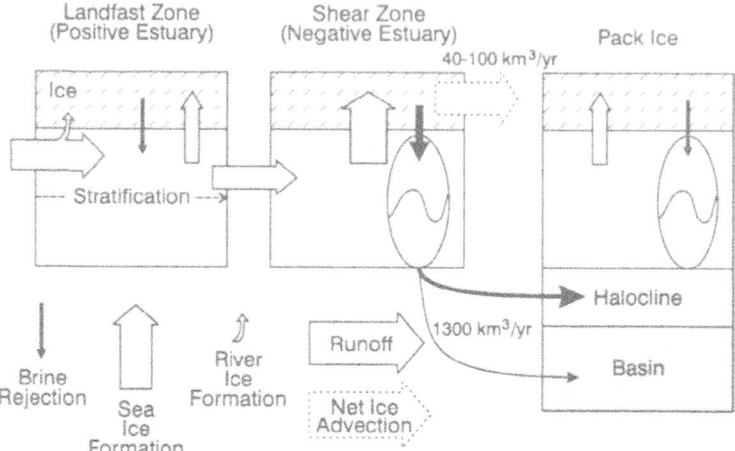

Figure 12. A schematic diagram showing the coupling in winter between the positive estuary where fresh water from the Mackenzie River dominates and the negative estuary where ice formation/brine production dominates. In the land-fast zone, salt rejected during sea ice growth is not able to overcome stratification supplied by the winter inflow and hence remains near the surface. In the shear zone, some of the salt rejected from the sea ice goes to removing stratification left over from the autumn producing a deepening mixed layer which is observed as a seasonal oscillation between stratification in summer and a polar mixed layer in winter. During favourable years when the shelf waters start with a relatively high salinity, penetrative convection can move brine to deeper layers of the interior ocean (halocline/basin) in which case there is a matching net export of brackish water in the form of sea ice (or ice melt) to balance the system. Under the pack ice the annual thaw/freeze cycle produces only a deepening mixed layer but no penetrative convection. In this proposed scheme, the penetrative convection in the negative estuary is controlled both by the amount of fresh water remaining at freeze-up and the amount of fresh water that leaks into the region during winter.

these two extremes. We may safely infer that today's ice climate allows the production of brine, at least intermittently on some shelves. However, it is not at all certain where the optimum for brine production is in relation to the present and whether projected changes would drive us toward or away from the optimum.

6. Future Desired Research

1. The perspectives developed here depend mainly on data from the Mackenzie shelf of the Beaufort Sea whose notable characteristics are a short residence time, a large freshwater yield and a net export of ice. Although it is clear that similar interplay between positive estuaries and mid-shelf flaw leads occurs almost everywhere in the Arctic, there are presently no data with which to evaluate this interplay

in settings with varied characteristics. Appropriate data must include sufficient tracers to separate the effects of runoff and ice (i.e., salinity plus $\delta^{18}O$), measurements of both ice and water phases, and studies that extend through an entire season or, better still, through several seasons. Such studies should be carried out especially on the Russian shelves where there are longer residence times, smaller yields and where large amounts of ice can be either exported or imported (Table 1).

2. Arctic circulation models need to incorporate shelves in a meaningful way, including at least: Distinction between seasons, distinction between near-shore and mid-shelf regions, coupling between the shelf regions throughout the year, and coupling of the shelf with the interior ocean. Without this, there is little hope that models will provide realistic predictions of the effects of climate change on the Arctic Ocean's freshwater balance.

References

1. Dyer, K.R. (1973) *Estuaries: A Physical Introduction*, Wiley-Interscience, New York.
2. Cameron, W.M. and Pritchard, D.W. (1963) Estuaries, in M.N. Hill (ed.), *The Sea*, John Wiley and Sons, New York, pp. 306–324.
3. Aagaard, K. and Carmack, E.C. (1989) The role of sea ice and other fresh water in the Arctic circulation, *Journal of Geophysical Research*, **94**, 14,485–14,498.
4. Vuglinsky, V.S. (1997) River water inflow to the Arctic Ocean – Conditions of formation, time variability and forecasts, in Aagaard, K., D. Hartmann, V. Kattsov, D. Martinson, R. Stewart and A. Weaver (eds.), *Polar Processes and Global Climate, Arctic Climate System Study*, WCRP-106, WMO/TD908, Rosario, Washington, pp. 275–276.
5. Macdonald, R.W., Paton, D.W., Carmack, E.C. and Omstedt, A. (1995) The fresh water budget and under-ice spreading of Mackenzie River water in the Canadian Beaufort Sea based on salinity and $^{18}O/^{16}O$ measurements in water and ice, *Journal of Geophysical Research*, **100**, 895–919.
6. Dean, K.G., Stringer, W.J., Ahlns, K., Searcy, C., Weingartner, T. (1994) The influence of river discharge on the thawing of sea ice, Mackenzie River Delta: Albedo and temperature analyses, *Polar Research*, **13**, 83–94.
7. Ingram, R.G. and Larouche, P. (1987) Variability of an under-ice river plume in Hudson Bay, *Journal of Geophysical Research*, **92**, 9541–9547.
8. Macdonald, R.W. and Carmack, E.C. (1991) The role of large-scale under-ice topography in separating estuary and ocean on an Arctic shelf, *Atmosphere–Ocean*, **29**, 37–53.
9. Pavlov, V.K. and Pfirman, S.L. (1995) Hydrographic structure and variability of the Kara Sea: Implications for pollutant distribution, *Deep-Sea Research*, **42**, 1369–1390.
10. Reimnitz, E., Dethleff, D. and Nürnberg, D. (1994) Contrasts in Arctic shelf sea-ice regimes and some implications: Beaufort Sea versus Laptev Sea, *Marine Geology*, **119**, 215–225.
11. Stirling, I.S. and Cleator, H. (1981) Polynyas in the Canadian Arctic, *Occasional Paper 45*, Canadian Wildlife Service, 70 pp.

12. Rigor, I. and Colony, R. (1997) Sea-ice production and transport of pollutants in the Laptev Sea, 1979–1993, *The Science of the Total Environment*, **202**, 89–110.

13. Duxbury, A.C. and Duxbury, A.B. (1994) *An Introduction to the World's Oceans*, Wm. C. Brown, Iowa.

14. Macdonald, R.W., Carmack, E.C. and Paton, D.W. (1999) Using the δ^{18}O composition in landfast ice as a record of arctic estuarine processes, *Marine Chemistry*, **65**, 3–24.

15. Antonov, V.S. (1978) The possible impact on the Arctic Ocean of the proposed transfer of water from the northern rivers of the USSR to the south, *Polar Geography*, **2**, 223–231.

16. Ingram, R.G. (1981) Characteristics of the Great Whale River plume, *Journal of Geophysical Research*, **86**, 2017–2023.

17. Bauch, D., Schlosser, P. and Fairbanks, R. (1995) Freshwater balance and the sources of deep and bottom waters in the Arctic Ocean inferred from the distribution of $H_2^{18}O$, *Progress in Oceanography*, **35**, 53–80.

18. Östlund, H.G. and Hut, G. (1984) Arctic Ocean water mass balance from isotope data, *Journal of Geophysical Research*, **89**, 6373–6381.

19. Pfirman, S.L., Kögeler, J. and Anselme, B. (1995) Coastal environments of the western Kara and eastern Barents Seas, *Deep-Sea Research*, **42**, 1391–1412.

20. Macdonald, R.W., Carmack, E.C., McLaughlin, F.A., Iseki, K., Macdonald, D.M. and O'Brien, M.O. (1989) Composition and modification of water masses in the Mackenzie Shelf Estuary, *Journal of Geophysical Research*, **94**, 18,057–18,070.

21. Dethleff, D. (1995) Sea ice and sediment export from the Laptev Sea flaw lead during 1991/92 winter season, in H. Kassens, D. Piepenburg, J. Thiede, L. Timokhov, H.-W. Hubberton and S.M. Priamikov (eds.), *Berichte zur Polarforschung*, pp. 78–93.

22. Omstedt, A., Carmack, E.C. and Macdonald, R.W. (1994) Modeling the seasonal cycle of salinity in the Mackenzie shelf/estuary, *Journal of Geophysical Research*, **99**, 10,011–10,021.

23. Melling, H. and Riedel, D.A. (1996) Development of seasonal pack ice in the Beaufort Sea during the winter of 1991–1992: A view from below, *Journal of Geophysical Research*, **101**, 11,975-11,991.

24. Melling, H. (1996) Water modification on arctic continental shelves: Seasonal cycle and interannual variation, In P. Lemke, L. Anderson, R. Barry and V. Vuglinsky. (eds.), *Proceedings of the ACSYS Conference on the Dynamics of the Arctic Climate System*, Göteborg, Sweden, pp. 78–82.

25. Colony, R. and Thorndike, A.S. (1985) Sea ice motion as a drunkard's walk, *Journal of Geophysical Research*, **90**, 965–975.

26. Carmack, E.C., Macdonald, R.W. and Papadakis, J.E. (1989) Water mass structure and boundaries in the Mackenzie shelf estuary, *Journal of Geophysical Research*, **94**, 18,043–18,055.

27. Anderson, D.L. (1961) Growth rate of sea ice, *Journal of Glaciology*, **3**, 1170–1172.

28. Maykut, G.A. (1986) The surface heat and mass balance, in N. Untersteiner (eds.), *The Geophysics of Sea Ice*, NATO ASI Series, Series B, Physics, pp. 418–463.

29. Melling, H. and Lewis, E.L. (1982) Shelf drainage flows in the Beaufort Sea and their effect on the Arctic Ocean pycnocline, *Deep-Sea Research*, **29**, 967–985.

30. Lawford, R.G. (1994) The hydroclimatology of north-flowing high latitude rivers, in P. Lemke, L. Anderson, R. Barry and V. Vuglinsky (eds.), *ACSYS Conference on the Dynamics of the Arctic Climate System*, World Meteorlogical Organization, Göteborg, Sweden, pp. 8–23.

31. Giovando, L.F. and Herlinveaux, R.H. (1981) A discussion of factors influencing dispersion of pollutants in the Beaufort Sea, *Pacific Marine Science Report*, **81-4**, Institute of Ocean Sciences, Sidney, B.C., 198 pp.

32. Macdonald, R.W., Wong, C.S. and Erikson, P.E. (1987) The distribution of nutrients in the southeastern Beaufort Sea: Implications for water circulation and primary

production, *Journal of Geophysical Research*, **92**, 2939–2952.

33. Aagaard, K., Coachman, L.K. and Carmack, E.C. (1981) On the halocline of the Arctic Ocean, *Deep-Sea Research*, **28**, 529-545.

34. Aagaard, K., Swift, J.H. and Carmack, E.C. (1985) Thermohaline circulation in the Arctic mediterranean seas, *Journal of Geophysical Research*, **90**, 4833–4846.

35. Martin, S., Munoz, E. and Drucker, R. (1992) The effect of severe storms on the ice cover of the Northern Tatarskiy Strait, *Journal of Geophysical Research*, **97**, 17,753–17,764.

36. Cavalieri, D.J. and Martin, S. (1994) The contribution of Alaskan, and Canadian coastal polynyas to the cold halocline layer of the Arctic Ocean, *Journal of Geophysical Research*, **99**, 18,343–18,362.

37. Gewarkiewicz, G.G. and Chapman, D.C. (1995) A numerical study of dense water formation and transport on a shallow sloping continental shelf, *Journal of Geophysical Research*, **100**, 4489–4507.

38. Melling, H. (1993) The formation of a haline shelf front in wintertime in an ice-covered Arctic sea, *Continental Shelf Research*, **13**, 1123–1147.

39. Melling, H. and Moore, R.M. (1995) Modification of halocline source waters during freezing on the Beaufort Sea shelf: Evidence from oxygen isotopes and dissolved nutrients, *Continental Shelf Research*, **15**, 89–113.

40. Swift, J.H., Jones, E.P., Aagaard, K., Carmack, E.C., Hingston, M., Macdonald, R.W., McLaughlin, F.A. and Perkin, R.G. (1997) Waters of the Makarov and Canada basins, *Deep-Sea Research*, **44**, 1503–1529.

41. Miller, J.R. and Russell, G.L. (1992) The impact of global warming on river runoff, *Journal of Geophysical Research*, **97**, 2757–2764.

42. Maxwell, J.B. and Barrie, L.A. (1989) Atmospheric and climatic change in the Arctic and Antarctic, *Ambio*, **18**, 42–49.

43. Cattle, H. (1985) Diverting Soviet rivers: Some possible repercussions for the Arctic Ocean, *Polar Record*, **22**, 485–498.

44. Barrie, L., Falck, E., Gregor, D., Iverson, T., Loeng, H., Macdonald, R., Pfirman, S., Skotvold, T. and Wartena, E. (1998) The influence of physical and chemical processes on contaminant transport into and within the Arctic, in D. Gregor, L. Barrie and H. Loeng (eds.), *The AMAP Assessment Report*, Chapter 3, pp. 25–116.

45. Hanzlick, D. and Aagaard, K. (1980) Freshwater and Atlantic water in the Kara Sea, *Journal of Geophysical Research*, **85**, 4937–4942.

46. Ivanov, V.V. (1994) River water inflow to the Arctic seas, in P. Lemke, L. Anderson, R. Barry and V. Vuglinsky (eds.), *ACSYS Conference on the Dynamics of the Arctic Climate System*, WMO, Göteborg, pp. 115–124.

47. Schlosser, P., Bauch, D., Fairbanks, R. and Bönisch, G. (1994) Arctic river-runoff: Mean residence time on the shelves and in the halocline, *Deep-Sea Research*, **41**, 1053–1068.

48. Aagaard, K. and Coachman, L.K. (1975) Toward an ice-free Arctic Ocean, *Eos*, **56**, 484–486.

SATELLITE VIEWS OF THE ARCTIC OCEAN FRESHWATER BALANCE

D.A. ROTHROCK[1], R. KWOK[2], and D. GROVES[1]

[1]*Applied Physics Laboratory, University of Washington, Seattle, Washington, 19195, USA*
[2]*Jet Propulsion Laboratory, 4800 Oak Grove Drive, Pasadena, California, 91109, USA*

1.   Introduction

The freshwater balance of the Arctic Ocean involves atmospheric moisture and its movement; precipitation onto and evaporation from the land and ocean surface; runoff from rivers into the ocean; the growth, movement and melt of sea ice; and the movement of freshwater within the ocean itself. Satellite observations have contributed to a surprisingly large portion of these topics, some quite directly and others rather indirectly. Satellites generally cannot see into sea ice or the ocean, so we are left to deduce what we can about the variables we wish to estimate from signals from the surface. On the other hand the atmosphere is seen throughout its depth by satellites, so we can even get some vertical resolution of where water resides in the atmosphere. In this chapter we treat only the Arctic Ocean and its surrounding seas, and not the ocean's drainage basin. For the most part, the satellite data record becomes quite valuable around 1979, and it keeps getting better with more sensors, better coverage, and better production of data sets applicable to the large-scale interests of this book.

We begin our consideration with ice extent and ice concentration, the variables with the most accepted and longest satellite record. From them, we can examine seasonal and longer-term changes of the horizontal coverage of ice. Next we face the issue of wanting to know the mass of sea ice, its thickness as well as its coverage. This subject is less developed, but we review promising new capabilities and later in the chapter some results about thickness deduced from concentration observations. This same use of concentration has been extended to some estimates of the ocean freshwater storage and budget. All satellite sensors that provide imagery have the potential to give us ice motion if ice 'features' can somehow be recognized at sequential times and "tracked"; we review ice motion data next. In particular, the ice mass balance of the Arctic Ocean can be described by specifying all sources and sinks of ice, and the ice flux through passages, especially Fram Strait, has been well studied with satellite ice motion data. The atmospheric moisture balance is quite amenable to satellite studies; moisture transports and the net exchange with the surface have been estimated. Because satellites do not see exactly what we want, the thread of the assimilation of satellite data into ice-ocean models runs through this chapter. This technique has much more potential for the book's subject than has yet been realized.

This is a book about geophysics, not remote sensing, so this chapter does not stress the latter. However, we have tried to provide a short review of the capability to

E.L. Lewis et al. (eds.), The Freshwater Budget of the Arctic Ocean, 409-451.

make each of these measurements and some of the uncertainties and pitfalls. Somewhat arbitrarily, we have put these reviews as introductions to some sections. Some readers may find them unnecessary or a distraction and can surely skip them. Excellent treatments of these issues are given by Massom [1] and Carsey [2].

We conclude our review with a summation of quantitative freshwater balance estimates from satellites, and a brief recapitulation of which variables are already well used for this problem and which variables hold unrealized potential.

2. Ice Coverage

2.1. MOTIVATION

The sea ice cover represents a sizable fraction of the freshwater storage in the Arctic Ocean (some 20% according to Aagaard and Carmack, [3]) and about a third of the freshwater export. This statement is from the point of view of arctic oceanographers, who define freshwater relative to a salinity of 34.8 psu, as developed in Sec. 3.1.1. Estimations of sea ice mass or volume are a crucial ingredient of the hydrology of the ocean. The mean state, the seasonal cycle, and change that may be occurring over many years are all of interest.

By far the most important record of sea ice comes from passive microwave radiometers. The record's importance stems primarily from its duration of over twenty years (and continuing) and its coverage, which is global except for a polar blind spot of no dire consequences. An example is shown in Figure 1. Ice concentration C tells us the fractional coverage by sea ice of an area from ten to several hundred kilometers across. The open water fraction is $1-C$. Because the radiometer can also differentiate thicker multiyear" ice (Figure 2) from "first-year" ice—ice that has not experienced a summer melt season and is therefore more saline—it gives us a weak proxy for ice thickness.

In particular, radiometers delineate the ice edge quite well. The ice *extent* is simply the integral over the included region of the area where C is greater than some cutoff, often taken as *15%*. This threshold is adopted to rid the data of incorrect readings of ice in the open ocean away from the sea ice cover. The region can be a portion of an ocean, a whole ocean, or the hemisphere.

$$Ice\ Extent = \iint_{region} \left\{ \begin{array}{l} 1,\ where\ C \geq 15\% \\ 0,\ otherwise \end{array} \right\} da \qquad (1)$$

The ice *area* is the integral over the region of the concentration itself; this definition excludes all open water from the sum. Ice area is smaller than ice extent.

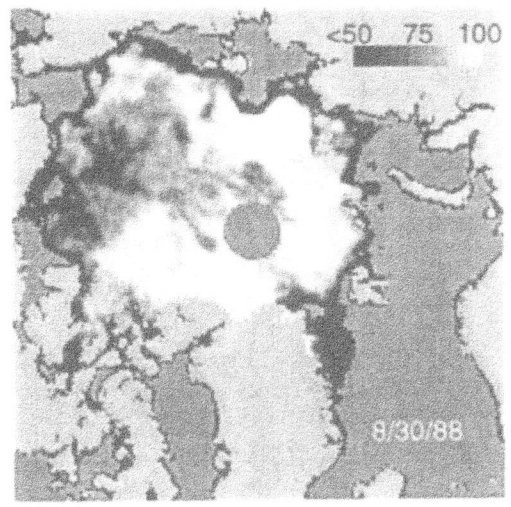

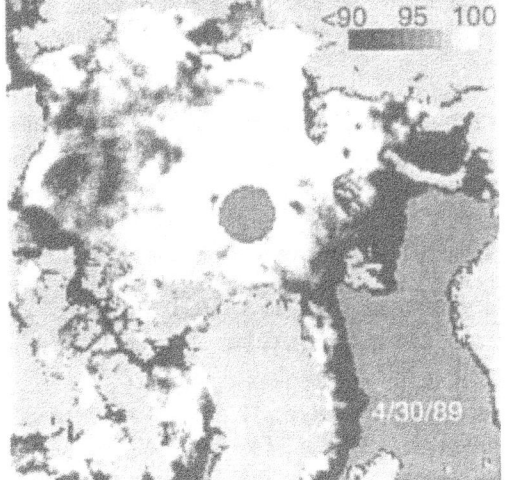

Figure 1. Microwave data of sea ice concentration in the Arctic Ocean. The panels represent the seasonal extremes of ice concentration: end of summer (top) and late winter (bottom). The grey scales are different. The ocean is shown as a uniform grey.

$$Ice\ area = \iint_{region} C \begin{cases} 1,\ where\ c \geq 15\% \\ 0,\ otherwise \end{cases} da \qquad (2)$$

Ice extent is a more robust variable than ice area because the 15% concentration contour is rather sharp whereas the integrand of (2) has more error than that of (1).

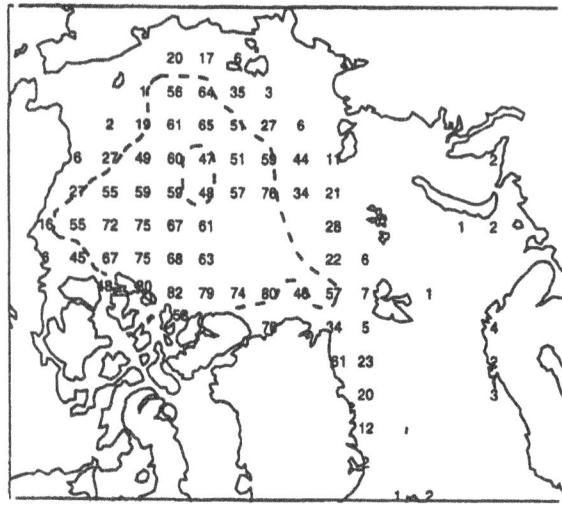

Figure 2. Multiyear ice concentration in percent for 31 December 1984, showing the 50% contour.

2.2 SATELLITE OBSERVATIONS OF CONCENTRATION

2.2.1 *Description of the Observations*

Passive microwave radiometers measure the radiation emitted from the surface and the atmosphere and reaching the satellite. By utilizing several frequencies between 6 and 85 GHz and with both horizontal and vertical polarization, they provide 7 to 10 signal "channels" of which two or three are independent of each other [4]. The first multichannel instrument, SMMR, the Scanning Multichannel Microwave Radiometer, was launched in 1978. Shortly before its demise in 1987, a new series was launched: SSMII, Special Sensor Microwave Imager; this series continues flying today. A single-channel instrument, ESMR, the Electronically Scanning Microwave Radiometer, flew on Nimbus5 from 1973 to 1976; its best application is in defining ice extent.

The multichannel sensor resolution varies between 12 and 150 km, smaller for higher frequencies. Data products are produced at 12 and 25 km grid spacing, which are ideal for a large-scale geophysical view. Coverage is excellent: global except pole-ward of 840 latitude for SMMR and 870 for SSMII. Because the atmosphere is fairly transparent at their frequencies, microwave sensors are "all-weather," providing data in darkness and through clouds.

Over sea ice the microwave signals are usually interpreted as concentrations of all ice and of only multiyear ice. The 15% concentration contour is usually taken as the ice edge. Estimates of the uncertainties in theses quantities vary~ ice extent, ±12 1cm; ice concentration, ± 6%, with possible biases of similar magnitude; and multiyear concentration, ±20%, again with biases of similar size. Further discussions of these issues are found in Chapters 4, 11, and 17 of Carsey [2].

Data sets are available through the National Snow and Ice Data Center (http://wwwnsidc.colorado.edu) for October 1978 to the present, for total, first-year, and multiyear ice concentrations. They are on a 25 x 25 km grid, daily and monthly, for both hemispheres. The passive microwave record is presented in three atlases: Gloersen *et al.* [5], Parkinson *et al.* [6], and Zwally *et al.* [7].

2.2.2 Difficulties in Interpretation

No observation is perfect, and passive microwave data have their share of problems. The first is that of mixed pixels. (A pixel is a "picture element," or roughly a footprint—the smallest resolved area. It is like the "point" of a "pointillist" painting.) The resolution of these sensors is a mixed blessing. Their footprint size of tens of kilometers makes for a manageably small data set, but it also means that each footprint contains signals from many different thicknesses of sea ice with different surface states and emissivities, and we must interpret this average signal from the footprint. This is done by assigning values (signatures) to what are assumed to be the "pure" constituents (water, first-year ice, multiyear ice) and interpreting intermediate values as mixtures. The assignment of the "pure type" signatures involved interpreting the wide scatter of data values. The dominant paradigm (Cavalieri *et al.,* [8]) is to assume that, say, multiyear ice has a very extreme value, so that no values can fall beyond it and imply greater than 100% multiyear ice (a physical impossibility). Figure 3a illustrates this viewpoint. An alternative is to assume that even "pure" multiyear ice has a distribution of values and to choose the middle of the distribution to represent the "pure" type (Thomas, [9]). The choice of the former method (Figure 3a) is thought by the present authors to cause a large negative bias in the multiyear concentrations in the data sets mentioned above, as illustrated in Figure 4. Figure 5 shows a comparison of ice concentrations estimated from the mixed S SM/I pixels and those estimated from much higher resolution synthetic aperture radar for which most pixels are not mixed, illustrating the negative bias in the passive microwave estimate (Kwok and Comiso, [10]).

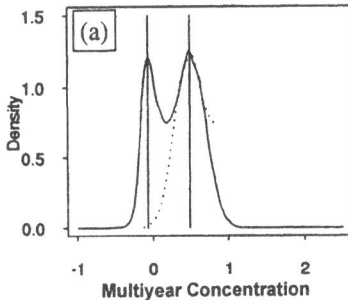 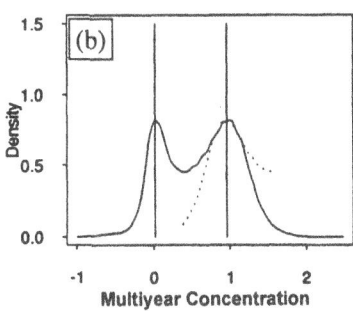

Figure 3. Distribution of multiyear concentrations for different choices of the signatures of pure multi-year ice, of first-year ice, and of open water. (after Thomas, [9])

In summer there are numerous complications, and the data have much less value. The winter sea ice emissivity at these frequencies is higher than that of water. When the snow cover gets wet, the emissivity jumps up, and the multiyear concentration incorrectly plummets. As summer wears on, the snow gives way to bare ice and emissivity falls somewhat. In July the surface gets melt ponds, which appear as water even though there may be several meters of ice beneath them.

414

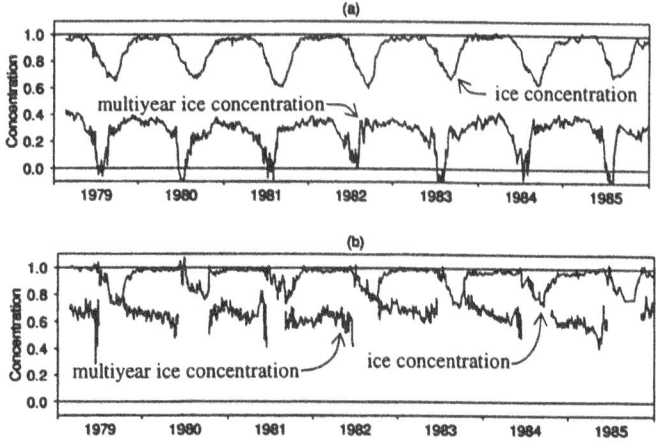

Figure 4. A seven year record of ice concentration and multiyear ice concentration using the two different methods suggested by Figures 2(a) and 2(b). (after Thomas, 1993)

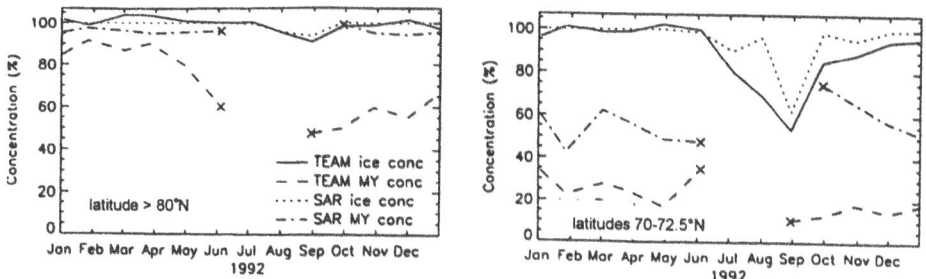

Figure 5. Comparison in two different latitude ranges of ice concentration estimated from passive microwav radiometer ("Team" algorithm) and from synthetic aperture radar (SAR). The passive microwave algorithm especially for multiyear ice, gives values substantially lower than those estimated from SAR. (After Kwok and Comiso, [10])

Ice concentration then is too low; it would be more realistically defined to exclude both open water *and* melt ponds. As the surface and ponds repeatedly refreeze and re-melt throughout summer, the emissivity and concentrations fluctuate. It is not until autumn when the surface finally refreezes for the winter that the signal is again reliable [9, 11].

Water vapor, clouds, and rain affect the signals, even though ice concentration algorithms attempt to minimize the effect [12]. In marginal ice zones, atmospheric water vapor and cloud liquid water raise total ice concentration by 10—15% and depress multiyear ice concentration by up to 50% according to Oelke [13]. Away from sea ice the data show spurious patches of ice unless a weather filter (an arbitrary threshold) is used [14].

To create a twenty-year record requires merging data from two different sensors. This requires careful intercalibration, particularly when one is looking for trends in the entire record [15].

2.3. VARIABILITY AND TRENDS IN EXTENT AND CONCENTRATION

The passive microwave record allows us to assess the variability of sea ice extent. Whatever their time scale, these variations affect the hydrologic cycle by changing the ice mass storage term, and they are connected with changes in ice production within and export from the Arctic Ocean. Figure 6 shows the seasonal cycle of ice extent; the ocean is completely covered from December through May, and the extent decreases throughout the summer, reaching a minimum in September. The autumn refreezing of the ocean is more rapid than the melt by a month.

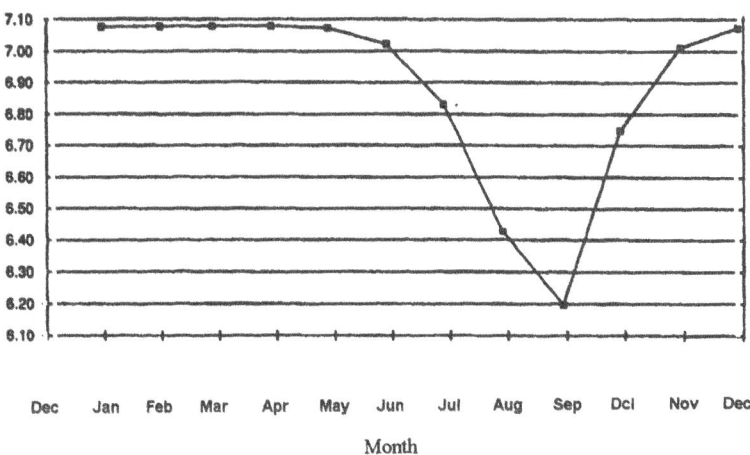

Figure 6. Average seasonal cycle of sea ice extent from SMMR, 1978—1987. (After Gloersen *et al.*, [5])

Ice concentration within the Arctic Ocean has striking interannual variations. Areas of low concentration during summer are especially important to the heat balance and to the subsequent evolution of the ice pack. Serreze *et alt.* [16] examined data and model output to explain the causes of the extreme summer of 1988, finding that strong spring divergence and fracturing, an unusually warm atmosphere and clear summer skies, and anomalous ocean heat could all have helped produce low ice concentrations. In fact, summer ice concentrations were abnormally low in 1990-1995; this is attributable to the reduced strength of the, arctic surface anticyclone during the years of a strong North Atlantic Oscillation index (Maslanik *et al.* [17]). Parkinson [18] analyzed the duration of the ice season during 1979—1986 and found trends in the spatial pattern; the eastern longitudes experienced shortening ice seasons, while western longitudes enjoyed lengthening ice seasons.

Although a study of the ESMR and SMMR records from 1973—1987 showed no trend [19], it did show which regions experience low variability and might be more telling of climate change because of their persistent ice cover. Table I shows results of a later analyses of the continuous SMMR and SSMI record from 1978 to 1995 which indicate a decrease in arctic ice extent—of about 4.5% over the 16.5 year record [15,20]. Figure 7 shows the record. Björgo *et al.* [15] take special caution to intercalibrate the brightness temperatures in the summer 1987 period of overlap and to readjust the ice concentration algorithm for each sensor. Other authors report similar findings [21].

TABLE 1. Trends in arctic sea ice extent (modified from [20]). These rates are derived from a 16.5-year record

Parameter	Rate of change	Rate of change (% I decade)
Ice extent	-0.33×10^6 km^2/ decade	-2.7
Ice area	-0.36×10^6 kin^2/ decade	-3.4
Ice concentration	-0.94% / decade	-1.1

2.4. USES OF EXTENT AND CONCENTRATION DATA WITH ICE MODELS

Because ice concentration does not tell us ice mass directly, it is natural to use ice concentration data in conjunction with an ice model. If modeled and observed ice concentrations agree, one is more willing to trust the model's other output, such as ice thickness. In a quite thorough intercomparison of model output with observations, Walsh *et al.* [22] examine ice extent finding the agreement shown in Figure 8. Tests of how modeled multiyear ice concentrations (Figure 9) compare with observations (Figure 2) in the middle of winter and in the middle of the Arctic Ocean show the modeled values to be somewhat higher than those observed [23]. This actually agrees with other evidence that satellite estimates of multiyear ice are biased low.

In the extreme one actually assimilates the concentration data into the model; because ice concentration data from satellites have excellent spatial and temporal coverage, assimilation is an attractive option. By assimilating ice concentration into a box model, Thomas *et al.* [24] estimated the concentrations of five classes (Figure 10). An approach for replacing modeled variables with observations where they exist is described by Maslanik and Maybee [25].

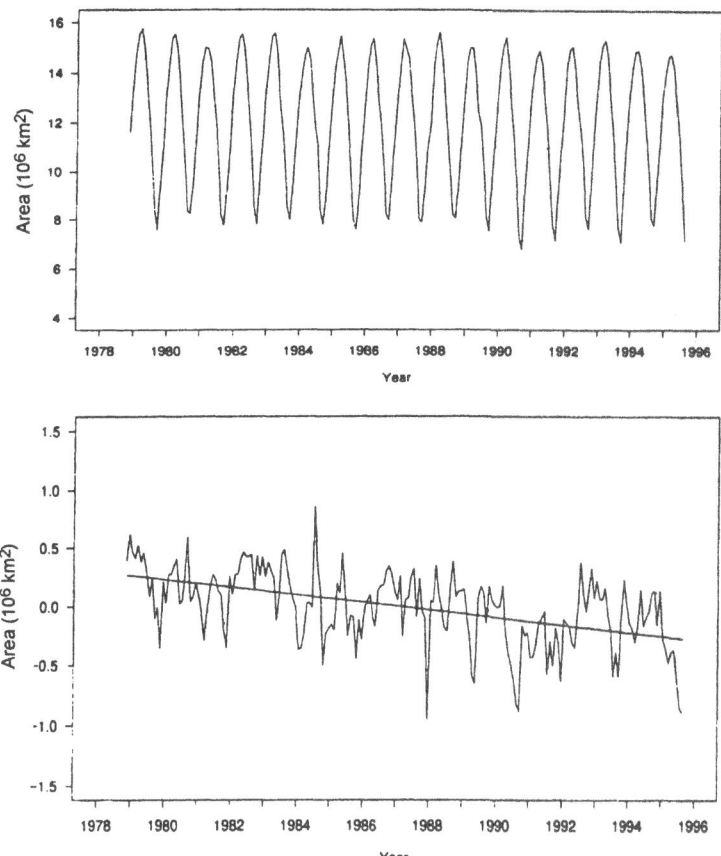

Figure 7. Monthly average arctic ice extent (top) and anomalies (bottom) for November 1978 to August 1995, with the estimated trend line. (after [15]).

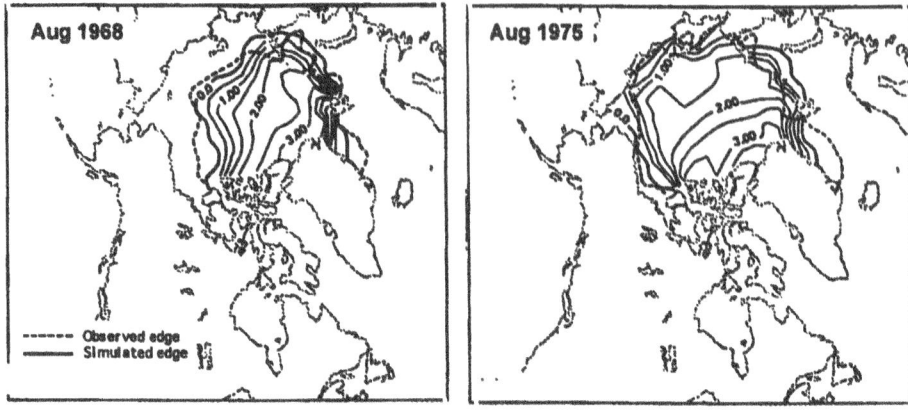

Figure 8. Simulated ice edge (bold solid lines) compared with observed ice edge (bold dashed lines) in two quite different Augusts. The model captures the variability very well. Thin solid lines show simulated ice thickness. (after [22]).

Figure 9. Modeled multiyear ice concentration (in percent) on 31 December 1984. The dashed contour is 50%. (after [23]).

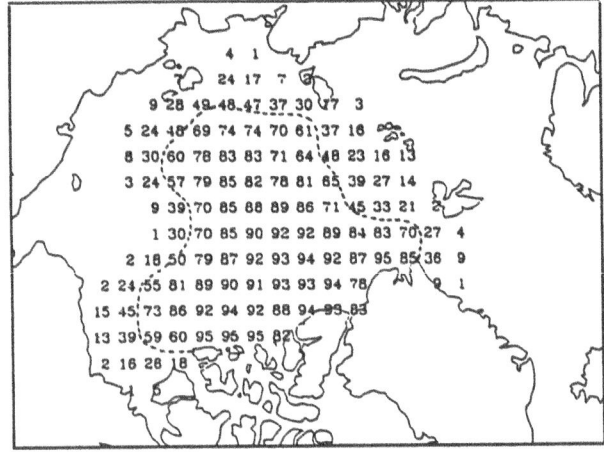

Figure 10.
Right: The seven cells of an ice model [24]. The Arctic Ocean here
consists of the seven cells in upper case letters; their combined
area is 6..92 x 10^6 km^2. Exchange between cells is given by buoys
with 10-day temporal resolution; the arrows show the mean
velocities.
Lower Left: One-year running mean of the concentration of four
ice types: ridged multiyear (RMY), level multiyear (LMY), ridged
first-year (RFY), level first-year (LFY), and open water (OW).
Lower Right: One-year running mean of the concentration in the
Canada Basin. Individual cells have much more variability than the
Arctic Ocean as a whole.

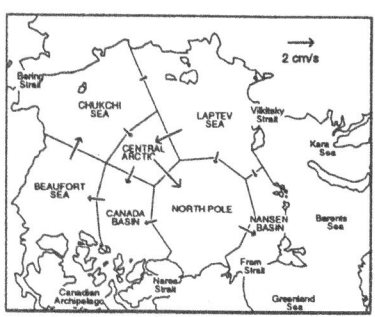

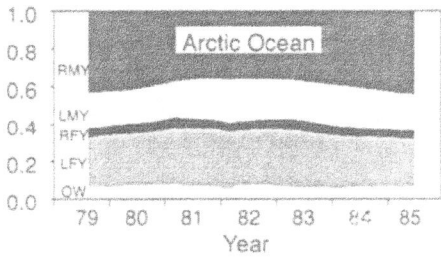

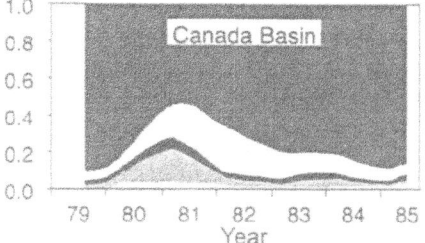

3. Ice and Ocean Freshwater Budget

3.1. MOTIVATION

Of course, to elucidate the hydrologic cycle, one wants to know the mass of sea
ice stored, produced, melted, or transported within and across the boundaries of the Arctic
Ocean. No satellite sensor has yet produced data sets on ice thickness that have been
applied to the subject of this book. However, some methods to estimate ice thickness have
been developed, and in Section 7 we will comment on their status. To say something
useful about ice mass, we can turn to models that contain thickness as a dependent
variable and are constrained by satellite observations.

Some additional definitions will be useful. One is that of the distribution of ice
thickness. Within an area of some tens of kilometers diameter, there are many different
ice thicknesses, and we describe these by a distribution function $g(h)$ defined by:

$$g(h)dh = \text{fraction of area covered by ice of}$$
$$\text{thickness } h \text{ to } h+dh \tag{3}$$

The cumulative distribution $G(h)$,

$$g(h) = \int_0^h g(h)dh \qquad (4)$$

= fraction of area covered by ice no thicker than h

is useful at times. Ice concentration is related to the thickness distribution: it is the fraction of area covered by ice thicker than some threshold thickness ε.

$$C = \int_\varepsilon^\infty g(h)dh \qquad (5)$$

An observational difficulty is that this threshold is not known exactly, and different techniques have differing capabilities even to define the threshold and certainly do not all have the same threshold, so some ambiguity is introduced.

The quantity we most want for hydrology is the mean ice thickness:

$$\overline{h} = \int_0^\infty hg(h)dh \qquad (6)$$

The average over all thicknesses including open water (zero thickness) within the defining area. Another mean we might call the "ice-only mean thickness" $\hat{h}$.

$$\hat{h} = \frac{1}{C} \int_\varepsilon^\infty hg(h)dh \qquad (7)$$

It is the mean over only the ice-covered portion $(h > \varepsilon)$ of the defining area. The two means are related by $\overline{h} = C\hat{h}$. The ice-only mean $\hat{h}$ is always greater than $\overline{h}$ because it excludes the open water area from the average. Care must be taken to be sure which mean is being discussed. We find $\overline{h}$ to be much more useful; for instance, to integrate the entire mass of ice, one can integrate $\overline{h}$ over some large region, but to find the same quantity from $\hat{h}$, one must also know C.

3.2   ICE THICKNESS FROM ASSIMILATED CONCENTRATION DATA

The concentration-assimilating box model of Thomas *et al.* computes the thickness distribution of ice in each cell on the 10-day resolution over seven years. There are actually four thickness distributions keeping track of level and ridged amounts of both first-year and multiyear ice. The distributions are illustrated in Figure 11. The mean thickness in each cell is compared with model and submarine observations in Figure 12. The results show a considerable temporal and regional variability of ice thickness, as shown in Figure 13.

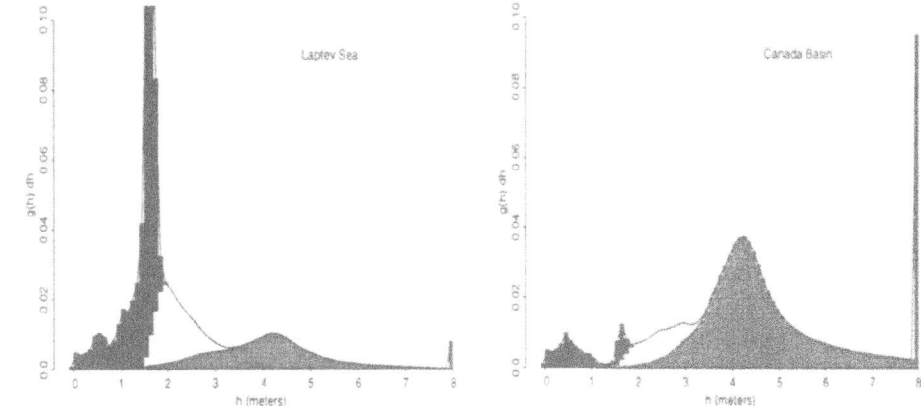

Figure 11. Seven-year mean thickness distributions in late May for two of the cells shown in Figure 10. Three distributions are shown here: darkest is level first-year ice, lightest is level multiyear ice, and intermediate is all ridged ice. The three are shown added together—that is, the outline is the total distribution. All ice thicker than eight meters is shown in the 8-m bin. The Laptey Sea shows a preponderance of level first-year ice; the Canada Basin on the other hand has mostly ridged ice, making it much thicker. (after Thomas *et at.,* 1996)

Figure 12. Winter ice thickness from several sources. (after [24]). T stands for the box model and concentration assimilation of Thomas *et at.* Two estimates are model results: FH is Flato and Hibler *[26]* and F is Flato [27]. The last, B, is from submarine estimates of Bourke and Garrett [28].

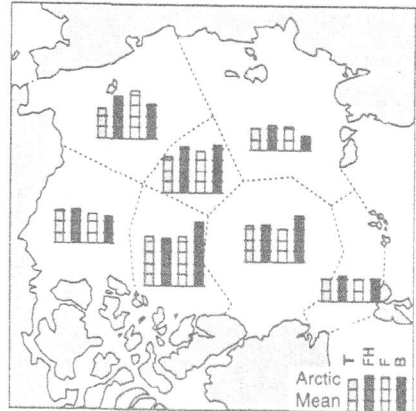

3.3. ICE AND OCEAN FRESHWATER STORAGE AND EXPORT

3.3.1 *Definitions*

In dealing with freshwater budgets, it is important to take care with definitions. We give a few useful numbers here and summarize them in Table 2. Salinities of water and ice and the volume density of ice can affect storage and flux terms by factors that are too large to

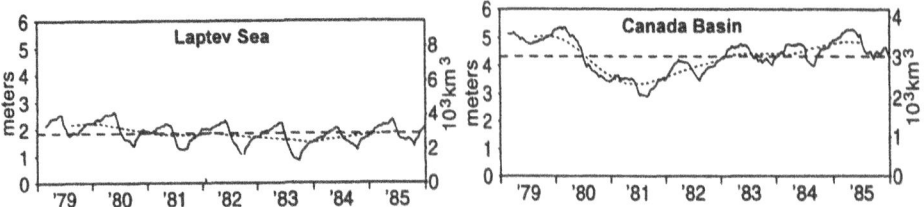

Figure 13. The seven year variation of mean ice thickness in two sample regions. The dashed curves are 1-year running means. Note the considerable difference in the mean, and the combination of a seasonal cycle of about a meter superimposed on dramatic interannual variations (after [24]).

be neglected as we review various estimates of the arctic freshwater budget. We multiply ice volume by $f_m = 1/1.12 = 0.89$ to convert to meltwater volume. To compute the freshwater equivalent of ocean water or ice meltwater with a salinity S, we multiply volume or volume flux by the factor:

$$fs = \{S_{ref} - S\}/\{S_{ref} - 0\} \qquad (8)$$

We use a reference salinity S_{ref} for the Arctic Ocean of 34.8 psu (following Steele *et al.* [29], who followed Aagaard and Carmack, [3]). For freshwater with $S = 0, f_s$ is unity, but as S increases toward S_{ref}, the freshwater equivalent decreases towards 0. We take average ice salinities from Thomas *et al.* [24]. Converting ice volume to its freshwater equivalent reduces the number by nearly 20%.

TABLE 2. Conversion table for converting ice volume to freshwater equivalent and yield. Use the numbers in the right-hand column to convert ice volume in ~ km³ to yield in meters

	Salinity (psu)	Water to FW equivalent (f_{Sv})	Ice to meltwater (f_m)	Ice volume to FW equivalent ($f_S \cdot f_m$)	Ice volume to FW yield ($f_S \times f_m/A$)
Arctic Ocean	2.4	0.93	0.89	0.83	0.120
Fram Strait	2.7	0.92	0.89	0.82	0.118

It is handy to think of a volume V or a volume flux Q as spread out over an ocean of Area A; this gives: es·

$$yield = V/A \ or \ Q/A \qquad (9)$$

made clear by the context. For the Arctic Ocean exclusive of the Kara, Barents, and GIN seas, two references (Thomas *et al*, [24]; Steele *et al.* [29]) use an area A of $6.92 \times 10^6 km^2$, which we use in this chapter. A standard reference on arctic freshwater balance [3] uses an area including the Kara and Barents seas of 9.55 x 10^6 km^2. One finds a variety of units; the following equivalency is useful:

$$\text{Volume flux: 1 Sverdrup} = 10^6 \text{ m}^3/\text{s} = 31.54 \text{ x } 10^3 \text{ km}^3/\text{yr} \qquad (10)$$

3.3.2. *Results*

In the box model of Thomas *et al.* [24] the seven-year mean ice thickness is 2.70 m with a freshwater yield of 2.24 m. This gives ice volume storage of 18.7 x 10^3 km^3 and a freshwater volume of 15.5x10^3 km^3. That model was extended to the upper 200 m of the Arctic Ocean by Steele *et al.* [29]. They found a seven-year mean freshwater storage of 7.65 m in the upper ocean, equivalent to 52.9 x 10^3 km^3.

The studies provide terms in the budget. Thomas *et al.* give a mean ice volume flux through Fram Strait of 1.9 x10^3 km^3/yr, and an equal net ice production. This value is somewhat lower than other estimates as will be seen in the broader discussion in Section 5.2. Steele *et al.* review the entire freshwater balance of the Arctic Ocean. Their estimates include several interesting variants from earlier estimates. In addition to the "low" Fram Strait flux just mentioned, they find a freshwater export through the Canadian archipelago of 1.23 x 10^3 km^3/yr (0.039 Sv), a third higher than the 0.92 x 10^3 km^3/yr (0.029 Sv) of [3], because it draws on shallower, fresher water. Their balance also includes a significant downwelling from the upper 200 m in the amount of 1.45 x 10^3 km^3/yr (0.046 Sv).

The seven-year time history of Arctic Ocean ice volume in Figure 14 shows an annual cycle of about 6 x10^3km^3/yr, about a third of the mean ice storage. The inter-annual variability is about half that. Much larger variability occurs on regional scales, as was shown in Figure 13. The variability in annual ice and oceanic outflows is shown in Figure *15*. By far the largest variations arise from Fram Strait export; annual ice outflow can vary by 100%. The high outflow in the early 80s corresponds to a period of diminished convection in the GIN Sea reported by Schlosser [30]. The ocean transports through Fram Strait and the Canadian archipelago vary out of phase, suggesting an alternating destabilization and convection in the Greenland Sea and the Labrador Sea. A picture of higher frequency variability is given in Figure 16. The ice export (advection) and

Figure 14. Time history of ice volume in the seven-cell box model of Thomas *et al.* *[24]*.

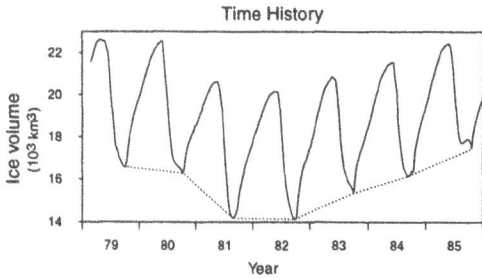

Figure 15. (left) Freshwater transport from the Arctic Ocean through Fram Strait and (right) the ocean export through Fram Strait and the Canadian archipelago (after [29]).

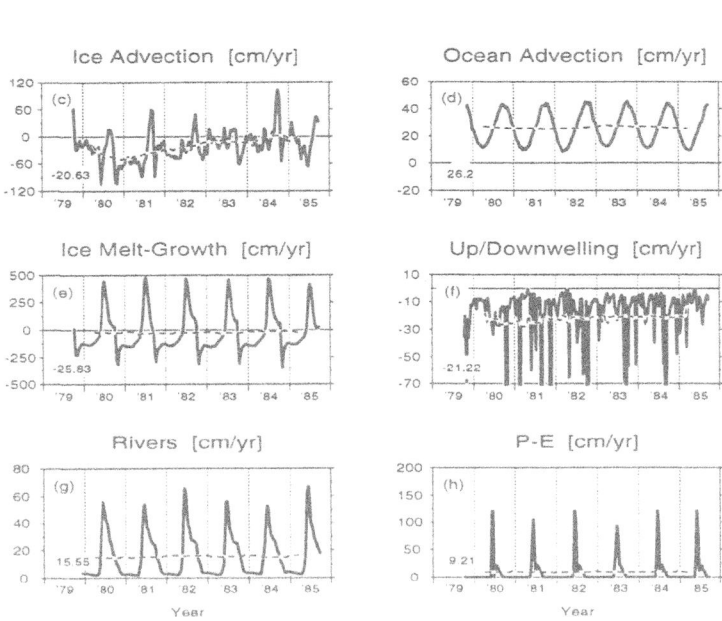

Figure 16. Time history of terms in the Arctic Ocean freshwater balance (after [29]). P-E is precipitation minus evaporation. All terms are yields in freshwater equivalents for an Arctic Ocean of area 6.92 x 10^6 km^2.

downwelling are seen to be variable on synoptic time scales; other terms are dominated by their annual cycle, which can consist of quite sharp spikes, as for the riverine input and the spring melt of accumulated snow on sea ice in the P-E panel of the figure.

4. Ice Circulation in the Arctic Ocean

4.1. MOTIVATION

Like any moving medium, an understanding of the sea ice mass balance requires that we describe where and how ice is produced, how it moves about the ocean surface, and where and how it melts. We can think of these advective motions as playing out over months and years. Motion plays a strong role, not only in advection but also in determining processes occurring as the ice moves along. Deformation, the spatial gradient of the velocity field, has a immense effect on thickness, ice production, and melt. Shearing and convergence cause ice to raft to twice its thickness and to ridge to many times its thickness thus increasing the storage of ice. Divergence and to some extent shearing cause the ice cover to pull apart and form leads, areas of open water. In winter, these are sites of rapid ice growth, and in summer they are sites of intense solar absorption because of the water's low albedo. These deformations are important on time scales more like days to weeks.

Ice modeling requires the best possible representation of ice motion in order to properly represent the ice mass balance. The more data we can bring to bear on modeling ice motion the better. Ice motion is also important as the driver of the ocean circulation. Tracking ice motion for this purpose is the satellite analogue of observing wind stress on the non-ice-covered ocean surface by means of a scatterometer.

4.2. SATELLITE OBSERVATIONS OF ICE MOTION

It is possible to recognize ice motion in sequential imagery from a wide variety of sensors. The smaller the spatial resolution, the better trackable features stand out. One of the most appealing is the Advanced Very High Resolution Radiometer (AVHRR), which has bands at visible, near infrared, and thermal infrared frequencies. The thermal band can be used during winter darkness. Its resolution is either 1 km or 5 kin, depending on the data product. Its only drawback is that clouds obscure the surface, moderately in winter and quite substantially in summer. Data products from this source for several years from the AVHRR Polar Pathfinder project are available through the National Snow and Ice Data Center (see http://www.nsidc.colorado.edu/NSIDC/index.html or http://polarbear. colorado.edu/).

It has recently been realized that even quite spatially coarse imagery can be successfully tracked, and this has led investigators to track first the 12-km imagery from 85 GHz SSM/I and then the 25-km imagery from 37 GHz SSM/I and SMMR, pushing the trackable record back to late 1978 [31,32,33]. An example is shown in Figure 17.

4.3. APPLICATIONS OF SATELLITE VELOCITY DATA

Because large-scale ice motion data sets from satellites are only now coming into pro-duction, we see so far only the beginnings of their application to the Arctic Ocean ice mass balance. An analysis of the response of the ice circulation patterns to changes in the

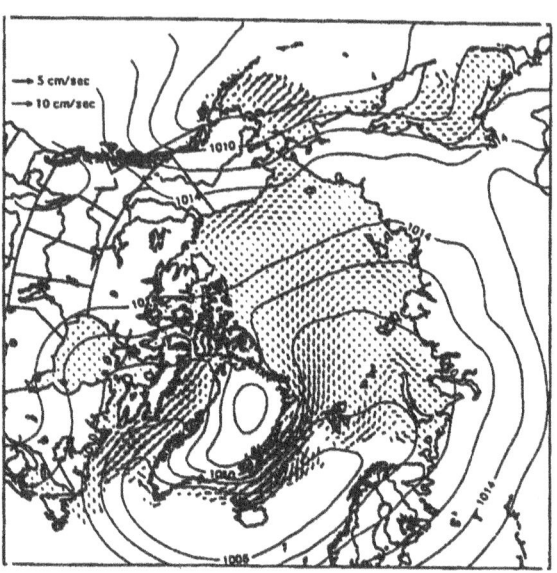

Figure 17. Mean sea ice motion for 1988-1994 from daily 85.5 GHz SSM/I data. The contours are mean sea level pressure in hPa from the NCEP reanalysis for the same years. To a rough approximation, ice motion follows isobars. (after [32]).

wind patterns as exemplified by atmospheric surface pressure shows striking changes in response to the variation known as the North Atlantic Oscillation [33]. Figure 18 shows how periods of low NAO index - a weakened Icelandic low - produce a very strong polar surface anticyclone and ice gyre in the Beaufort Sea. Periods of high NAO index substantially weaken the anticyclone and nearly eradicate the gyre. Even the mean field shows a less developed gyre and a shifted mean flow pattern.

Fields of motion data are clearly important for validating ice models. Steps in this direction have been taken by Fowler *et at.* [34]. Assimilation of these data sets has been started by Maslanik and Maybee [25] and Meier *et al.* [35].

The more regional application of these data sets to the ice fluxes at boundaries of the Arctic Basin is the subject of a later section.

5. Ice Fluxes at Straits

5.1. MOTIVATION AND DEFINITIONS

About 14% of the ice mass in the Arctic Ocean is exported each year through Fram Strait. Ice flows through other passages and straits in lesser amounts. Satellites have long been used to give quantitative estimates of the area flux flowing through Fram Strait, of the profile of velocity across the strait, and the concentration of ice within the strait. The export of ice from the Arctic Ocean into the Greenland-Iceland-Norwegian seas and through the Canadian archipelago into the Labrador Sea is thought to play a major role in controlling where and how much vertical convection can occur in these two basins. This convection is a link in the thermohaline circulation of the global ocean.

Figure 18. Mean ice motion vectors from passive microwave imagery and surface pressure contours for October through May, 1978—1996.
(a) Mean over periods when the NAO index is greater than +1.
(b) Mean over periods when the NAO index is less than —L.
(c) The difference between the two: (a)—(b). The contour interval is 1 mbar. (After Kwok [36]).

The fundamental quantity we want as a flux measure across a boundary segment is the ice volume flux:

$$Volume\ flux(t) = \int \overline{h}(x, t)u(x, t)\,dx \tag{11}$$

or, if one were using the ice-only mean thickness of Eq. (7),

$$\int_{Across\ strait} C(x, t)\hat{h}(x, t)u(x, t)\,dx \qquad \hat{h}\ \text{excludes open water} \tag{12}$$

We prefer the first expression since it is independent of any definition or observation of concentration. Satellite motion data provide us directly with ice area flux:

$$Area\ flux(t) = \int_{Across\ strait} u(x,t)\,dx \qquad (13)$$

or if one wishes to exclude open water from the definition,

$$Ice\ concentration\ flux(t) = \int_{Across\ strait} C(x,t)u(x,t)\,dx \qquad (14)$$

We do not make any use of the latter definition here.

A summary of observations of area and volume flux through Fram Strait is presented in Table 3; this is a complicated table, and the various columns are discussed in turn in the following sections.

5.2. ICE FLUX THROUGH FRAM STRAIT

5.2.1   Area Flux

We begin with those estimates that do not use satellite data. An early estimate of the mean area flux through Fram Strait by Vowinckel [37] was made from surface pressure data; his area flux estimate is 0.90 units. The unit of area flux in this section is $10^6\ km^2$ /yr. A later estimate of ice concentration flux by Englebretson and Walsh [38], also with pressure data, is some 23% lower. Ice concentration cannot account for the difference since concentrations in Fram Strait are about 90%. As sufficient buoy drift data accumulated two estimates of the flux were made: an area flux of 1.21 units by Vinje and Finnekåsa [39] and ice concentration flux of 0.84 units by Moritz[40], some 30% lower.

Since the late 1970s the movement of ice through the strait has been estimated by satellite with visible and later with synthetic aperture radar imagery by Vinje and coworkers. The earliest estimate by Vinje [41] of 0.75 units is from a rather short data record. Later longer records revise this estimate slightly upwards. A record of two years' length was taken from AVHRR by Martin [[42]; see also Martin and Wadhams, [43]. A typical day's motion is shown in Figure 19. The two-year mean area fluxes and imputed volume fluxes are given by season in Table 4. Note that velocities are strongest in early winter, strong in winter, drop off somewhat in spring, and are quite weak in summer. Because the data period is one of high NAO index, which strengthens outflow, one might expect this estimate to be above longer-term estimates, but it is not.

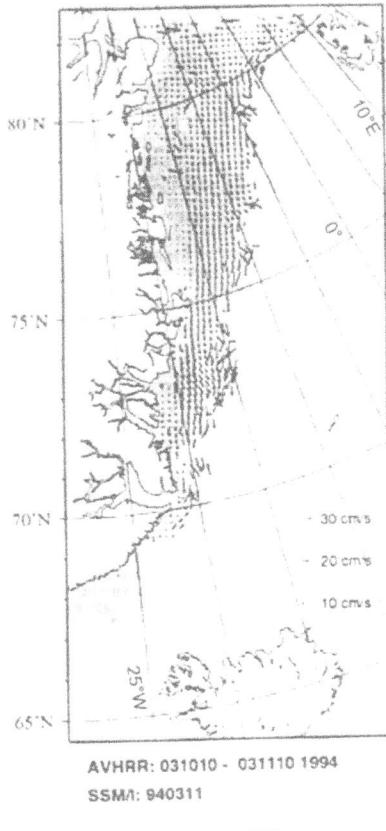

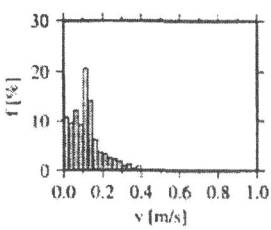

AVHRR: 031010 - 031110 1994
SSM/I: 940311

Figure 19. (left) An example of ice drift through Frani Strait from AVHRR images on March 10 and 11, 1994. The ice concentration is shown by the grey scale with the scale shown at the bottom. (right) A histogram of drift speed, in percent. (after [42])

Another analysis of Fram Strait flux has been made from an 18-year record of daily velocities tracked from SMMR and SSM/I [44]. The tracked data cover the eight winter months October through May; summer ice motions are inferred from surface pressure data to complete the annual estimates. The mean seasonal cycle is shown in Figure 20. Two 3-month means from this figure can be compared (in parentheses) with values in the second column of Table 4. The data in the figure show a peak in the deep of winter, different from the early winter peak in the table, but the different periods of the two data sets could well account for the difference.

TABLE 3. Observational estimates of the ice flux at Fram Strait. The units are such that area flux multiplied by mean thickness in meters gives volume flux; the "inferred ice thickness" is computed simply as the ratio of mean volume flux to mean area flux. When interannual variability was estimated, the range of annual means is given on a second line in parentheses. The area flux is the integral of velocity across che flux gate unless otherwise indicated. The volume flux is ice volume and in not converted Co freshwater equivalent, buc see Sec.3.3. Uses of satellite data are underlined. To convert volume flux in 10^3 km^3/yr to Sverdrups, divide by 31.5. Values in boldface are those in which we place the most confidence for estimating the long-term mean as explained in the text.

Source	Ice velocity	Ice state	Area flux - ∫udx (10^6 km^2/yr)	Ice volume flux (10^3 km^3/yr)	(Sv)	Inferred ice thickness (m)	Flux gate (°N)	Duration of record
Vowinckel [37]	Pressure difference + steady current	Complete ice coverage, 3 m	0.90	3.1	0.10	3.4		
Englebretson and Walsh [38]	Pressure difference + steady current	NPOC ice concentration maps, some with passive microwave data	∫ Cudx = **0.69**				79-80	30 years: 1953—1984
Vinje [41]	LANDSAT tracking		0.75					2 months: Apr—May 1976
Wadhams [45]	Apr-May76 (Vinje, [41])	Submarine, Apr-May79	1.10	40	0.13	3.6	80	2 months: Velocity: Apr—May 1976 Thickness: Apr—May 1979
Vinje and Finnekåsa [39, Table 2]	Buoy drift	Concentration from ice charts, in situ thickness, draft from submarines	**1.28**	**5.0**	0.16	4.2	81	Motion, 9 years: 1976—1984
Moritz [40]	Buoy and station drift	Concentration from charts with some passive microwave data	∫Cudx = 0.84					10 years: 1965, 1976—1984

*FWE is "freshwater equivalent." 1MW is "ice meltwater." (Aagaard and Carmack use 4 psu for ice salinity, and speak only of a salinity correction in converting ice to FWE)

TABLE 3. (cont.)

Source	Ice velocity	Ice state	Area flux ∫udx (10^{6}km^2/yr)	Ice volume flux (10^{6}km^3/yr)	(Sv)	Inferred ice thickness (m)	Flux gate (^{o}N)	Duration of record
Aagaard and Carmack [3]	0.16 Sv [Vinje & Finnekåsa, [39]] less 006 Sv [Untersteiner, 1988]			3.40 (FWE*=2.79)	0.11 (IMW* = 0.10		81	
Thomas and Rothrock [46]	Buoy drift (Moritz, [40])	SMMR total and multi-year concentrations	∫Cudx = 0.83 (0.50-1.21)				81	7 years: 1979-1985
Thomas et al. [24]	Buoy drift	SMMR ice concentration with ice thickness model	0.97	1.9 (1.1-3.0)	0.060 (0.03-0.10)	2.0	81	7 years: 1979-1985
Martin [42]	AVHRR tracking	2.2 - 3.0 m assumed	0.71	1.6-2.1	0.05-0.07	2.2-10	79-80	2 years: 1993-1994
Vinje et al. [47]	Buoy drift, SAR tracking (93-94) and pressure difference	Ice draft from moored sonars	1.10 (0.82-1.53)	2.9 (2.1-4.7)	0.088 (0.07-0.15)	2.5	79	6 years: 1990-1996
Kwok and Rothrock [44]	SMMR, SSM/I tracking; pressure diff, in summer	Vinjeerat [1998]	0.92 (0.72-1.12)	2.4 (1.8-3.3)	0.076 0.06-0.10)	2.6	81	Area, 19 years: 1978-1996 Volume, 5 years: 1990-1995

*FINE is "freshwater equivalent." 1MW is "ice meltwater." (Aagaard and Carmack use 4 psu for ice salinity, and speak only of a salinity correction in convening ice to FWE)

TABLE 4. Seasonal sea ice area and ice volume fluxes across 79°N (computed from Martin, 1996). Values are means for 1993 and 1994. The values in parentheses are taken for comparison from Figure 20.

	Area flux (10^6 km^2/yr)	Vol. flux if h=2.2m (10^6 km^3/yr)	Vol. flux if h=3.0m (10^6 km^3/yr)
Jan—Mar	0.85 (1.10)	1.9	2.5
Apr—Jun	0.66	1.5	2.0
Jul—Sep	0.13	0.3	0.4
Oct—Dec	1.22 (1.02)	2.7	3.7
Ann. avg.	0.71	1.6	2.1

The area flux for the 18-year period is shown in Figure 21. The mean flux is 0.92 units. One can discern a slight increase during the latter half of the Oct—May record. During the deep winter (December through March) the flux is quite strongly correlated with the North Atlantic Oscillation index, and it is during the high NAO index of the late 1980s and early 1990s that these winter fluxes through Fram Strait are stronger than normal [44].

Those estimates of mean area flux from a record of two years or longer from satellite tracking or buoys are shown bold in the table; their average is 1.0 x 106 ~2 yr~¹ with a standard deviation of 10% which is a reasonable estimate of the uncertainty in the mean. The area flux clearly varies interannually by 150%, and over a seasonal cycle of a factor of 2.4 [47, 44]. The effort at more direct measurements since 1964 has not changed Vowinckel's estimate of the mean by much but has certainly pinned down the uncertainty and the variability.

Figure 20. Monthly mean area flux for the years 1978 to 1996 from passive microwave tracking (after [44]).

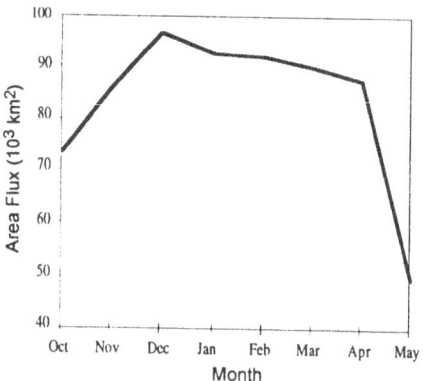

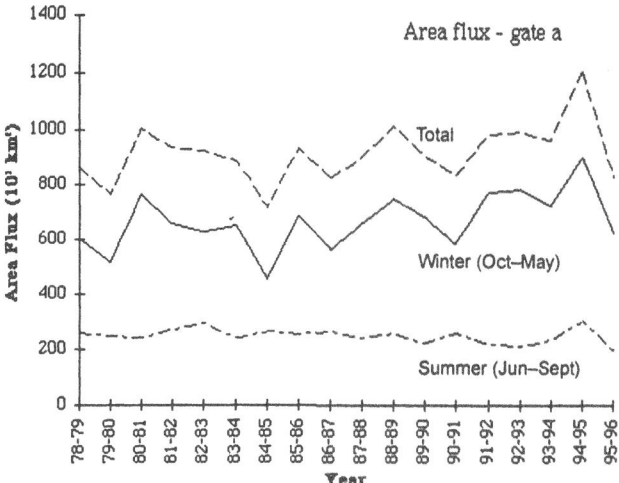

Area flux - gate a

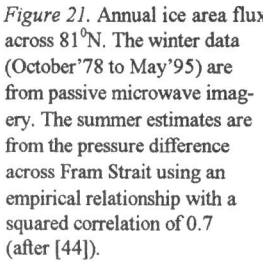

Figure 21. Annual ice area flux across 81°N. The winter data (October'78 to May'95) are from passive microwave imagery. The summer estimates are from the pressure difference across Fram Strait using an empirical relationship with a squared correlation of 0.7 (after [44]).

5.2.2 Velocity Profiles

Figure 22 shows velocity profiles across Fram Strait determined by buoys and by satellite tracking. The two profiles of Martin show the large seasonal variability, but the shortness of those records means they are not representative of a mean profile. The reversal or backwater near Greenland shown in Martin's summer profile is noted in other sources, such as Vinje and Finnekåsa [[39]. All but Moritz's profiles show maxima right at the shelf break on the Greenland side. Monthly mean profiles across 81 °N are shown in [44].

The profiles show why sampling this flux is challenging and why this quantity is elusive. Ice velocity has strong gradients, not only across the strait as illustrated but also in the north—south axis of flow [41]. Kwok and Rothrock [44] report that the area flux increases by 16% between their two flux gates at about 81°N and 79°N. The line segment representing the strait, or as in Table 3 "flux gate," is chosen differently depending on the purpose or simply on the coordinate system of the investigation. The way one averages the data whether curve fitting or binning determines the shape of the profile. One cannot use data only when the ice cover is present or only when a buoy drifts through without biasing the profile, especially at the eastern end. When no ice is present on the flux gate, the ice velocity of zero must be averaged with those periods when a substantial ice drift occurs (see Moritz's treatment with "bogus" data). Because of the strong gradients in both motion and thickness in this region, coincidence of measurements is extremely important in flux calculations.

5.2.3 Ice Volume Flux

To extend the area fluxes to estimates of volume flux has been a challenge. The early estimates using submarine ice draft data and *in situ* thickness from drilling give ice volume fluxes of 4.0-9.1 units by Wadhams [45] and *5.0* units by Vinje and Finnekåsa [39].

434

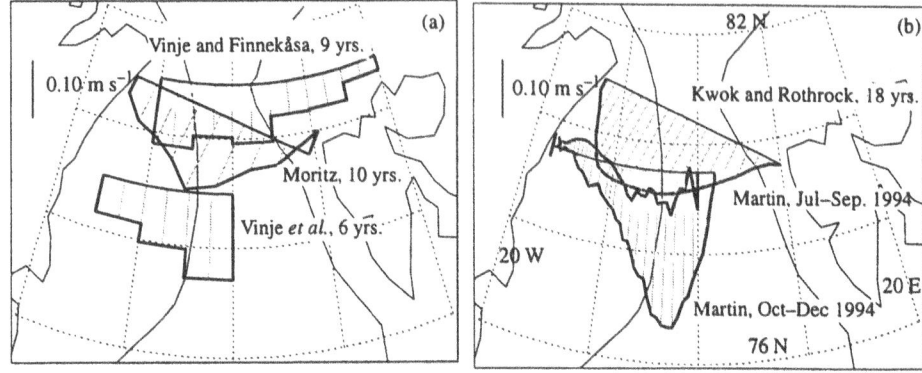

Figure 22. Velocity profiles in Fram Strait (a) from buoy and mixed observations, (b) from satellite observations. The 1000-in isobath is shown, with Greenland on the left and Spitzbergen on the right. Each graphic contains a velocity scale.

The unit of ice volume flux in this section is 10^3 km^3 /yr. Results discussed here are summarized in Table 3. Aagaard and Carmack revised these numbers downward to 3.4 units to allow for melting of ice by the West Spitzbergen current. When they assimilated concentration data, Thomas *et al.* [24] obtained the much lower value of 1.9 units. The effective thickness of their export was only 2.0 m, down from 4 m of earlier estimates; more recent thickness estimates have been about 2.5 m.

Vinje *et al.* [47] used a 6-year record of ice thickness from moored sonars to estimate an ice volume flux of 2.9 units. Kwok and Rothrock [44] combined those ice thickness estimates with motions from passive microwave imagery to obtain a 5-year mean volume flux of 2.4 units; their area flux and volume flux estimates are about 20% lower than those of Vinje and Finnekåsa. The four summer months contribute about 24% of their annual flux estimates.

What is the best estimate of the mean? Estimates range from 1.6 to 5.0 x 10^3 km^3 /yr. The uncertainty comes primarily from thickness. The inferred mean thickness of the export is from 2 to 4 m, a range of 100%. This range does not seem to correspond to the latitude of the flux gate. Note that values published earlier (higher rows in the table) are greater and are based on thickness estimates from the 1970s. The lower values depend on the thickness record in the 1990s. It is quite possible that the thickness varied between these decades. The early 1990s with their high NAO index would draw relatively thin ice from the Eastern Arctic through Fram Strait, while the earlier period might have drawn thicker ice from north of Greenland. Attributing the most weight to the three bold numbers for ice volume flux in Table 3, the best estimate can be taken as their average, or 3.4 ± 1.5 x 10^3 km^3 /yr.

A multiyear record of volume flux is patched together in Figure 23. One sees an increased export of ice in the 1990s over what was estimated for the early 1980s. The range of annual values in the figure and in Table 3 is from a low of 1.1 units in 1984 to a high of 4.7 units in 1994—1995, and we take this range of a factor of 4 to indicate the

Figure 23. Annual volume flux through Fram Strait, seven years from a box model assimilating ice concentration on the left (after Thomas *et al.* [24]), and five years tracked from microwave tracked motions on the right (after [44]).

interannual variability, not observational error. This interannual variability certainly contributes to the range of estimates of the mean.

Of the many tests one might make of ice model behavior, the mean and the variations of area and volume fluxes at Fram Strait seem to rank high. Table 5 gives modeled fluxes from the literature. All but one model have two categories of sea ice: open water and level ice of a variable thickness. The exception, Zhang *et al.* [[48], uses a 12-category thickness distribution ice model that explicitly simulates ice ridging and tends to produce thicker ice than 2-category models; thicker ice tends to move more slowly, giving a lower area flux but a higher volume flux than 2-category models.

TABLE 5. Model estimates of the ice flux at Fram Strait. The units are such that area flux multiplied by mean thickness in meters gives volume flux. The "inferred ice thickness" is computed simply as the ratio of mean volume flux to mean area flux. When interannual variability has been estimated, the range is given on a second line in parentheses.

Fram Strait ice flux	Area flux $10^6 \, km^2/yr$	Volume flux	Inferred ice thickness (in)	Period
Walsh et al. [22]	0.63	1.4 (0.6-2.2)	2.2	195 1-1980
Flato and Hibler [44]		1.9		1981-1983
Hakkinen [50]		2.0 (1.1-3.6)		1955-1975
Harder et al. [51]		2.7 (2.0-3.8)		1986-1992
Zhang et al. [52]	1.13 (0.78-1.34)	2.6 (0.8-4.5)	2.3	1979-1985
Zhang et al. [48]	0.87 (0.62-1.58)	3.3 (2.0-5.7)	3.8	1979-1996

436

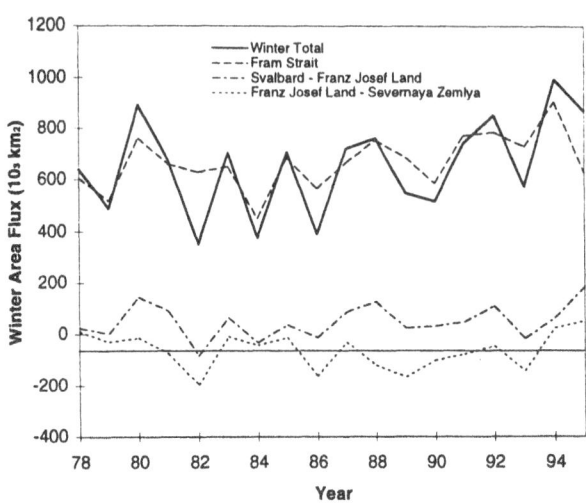

Figure 24. Ice area flux between the Arctic Ocean and the Atlantic Ocean for winter: October through May. These are tracked over 2-day intervals from passive microwave imagery following Kwok *et al.* [33]. Export from the Arctic Ocean is shown positive.

5.3.   ICE FLUX THROUGH OTHER STRAITS

There are wide passages between the Arctic Ocean and the Barents and Kara seas. The area flux through them is shown in Figure 24 to be a small fraction of that through Fram Strait. The 18-year mean winter ice export from the Arctic Ocean is 0.047×10^6 km^2 between Svalbard and Franz Josef Land. Approximately 0.065×10^6 km^2 of sea ice is imported into the Arctic Ocean between Franz Josef Land and Severnaya Zemlya. Combined, and assuming ice no thicker than 2 m, and making no allowance for the four summer months, this gives the very small net import to the Arctic Ocean of no more than 0.04×10^3 km^3 /yr.

The variability of the ice flux through these passages is high and negatively correlated with the NAO [Kwok, submitted]. High ice import is associated with positive phases of NAO. The large-scale cyclonic pattern of atmospheric circulation due to the Icelandic Low (see Figure 18) tends to reduce ice export between Svalbard and Franz Josef Land, and enhance ice import between Franz Josef Land and Severnaya Zemlya.

The flux of ice and water through the Canadian archipelago is treated in another chapter [Melling, this volume].

6. Atmospheric Moisture Balance

6.1. MOTIVATION

The freshwater budget of the Arctic surface is strongly influenced by atmospheric processes. Winds transport moisture from lower latitudes to the Arctic where much of it condenses and precipitates as snow and rain. Studies using rawinsondes around the perimeter of the Arctic Basin reveal that penetrating cyclones account for around 85% of the poleward moisture transport into the Arctic [53]. Although synoptic activity is stronger in the winter, summertime storms are warmer and contain more moisture, resulting in summertime moisture transport peaks. Freshwater also evaporates from the ice as well as the ocean surface through cracks in the ice pack. These freshwater fluxes are large during certain times of the year and are strongly influenced by atmospheric conditions.

Precipitation, evaporation, and atmospheric moisture influence the Arctic in a variety of ways. Precipitation is a direct flux of freshwater to the surface of the ice pack and eventually the ocean. Snowfall maintains the high albedo of sea ice. By supporting part of the temperature difference between the atmosphere and the ocean, snow accumulation allows thinner sea ice to exist in equilibrium. Evaporation increases the salinity of the ocean surface and can be a significant local source of moisture to the atmosphere. During the winter, when the atmosphere is cold and dry, evaporation from the ocean surface often dominates the atmospheric moisture budget. Atmospheric moisture influences the surface energy balance by absorbing and re-emitting longwave radiation. An understanding of the atmospheric moisture budget is necessary to quantify these fluxes and effects and to predict arctic change and variability in the future. Data from satellites can provide historical daily atmospheric fields of moisture and temperature over the Arctic, enabling the study of a wide variety of moisture related topics. Here we discuss only the atmospheric moisture budget.

6.2. SATELLITE OBSERVATIONS OF ATMOSPHERIC MOISTURE

The TIROS-N Operational Vertical Sounder (TOVS) has flown on NOAA polar orbiting satellites from 1979 to the present. The orbits cross the poles about 14 times per day with a swath approximately 2200 km wide, providing consistent daily coverage over the arctic region. Two of the TOVS instruments are used to retrieve temperature and vapor profiles of the atmosphere as well as surface characteristics.

A High-Resolution Infrared Radiation Sounder (HIRS-2) measures radiation in 19 infrared bands (3.7-15 μm) and one visible band (0.7 μm). The Microwave Sounding Unit (MSU) measures radiation in four bands around 55 GHz. The absorption and re-emission of radiation by well-mixed atmospheric gasses - namely carbon dioxide, nitrous oxide, and oxygen - peak at different altitudes for the different bands of TOVS. Therefore, weighting functions of height for each radiation band can be constructed and used with inverse radiative transfer theory to convert the measured radiances into atmospheric temperature profiles. Vapor profiles are retrieved in a similar manner using TOVS channels sensitive to water vapor.

The Improved Initialization Inversion (31) method, developed at Laboratoire de Mdt~orologie Dynamique du CNRS, in Palaiseau, France, is a physico-statistical retrieval process used to convert retrieved radiances into atmospheric geophysical quantities [54]. It first uses pattern recognition to match the measured radiances with the most appropriate temperature and vapor profile contained in the TOVS Initial Guess Retrieval (TIGR) climate data set, which is comprised of approximately 1800 representative atmospheric profiles and their modeled radiative emission signature. For accurate retrievals over sea ice, separate steps developed by Francis [55] are used to discriminate between cloud tops and the sea ice surface. The final retrieval is then calculated by minimizing the differences in the brightness temperatures of the measured sounding and the first guess profile [56]. This method returns a variety of geophysical variables including temperature data at 10 levels (50, 70, 100, 300, 400, 500, 600, 700, 850, 900 mb, and the surface) and vapor profiles as precipitable water averaged over layers (300-400 mb, 400-500 mb, 500-700 mb, 700-850 mb, and 850-1000 mb). The rms errors for the atmospheric and surface skin temperatures have been shown to be less than 3K and the layer precipitable water is accurate to 20% [57]. The data are archived as the TOVS Polar Pathfinder (Path-P) data set at the National Snow and Ice Data Center (http://www-nsidc.colorado.edu/NSIDC/index.html) and on a 100 km grid and spanning the years 1979-1996.

6.3. ATMOSPHERIC MOISTURE BUDGET FROM SATELLITE DATA

6.3.1 *Methodology*

To construct the complete atmospheric moisture budget one must compute the transport and convergence or divergence of atmospheric moisture. The atmosphere is modeled as a horizontal grid of vertical columns each with a finite number of layers with the moisture budget for each column described by Equation 15:

$$P - E = \sum_{L=0}^{top} \left(-\nabla \cdot (Q_L \nabla_L) \frac{\partial Q_L}{\partial t} \right) \tag{15}$$

where P is precipitation, E is evaporation or sublimation, Q is precipitable water, V is the horizontal wind vector, and L signifies a particular horizontal atmospheric layer. In this model, the horizontal moisture flux convergence minus the change in moisture over a particular time period is the amount of vertical moisture transport in or out of the layer. Cloud storage is considered negligible. Summing all layers over a grid box yields the net transport in or out of the atmospheric column. Since moisture transport across the tropopause is negligible, this transport is the net precipitation to the surface or P-E. Groves [58] uses 6-hourly NCEP/NCAR Reanalysis upper-level winds with TOVS temperature and moisture profiles to calculate the right hand side of Equation 15.

Evaporation is estimated over ice-free ocean surfaces, following usual practice, as the product of the near surface vapor deficit times the near-surface wind speed times a transfer coefficient which parameterizes the stability of the boundary layer and is empirically determined. Over the Arctic Ocean, the presence of sea ice necessitates a somewhat different method. There, evaporation is estimated by [58]:

$$E = \rho_{air} U_{10} [C_i(q_j - q_{air}) Ice + C_w(q_w + q_{air})(1 - Ice)] \tag{16}$$

where Ice is the sea ice concentration estimated by passive microwave data [59]. The 10-meter wind speed is calculated by modifying the surface geostrophic wind according to the TOVS-retrieved atmospheric stability. TOVS surface temperature and low level moisture and temperature retrievals are used with climatological relative humidity profiles (from rawinsondes) to estimate the near-surface vapor deficits ($q_{i,w}$ - q_{air}). The transfer coefficients ($C_{1,w}$) vary seasonally but not spatially and are chosen to tune the monthly climatology of E to independent monthly evaporation climatologies [60].

Monthly precipitation fields are computed as the sum of net precipitation and evaporation—that is, as the sum of the right-hand sides of Equations 15 and 16. These fields are completely independent of any parameterized precipitation processes and can serve as good tests of such parameterizations in models.

6.3.2 Calculated Arctic Moisture Budget

Table 6 shows the annual mean climatological moisture flux convergence, evaporation, and precipitation for two regions north of 65°N calculated from 17 years of data (1980-1996). One region is the Arctic Ocean domain used by Steele *et al.* [19961 and consists of the entire Arctic Basin except for the Barents and Kara seas. The other consists of the Barents, Kara, and GIN seas (see Figure 10). The values in parenthesis are estimates from other sources as described in the caption. Note that the results from this methodology compare well to independent estimates of P-E and precipitation. For the Central Arctic region, despite the presence of sea ice cover, evaporation and horizontal moisture flux convergence provide nearly equal amounts of moisture for precipitation. In mostly ice-free seas (region 2), evaporation is more than twice as large as flux convergence.

TABLE 6. Annual climatologies of the atmospheric moisture budget for two regions. The methodology is described in the text. The domains are as shown in Figure 10.

	Arctic Ocean (cm/year)	GIN, Barents, & Kara Seas (cm/year)
Flux Convergence (P-E) *	10.2 (9.O)*	18.1
Evaporation	9.8	44.2
Precipitation **	20.0 (20.0)**	62.3 (57 5)**

* Value in parenthesis is from [61].

**Values in parenthesis are from an improved Legates/Willmott climatology (from M. Serreze, 1998, personal communication).

The 10.2 cm annual net precipitation converts to a freshwater input rate of 0.022 Sverdrups (1 Sv = 10^6 m^3 s^{-1}), about 1/7 of the annual source of freshwater to the Arctic Ocean system, the other sources being river runoff and inflow from the Bering Sea. Although this may appear to be an insignificant amount, the timing of the deposition of this source makes it an important flux. Steele *et al.* [29] assume that September to June snowfall (P-E) all melts and drains into the ocean at the beginning of the summer (as defined by surface temperature and sea ice concentration). Assuming that the melt all occurs in one month, the freshwater flux from the snowmelt is 0.17 Sv on average over the Arctic—a dominant source of freshwater for that month.

Figure 25 shows the spatial distribution of the climatological annual precipitation (1980-1996) over the polar region and for comparison the Legates/Willmott climatological precipitation (derived from buoy and ice station reports). The difference, of course, is that behind the satellite climatology are daily and monthly fields and a record of interannual and regional variability.

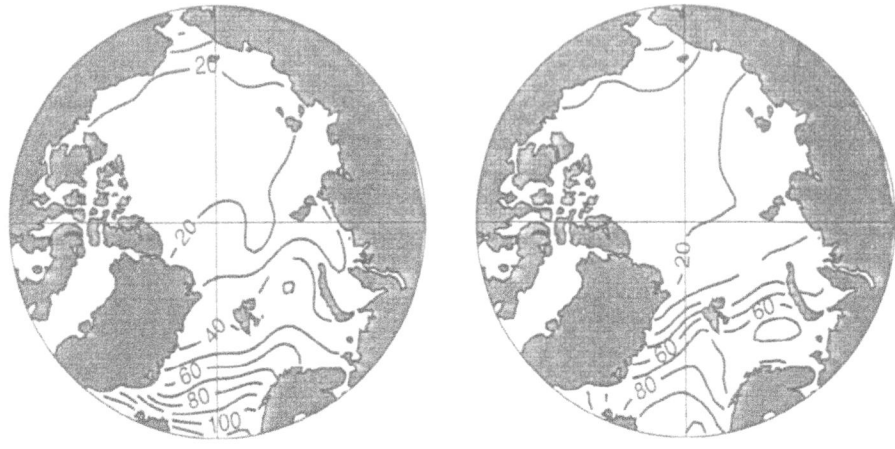

Figure 25. (left) Climatological annual precipitation in cm/year computed from 1980—1996 using Equation 15. (right) Climatological annual precipitation in cm/year from an improved version of the Legates/Willmott data set (Serreze, 1998, personal communication).

Figure 26 shows the climatological seasonal cycles of the different components of the moisture budget for the two regions discussed above. There are striking differences between the moisture budgets for these two areas. Over the central Arctic, vapor flux convergence is at its minimum during the winter and maximum during the summer. This reflects the fact that even though cyclone intensity is stronger during the winter, cold winter air limits the moisture transported and subsequently deposited into the Arctic. Over ice, evaporation is limited in winter by low concentrations of open water but increases in the springtime as the surface temperature rises, leveling off in May and June when the moisture of the overlying air increases and limits evaporation. Over ice free regions (Figure 26, right), flux convergence is similar throughout the year, and evaporation peaks in the wintertime owing to the large temperature contrast between the warm ocean and cold atmosphere.

The high spatial and temporal resolution of satellite data permits the computation of all the moisture budget quantities on a synoptic timescale from 1979 to the present, at a basin-wide spatial resolution. There is substantial regional and interannual variability of these quantities which remains to be utilized in sea ice and moisture budget modeling efforts.

7. Satellite Data on the Horizon

To this point, we have treated work that has been accomplished by applying satellite data sources to topics relevant to the arctic freshwater balance. There are new methods providing new data sets that will soon be able to make contributions to these topics.

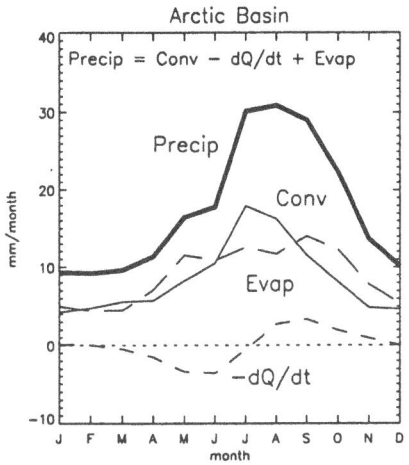
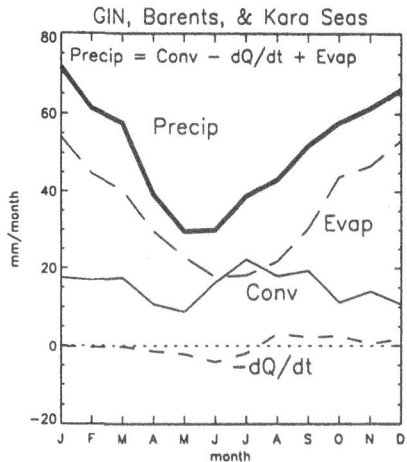

Figure 26. (left) Climatological seasonal cycle of the moisture budget components. (right) Climatological seasonal cycle of the moisture budget components for the Barents, Kara, and GIN seas. Note that the vertical scales differ by a factor of two.

7.1. ICE CONCENTRATION FROM SCATTEROMETERS

Scatterometers measure the radiation backscattered from the ice cover at a range of incidence and azimuth angles. The scatterometer data at *5.3* GHz (from the European Earth Remote Sensing Scatterometer, EScat) and 14 GHz (from the NASA Scatterometer, NSCAT) are relatively free of weather effects that plague sea ice type retrievals from passive microwave data. It has been shown that the backscatter from multiyear ice, first-year ice, and open water has intensities, incidence angle and azimuth angle dependence that are stable and can be used as discrimination factors between sea ice types [62, 63]. Some investigators have moved toward blending the active and passive microwave data sets to eliminate some of the shortcomings of the passive microwave retrieval procedures [64]. There is considerable work to be done to validate and develop these data sets. Gridded EScat backscatter fields, from August 1991 through May 1996, on a polar projection are available through the French Processing and Archiving Facility (http://www.ifremer.fr/cersat).

7.2 ICE MOTION

In Sections 4 and 5 we saw that potent data sets of ice motion are just now coming into use. They have the temporal and spatial coverage and resolution to contribute strongly to the subject of this book and indeed are already providing new results. Below we discuss two additional sources of data that should become important in the near future.

A review of motion data and a discussion of their errors is given by Holt *et al.* [65]; the passive microwave tracking is newer; its errors are discussed by Kwok *et al.* [33] and Meier *et al.* [35].

7.2.1   *Ice Motion from Scatterometers*

Ice motion fields have also been derived from sequential NASA Scatterometer (NSCAT) data [66]. This scatterometer has rather coarse spatial resolution (~25 km), but it is unaffected by cloud cover and is capable of providing basin-scale coverage on a daily basis. The quality of the NSCAT motion fields (~5 km uncertainty) is comparable to that derived from the 85 GHz SSM/I data. There was hope that features in longer wavelength active microwave imagery would be better suited for ice tracking during the summer when passive microwave imagery is affected by atmospheric water vapor and surface conditions. Unfortunately, the spacecraft failed after less than a year of operation between September 1996 and June 1997, not long enough to provide an ice motion record that is complementary to the other data sets discussed above. QuikSCAT, to be launched in May 1999, is an NSCAT replacement, and there are plans to produce ice motion measurements from QuikSCAT data on a routine basis.

7.2.2   *Ice Motion from Synthetic Aperture Radar*

The high resolution and all weather capability of synthetic aperture radar have made it amenable to tracking. The quality of the ice motion from sequential SAR imagery is comparable to that from drifting buoys. Discontinuities in the motion fields can be measured by dense sampling of ice displacements in SAR data. The recent application to the large coverage lower resolution SAR imagery from RADARSAT will allow this sensor to contribute strongly to the issue of large-scale ice balance and thereby to the subject of this book. Figure 27 shows the tracked deformation around the SHEBA (Surface Heat Budget of the Arctic) ice camp. Note that deformation is concentrated along distinct slip lines; one sees two strong shears on the diagonal and several weaker ones running across the patch.

7.3. ICE THICKNESS

All the techniques discussed here are in some stage of development and have not been widely enough applied to contribute to questions of large-scale hydrologic balance. But each has promise to make contributions in the next decade.

Satellite altimetry offers the most direct measurement of ice thickness, by observing the ice freeboard above the ocean surface and using empirical relations or the isostatic assumption to infer thickness. Other satellite estimates of thickness are inferred from brightness temperature, backscatter, physical surface temperature, kinematic history, and the like. The topic has been reviewed recently in several chapters of Carsey [2] and at a workshop [67].

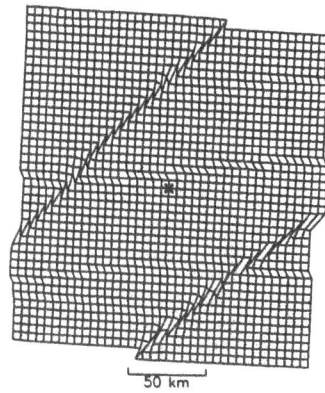

Figure 27. Ice deformation near the SHEBA camp (x). A regular 5 km by 5km grid on 22 May 1998 had deformed to this grid three days later.

7.3.1   *Ice Thickness from Altimetry*

To estimate ice thickness, altimetry has to overcome the handicap of observing only the 11% of the ice that protrudes above the ocean surface. In effect, altimetric errors are magnified by a factor of ten when applied to estimating ice thickness. Efforts to reduce the errors to usefully low levels have been pursued by Peacock *et al.* [68] who have reanalyzed waveforms to eliminate pulse blurring. By intercomparing pulses for collinear tracks they find differences with a standard deviation of about 7 cm, which could be reduced by pulse averaging. This uncertainty is to be compared with a typical freeboard of about 30 cm. Comparisons of altimetric ice thickness estimates with observations of ice draft from moored and submarine-mounted upward-looking sonars are underway. The impact of this method for climate and sea ice studies would be enormous if continual long-term direct observations of ice freeboard and thence thickness can be realized.

7.3.2   *Thin Ice Thickness from Active and Passive Microwave Sensors*

When no multiyear ice is present, passive microwave brightness temperature bears some relationship to actual ice thickness, as shown in Figure 28. When no multiyear ice is present, the thickness of thin ice can be estimated to about ±30%. This method has been applied primarily in the Bering Sea [69] but should have equal applicability in regions of the Arctic Ocean that can be determined, perhaps kinematically, to have no multiyear ice.

Synthetic aperture radar has the advantage of much finer spatial resolution compared with passive instruments, so that individual pixels tend to consist of a single ice type. However, there are many different surface states that give a complete range of backscatter. Furthermore, different ice types can produce the same backscatter for a single channel instrument as seen in Figure 29. Observations at several frequencies can improve the ability to resolve thickness. The figure does not illustrate the fact that the backscatter from open water, whose fraction is important to the ice mass balance and air—sea heat exchange, can range from —20 dB for calm water to +5 or 10 dB in strong winds. A useful review of this topic is found in [70].

Figure 28. Thin ice types as seen in passive microwave brightness temperatures. In standard algorithms, these types are treated as mixtures of open water and first-year ice. (after [70])

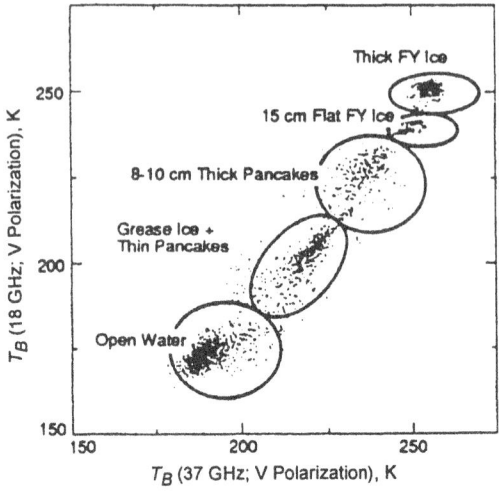

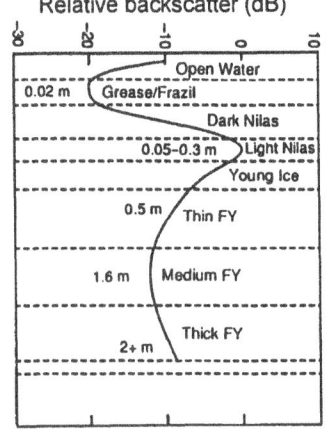

Relative backscatter (dB)

Figure 29. Radar backscatter σ_0 for several types of first-year ice. The backscatter is for combined C- and X-bands VV and HH polarizations ([70]). Note that different thicknesses can have the same backscatter, making it difficult to distinguish ice thickness uniquely from a single measurement. The backscatter of open water can range from -20 dB for very still water to +5 or 10dB for windy conditions. The high backscatter of nilas can result from "frost flowers," crystalline protrusions that form when salt is extruded from the growing and cooling thin ice. Observations at several frequencies can improve the ability to resolve thickness.

7.3.3   Ice Thickness from Kinematics—RGPS

The spatial distribution of thin ice can be determined kinematically as is done in the RADARSAT Geophysical Processor System (RGPS, [71]). In this system, a large number of Lagrangian elements on the ice cover are followed over the course of an ice growth season. The thin ice distribution is derived from the time series of area change of each element. A new category of thin ice is created with each increase in area. The history of freezing-degree-days of a thin ice category is used to estimate its thickness. Decreases in area are interpreted as ridging events that consume the thinnest ice available in an element.

In the process, the ice volume created over a season is recorded in the thin ice thickness distribution and in the volume of newly formed ridges. Understanding of radar backscatter is not utilized in this scheme. Figure 30 shows the fractional coverage of the ice cover by thin ice and ridged ice on December 21, 1996, derived from sequential RADARSAT data. The production of these data sets began in January 1999, and the products are available through the RGPS Web site (http://www.radar.jpl.nasa.gov/rgps) and the Alaska SAR Facility (http://www.asf.alaska.edu/).

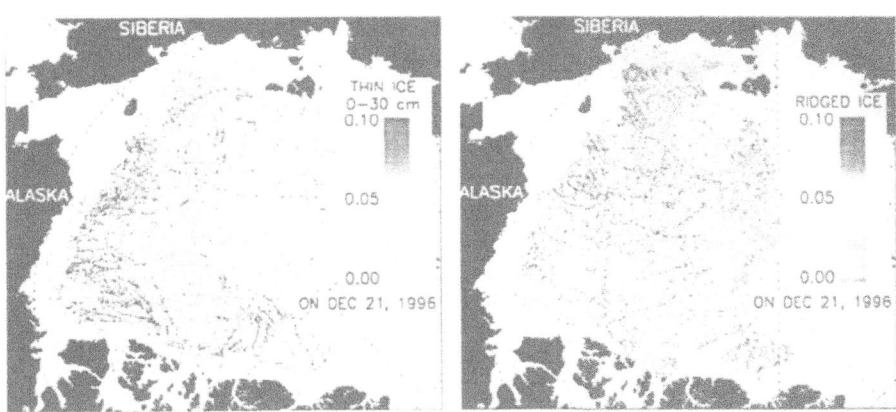

Figure 30. Concentrations of thin ice and of ridged ice derived from the RADARSAT Geophysical Processor System.

7.3.4   *Thin ice Thickness from Thermal Imagery*

The thicker ice is, the closer its surface temperature is to the air surface temperature, which in winter is on the order of -30^0C. Thin ice on the other hand is closer to the freezing point at the bottom of the ice sheet (-1.8^0C). Ice up to about one meter of thickness is observably warmer than the surrounding thick ice. The possibility of using thermal imagery to quantify the thickness of thin ice was pursued as early as 1971 by LeSehack *et al.* [72] and more recently by several investigators. The comparison in Figure 31 of ice thickness estimated from thermal signals with thickness observed by moored upward looking sonar shows that ice covers with quite different amounts of thin ice can be quantified by this method. Its drawback is that it requires moderately high 1 km resolution of AVHRR or finer in order to resolve the thin ice in leads. The method has been utilized more for comparisons with other thin ice estimates; say from microwave sensors, than for producing large data sets.

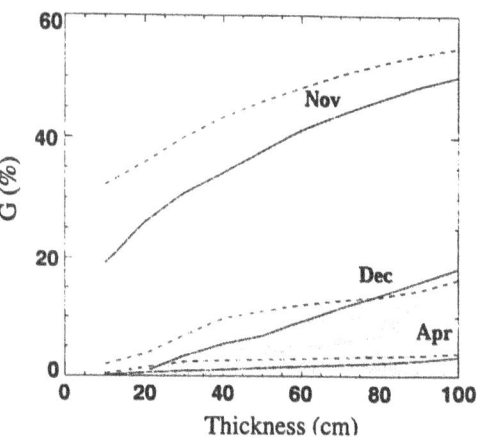

Figure 31. Cumulative ice thickness distributions estimated from AHVRR imagery (solid lines) and from moored sonars (dashed lines). Each of the three pairs of solid and dashed lines represents a comparison of different ice conditions. The grey areas are estimated error bars for the solid lines. (after [73])

8. Satellites Contributions to the Arctic Ocean Freshwater Budget (AOFWB)

8.1. GEOPHYSICAL RESULTS

Analyses of satellite data have elucidated a great deal about the freshwater balance of the Arctic Ocean. In Tables 7 and 8 we summarize some of the important results. The storage terms estimated from satellite data are listed in Table 7. The ocean stores the largest amount of freshwater, equivalent to nearly 8 m, over three quarters of the total of about 10 m. The rest is ice. The cold atmosphere contains a mere half centimeter (compared with about 5 cm in the tropical atmosphere). The annual cycles of moisture in ice and ocean are out of phase since they swap water back and forth during a freeze-thaw cycle, with runoff and melting snow adding to the ocean's summer maximum.

TABLE 7. Storage of freshwater in the Arctic Ocean as estimated with satellite data. Values are in freshwater equivalent relative to a salinity of 34.8 psu. Actual ice volume and thickness are given in parentheses. Conversions are performed as described in Sec. 3.3.1 above. The values refer to the Arctic Ocean in Figure 10,

	Source	Yield (m)	Volume (10^3 km^3)			
		Mean	Mean	Avg. ann. max.	Avg. ann. min.
Sea ice	Thomas et al. 1996	224 (2.70)	15.0 (18.7)	17.0 in May (21.3)	12.2 in Sept. (15.3)
Ocean	Steele et al. 1996	7.65	52.9	56.7 in Sept.	50.5 in May
Atmosphere	Sec. 6.3	0.005	0.033	0.077 in July	0.012 in Jan.
Total		9.90	54		

which excludes the Barents, Kara, and GIN seas and has an area of 6.92×10^6 km^2.

With their ability to track ice, satellites have provided long term records of the ice circulation and area flux through important passages. As an estimate of the area flux through Fram Strait, a quantity ideally suited to satellite observation, we have settled here on the mean value of 1.00 ($\pm$ 0.1) $\times 10^6$ km^2/yr. Table 8 shows volume fluxes. Sea ice is a major term in the AOFWB, somewhat larger than the combined ocean freshwater export. The exchange of ice with the Kara and Barents seas is found to be two orders of magnitude smaller. The atmosphere is a substantial source of AOFW. The ice export is a surprisingly high 19% of the ice storage.

TABLE 8. Transport of freshwater in the Arctic Ocean as estimated with satellite data. Values are in freshwater equivalent relative to a salinity of 34.8 psu. Sea ice volume transport and yield are in parentheses.

	Source	Transport (10^3 km^3/ yr)	Yield (cm /yr)
FW equivalent of ice export through Fram Strait (Ice volume flux)	Sec. 5.2.3	2.8 (3.4)	40 (49)
FW equivalent of ice export to Kara and Barents seas (Ice volume flux)	Sec. 5.3	—0.03 (—0.04)	—0.4 (—0.6)
FW equivalent of ocean export through Fram Strait	[29]	0.76	11
FW equivalent of ocean export through Canadian archipelago	[29]	1.13	16
Atmos'c advection into Arctic Ocean = P-E	Sec. 6.3.2	0.69	10

The residence time of water in each medium can be computed as the storage divided by the net export. For the ice the lifetime is about 8 years, for the ocean about 30 years. Atmospheric moisture has minuscule storage and yet a transport comparable to those of the ice and ocean because of the short synoptic lifetime of atmospheric moisture, roughly 20 days.

By providing observations year in and year out as reviewed in several sections of this chapter, satellite observations allow us to pin down seasonal cycles, regional variations, and interannual changes, and look for possible trends. Seasonal cycles and trends of ice extent are the most direct examples, including the quite significant finding that ice extent is declining by about 0.03 x 106 km^2 yr^1, or about 3% per decade. This decline has import for global climate far beyond the freshwater balance. Less direct methods have used concentration data to provide seasonal and interannual variability of ice and ocean freshwater storage.

The ocean is, of course, less well observed from space than ice. Nonetheless documenting from satellites the best ocean surface forcing in the guise of ice drift and surface freshwater forcing provides a necessary ingredient for modeling ocean behavior realistically.

8.2. UTILIZATION OF SATELLITE DATA

Many applications of satellite data to the Arctic Ocean freshwater balance are in an early stage. An initial outline of this chapter included many potential topics for which long records of satellite data exist but on which little work has yet been published. Applying satellite data to this topic is a scientific opportunity ripe for exploitation.

Some data are being quite well utilized, or at least we know how to make useful geophysical products from the satellite signal. The ice concentration record from passive microwave observations is the best example. Even with mature products such as these there is always room for improvement: it would seem that a reanalysis of these data with more effort to remove or even perhaps utilize the atmospheric moisture signal would be warranted, as would another look at summer melt ponds and how they should be acknowledged and perhaps quantified in a new summer data product.

Other data are just becoming available but have not yet been applied to this problem. Ice motion data stand out in this category. The new two-decade records of motion from AVHRR and passive microwave tracking will do much to improve our representations of the ice mass balance and the spatial patterns of freshwater forcing of the ocean by ice growth and melt. New motion products from scatterometers and SAR will play important roles too. Atmospheric moisture, its transport, precipitation, and evaporation are another subject area with new 20-year records needing exploitation. Realistic snow precipitation patterns with interannual variability are needed by ice modelers and are now available. How is climate change in the Arctic reflected in atmospheric moisture, in precipitation, and then in cloudiness and the heat balance?

Still other data are at a very immature stage and may contribute to this subject only after considerably more work at interpreting the satellite signal. Thickness estimates from space are the most challenging observations. Several upcoming methods were reviewed here. It is impossible to say now what may be their eventual contribution to AOFWB.

All of these data sets stand to be utilized to their fullest when used in conjunction with models. This subject is in its infancy. Think of what numerical weather prediction models do with observations and think of the infrastructure it takes to merge data with models. Assimilation will surely be the fastest growing area of arctic air-ice-ocean research in the coming years.

9. References

1. Massom, R. (1991) *Satellite Remote Sensing of Polar Regions: Applications, Limitations and Data Availability,* Bellhaven Press, London.
2. Carsey, F.D. (1992) *Microwave Remote Sensing of Sea Ice,* American Geophysical Union, Washington, D.C.
3. Aagaard, K. and Carmack. EC. (1989) The role of sea ice and other fresh water in Arctic circulation, *J. Geophys. Res.* **94**(C 10), 14,485—14,498.
4. Rothrock, DA., Thomas, D.R., and Thorndike, A.S. (1988) Principal component analysis of satellite passive microwave data over sea ice, *J. Geophys. Res.* **93**(C3), 232 1—2332.
5. Gloersen, P., Campbell, W.J., Cavaliers, D.J., Comsso, J.C., Parkinson, CL., Zwally, H.J. (1992) *Arctic and Antarctic Sea Ice, 1978—1 987: Satellite Passive-Microwave Observations and Analysis,* NASA SP-51 1, Washington, D.C.
6. Parkinson, CL., Comiso, J.C., Zwally, H.J., Cavalieri, D.i., Gloersen, P., and Campbell, W.J. (1987) *Arctic Sea Ice, 1973—1976: Satellite Passive-Microwave Observations,* NASA SP-489, National Aeronautics and Space Administration, Washington, D.C.
7. Zwally, H.J., Comiso, J.C., Parkinson, C.L., Campbell, W.J., Carsey, F.D., and Gloersen, P, (1983) *Antarctic Sea Ice, 1973-1976: Satellite Passive-Microwave Observations,* NASAA SP-459, Washington, D.C.
8. Cavalieri, D.J., Gloersen, P., and Campbell, W.J. (1984) Determination of sea ice parameters with the Nimbus 7 SMMR, *J. Geophys. Res.* **89**(D4), 5355—5369.
9. Thomas, D.R. (1993) Arctic ice signatures for passive microwave algorithms, *J. Geophys. Res.* **98**(C6), 10,037—10,052.

10. Kwok, R. and Comiso, J. (1997) The perennial ice cover of the Beaufort Sea from active- and passive-microwave observations, *Ann. Glaciol.* **25**, 376-381.

11. Comiso, J.C. and Kwok, R. (1996) Surface and radiative characteristics of the summer Arctic sea ice cover from multisensor satellite observations, *J. Geophys. Res.* **1O1**(C12), 28,397—28,416.

12. Maslanik. J. A. (1992) Effects of weather on the retrieval of sea ice concentration and ice type from passive microwave data. *Int. J. of Rem. Sens.* **13**(1). 37—54.

13. Oelke, C. (1997) Atmospheric signatures in sea-ice concentration estimates from passive microwaves: Modelled and observed, *Int. J. Rem. Sens.* **18**, 1113—1136.

14. Cavalieri, D.J., St. Germain, KM., and Swift, CT. (1995) Reduction of weather effects in the calculation of sea ice concentrations with the PMSP SSM/I, *J. Glaciol.* **41**(139), 455—464.

15. Bjørgo, E., Johannessen, OM., and Miles, M.W. (1997) Analysis of merged SMMR-SSMI time series of Arctic and Antarctic sea ice parameters, *Geophys. Res. Lett.* **24**(4), 413—416.

16. Serreze, MC., Maslanik, JA., Preller, R.H., and Barry, R.G. (1990) Sea ice concentrations in the Canadian Basin during 1988: Comparisons with other years and evidence of multiple forcing mechanisms, *J. Geophys. Res.* **95**(C12), 22,253—22,267.

17. Maslanik, JA., Serreze, M.C., and Barry, R.G. (1996) Recent decreases in Arctic summer ice cover and linkages to atmospheric circulation, *Geophys. Res. Lett.* **23**(13), 1677—1680.

18. Parkinson, C. (1992) Spatial patterns of increases and decreases in the length of the sea ice season in the north polar region, 1979—1986, *J. Geophys. Res.* **97**(C9), 14,377—14,388.

19. Parkinson, C.L. (1991) Interannual variability of the spatial distribution of sea ice in the north polar region, *1. Geophys. Res.* **96**(C3) 4792—4801.

20. Johannessen, OM., Miles, MW., and Bjørgo, E. (1996) Global sea ice monitoring from microwave satellites, *Remote Sensing for Sustainable Future, Volume II*, Proceedings from the 1996 International Geoscience and Remote Sensing Symposium, 932—934.

21. Cavalieri, D.J., Gloersen, P., Parkinson, C.L., Comiso, J.C., and Zwally, H.J. (1997a). Observed hemispheric asymmetry in global sea ice changes. *Science* **278**, 1104—1108.

22. Walsh, J.E., Hibler, W.D., and Ross, B. (1985) Numerical simulation of northern hemisphere sea ice variability, 1951—1980, *J. Geophys. Res.* **90**(C3), 4847—4865.

23. Walsh, J.E. and Zwally, H.J. (1990) Multiyear sea ice in the Arctic: Model- and satellite-derived, *J. Geophys. Res.* **95**(C7), 11,613—1 1,628.

24. Thomas, D., Martin, S., Rothrock, D., and Steele, M. (1996) Assimilating satellite concentration data into an Arctic sea ice mass balance model, 1979—1985, *J. Geophys. Res.* **101**(C9), 20,849—20,868.

25. Maslanik, J.A. and Maybee, H. (1994) Assimilating remotely-sensed data into a dynamic-thermodynamic sea ice model, *Surface and Atmospheric Remote Sensing: Technologies, Data Analysis, and Interpretation; Volume III*, Proceedings of 1994 International Geoscience and Remote Sensing Symposium, Pasadena, CA, 1306-1308.

26. Flato, G.M. and Hibler, W.D. (1995) Ridging and strength in modeling the thickness distribution of arctic sea ice, *J. Geophys. Res.* **100**(C9), 18,611—18,626.

27. Flato, G.M. (1995) Spatial and temporal variability of Arctic ice thickness, *Ann. Glaciol.* **21**, 323—329.

28. Bourke, R.H. and R.P. Garrett (1987) Sea ice thickness distribution in the Arctic Ocean, in *Cold Regions Science and Technology* Vol. 13, Elsevier Sci., New York, pp. 259—280.

29. Steele, M., Thomas, D., Rothrock, D., and Martin, 5. (1996) A simple model study of the Arctic Ocean freshwater balance, 1979—1985, *J. Geophys. Res.* **101**(C9), 20,833—20,848.

30. Schlosser, P., Bonisch, G., Rhein, M., and Bayer, R. (1991) Reduction of deep.water formation in the Greenland Sea during the 1980's: Evidence from tracer data, *Science,* 251, 1054-1056.

31. Agnew, L.. Le., H., and Hirose. T. (1997) Estimation of large scale sea ice motion from *SSMII 85.5* GHz imagery, *Ann. Glaciol.* **25**. 305—311.

32. Emery, W.J., Fowler, C.W., and Maslanik, JA. (1997) Satellite-derived maps of Arctic and Antarctic sea ice motion: 1988 to 1994, *Geophys. Res. Lett.* **24**(8), 897—900.

33. Kwok, R., Schweiger, A., Rothrock, DA., Pang, S., and Kottmeier, C. (1998) Sea ice motion from satellite passive microwave data assessed with ERS and buoy motions, *J. Geophys. Res.* **103**, 8191—8214.

450

34. Fowler, C., Maslanik, J.A., and Emery, W.J. (1994) Intercomparison of observed and simulated ice motion for a one year time series, *Surface and Atmospheric Remote Sensing: Technologies, Data Analysis, and Interpretation: Volume III*, Proceedings of 1994 International Geoscience and Remote Sensing Symposium, Pasadena, CA, 1303—1305.

35. Meier, W.N., Maslanik, JA., and Fowler, C.W. (1999) Error analysis and assimilation of remotely-sensed ice motion within an arctic sea ice model, *J. Geophys. Res.* in press.

36. Kwok, R. (1999) Recent changes in the Arctic Ocean sea ice motion associated with the North Atlantic Oscillation, *Geophys. Res. Lett.,* submitted.

37. Vowinekel, E. (1964) Ice transport in the East Greenland Current and its causes, *Arctic* **17**, 111—119.

38. Englebretson, RE. and Walsh, J.E. (1989) Fram Strait ice flux calculations and associated arctic ice conditions, *Geojournal* **18(1)**, 61—67.

39. Vinje. T. and Finnekåsa, Ø. (1986) *The ice transport through the Fram Strait,* Norsk Polarinstitutt, Oslo, Skrifter Nr. **186**. 39 pp.

40. Moritz, R. (1988) The Ice Budget of the Greenland Sea, Technical Report, *APL-UWTR 8812.* Applied Physics Laboratory, University of Washington, Seattle.

41. Vinje, T. (1982) The drift pattern of sea ice in the Arctic with particular reference to the Atlantic approach, in L. Rey, L. and B. Stonehouse (eds.), *The Arctic Ocean.*, Macmillan, London.

42. Martin, T. (1996) Sea ice drift in the East Greenland Current, *Proceedings of the 4th Symposium on Remote Sensing in Polar Environments*, Lyngby, Denmark, 10 1—105.

43. Martin, T. and Wadhams, P. (1999) Sea-ice flux in the East Greenland Current, *Deep.Sea Res.*, Part II., **46**, 1063—1082.

44. Kwok, R. and Rothrock, D.A. (1999) Variability of Fram Strait ice flux and North Atlantic Oscillation, *1. Geophys. Res.* **104**(C3), 5177—5190.

45. Wadhams. P. (1983) Sea ice thickness distribution in Fram Strait, *Nature* **305**, 108—111.

46. Thomas, D.R. and Rothrock, D.A. (1993) The Arctic Ocean ice balance: A Kalman smoother estimate, J. Geophys. Res. **98**(C6) 10,053-10,067.

47. Vinje, T., Norland, N., and Kvambekk, A, (1998) Monitoring ice thickness in Fram Strait, *1. Geophys. Res.* **103**(C5), 10,437—10,449.

48. Zhang, J., Rothrock, D.A., and Steele, M. (1998b) Recent changes in arctic sea ice: The interplay between ice dynamics and thermodynamics, submitted to *J. Climate.*

49. Flato, G.M. and Hibler, W.D. (1992) Modeling pack ice as a cavitating fluid, *J. Phys. Oceanog.* **22**, 626—651.

50. Häkkinen, 5. (1993) An arctic source of the Great Salinity Anomaly: A simulation of the Arctic ice-ocean system for 1955—1975, *J. Geophys. Res.* **98**(C9), 16,397—16,410.

51. Harder, M., Lemke, P., and Hilmer, M., (1998) Simulation of sea ice transport through Fram Strait: Natural variability and sensitivity to forcing, *J. Geophys. Res.* **103**. 5595—5606.

52. Zhang, J., Hibler, W.D., Steele, M., and Rothrock, DA. (1998a) Arctic ice-ocean modeling with and without climate restoring, *J. Phys. Oceanogr.* **28**, 191—217.

53. Overland, J.E., Turet, P., and Gort, A.H. (1996) Regional variations of moist static energy flux into the Arctic, *J. Clinsate* **9**(1), 54—65.

54. Chédin, A., Scott, NA., Wahiche, C., and Moulinier, P. (1985) The improved initialization inversion method: A high resolution physical method for temperature retrievals from satellites of the TIROS-N series, *J.Chin. Appl. Meterol.* **24**, 128—143.

55. Francis. JA. (1994) Improvements to TOVS retrievals over sea ice and applications to estimating Arctic energy fluxes. *J. Geophy Res.* **99**(D5), 10,395—10,408.

56. Köpken, C., Heinemann, G., Chédin, A., Claud, C., and Scott, N. A. (1995) Assessment of the quality of TOVS retrievals obtained with the 31 algorithm for Antarctic conditions. *J. Geophys. Res.* **101(D3)**, 5143—5 158.

57. Francis, J.A. and Schweiger, A.J. (1999) The NASA/NOAA TOVS Polar Pathfinder: 18 years of Arctic data, *Proceedings of the SthAMS Conference on Polar Meteorology and Oceanography*. Dallas, TX, 62—65.

58. Groves, D.G. (1999) A new moisture budget of the Arctic atmosphere derived from 19 years of daily TOVS satellite moisture retrievals and NCEP reanalysis winds, *Proceedings of the 5th AMS Conference on Polar Meteorology and Oceanography*, Dallas, TX, 107—112.

59. Cavalieri, D. J., Parkinson, C.L., Gloersen, P., and Zwally, H.J. (1997b) Arctic and Antarctic sea ice concentrations from multichannel passive-microwave satellite data sets: October 1978 to September 1995, *User's Guide, NASA Technical Memorandum 104647*, p. 17.

60. Briazgin, N.N., Burova, L.P., and Khrol. V.P. (1996) *Atlas of the water balance of the northern polar area* (in Russian), Arctic and Antarctic Research Institute.

61. Vowinkel, E. and S. Orvig (1970) The Climate of the North Polar Basin, World Survey of Climatology, Vol. 14 Climates, of the Polar Regions. S. Orvig (ed.) Elsevier, 129-252.

62. Gohin, F. and Cavanie, A. (1994) A first try at identification of sea ice using the three beam scatterometer of ERS-1, *ml. J. Remote Sens.* **15**(6), 1221—1228.

63. Grandell, J., Johannessen, J.A., and Hallikainen, MT. (1999) Development of a synergistic sea ice retrieval method for the ERS-1 AMI wind scatterometer and SSMII Radiometer, *IEEE Trans. Geosci. And Remote Sens.* **37**(2), 668—679.

64. Yueh, S. and Kwok, R. (1998) Arctic sea ice extent and melt onset from NSCAT observations, *Geophys. Res. Lett.* 25(23), 4369-4372.

65. Holt, B., Rothrock, DA., and Kwok, R. (1992) Determination of sea ice motion from satellite images, in F.D. Carsey (ed.), *Microwave Remote Sensing of Sea Ice*, American Geophysical Union, Washington, DC, pp. 343—354.

66. Liu, A., Zhao, Y., and Wu, S.Y. (1999) Arctic sea ice drift from NSCAT and SSM/I data, *J. Geophys. Res.*, in press.

67. Thorndike, A.S., Parkinson, C., and Rothrock, D.A. (1992) *Report of the Sea Ice Thickness Workshop*, 19—21 November 1991, New Carrollton, Maryland, Applied Physics Laboratory, University of Washington, Seattle, Washington.

68. Peacock, N.R., Laxon, S.W., Maslowski, W., Winebrenner, D.P., and Arthern, R.J. (1998) Geophysical signatures from precise altimetric height measurements in the Arctic Ocean, *Proceedings from the 1998 International Geoscience and Remote Sensing Symposium*, 1964—1966.

69. Markus, T. and Cavalieri, D. J. (1996) Comparison of open water and thin ice areas derived from satellite passive microwave data with aircraft measurements and satellite infrared data in the Bering Sea, *Remote Sensing for Sustainable Future, Volume III*, Proceedings from the 1996 International Geoscience and Remote Sensing Symposium, 1529—1531.

70. Grenfell, T.C., Cavaliers, D.J., Comiso, J.C., Drinkwater, M.R., Onstott, R.G., Rubinstein, I., Steffen, K., and Winebrenner. D.P. (1992) Considerations for microwave remote sensing of thin sea ice, in F.D. Carsey (ed.), *Microwave Remote Sensing of Sea Ice*, American Geophysical Union. Washington. DC. pp. 291—300.

71. Kwok, R., Rothrock, DA., Stern, H.L., and Cunningham, G.F. (1995) Determination of the age distribution of sea ice from Lagrangian observations of ice motion. *IEEE Trans. Geosci. Remote Sens.* **33**(2), 392—400.

72. LeSchack, L.A., Hibler, W.D., and Morse, F.H. (1971) Automatic processing of arctic pack ice data obtained by means of submarine sonar and other remote sensing techniques, in J.B. Lomax (ed.), *Propagation Limitations in Remote Sensing. 17th Symp. Electromag. Wave Prop. Panel of AGARD*, Colorado Springs, USA, 5/1—19.

73. Yu,Y and Rothrock, DA. (1996) Thin ice thickness from satellite thermal imagery, *J. Geophys. Res.* **101**(C1O), 25,753—25,766.

TRACER STUDIES OF THE ARCTIC FRESHWATER BUDGET

P. SCHLOSSER[1,2], B. EKWURZEL[1], S. KHATIWALA[1,3],
B. NEWTON[1,3], W. MASLOWSKI[4], S. PFIRMAN[5,1]

[1] *Lamont-Doherty Earth Observatory of Columbia University*
   Palisades, NY 10964
[2] *Department of Earth and Environmental Engineering*
   Columbia University, New York, NY 10027
[3] *Department of Earth and Environmental Sciences*
   Columbia University, New York, NY 10027
[4] *Wieslaw Maslowski, Department of Oceanography*
   Naval Postgraduate School, Monterey, CA 93943
[5] *Department of Environmental Science, Barnard College*
   Columbia University, New York, NY 10027

1. Introduction

The freshwater lens covering the surface of the Arctic Ocean is roughly 50 to 150 meters thick and consists of river runoff, sea-ice meltwater, and low-salinity water of Pacific origin imported through Bering Strait. Whereas salinity data provide us with a good picture of the distribution and variability of the total freshwater contained in the Arctic Ocean, they cannot, in general, distinguish between the individual freshwater components. To obtain this information, measurements of additional water mass properties have to be performed.

The most promising variables for visualization and quantification of the individual freshwater components in the Arctic Ocean are the stable isotopes of water $^1H_2^{18}O$ and $^1H^2H^{16}O$ (HDO), dissolved inorganic carbonate, dissolved barium (Ba), and nutrients. These tracers have different sources and sinks in the Arctic Ocean. Therefore, they can be used, in combination, to identify the sources of the freshwater. Additionally, transient tracers allow us to determine the mean residence time of the freshwater. They can also be used to calibrate models used in studies of the Arctic freshwater balance.

In this chapter, we briefly outline the basic principles of the tracer method for studies of the Arctic freshwater balance, present results from field studies, and discuss a strategy for incorporating the tracer data into model studies. We briefly introduce stable isotopes of water, dissolved carbon, and barium as freshwater tracers, and we focus our discussion on the application of the stable isotopes of water because the spatial coverage of stable isotope data is better than that for other tracers and their integration into models

453

E.L. Lewis et al. (eds.), The Freshwater Budget of the Arctic Ocean, 453-478.
© 2000 *Kluwer Academic Publishers.*

is further along. However, the ultimate goal of tracer studies is to use them in an integrated fashion.

2. Principles

All components contributing to the freshwater of the Arctic Ocean by definition have salinities lower than that of the Atlantic water which is used here as reference salinity. River runoff and sea-ice meltwater have salinities of 0 and between 3 and 5, respectively. Although much saltier than river runoff and sea-ice meltwater, the

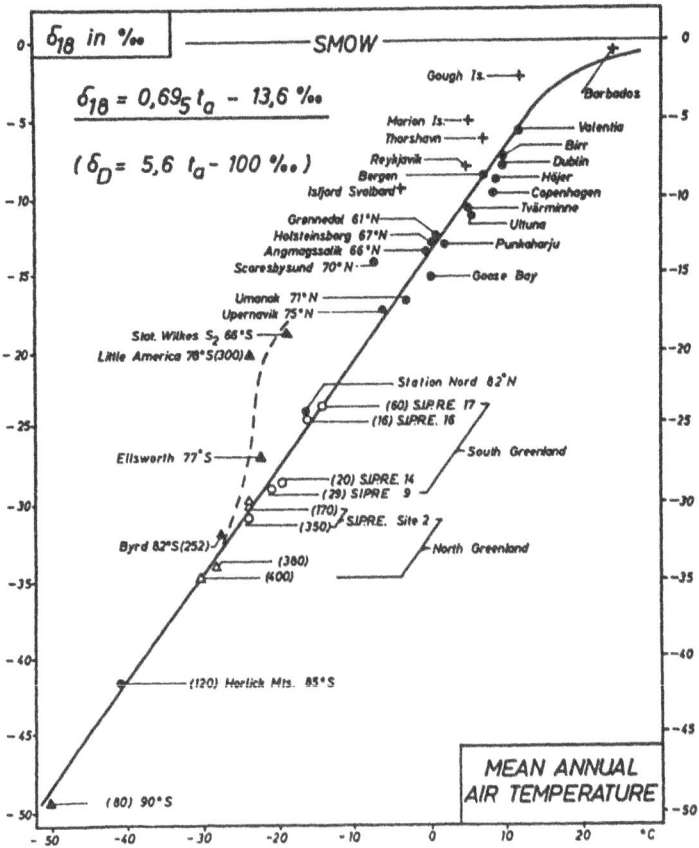

Figure 1: Plot of $\delta^{18}O$ for sites from a variety of latitudes (from Dansgaard, 1964). The $\delta^{18}O$ values are decreasing with increasing latitude (decreasing temperature). Typical values for high northern latitude sites are approximately –20‰.

Pacific inflow through Bering Strait with its average salinity of approximately 33 psu, about 2 psu lower than Atlantic water, represents a negative salt flux or a freshwater source. Salinity data have been used in past studies to estimate fluxes of the individual freshwater components into and out of the Arctic Ocean. However, without quantifying the separate components, gross assumptions must be made which introduce large uncertainties. Thus, it is difficult to assess the distribution and mean residence times of the individual freshwater components within the upper layers of the Arctic Ocean from salinity data alone. There is a need for other independent tracers in order to match the number of equations with the number of unknowns.

2.1. STABLE ISOTOPES OF WATER

It was recognized early on that stable isotopes of water ($^1H_2^{18}O$ and $^1H^2H^{16}O$ or HDO) have distinct signals in natural waters (e.g., [1]). These signals are caused by isotope fractionation during evaporation and precipitation. Fractionation of stable isotopes in natural waters is primarily linked to the temperature at which atmospheric water vapor condenses and precipitates. As air masses move from low to high latitudes, the heavy isotopes are precipitated preferentially, leaving behind isotopically depleted water vapor. Therefore, precipitation and consequently runoff in high latitudes typically have lower

Table 1. Weighted Mean River $\delta^{18}O$

River	Mean annual discharge ($km^3 yr^{-1}$)	$\delta^{18}O$ (‰)
Severnaya Dvina	99[a]	-13.3[c]
Pechora	108[a]	-14.4[c]
Ob + Pur + Taz	458[b]	-16.1[d]
Yenisey	573[b]	-17.2[d]
Olenek	30[b]	-20.4[d]
Lena	536[b]	-19.4[d]
Yana	31[b]	-21.1[d]
Indigirka	49[b]	-23.8[c]
Kolyma	107[b]	-22.4[c]
Mackenzie	340[a]	-20.3[e]
Combined discharge	2331	
$\delta^{18}O$ weighted mean		-18.13

[a] [53]
[b] [52]
[c] [5]
[d] [3]
[e] [54]

$H_2^{18}O/H_2^{16}O$ ratios (and HDO/H_2O ratios) than waters in low latitudes (e.g., [2]; Fig. 1). This feature can be used to identify the freshwater components in the Arctic Ocean originating from meteoric water sources (precipitation and river runoff). The $H_2^{18}O/H_2^{16}O$ ratios of waters in the Arctic rivers cover a range from about –13 to -24‰ (Table 1). The average value of all rivers discharging into the Arctic Ocean is about –

18‰ [3]. There are distinct differences between the $H_2^{18}O/H_2^{16}O$ ratios of individual rivers (e.g., -15‰ in the Pechora River [4] versus -22‰ in the Yukon [5]). However, these differences are not large enough to be useful for quantification of the freshwater contributions from individual rivers. There is a need for other independent tracers to match the number of equations with the number of unknowns.

2.2. BARIUM

A promising tracer that is used for separation of the freshwater contributions from Eurasian and American rivers is barium (Ba). The following discussion is based upon data presented in Falkner *et al.* [6] and Guay and Falkner [7]. Ba is delivered to the surface ocean through terrestrial runoff; it is scavenged biologically and ultimately sequestered in sediments. Dissolution at the sedimentary interface and upwelling partially recycles the scavenged fraction in shallow seas. Wherever the fluvial inputs dominate over biological sinks and sedimentary sources, Ba may be a useful tracer for river runoff. In their survey of Arctic rivers, Falkner and Guay find that barium concentrations in North American rivers are significantly higher than those in Eurasian rivers. The Mackenzie, by far the largest North American freshwater source, delivers an effective concentration of 520 nmol l^{-1}, whereas the Eurasian rivers carry between 5 and 20 percent of that concentration. Water flowing through Bering Strait has intermediate Ba levels, due to the influence of North American rivers that discharge into the North Pacific (e.g. the Yukon River) and upwelling in the Gulf of Anadyr. However, as a result of biological activity, there is strong variability in the inflow region. The ambient, mostly Atlantic waters contribute a small, but potentially problematic quantity of barium.

Guay and Falkner [7] constrain the non-fluvial sources and sinks well enough to conclude that North American runoff dominates the Canadian Basin, while Siberian rivers are the source for the rest of the Arctic Ocean. They also use Ba, together with salinity, to quantify the penetration of river runoff into distinct water masses comprising the halocline over the open Arctic Ocean. However, the assignment of quantitative fractions to the North American and Eurasian sources cannot be accomplished based on Ba alone. The non-fluvial contributions present a 3- (or 4-) component mixing problem. Ba will have to be combined with additional tracers to yield more precise results.

2.3. DISSOLVED CARBONATE

Arctic river drainage basins contain high total organic carbon (TOC), total carbonate (C_T), and total Alkalinity (A_T) [8,9]. This occurs through organic matter decay that releases carbon dioxide, which then reacts with different minerals as shown by the simplified reaction in [9]:

$$MeCO_3(s) + CH_2O(org) + O_2 \oslash Me^{2+} + 2HCO_3^-$$ (1)

Hence this addition of total carbonate can also be used to trace river runoff on Arctic shelf seas and the interior ocean basin. As Anderson *et al.* [8] point out, one can obtain a qualitative separation of the river runoff, but if river runoff were to mix with Atlantic water to produce a salinity of 31.5, the resulting normalized total alkalinity would be close to the Pacific inflow value. To eliminate this problem, they plot normalized total alkalinity and silicate concentrations for Arctic tracer data collected in 1991 with lines depicting the mixing relationships between all the water masses [10].

2.4. NUTRIENTS AND PO$_4$*

As pointed out before (e.g., [11]), the H$_2^{18}$O/H$_2^{16}$O ratio is a good tracer for contributions of river runoff and meteoric water to the water masses of the Arctic Ocean. In order to separate the freshwater component associated with the Pacific inflow, an additional tracer is required. Studies using silicate were able to define part of the Pacific inflow [11]. However, the surface water signal is not well resolved by this tracer. Jones et al. [12] used the marked difference in nutrient concentrations between Atlantic water and Pacific water to separate these water masses in Arctic Ocean surface waters. They found linear relationships between nitrate and phosphate based on St. Anna Trough data, which they use for Atlantic Water inflow, and Chukchi shelf slope data, which they use for the Pacific water relationship. Thus, for each surface nutrient data point within the upper 30 m, Jones et al. [12] created a map of the frontal position dividing Pacific water dominated versus Atlantic dominated surface waters. For their calculations, they assume precipitation, runoff, and sea-ice meltwater to have the same N-P relationship as Atlantic water.

In order to quantify all three principal freshwater sources, an alternate approach was introduced by Ekwurzel [3]. In this approach, PO$_4$*, a combination of phosphate and oxygen with a behavior close to that of a conservative tracer [13], is used in combination with salinity and the H$_2^{18}$O/H$_2^{16}$O ratio to quantify the contribution of Pacific inflow through Bering Strait. The advantage of this method is that all four water masses (Atlantic Water, Pacific Inflow, sea ice meltwater, and meteoric water) can be quantified.

3. Available data

The first studies using stable isotopes for separation of the components of the Arctic freshwater balance were performed several decades ago (e.g., [14,15]). In the framework of his tracer studies of the Arctic Ocean, Östlund started to more or less routinely measure H$_2^{18}$O/H$_2^{16}$O ratios of the water samples he collected for tritium measurements (e.g., [16,17,18]). Most of these data are summarized in Östlund and Grall [19]. Building upon this work, water samples for measurement of the H$_2^{18}$O/H$_2^{16}$O ratio were collected extensively during the 1980s and 1990s (e.g., [20,11,4,3]). The presently available data sets are sufficiently large to yield contoured fields of the H$_2^{18}$O/H$_2^{16}$O ratio in the surface waters of the Arctic Ocean (e.g., [21]). There are more than 1100 data points available for the upper 15 meters of the water column in the Arctic Ocean (including the shelf seas and river deltas).

The Ba data set is smaller than the H$_2^{18}$O/H$_2^{16}$O data set, although it has grown rapidly during the 1990s (e.g., [6,7]). The available data provide high-resolution vertical sections along ship tracks, as well as partial maps of the Ba distribution on a variety of isopycnals in the near-surface waters of the Arctic Ocean [7].

458

4. Visualization of the individual freshwater components

To demonstrate the application of the multi-tracer approach to studies of the Arctic freshwater balance, we use a trans-Arctic section consisting of three segments

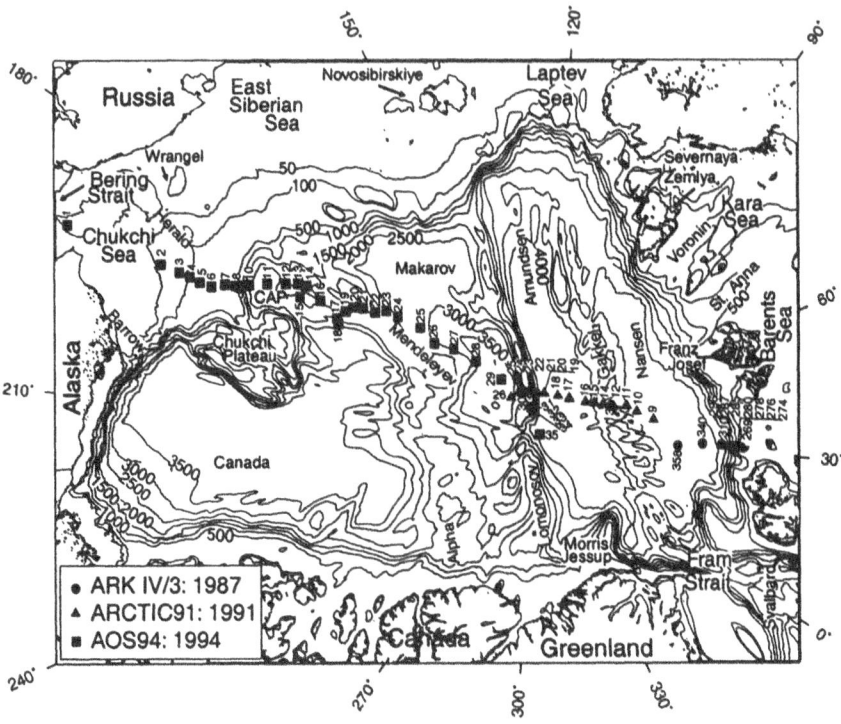

Figure 2: Geographic locations of the stations used to construct the Transarctic section plotted in Fig. 3.

occupied in 1987, 1991, and 1994 (Fig. 2). All three segments of the section were occupied during the summer. Therefore, the resulting sections represent the freshwater distribution during the summer season. This section has to be interpreted with caution due to the lack of synopticity during a time of changing water mass distribution and circulation patterns (e.g., [22]). However, the main features of the distribution of freshwater are clearly visible in this section.

4.1. SALINITY

The salinity distribution (Fig. 3a) already indicates that the bulk of the freshwater is concentrated in the top ca. 100 to 200 meters of the water column. Freshwater concentrations decrease below the surface mixed layer. Values range from above 34 close to the Barents Sea to below 33 in the Canadian Basin. The core of the Atlantic-derived water is located at ca. 300 to 500 meters depth and the freshwater contributions to this water mass are negligible in terms of the overall freshwater inventories of the upper few hundred meters of the water column (Fig. 3a). However, the salinity distribution does not provide straightforward

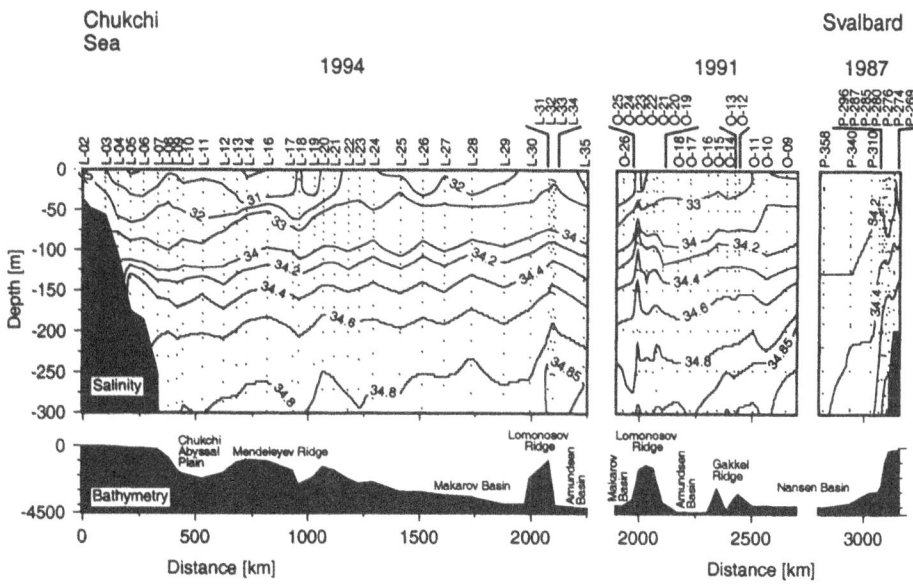

Figure 3a: Distribution of salinity along a transArctic section (for geographical position of the stations, see Fig. 2; note changes in contour intervals). The section has been constructed using data from three icebreaker cruises to the Arctic (Polarstern cruise 1987, Arctic 91 expedition, and Arctic Ocean Section 94). The data were provided by the Ocean Data facility of the Scripps Institution of Oceanography.

information on the individual freshwater components contained in the surface waters. This information can be gained by solving equations (2) through (5) described below.

4.2. $H_2^{18}O/H_2^{16}O$ AND PO_4^*

Adding the $H_2^{18}O/H_2^{16}O$ ratio to the salinity data (Fig. 3b) already reveals some additional qualitative information on the distribution of the freshwater sources, specifically the river runoff. The $H_2^{18}O/H_2^{16}O$ ratios in the surface waters are low in the presence of river runoff. Most of the river runoff is found in the upper 50 to 100 meters of the water column. This is reflected by $\delta^{18}O$ values significantly below that of Atlantic-derived water ($\delta^{18}O$: ca. 0.3 ‰). In contrast to the salinity distribution, there is a sharp gradient in the 1987 and 1991 $\delta^{18}O$ section over the Gakkel Ridge, indicating a transition from a region influenced by sea-ice meltwater (south of the ridge) to one influenced by river runoff.

To quantify the individual freshwater components contained in the surface waters of the Arctic Ocean, we added PO_4^* as a third tracer instead of silicate because it is more conservative and it tracks all waters with origin in the Pacific inflow region.

Silicate seems to account mainly for Bering Strait water entering the Chukchi Shelf during winter.

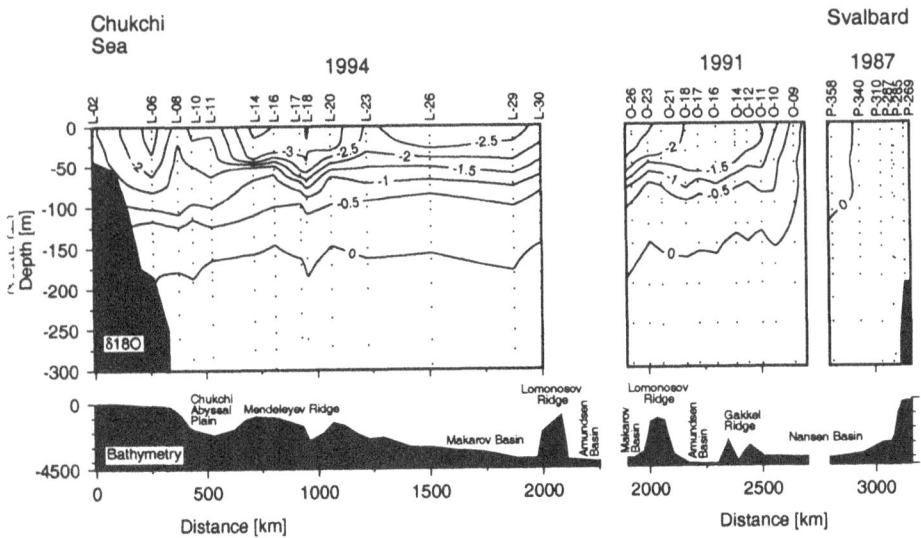

Figure 3b: same as 3a except for $\delta^{18}O$. From Ekwurzel [3].

The following balance equations are being used (e.g., [17,20,11,3]):

$$f_a + f_r + f_i + f_p = 1 \tag{2}$$
$$f_aS_a + f_rS_r + f_iS_i + f_pS_p = S_m \tag{3}$$
$$f_aO_a + f_rO_r + f_iO_i + f_pO_p = O_m \tag{4}$$
$$f_aX_a + f_rX_r + f_i X_i + f_pX_p = X_m \tag{5}$$

where f_a, f_r, f_i, and f_p are the fractions of Atlantic-derived water, river runoff, sea-ice meltwater, and Pacific Water, respectively. S_x and O_x are the corresponding salinities and $H_2^{18}O/H_2^{16}O$ ratios. X_x is a variable that has to be measured in addition to salinity and the $H_2^{18}O/H_2^{16}O$ ratio in order to quantify the fraction of Pacific water in areas where such water is present (so far, we have used both silicate and PO_4^* for this purpose). S_m, O_m, and X_m are the observed values.

Solution of the balance equations requires knowledge of the salinity and tracer values of the end members (Table 2), as well as their measured values. Selection of the end members has been discussed in detail in the literature (e.g., [17,20,11,4,3]).

Table 2. Values for end members used in solution of equations (2) to (5) ([3])

	Salinity	$\delta^{18}O$ (‰)	PO$_4$* (μmol kg^{-1})
Atlantic water	35 ± 0.05	0.3 ± 0.1	0.7 ± 0.05
River runoff	0	-18 ± 2	0.1 ± 0.1
Sea ice	4 ± 1	surface + (2.6 ± 0.1)*	0.4 ± 0.2
Pacific water	32.7 ± 1	-1.1 ± 0.2	2.4 ± 0.3

* For each station this value is set as the surface $\delta^{18}O$ measurement plus the maximum equilibrium fractionation factor [55,17,56].

4.3. RIVER-RUNOFF DISTRIBUTION

The solutions of equations (2) through (5) clearly delineate the distribution of river-runoff with high concentrations in the upper 50 to 100 meters (highest fractions: >14%; Fig. 3c). The 10% isoline is located at roughly 50 meters depth in the Canadian Basin portion of the section. In the Eurasian Basin, the highest river-runoff fractions are about 8 to 10 % in the upper 50 meters of the Amundsen Basin. In the Nansen Basin, they drop from from about 6 to 8% over the Gakkel Ridge to practically zero at the Barents Sea slope. The 1% isoline generally is located at about 200 meters depth, except in the Nansen Basin where it slopes upward and outcrops close to the Barents Sea.

The calculated percent fractions (Fig. 3c) represent the fraction of river-runoff originally added to the water column, i.e., before sea-ice formation. To obtain the real river-runoff fraction, the sea-ice meltwater fraction (see below and Fig. 3d) has to be added to the river-runoff fraction. This procedure provides a fairly accurate estimate of the river-runoff fraction actually contained in the upper waters for regions without significant Pacific freshwater input. In regions with large Pacific freshwater input, this number provides an upper limit of the river-runoff fraction.

The river-runoff fractions can be integrated between the Atlantic core where their values are practically zero and the surface to obtain the total inventory of river-runoff water at a given site. These values are equivalent to the thickness of a layer of pure river-runoff (S=0). Along our section, we obtain inventories ranging from values below 1 meter close to the Barents Sea to more than 14 meters over the Mendeleyev Ridge. The maximum number has to be reduced by about 5 meters in order to obtain the fraction of river-runoff residing in the water column. This means that in the Canadian Basin, the river-runoff inventory is approximately 8 to 10 meters. There is a sharp front over the Gakkel Ridge with inventories decreasing to values close to zero in the Nansen Basin (especially the southern Nansen Basin).

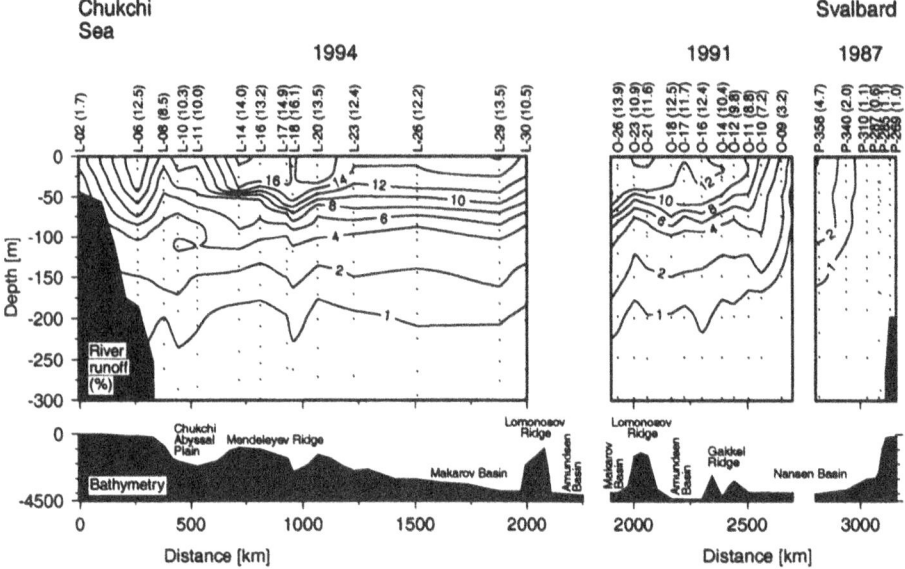

Figure 3c: same as 3a except for the fraction of river-runoff. The inventories of river-runoff are indicated above the location of the stations (for explanation, see text).

4.4. SEA-ICE MELTWATER DISTRIBUTION

The main patterns in the distribution of sea-ice meltwater are similar to those of the river-runoff. Negative fractions indicate that freshwater was extracted (cold distillation) from the water column to form sea-ice, positive numbers indicate addition of sea-ice meltwater. The upper 50 meters of the water column in the Canadian Basin have sea-ice meltwater fractions of roughly −5%, i.e., about 5% of the water was used to form sea-ice. The 0% isoline is located close to 200 meters depth throughout the section. There is a sharp front over the Gakkel Ridge with lower values to the south (Nansen Basin). Over the central Nansen Basin, the fractions turn from negative values to positive values indicating the transition from a sea-ice formation signal to a sea-ice meltwater signal.

Calculating inventories in the same way as done for the river-runoff, we obtain signals equivalent to the formation of about 3 to 5 meters of sea-ice between the Gakkel and Mendeleyev ridges, whereas we estimate addition of 1.5 meters of sea-ice meltwater to the southern Nansen Basin. There are a few stations over the Chukchi Plateau that have slightly positive sea-ice meltwater fractions indicative of addition of sea-ice meltwater. Whereas we do not expect the values over the central Basin to change much seasonally, the values in the southern Nansen Basin and the Chukchi Rise most likely will have a strong seasonal signal (sea-ice formation during winter, and melt during summer).

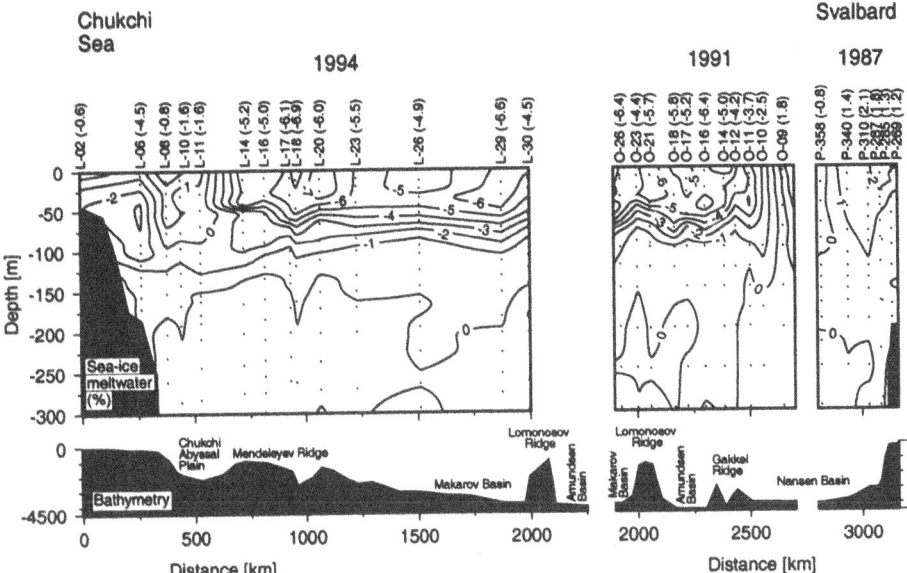

Figure 3d: same as 3a except for the fraction of sea-ice meltwater. Negative numbers for fractions mean that sea-ice has been formed from the water column, positive fractions indicate addition of sea-ice meltwater. The inventories above the station locations indicate the amount of sea-ice formed from (negative signs) or added to (positive signs) the water column.

4.5. PACIFIC COMPONENT

Finally, the distribution of Pacific water was calculated. Because in our section the influence of the Pacific water is practically restricted to the Canadian Basin, we use a three component mass balance for most of the Eurasian Basin and a four component balance for the rest of the section (Fig. 4). Using the end members listed in Table 2, the fractions of river runoff, sea-ice meltwater, and Pacific water could be calculated and contoured along the section.

464

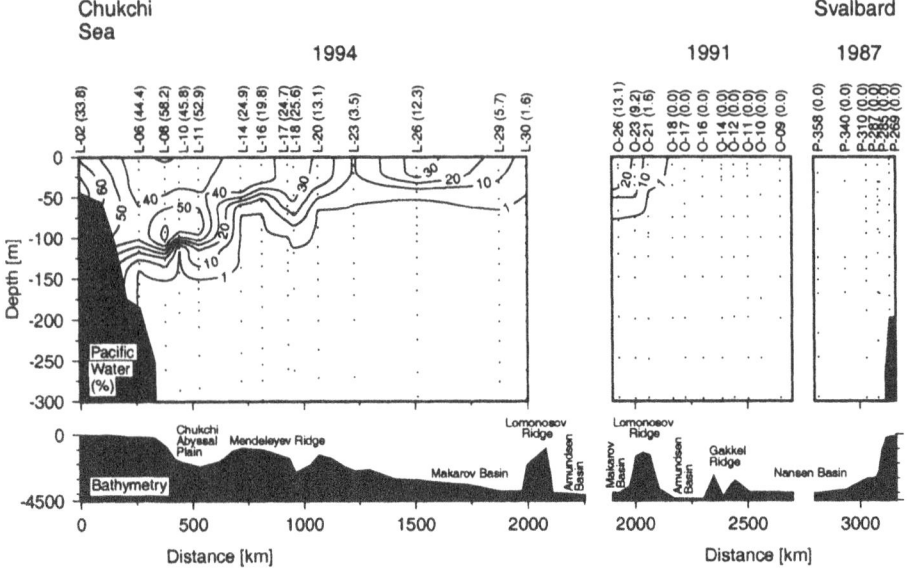

Figure 3e: same as 3 a except for the fraction of Pacific Water

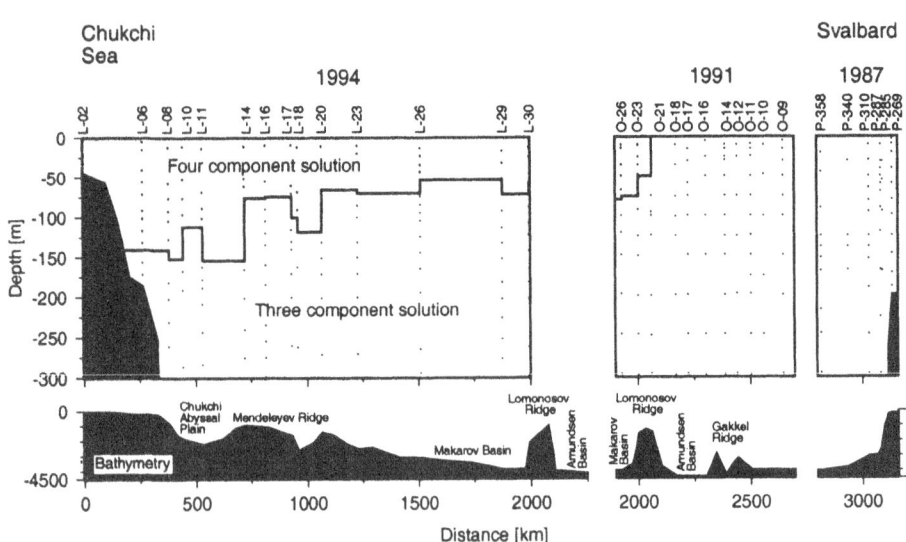

Figure 4: Plot indicating where either the three-component mass balance was used to
calculate the fractions of river-runoff and sea-ice meltwater, or where the
four-component balance was applied to determine the river-runoff, sea-ice meltwater,
and Pacific-derived freshwater components, respectively.

In 1994, the bulk of the Pacific Water was confined to the area south of the
Mendeleyev Ridge in our section (Fig. 3e). The highest fractions (above 50%) were
found in the southern Canadian Basin at depth close to about 75 meters. They dropped to

values below 1% (practically below the detection limit) at a depth of 150 to 200 meters. Above the maximum (75 meters), the Pacific Water fractions decreased to about 30 to 40% near the surface. In the Makarov Basin, the Pacific Water was confined to the upper 50 meters with maximum values of about 10 to 30% (1994). In 1991, the Pacific Water tongue reached across the Lomonosov Ridge into the Eurasian Basin (about 10% in the upper 50 meters over the Lomonosov Ridge). As described by Morison *et al.* [22], Carmack *et al.* [23] and McLaughlin *et al.* [24] and quantified by Ekwurzel [3], Pacific Water retreated during the late 1980s and early 1990s. Our observations fell exactly into this period. This means that some of the features observed in our three sections have to be interpreted in the context of a changing water mass distribution in the Arctic Ocean.

5. $\delta^{18}O$ distribution in the surface waters of the Arctic Ocean and the GIN seas

As described above, $\delta^{18}O$ is an excellent qualitative indicator of river runoff and meteoric water. The $\delta^{18}O$ distribution in the surface waters of the Arctic Ocean provides information on the strength of the river runoff signal. If combined with salinity and PO_4*, the distributions of sea-ice meltwater and Pacific water can also be reconstructed. Additionally, these maps can be used as calibration tools in model simulations (see below).

The bulk of the available $\delta^{18}O$ data (published and unpublished) have been compiled into a comprehensive data base. The surface $\delta^{18}O$ map constructed from this data base (Fig. 5a) was presented by Schlosser *et al.* [21]. It is a first step towards a more complete utilization of salinity, $\delta^{18}O$, and PO_4* data in studies of the sources and pathways of the individual freshwater components contained in Arctic surface waters. The data set includes data collected over a period of several decades. In view of the large changes recently observed in the Arctic Ocean, the $\delta^{18}O$ distribution has to be interpreted carefully. Only first-order features are sufficiently robust to be useful in straightforward interpretation of the data. Details of the $\delta^{18}O$ field at specific sites have to be evaluated for the years in which the samples were collected.

Many features of the $\delta^{18}O$ distribution (Fig. 5a) are similar to those of the salinity distribution (Fig. 5b). This is due to the fact that most of the freshwater contained in the surface layers of the Arctic Ocean is contributed by river runoff with its very low $\delta^{18}O$ signal (ca. –18‰) and Pacific inflow with a $\delta^{18}O$ value (ca. –1 ‰) that is significantly below that of Atlantic-derived water (ca. 0.3 ‰).

466

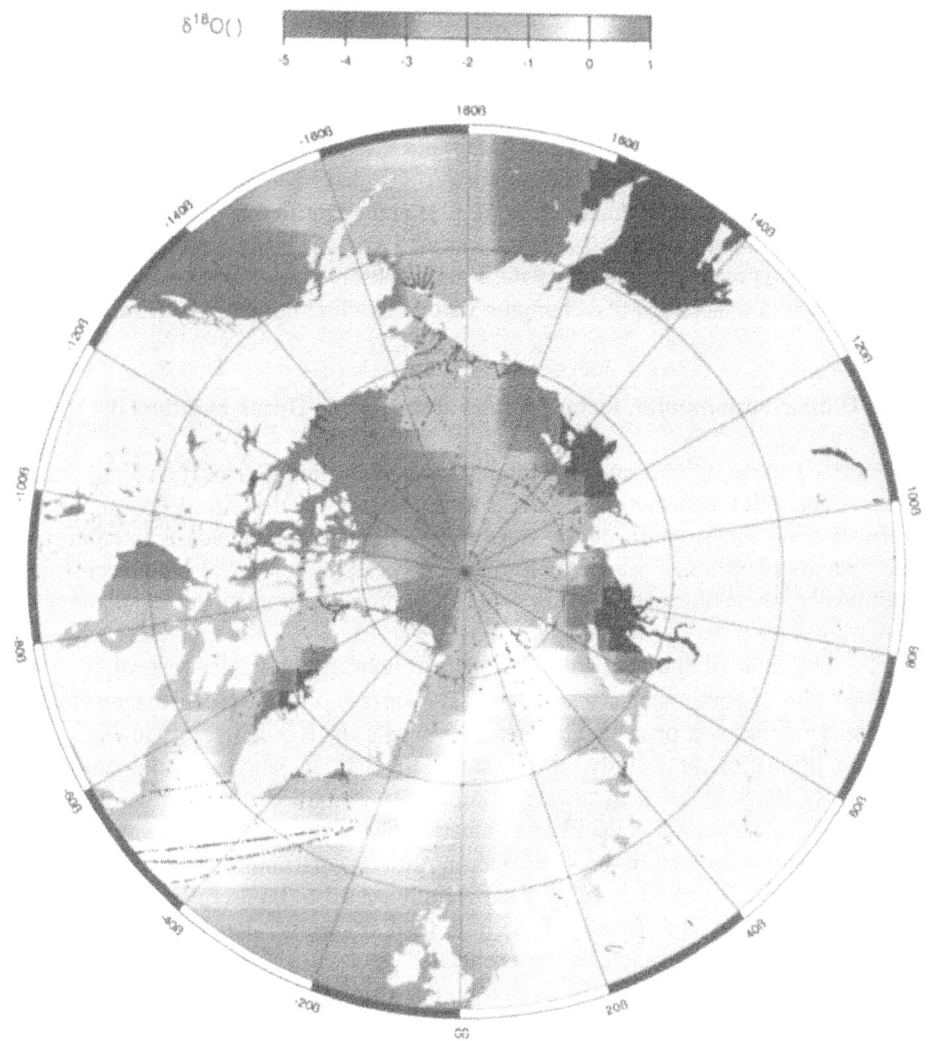

Figure 5a: Distribution of $\delta^{18}O$ in the top 15 meters of the water column. For discussion, see text. Black shading near the coast of Greenland and Siberia indicates values < -5‰.

(For a colour version of this figure, see p. 616)

The highest $\delta^{18}O$ values (0 to +1‰) are observed in the North Atlantic Current and over the Barents shelf where almost pure Atlantic water enters the Arctic Ocean (e.g., [25,26,27]), and there is minimal admixture of river water (e.g., [11]). The lowest $\delta^{18}O$ values are observed in and close to the river deltas, in regions where there is a high fraction of river water. River influence is especially pronounced in the Kara Sea (Ob and Yenisey rivers), the Laptev Sea (Lena River), and close to the Mackenzie delta in the southern Canada Basin (Fig. 5a). The salinity distribution (Fig. 5b) also shows the river plumes as areas with low salinities, although the Mackenzie River plume signature is not as pronounced in salinity as it is in $\delta^{18}O$. The Mackenzie influence appears to extend all the way to the North Pole in the $\delta^{18}O$ field.

There is a sharp front in the $\delta^{18}O$ distribution at ca. 155°W between surface waters influenced by the Mackenzie and waters flowing through Bering Strait. Bering Strait water can be distinguished by its relatively low salinity (about 32) and moderate $\delta^{18}O$ values ($\approx$-1‰; see also Cooper et al., [28]). The surface waters in the interior of the Arctic Ocean are relatively homogeneous in salinity and $\delta^{18}O$. Slightly higher values are observed in the Eurasian Basin due to its closer proximity to the high-salinity Atlantic inflow compared to the Canadian Basin which is supplied by low-salinity Bering Strait water. There seems to be flow of low $\delta^{18}O$ water from the East Siberian shelf into the Canadian Basin at ca. 160°E which could be contributed by river-runoff leaving the shelf and exchanging with the interior basin However, this feature is not observed in the salinity distribution.

The outflow of river water from the Arctic Ocean along the East Greenland coast and through the Canadian Archipelago is reflected in low salinities ($\approx$32 in the East Greenland Current, Kane Basin and Lancaster Sound) and $\delta^{18}O$ values ($\approx$-2 ‰). The $\delta^{18}O$ field suggests a continuous flow from the East Greenland Current into the West Greenland Current. To the west, there is a clear increase in $\delta^{18}O$ from the central Arctic (-4‰) through the Archipelago to Baffin Bay (-1‰). The corresponding salinity gradient is not as pronounced. In Baffin Bay, the West Greenland Current follows a cyclonic path and combines with the Canadian throughflow, exiting the study area to the south along the coast of Labrador.

The data indicate a local addition of glacial meltwater in Baffin Bay originating from the West Greenland coast. Along the west coast of Greenland, Bedard et al. [29] measured glacial meltwater inflow ranging from -17‰ to -21.7‰ in Cornelius Grinnell Fjord (63°10'N) and Qaumarujuk Fjord (71°09'N), respectively. The largest source of icebergs in the world is Ilulissat glacier, which discharges about 20 million tons of ice per day (ca. $2.3*10^{-4}$ Sv) into Disko Bay. This glacier, plus the ca. 20 glaciers to the north along the coast, are the main producers of icebergs in the North Atlantic.

In the Labrador and the Norwegian/Greenland seas, the main features of the salinity distribution and the $\delta^{18}O$ field are similar. High salinity, high $\delta^{18}O$ water flow toward the Arctic Ocean in the Norwegian Current, and low-salinity, low-$\delta^{18}O$ water exits the Arctic Ocean in the East Greenland Current and through the Canadian Archipelago.

Each summer there is melting of all of the 20-40 cm of snow deposited on the sea-ice, as well as the upper 30-70 cm of the ice itself [30,31]. Because most samples used in this study were collected in the summer, this could bias the surface $\delta^{18}O$ and salinity fields toward lower levels than the annual average. Isotopic profiles of cores from drifting multiyear Arctic sea ice indicate that ice formed during summer is usually more depleted than ice formed during winter due to snow melt admixed with surface ocean water [14,32,33,34].

468

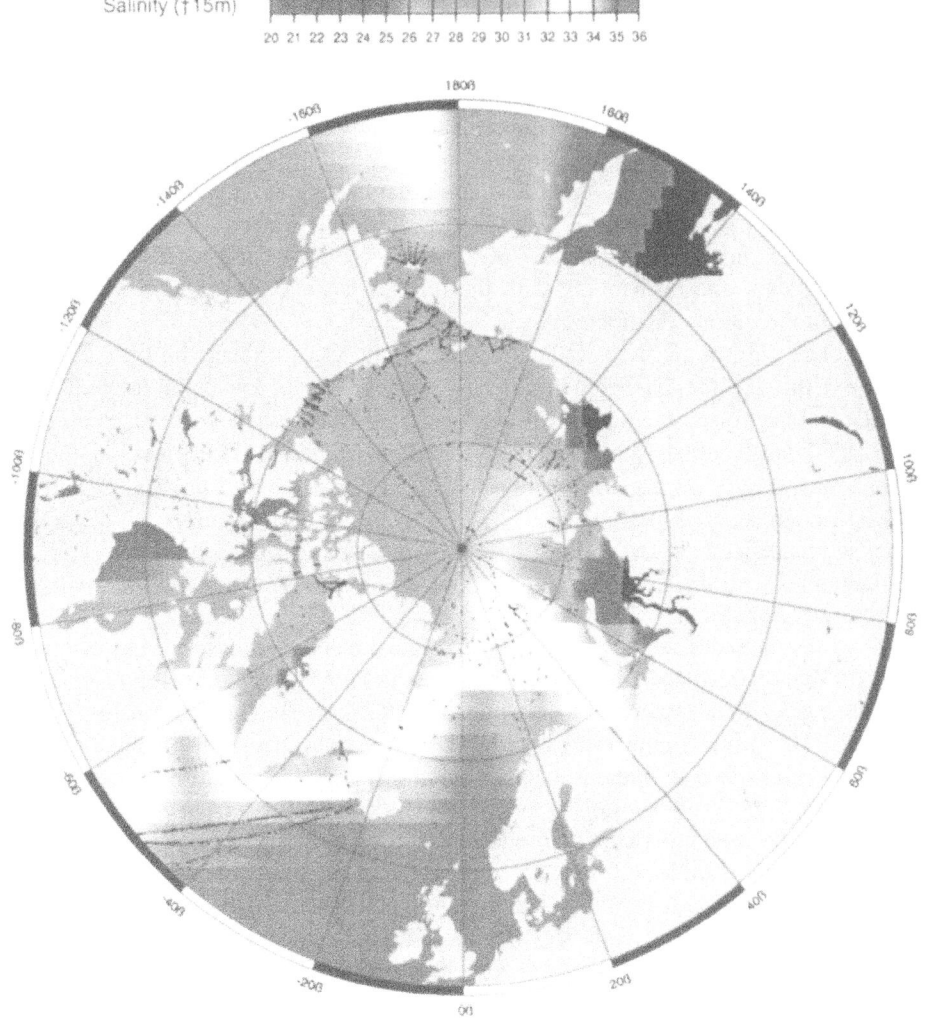

Salinity (†15m)

20 21 22 23 24 25 26 27 28 29 30 31 32 33 34 35 36

Figure 5b: Map of the salinity of the top 15 meters of the water column using the same stations as in Fig. 5a. For discussion, see text.

(For a colour version of this figure, see p. 617)

The $\delta^{18}O$ distribution discussed here has been compiled using data covering a period of at least two decades. During this time, studies performed in the Nordic seas and the Arctic Ocean revealed strong evidence for variability on a variety of time scales. This fact has to be taken into account when interpreting the $\delta^{18}O$ field. Details of the interpretation of the temporal and spatial variability is beyond the scope of this contribution.

6. Use of isotope data for model calibration

6.1. GENERAL CONSIDERATIONS

In the upper layers of the Arctic Ocean, the freshwater distribution exerts a controlling influence over the density field, and therefore over any buoyancy-driven currents. Modeling experiments with both idealized and realistic boundary conditions have been performed to try to segregate the dynamical influences of sea ice, runoff, bathymetry and winds on the surface circulation. (see, e.g., [35], for a review of early modeling work and [36,37,38], among others, for more recent advances). However, even qualitative understanding of the relative importance of these different forces has been elusive.

Partly, progress has been slow because the data do not sufficiently constrain the model results. With only limited observations, it is not obvious which model results are right and which are wrong, and it is difficult to diagnose the sources of errors. The inclusion of freshwater tracers into circulation models promises to improve both their diagnosis and calibration, as well as to provide insight into the processes controlling freshwater fluxes.

By adding artificial dyes to the rivers which flow into a particular marginal sea, model output manifests, at first glance, the interactions between coastal and shelf circulation, and between the shelf-seas and the ocean basins. Since the shelf-seas are the places where most water mass transformation takes place, these same tracers can be used to follow the circulation of the newly (trans)formed water masses. The tracer distributions in the model (as in the real ocean) act as integrators of formation, transport and mixing of water masses. Thus, one advantage of freshwater tracers in the Arctic is that they highlight dynamically important effects of the model physics, making it simple and immediate to diagnose obvious problems.

With the availability of the growing body of observations described above, freshwater tracers also add an important constraint to the model-data comparison. Circulation patterns may seem plausible if they are in the correct direction, and the velocities fall within the observed range. The successful transport of appropriate fractions of different source waters throughout the model domain poses a more stringent test, one that is more directly relevant to questions involving freshwater relevant to climate studies. Starting with artificial dyes added to the individual rivers, comparison with conservative tracers is straightforward. For example, at any model point, and at any point in time, we can calculate the individual river contributions, as well as the Atlantic and Pacific water components. The simulated $\delta^{18}O$ field is simply a weighted sum of the $\delta^{18}O$ values of these water types. Similarly, if we would dye the Eurasian and North American rivers with two different dyes, we could use their fractions in the model output to calculate a (model) Ba field.

6.2. INITIAL RESULTS

Our own approach has been to introduce an artificial tracer for each of the major rivers, one for Bering Strait Inflow, and one for Atlantic waters moving north of (approximately) Denmark Straight into a high-resolution, basin-scale GCM. The ocean model is of the Bryan-Cox-Semtner type (see, e.g., [39], which has been reformulated with a free surface by Killworth, *et al.* [40]. The numerical scheme is by Dukowicz and Smith [41]. The sea-ice component is a reformulation of Hibler's model with a linear thermal profile and plastic-viscous rheology, by Hunke and Zhang [42]. The model is forced with realistic atmospheric fields from the ECMWF reanalysis and operational analyses between 1979 and 1997. We have used the $\delta^{18}O$ fields from several icebreaker cruises [11,3, 20] to compare the model freshwater distribution to observations. In our case, the salinity fields would have been very difficult to diagnose, since our GCM still has a weak salinity relaxation term. The comparison of modeled and observed river runoff distributions instantly highlighted model deficiencies that had escaped detection from studying only the salinity, temperature and velocity fields. Furthermore, a comparison of the river tracer distributions and salinity field facilitated the diagnosis of the model problems. A subtle combination of grid-scales, vertical resolution and adjustments to avoid numerical effects had limited the cross-shelf transport in spite of "realistic" currents throughout the model run. Finding and correcting such errors is key to using models such as ours to answer questions related to halocline maintenance by shelf-water injections, the fate of river- or shelf-deposited contaminants, and changes in the salinity of Arctic outflows, among others.

Where the model results can be validated, the inclusion of water-mass tracers can clarify the model dynamics in more detail. For example, in our simulations, we delineate the pathways of the fraction of river water that is detrained off the shelf across the Arctic (e.g., Fig. 6a). Comparison of the river-runoff distributions simulated for 1979 and 1993 (Fig. 6b) demonstrates that we are able to reproduce the shift in the river-runoff distribution observed over the past decade. In both our river-runoff and our Atlantic water distributions, we see the eastward shift in property fronts, which Morison *et al.* [22] and Steele and Boyd [43] documented from observations. We can use the model to determine the relative importance of buoyancy and wind forcing in shifting the fronts.

The integration of river runoff into large scale circulation models is still at an early stage, and the task of modeling coastal, shelf and deepwater processes in a single domain (with finite computing resources) is not simple. We expect that the current efforts will unmask problems, as well as solutions. As we work through the initial problems, we expect the models to become more sophisticated. For example, an obvious enhancement would be the addition of a sea-ice meltwater tracer. Besides its dynamical importance, sea-ice meltwater is important in studies of the fate of contaminants in the Arctic. Together with the addition of a tracer of the sea-ice component of coupled ice-ocean models, this would also close a serious gap in the current modeling of tracer tansport.

Forcing fields could also be improved and experimented with. Typically the current models use a 'climatological' annual cycle for runoff. However, there are interesting questions that can be posed by varying the strength of the runoff signal. For example, shifts in storm tracks and moist static energy flux [44] suggest that climate change in the Arctic is expressed, in part, by a shift of the longitudinal distribution of precipitation. This poses the question if a shift of river-runoff from the Eurasian to the North American rivers would have an impact on the salinity of the outflow through the Davis and Denmark Straits. It has also been suggested that low latitude warming will lead to high-latitude precipitation and runoff. High latitude

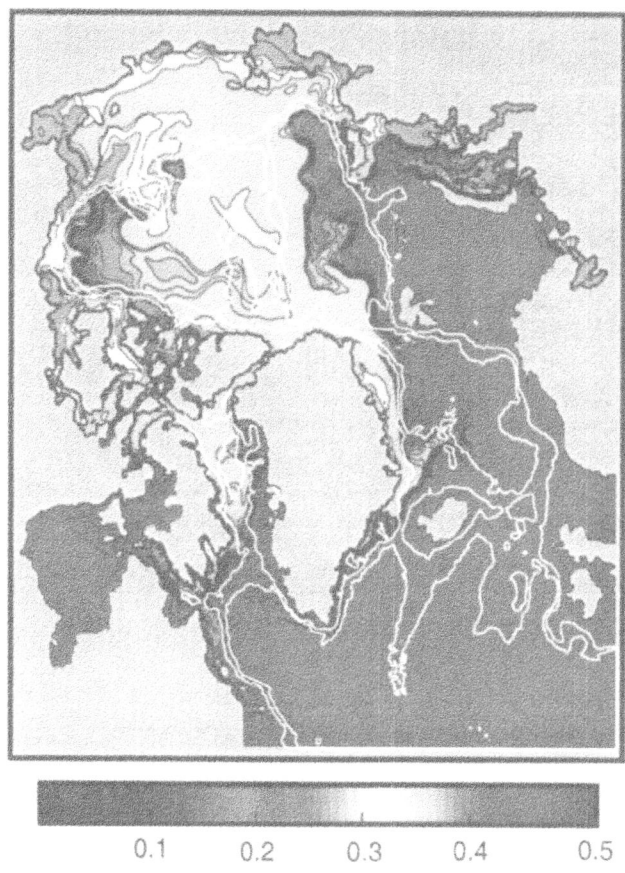

Figure 6a: Results of the simulation of the Arctic freshwater distribution in a high-resolution GCM. Displayed are the fractions of freshwater.

(For a colour version of this figure, see p. 618)

warming, on the other hand, may tend to shift the freshwater source from river runoff towards snowfall over the Arctic Ocean (Mark Stieglitz, pers. comm.). Experiments with such scenarios require sophisticated tracking of freshwater from its source to its ultimate destination in the Arctic outflows.

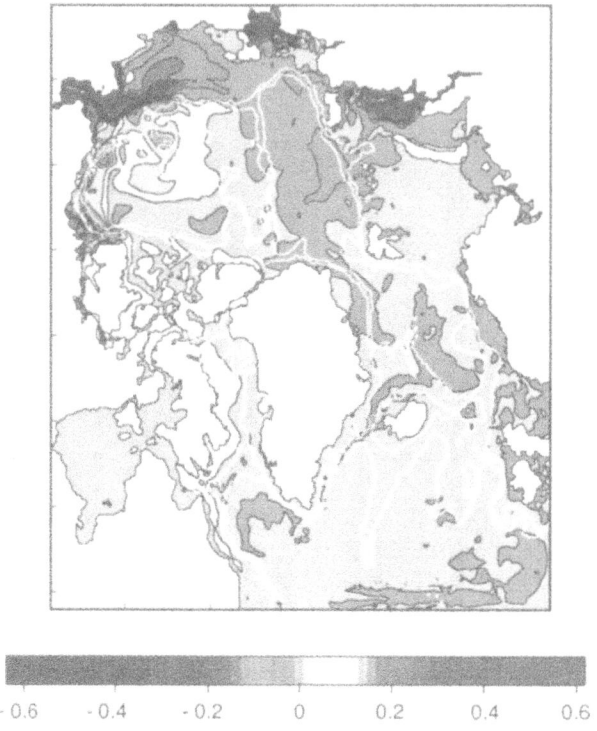

Figure 6b: Difference in freshwater fraction distribution between 1993 and 1979, as simulated by a high-resolution GCM.

(For a colour version of this figure, see p. 618)

7. Mean residence time of the freshwater

The mean residence time of the water masses in the Arctic Ocean can be determined by using a combination of tracers. For the surface waters, i.e., the waters that contain the bulk of the freshwater, we typically use tritium/^{3}He ages and CFC data (e.g., [45,46,47,48,49]). Here we present the principal distribution of tracer ages in the surface waters of the Arctic on the basis of the 1996 section of the US nuclear submarine POGY in the Canadian Basin.

The main feature of the tritium distribution in the upper waters of the Canadian Basin is a lens of water (depth ≤ 50 m) with high tritium concentrations (>5 TU) indicating the presence of river runoff (Fig. 8a). This feature is confirmed by the presence of low δ^{18}O values ([21]; their Fig. 6b). Below the surface mixed layer, tritium concentrations decrease through the halocline (50 to 250 m depth) to about 1.5 to 2 TU in the Atlantic layer. Intermediate waters underlying the Atlantic layer (500 to m depth) have variable tritium concentrations reflecting the different mean transit times of these waters from the boundary into the interior of the Canadian Basin (for details of the intermediate water circulation, see e.g., [50]).

Figure 7: Geographical locations of the stations used in Fig. 8.

The distribution of δ^3He (Fig. 8b) shows a high degree of variability along the POGY section. The most significant features are the buildup of tritiogenic ^{3}He in the halocline (tritiocline; for explanation, see, e.g., [48]), as well as the lateral gradients in the δ^3He field (lower concentrations at stations 36 and 37). Maximum δ^3He values of close to 25% were observed between 100 and 150 meters depth. These values are comparable to δ^3He data obtained from other sections across the central Arctic Ocean (e.g., [48,3]). The surface values vary from close to 0 to about 3 to 5 %, i.e., significantly above the solubility equilibrium with the atmosphere (-1.8%). This feature reflects reduced gas exchange through the perennial sea ice cover (e.g., 46,48,47]).

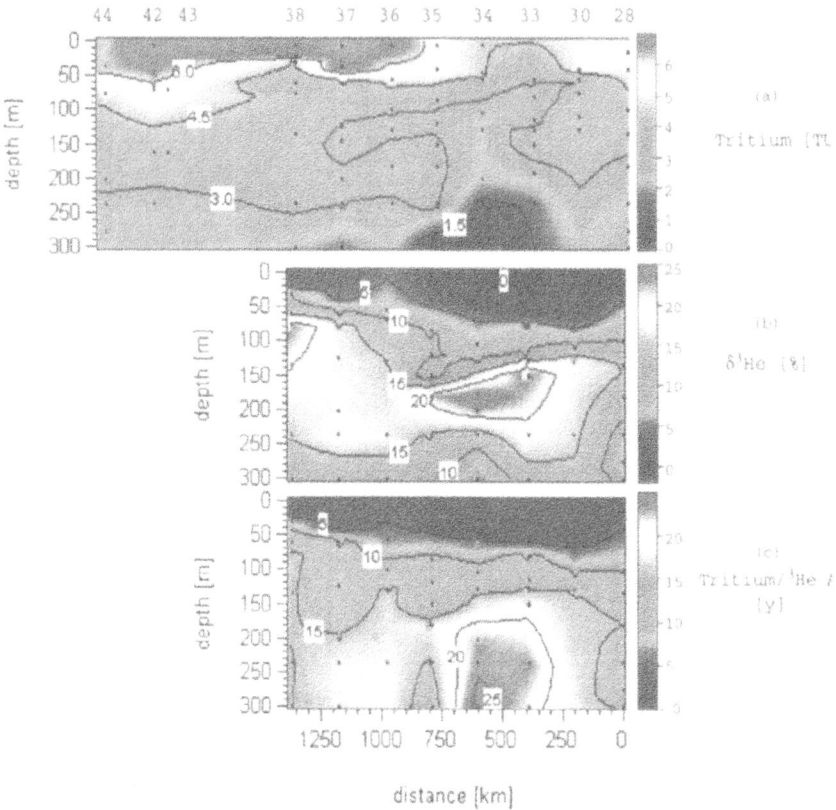

Figure 8: (a) Tritium section along a portion of the cruise track of the US nuclear submarine POGY (1996). (b) same as (a) except for δ³He. (c) same as (a) except for the tritium/³He age.

(For a colour version of this figure, see p. 619)

The combination of tritium and ³He allows us to calculate the tritium/³He age (Fig. 8c). This age field shows a fairly smooth distribution, i.e., some of the variability in the tritium and δ³He fields is compensated for in the tritium/³He age field. Mean residence times of the surface waters range from about 2 to 5 years (Fig. 8c). They have to be considered as lower boundaries because some of the tritiogenic ³He might have escaped through leads in the ice cover. The mean residence times of the halocline waters vary from about 5 to 15 years. As pointed out by Wallace *et al.* [47], tritium/³He ages between modern (0) and about 10 to 15 years should be close to the 'real' mean residence times of the waters in the surface layer. Higher tritium/³He ages are affected significantly by mixing and have to be interpreted by using models correcting for the non-linearity of the tritium/³He age with respect to mixing. Although based on different tritium and δ³He fields (transient nature of these tracers), the resulting tritium/³He ages agree with previous estimates from other regions of the Arctic Ocean (Eurasian Basin: [46,48]; Canadian Basin: [3]). They indicate that freshwater added to the surface layers will be flushed out of the system on time scales of a few years to less than one decade. There seems to be little lateral variation in the mean residence time of the surface waters in the Arctic Ocean.

The mean residence times derived from tritium/^{3}He data agree well with those derived from CFC data [51]. They are lower in the near surface waters (top 100 to 200 meters) if compared to mean residence times derived from tritium data alone [16] or tritium and δ^{18}O data [17].

8. Future research priorities

Tracer studies add significantly to the understanding of the Arctic freshwater balance. The most significant contributions can be expected in terms of (1) separation of the individual sources contributing to the overall freshwater balance of the Arctic Ocean and its variability, (2) visualization of the freshwater components along sections or on isopycnals, (3) calculation of the inventories of the individual freshwater components at any given location, (4) estimation of mean residence times of the freshwater in the Arctic ocean, and (5) calibration of circulation models used to simulate the freshwater balance.

In order to use the tracer method most efficiently, the following tasks have to be completed:

- Close the gaps in the first, multi-tracer survey of the Arctic Ocean basins. For many sections we do not have adequate tracer coverage (no or incomplete tracer data sets).
- Establish sites for long-term observations to place specific sections into climatological context.
- Repeat surveys of tracer fields in a quasi-synoptic fashion at time intervals to be determined by evaluation of the existing data sets.
- Fully utilize the existing data in model studies.
- Combine estimates of the mean currents in Fram Strait and the Canadian throughflow and tracer data to arrive at estimates of freshwater fluxes out of the Arctic Ocean (e.g., [11])

Acknowledgements

This work was supported by the National Science Foundation through grants by the national Science Foundation (grants OPP 08924 and OPP 95-30795 and OPP 95-30795), and the Office of Naval Research (grants N00014-90-J-1362, N00012-94-I-0809, N00014-93-I-0830, and N00014-93-30795). The W.M. Keck Foundation contributed significantly to the establishment of the L-DEO tritium/^{3}He laboratory. L-DEO contribution no. 6022.

476

9. References

1. Craig, H. and Gordon, L.I. (1965) Deuterium and oxygen 18 variations in the ocean and the marine atmosphere, in: E. Tongiorgi (ed), *Stable isotopes in oceanographic studies and Paleotemperatures*. Spoleto Conf. Proc., Cons. Naz. Ric., V. Lischi Figli, pp. 9-130.
2. Dansgaard, W. (1964) Stable isotopes in precipitation. *Tellus*, **XVI**, 436-468.
3. Ekwurzel, B. (1998) Circulation and mean residence times in the Arctic Ocean derived from tritium, helium, and oxygen-18 tracers, Ph.D. thesis, Columbia University, New York.
4. Frank, M. (1996) Spurenstoffuntersuchungen zur Zirkulation im Eurasischen Becken des Nordpolarmeeres, Ph.D. thesis, Ruprecht Karls Universität Heidelberg, Heidelberg.
5. Létolle, R., Martin, J.M., Thomas, A.J., Gordeev, V.V., Gusarova, V., and Sidorov, I.S. (1993) ^{18}O Abundance and dissolved silicate in the Lena delta and Laptev Sea (Russia), *Marine Chem.*, **43**, 47-64.
6. Falkner, K.K., Macdonald, R.W., Carmack, E.C., and Weingartner, T. (1994) The potential of barium as a tracer of Arctic water masses, in O.M. Johannessen, R.D. Muench, and J.E. Overland (eds.), *The polar oceans and their role in shaping the global environment*, pp. 63-76, American Geophysical Union, Washington DC.
7. Guay, C.K. and Falkner, K.K. (1997) Barium as a tracer of Arctic halocline and river waters, *Deep Sea Res.* **44**, 1543-1569.
8. Anderson, L.G., Olsson, K., and Skoog, A. (1994) Distribution of dissolved inorganic and organic carbon in the Eurasian Basin of the Arctic Ocean, in O.M. Johannessen, R.D. Muench, and J.E. Overland (eds) *The polar oceans and their role in shaping the global environment*, pp. 255-262, American Geophysical Union, Washington.
9. Anderson, L.G. (1995) Chemical oceanography of the Arctic and its shelf seas, in W.O. Smith. Jr., and J.M. Grebmeier (eds.) *Arctic oceanography: marginal ice zones and continental shelves*, pp. 183-202, American Geophysical Union, Washington, DC.
10. Anderson, L.G., Björk, B., Holby, O., Jones, E.P., Kattner, G., Koltermann, K.P., Liljeblad, B., Lindegren, R., Rudels, B., and Swift, J. (1994) Water masses and circulation in the Eurasian Basin: Results from the Oden 91 expedition, *J. Geophys. Res.*, **99** (C2), 3273-3283,
11. Bauch, D., Schlosser, P., and Fairbanks, R.G. (1995) Freshwater balance and the sources of deep and bottom waters in the Arctic Ocean inferred from the distribution of $H_2^{18}O$, *Progress in Oceanography*, **35**, 53-80.
12. Jones, E.P., Anderson, L.G., and Swift, J.H. (1998) Distribution of Atlantic and Pacific waters in the upper Arctic Ocean: Implications for circulation, *Geophys. Res. Lett.*, **25**, 765-768.
13. Broecker, W.S., Peacock, S.L., Walker, S., Weiss, R.F., Fahrbach, E., Schroeder, M., Mikolajewicz, U., Heinze, C., Key, R., Peng, T.-H., and Rubin, S. (1998) How much deep water is formed in the Southern Ocean?, *J. Geophys. Res.*, **103**, 15,833-15,843.
14. Friedman, I., Schoen, B., and Harris, J. (1961) The deuterium concentration in Arctic sea ice, *J. Geophys. Res.*, **66**, 1861-1864.
15. Redfield, A.C., and Friedman, I. (1961) The effect of meteoric water, melt water and brine on the composition of Polar Sea water and of the deep waters of the ocean, *Deep-Sea Res.*, **16**, 197-214.
16. Östlund, H.G. (1982) The residence time of the freshwater component in the Arctic Ocean, *J. Geophys. Res.*, **87**, 2035-2043.
17. Östlund, G.H. and Hut, G. (1984) Arctic Ocean water mass balance from isotope data, *J. Geophys. Res.*, **89**, 6373-6381.
18. Östlund, H.G., G. Possnert, G., J.H. Swift, J.H. (1987) Ventilation rate of the deep Arctic Ocean from carbon 14 data, *J. Geophys. Res.*, **92**, 3769-3777.
19. Östlund, H.G. and Grall, C. (1993) Arctic Tritium: 1973-1991. *Tritium Laboratory data report no. 19*, University of Miami, Rosenstiel School of Marine and Atmospheric Science, Miami, Florida, 33149.
20. Schlosser, P., Bauch, D., Fairbanks, R., and Bönisch, G., (1994). Arctic river-runoff: mean residence time on the shelves and in the halocline, *Deep-Sea Research*, **41**, 1053-1068.
21. Schlosser, P., Bayer, R., Bönisch, G., Cooper, L., Ekwurzel, B., Jenkins, W.J., Khatiwala, S., Pfirman, S., and Smethie, W.M., Jr. (1999) Pathways and mean residence times of dissolved pollutants in the ocean derived from transient tracers and stable isotopes. *Science of the Total Environment*, **237/238**, 15-30.
22. Morison, J.H, Steele, M., and Andersen, R., (1998) Hydrography of the upper Arctic Ocean measured from the nuclear submarine USS Pargo, Deep Sea Res. **45**, 15-38.
23. Carmack, E.C., Macdonald, R.K.W., Perkin, R.G., McLaughlin, F.A., and Pearson, R.J., (1995) Evidence for warming of Atlantic Water in the Southern Canadian Basin of the Arctic Ocean, Results from the Larsen-93 Expedition, *Geophys. Res. Letters* **22**, 1061-1064.
24. McLaughlin, F.A., Carmack, E.C., and Macdonald, R.W. (1996) Physical and geochemical properties across the Atlantic/Pacific water mass front in the southern Canadian Basin, *J. Geophys., Res.* **101**, 1183-1197, 1996.

25. Loeng, H (1991) Features of the physical oceanographic conditions of the BarentsSea, 1981, *Polar Research* **10**, 5-18.

26. Pfirman, S.L., Bauch, D., and Gammelsrød, T. (1994) The Northern Barents Sea: Water mass distribution and modification, in O.M. Johannessen, R.D. Muench, and J.E. Overland (eds) *The polar oceans and their role in shaping the global environment*, AGU Geophysical Monograph no. 85, pp. 77-94.

27. Schauer, U., Muench, R.D., Rudels, B., and Timokhov, L. (1997) Impact of eastern Arctic shelf waters on the Nansen Basin intermediate layers. *J. Geophys. Res.* **102**, 3371-3382.

28. Cooper, L.W., Whitledge, T.E., and Grebmeier, J.M. (1997) The nutrient, salinity, and stable oxygen isotope composition of Bering and Chukchi Seas waters in and near the Bering Strait, *J. Geophys. Res.* **102**, 12,563-12,573.

29. Bedard, P., Hillaire-Marcel, C., and Page, P. (1981) ^{18}O modeling of freshwater inputs in Baffin Bay and Canadian Arctic coastal waters, Nature **293**, 287-289.

30. Untersteiner, N. (1961) On the mass and heat budget of Arctic Sea Ice, *Arch. Meteor. Geophys. Bioklim.* **A12**, 151-182.

31. Romanov, I.P. (1992) The ice cover of the Arctic Ocean. AARI, St. Petersburg, 211 pp.

32. Mel'nikov, I.A. and Lobyshev, V.I. (1985) Fractionation of ^{18}O in the snow and ice cover of the central Arctic Basin. *Oceanology*, **25**(2), 181-184.

33. Pfirman, S.L., Lange, M., Wollenburg, I., and Schlosser, P. (1990) Sea ice characteristics and the role of sediment inclusions in deep sea deposition: Arctic-Antarctic comparisons, in: U. Bleil and J. Thiede (eds.), *Geologic History of the Polar Oceans: Arctic versus Antarctic*, NATO ASI Series C, Kluwer Academic Publishers, Dordrecht, pp 187-211.

34. Lobyshev, V.J., Mel'nikov, I.A., Yesikov, A.D., and V.V. Nechayev (1984) Study of the oxygen isotope composition of Arctic drift ice in connection with activation of the growth of phytoplankton at the boundary of the melting ice. *Biophysics*, **29**(5) 912-917.

35. Hakkinen, S. (1990) Models and their applications to polar oceanography, in W.O. Smitjh (ed.), *Polar Oceanogrphy, Part 1, Physical Science*, Academic Press, San Diego, CA., pp. 225-384.

36. Holland, D.M, Mysak, L.A., and Oberhuber, J.M. (1996) An investigation of the general circulation of the Arctic Ocean using an isopycnal model, *Tellus* **48A**, 138-157.

37. Proshutinsky, A.Y. and Johnson, M.A. (1997) Two circulation regimes of the wind-driven Arctic Ocean. *J. Geophys. Res.* **102**,12,493-12,514.

38. Weatherly, J.W. and Walsh, J.E. (1996) The effects of precipitation and a river runoff in a coupled ice-ocean model of the Arctic. *Climate Dynamics* **12**, 785-798.

39. Semtner, A.J. and Chervin, R.M. (1988) A simulation of the global ocean circulation with resolved eddies, *J. Geophys. Res.*, **93**, 15,502-15,522.

40. Killworth, P.D., Stainforth, D., Webb, D.J. and Paterson, S.M. (1991) The development of a free-surface Bryan-Cox-Semtner ocean model. *J. Phys. Oceanography* **21**, 1333-1348.

41. Dukowicz, J.K. and Smith, R.D. (1994) Implicit free-surface method for the Bryan-Cox-Semtner ocean model, *J. Geophys. Res.* **99**, 7991-8014.

42. Hunke, E.C. and Zhang, Y. (1999) A comparison of sea ice dynamics models at high resolution, *Mon. Weather Rev.* **127**, 396-408

43. Steele, M. and Boyd, T. (1998) Retreat of the cold halocline layer in the Arctic Ocean, *J. Geophys. Res.* **103**, 10,419-10,435.

44. Maslanik, J.A., Serreze, M.C., and Barry, R.G. (1996) Recent decreases in Arctic summer ice cover and linkages to atmospheric circulation anomalies. *Geophys. Res. Letters* **23**, 1677-1680.

45. Wallace, D.W.R., Moore, R.M., and Jones, E.P. (1987) Ventilation of the Arctic Ocean cold halocline: rates of diapycnal and isopycnal transport, oxygen utilization, and primary production inferred using chlorofluoromethane distributions, *Deep Sea Res.* **34**, 1957-1979.

46. Schlosser, P., Bönisch, G., Kromer, B., Münnich, K.O. and Koltermann, K.P. (1990) Ventilation rates of the waters in the Nansen Basin of the Arctic Ocean derived from a multi-tracer approach, *Journal of Geophysical Research*, **95**, 3265-3272.

47. Wallace, D.W.R., Schlosser, P., Krysell, M., and Bönisch, G. (1992) Halocarbon ratio and tritium/^{3}He dating of water masses in the Nansen Basin, Arctic Ocean, *Deep-Sea Res.* **39**, 435-458.

48. Schlosser, P., Bönisch, G., Kromer, B., Loosli, H.H., Bühler, B., Bayer, R., Bonani, G., and Koltermann, K.P. (1995) Mid 1980s distribution of tritium, ^{3}He, ^{14}C and ^{39}Ar in the Greenland/Norwegian seas and the Nansen Basin of the Arctic Ocean, *Progress in Oceanography* **35**, 1-28.

49. Frank, M., Smethie, W.M., Jr., and Bayer, R. (1998) Investigation of subsurface water flow along the continental margin of the Eurasian basin using the transient tracers tritium, ^{3}He, and CFCs, *J. Geophys. Res.*, **103**, 30,773-30,792.

50. Smethie, W.M. Jr., Schlosser, P., Hopkins, T.S., and Bönisch, G. (2000). Renewal and circulation of intermediate waters in the Canadian Basin observed on the SCICEX-96 cruise, *J Geophys. Res.*, **105**, 1105-1121..

51. Wallace, D.W.R., and Moore, R.M. (1985) Vertical profiles of CCL3F (F-11) and CCl2F2 (F-12) in the central Arctic Ocean basin. *J. Geophys. Res.*, **90**, 1155-1166.

52. Pavlov, V.K., Timokhov, L.A., Baskakov, G.A., Kulakov, M.Y., Kurazhov, V.K., Pavlov, P.V., Pivovarov, S.V., and Stanovoy, V.V. (1996) Hydrometeorological regime of the Kara, Laptev, and East-Siberian Seas, Applied Physics Laboratory, University of Washington, Seattle. 179 pp.

53. Becker, P. (1995) The effect of Arctic river hydrological cycles on Arctic Ocean circulation, Ph.D. thesis, Old Dominion University, Norfolk, VA.

54. Macdonald, R.W., Carmack, E.C., McLaughlin, F.A., Iseki, K., Macdonald, D.M., and O'Brien, M.C. (1989) Composition and modification of water masses in the Mackenzie shelf estuary, *J. Geophys. Res.* **94**, 18,057-18,070.

55. Macdonald, R.W., Paton, D.W., Carmack, E.C., and Omstedt, A. (1995) The freshwater budget and under-ice spreading of the Mackenzie River water in the Canadian Beaufort Sea based on salinity and $^{18}O/^{16}O$ measurements in water and ice, *J. Geophys. Res.* **100**, 895-919.

56. Eicken, H. (1998) Deriving modes and rates of ice growth in the Weddell Sea from microstructural, salinity and stable-isotope data, in *Antarctic sea ice: Physical processes, interactions and variability*, pp. 89-122, American Geophysical Union, Washington, DC.

EXCHANGES OF FRESHWATER THROUGH THE SHALLOW STRAITS OF THE NORTH AMERICAN ARCTIC

HUMFREY MELLING
Fisheries and Oceans Canada
Institute of Ocean Sciences, P.O. Box 6000, Sidney, B.C.
Canada, V8L 4B2
Internet: mellingh@dfo-mpo.gc.ca

1. Introduction

The shallow straits of the North American Arctic include Bering Strait, which connects the Pacific Ocean to the Arctic Ocean, and the myriad channels of the Canadian Arctic Archipelago, which connect the Arctic Ocean to the Atlantic. A net flux of freshwater passes from the North Pacific into the North Atlantic via these straits (plus Fram Strait), and thereby stabilizes a global imbalance in precipitation less evaporation between the two major ocean basins [0, 2]. At the same time, the fluctuating outflows of fresh Arctic waters into the Greenland Sea, via Fram Strait, and into the Labrador Sea, via the Canadian Arctic Archipelago, may control the occurrence of deep ocean convection in these areas [3, 4, 5, 6]. For these reasons, the freshwater fluxes through the shallow straits of the North American Arctic are important within the global climate system.

To meet the needs of climate prediction, we need (a) to estimate to known accuracy the magnitudes of the present fluxes of freshwater through these straits, and their seasonal and interannual variations, and (b) to develop an understanding of the forcing of and controls on these fluxes.

It is important to clarify the meaning of the term *freshwater flux*. Formally, the freshwater flux is the rate of transport of seawater mass, minus the rate of transport of the mass of salt contained within it. More intuitive, if less rigorous, is the *freshwater anomaly flux*. This is the volume flux of zero-salinity water that must be combined with a volume flux of reference-salinity water to yield the flux of seawater of the salinity observed. Imprecision associated with use of volume rather than mass in this definition is negligible relative to the inaccuracy of present estimates. Here we use *freshwater flux* to denote the latter concept. The choice of reference salinity is arbitrary. Here we use the convenient whole-number value 35, which approximates the maximum salinity within the Arctic basins. This reference yields a freshwater flux at typical Arctic surface-water salinity that is about 8% higher than the 34.8-reference used elsewhere. The difference is insignificant relative to present uncertainty in these fluxes.

Past efforts to observe flows in the shallow straits of the North American Arctic have rightly concentrated on determining the fluxes of water volume, and their variations. Even the accuracy of volume-flux estimates is low, since few data are available. Measuring the fluxes of freshwater is a much greater challenge. According to

479

E.L. Lewis et al. (eds.), The Freshwater Budget of the Arctic Ocean, 479-502.
© 2000 *Kluwer Academic Publishers.*

the most recent estimates, by Aagaard and Carmack [3], the principal oceanic influx of freshwater (adjusted to reference salinity 35) to the Arctic is through Bering Strait (1800 km^3 y^{-1}). The major outflow is through Fram Strait (2800 km^3 y^{-1}) as ice, with an additional 1170 km^3 y^{-1} as water. These authors assess the freshwater outflow through the Canadian Arctic to be about the same as Fram Strait (1230 km^3 y^{-1}), but the ice outflow to be smaller (155 km^3 y^{-1}). Ingram and Prinsenberg [7] derive a much larger value (7100 km^3 y^{-1}) for the Canadian freshwater outflow. Estimates of the net export south through Davis Strait into the Labrador Sea are intermediate, 3100 km^3 y^{-1} as water and 1100 km^3 y^{-1} as ice [8]. Clearly, we are far from understanding the oceanic fluxes of freshwater in the Arctic.

This paper will illustrate the observational basis of these preliminary estimates, and propose a more sophisticated context to guide future consideration of this issue.

2. Freshwater Flux: What do we need to know?

The magnitude of the freshwater flux is the fundamental variable of interest. A freshwater budget can be compiled if the flux is known for all openings to the Arctic. However, for understanding the dynamics of freshwater flows, and their impacts on stratification and convection in the Pacific, Arctic and Atlantic Oceans, this knowledge is not enough. It is also important to know: 1) What is the salinity range of waters which constitute the freshwater flux at its source? 2) What is the salinity spectrum of freshwater flux at delivery? 4) What is the frequency co-spectrum of the flux? 4) Is the flux carried by a simple through-flow, or is it the net result of much larger transports by counter-flowing streams? 5) How much of the flux is carried as pack ice?

The moving pack ice and seawater that carry the freshwater flux encompass a range of values in salinity. Hence, the volume flux has a salinity spectrum (Figure 1). The total freshwater flux is the integral over salinity of this spectrum, weighted by freshwater anomaly. The effects of a freshwater flux that is carried by a homogeneous water mass of zero salinity are different from those if the same flux involves a much larger volume of water of salinity near 35. It makes a difference if the flux is taken from the surface of the source reservoir, or if it is withdrawn selectively over a range of depths. It makes a difference if the transported water is stratified, so that baroclinic effects are relevant, or if it is homogeneous.

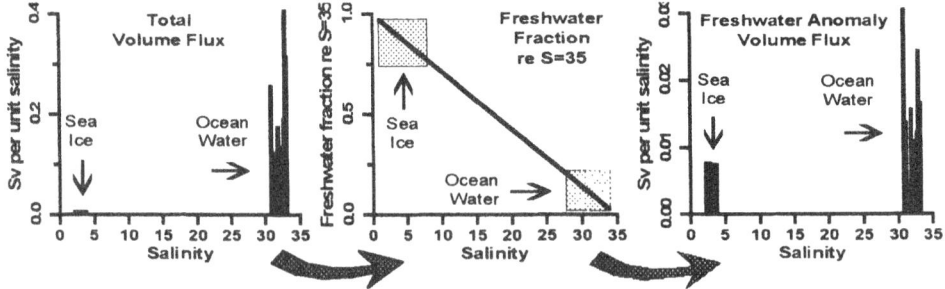

Figure 1. Salinity spectrum of the freshwater anomaly flux. This schematic illustrates the relationship between freshwater-anomaly flux and the volume flux, both expressed as functions of the salinity of the ice and seawater transported.

It makes a difference if the flux is delivered at the surface of the receiving reservoir, or if it interleaves at depth. The salinity spectrum of the freshwater flux is important, because it determines the impact of the freshwater flux on the stratification of both reservoirs. The spectrum may change during transit between reservoirs because of mixing and because of freezing and melting, which exchange freshwater between the solid and liquid components of the oceanic transport.

Since the freshwater flux is the covariance of volume flux and salinity, both of which fluctuate at periods that range between semi-diurnal and multi-annual, it has a frequency spectrum that, in general, will not peak at zero frequency. Therefore, the freshwater flux cannot be calculated simply as the product of long-term mean values of volume flux and freshwater anomaly. Unfortunately, this simplified method of calculation has commonly been used when estimating freshwater fluxes. A thorough knowledge of the co-varying time-series of current and salinity is necessary to compute freshwater flux correctly.

Counter-flowing currents commonly exist simultaneously in straits. Fram Strait, where the inflowing Atlantic water on the eastern side co-exists with out-flowing Arctic waters on the western side, is a well-known example. Counter-flows may also be separated vertically, as they are in an estuary. Both vertically and horizontally separated counter-flows occur throughout the Canadian Arctic Archipelago. A third form of counter-flow has the ocean and the pack ice as the opposing streams, as in the "conveyor-belt" circulation of the Bering Sea described by Pease [9]. In all such situations, there is an exchange of water between the reservoirs, not simply a transfer from one, regarded as source, to another as sink. The fluxes in each direction, as well as the net flux, have significance in the context of climate. Thus, knowledge of the spatial co-variation of current and salinity over the measurement section is needed for this calculation. The existence of counter-flows in all the channels that connect the Arctic to the rest of the world's ocean complicates the measurement and modeling of the freshwater fluxes through them.

The previous discussion can be summarized within an integral expression for the freshwater flux, Q, through a channel cross-section of width W and depth $H(y)$, where $y \in (0, W)$:

$$Q = \frac{1}{T} \int_0^T \int_0^W \int_{-H(y)}^{h(y)} u(y,z,t) \cdot \left(1 - S(y,z,t)/S_{ref}\right) \cdot dz \, dy \, dt \qquad (1)$$

Here y and z are the cross-channel and vertical coordinates, t is time, $u(y,z,t)$ and $S(y,z,t)$ are the fields of along-channel flow speed and salinity over the cross-section, $h(y)$ is the ice freeboard, and T is the period of integration. The vertical integral extends from the seafloor to the top of the (snow-covered) ice. In this application, T should be at least one year because seasonal cycles in salinity, ice and current in the Arctic are large.

Systematic observations of the freshwater flux as ice have been acquired only in Fram Strait. Here ice dominates the total freshwater transport, for three reasons. It is predominantly old ice of high average thickness. Its freshwater anomaly is large because its salinity is low (~5). It is the swiftest component of the water column because it rides a baroclinic, surface-intensified buoyancy current. For the same reasons, a potentially important freshwater transport as ice through other straits should not be overlooked. For example, within the Canadian Arctic, an 8-month drift of 3-m ice through the major channels at a modest 16 cm s^{-1} could transport as much freshwater annually as the ocean (1230 km^3).

3. Aspects of Freshwater Flux through Straits

Although the freshwater fluxes through Arctic straits may perhaps be measured without theoretical comprehension, the need to parameterize and model the fluxes in a changing climate requires an understanding of their forcing and of the controls upon them. Notwithstanding the geometrical complexity of real straits, idealized theoretical models of channel flow in rotating frames of reference will provide ample challenge for some time to come.

Relevant aspects of channel flows are the channel geometry, the stratification and width of the flow, mixing, flow forcing and control, re-circulation, and the dynamics of pack-ice drift along channels.

The width, depth and length of a channel are its basic geometrical parameters. For greater realism, constrictions in the channel cross-section, and curvature of its walls must be introduced. For example, Bering Strait more closely resembles an opening in a barrier than a strait since it widens rapidly to north and south of its narrowest section. Changes in either width or depth (at a sill) create constrictions in a channel flow. Constrictions impose hydraulic controls on the speed of flow and may introduce hydraulic re-adjustments in flow that promote mixing. In a stratified flow, sills can induce a vertical separation of the inflow and thus a selective withdrawal of fluid from the source reservoir. Sharp channel curvature may induce separation of the flow from the channel wall in the horizontal plane and consequent re-circulation and mixing.

The density stratification of channel flow influences its width (through baroclinic adjustment), its susceptibility to hydraulic and baroclinic instability, and the selective withdrawal of the flow from its source reservoir. The salinity spectrum of the freshwater flux is directly related to stratification because density is dominated by salinity at the low temperature of Arctic seawater.

The width of a channel flow in a rotating frame is dependent on its stratification. An unstratified flow has a width comparable to the Rossby radius, dependent upon the gravitational acceleration, channel depth H and Coriolis parameter. For Bering Strait (50 m deep) this scale is 170 km, far exceeding the width of the strait. Stratified flows are narrower, scaling with the internal Rossby R_{oi}, which in Arctic seas is only 5-10 km. The width of the upper layer flow, given by $(c/u)R_{oi}$ (with $c^2 \sim g'h$ and u the speed of the surface current) is typically less than the width of the important Arctic straits [10].

Mixing in sea straits is driven by turbulence. The predominant source of turbulence kinetic energy is the flow itself, via frictional interaction with the seabed and mesoscale instabilities of the current. The average rate of mixing is proportional to u^3/H. This variable increases by several orders of magnitude in the narrow, shallow cross-sections of sea straits, where current accelerates to maintain volume flux. Although fluctuating flows in the tidal band have little direct role in carrying the freshwater flux, they are important sources of energy for mixing. The 'inertial' latitudes are a relevant curiosity of polar ocean dynamics [11]. At these latitudes (71.0°, 74.5° and 85.7°, for the N$_2$, M$_2$ and S$_2$ tides), semi-diurnal tidal constituents resonate with inertial waves, causing a thickening of the Ekman layer of the clockwise-rotary tidal component. The effect is important in the shallow straits of central Canadian Arctic near 75° N [12]. As channel flows become more homogeneous through mixing, their baroclinicity is reduced. Therefore, mixing weakens internal hydraulic effects, strengthens the barotropic flow and narrows the salinity spectrum of the freshwater anomaly flux [13]. Mixing drives the estuarine circulation, by forcing the entrainment of underlying waters into the outflow.

Flows in sea straits are forced over a wide range of frequencies. At short period (hours to months), where tides and storms are active, forcing is relatively well understood. In the Arctic, the annual cycle in ice cover (strong and compact in winter) imposes an annual modulation on direct atmospheric forcing by wind. At long period, poorly understood steric and eddy-topographic effects [14] dominate. The flow through the Arctic from Pacific to Atlantic is generally attributed to the higher sea level of the Pacific [0, 2]. This arises from the lower salinity of Pacific waters, subject to an assumed level of no motion between the ocean basins. Relative to the 1100 dbar level, the surface of the Atlantic is thought to be 0.65 m lower than the Pacific and 0.15 m lower than the Arctic [0]. The drop across the Canadian Archipelago from the Arctic to Baffin Bay has been independently estimated as 0.3 m relative to 250 dbar [15]. The discrepancy is indicative of uncertainty in the magnitude of steric forcing.

The principal controls on channel flow in a rotating frame of reference are friction and geostrophy. Frictional drag is applied at the surface, in addition to the seafloor, when Arctic straits are capped by fast ice. Therefore, in winter a pipe flow supplants channel flow within the Canadian Arctic Archipelago. If the seafloor and the ice canopy are relatively smooth, then flat-plate boundary layers develop, and drag has a quadratic dependence on flow speed. Note, however, that if only sub-diurnal flow is considered explicitly, the drag coefficient must be increased to parameterize the high drag implicit in strong tidal flows. In reality, the seafloor and the ice cover of Arctic straits are topographically rough. Thus, flat-plate drag must be augmented by form drag, and in stratified flows, by internal hydraulic effects [16].

The geostrophic control of channel flow is a complex topic [17, 18, 19, 20]. In simple terms, a barotropic channel flow is limited to the speed at which the cross-channel sea-level difference established through geostrophic adjustment equals the along-channel difference that drives the flow. The volume flux driven by a sea-level difference h through a uniform channel of depth H is ghH/f at control. Solutions for realistic channel flows are difficult [21, 22]. A real strait may have several control points imposed by constrictions in width and in depth. Moreover, frictional and rotational controls are strongly coupled via Ekman pumping and do not operate independently [23].

Re-circulation refers to the return of water to its source following its passage partway along the exit channel. Re-circulation will occur wherever counter-flows coexist in straits, which is throughout the Arctic. Turbulent mixing and flow instability cause exchange of water between the inflow and the outflow. Modified water can also re-circulate back to its source via a temporary reversal of the direction of flow in a channel. This occurs frequently in Bering Strait. Or, as also occurs in Bering Strait, seawater moving in one direction may freeze to ice moving in the other, be carried to the source reservoir, melt and be returned via the oceanic flow. Because re-circulation blurs the distinction between source and sink, and depends on details of mixing and mesoscale circulation, it is a serious impediment to understanding of the fluxes through straits.

A unique aspect of polar channel flows is their seasonally varying cover of pack ice. When this consists of small floes at moderate concentration, its main effect on flow is to alter the stress exerted by wind on the ocean surface. However, when large thick floes are present at high concentration, they can jam within the channel [24, 25]. As ice drift continues downstream of the blockage, an arch becomes evident marking the boundary between open water and fast ice. The formation of ice arches on scales of 10-100 km can be simulated using both granular flow [24] and rigid-plastic rheological models of pack ice [25]. The most important variables are the channel width and geometry (converging or diverging walls), the magnitudes and directions of the wind and current stresses, and the mechanical properties of the ice. Empirical methods have been developed to predict arching of ice in Bering Strait [26]. However, an inability to predict the mechanical properties of ice (angle of internal friction, cohesive and compressive strength) on a geophysical scale, and the likelihood that arching is stochastic, hinder a general understanding of this phenomenon. In addition to an obvious effect on the ice flux through straits, a fast-ice canopy reduces the oceanic flux by imposing additional drag at the upper boundary of the flow.

4. Flow through Bering Strait

Bering Strait occupies the gap between Asia and America and joins the Bering and Chukchi seas. It is 85 km wide, but two islands within the strait obstruct 9 km of the cross-section. The width of Bering Strait greatly exceeds its length. Its greatest depth is 60 m, but sills of 45-m depth are found within 100 km to both the north and the south (Figure 2).

In relation to the aspects of channel flow discussed above, how well do we know the flow through Bering Strait?

Surprisingly few hydrographic cross-sections of Bering Strait have been published in the open literature. Coachman and Aagaard [27] present a section acquired in August 1964. Below 20-m depth, the salinity increases across the strait from 31.5 near Cape Prince of Wales to 33. Above 20 m, waters east of Diomede Island are 2.5 kg m^{-3} lower in density than those below, and those within 10 km of Cape Dezhneva are less dense by 0.3 kg m^{-3}. A cross-section from October 1981 also exhibits a low-density boundary flow along the eastern shore and a cross-strait salinity gradient at depth [28]. These features are associated with the Alaskan coastal current, carrying outflow from the Yukon River (300 km^3 y^{-1}) north, and with the south-flowing Siberian coastal current driven by outflow from Siberian rivers. Sections measured in September-October in 1992 [29] and in 1993 [30] reveal similar hydrographic structure. Since the salinity deficit (relative to 35) varies by 2-3 times across these sections, the Bering Strait through-flow is clearly not homogeneous in the context of freshwater flux.

The stratification evident in these observations permits baroclinic effects within Bering Strait, namely vertical shear and flow variation on small scales. From quasi-synoptic observations, via anchor station in 1964 [27], and ship-mounted acoustic Doppler current profiler (ADCP) in 1993 [28, 30], large (up to 3 times) cross-strait

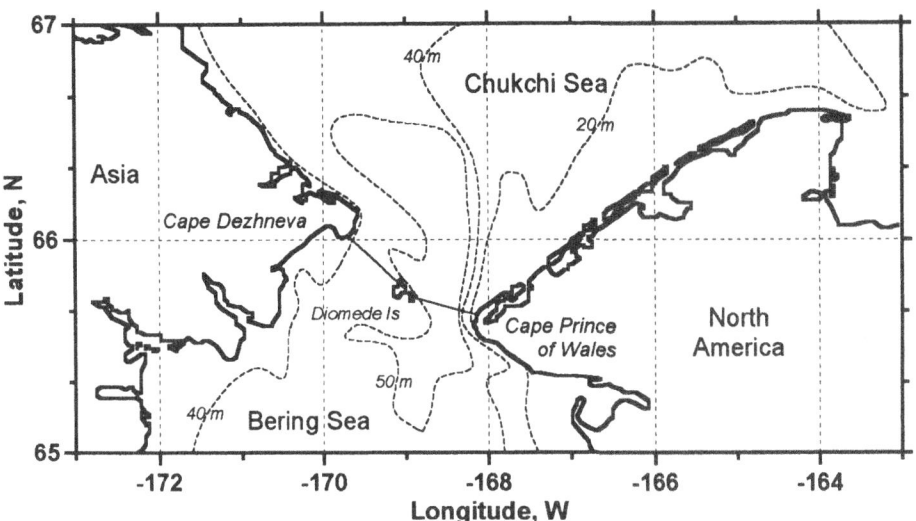

Figure 2. Bering Strait. A fine line marks the reference cross-section.

variations in flow speed have indeed been reported. Even over longer intervals, flows at different depths may differ by similar ratios, as illustrated by monthly mean current profiles for 1990-91, calculated from moored ADCP data [28]. The existence of significant structure in the cross-section of flow through Bering Strait contradicts the conventional wisdom of a negligible baroclinic influence on this flow [27, 28, 31].

Sub-inertial currents in Bering Strait can exceed 100 cm s^{-1}, but average speeds are much lower (20-40 cm s^{-1}) [28, 32]. Tidal flow is also weak (< 20 cm s^{-1}) [31, 33]. Therefore, the mixing coefficient (u^3/H) is relatively small (< 0.5 x 10^{-3} m^3 s^{-1}). This, in combination with the short length of the strait, implies that frictional drag is probably unimportant as a control on the flow. Water-mass modification by mixing within the strait itself is probably negligible for the same reasons. Despite this, the properties of Pacific waters that reach the Arctic having traversed the intervening 1900 km of very shallow continental shelf in the Bering and Chukchi seas are undoubtedly substantially changed from those at source.

Current in Bering Strait has been measured sporadically since 1932 [27]. Until the 1970's, observations were of short duration. Although not ideal for studying the forcing of flow, short records have revealed strong variations (at times, reversals) in current at intervals of days that were correlated with changes in air pressure and wind on a regional scale [34]. Longer time series acquired using internally recording instruments were first acquired in 1976-77 [35]. Subsequent observations by this means have been reported for 1981-82 [31], 1984-85 [32, 36] and 1990-94 [28]. These measurements have confirmed the strong forcing of Bering Strait through-flow by the regional atmospheric circulation: day-to-day fluctuations are driven by passing storms, and the strong annual cycle by seasonal change in the locations and intensities of the Aleutian low and Siberian high. The most recent version [36] of the regression of oceanic volume flux against wind speed during 1976-77 is illustrated in Figure 3 (earlier versions in [31, 35] contain errors). Monthly means are plotted. The wind speed is the along-strait (192°

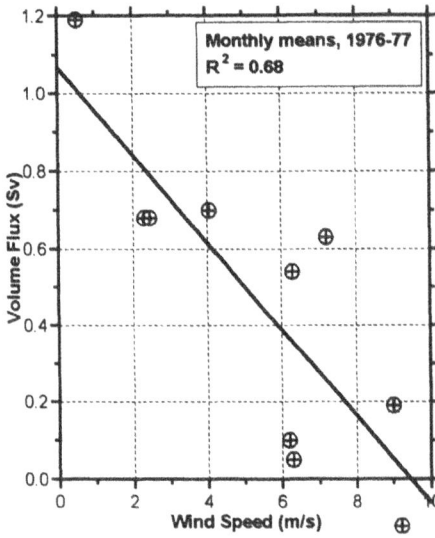

Figure 3. Linear regression of estimated northward volume flux through Bering Strait against wind (adapted from [31])

heading) component of the surface geostrophic wind scaled by 0.7 and rotated anti-clockwise by 15°. The flux has been estimated from measurements by six near-bottom current meters widely separated along a 500-km arc across the Chukchi Sea 350 km north of Bering Strait, and is undoubtedly an approximation. Nonetheless, the figure illustrates the retarding effect of wind on the 1-Sv northward flux which persists when along-strait wind is zero. The linear regression on wind (8 degrees of freedom) accounts for about two-thirds of the variance in monthly average volume flux. The flux for zero wind is that driven by the global steric gradients discussed in Section 3. Numerical models of Bering Strait support the attribution of forcing to steric and atmospheric influences [37, 38]. Unfortunately, we know little about the sea-level difference between the Pacific and the Arctic. The present estimate of its average value is crude [0], and nothing is known of its fluctuations, or of how it might respond to changing climate.

As already noted, friction is unlikely to be an important control on the flow through Bering Strait. However, a rotational hydraulic control on steady barotropic flow through the strait has been demonstrated via numerical modeling. At values of the Pacific-Arctic sea-level difference that exceed 0.4 m, the volume flux is limited to just over 1 Sv [37]. With the imposition over the domain of a (rather high) 0.19 N m^{-2} wind stress on a 234° heading, the modeled flux decreases to the average of values observed. Thus countervailing wind is also an important control on the flow through Bering Strait. However, a detailed quantitative understanding of these controls under varying conditions of stratification and ice cover has yet to be presented.

Two modes of re-circulation operate in Bering Strait, flow reversal and the ice-ocean conveyor belt. The direction of water movement reverses frequently [28, 31, 35, 36]. Thus water which has moved northward through the strait may return southward at a later time, with properties modified by mixing and other processes in the southern Chukchi Sea. Reversals can achieve a net property flux even if the net volume flux is zero. Large (60,000 km^2) areas of ice grown in the southern Chukchi Sea are driven southward through Bering Strait by storm winds at intervals during winter [26, 39]. Sometimes, this ice returns north with a change in wind [40, 41]. Using satellite imagery over an 11-year period, Kozo et al. [26] have determined that, on average, ice is returned to the Chukchi Sea about half as often as it is exported, and that the area of ice returned is generally smaller. Thus, most often, the ice continues to drift southward to melt in the southern Bering Sea [9, 42], thereby achieving an export of freshwater at low salinity from the Arctic. The melt water subsequently mixes with Bering Sea waters, and may join the northward flow and eventually re-enter the Chukchi Sea as part of a much larger volume flux at higher salinity. The conveyor-belt re-circulation in this region thus alters the salinity spectrum and timing of the freshwater flux. The volume of ice involved annually in this re-circulation is not known.

The ice season in Bering Strait typically begins in mid November and ends in late May. In early winter, the pack ice is thin and flows relatively freely through the strait in response to wind and current. The principal dynamic effect of the pack is to enhance the transfer of momentum from the atmosphere to the ice by increasing surface roughness. As the ice becomes thicker and more compact, it may jam up to form vast arches across the southern Chukchi Sea that resists southward forcing by wind [43]. Single arches form frequently, but typically collapse within 1-2 days, usually in response to strengthening northerly winds, to spawn the large movements of ice through Bering Strait noted in the preceding paragraph. Double arches, which separately span the east and west channels of Bering Strait, are less common: Torgerson and Stringer [44] spotted seven occurrences in cloud-free satellite imagery during 1974-85. However, double arches are more stable than single arches, and can prevent the southward drift of ice (and enhance the frictional drag on oceanic flows) for up to a month. Clearly, in the winter months, the dynamics of seawater and ice fluxes through Bering Strait are strongly coupled, but this coupling is poorly understood. We cannot presently predict how through-flow would change were Bering Strait to become ice-free in winter in response to warming climate. In the past, studies of ice and ocean in Bering Strait have proceeded independently. Future research must be integrated, particularly for freshwater flux, of which sea ice constitutes a fraction disproportionate to its volume.

In North America, our best estimates of volume flux (which excludes ice) and its long-term variations are derived from a flow measurement at one point in Bering Strait, and a sequence of assumptions: that volume flux through a 500 km section across the Chukchi was accurately calculated from measurements by six near-bottom current meters in 1976-77; that a time-invariant scaling of surface geostrophic wind provides a realistic measure of atmospheric forcing over the ocean; that linear regressions of these ten values of monthly mean volume flux against a monthly mean scaled geostrophic wind component, and against a monthly mean near-bottom flow in Bering Strait, are statistically robust; that these regression curves are generally valid for this flow at other times; that currents in Bering Strait are horizontally and vertically uniform. The paper by Roach *et al.* [28] summarizes current knowledge: a long-term mean transport of 0.83 Sv (25,200 km^3 y^{-1}), uncertain by an estimated 30%, an annual variation of 0.5 Sv amplitude peaking in July, and inter-annual variations over ±0.3 Sv.

This paper also presents long time series of salinity and temperature from the recording instruments near the seafloor. In salinity, the series contains strong fluctuations on synoptic times scales with amplitude as large as 1, superimposed upon an annual cycle of comparable amplitude, highest in late March and lowest in late September. Differences between values in the east and west channels, and between different years in the same channel can be as large as 1. The highest value measured was close to 34.8. These series contain useful insights for the measurement of freshwater flux. First, the fluctuations in salinity (29.3-34.8) represent a 30-fold change in freshwater anomaly. Second, the synoptic variability in salinity, undoubtedly coherent with fluctuations in the along-channel flow, may contribute significantly to the freshwater anomaly flux.

Third, the salinity is not uniform across Bering Strait, even close to the seafloor in winter. In summer, the season that dominates the freshwater flux because of high flow and low salinity, waters in the strait are less homogeneous.

The Russian method of estimating flux through Bering Strait has been similar to the American, but is based on a much longer series of systematic measurements. An observational station has been operated on Ratmanova (Big Diomede) Island since 1941. For the same political reasons restricting Americans to the eastern channel, Russians have worked predominately in the western channel. Russian observations, dating from the 1930's, are summarized by Fedorov and Yankina [45]. All the work reported is derived from occasional direct measurements of current, in combination with regression equations based on systematic observations of more easily acquired data, such as wind, sea level and air pressure. For example, values between 2.16 and 2.87 have been used as the ratio of total through-flow to that through the western channel. With the exception of those data acquired from the anchored hydrologic station *Naukan* during 1940-42 [46], few observations have been made in the strait during wintertime. The work thus resembles American work prior to the first use of long-term moorings in 1976, although details are not available.

The most recent estimate of mean annual flux by Russian scientists is 30,000 $km^3 y^{-1}$ [45], 20% above the present American value. Their estimates of monthly average flows vary by more than three times between a low in February and a high in August. Vorob'yev and Tiguntsev [46] identify a decadal variation in this flux. In the early 1960's, this contributed a +15% anomaly in flow to the Arctic, which correlated with a drop of 0.04 m in Arctic sea level. Because Russian scientists (as also Americans) have assumed a constant steric influence in calculating through-flow, the correlation can only be attributed to a common meteorological influence. It is plausible, however, that the change in Arctic sea level could be indicative of a decadal variation in the steric forcing of flow through Bering Strait.

Fedorova has also published estimates of the flux of salt through Bering Strait [47]. Occasional observations (in summer) were used to establish a linear relation between the average salinity in the eastern channel and that in the western. Then, a linear relation was determined between 112 monthly means of the average salinity in the western channel and the surface salinity measured at Ratmanova Island. The salt flux was calculated by multiplying the monthly mean salinity in the strait deduced via these regressions from observations at Ratmanova Island, and the monthly mean volume flux estimated in [45]. The resulting long-term (1941-1965) average salt flux is 9.59 x 10^{14} kg y^{-1} (erroneously written as 9.59 x 10^{11} kg y^{-1} in [45]) with inter-annual variations of ±15%, and a pronounced annual cycle. This value is equivalent to a freshwater flux of 2570 $km^3 y^{-1}$ (0.08 Sv) and exceeds that in [3] by more than 40%. Since the Russian value is built upon much the same assumptions as the American, this discrepancy can be considered a measure of the precision of such estimates. Accuracy is probably less good, since neither estimate includes flux as sea ice, or flux via the covariance of current and salinity at periods shorter than two months.

In summary, we have developed some capability to measure the volume flux through Bering Strait, but at present do not have observations that would allow the calculation of freshwater flux by evaluating the space and time integral of freshwater-weighted transport. We are still far from answering the five questions concerning freshwater flux posed in Section 2.

5. Flow through the Canadian Arctic Archipelago

The Canadian Arctic Archipelago occupies 2.5×10^6 km^2 of continental shelf that comprise 20% of the Arctic Ocean. Its many channels have been deepened by glacial action to form network of basins as deep as 600 m, separated by sills (Figure 4). Deep (365-440 m) sills at the western margin of the continental shelf are the first impediment to inflow from the Canada Basin, but the shallowest sills are in the central and southern parts. The complex topography of the Archipelago has a strong influence on ocean circulation. From the standpoint of flux measurement, there is an optimal set of relatively narrow, shallow straits through which all flow must pass (circled areas in Figure 4): Bellot Strait, Barrow Strait (east of Peel Sound), Wellington Channel, Cardigan Strait, Hell Gate and Kennedy Channel. However, we judge Bellot Strait to be unimportant because it has such a small cross-section at the sill (< 24 m deep, 1.9 km wide).

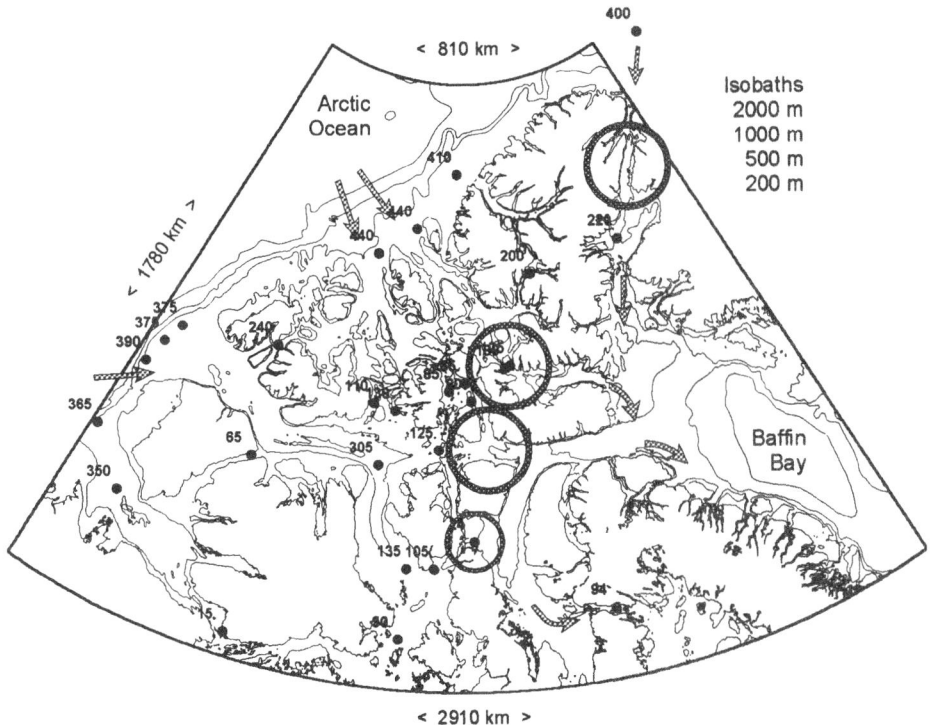

Figure 4. The Canadian Arctic Archipelago. Principal sills are marked by small circles, and their depths indicated in meters. Larger circles mark a group of relatively shallow, narrow passages that must carry all through-flow. From south to north these are: Bellot Strait, Barrow Strait/Wellington Channel, Cardigan Strait/Hell Gate, Kennedy Channel

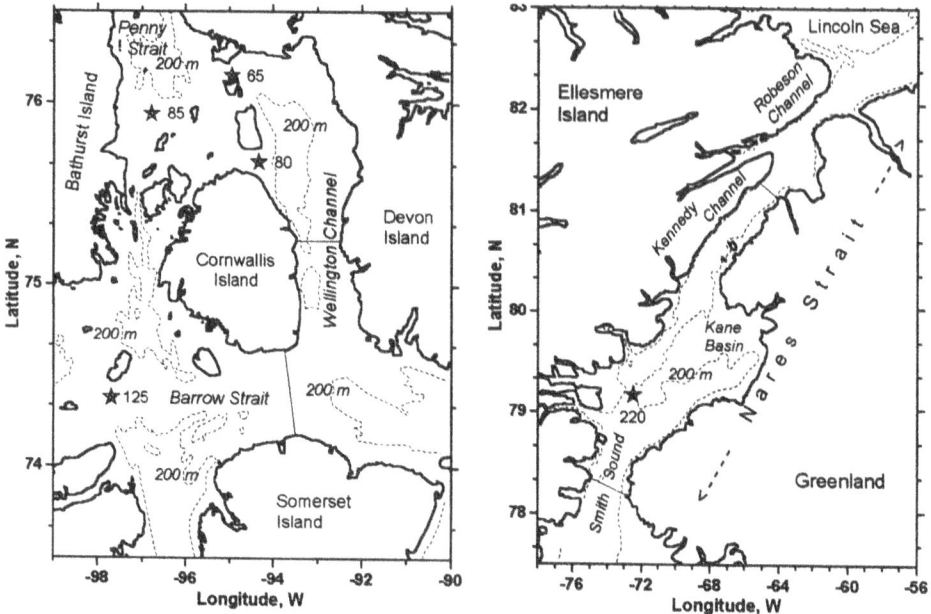

Figure 5. Maps (above and below) showing key flow constrictions (fine lines) within the Canadian Archipelago. Sills are marked by stars with depth in meters. Map scales differ.

Fine lines on the maps of Figure 5 indicate crossings of these straits that are expedient for flux measurement. In several instances, the expedient crossing is the narrowest, not that at the sill, because the measurement of flux through a narrow deep section is more tractable with present technology than through a shallow wide one. The cross-sections are depicted on a common scale for comparison with Bering Strait in Figure 6, and the properties of all sections, are summarized in Table 1. The combined width of these cross-sections is 1.6 times that of Bering Strait, their area is 7.7 times, and their average depth is 4.7 times. A flux of 1 Sv through these sections can be accomplished by an average flow of 4 cm s^{-1}, whereas in Bering Strait the same flux requires a 32 cm s^{-1} current. A significant freshwater flux (0.05 Sv) can be carried through the Archipelago by 2.5-m ice drifting at 16 cm s^{-1}. In Bering Strait, typical 0.5-m ice must drift at 130 cm s^{-1} for the same flux. Water passing through the Archipelago travels a lesser distance between ocean basins than that passing through Bering Strait, but moves through narrow channels, not across broad shelves (Table 1).

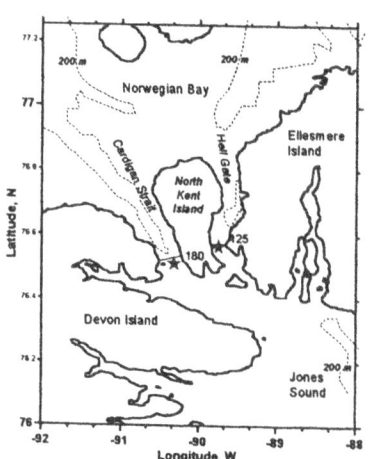

Is the sill in Davis Strait, at the southern end of Baffin Bay, a better place to monitor the flow through the Archipelago on a single section? Davis Strait is certainly easier to reach than the Archipelago, but its sill is wider (360 km) and deeper (650 m). Thus the strait is much larger in section (140 x 10^6 m^2), and the difficulty of

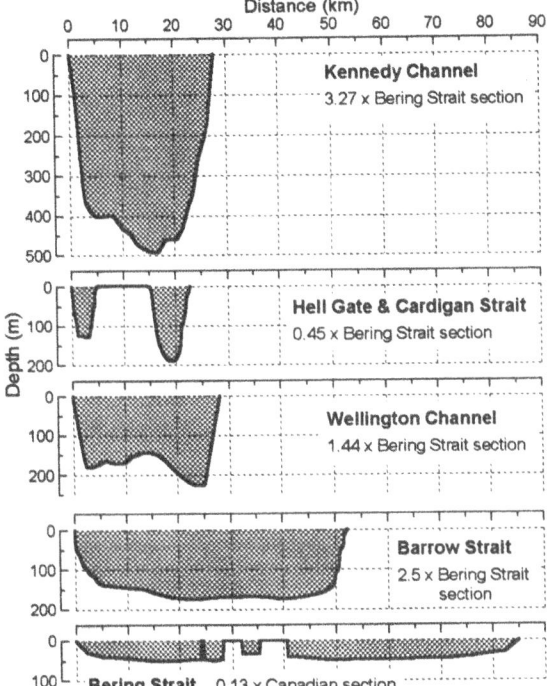

Figure 6. Key channel sections on a common scale for the Canadian Archipelago, and for Bering Strait. Sections are drawn as viewed from the Arctic Ocean.

measuring oceanic fluxes is correspondingly greater. As within the Archipelago, there are strong counter-flows within Davis Strait that complicate flux measurement: on the eastern side is the north flowing West Greenland Current, and on the western side, the Baffin Current. Within the latter, icebergs drifting from northern Baffin Bay to the Grand Banks threaten the survival of instruments moored within 300 m of the surface.

The pack ice within the Archipelago is impenetrable in winter, and a challenge to ice-breaking ships in summer. Thus, oceanographic understanding of this complex region is patchy and inadequate. It is humbling to note that the earliest geostrophic calculations of the volume flux through the Archipelago, based on bottle casts, are still frequently cited. We discuss the limited knowledge of the flow through the Canadian Arctic Archipelago with reference to those aspects of channel flow summarized in Section 3.

Hydrographic sections measured both during winter and summer demonstrate that flows through the Canadian Arctic are stratified year-round, that cross-flow scales are small and that baroclinic effects are important [48-55]. Representative conditions in summer and winter are illustrated by sections in Barrow Strait [56, 57]. Isohaline surfaces rise towards the north in both seasons. In summer, when stratification is strong (2.5 kg m^{-3}), isohalines near the seafloor at 150 m are level. In winter, the stratification is weaker (1 kg m^{-3}) and isohalines flatten little with depth, implying that less of the volume flux is carried by the baroclinic mode. This deduction is consistent with direct observations of the seasonal variation in vertical shear of the flow in Barrow Strait [57]. The deficit in salinity varies by almost a factor of two across the winter section (from 1.5 to 3), and by three across that in summer (from 1.5 to 5). The Canadian Archipelago through-flow is clearly not homogeneous in the context of freshwater flux.

The most saline flow into the Archipelago at the deep sills bordering the Beaufort Sea is 34.85 [13, 52]. At the sill in the deepest passage (Nares Strait), it is 34.4 [51]. At the shallower sill in Barrow Strait, it is 33.2 [49, 52, 57]. Thus, flows via different paths within the Canadian Arctic Archipelago contribute freshwater anomaly fluxes with different salinity spectra to the Baffin Current. No channel is sufficiently deep to permit the passage of waters of Atlantic origin (S > 34.65).

Table 1. Properties of the key channel sections in the North American Arctic. The length is the distance between the 1000-m isobaths along the shortest path that includes the channel

Channel	Section Area m^2	Area As fraction	Section Width km	Width as fraction	Mean Depth m	Length km
Kennedy Channel	10,370,000	42.7%	27.7	22.6%	374	1250
Hell Gate & Cardigan Strait	1,420,000	5.8%	12.4	10.1%	115	1200
Wellington Channel	4,550,000	18.7%	28.3	23.1%	160	1450
Barrow Strait	7,930,000	32.6%	52.3	42.6%	152	1700
Bellot Strait	31,600	0.1%	1.9	1.5%	17	-
Sum of Sections	24,301,600	100%	122.6	100%	198	1400

Direct measurements of flow within the Canadian Arctic Archipelago are sparse: only 31 records exist that exceed 250 days in duration [53, 57-67]. Our poor knowledge reflects the expense and difficulty of working in this vast area with its heavy year-round ice cover, and its lack of a geomagnetic reference for flow direction (because of proximity to the north geomagnetic pole). Excluding the Baffin Current across the mouth of Lancaster Sound [63-65], the strongest average flows towards the Atlantic, of order 20-30 cm s^{-1}, have been observed in Barrow and Penny Straits and in Byam Martin, Wellington and Robeson Channels. Strongest currents flow near the surface along the right side of these channels (looking towards Baffin Bay). A counter current flows along the opposite side of each channel, although in narrow sections this counter flow of denser water may not extend to the surface [52, 53, 66, 67]. The result is a cyclonic circulation in channels throughout the Archipelago. Current speed varies seasonally: flows toward the Atlantic tend to be strongest in summer [57], but so also do those away from it [67]. There are insufficient data to establish whether this intra-annual variation implies a corresponding seasonal cycle in net through-flow, or whether it is connotes simply a strengthening of the cyclonic circulation within channels in summer, with little effect on net flux.

Tidal currents are strong (50-150 cm s^{-1}) in the vicinity of the key sections mapped in Figure 5 [12, 53, 57, 62, 63], and the channels are shallow. The mixing coefficient (u^3/H) is large $(0.5 - 50 \times 10^{-3} \ m^3 \ s^{-1})$. In the course of their transit over several hundred kilometers in these areas, through-flowing waters are significantly modified by tidally driven mixing. The water column is more homogeneous in the central Archipelago than in the 'reservoirs' to the west and to the east, so that the surface density has a local maximum there. The effect is most pronounced on waters flowing from the northwest to the southeast around Cornwallis Island [52]. Along the deeper passages (Barrow Strait, Nares Strait), the impact of mixing is less dramatic [51, 52].

As in the case of Bering Strait, it is plausible, but not proven, that a difference in sea level between basins drives the flow through the Canadian Arctic. Independent estimates of this difference, based on steric anomalies, are 0.15 m [0] and 0.3 m [15].

Because an appreciable fraction of the flow eastward through the Canadian Archipelago is carried in the baroclinic mode, the width of currents scales with the internal Rossby radius. The channels of the Archipelago are typically much wider than the baroclinic currents. Thus Arctic water meets a reservoir of Baffin Bay water well within the Archipelago. Here, strong mixing driven by tidal energy entrains water from the Baffin reservoir into the Arctic through-flow, thereby setting up an estuarine circulation. Because this 'estuary' is rotational, the Arctic outflow and compensating inflow from Baffin Bay are separated primarily in the horizontal plane, not the vertical plane as in the familiar fluvial estuary.

Mixing also contributes to secondary circulation by reducing the stratification of the through-flow relative to that of the source and sink reservoirs. This creates an internal pressure gradient along the channel that drives a convergence of less dense surface water towards the central sill from both ends of the channel.

Clearly, the Canadian Arctic Archipelago cannot be regarded as a network of channels that simply convey water from the Arctic to the Atlantic. Instead, water is exchanged between these reservoirs in both directions (although the net flux of volume is towards the Atlantic), may be modified by mixing, freezing and melting during the months spent within these channels and may ultimately be re-circulated back to its source. In contrast to Bering Strait, where re-circulation is usually dependent upon temporal reversals in flow direction, that within the Canadian Arctic is implicit in the spatial pattern of the circulation and the strength of tidally forced mixing and entrainment. The important net fluxes of volume and fresh water must, therefore, be calculated as the differences between the much larger fluxes that flow in opposite directions through different parts of the same cross-section.

Friction is likely to have a greater importance in controlling the flow through the Canadian Archipelago than that through Bering Strait, because the channels are long, relatively shallow and tidally energetic. Frictional drag will be doubled in winter, when pack ice forms a static canopy over them. In contrast, wind stress is likely to have lesser importance, both because the channels and the reservoirs are deeper and ice covered, and because the reservoirs are widely separated. Rotational hydraulic controls on through-flow at various locations, coupled to friction through Ekman effects and varying with hydrographic conditions, are highly probable. However, even a preliminary understanding of the controls on flow through the Archipelago lies in the future. We cannot presently predict how through-flow would change were the ice cover of the

494

Canadian Arctic Archipelago to remain mobile throughout the winter in response to warming climate.

Estimates of the flux of volume through the Canadian Arctic Archipelago are few and outdated. Those most frequently cited [15, 49] are derived from hydrographic data using the geostrophic method, which requires the assumption of a depth at which there is no flow. They must be regarded as preliminary, because the existence of a depth of no motion in channel flows is unlikely, and because only summertime hydrographic data were used. Collin [49] summarized calculations for Lancaster, Jones and Smith Sounds using data from 1928, 1954 and 1957: estimates for the combined flux through the three channels ranged from 22,000 to 54,000 km^3 y^{-1} (0.7-1.7 Sv). Muench [15] used a section across Baffin Bay to capture the entire outflow (and northward flux via the West Greenland Current).

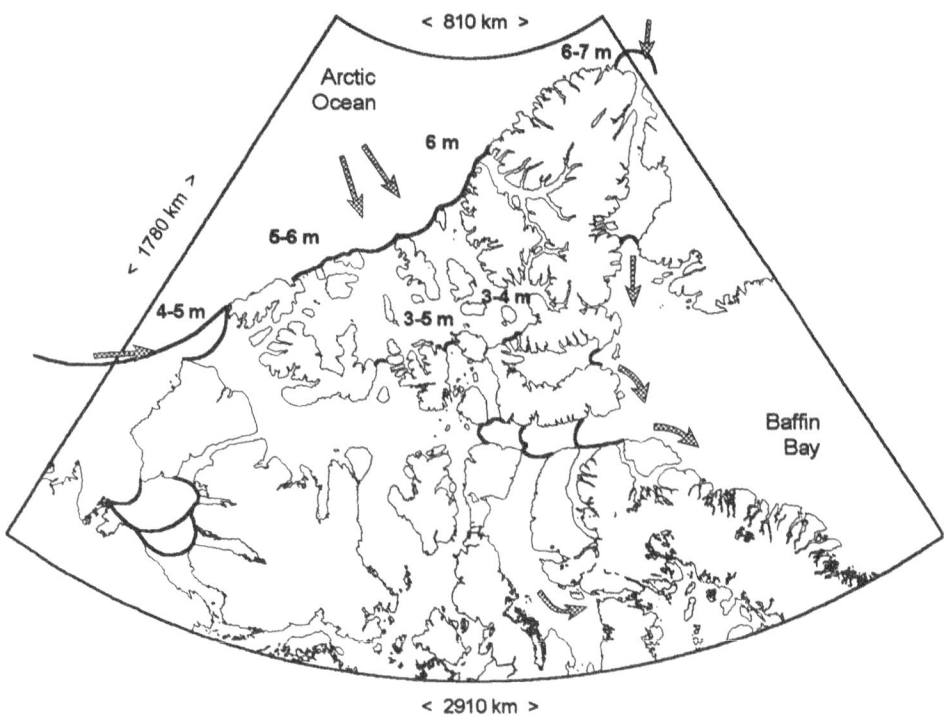

Figure 7. Locations of stable ice arches in the Canadian Archipelago in late winter (thick curved lines) and average ice thickness (m).

The net southward flux for four summers in the 1960's varied between 47,000 and 85,000 km^3 y^{-1} (1.5-2.7 Sv). The only alternative direct estimates are also based on dynamic calculations, but used measurements of current (usually near the surface) in lieu of a no-motion assumption at depth [57, 59, 67]. Typically, these measurements were not sufficiently dense in space to resolve the baroclinic features of the flows. In addition, for logistic reasons, they were made in winter and at locations with stable ice in the central and northern reaches of the Archipelago. Flux values from this hybrid method are 22,000 km^3 y^{-1} (0.7 Sv) through Nares Strait in 1972 [59], 16,000 km^3 y^{-1} (0.5 Sv) through Barrow Strait in 1981 and 1984 [57, 67], and 9,000 km^3 y^{-1} (0.3 Sv) through Wellington Channel in 1984. These values match earlier estimates for Smith and Lancaster Sounds within the calculated ±50% uncertainty. Estimates of total flux through the Archipelago have been derived using salt and heat balances for Baffin Bay, in one instance [68], and the Arctic Ocean in the other [69]. These yield values lower than those via conventional means, 22,000 and 17,000 km^3 y^{-1}, respectively (0.7 & 0.54 Sv), but address only baroclinic fluxes.

Pack ice is a year-round presence in the Canadian Arctic Archipelago. Although southern and eastern regions may clear wholly or in part by late summer, ice concentrations in the west and north are always high. Between late January and late July, the ice is immobilized in most channels by stable arches [70, 71] (Figure 7), but in Nares Strait ice frequently drifts from the Arctic Ocean throughout the year: 1962-63 [70, 73], 1989-90 [74, 75], 1992-95 [76], 1996-97 and 1997-98. Consequently, the flux of freshwater as ice through the Archipelago has a strong seasonal cycle. Heavily deformed floes from the perimeter of the Arctic Ocean pack probably supply the ice flux. Only three tracked floes have passed from the Canada Basin through the Archipelago during the two decades of the International Arctic Buoy Program, presumably because few buoys survive passage through this dynamic interfacial zone. Observations of ice drift within the Canadian Archipelago are scarce and unsystematic. Typical values for mobile pack are 10-20 cm s^{-1} in Nares Strait, 10-15 cm s^{-1} in the eastern and central channels of the Archipelago, and 5-10 cm s^{-1} in the west and north [73, 71-78]. Ice drifting at 6 cm s^{-1} will travel 1000 km in six months.

Knowledge of ice thickness within the Canadian Archipelago is poor (Figure 7). Submarine sonar has revealed that a band of very thick ice (4-7 m) extends 100-200 km from the coast to the west and north of the Archipelago [79]. Observations in M'Clure Strait by submarine sonar [80], and among the Queen Elizabeth Islands by extensive drilling during the 1970's [81-1] show a 3-5 m ice thickness to be typical in these areas. Assuming that ice of 2.5-m thickness drifts through the key sections within the Archipelago at 10 cm s^{-1} for 6 months of the year, the flux of freshwater (as ice) is 480 km^3 y^{-1} (0.03 Sv for 6 months). This is almost 20% of the ice export through Fram Strait.

Because ice makes a large contribution to the freshwater flux, and because an sizeable fraction of the water-column flux is associated with the baroclinic mode, the freshwater flux through the Canadian Arctic Archipelago is carried very close to the surface (Figure 8). Data from Barrow Strait [57] have been used to calculate that half of the freshwater flux occurs in the upper 10 m in summer and in the upper 44 m in winter (when ice is stationary). Instruments that might be deployed near the surface to measure current and salinity are vulnerable to destruction by drifting ice and icebergs.

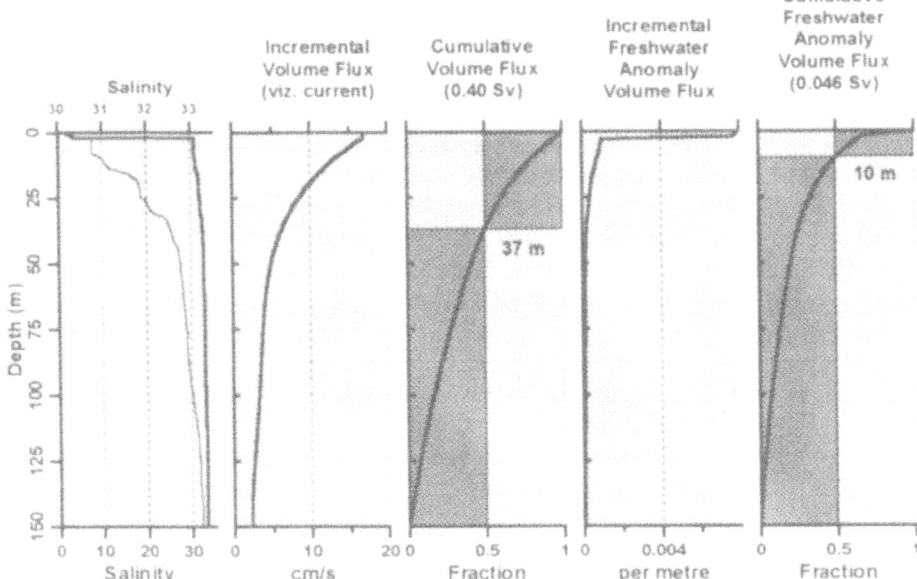

Figure 8. Idealized calculation of the depth dependence of fluxes through Barrow Strait in summer, when both the ocean and the ice cover contribute to freshwater flux. Note that half the volume flux occurs above 37 m, and half the freshwater flux above 10 m.

Consequently, it would be very difficult to measure the freshwater flux through the Archipelago . with existing technology.

A quantitative discussion of the freshwater flux through the Canadian Arctic Archipelago is clearly premature. First, the flux of volume has not been measured to established accuracy in any channel. Second, measurements of the covariance between freshwater anomaly and current that fully resolve the freshwater flux (including ice) across even a single channel section have yet to be made. Approximate estimates of the flux of salt through the Archipelago have appeared in the literature [3, 57, 59], but the wide variation (32-34) in the assumed salinity of the through-flow for these estimates is indicative of their large uncertainty. The range in value is equivalent to a three-fold variation in freshwater anomaly and therefore also in freshwater flux!

6. Challenges

To measure and model the fluxes of freshwater through the shallow straits of the North American Arctic is a challenging undertaking. Attaining these goals will engage Arctic oceanographers for many years to come. In this pursuit, it is important to acknowledge that in the context of freshwater flux pack ice is an integral part of the ocean. Because there is a strong and variable dynamical interdependence between the fluxes of seawater and sea ice, the study of oceanic flows must be integrated with those of ice. This has not been the case in the past.

Table 2. Comparison of shallow straits in the North American Arctic

Canadian Arctic Archipelago	Bering Strait
Complex geometry of interconnecting channels	Simple geometry
Not so shallow (> 100 m)	Very shallow (<50 m)
Prominent sills	No sills
Strongly stratified in most areas	Weak stratification
Baroclinic effects important	Barotropic flow dominant
Channel width >> flow width	Channel width < flow width
Year-round ice cover (12 months)	Seasonal ice cover (4 months)
Ice immobile for up to half the year	Ice usually mobile
Thick ice (3-4 m)	Thin winter ice (< 1 m)
Combined minimum channel width, 27% of Fram Strait	Minimum channel width, 75.8 km
Combined channel cross-section at minimum width, 24.3×10^6 m^2	Channel cross-section at minimum width, 3.17×10^6 m^2
Tidal currents are strong (50-150 cm/s)	Tidal currents are weak (20-30 cm/s)
$u^3/L \sim 0.5 - 50 \times 10^{-3}$ m^3/s	$u^3/L \leq 0.5 \times 10^{-3}$ m^3/s

On the theoretical side, advances are necessary in the understanding, parameterization and modeling of:

- Differences in sea level between ocean basins
- Rotating flow in channels of realistic geometry
- Boundary stress at the seafloor and the ice canopy in tidal channels
- Buoyant boundary flow through a maze of 'wide' interconnected channels
- Lagrangian aspects of mixing in channels
- Flow of pack-ice in channels of realistic geometry
- Stable ice-arch formation in channels of realistic geometry
- Dynamical interactions between the flows of water and pack ice in channels

On the technical/logistic side, solutions are required to problems of:

- Measuring current direction in the vicinity of the geomagnetic pole in the Canadian Arctic Archipelago
- Gaining access to remote, ice choked areas of the Canadian Arctic in a reliable and cost-effective manner
- Recovering moorings reliably through the very thick multi-year ice of the Canadian Arctic Archipelago
- Measuring the volume flux of a narrow, meandering baroclinic flow in a wide channel
- Measuring the vertical profiles of current and salinity within 40 m of the ice canopy

498

In order to derive robust estimates of the freshwater fluxes through the shallow straits of the North American Arctic, there is a great need for more observations. Some, such as those required to calculate volume flux, are practical even beneath moving ice using Doppler sonar. However, investigators must position such instruments carefully in order to measure the strong currents in the low density surface layer that is only 5-10 m thick in summer. Also, many locations across a section must be instrumented in order to resolve the baroclinic features of the flow. Other observations, such as those in remote areas of the Canadian Archipelago and those required to calculate freshwater flux, must await technological advances. In the case of the latter, measurements of salinity are particularly needed within 40 m of the surface, where the freshwater anomaly is largest and the risk to *in situ* salinity recorders from moving ice is extreme. Without such data, it will not be possible to evaluate the freshwater flux correctly as a covariance (Section 2). Even with suitable new technology, many (10 or more) moorings will be needed across a passage such as Barrow Strait in order to resolve the small scales of flow.

It should be very clear from this discussion that the two pathways for freshwater transport through the North American Arctic, Bering Strait and the Canadian Arctic Archipelago, have different characteristics (Table 2). Thus, assumptions that simplify approaches to measurement and modeling in one area may not be valid in the other. For this reason, it is probable that real progress in measuring fluxes through Bering Strait may be possible over the next few years. In the Canadian Arctic, progress will be much slower.

The first priority in future work must be the acquisition of volume-flux values for both Bering Strait and the Canadian Arctic Archipelago with substantiated limits of confidence. We also require an improved understanding of the forcing of these flows. Subsequently, research might progress to the acquisition of observations sufficient to quantify the freshwater flux through Bering Strait, and its interannual variations. At the same time, technological solutions to the problems of measurement in the Canadian Arctic should be sought. Ultimately, successful new technology should be used to measure the freshwater flux through the Canadian Arctic. This new technology could also be used for the same purpose beneath the heavy ice of the East Greenland Current in western Fram Strait.

It should not be forgotten that the ultimate goal of much research into the dynamics of the global climate system is to predict the response to change. The emphasis here has been on the more modest objective of measuring and understanding present conditions. Until this objective is achieved, prediction of the changes in freshwater fluxes through the North American Arctic that may accompany dramatic change in climate is beyond our reach.

7. References

1. Stigebrandt, A. (1984) The North Pacific: a global-scale estuary. *Journal of Physical Oceanography* **14**, 464-470.
2. Wijffels, S.E., R.W. Schmitt, H.L. Bryden and A. Stigebrandt (1992) Transport of freshwater by the oceans. *Journal of Physical Oceanography* **22**, 155-162.
3. Aagaard, K. and E.C. Carmack. (1989) The role of sea ice and other fresh water in the Arctic circulation. *Journal of Geophysical Research* **94**, 14,485-14,498.
4. Shaffer, G. and J. Bendtsen (1994) Role of the Bering Strait in controlling North Atlantic ocean circulation and climate. *Nature* **367**, 354-357.

5. Goosse, H., T. Fichefet and J.-M. Campin (1997) The effects of the water flow through the Canadian Archipelago in a global ice-ocean model. *Geophysical Research Letters* **24**, 1507-1510.

6. Goosse, H., J.M. Campin, T. Fichefet and E. Deleersnijder (1997) Sensitivity of a global ice-ocean model to the Bering Strait throughflow. *Climate Dynamics* **13**, 349-358.

7. Ingram, R.G. and S. Prinsenberg (1998) Coastal oceanography of Hudson Bay and surrounding eastern Canadian Arctic waters. *The Sea, Volume 11, The Global Coastal Ocean*, A.R. Robinson & K.H. Brink (editors), John Wiley and Sons Inc., New York, 835-861.

8. Loder, J.W., B. Petrie and G. Gawarkiewicz (1998) The coastal ocean off northeastern North America: a large-scale view. *The Sea, Volume 11, The Global Coastal Ocean*, A.R. Robinson & K.H. Brink (editors), John Wiley and Sons Inc., New York, 105-133.

9. Pease, C.H. (1980) Eastern Bering Sea ice processes. *Monthly Weather Review* **108**, 2015-2023.

10. Leblond, P.H. (1980) On the surface circulation in some channels of the Canadian Arctic Archipelago. *Arctic* **33**, 189-197.

11. Furevik, T. and A. Foldvik (1996) Stability at M2 critical latitude in the Barents Sea. *Journal of Geophysical Research* **101**, 8823-8838.

12. Stronach, J.A., J.A. Helbig, S.S. Salvador, H. Melling and R.A. Lake (1987) Tidal elevations and tidal currents in the Northwest Passage. *Canadian Technical Report of Hydrography and Ocean Sciences No 97*, Institute of Ocean Sciences, Sidney, Canada, V8L 4B2. 329 pp.

13. Melling, H., R.A. Lake, D.R. Topham and D.B. Fissel (1984) Oceanic thermal structure in the western Canadian Arctic. *Continental Shelf Research* **3**, 233-258.

14. Holloway, G. (1992) Representing topographic stress for large-scale ocean models. *Journal of Physical Oceanography* **22**, 1033-1046.

15. Muench, R.D. (1971) The physical oceanography of the northern Baffin Bay region. *North Water Project Scientific Report No. 1*, Arctic Institute of North America, University of Calgary, Calgary, Canada, 150 pp.

16. Pite, H.D., D.R. Topham and B.J. van Hardenberg (1995) Laboratory measurements of the drag force on a family of two-dimensional ice keel models in a two-layer flow. *Journal of Physical Oceanography* **25**, 3008-3031.

17. Whitehead, J.A., A. Leetma and R.A.Knox (1974) Rotating hydraulics of strait and sill flows. *Geophysical Fluid Dynamics* **6**, 101-125.

18. Sambuco, E. and J.A. Whitehead (1976) Hydraulic control by a wide weir in a rotating fluid. *Journal of Fluid Mechanics* **73**, 521-528.

19. Toulany, B., and C. Garrett (1984) Geostrophic control of fluctuating barotropic flow through straits. *Journal of Physical Oceanography* **14**, 649-655.

20. Whitehead, J.A. (1986) flow of a homogeneous rotating fluid through straits. *Geophysical and Astrophysical Fluid Dynamics* **36**, 187-205

21. Killworth, P.D. (1995) Hydraulic control and maximal flow in rotating stratified hydraulics. *Deep-Sea Research* **42**, 859-871.

22. Whitehead, J.A. (1998) Topographic control of oceanic flows in deep passages and straits. *Reviews of Geophysics* **36**, 423-440.

23. Johnson, G.C. and D.R. Ohlsen (1994) Frictionally modified rotating hydraulic channel exchange and ocean outflows. *Journal of Physical Oceanography* **24**, 66-78.

24. Sodhi D.S. (1977) Ice arching and the drift of pack ice through restricted channels. *CRREL Report No 77-18*, U.S. Army Cold Regions Research and Engineering Laboratory, Hanover, N.H. Available NTIS. 11 pp.

25. Pritchard, R.S., R.W. Reimer and M.D. Coon (1979) Ice flow through straits. In *Proceedings of POAC'79*, Vol. **3**, Norwegian Institute of Technology, Trondheim, Norway. 61-74.

26. Kozo, T.L., W.J. Stringer and L.J. Torgerson (1987) Mesoscale nowcasting of sea ice movement through the Bering Strait with a description of major driving forces. *Monthly Weather Review* **115**, 193-207.

27. Coachman, L.K. and K. Aagaard (1966) On the water exchange through Bering Strait. *Limnology and Oceanography* **11**, 44-59.

28. Roach, R.T., K. Aagaard, C.H. Pease, S.A. Salo, T. Weingartner, V. Pavlov and M. Kulakov (1995) Direct measurements of transport and water properties through the Bering Strait. *Journal of Geophysical Research* **100**, 18443-18457.

29. Falkner, K.K., R.W. Macdonald, E.C. Carmack and T. Weingartner (1994) The potential of barium as a tracer of Arctic water masses. In "The Polar Oceans and their Role in Shaping the Global Environment", *Geophysical Monograph* **85**, 63-76. American Geophysical Union.

500

30. Weingartner, T.J., S. Danielson, Y. Sasaki, V. Pavlov and M. Kulakov (1999) The Siberian Coastal Current: a wind and buoyancy forced Arctic coastal current. *Journal of Geophysical Research* (in press).
31. Aagaard, K., A.T. Roach and Schumacher (1985) On the wind-driven variability of the flow through Bering Strait. *Journal of Geophysical Research* **90**, 7213-7221.
32. Muench R.D., J.D. Schumacher and S.A. Salo (1988) Winter currents and hydrographic conditions on the Northern Central Bering Sea Shelf. *Journal of Geophysical Research* **93**, 516-526.
33. Mofjeld, H.O. (1986) Observed tides on the northeast Bering Sea shelf. *Journal of Geophysical Research* **91**, 2593-2606.
34. Coachman, L.K., K. Aagaard and R.B. Tripp (1975) *Bering Strait: The Regional Physical Oceanography*. University of Washington Press, Seattle, 172 pp.
35. Coachman, L.K. and K. Aagaard (1981) Re-evaluation of water transports in the vicinity of Bering Strait. In *The Eastern Bering Sea Shelf: Oceanography and Resources, Vol.* **1**. D.W. Hood and J.A. Calder (ed.), University of Washington Press, Seattle, 95-110.
36. Coachman, L.K. and K. Aagaard (1988) Transports through Bering Strait: Annual and interannual variability. *Journal of Geophysical Research*, **93**, 15535-15539.
37. Overland, J.E. and A.T. Roach (1987) Northward flow in the Bering and Chukchi Seas. *Journal of Geophysical Research* **92**, 7097-7105.
38. Spaulding, M., T. Isaji, D. Mendelsohn and A.C. Turner (1987) The numerical simulation of wind-driven flow through the Bering Strait. *Journal of Physical Oceanography* **17**, 1799-1816.
39. Ahlnas, K. and G. Wendler (1979) Sea-ice observations by satellite in the Bering Chukchi and Beaufort Seas. In *"Proceedings of POAC'79"*, Vol. 1, Norwegian Institute of Technology, Trondheim, Norway. 313-329.
40. Muench, R.D. and K. Ahlnas (1976) Ice movement and distribution in the Bering Sea from March to June 1974. *Journal of Geophysical Research* **81**, 4467-4476.
41. Pease, C.H. and S.A. Salo (1987) Sea ice drift near Bering Strait during 1982. *Journal of Geophysical Research* **92**, 7107-7126.
42. Pease, C.H. (1988) Meridional heat transport by the ice and ocean in the western Arctic. In *Proceedings of the Second Conference on Polar Meteorology and Oceanography*, March 29-31, 1988, Madison. American Meteorological Society, Boston, Ma.
43. Kovacs, A., D.S. Sodhi and G.F.N. Cox (1982) Bering Strait sea ice and the Fairway Rock ice foot. *CRREL Report No.* **82-31**, 44 pp. Available U.S. Army CRREL, Hanover, N.H.
44. Torgerson, L.J. and W.J. Stringer (1985) Observations of double arch formation in the Bering Strait. *Geophysical Research Letters* **12**, 677-680.
45. Fedorova, Z.P. and A.S. Yankina (1964) The passage of Pacific Ocean water through the Bering Strait into the Chukchi Sea. *Deep-Sea Research* **11**, 427-434.
46. Vorob'yev, V.N. and L.A. Tiguntsev (1974) Long-period variations in water discharge through Bering Strait and of the mean sea-level of the Arctic seas. *Oceanology* **14** (English translation), 32-36.
47. Fedorova, Z.P. (1968) Salt transfer through the Bering Strait into the Chukchi Sea. *Oceanology* **8** (English translation), 37-41.
48. Bailey, W.G. (1957) Oceanographic features of the Canadian archipelago. *Journal of the Fisheries and Research Board of Canada* **14**, 731-769.
49. Collin, A.E. (1962) Oceanographic observations in the Canadian Arctic and the adjacent Arctic Ocean. *Arctic* **15**, 194-201.
50. Collin, A.E. (1963) Waters of the Canadian Arctic Archipelago. In *'Proceedings of the Arctic Basin Symposium'*, Washington D.C., October 1962. Arctic Institute of North America, 312 pp.
51. Herlinveaux, R.H. and H.E. Sadler (1973) Oceanographic Features of Nares Strait in August 1971. Unpublished manuscript. Institute of Ocean Sciences, Sidney, Canada, V8L 4B2.104 pp.
52. De Lange Boom, B.R., H. Melling and R.A. Lake (1987) Late winter hydrography of the Northwest Passage: 1982, 1983 and 1984. *Canadian Technical Report of Hydrography and Ocean Sciences No* **79**, Institute of Ocean Sciences, Sidney, Canada, V8L 4B2. 165 pp.
53. Prinsenberg, S.J. and E.B. Bennett (1989) Transport between Peel Sound and Barrow Strait in the Canadian Arctic. *Continental Shelf Research* **9**, 427-444.
54. Addison, V.G. (1987) The physical oceanography of the Northern Baffin Bay-Nares Strait region. M.Sc. thesis, Naval Postgraduate School, Monterey, CA. 99 pp.
55. Bourke, R.H., V.G. Addison and R.G. Paquette (1989) Oceanography of Nares Strait and northern Baffin Bay in 1986 with emphasis on deep and bottom water formation. *Journal of Geophysical Research* **94**, 8289-8302.

56. Jones, E.P. and A.R. Coote (1980) Nutrient distributions in the Canadian Arctic Archipelago: Indicators of summer water-mass and flow characteristics. *Canadian Journal of Fisheries and Aquatic Sciences* **37**, 589-599.

57. Prinsenberg, S.J. and E.B. Bennett (1987) Mixing and transports in Barrow Strait, the central part of the Northwest Passage. *Continental Shelf Research* **7**, 913-935.

58. Barber, F.G. (1965) Current observations in Fury and Hecla Strait. *Journal of the Fisheries and Research Board of Canada* **22**, 225-229.

59. Sadler, H.E. (1976) Water, heat and salt transports through Nares Strait, Ellesmere Island. *Journal of the Fisheries and Research Board of Canada* **33**, 2286-2295.

60. Fissel, D.B. and J.R Marko (1978) A surface current study of eastern Parry Channel, NWT, Summer 1977. *Canadian Contractor Report of Hydrography and Ocean Sciences No* **78-4**, Institute of Ocean Sciences, Sidney, Canada, V8L 4B2. 66 pp.

61. Godin, G. (1979) Currents in Robeson Channel, Nares Strait. *Marine Geodesy* **2**, 351-364.

62. Sadler, H.E. (1982) Water flow into Foxe Basin through Fury and Hecla Strait. *Le Naturaliste Canadien* **109**, 701-707.

63. Fissel, D.B., D.D. Lemon and J.R. Birch (1982) Major features of the summer near-surface circulation of western Baffin Bay, 1978 and 1979. *Arctic* **35**, 180-200.

64. Fissel, D.B. (1982) Tidal currents and inertial oscillations in northwestern Baffin Bay. *Arctic* **35**, 201-210.

65. Lemon, D.D. and D.B. Fissel (1982) Seasonal variations in currents and water properties in northwestern Baffin Bay, 1978 and 1979. *Arctic* **35**, 211-218.

66. Greisman, P and R.A. Lake (1984) Current observations in the channels of the Canadian Arctic Archipelago adjacent to Bathurst Island. *Pacific Marine Science Report No* **78-23**, Institute of Ocean Sciences, Sidney, Canada, V8L 4B2. 127 pp.

67. Fissel, D.B., J.R. Birch, H. Melling and R.A. Lake (1988) Non-tidal flows in the Northwest Passage. *Canadian Technical Report of Hydrography and Ocean Sciences No* **98**, Institute of Ocean Sciences, Sidney, Canada, V8L 4B2., 143 pp.

68. Rudels, B. (1986) The outflow of polar water through the Arctic Archipelago and the oceanographic conditions in Baffin Bay. *Polar Research* **4**, 161-180.

69. Steele, M., D. Thomas, D. Rothrock and S. Martin (1996) A simple model of the Arctic Ocean freshwater balance, 1979-1985. *Journal of Geophysical Research* **101**, 20833-20848.

70. Dunbar, M. (1969) The geographical position of the North Water. *Arctic* **22(4)**, 438-441.

71. Marko, J.R. (1978) A Satellite Imagery Study of Eastern Parry Channel. *Contractor Report Series No.* **78-5**, 134 pp. Institute of Ocean Sciences, Sidney, Canada, V8L 4B2

72. Nutt, D.C. and L.K. Coachman (1963) A note on ice island WH-5. *Arctic* **16**, 204-206.

73. Nutt, D.C. (1966) The drift of ice island WH-5. *Arctic* **19**, 244-262

74. Kozo T.L. (1991) The hybrid polynya at the northern end of Nares Strait. *Geophysical Research Letters* **18**, 2059-2062.

75. Kozo T.L., R.W. Fett, L.D. Farmer and D.S. Sodhi (1992) Clues to *deformation features in coastal sea ice. EOS Transactions of the American Geophysical Union* **73**, 385-389.

76. Agnew, T.A. (1998) Drainage of multi-year sea ice from the Lincoln Sea. *CMOS Bulletin* **26**, 101-103.

77. Marko, J.R. (1977) A Satellite-based Study of Sea Ice Dynamics in the Central Canadian Arctic Archipelago. *Contractor Report Series No.* **77-4**, 106 pp. Institute of Ocean Sciences, Sidney, Canada, V8L 4B2

78. Jeffries, M.O. and M.A. Shaw (1993) The drift of ice islands from the Arctic Ocean into the channels of the Canadian Arctic Archipelago: the history of Hobson's Choice Ice Island. *Polar Record* **29**, 305-312.

79. Bourke, R.H. and R.P. Garrett (1987) Sea ice thickness distribution in the Arctic Ocean. *Cold Regions Science and Technology* **13**, 259-280.

80. McLaren, A.S., P. Wadhams and R. Weintraub (1984) The sea ice topography of M'Clure Strait in winter and summer of 1960 from submarine profiles. *Arctic* **37**, 110-120.

81. Wetzel, V.F. (1976) Statistical Study of Late Winter Ice Thickness in the Arctic Islands (1971-1975). *APOA Report No.* **96**, Arctic Petroleum Operators Association. Available from Pallister Resources, Calgary, Alberta. 9 pp + append.

82. Wetzel, V.F. (1978) Statistical Study of Late Winter Ice Thickness in the Queen Elizabeth Islands (1977). *APOA Report No.* **142**, Arctic Petroleum Operators Association. Available from Pallister Resources, Calgary, Alberta. 15 pp + append.

83. Wetzel, V.F. (1981) Statistical Study of Late Winter Ice Thickness in the Queen Elizabeth Islands (1978-80). *APOA Report No.* **174**, Arctic Petroleum Operators Asociation. Available from Pallister Resources, Calgary, Alberta. 29 pp + append.

THE TRANSFORMATIONS OF ATLANTIC WATER IN THE ARCTIC OCEAN AND THEIR SIGNIFICANCE FOR THE FRESHWATER BUDGET

BERT RUDELS[1] AND HANS J. FRIEDRICH[2]
[1]Finnish Institute of Marine Research
PL 33, FIN-00931 Helsinki, Finland
[2]Institut für Meereskunde, Universität Hamburg,
Troplowitzstraße 7, D-22529 Hamburg, Germany

1. Introduction

The Arctic Ocean comprises the inner part of the Arctic Mediterranean Sea, which also includes the Nordic Seas. The Arctic Ocean constitutes the northernmost extension of the North Atlantic, and – through the shallow and narrow Bering Strait – it is connected to the North Pacific. The Arctic Ocean thus forms a direct link between the two "opposite poles" of the world ocean – the main pool of low salinity water (Pacific) and the main pool of high salinity water (Atlantic). The transport through the Bering Strait is predominantly northward and brings Pacific water into the Arctic Ocean and ultimately to the North Atlantic. Because of its lower salinity and density the Bering Strait inflow is confined to the upper 150-200m and is mainly restricted to the North American side of the Arctic Ocean. Most of the Pacific water enters the North Atlantic through the Canadian Arctic Archipelago, where it constitutes the bulk of the outflow. Analysis of nutrient data shows that some Pacific water is also present in Fram Strait; in the low salinity outflow of the East Greenland Current [1] as well as in the deep waters [2]. Here, however, its contribution is small and, to the first order, the Arctic Ocean can be viewed as the inner part of an extensive North Atlantic bay which – because of the strong river run-off – acts as a large estuary of truly global scale. The great rivers enter the Arctic Ocean at the wide shelf seas which comprise more than 1/3 of the 9 $10^{12}m^2$ large surface area of the Arctic Ocean.

The water masses of the Arctic Ocean are then essentially created from two source waters – Atlantic Water and freshwater, deriving from run-off and from net precipitation. The freshwater content of the different water masses is therefore estimated relative to the salinity of the Atlantic Water, here set to 35. It should be kept in mind that all relative estimates of freshwater content are ambiguous. A stricter approach would be to work with the absolute freshwater content, being defined as the mass of sea water minus the mass of dissolved salt [3].

The shelf seas are the areas where the main water mass formation occurs. Transformations are controlled by a combination of the harsh, seasonally varying

E.L. Lewis et al. (eds.), The Freshwater Budget of the Arctic Ocean, 503-532.
© 2000 *Kluwer Academic Publishers.*

climate, causing freezing and melting, the seasonal cycle of the river run-off, and the shallow, varying bottom topography. The Pacific inflow is regarded as an independent surface wedge along the American continent, bulging out into the Canada Basin at the Beaufort Gyre and exiting through the Canadian Arctic Archipelago. The interactions between the two regimes are not considered, but in the context of freshwater and salt balances for the Arctic Ocean the Pacific water can be viewed as an additional low salinity shelf input attaining its salinity without interacting with Atlantic Water.

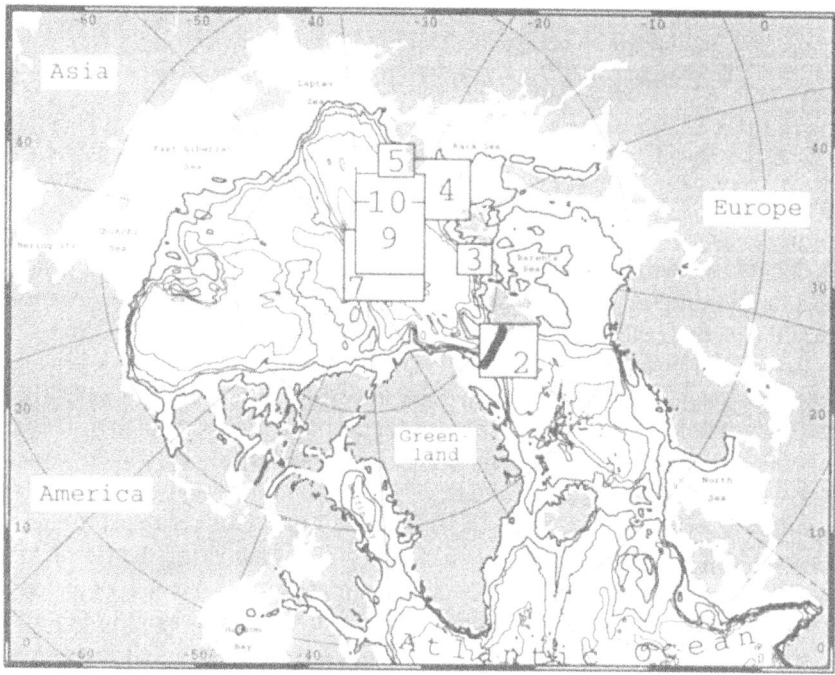

Figure 1.   Map of Arctic Mediterranean Sea. Numbered frames indicate the areas with CTD-observations discussed in the context of corresponding figures. Note: we choose this particular map, because it includes the area of investigation and the source region of our main water mass, the Atlantic Water, reducing simultaneously the distortion of more common configurations. It is a Mercator projection after the positive pole has been moved to 115°E;30°N (China). The zero meridian is taken to pass through Greenwich too, and we count 'longitude' negative to the left. The topography has been developed from ETOPO 5 [4] with the full resolution of 5'. Here the isolines are shown for 200, 1000, 2000, 3000, 4000 and 5000m with lines of decreasing thickness. The co-ordinates shown on the frame are the co-ordinates given on the blow-up maps in the figures below.

We first review the exchanges between the Arctic Ocean and the Nordic Seas, which serve as an anteroom between the Arctic Ocean and the North Atlantic. The circulation and water mass transformations in the Arctic Ocean are then examined. The distribution of the freshwater within the water column is discussed and the exchanges and residence times for the different layers are estimated as functions of the freshwater input. Temporal changes will be considered as far as existing observations so permit. The location of the examined stations and sections are shown on the map in Figure 1 and station information is given in an appendix.

2. Exchanges

Atlantic Water enters the Arctic Mediterranean, as the North Atlantic sub polar gyre splits south of the Greenland-Scotland Ridge. One part continues westward south of the ridge, while the rest flows northward into the Norwegian Sea as the Norwegian Atlantic Current. As it reaches the latitude of the Barents Sea the current again bifurcates. One part flows northward as the West Spitsbergen Current and the other part turns east and enters, together with the Norwegian Coastal Current, the Barents Sea. The West Spitsbergen Current comprises several current bands and as it approaches Fram Strait parts of the current recirculate toward the west, join the East Greenland Current and flow southward along the Greenland continental slope. What remains of the West Spitsbergen Current enters the Arctic Ocean. The major inflow occurs north of Svalbard and continues eastward as a boundary current along the continental slope. Another, weaker flow goes around the Yermak Plateau, turning eastward further to the north [5].

That Atlantic Water, here identified by temperatures above 0°C, is present in the Arctic Ocean at depths between 200 and 600-700m was established by Nansen during the Fram expedition [6]. He assumed that this warm layer derived from an inflow of Atlantic Water through Fram Strait. The outflow of sea ice and low salinity polar water through Fram Strait was known earlier [7], as was the westward recirculation of the West Spitsbergen Current [8] and the existence of warm Atlantic Water below the Polar Water of the East Greenland Current [9]. This is now often called the Return Atlantic Current [10].

As the knowledge of the bathymetry of the Arctic Ocean grew, it became clear that Fram Strait is a deep (~2500m) passage connecting the Arctic Ocean to the Nordic Seas allowing deep exchanges, and especially a deep outflow, to occur. The potential temperature and salinity sections in Figure 2 reveal large differences between the eastern, inflow, and western, outflow, side of the strait. In particular, the -0.5°C and +0.5 °C isotherms are much wider apart at the western continental slope than to the east. This implies that the waters in this temperature range are mainly created in the Arctic Ocean and exit through Fram Strait. The Arctic Intermediate Water formed in the Greenland Sea [11] can be identified as a salinity minimum at ~1000m in the central and eastern part of the strait. However, the minimum does not extend to the Svalbard continental slope and lies outside the deep boundary flow suggested by the rise of the -1 °C isotherm. Deep waters, rather than intermediate water, from the Nordic Seas would then be expected to add to the inflow of Atlantic Water to the Arctic Ocean.

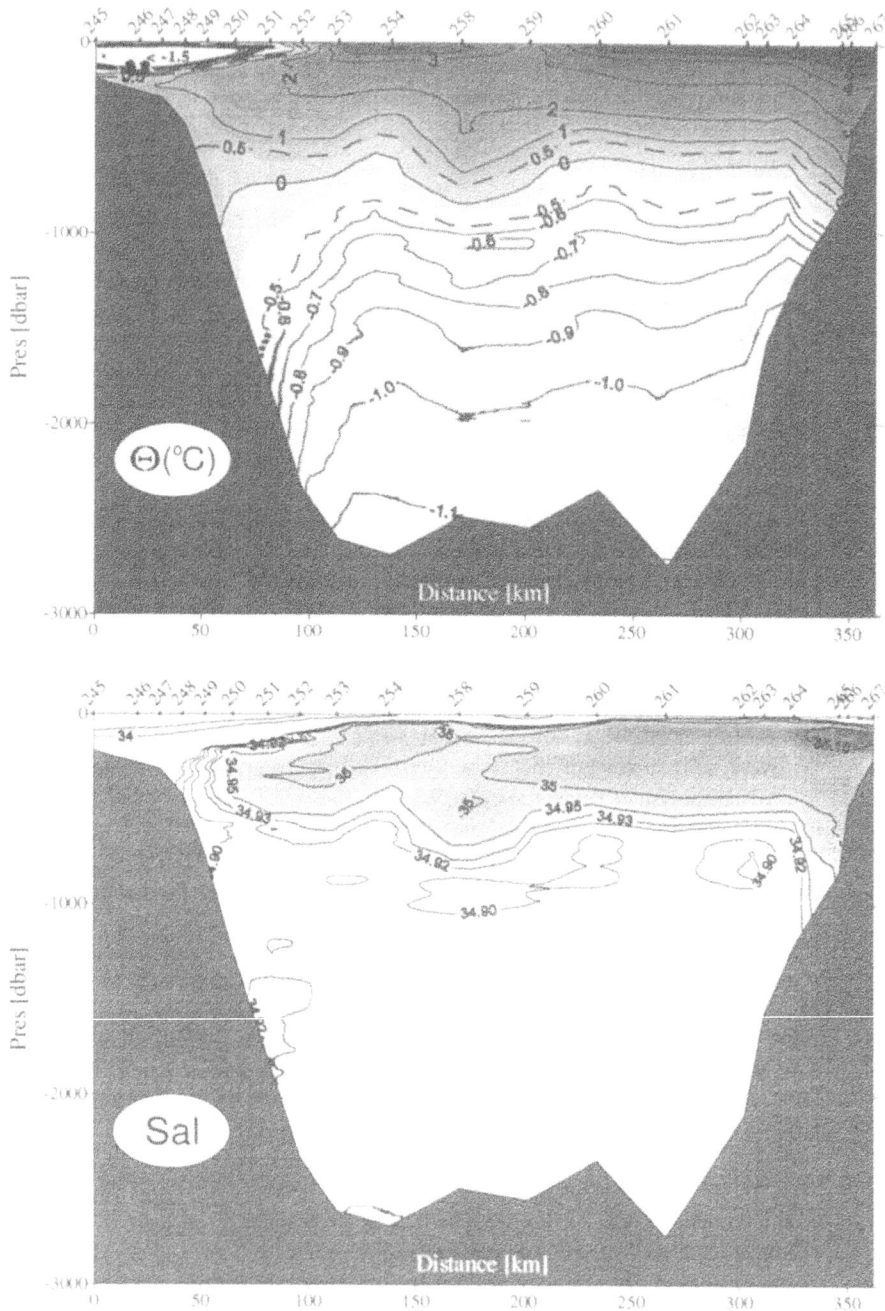

Figure 2. Sections potential temperature and salinity across Fram Strait taken along 78°55'N by RV Lance in August 1984 (Replotted from [13]).

The estimates of the Atlantic inflow have varied considerably during the last 30 years. The transports given in the first edition of the introductory text *Descriptive Physical Oceanography* [12] were 3.5 Sv entering the Arctic Ocean from the Nordic Seas and 3.7 Sv leaving in the East Greenland Current. The inflow through the Bering Strait and outflow through the Canadian Arctic Archipelago were there estimated to 1 Sv and 0.8 Sv respectively (1 Sv = $10^6 m^3 s^{-1}$). No separation of the Atlantic inflow between Fram Strait and the Barents Sea was made. These estimates were most likely derived from budget considerations and geostrophic computations.

In the 1970s the first transport estimates using direct current measurements became available. Aagaard and Greisman [14] proposed a new budget based upon the measurements of the northward flow of the West Spitsbergen Current, and with the assumptions that the exchanges through Fram Strait were in balance, and that all other transports were decidedly smaller. They obtained an inflow of 7.1 Sv Atlantic Water in the West Spitsbergen Current and a corresponding outflow of 1.8 Sv Polar water and 5.3 Sv modified Atlantic Water in the East Greenland Current. For the other passages their estimates were: Canadian Arctic Archipelago, -2.1 Sv (– = out of the Arctic Ocean).; Bering Strait 1.5 Sv; Barents Sea 0.7 Sv. These budgets are now quoted directly, or given slightly changed, in introductory and reference literature [15, 16].

Continued current measurements west of Svalbard indicated that the West Spitsbergen Current is highly variable, in space as well as in time, and the estimated inflow showed annual as well as seasonal variations [17]. In 1983 and 1984 the Marginal Ice Zone Experiment was launched in Fram Strait. Although the focus of the MIZEX programme was on air-sea-ice interaction and smaller scale mixing processes, it considerably increased the knowledge of the large scale circulation in Fram Strait, especially showing the presence of eddies and recirculation branches [18]. This recirculation occurring in Fram Strait was not considered in [14]. When complete hydrographic sections across the entire strait became more common, making the geostrophic calculations more reliable, and year long current meter records became available not only from the inflow side west of Svalbard but also from the East Greenland Current, the estimates of the volume really entering the Arctic Ocean again became smaller [10, 19, 20, 21].

Transport estimates based upon geostrophic computations tend to be smaller than those obtained from moored current meters, perhaps because of the neglect of a possible, but undetermined, barotropic part of the velocity. To deal with this problem Rudels [21] minimised the geostrophic transports in Fram Strait while requiring mass and salt balances for the entire Arctic Ocean and introducing additional constraints on the deep circulation. The exchanges through the other passages were assumed known from literature or were estimated separately. Houssais et al. [22] extended this approach by only requiring balances of mass and salt for the geostrophic transports in different layers between two hydrographic sections in Fram Strait. Additional information from the other passages was then not needed. They found a net outflow from the Arctic Ocean and only a small flow of Atlantic Water into the Arctic Ocean.

It has recently been appreciated that the Atlantic Water entering the Arctic Ocean over the Barents Sea might be as large or larger than that through Fram Strait [21, 23,

24]. A large inflow has also been substantiated by modelling work [25]. This is a return to the view held in the beginning of the century, when most of the Atlantic Water entering the Arctic Ocean was assumed to pass over the Barents Sea [26]. A strong inflow over the Barents Sea is also consistent with a net outflow in Fram Strait.

Considering the range of the different "best" estimate for the transports between the Nordic Seas and the Arctic Ocean proposed in the last decades, and taking into account the ongoing ACSYS programme and that a major effort to determine the exchanges (and their variability) through Fram Strait and the Barents Sea is now under way within the EC Mast III programme VEINS, it may easily be that transports, stated here and now, soon will become obsolete. However, it would be cowardly, in a work like this, not to volunteer an estimate. In our opinion the exchanges through the four major passages are: Barents Sea, 2 Sv [23, 24, 27]; Bering Strait, 0.8 Sv [28]; The Canadian Arctic Archipelago, -1 Sv [29]; Fram Strait; the East Greenland Current, -3 to -3.5 Sv [19, 21]; the West Spitsbergen Current, 1 - 1.5 Sv [10, 21, 27]. For the inflow in the West Spitsbergen Current we assume that it consists of Atlantic Water and that the inflow of intermediate and deep waters is small and can be neglected. However, it should be mentioned that the tracer constrained models from Heinze et al. [30], Smethie et al. [31] and the estimates by Rudels [32], and used in [21], indicated a deep inflow of 1 Sv. This is discussed further below (section 4). An inflow of deep water is not likely to interact strongly with the Arctic Ocean freshwater and may be ignored on these grounds. The Barents Sea inflow comprises both Atlantic Water and water from the Norwegian Coastal Current. To properly consider the waters constituting the outflow in the East Greenland Current, we first have to examine the circulation and the transformations of the Atlantic Water within the Arctic Ocean.

3. Transformations

As Atlantic Water enters the Arctic Ocean, it interacts with freshwater in different forms; sea ice, river run-off, evaporation and precipitation, and in different settings; in the deep basins, at the continental slope and on the shelves, and subjected to different forcing; wind stress, cooling, freezing and melting. Furthermore, the two inflow branches, following separate paths, interact differently.

The part of the West Spitsbergen Current that turns eastward encounters sea ice north of Svalbard. The sensible heat of the warm Atlantic Water induces melting and the upper part of the inflowing water column becomes transformed into a less saline layer which then covers the warm, saline core of the Fram Strait branch. The Atlantic Water does not "dive" beneath the low salinity surface water of the Arctic Ocean, as is occasionally stated. Its upper part interacts with sea ice and loses its warm, saline characteristics. The now less saline surface layer follows the warm Atlantic core eastward as a part of the boundary current along the Eurasian continental slope. In winter the surface layer becomes homogenised down to the thermocline by convection caused by brine rejection during freezing. A deep, comparably saline, winter mixed layer, but no cold halocline, is thus present in the southern Nansen Basin. In summer seasonal ice melt creates a low salinity surface layer which is separated from the Atlantic Layer by a temperature minimum, marking the depth of the last winter convection. Here the intermediate depth water in the Arctic Ocean with temperatures >0°C will be denoted the Atlantic Layer. In winter the melt water is removed by freezing and convection again reaches the thermocline [33].

A strong input of low salinity shelf water to the interior of the Arctic Ocean does not take place until as far east as the Laptev Sea. Once this input occurs, the salinity of the surface water becomes reduced. Winter convection no longer reaches down to the thermocline but stops shallower than the temperature minimum of the convection occurring the previous winter. The air-sea-ice exchanges become decoupled from the Atlantic Layer, and the deeper part of the older winter mixed layer becomes a cold halocline, which further insulates the Atlantic Layer.

The stream entering the Barents Sea separates into two paths. One turns northward west of the Central Bank into the Hopen Trench, where it again splits. One part returns along the slope of the Svalbard Bank to the Norwegian Sea, a small part crosses the sill between Edgeöya and the Grand Bank into the northern Barents Sea, and the third part flows eastward, north of the Central Bank. The rest of the Atlantic Water flows, together with the water of the Norwegian Coastal Current, directly to the eastern Barents Sea [34].

The river input to the Barents Sea is small, and the main freshwater source is the Norwegian Coastal Current which transports the run-off from the Baltic Sea and from Scandinavia into the Barents Sea. In winter the Norwegian Coastal Current is cooled to freezing temperature and ice forms in the south-east part of the Barents Sea and on the shallow banks west of Novaya Zemlya. The rejection of brine and the accumulation of cold, saline water at the bottom increase the density of the water over the banks throughout the winter [35, 36]. As dense water leaves the banks, sinking into the deeper depressions, it interacts with the Atlantic Water. Some enters the Atlantic core, while denser fractions reach the bottom of the depressions and, by entraining Atlantic Water underway, become warmer. Ice formation and subsequent melting, brine rejection and convection thus cause a separation of the inflowing water into low density surface, and high density deep and bottom waters [35].

The high density part continues into the Arctic Ocean, predominantly by entering the Kara Sea between Novaya Zemlya and Franz Josef Land and then sinking down the St. Anna Trough [24, 35, 37, 38]. Some dense water is also found to enter the Arctic Ocean directly from the Barents Sea through the Victoria Channel [32], but this water could have formed around Franz Josef Land (see below). Dense water formation also occurs at the Central Bank [34, 39], but the dense water created here mostly returns toward the Norwegian Sea, sinking down the Bear Island Channel and joining the deeper levels of the West Spitsbergen Current [23, 39]. It may eventually reach the Arctic Ocean via Fram Strait.

The low salinity, low density surface water, created by ice melt, may continue eastward into the Kara Sea either north or south of Novaya Zemlya. However, it also supplies the low salinity surface water in the northern Barents Sea [35]. This less saline water is normally denoted Arctic Water, when Barents Sea water masses are discussed

510

[34]. Some Arctic Water flows south-westward and enters the Norwegian Sea south of Björnöya, where it joins the West Spitsbergen Current and moves north. The same happens with the East Spitsbergen Current once it flows around the southern cape of Svalbard and enters Fram Strait. The freshwater, which has been extracted by ice formation and subsequently released by ice melt in the Barents Sea thus eventually ends up in the Arctic Ocean, either directly or via the West Spitsbergen Current.

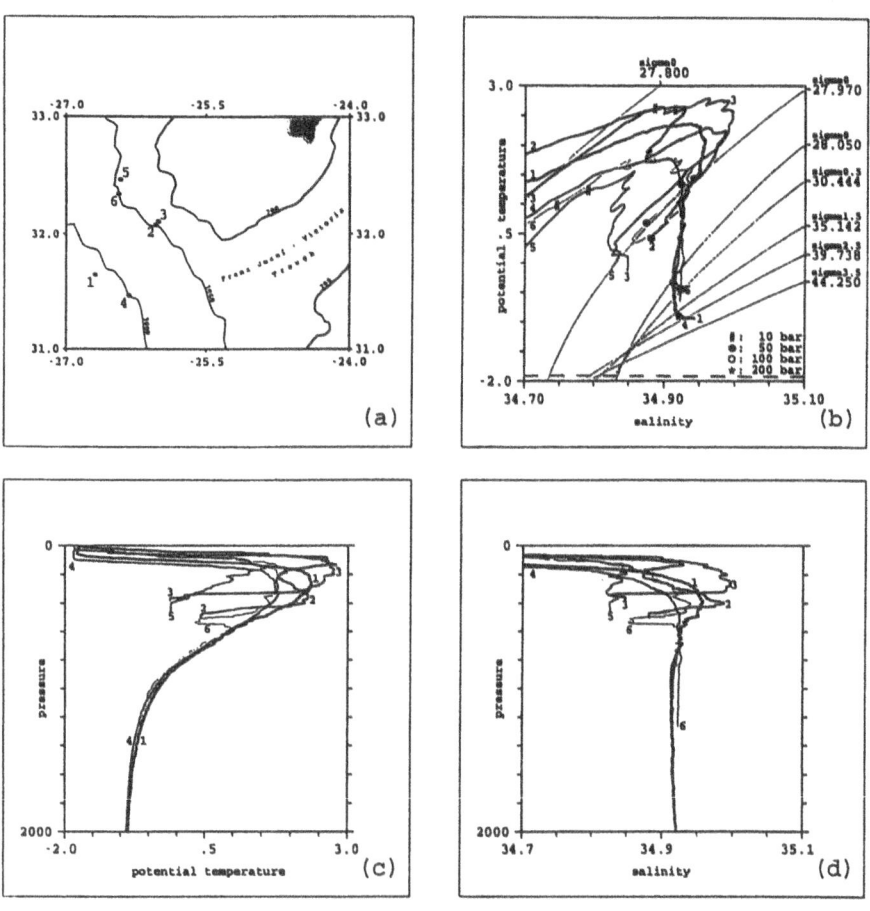

Figure 3.   Profiles of potential temperature and salinity and Θ-S curves from stations north of the Barents Sea taken by HMS Ymer in 1980 and RV Polarstern in 1993. For station information see appendix.

The upper layer in the northern Barents Sea is in winter too saline to supply the winter mixed layer of the Arctic Ocean. If it flows north, it will enter the water column in the thermocline or in the Atlantic Layer core. At some occasions the density of the Barents Sea outflow to the north has increased so much that it enters still deeper, below the temperature maximum of the Atlantic Layer. Figure 3 shows dense shelf water found at the bottom at stations 3 and 5 intruding into the Arctic Ocean water column at the continental slope, stations 2 and 6, north-west of Franz Josef Land. Such dense slope plumes must involve convection to the bottom of the shelf and the accumulation of brine enriched water during the winter. The shallow shelves north of Svalbard and especially around Franz Josef Land, where in winter persistent lee polynyas are present [40], may be possible source areas, as well as the banks west of Novaya Zemlya. The stations shown in Figure 3 were taken with 13 years interval between expeditions, which suggests a permanence of the shelf outflow and implies that it is entering the Arctic Ocean from a localised source, most likely the Victoria Channel. The differences in temperature and salinity in the Atlantic Layer between 1980 and 1993 are also evident. The higher temperatures in 1993 show the recently reported warmer Atlantic Layer [41] and 1980 was the year, when the great salinity anomaly reached the Arctic Ocean [42], which could explain the low salinities observed that year.

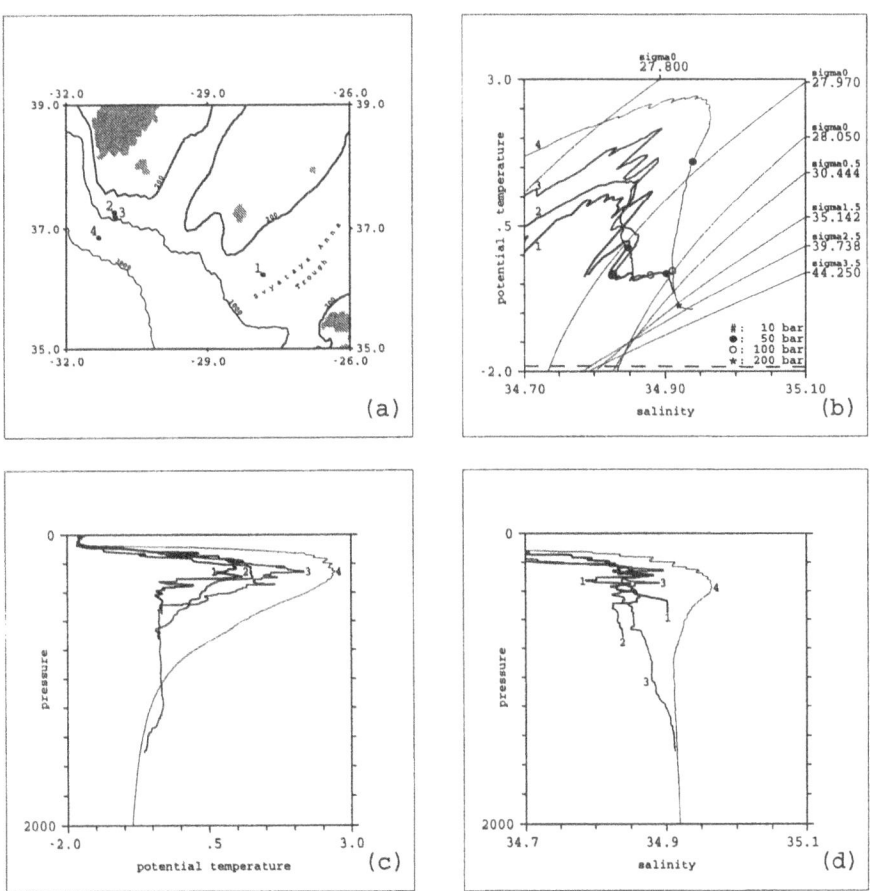

Figure 4.   Profiles of potential temperature and salinity and Θ-S curves from stations in the St Anna Trough and north of the Kara Sea taken by RV Polarstern in 1996. For station information see appendix.

The main outflow of the denser fractions of the Barents Sea branch, however, occurs between Novaya Zemlya and Franz Josef Land and the waters continue down the St. Anna Trough. The Atlantic Water has become significantly colder and denser but also less saline. It sinks into and beneath the Atlantic Water of the Fram Strait branch, as they meet in the St. Anna Trough [43] (see also Figure 4), and can be identified as a 1000m thick cold, low salinity wedge at the continental slope east of the St. Anna Trough [38] (see also Figures 4 and 5). A comparison between Figure 3 and Figures 4 and 5 shows that the St. Anna Trough outflow displaces the basin water column from the slope rather than intrudes at its density level. The outflow in the trough is stratified (see station 1 in figure 4) and the less saline, and less dense, part stays high on the slope while the more saline and denser bottom water slides further down the slope (Figures 4 and 5).

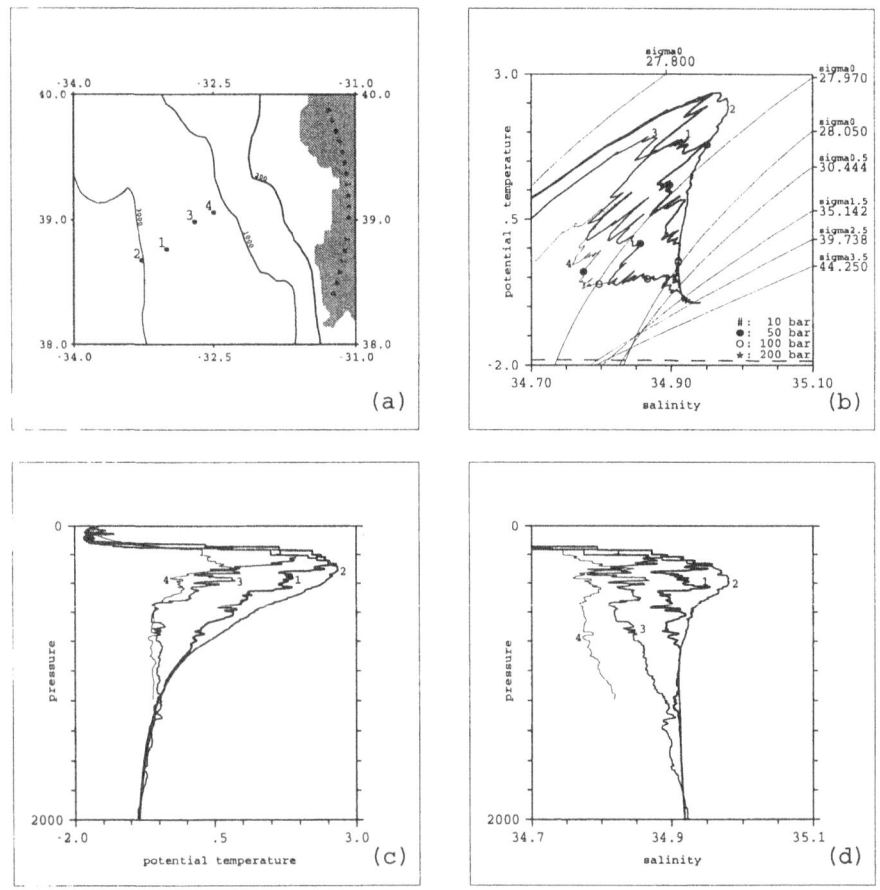

Figure 5. Profiles of potential temperature and salinity and Θ-S curves from stations north of Severnaya Zemlya taken by RV Polarstern in 1995. For station information see appendix.

The Laptev and East Siberian Seas supply the Polar Mixed Layer, while the contributions from the Barents and Kara Seas sink deeper into the water column. However, once the halocline has been formed by the reduction of the depth of the winter convection in the interior of the basins, it has a finite thickness and density range and can increase its thickness by absorbing water convecting from the shelves further to the east [33]. Some of these inputs are provided by shelf water of Pacific origin, which also enters the shelves from behind and covers the Chukchi Sea and the Alaskan shelf and incorporates the run-off of the large North American rivers, the Yukon and the Mackenzie.

In contrast to the Fram Strait inflow the Barents Sea branch does not transport much sensible heat to the Arctic Ocean. Most, if not all, its heat has been lost to the atmosphere in the Barents Sea. However, the St. Anna inflow is one of the few instances, as are, to a lesser extent, the denser plumes leaving the Barents Sea, where freshwater is brought directly from the shelves into the deeper part of the water column of the Arctic Ocean. This occurs because the deep St. Anna Trough allows the cold, dense, less saline Barents Sea branch to enter, with little entrainment, beneath the warm core of the Fram Strait branch and transfer freshwater into the deep.

The presence of inversions in the salinity and temperature profiles in Figures 4 and 5 implies lateral and isopycnal mixing between the two inflow branches. A high variability was observed in the temperature and salinity profiles at stations where multiple casts have been taken [44], indicating sharp fronts between the water masses and/or the presence of smaller scale eddies. The inversions also show the existence of layering structures and small-scale mixing processes such as double-diffusive convection and frontal interleaving (Figure 5). The layering structures can be followed throughout the Eurasian Basin and indicate that a large part of the Atlantic inflow recirculates in the Eurasian Basin (see Figure 5 in [37]). The considerably lower temperature of the temperature maximum of the Atlantic layer east of Severnaya Zemlya (see Figure 6 in [38]) even implies that a substantial fraction of the Fram Strait branch may recirculate within the Nansen Basin [44].

The less saline part of the Barents Sea branch that enters the Kara Sea acts as a receptacle for the run-off from the rivers Ob and Yenisey and constitutes the bulk of the low salinity shelf water which eventually reaches the Arctic Ocean. Most of the low salinity water of the Kara Sea passes through the Vilkiltskij Strait into the Laptev Sea, where it incorporates the run-off from the Lena, Khatanga and Yana rivers, before it enters the Arctic Ocean close to the Lomonosov Ridge. Some, probably annually varying, fraction stays on the shelf and continues eastward into the East Siberian Sea. The Barents Sea branch thus supplies the oceanic part of the shelf water on the Eurasian shelf as far east as the East Siberian Sea through the backdoors formed by the straits between the islands and the continent. There is no need for a return flow of saline water from the Arctic Ocean across the shelf-break to compensate for the outflow of low salinity shelf water.

514

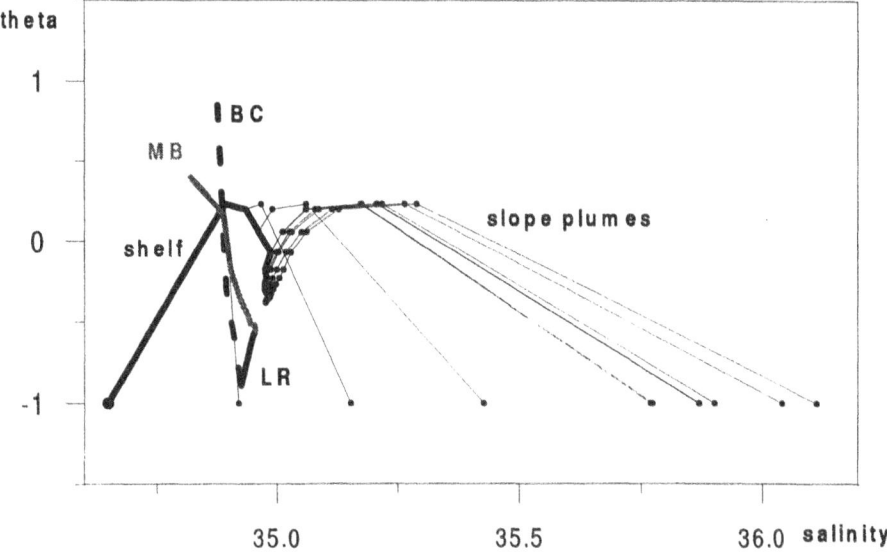

Figure 6.   The evolution of the Θ-S characteristics of the shelf-slope plumes. "BC" is the Θ-S curve for the boundary current, "MB" for the Makarov Basin. "slope plumes" shows the evolution of the Θ-S properties of individual plumes sinking to different depth. The small dots indicate the initial characteristics of the plumes. The starting point has been chosen to 200m and the initial temperature is assumed to have risen from the freezing point to -1 °C due to entrainment of ambient water. "Shelf" indicates the Θ-S curve connecting the characteristics of the plumes when they have reached their final depths and merge into the boundary current. "LR" indicates the final Θ-S characteristics of plumes of Amundsen Basin water sinking down the Lomonosov Ridge.

To penetrate below the halocline, the initial salinity of the shelf water must be high. No freshwater is therefore exported, by this process, directly from the shelves into the deep. However, as the boundary plumes sink down the slope, they entrain less saline ambient water and redistribute it downwards. The evolution of the properties of idealised shelf-slope plumes in the Canadian Basin as they sink, reaching different terminal depth in the water column, is shown in Figure 6, using the information given in [2]. The salinities of the plumes must be above 35 when they pass 200m, if they are to reach deeper than the temperature maximum in the Atlantic Layer. The input of freshwater from the shelves to the deeper layers is thus small. However, Figure 6 also indicates that the deeper layers in the Canadian Basin are colder than they should be, if they were only ventilated by the slope convection. The colder, less saline characteristics are due to Amundsen Basin deep water which passes through gaps in the central Lomonosov Ridge and sinks down the ridge into the deeper Makarov Basin [2]. The entrainment of the deep water bends the Θ-S curves of the deep sinking shelf-slope plumes toward colder and less saline values (Figure 6). The conditions are different in the Eurasian Basin east of the St. Anna Trough. There an almost undiluted Barents Sea branch provides low salinity ambient water and plumes sinking through this layer will induce a larger transport of freshwater towards the deeper layers.

4. Freshwater storage and export

The Atlantic Water acts as a receptacle for the freshwater. In this process the Atlantic Water becomes transformed into the different water masses, which constitute the stratified Arctic Ocean. In a very simplified picture of the Arctic Ocean water column, the following, most conspicuous water masses may be identified. The polar surface water includes the shelf derived, and basin modified, less saline Polar Mixed Layer, and the underlying halocline, partly created by ice melt and winter homogenisation in the deep Eurasian Basin and augmented by shelf outflows east of the Laptev Sea. The Atlantic Layer is, in the Eurasian Basin, predominantly comprised by Fram Strait branch water while comparably larger fractions of Barents Sea branch water and shelf inputs are present in the Canadian Basin. In the intermediate water depth range the water column mostly reveals the presence of Barents Sea branch water, which is strongly modified by slope convection and shelf water input in the Canadian Basin. Finally the deep waters are supplied mostly by the Barents Sea inflow but are, as well as the intermediate depth waters, augmented through slope convection by dense shelf water and vertically redistributed ambient waters. A more comprehensive water mass classification, aimed at the Fram Strait water mass distribution but showing its relation to Arctic Ocean and the Greenland Sea water masses, is given elsewhere [45, 46].

To obtain volume estimates and transport rates we conceive a simple box model of the Arctic Ocean. The deep basins of the Arctic Ocean comprise an area of about 4.5 $10^{12}m^2$ with the Eurasian Basin making up 40%, the Canadian Basin 60% [47]. Because of the steep continental slope this area can be taken as representative down to 2000m.

According to the classical literature [48] the Polar Mixed Layer is about 50m deep with an average salinity of 32.7. However, this is over the entire Arctic Ocean, including the part dominated by the Bering Strait inflow. We only consider the part influenced by the Atlantic inflow, where the salinity is higher and examining Table 1 and Figures 4, 5 and 6 given in [33] we adopt 33.2 for the salinity of the Atlantic derived part of the Polar Mixed Layer. The thickness and the salinity of the halocline varies throughout the basin. Its mean salinity is here set to 34.3 and it is taken to extend from 50 to 200m. A part of the thermocline is then also included. The upper waters in the Canadian Basin are partly derived from the Pacific inflow through Bering Strait. We take the Atlantic derived water in the Polar Mixed Layer to be confined within an area of 2.5 $10^{12}m^2$, and the Atlantic derived halocline to extend over 3.0 $10^{12}m^2$. The Atlantic Layer ($\Theta > 0°C$) reaches across the entire Arctic Ocean. It is found at deeper levels in the Canadian Basin, but its thickness hardly varies across the Arctic Ocean. The Atlantic Layer is taken to occupy the depth range between 200 and 700m. The intermediate depth range below 700m is dominated by the Barents Sea inflow, and it is assumed to extend down to 1700m, the sill depth for the boundary current as it crosses the Lomonosov Ridge into the Canadian Basin [37, 49]. As deep water we consider the depth range 1700-2500m, the upper limit being at the sill depth of the Lomonosov Ridge and the lower limit corresponding to the sill depth in Fram

Strait. Below this level the area of the deep basins decreases rapidly and the volume of the bottom water below 2500m is about half that assumed for the deep water (Table 1).

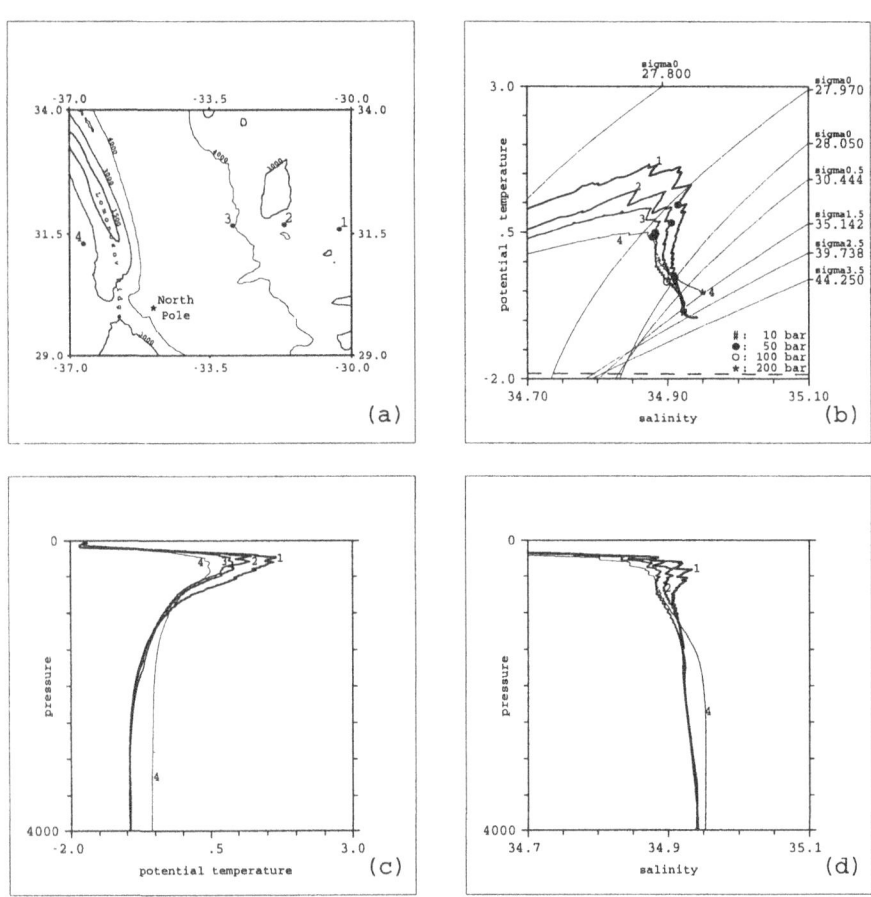

Figure 7. Profiles of potential temperature and salinity and Θ-S curves from stations in the Nansen, Amundsen and Makarov basins used to estimate the average water mass characteristics. For station information see appendix.

To determine salinities representative for the deeper layers we examine the profiles from stations in the interior of the Nansen, Amundsen and Makarov Basins (Figure 7). These indicate that a mean salinity of 34.9 is reasonable for the Atlantic Layer and 34.91 for the intermediate depth waters, if all basins are considered. For the deep and bottom (> 2500m) waters we choose mean salinities of 34.935 and 34.945 respectively. The amount of freshwater (relative to 35) in each layer can then be determined (Table 1). It should be kept in mind that the depths, areas and salinities chosen here, although realistic, do not represent any absolute truth. They are used for illustrative purpose.

For a comparison we also estimate the storage of freshwater as ice. The largest sea ice production occurs over the shelves which are ice free in the beginning of the winter, allowing for a rapid growth rate. Ice is exported from the shelves but the ice thickness continues to grow in the interior of the Arctic Ocean, increasing the salinity of the Polar Mixed Layer above that of the injected shelf water. The mean ice thickness in the central Arctic Ocean is commonly given as 3m. Assuming an extent of the ice cover to be $7 \cdot 10^{12} m^2$, which is intermediate between the total and the basin area of the Arctic Ocean, the freshwater storage in the ice cover becomes larger than that in the upper 200m but only about 1/2 of the freshwater storage in the Atlantic derived waters of the Arctic Ocean (Table 1).

TABLE 1. Thickness, area and volume, and freshwater content in m and volume for the different layers.

depth interval m.	area $10^{12} m^2$.	volume $10^{12} m^3$.	salinity	freshwater, thickness m.	freshwater, volume $10^{12} m^3$.
+3 (ice)	7	21	0	3.00	21
0-50	2.5	125	33.2	2.57	6.43
50-200	3.0	450	34.3	3.00	9.00
200-700	4.5	2250	34.90	1.43	6.43
700-1700	4.5	4500	34.91	2.57	11.57
1700-2500	4.0	2000	34.935	1.49	5.94
>2500		1500	34.945		2.36

We define the residence time as volume of layer divided by export. With the limited amount of information used here we cannot estimate residence times separately for all layers introduced in Table 1. However, if the Polar Mixed Layer and the halocline are combined into polar surface water and the Atlantic and the intermediate depth waters are considered as one water mass, we can find the export of volume, w_1 and w_2 and freshwater, fw_1 and fw_2, in these two layers for different inflow rates of Atlantic Water, aw, as functions of the freshwater input to the water column, fw.

The freshwater input to the Arctic Ocean from river run-off and net precipitation is about 0.13 Sv [27]. An additional 0.01 Sv is added by the Norwegian Coastal Current. The freshwater provided by the Bering Strait inflow is about 0.06 Sv (relative to 35), while the corresponding freshwater export through the Canadian Arctic Archipelago is 0.05 Sv using the transport estimate given in [29]. If the Pacific water is assumed to exit primarily through the Canadian Arctic Archipelago, the Bering Strait only contributes 0.01 Sv to the freshwater in the Atlantic domain. The freshwater exported through Fram Strait, as ice or in the water column, then becomes 0.15 Sv. Of the Atlantic water entering the Arctic Ocean about 0.2 Sv leaves through the Canadian Arctic Archipelago [21], and some will supply the deep (>1700m) layers [2]. Using the transport estimates proposed in section 2, 3 Sv of Atlantic Water must absorb the freshwater added to the water column. However, the inflow of Atlantic Water through Fram Strait is often assumed to be larger than the 1-1.5 Sv proposed here and we also consider an inflow of 3 Sv through Fram Strait making a total inflow of 5 Sv.

If a steady state is assumed, the outflow of freshwater and Atlantic Water in the two layers must balance the input of freshwater and Atlantic Water and the balance equations for the two layers become;

$$fw_1 + fw_2 = fw = \frac{FW_1}{T_1} + \frac{FW_2}{T_2}$$

$$aw_1 + aw_2 = aw = \frac{AW_1}{T_1} + \frac{AW_2}{T_2}$$

where $FW_{1,2}$ are the volumes of freshwater and $AW_{1,2}$ the volumes of Atlantic Water and T_1 and T_2 are the residence times of the water in the upper (1) and lower (2) layer respectively. The outflows in the two layers become functions of the freshwater input and when fw increases from 0.01 Sv to 0.14 Sv the exports in the two layers vary inversely. A higher freshwater input leads to a larger transport in the upper and a smaller transport in the lower layer (Figure 8a). For the inflow of 3 Sv of Atlantic Water not more than 0.08 Sv of freshwater can be exported, all in the upper layer (Figure 8b). For the higher inflow the corresponding limit is 0.12 Sv. Such a large outflow of polar surface water in the East Greenland Current is contrary to most estimates, which lie between 1 Sv and 2 Sv and often closer to 1 Sv. 1 Sv corresponds to a freshwater export of 0.03 - 0.04 Sv and 0.04 - 0.05 Sv for aw equal 3 Sv and 5 Sv respectively. For the case with aw = 3 Sv and fw = 0.03 Sv the freshwater export is about 5 times as large in the upper than in the lower layer. For aw = 5 Sv the ratio of the transports is about 2:1 (Figure 8b). To have a freshwater export in the water column as large as the ice export would require an outflow in the upper layer of 3 Sv which appears much too high (Figure 8).

The residence time for the water in the lower layer is about 100 years for the low inflow case with fw = 0.03 Sv. This is reasonable for an Atlantic inflow of 3 Sv. For higher inflow of Atlantic Water a more rapid renewal of the lower layer is expected, which again points toward an input of freshwater < 0.05 Sv. The residence time in the upper layer then lies between 15 and 20 years for both inflow situations. This is high compared with the 10 years commonly given [50, 51]. The residence time cannot be much lower though, because not enough freshwater is available. We have not separated the Polar Mixed Layer and the halocline, and the Polar Mixed Layer may be ventilated more rapidly. On the other hand, some Tritium/Helium-3 observations indicate that the upper part of the halocline may be renewed fairly slowly [52]. As a comparison we may note that the residence time for the water supplied by the Bering Strait inflow to the Pacific dominated part of the Arctic Ocean ($325 \; 10^{12} m^3$) is about 10 years.

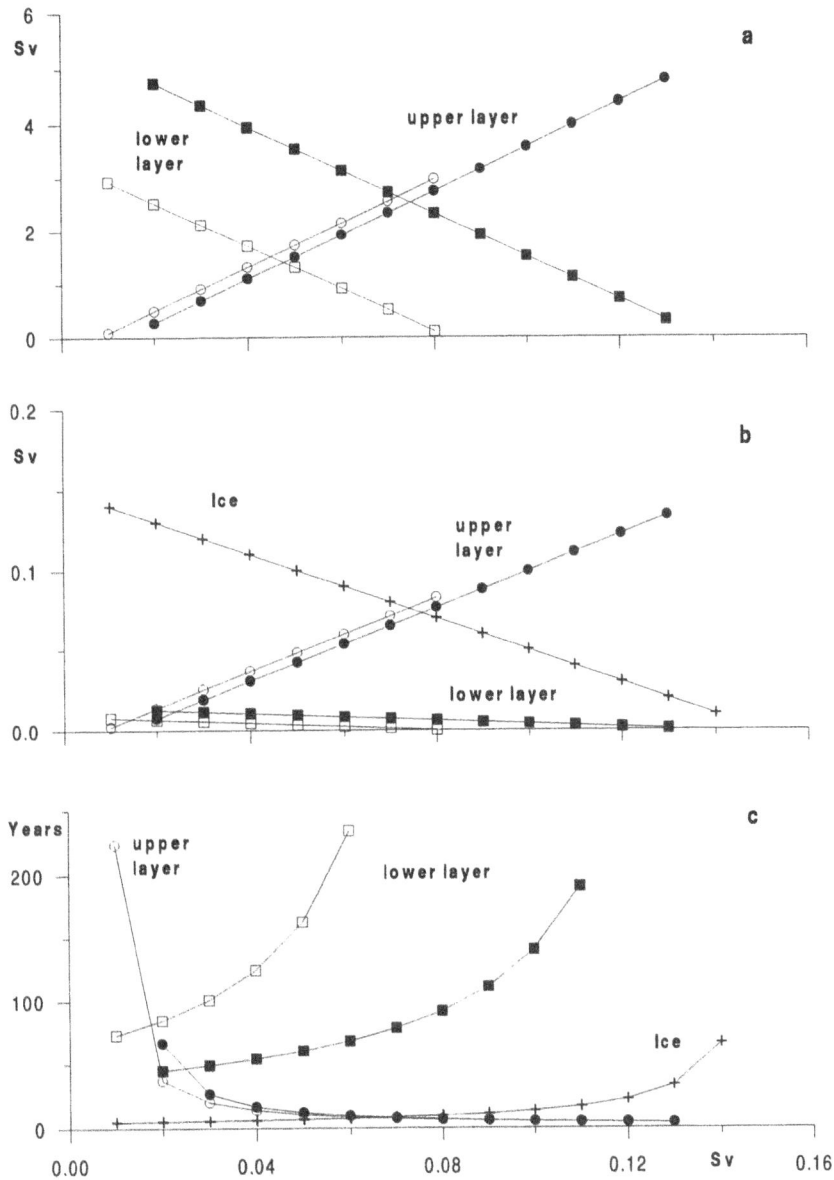

Figure 8. a) The transports of water in the idealised upper and lower layers exiting through Fram Strait as functions of the freshwater input to the water column. b) The freshwater export in the two layers and in the ice export as functions of the freshwater input to the water column. c) The residence time in the Arctic Ocean for the waters in the two layers and for the water exported as ice as functions of the freshwater input to the water column. Crosses indicate ice export, circles fluxes in the upper, squares fluxes in the lower layer. Open symbols refer to an Atlantic inflow of 3 Sv, filled symbols refer to an inflow of 5 Sv.

It is thus difficult to export more freshwater in the water column without really changing the characteristics of the water masses. The ice cover and the ice export is much less constrained and can buffer and export changes in the freshwater input. The ice export figure of 0.1 Sv, which has been around for a long time [53], fits well with the freshwater export in the water column and the freshwater input. It should be noted that seasonal ice melt is not considered as freshwater stored in the water column but is included in the ice export. The estimates thus overestimate the ice export and underestimate the freshwater export in the liquid phase. When measurements of the exports are made in Fram Strait, the ice export is observed over the entire year [54], while the hydrographic observations are mainly made in summer when the seasonal melt water input is at its maximum. Freshwater fluxes based on observations in Fram Strait may then overestimate the export in the water column as well as the total freshwater transport.

If the fraction of freshwater exported as ice should increase the salinity of the upper layer becomes higher, or the outflow of the upper layer decreases and that in the lower layer increases. The circulation would change towards one with more vigorous exchanges in the Atlantic and intermediate layer and weaker fluxes of polar surface water. This is consistent with a more active slope convection and a stronger ventilation of the deeper layers which would result from larger ice formation. If the changes were such as to decrease the ice export, the outflow of polar surface water becomes larger and the Atlantic and intermediate water circulation weakens.

We have ignored exchanges occurring deeper than 1700m, which are expected to be less than 0.5 Sv and with a small freshwater fraction [55]. In this context it is appropriate to discuss our neglect of the inflow of deep water to the Arctic Ocean through Fram Strait. An inflow of Greenland Sea Deep Water was earlier thought to supply the deep waters of the Arctic Ocean as well as the Norwegian Sea [26, 56]. The colder deep water in the Eurasian Basin as compared to the Canadian Basin was assumed to be due to the blocking of the coldest, densest deep water from the Greenland Sea by the Lomonosov Ridge, preventing it from reaching the Canadian Basin [57]. However, it has been found that, at least at present, the inflow of deep water from the Nordic Seas to the Arctic Ocean is weak. Most of it mixes with the outflowing Arctic Ocean deep waters in Fram Strait [22], and the deep water from the Nordic Seas is only detectable in temperature and salinity close to the Yermak Plateau [2]. Moreover, the slope convection in the Canadian Basin, penetrating through the Atlantic Layer of the boundary current, has been shown to generate differences in temperature and salinity between the Canadian and Eurasian Basin water columns above the sill depth of the Lomonosov Ridge, which are in agreement with observations [58].

The convection in the Greenland Sea mixes freshwater from the surface layer into deeper waters, which partly originate from the Arctic Ocean, making them fresher. If some of these intermediate and deep waters are transported into the Arctic Ocean through Fram Strait, they constitute a deep, freshwater import. However, most of the freshwater convected downward in the Greenland Sea derives from sea ice and polar surface water which have entered the Greenland Sea gyre from the Arctic Ocean. The convection in the Greenland Sea can then be viewed as a last redistribution of the freshwater of the Arctic Ocean from the surface into the deep.

5. Temporal Changes

The temperature of the Atlantic Layer in the boundary current at the continental slope in the Arctic Ocean has recently been observed to be considerably higher than in the historical data [41, 59]. Later observations and analysis have supported these observations [38, 49, 60, 61, 62]. Whether the change is due to a large but natural fluctuation, or if it indicates a persistent trend, is presently not known.

Other deviations from the established picture of the water masses in the Arctic Ocean have been observed. The low salinity upper layers of the Canadian Basin and the Beaufort Gyre, deriving from the Pacific inflow, have recently been located closer to the North American continent and been more confined to the Canada Basin than previously observed. Finally, and perhaps most striking, the Polar Mixed Layer in the interior of the Eurasian Basin is reported to be more saline than in the historical data, 34.3 as compared to 32.7 (or the 33.2 used here), and to extend much deeper, 140m as compared to 50m, reaching down to the thermocline at the top of the Atlantic Layer [63, 64, 65, 66] (see also Figure 9). No cold halocline was present, or in the view of expounded in [33], a winter mixed layer but no overlying, protective, shelf derived surface water was found. These observations indicate that the Arctic Ocean is not a quiescent environment remaining in a quasi-steady state but exhibits as large a variability as other regions in the ocean.

The hydrographic conditions in the Arctic Ocean are determined locally by the varying wind fields and by the sea surface heat loss, and advectivly by the inflows from the North Atlantic and the North Pacific as well as from the river run-off. The increase in temperature of the Atlantic Layer well into the Canadian Basin has led to the suggestion that a shift has occurred in the boundary between the domains dominated by Pacific and Atlantic derived waters respectively. Previously the boundary was believed to run along the Lomonosov Ridge. Now it should have moved to the Mendeleyev Ridge [49, 59, 67].

The higher temperatures of the Atlantic Layer probably derive from a warmer, but perhaps also a stronger inflow of Atlantic Water through Fram Strait. A possible cause for higher temperatures is a smaller heat loss in the Norwegian Sea in winter [61]. However, higher temperatures in the Atlantic Layer would not, by themselves, lead to a shift in the boundary between the two domains, especially since the Atlantic Layer is always present in the Canadian Basin below the Pacific waters.

The most likely explanation for the boundary shift is a change in the atmospheric circulation pattern. When the anti-cyclonic circulation over the Canadian Basin is strong, the Pacific Water is brought from the shelf and slope into the Beaufort gyre and extends far into the Canada Basin. When the anti-cyclonic circulation is weaker, as it has been in recent years, the Pacific water becomes restricted to the area close to the North American continent. A weaker anti-cyclonic circulation also allows the cyclone paths from the North Atlantic to reach far into the Arctic. The wind field could then drive much of the river run-off from Ob, Yenisey and Lena eastward from the Laptev Sea. The upper layers of the Makarov Basin then receive an increased amount of Atlantic derived shelf water which normally enters the Eurasian Basin at the Laptev Sea shelf break. This would result in a shift in the boundary between the two regimes similar to the one observed [66, 68].

As the Eurasian Basin receives less low salinity shelf water, the deep winter mixed layer, initially derived from ice melt north of Svalbard and subsequent winter homogenisation in the southern Nansen Basin (see above), does not become overrun by shelf water of lower salinity. No halocline is formed, or rather, the halocline extends to the surface, and local convection can reach the thermocline and the Atlantic Layer. A reduction of the shelf outflow also decreases the water volume above the Atlantic Layer and its upper boundary is likely to shift upward relative to its position in the historical records, making the temperature increase appear even more drastic.

The boundary between the Pacific and Atlantic domains is leaky. Surface waters may be transferred horizontally between the two domains, and in the deep sufficiently dense shelf water could sink out of the Pacific water into the Atlantic derived halocline [69], into the Atlantic Layer, and occasionally still deeper. Some deeper lying water could be brought onto the shelves also in the Pacific domain. In the mass balance for the different layers of the Canadian Basin computed in [37] the changes in the characteristics of the boundary current, necessary to create the water mass characteristics of the Canadian Basin water column, are accomplished, if dense water corresponding to 10% of the volume of the boundary current, crossing the Lomonosov Ridge, convects from the shelves. Assuming that all this shelf water is of Pacific origin and that the strength of the boundary current ranges between 0.5 Sv and 1 Sv, then 0.05-0.1 Sv leaves the Pacific and enters the Atlantic domain. This is 1/10 of the Bering Strait inflow.

The present atmospheric circulation pattern leads to a deeper and more confined Pacific water lens. The shelf water would be less saline, requiring more ice formation to attain the high salinity needed for deep reaching slope convection and vertical exchanges may be inhibited. The thicker intervening layer more strongly dilutes the sinking plumes, removing their density excess. When less shelf water enters the Atlantic Layer the cooling through slope convection is reduced. If this atmospheric circulation regime prevails over a longer period (decades), the Atlantic Layer in the Canada Basin will also become warmer.

Being at the surface, the distribution of the Pacific waters is strongly influenced by the atmospheric forcing, but the deeper Atlantic Layer should be less affected. Its circulation pattern would be the same as before, but, because of the higher core temperatures, it is now possible to follow the paths of the Atlantic inflow and observe the slope-basin interactions. The boundary current bifurcates at all major ridges, bringing waters from the slope into the interior of the basins. The interior changes more slowly and sub-surface fronts are observed in temperature and salinity [49]. Because of the changes in the inflow characteristics the difference in Θ-S properties between the old interior water column and the boundary current has become larger.

This may increase the occurrence of meso- and small-scale mixing processes e.g. double-diffusive convection [61].

Below 200m the water column in the Canadian Basin is created by waters in the boundary current, which have become transformed by slope convection in the Canadian Basin. Because of the changes of the inflow properties of both the Fram Strait and in the Barents Sea branches (see below) the interactions between slope convection and the boundary current now result in a water column with characteristics different from those of the interior of the Canadian Basin. Assuming, as above, the strength of the boundary current crossing the Lomonosov Ridge to be between 0.5 Sv and 1 Sv, about 200-100 years are needed to completely replace the water mass between 200 and 1700m in the Canadian Basin with waters having the new properties. After this the slope convection will again transform the inflowing boundary current into waters with characteristics matching the Canadian Basin water column [44]. Variations in the properties of the boundary current are likely to occur before this has happened, and an equilibrium may not be, and may never have been, reached.

In Fram Strait a change appears to have taken place between the 1980s and the present. In the 1980s (Figure 2) warm Atlantic Water was present almost over the entire cross section, implying a strong recirculation at and north of 79°N. Now warm Atlantic Water only extends over 3/4 of the strait, leaving a much freer passage for the water of the Atlantic Layer to exit the Arctic Ocean [70]. One interpretation, which would also be consistent with other recently reported changes, is that the Atlantic Water which now crosses 79°N no longer recirculates in the vicinity of the strait but joins the boundary current and enters the Arctic Ocean. More Atlantic Water then participates in the circulation gyres in the different basins and the ventilation of the Atlantic Layer would become more rapid [70]. This would agree with increased temperatures in the Atlantic Layer. It could also imply a larger inflow of Atlantic Water to the Arctic Ocean than in the 1980s, and the exchanges may now be better described by the high exchange case (5 Sv) considered above (section 4). The low exchange case (3 Sv) would then correspond to the situation holding in the 1980s.

The cause for such a change, if the interpretation made above is correct, is not known. However, a weaker recirculation in Fram Strait intensifies the communication between the Nordic Seas and the Arctic Ocean and ties the two main areas of water mass formation in the Arctic Mediterranean more closely together. Since the deep water production in the Greenland Sea is reported to be reduced [71], at the same time as warmer Atlantic Water is observed in the Arctic Ocean, one interpretation would be that deep convection in the Greenland Sea transforms more Atlantic Water and induces a stronger recirculation in Fram Strait, bringing the Atlantic Water towards the convecting Greenland Sea. Instead the created Greenland Sea Deep Water penetrates northward into the Arctic Ocean as visualised in [36] and [56] in contrast to the present day situation with Arctic Ocean deep waters gradually filling up the Greenland Sea [58]. Such an interpretation, although not inconsistent with the present meteorological situation, must, however, be considered as highly speculative.

The most important reported change is associated with the deep, saline mixed layer in the Eurasian Basin and the possibility of a direct heat transfer from the Atlantic

Layer to the sea surface. It is therefore relevant to draw attention to different aspects of the interactions between the Atlantic Water and freshwater and their relation to the Arctic climate. The Atlantic inflow through Fram Strait transports a large amount of heat into the Arctic, and the Atlantic Layer contains enough heat to melt 20m of ice [72]. If the freshwater input were reduced, the stratification of the upper layers becomes weaker and the shelf water formed in winter would sink into or below the Atlantic Layer. The heat transport from the Atlantic Layer to the sea surface due to turbulent entrainment into the mixed layer would then be stronger and the ice cover could be significantly reduced and even disappear [72]. However, later work has shown that the freshwater added by the Bering Strait inflow would supply a buffer which prevents this from happening [73]. Also a small reduction of the ice thickness by melting would increase the stability sufficiently to protect the ice cover from further melting [73, 74]. Furthermore, the main cooling of the Atlantic Layer occurs through the incorporation of cold, shelf derived plumes, and the largest heat loss of the Atlantic Layer is due to the downward re-distribution of warmer water by deeper sinking entraining plumes. These effects of the slope convection would increase with less freshwater present on the shelves.

The presently most discussed scenario is one with increasing, not decreasing, freshwater input. A larger freshwater input to the Arctic Mediterranean would reduce the convection and thus the deep water formation in the Greenland Sea [27], but possibly also in the Arctic Ocean. The compensating inflow of Atlantic Water to the Arctic Mediterranean may then be reduced, and as less sensible heat is advected toward the north, the ice formation would increase. If the brine release and density increase connected with the ice formation on the shelves do not overcome the density decrease caused by the larger run-off, the northern limb of the thermohaline circulation will remain weak. This may lead to a regionally colder climate, especially in north-western Europe. Such a change could follow from increased precipitation caused by a globally warmer climate.

Returning now to the deep, saline mixed layer, we note that in a Θ-S diagram the thermocline does not form a straight mixing line but is curved, having a shape similar to an isopycnal (Figure 9, see also [64, 65]). This implies that winter convection caused by brine rejection penetrates to a density level within the thermocline and incorporates its upper part into the mixed layer, and that convection is the main mechanism for transporting heat upwards. Only a limited amount of heat can be brought into the mixed layer, and subsequently to the sea surface, by this process. A large flux of sensible heat from below reduces the ice formation, and the convection, and thus stops the heat flux from the Atlantic Layer into the mixed layer.

At the continental slope the Θ-S curves of the thermocline are straight (Figure 9). This indicates turbulent mixing between the mixed layer and the Atlantic Layer, and implies that the entrainment of warmer water from below into the mixed layer is mainly due to mechanically generated turbulence and does not depend upon the freezing rate. The upward heat flux then continues also when the ice formation is reduced. The upward flux of salt from the Atlantic Layer cannot compensate for the reduced brine release and the salinity of the mixed layer is lower at the slope than in the interior (Figure 9). However, no intermediate halocline is normally present at the Eurasian continental slope west of the Laptev Sea, and the situation here would not be different from what it was before. The vertical heat flux may increase because warmer Atlantic Water is entrained, but a slight decrease in the ice formation reduces the salinity of the mixed layer and would increase the stability at its base sufficiently to counteract this effect. The upward heat flux would then not be much larger, when the winter mixed layer reaches down to the Atlantic layer, compared with the classic Arctic situation with a shallow mixed layer and an intermediate halocline (Figure 9, see also [65]). The ice cover is likely to remain.

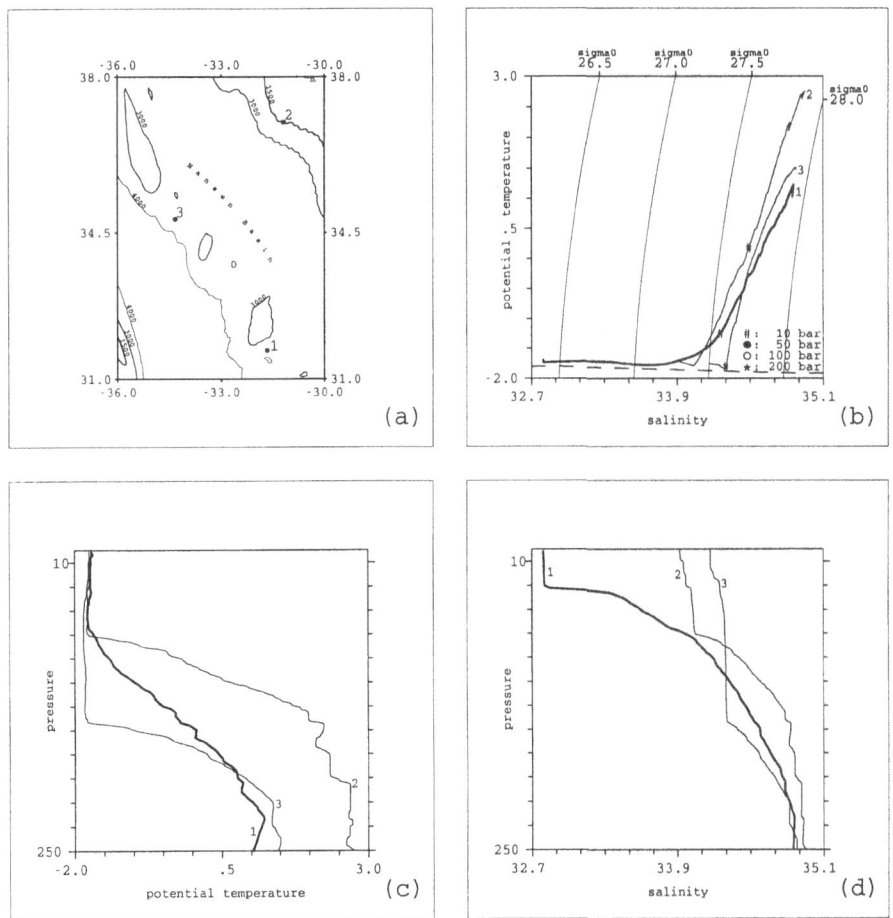

Figure 9.   Profiles of potential temperature and salinity and Θ-S curves from stations in the Eurasian Basin for the upper 250m showing the change in the mixed layer between 1991 (IB Oden) and 1996 (RV Polarstem). For station information see appendix.

526

During the same period as the freshwater content in the upper layers in the Eurasian Basin has been reduced, the salinity of the water injected down the St. Anna Trough has decreased, implying a larger content of freshwater. This change may be due to a more extensive rainfall over the coast of Norway [75], adding freshwater to the Norwegian Coastal Current, which could change the characteristics of the St. Anna Trough outflow [76].

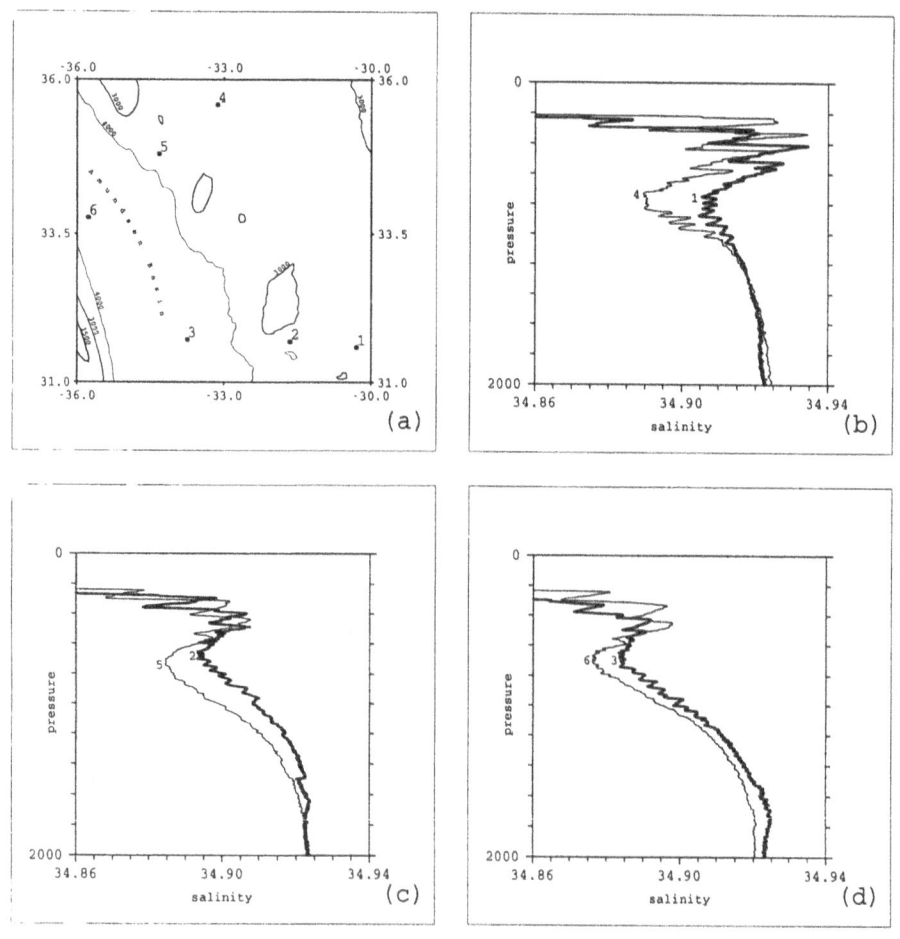

Figure 10.   Salinity profiles from stations in the Nansen Basin, the Nansen-Gakkel Ridge and the Amundsen Basin showing the difference in freshwater content in the intermediate layers between 1991 and 1996. For station information see appendix.

Another, but perhaps more farfetched possibility, is that less of the run-off from Ob and Yenisey is transported through the Vilkiltskij Strait into the Laptev Sea and more enters the Arctic Ocean directly across the Kara Sea shelf break. Some shelf water in the Kara Sea becomes in winter dense enough to sink down the St. Anna Trough and join the deep outflow, and the salinity of this contribution would vary with the salinity of the shelf water. Such a change could have several causes, e.g. different wind forcing or stronger ice formation in the Kara Sea. Sea ice formed in the Kara Sea could block the passage for top layers through the Vilkiltskij Strait allowing only the deeper part, containing much less freshwater, to pass into the Laptev Sea. If some sea ice formed in the Kara Sea drifts into and melts in the Barents Sea, the freshwater would eventually become incorporated in the dense Barents Sea outflow the following winter making it less saline. A stronger ice export from the Kara Sea to the Barents Sea in 1992/93 as compared with 1987 has indeed been observed [77]. Freshwater could then be redirected from the polar surface water to the intermediate waters by shifting the shelf-basin outflow between the Laptev and the Kara Seas, and would couple high salinity in the mixed layer with lower salinity of the St Anna Trough outflow.

To compare the freshwater decrease in the upper layers with the freshwater increase in the intermediate depth range we must subtract the freshwater added by seasonal ice melt, which causes the salinity of the upper part to be lower than the salinity at the temperature minimum (33.7 in Figure 9). Assuming a basic salinity of 34.3 in the halocline [33] this gives a freshwater storage of 1.9m in 1991 compared to 0.01m in 1996 for the stations in Figure 9. In the intermediate layers the salinity between 400 and 1000-1200m was ~0.02 lower in 1996 (Figure 10), which corresponds to a freshwater increase of 0.35m. Approximately 20% of the freshwater reduction in the upper layer could then be accounted for by the increased freshwater content at intermediate levels (Figure 10).

The main cause for the higher salinities in the upper layers (80%) must then be the shift of the shelf water input from the Laptev Sea eastward into the East Siberian Sea and further into the Canadian Basin. The two salinity changes, the increase in the mixed layer and the decrease in the intermediate layer, are then largely uncorrelated, in a strict oceanographic sense, if not in a wider meteorological, and climatological, sense. Changes in the wind fields could move the run-off from the Laptev Sea to the East Siberian Sea and then into the Canadian Basin. The same meteorological conditions could also cause higher precipitation in Scandinavia leading to lower salinities in the Norwegian Coastal Current and thus to lower salinities of the Barents Sea branch in the St. Anna Trough. A direct coupling to the run-off from Ob and Yenisey is then not required.

6. Causes for the changes – the North Atlantic Oscillation?

A shift in the atmospheric circulation pattern in agreement with the reported changes from the Arctic Ocean can be related to the variations of large scale air pressure distributions in the North Atlantic or the North Atlantic Oscillation (NAO) [78]. The NAO is defined as the variations in the pressure difference between the Azores high pressure cell and the Icelandic low pressure cell (the NAO index) and has recently been invoked to explain the shift in the convective activity in the North Atlantic between the Greenland Sea and the Labrador Sea [79]. In years with high NAO index the low pressure systems from the North Atlantic cause a higher cyclonic activity in the Arctic Mediterranean and bring warmer, moist air to north-western Europe. The deep convection in the Greenland Sea is reduced, while cold air brought down over the Labrador Sea leads to strong and deep convection there. When the index is low the high pressure system over Greenland brings cold air over the Greenland Sea. The cooling and the deep convection increase in the Greenland Sea. At the same time warmer air from the south is carried over the Labrador Sea and the deep convection there closes down.

The changes in the Arctic Ocean follow a pattern which could partly be explained by the same shift. Decreasing surface heat loss in the Norwegian Sea leads to warmer Atlantic Water entering through Fram Strait. High cyclonic activity increases the rainfall over north-western Europe causing a freshening of primarily the Barents Sea inflow. A weakening of the Arctic high allows the cyclones to penetrate deep into the Arctic and the Pacific wedge would be confined to the Canada Basin and the American continent. The shelf water normally entering the Eurasian Basin is then deflected toward the East Siberian Sea and ultimately to the Canadian Basin.

Incidentally, the forcing of the wedge of Pacific water close to the North American continental shelf by the prevailing wind fields may lead to a larger export of Pacific water not only through the Canadian Arctic Archipelago but also through Fram Strait by the creation of a buoyancy driven boundary current along the shelf. Pacific water has recently been followed into the Eurasian Basin and to Fram Strait [1]. The strong pulses of freshwater exports which would be associated with such a boundary current could explain occasional, high transports of freshwater in the water column which are unlikely if only waters of the Atlantic domain are involved. Since the flow through the Canadian Arctic Archipelago in this scenario also would increase, it means that the volume of Pacific water in the Arctic Ocean would be reduced.

The NAO is believed forced primarily by conditions holding at mid latitudes. If the NAO exerts a dominating influence over the conditions in the Arctic Mediterranean we would expect a return toward more "familiar" conditions in a few years.

7. Recommendations

I: Establish long-term monitoring of the inflow of Atlantic Water by deploying current meter arrays in Fram Strait and between Franz Josef Land and Novaya Zemlya.

II: Re-examine existing hydrographic data to determine interannual and longer term variability.

III: By hydrographic observations and modelling work, try to establish if any connection exists between the intensity of the deep convection in the Greenland Sea, and the relative strength of the inflow of Atlantic and deep water to the Arctic Ocean, and recirculation of Atlantic Water in Fram Strait.

Acknowledgement: This work has been supported by the European Commission MAST III Programme: Variability of Exchanges in the Nordic Seas (VEINS), through contract MAS3-CT96-0070.

Appendix

Table giving the station information for stations shown in the figures.

Figure/#	ship	station	position	date
02/*	Lance	all	section ~78.92°N	August 1984
03/1	Polarstern	019	82.755°N; 040.207°E	18.08.1993
03/2		025	82.125°N; 042.565°E	21.08.1993
03/3		026	82.095°N; 042.725°E	21.08.1993
03/4	Ymer	048	82.482°N; 038.393°E	25.07.1980
03/5		205	82.333°N; 045.933°E	17.09.1980
03/6		206	82.383°N; 045.083°E	17.09.1980
04/1	Polarstern	011	81.393°N; 073.313°E	26. 07. 1996
04/2		032	82.040°N; 091.833°E	01.08.1996
04/3		034	82.092°N; 091.502°E	02.08.1996
04/4		037	82.517°N; 092.297°E	02.08.1996
05/1	Polarstern	025	81.098°N; 105.357°E	07.08.1995
05/2		027	81.222°N; 106.577°E	08.08.1995
05/3		029	80.835°N; 104.168°E	10.08.1995
05/4		031	80.727°N; 103.302°E	11.08.1995
07/1	Oden	010	85.730°N; 048.052°E	25.08.1991
07/2		013	86.737°N; 056.975°E	27.08.1991
07/3		016	87.603°N; 069.745°E	29.08.1991
07/4		026	88.015°N; 163.603°E	04.09.1991
09/1	Oden	013	86.737°N; 056.975°E	27.08.1991
09/2	Polarstern	036	82.333°N; 091.997°E	02.08.1996
09/3		050	85.168°N; 109.288°E	07.08.1996
10/1	Oden	010	85.730°N; 048.052°E	25.08.1991
10/2		013	86.737°N; 056.975°E	27.08.1991
10/3		017	88.053°N; 085.055°E	30.08.1991
10/4	Polarstern	043	84.202°N; 100.533°E	05.08.1996
10/5		050	85.168°N; 109.288°E	07.08.1996
10/6		056	86.160°N; 125.813°E	09.08.1996

References:

1   Jones, E.P., Rudels, B. and Anderson, L.G. (1995) Deep Waters of the Arctic Ocean: Origins and Circulation, *Deep-Sea Res.*, 42, 737-760.
2   Jones, E.P., Anderson, L.G. and Swift, J.H. (1998) Distribution of Atlantic and Pacific Water in the upper Arctic Ocean: implications for the circulation, *Geophys. Res. Letters*, 25, 765-768.
3   Wijffels, S., Schmitt, R.W., Bryden H.L. and Stigebrandt, A. (1992) On the transport of freshwater by the oceans. *J. Phys. Oceanogr.*, 22, 155-162.
4   Hirtzler, J.R. (1985) Relief of the surface of the earth. Rep. MGG-2, National Geographic Data Center, Boulder, Colorado.
5   Perkin, R.G., and Lewis, E.L. (1984) Mixing in the West Spitsbergen Current, *J. Phys. Oceanogr.*, 14, 1315-1325.

530

6 Nansen, F. (1902) *Oceanography of the North Polar Basin. The Norwegian North Polar Expedition 1893-96.* Scientific Results III (9), 427 pp.

7 Petermann, A. (1965) Der Nordpol und Südpol, die Wichtigkeit ihrer Erforschung in geographischer und kulturhistorischer Beziehung. Mit Bemerkungen über die Strömungen der Polar-Meere. *Pet. Mitt.* pp 146-160, Gotha.

8 Mohn, H. (1887) *Den Norske Nordhavsekspedition 1876-1878. Bd II. Nordhavets Dybder, Temperatur og Strømningar,* Christiania.

9 Ryder, C. 1891-1892, Tidligere Ekspeditioner til Grønlands Østkyst nordfor 66° Nr. Br., *Geogr. Tids.* Bd. 11, pp. 62-107, København.

10 Bourke, R.H., Weigel, A.M. and Paquette, R.G. (1988) The westward turning branch of the West Spitsbergen Current, *J. Geophys. Res.* 93, 14065-14077.

11 Carmack, E.C. (1990) Large-Scale Physical Oceanography of Polar Oceans, in W.O. Smith Jr. (ed.) *Polar Oceanography, Part A,* Academic Press, San Diego, California, pp. 171-212.

12 Pickard, G.L. (1963) *Descriptive Physical Oceanography, an introduction.,* Pergamon Press, Oxford. 200pp.

13 Dobberphul, F. (1992) Abschätzung der absoluten Wassenmassestransporte durch die Framstrasse aus geostrophischen Berechnungen und direkten Strömungsmessungen, Diplomarbeit, Universität Hamburg, 92pp.

14 Aagaard, K. and Greisman, P. (1975) Towards new mass and heat budgets for the Arctic Ocean, *J. Geophys. Res.,* 80, 3821-3827.

15 Pickard, G. L. and Emery, W.J. (1990) *Descriptive Physical Oceanography, an introduction.,* Pergamon Press, Oxford. 320pp.

16 Tomczak, M and Godfrey, J.S. (1994) *Regional Oceanography: an Introduction.,* Pergamon, Oxford. 422pp.

17 Hanzlick, D.J. (1983) The West Spitsbergen Current: Transport, forcing and variability, Ph.D. thesis, University of Washington, 127 pp.

18 Quadfasel, D., Gascard, J.-C. and Koltermann, K.P. (1987) Large-scale oceanography in Fram Strait during the 1984 Marginal Ice Zone experiment. *J. Geophys. Res.,* 92, 6719-6728.

19 Foldvik, A., Aagaard, K. and Törresen, T. (1988) On the velocity field of the East Greenland Current. *Deep-Sea Res.,* 35, 1335-1354.

20 Jónsson. S and Foldvik, A., 1992. The transport and circulation in Fram Strait. *ICEM, C.M.* 1992/C.10 Hydrography Committee, 10 pp.

21 Rudels, B., (1987) On the mass balance of the Polar Ocean, with special emphasis on the Fram Strait. *Norsk Polarinstitutt Skrifter* 188. 53pp.

22 Houssais, M.-N., Rudels, B., Friedrich, H. and Quadfasel, D. (1995) Exchanges through Fram Strait, Extended Abstract, Nordic Seas Symposium, Hamburg 7/3-9/3 1995, 87-92.

23 Blindheim, J. (1989) Cascading of Barents Sea bottom water into the Norwegian Sea, *Rapp. P.-v. Reun. Cons. int. Explor. Mer,* 188, 49-58.

24 Loeng, H., Ozhigin, V., Ådlandsvik, B. and Sagen, H. (1993) Current Measurements in the north-eastern Barents Sea. ICES C.M. 1993/C:41, Hydrographic Committee, 22 pp.

25 Harms, I.H. (1997) Water mass transformations in the Barents Sea - application of the Hamburg Shelf Ocean Model (HamSOM), *ICES J. of Marine Science,* 54, 351-365.

26 Helland-Hansen, B. and Nansen, F. (1909) *The Norwegian Sea. Its physical oceanography based upon the Norwegian researches 1900-1904.* Rep. on Norw. Fishery and Marine Investigations II(1), Kristiania.

27 Aagaard, K. and Carmack., E.C. (1989) The role of sea ice and other fresh water in the Arctic circulation, *J. Geophys. Res.,* 94, 14485-14498.

28 Coachman, L.K. and Aagaard., K. (1988) Transports through Bering Strait: Annual and interannual variability, *J. Geophys. Res.* 93, 15535-15539.

29 Rudels, B. (1986) The outflow of Polar Water through the Arctic Archipelago and the oceanographic conditions in Baffin Bay. *Polar Res.* 4 n.s., 161-180.

30 Heinze, Ch., Schlosser, P., Koltermann, K.-P. and Meincke, J. (1990) A tracer study of the deep water renewal in the European polar seas, *Deep-Sea Res.* 37, 1425-1453.

31 Smethie, W.M., Chipman, D.W., Swift J.H. and Koltermann, K.-P. (1988) Chlorofluoromethanes in the Arctic Mediterranean seas: evidence for formation of bottom water in the Eurasian Basin and the deep-water exchange through Fram Strait, *Deep-Sea Res.,* 35, 347-369.

32 Rudels, B. (1986) The Θ-S relations in the northern seas: Implications for the deep circulation, *Polar Res.,* 4 n.s. 133-159.

33 Rudels, B., Anderson, L.G. and Jones, E.P. (1996) Formation and evolution of the surface mixed layer and the halocline of the Arctic Ocean, *J. Geophys. Res.* 101, 8807-8821.

34 Loeng, H. (1991) Features of the physical oceanographic conditions in the central parts of the Barents Sea, *Polar Res.* 10, 5-18.

35 Midttun, L. (1985) Formation of dense bottom water in the Barents Sea, *Deep-Sea Res.*, 32, 1233-1241.

36 Nansen, F. (1906) Northern Waters. Captain Roald Amundsen's oceanographic observations the Arctic seas in 1901. *Vid-selskap Skrifter I, Mat.-Naturv. kl.* Dybvad Christiania 1 (3) 145 pp.

37 Rudels, B., Jones, E.P., Anderson, L.G. and Kattner, G. (1994) On the intermediate depth waters of the Arctic Ocean, in O.M. Johannessen, R.D. Muench and J.E. Overland (eds.), *The role of the Polar Oceans in Shaping the Global Climate*, American Geophysical Union, Washington, pp. 33-46.

38 Schauer, U., Muench, R.D. Rudels, B. and Timokhov, L. (1997) Impact of eastern Arctic shelf water on the Nansen Basin intermediate layers, *J. Geophys. Res.* 102, 3371-3382.

39 Quadfasel, D., Rudels, B. and Selchow, S. (1992) The Central Bank vortex in the Barents Sea: water mass transformation and circulation, *ICES mar. Sci Symp.*, 195, 40-51.

40 Martin, S and Cavalieri, D.J. (1989) Contributions of the Siberian shelf polynyas to the Arctic Ocean intermediate and deep water, *J. Geophys. Res.*, 94, 12725-12738.

41 Quadfasel, D., Sy, A., Wells, D. and Tunik, A. (1991) Warming in the Arctic, *Nature*, 350, 385.

42 Dickson, R.R., Meincke, J., Malmberg, S.-A. and Lee, A.J. (1988) The " Great Salinity Anomaly" in the Northern North Atlantic 1968-1982, *Progress in Oceanogr.*, 20, 103-151.

43 Hanzlick, D. and Aagaard, K. (1980) Freshwater and Atlantic Water in the Kara Sea, *J. Geophys. Res.*, 85, 4937-4942.

44 Rudels, B., Muench, R.D., Gunn, J. Schauer, U. and Friedrich, H.J (2000) Evolution of the Arctic Ocean boundary current north of the Siberian shelves. *J. Marine Systems*, (accepted).

45 Friedrich, H., Houssais, M.-N., Quadfasel, D. and Rudels, B. (1995) On Fram Strait Water Masses. Extended Abstract, Nordic Seas Symposium, Hamburg, 7/3-9/3 1995, 69-72.

46 Rudels, B., Friedrich, H.J. and Quadfasel, D. (1999) The Arctic Circumpolar boundary current. *Deep-Sea Research II*, 46, 1023-1062.

47 Aagaard, K., Swift, J.H. and Carmack, E.C. (1985) Thermohaline circulation in the Arctic Mediterranean Seas, *J. Geophys. Res.* 90, 4833-4846.

48 Coachman, L.K. and Aagaard, K. (1974) Physical Oceanography of the Arctic and Sub-Arctic Seas, in . Y. Herman (ed.), *Marine Geology and Oceanography of the Arctic Ocean*, Springer, New York, pp. 1-72.

49 Swift, J.H.; Jones, E.P., Carmack, E.C., Hingston, M., Macdonald, R.W., McLaughlin, F.A. & Perkin. R.G. (1997) Waters of the Makarov and Canada Basins, *Deep-Sea Research II*, 44, 1503-1529.

50 Östlund, H.G. and Hut, G. (1984) Arctic Ocean water mass balance from isotope data, *J. Geophys. Res*, 89, 6373-6381.

51 Schlosser, P., Bauch, D., R. Fairbanks, R. and Bönisch, G. (1994) Arctic river-runoff: mean residence time on the shelves and in the halocline, *Deep-Sea Res.*, 41, 1053-1068.

52 Bauch, D., Schlosser, P. and Fairbanks, R.G. (1995) Freshwater balance and sources of deep and bottom waters in the Arctic Ocean inferred from the distribution of H_2O^{18}. *Progress in Oceanogr.* 35, 53-80.

53 Fletcher, J.O. (1965) The heat budget of the Arctic Basin and its relation to climate, the Rand Corporation R. 444-PR, Santa Monica, California, 179 pp.

54 Vinje, T. Nordlund, N. and Kvambekk, Å. (1998) Monitoring ice thickness in Fram Strait. *J. Geophys. Res.*, 103, 10437-10449.

55 Anderson, L.G., Jones, E.P. and Rudels, B. (1999) Ventilation of the Arctic Ocean estimated by a plume entrainment model constrained by CFCs, *J. Geophys. Res.*, 104, 13423-13429.

56 Kiilerich, A. (1945) On the hydrography of the Greenland Sea, *Medd. om Grönland*, 144 (2) 63 pp.

57 Worthington, L.V. (1953) Oceanographic results of Project Skijump I and II, *Trans. Am. Geophys. Union*, 34(4), 543-551.

58 Meincke, J., Rudels, B. and Friedrich, H. (1997) The Arctic Ocean -Nordic Seas Thermohaline System, *ICES J. of Marine Science*, 54, 283-299.

59 Carmack, E.C., Macdonald, R.W., Perkin, R.G., McLaughlin, F.A. and Pearson, R.J. (1995) Evidence for warming of Atlantic Water in the southern Canadian Basin of the Arctic Ocean: Results from the Larsen-93 expedition, *Geophys. Res. Letters*, 22, 1061-1064.

60 Aagaard, K., Barrie, L.A., Carmack, E.C., Garrity, C., Jones, E.P., Lubin, D., Macdonald, R.W., Swift, J.H., Tucker, W.B., Wheeler, P.A. and Whritner, R.H. (1996) U.S., Canadian Researchers Explore the Arctic Ocean, *Eos*, 177, 209, 213.

532

61 Carmack, E.C., Aagaard, K., Swift, J.H., Macdonald, R.W., McLaughlin, F.M., Jones, E.P., Perkin, R.G., Smith, J.N., Ellis, K.M. and Killius, L.R. (1997) Changes in temperature and tracer distributions within the Arctic Ocean: Results from the 1994 Arctic Ocean section, *Deep-Sea Research II*, 44, 1487-1502.

62 Grotefendt, K., Logemann, K., Quadfasel, and D. Ronski, S. (1998) Is the Arctic Ocean warming? *J. Geophys. Res.*, 103,27,679-27,688.

63 Alekseev, G.V., Bulatov, L.V., Zakharov, V.F. and Ivanov. V. (1998) Interannual changes of water temperature and salinity in the context of recent icebreaker observations to the north of the Kara and the Laptev Seas. Proceedings of the ACSYS Conference on Polar processes and Global climate, Rosario, Orcas Island, Washington, USA, 3-6 Nov. 1997, WCRP-106, 15-17.

64 Rudels, B., Schauer, U. and Muench, R.D. (1998) Deep saline mixed layer and temporal variations in the Eurasian Basin water column, Proceedings of the ACSYS Conference on Polar processes and Global climate, Rosario, Orcas Island, Washington, USA, 3-6 Nov. 1997, WCRP-106, 220-222.

65 Schauer , U., Rudels, B., Jones, E.P., Anderson, L.G. Muench, R.D. Björk, G., Swift, J.H., Ivanov, V. and Larsson, A.-M. (2000) Water mass distribution in the Eurasian Basin and connections with the Canadian Basin: Results from ACSYS-96, *J. Geophys. Res.* (submitted).

66 Steele, M. and Boyd, T. (1998) Retreat of the cold halocline layer in the Arctic Ocean, *J. Geophys. Res.*, 103, 10419-10435.

67 McLaughlin, F.A., Carmack, E.C., Macdonald, R.W. and Bishop, J.K.B. (1996) Physical and geochemical properties across the Atlantic/Pacific water mass from in the southern Canadian Basin. *J. Geophys. Res.*, 101, 1183-1197.

68 Proshutinsky. A.Y. and Johnson, M.A. (1997) Two circulation regimes of the winddriven Arctic Ocean, *J. Geophys. Res.*, 102, 12493-12514.

69 Melling, H and Lewis, E.L. (1982) Shelf drainage flows in the Beaufort Sea and their effect on the Arctic Ocean pycnocline, *Deep-Sea Res.*, 29, 967-985.

70 Rudels, B., Meyer, R, Fahrbach, E, Ivanov, V, Quadfasel, D., Schauer, U, Tverberg, V., Woodgate, R.A. and Østerhus, S. (2000) Water mass distribution in Fram Strait and over the Yermak Plateau in summer 1997. *Annales Geophysicae*, (submitted).

71 Schlosser, P., Böhnisch, G., Rhein M. and Bayer, R. (1991) Reduction of deep water formation in the Greenland Sea during the 1980s: Evidence from tracer data, *Science*, 251, 1054-1056.

72 Aagaard, K. and Coachman, L.K. (1975) Toward an ice-free Arctic Ocean, *Eos*, 56, 484-486.

73 Stigebrandt, A. (1981) A model for the thickness and salinity of the upper layers of the Arctic Ocean and the relation between the ice thickness and some external parameters, *J. Phys. Oceanogr.*, 11, 1407-1422.

74 Rudels, B. (1989) The formation of Polar Surface Water, the ice export and the exchanges through the Fram Strait, *Progress in Oceanogr.*, 22, 205-248.

75 Cayan, D.R. and Reverdin, G. (1994) Monthly precipitation and evaporation variability estimated over the North Atlantic and the North Pacific, in Atlantic Climate Change Program: Proceedings of PI's meeting, May 9-11, 1994, 28-32.

76 Gerdes, R and Köberle, C. (1999) Numerical simulations of salinity anomaly propagation in the Nordic Seas and the Arctic Ocean, *Polar Res.*, (in press).

77 Martin, T. and E. Augstein (1998) Large-Scale Drift of Arctic Sea Ice Retrieved from Passive Microwave Satellite Data, *J. Geophys. Res.* (submitted)

78 Dickson R.R., Osborn, T.J., Hurrell, J.W., Meincke, J., Blindheim, J., Ådlandsvik, B., Vinje, T., Alekseev; G. and Maslowski, W. (1998) The Arctic Ocean Response to the North Atlantic Oscillation, *Journal of Climate*, (submitted).

79 Dickson, R.R., Lazier, J., Meincke, J. Rhines, P and Swift, J.H. (1996) Long-term co-ordinated changes in the convective activity of the North Atlantic, *Progress in Oceanogr.*, 38, 241-295.

MODELLING THE VARIABILITY OF EXCHANGES BETWEEN THE ARCTIC OCEAN AND THE NORDIC SEAS

Rüdiger Gerdes
Alfred-Wegener-Institut für Polar und Meeresforschung
Columbusstraße
D-27568 Bremerhaven
Germany

1. Introduction

A large scale warming in the interior of the Arctic indicates an increased influence of Atlantic Water [1,2,3]. The front between water masses of Atlantic and Pacific character, measured by hydrographic and geochemical properties in the southern Canadian Basin, has apparently shifted from a position over the Lomonosov Ridge to a more eastern position over the Mendeleyev Ridge[4]. Analysis of recent cruise data and historical data by Steele and Boyd [5] and Grotefendt et al. [6] indicate a wide spread warming in the Atlantic layer of the Eurasian Basin. They attribute the warming to an increased inflow of Atlantic water through Fram Strait and the Barents Sea.

Comparing results from two expeditions into the Eurasian Basin in 1991 and 1996, Schauer et al. [7] identify a warming from 1991 to 1996 in the core of the Atlantic Water at around 200m and reduced salinity in a layer between 500m and 1200m depth in the Amundsen Basin. The latter is attributed to an increased outflow from the Barents Sea that feeds into a boundary current along the rim of the Eurasian Basin [8,9]. This view is consistent with a numerical experiment by Gerdes and Köberle [10] (hereafter GK) in which an anomaly in surface fresh water flux was placed off the coast of Norway. There the largest fluctuations in precipitation occur during cycles of the North Atlantic Oscillation [11]. Schauer et al. [7] confirm the disappearance of the Arctic halocline in the central Eurasian Basin during recent years that was detected by Steele and Boyd [5]. The changes in the halocline of the Amundsen Basin have consequences for the air-sea fluxes and ice conditions in that area as the relatively warm Atlantic layer is more readily accessible to melt ice and warm the overlying atmosphere. It indicates a possible change in the fresh water pathways in the Arctic Ocean with conceivable consequences for its fresh water export to the Nordic Seas and the deep-water formation areas of the North Atlantic in general.

It is crucial to determine whether these changes are manifestations of a long term trend towards a different state of the Arctic Ocean or are a phase of long term variability in order to assess the function of the Arctic as an early indicator of global warming and to estimate possible impacts of global climatic change. Dickson et al. [11] suggest that the long-term atmospheric trend and the quasi-decadal variability in the North Atlantic Oscillation (NAO) combine in recent years to an extreme state of the Arctic climate system.

E.L. Lewis et al. (eds.), The Freshwater Budget of the Arctic Ocean, 533-547.
© 2000 *Kluwer Academic Publishers.*

Model experiments are a tool to investigate the relations between climate system components, their individual responses to external forcing, feedbacks, and internal modes of variability. Here I want to present model experiments that aim at clarifying these dynamical relationships and discuss modeling factors that affect the water mass exchange between the Arctic Ocean and the Nordic Seas. In the following section I shall first discuss model results for the variability of the oceanic fresh water export of the Arctic Ocean through Fram Strait. The variability and the different model results for the mean export can only be understood by addressing the interior Arctic Ocean variability that leads to fresh water transport fluctuations. This will be the topic of the third section while the fourth section presents model estimates for the Atlantic inflow through Fram Strait and the Barents Sea and possible consequences of changes in these inflows. The final section gives a summary and lists some open questions relating to modelling the Arctic Ocean - Nordic Seas exchanges.

2. Variability of the oceanic fresh water transport through Fram Strait

Ocean-sea ice models driven with realistic atmospheric forcing exhibit considerable fluctuations in the strength and the properties of boundary currents leading from the Arctic into the Nordic Seas and the sub polar Atlantic. Fig. 1 shows an example from the GK [10] model that was forced with ECMWF surface fields for the period 1986-1992. The section cuts through the Labrador Sea and the Irminger Sea at approximately 63°N, showing the signatures of the Labrador Current, the West Greenland Current and the East Greenland Current (EGC). The salinity in these currents varies from year to year with e.g. a minimum salinity in the EGC of 34.0 in 1986 and again in 1992 compared to 34.5 in 1988. This pronounced interannual variability is especially significant because the boundary currents represent the pathway of Arctic Ocean anomalies to lower latitudes where they can influence deep water mass formation and, through changes in the ocean-atmosphere heat flux, the atmospheric circulation.

The variability in the model is caused by fluctuations of fresh water export from the Arctic. The export occurs as sea ice and liquid water. In the annual mean, sea ice forms in the Arctic and melts predominantly in the marginal ice zone of the EGC where melt rates reach several meters per year [12]. Considerable fluctuations in the ice export through Fram Strait have been observed and the time series have been extended using empirical relationships [13,11]. Especially large transports occur in 1983, 1989, and 1995. A result from a numerical ocean-sea ice model is shown in Fig. 2 (C. Köberle, pers. communication): The average of 0.114 Sv (or 3600 km^3/yr) is in good agreement with the estimated mean of 3372 km^3/yr for the years 1990 through 1994 [13]. Superimposed is a long term variability with maximum ice transport of 0.16 Sv in 1982, a minimum of 0.09 Sv in 1986, and another maximum of 0.19 Sv in 1989. Very similar results were found in the coupled model of Zhang et al. [14] (hereafter ZRS) that was forced with daily data for the period 1979 through 1996. In this model maxima occur in 1981, 1989 and an especially large export in 1995 (Zhang, pers. comm.). The time dependence is in good agreement with observations and the reconstructed time series of Vinje et al. [13]. Maximum ice export happens in years when the NAO index is high. There seems to be little or no time lag between the forcing and the response of the sea ice in the two models for the simulated time period.

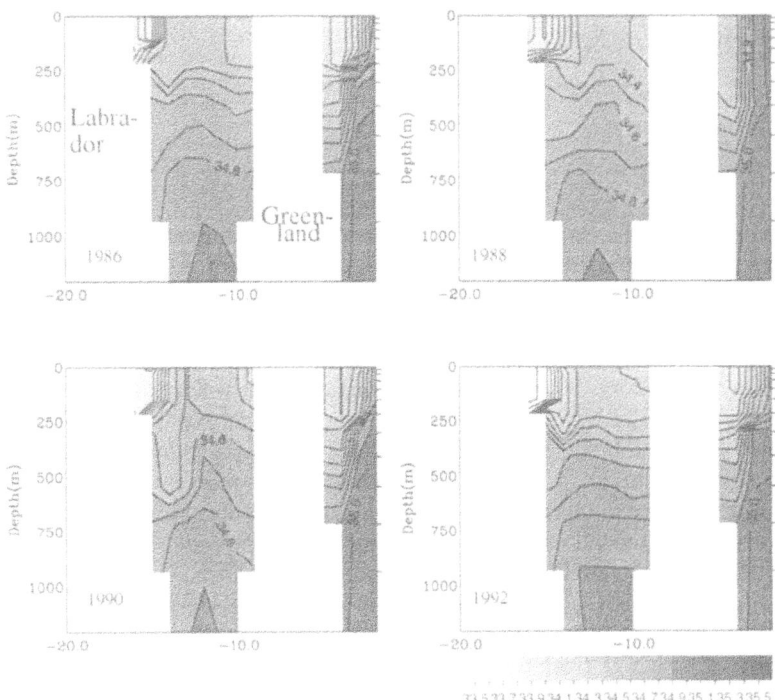

Figure 1. Salinity section through the Labrador Sea and the Irminger Sea at approximately 63°N looking north. It shows the signatures of the Labrador Current, the West Greenland Current and the East Greenland Current (EGC). The results for four years are taken from the model of Gerdes and Köberle [10] that in this case was forced with atmospheric surface fields from the ECMWF analysis for the period 1986-1992. Contour interval is 0.1.

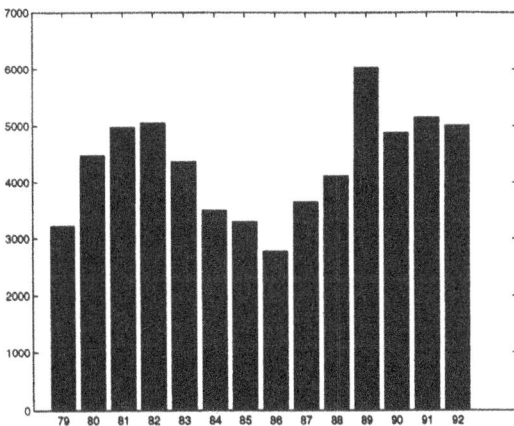

Figure 2. Time series of southward sea ice transport in Fram Strait from the GK [10] model. The forcing in this case was taken from the ECMWF reanalysis data set. Units are km³/year.

The oceanic fresh water transport can be calculated relative to a prescribed reference salinity S_o (here $S_o=34.8$) as:

$$FWT_{ocean} = \int\limits_{l_E}^{l_W} \left(\int\limits_{-z_0}^{0} \frac{S_0 - S}{S_0} dz \right) u_n dl \qquad (1)$$

where l_E and l_W denote the eastern and western boundaries of Fram Strait, respectively, z_o is the thickness of the fresh layer near the surface (here taken as 200m), and u_n is the velocity component perpendicular to the integration path with its positive direction defined as out of the Arctic Ocean. The results from the GK [10] and ZRS [14] models are shown in Fig. 3. Amplitude and phase of the variability of the oceanic fresh water export are similar in the models. The phase of the oceanic fresh water transports is slightly delayed compared to the NAO index (Fig. 4), suggesting an oceanic response time of perhaps 1 to 2 years that is somewhat slower than for the sea ice component. For instance, the ZRS model simulates an especially large oceanic fresh water export during 1995 and 1996, while the sea ice exhibits an extremely high export in 1995, almost collapsing in the following year. Both models show an increasing trend that could be caused by a still existing imbalance in the model or the increasing cyclonic tendency of the wind field over the integration period [15]. A similar analysis of oceanic fresh water transport through Fram Strait has been made for a simulation of the 1960-1975 period by Häkkinen [16]. Her integration shows high fresh water export during the 1960s with a maximum anomaly of more than 400 km³/yr additional southward fresh water transport in 1966 (Fig. 5), immediately preceding the Great Salinity Anomaly in the Nordic Seas and the Sub polar North Atlantic [17]. This is remarkable since the whole period was marked by a consistently low NAO index and the results thus differ from the simulations for the more recent period, suggesting that different oceanic processes are important during different periods.

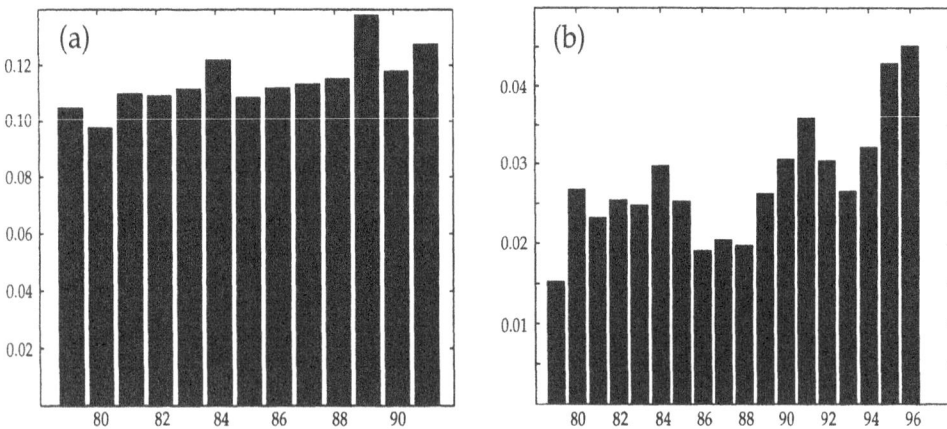

Figure 3. Oceanic fresh water transport in the upper 200m of Fram Strait using a reference salinity of 34.8. Positive values indicate transport out of the Arctic Ocean. Results are taken from a) the GK [10] model forced with ECMWF reanalysis data, and b) the ZRS [14] model. Units are Sv (1Sv= $10^6 m^3 s^{-1}$) with 0.1Sv=3155km³/year. Note the different scales.

Depending on the mean circulation in the Arctic Ocean, models can produce very different estimates of mean oceanic fresh water transport through Fram Strait. The GK [10] model predicts an average oceanic export of 3600 km³/yr while the ZRS [14] model predicts 874 km³/ yr for a similar integration period. Low salinity water arrives at Fram Strait from the interior Arctic with an oceanic transpolar drift that is aligned with the Mendeleyev-Alpha Ridge in the GK model (see Gerdes and Schauer [18], their Fig. 9) . Thus the average salinity of the Arctic Ocean outflow at Fram Strait is around 33.5 in the upper 200 m. Together with a surface intensified current with 3 Sv southward transport this results in the large oceanic fresh water export in that model. The near surface circulation in the ZRS model on the other hand contains a westward current through the Nansen Basin (see Fig. 8 below) that carries relatively saline water toward Fram Strait. Although the ZRS prediction is similar to the estimate of Aagaard and Carmack [19] one should note that their estimate is based on an outflow salinity of 33.7 and a volume transport of 1 Sv in the EGC. These low salinities are probably due to a large contribution of Pacific type water from the Canadian Basin [20].

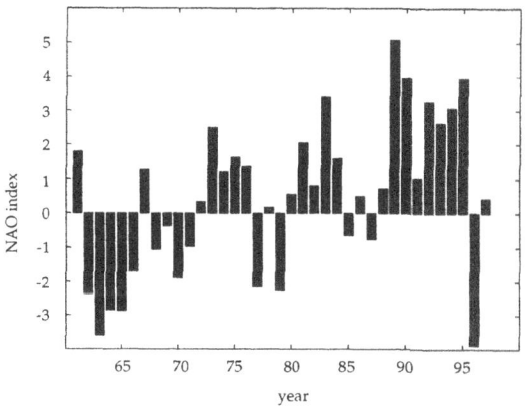

Figure 4.   North Atlantic Oscillation index for the months December through March. The index is based on the difference of normalized sea level pressures between Lisbon and Iceland. The pressure anomalies were normalized by the long time standard deviation of the individual time series (compare [21]).

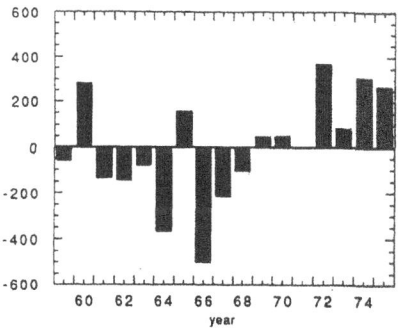

Figure 5.   Oceanic fresh water transport anomalies through Fram Strait for the 1960-1975 period from the model of Häkkinen [16]. Only transports in the upper 30m are integrated. Units are km³/ year. Negative values indicate southward transport.

538

3. Arctic Ocean variability

To understand oceanic fresh water transport variability in Fram Strait we have to consider large-scale current and water mass variability in the interior of the Arctic Ocean. Häkkinen [16] modelled high oceanic fresh water export during a period of low NAO index in contrast to the GK [10] and ZRS [14] models. Events of high fresh water export in her model are not due to enhanced volume flux because that shows little variability during the period of highest fresh water export from the Arctic Ocean (her Fig.10c). On the other hand, several low salinity anomalies could be traced that originated in the East Siberian Sea and then propagated westward to the Laptev Sea, north of Franz-Josef-Land, and finally appearing in Fram Strait, some 3-4 years later (Fig. 6). Since run-off and precipitation have no year-to-year variability in this simulation, the salt anomalies result from wind forced interannual variations in sea ice melt and formation. For this kind of variability one cannot expect a simple correlation with the atmospheric forcing and the signal in Fram Strait. A lagged correlation or, because of combined changes in the circulation and in the source region, a more complicated relationship is to be expected.

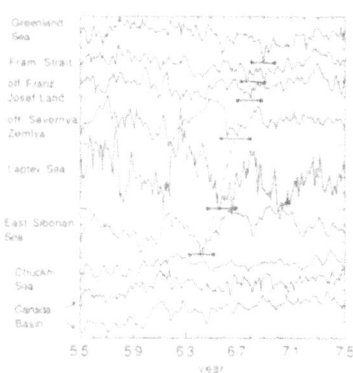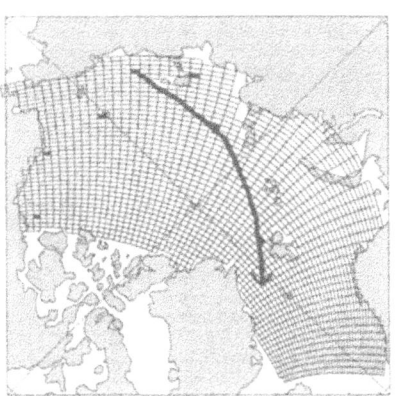

Figure 6. Propagation of salinity anomalies in the Häkkinen [16] model. The path of the anomaly originating 1964 in the East Siberian Sea is highlighted.

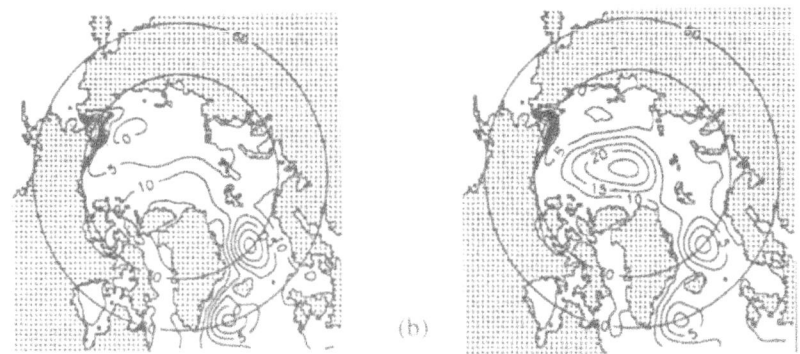

Figure 7. Mean annual surface height (in cm) from the Proshutinsky and Johnson [22] model for (a) 1974 and (b) 1977.

With a simple wind driven model Proshutinsky and Johnson [22] investigate the response of a barotropic model to wind forcing derived from the National Meteorological Center daily surface pressure data. The high NAO index year 1974 (Fig. 7a) for instance shows weak cyclonic flow in the interior Arctic and a strong cyclonic circulation in the Nordic Seas. For the low index year 1977 (Fig. 7b) Proshutinsky and Johnson find strong anticyclonic circulation in the Arctic and a weak circulation in the Nordic Seas. The response seems local to the basins with little changes in Fram Strait. This could be a result of the simplifications in the model. The results are broadly consistent with the sea surface height composites that Dickson et al. [11] constructed for different NAO index periods from hydrographic data. During low NAO index conditions surface height in the Arctic Ocean indicates a broad and weak transpolar drift and a relatively large and intense Beaufort Gyre. The cyclonic gyre in the Nordic Seas is weak. During high NAO index conditions the Beaufort Gyre is displaced toward Alaska and the transpolar drift is intense and focused toward Fram Strait. The cyclonic circulation in the Nordic Seas is strong. This suggests intensified exchanges between the Arctic Ocean and the Nordic Seas during high index years. However, there is no data available for a similar analysis in Fram Strait itself. Neither the simple model of Proshutinsky and Johnson [22] nor the surface height field of Dickson et al. [11] thus gives sufficient information about the exchanges between the Arctic Ocean and the Nordic Seas through Fram Strait.

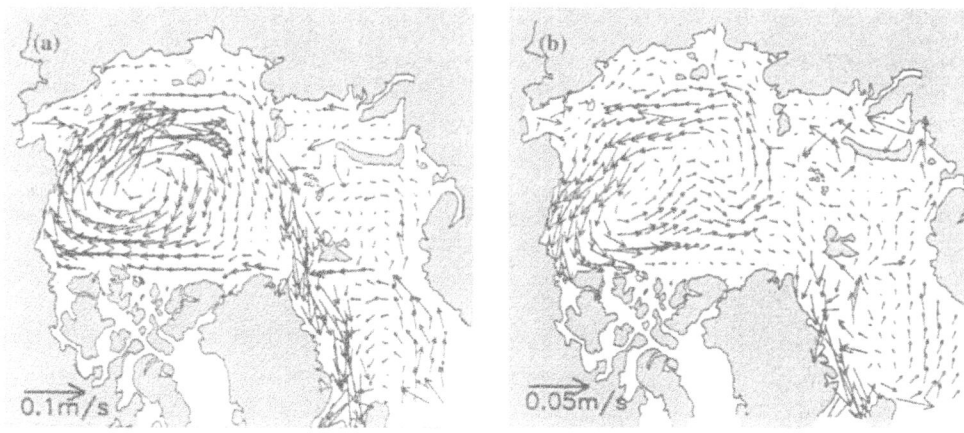

Figure 8.   (a) 1979-88 mean ocean velocities at the surface from the ZRS [14] model,
(b) differences between 1989-96 mean velocities and 1979-1988 mean velocities

A general circulation model can be used to investigate the relationship between interior Arctic response to the large scale atmospheric forcing and the transport fluctuations in Fram Strait. Zhang et al. [14] simulate the periods 1979-1988 and 1989-1996 using a 40 km resolution, 21 level sea ice-ocean general circulation model. Forcing data is taken from the International Arctic Buoy Program and the NCEP analysis. A generally more cyclonic wind field characterizes the second period over the Nordic Seas and the Arctic Ocean. The resulting ocean surface circulation is shown in Fig. 8. During the more recent period a general cyclonic tendency over the whole domain is visible, weakening the anticyclonic circulation in the Canadian Basin and strengthening the cyclonic circulation in the Nordic Seas. This response is as expected from data analysis

540

and simple models, although the reduced transpolar drift during high index years is in contrast to Dickson et al.'s [11] picture derived from surface height data. The response to the more cyclonic wind forcing during 1989 through 1996 compared to 1979 through 1988 consists of cyclonic circulation cells in the Arctic Ocean and the Nordic Seas (Fig. 8b). In addition, there is enhanced communication between the basins through the Barents Sea and Fram Strait. The source for the waters making up the enhanced Fram Strait outflow is apparently the Canadian Basin whereas a transpolar drift from the Eurasian Basin dominates the mean circulation. The anomalies thus exhibit a behaviour that differs from the mean circulation. If one subtracts the local recirculations in the Arctic Ocean and the Nordic Seas, the remaining pattern is a cyclonic loop along the margin of the domain, from the Norwegian Sea, the Barents Sea, along the Siberian shelf seas and the Canadian Arctic to Fram Strait. The situation is similar at deeper levels, e.g. the 500 m level shown in [14] (their Fig. 2). The anomalous cyclonic circulation loop exists in a similar fashion in the GK [10] model. In both models it appears that this loop is switched on almost immediately, resulting in a response of the Arctic Ocean-Nordic Sea exchanges that is in phase with the atmospheric forcing (see above, Figs. 3 and 4).

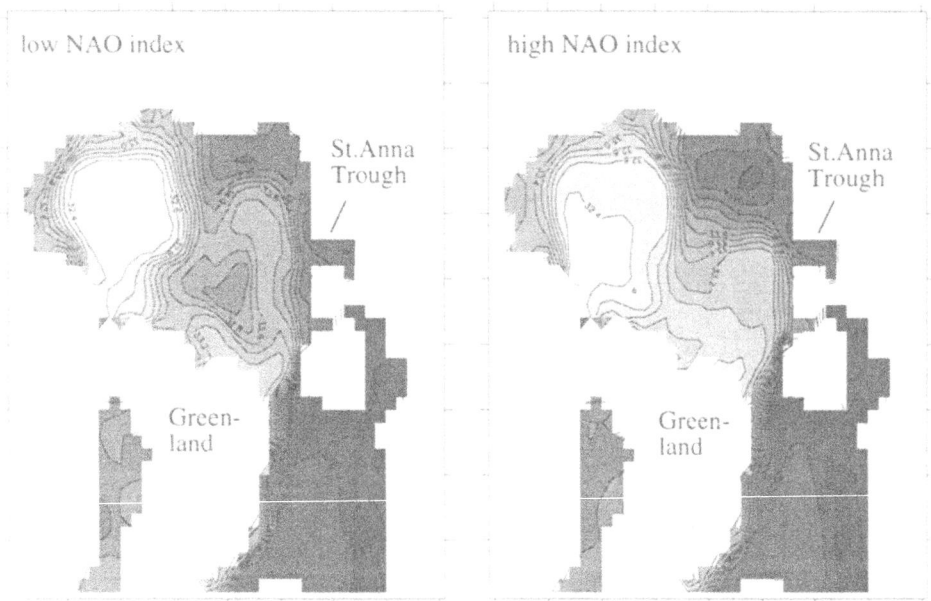

Figure 9. Annual mean salinity distribution at 211m from the GK [10] model forced with ECMWF reanalysis data, (a) 1985, (b) 1989

The differences between the Häkkinen [16] experiment for the forcing period 1960-1975 and the ZRS [14] and GK [10] models for more recent forcing periods could be due to the different forcing periods or properties of the different models used. Another possibility is an interplay of circulation changes and water mass properties, i.e. the correlation of velocity and salinity anomalies at Fram Strait, which might vary over long time scales. The example from the GK model (Fig. 9) shows the salinity distribution at 200 m depth for 1985 (a low NAO index year with small Fram Strait fresh water export) and 1989 (a high NAO index year). During the high index year, the salinity distribution is characterized by a strong outflow from the Barents Sea that generates high salinities in the eastern Eurasian Basin. A tongue of saline water can be seen flowing along the Lomonosov Ridge toward Fram Strait. Note that at 200 m the Barents Sea outflow is saline compared to the ambient water while it is relatively fresh at deeper levels [18]. Closer to Fram Strait, the salinity of the out flowing water is relatively low during the high NAO index year such that high transport and low salinities combine in a large oceanic fresh water export. In the low index year, outflow from the Barents Sea is relatively weak and salinity in the eastern Eurasian Basin correspondingly low. The remnant of previous strong outflow from the Barents Sea is still present in the form of a lens of saline water near the North Pole. This contributes to relatively high salinities at Fram Strait and also appears to block the flow of fresh water from the Canadian Basin toward Fram Strait. The travel time from the Barents Sea to Fram Strait along the rim of the Eurasian Basin is approximately ten years, comparable to the period between high inflow events during the last 30 years. Changes in the periodicity in the forcing - as the long term changes in the NAO index (Fig. 4) indeed suggest - and changes in the forcing patterns can interrupt this relationship and result in different behaviour of the oceanic fresh water transport through Fram Strait. This may explain the apparent shift in regimes between the 1960s and 70s considered in Häkkinen's [16] experiment and the 1980s and 90s simulated in ZRS [14] and GK [10].

4. Atlantic inflow through Fram Strait and Barents Sea

The Atlantic inflow into the Arctic Ocean takes two routes, the West Spitsbergen Current (WSC) through Fram Strait and a flow from the Norwegian Sea into the Barents Sea. The division in Barents Sea and Fram Strait inflows is important because the two branches influence the interior Arctic Ocean hydrography and circulation in different ways. Fram Strait inflow supports a layer of warm and saline water in the Arctic Ocean; fluctuations in this inflow will affect the extent and the properties of this layer [2,3,6]. The Barents Sea branch, on the other hand, feeds a cold and relatively fresh boundary current at intermediate depths that begins at the St. Anna Trough and continues along the rim of the Eurasian Basin [8,9,23]. Anomalies generated in the Nordic Seas or further upstream in the Atlantic will experience very different fates whether they take the Barents Sea or the Fram Strait branch because water mass transformation differs for the two branches. Signals entering Fram Strait become protected from further modification through exchanges with the atmosphere by the halocline that exists near the surface north of Fram Strait. Recent observations reported by Steele and Boyd [5] and Rudels et al. [24] indicate a retreat of the cold halocline layer from the western Eurasian Basin that could lead to a change in these conditions. However, so far warm and saline conditions in the inflow will survive until far into the Arctic proper. Following the path of the WSC into the Arctic the surface heat flux in the AWI (Alfred-Wegener-Institut) high-resolution sea ice-ocean model (Fig. 10) diminishes quickly. Large oceanic heat losses, on the other hand, take place in the Barents Sea. Once entering the shallow Barents Sea the water is exposed to strong cooling that results in convective overturning. Convection is maintained in spite of fresh water fluxes from melting of ice that enters the region from the Nansen Basin [12]. A warm and saline signal entering the Barents Sea is thus transformed into a cold and less saline signal. Close to the inflow into the Barents Sea, temperature fluctuations in phase with upstream signals in the Norwegian Sea and also in phase with the NAO wind

542

anomalies have been observed (see [6] for an overview). However, the water that exits from the Barents Sea into the Arctic Ocean at the St. Anna Trough is close to the freezing point, i.e. the original temperature anomaly has been lost. The temperature distribution at 300 m depth for the AWI high-resolution model (Fig. 11) shows this transformation. Atlantic water with 7°C in the Norwegian Current enters the Barents Sea trough the Bear Island Trough. Water flowing back out of the Bear Island Trough is already much colder and contributes to a drop in core temperature of the Atlantic water that continues toward Fram Strait. The outflow through the St. Anna Trough is close to the freezing point, considerably lower than the Fram Strait branch at that position. Due to the contributions from the Norwegian Coastal Current and melt water, this outflow is also less saline than the Fram Strait branch. An increase of the flow through the Barents Sea, as would be expected during high NAO index atmospheric conditions, thus yields the counterintuitive result of an increase in cold and fresh water inflow into the Arctic Ocean. Note that this is not the case for shallower levels where the flow into the Arctic Ocean is saline compared to the ambient water (see Fig. 9).

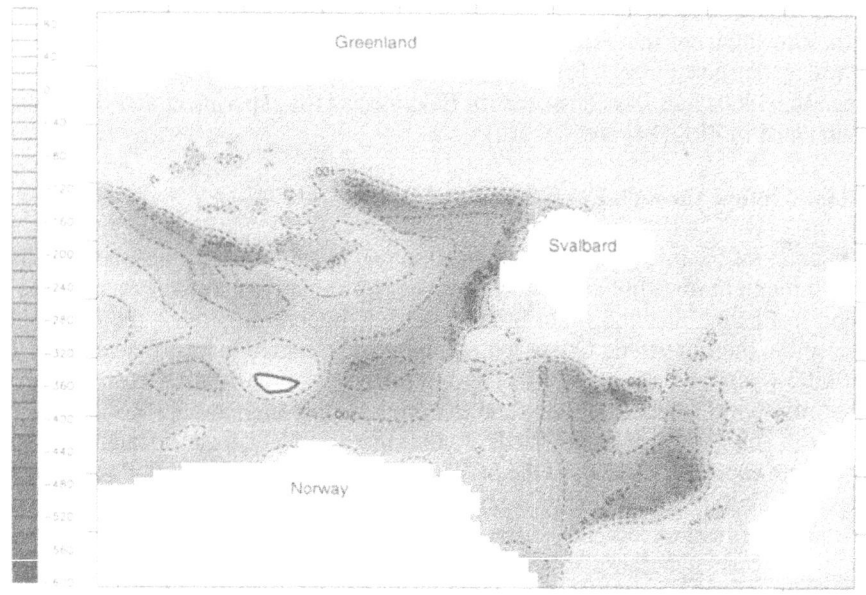

Figure 10. Surface heat flux (in W/m^2) for January 1982 (ECMWF re-analysis forcing) in the AWI high-resolution ocean-sea ice model.

There is a wide range of observational estimates for the Atlantic inflows into the Arctic Ocean (see [25] for a compilation). A re-evaluation of the available observational evidence indicates that the Barents Sea inflow is a major pathway for Atlantic water, at least as strong as the Fram Strait inflow [26]. Numerical models also show a wide range of transport estimates. Aukrust and Oberhuber [27] simulate a transport of 1 Sv in the West Spitzbergen Current (WSC) and an outflow of slightly above 3 Sv with the EGC. The oceanic fresh water export through Fram Strait of $4282km^3/yr$ is among the highest estimates in numerical models. A similarly high export is only found in the Hadley Center climate model [28] (4030km³/ yr) and in the GK model (3600km³/yr)[1]. In a rather coarse resolution model, Holland et al. [29] find an inflow of 2.5 Sv through the Barents Sea and almost no northward transport through Fram Strait. In a high-resolution version of the Oberhuber [30] model, Karcher and Oberhuber (1998, Modeling the ventilation of the upper and intermediate water of the Arctic Ocean with an isopycnic model, private communication) find 2.6 Sv inflow through the Barents Sea and 1.7 Sv through Fram Strait.

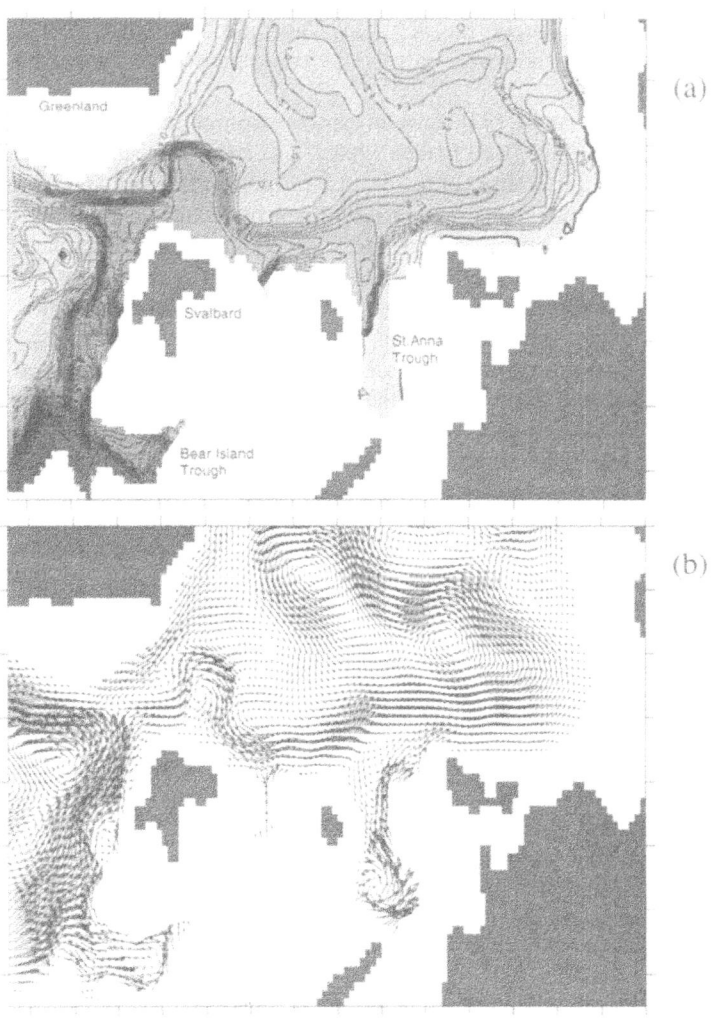

Figure 11.   Potential temperature (a) and circulation (b) at 300 m depth in the AWI high-resolution ocean-sea ice model.

The fresh water export through Fram Strait in this model is 1270km^3/yr. At the end of a simulation of ten cycles of the 1979-1985 forcing period Zhang et al. [31] produce an inflow of 0.8Sv through the Barents Sea while 1.2Sv enter the Arctic Ocean through Fram Strait. Using daily surface forcing ZRS find an increasing Barents Sea inflow, starting with 0.6 Sv in 1979 and ending with 1.6 Sv in 1996 while the Fram Strait inflow remains more or less constant at 2.5 Sv. In the standard case of Gerdes and Schauer [18] the inflow through the Barents Sea is 3.2Sv, more than three times larger than through Fram Strait. The very high-resolution model of Maslowski et al. [32] shows a net inflow into the Barents Sea of 2.2 Sv. This model contains a WSC that carries 3.5 Sv towards Fram Strait, however, only 1.5 Sv continue with the boundary current into the Arctic Ocean proper.

Maslowski et al. [32] show an especially pronounced example of recirculation of Atlantic Water in Fram Strait in their Plate I. The local recirculation of Atlantic Water is a problem that makes estimates of the Fram Strait inflow difficult and hard to compare. The exact position for the transport estimate influences the results such that spatial variations in hydrographic sections or current meter arrays may obscure natural fluctuations. The amount of recirculation in comparison to the through flow and the variability of this ratio will be important for the variability of the oceanic fresh water transport.

There are several possible reasons for the differences between models. Differences in parameterizations of subgridscale processes and in spatial resolution of the model are especially important for the local details of the circulation in the troughs and straits. Net exchanges between the Arctic Ocean and the Nordic Seas might be less affected because they are presumably determined by large scale forcing functions. Whether this is true or not is, as in many other cases in oceanography, an open and very important question because the answer will determine the numerical requirements for models of the high latitude Atlantic. Numerical properties of models are different, however, without direct intercomparisons it is hard to judge the effects of different numerics. Only a few such direct intercomparisons have been made, regarding vertical resolution and the advection scheme for temperature and salinity [18] as well as subgrid scale parameterizations like the "neptune" effect [33].

It is known that upstream conditions have a considerable influence on the simulated Atlantic inflow into the Arctic Ocean. Mellor and Häkkinen [34] report model sensitivity with respect to inflow boundary conditions in the Nordic Seas in their regional model of the Arctic. With an inflow of 7 Sv uniformly distributed over the boundary between Iceland and Norway they find almost no inflow through Fram Strait at 500 m depth and too cool Atlantic water in the interior of the Arctic Ocean. With the inflow velocity profile modified such that the transport occurs in a strong barotropic current near the Norwegian coast, the flow through Fram Strait is much enhanced. Such sensitivities suggest the possible effect of remote forcing of long term fluctuations in the Fram Strait transport.

The stronger inflow via the Barents Sea branch in recent model results could also be due to the atmospheric forcing that was used to drive the models. As pointed out earlier, the wind stress over the Arctic Ocean has changed considerably during recent decades [15,35]. The cyclonic tendency in wind forcing tends to emphasise the Barents Sea branch as its increasing transport during the ZRS simulation for the 1979-1996 forcing period indicates. Forcing with earlier, more anticyclonic wind fields will yield a relatively low inflow through the Barents Sea.

5. Future research priorities

Most models agree that the mean oceanic fresh water transport is considerably lower than the mean ice transport through Fram Strait. This is even more so for the transport fluctuations where serious changes in sea ice transport have been observed and modelled for recent decades. This emphasises the role of sea ice for the variability and stability of the large-scale circulation because deep-water formation and thus the thermohaline circulation are susceptible to changes in the fresh water supply from the Arctic.

A few models, however, give much larger mean oceanic fresh water transports [10, 27, 28] while exhibiting similar variability. The large scale Arctic Ocean circulation delivers relatively fresh water from the Canadian Basin at Fram Strait in these models. The interior pathway in these models is more consistent with recent observations of an eastward shift of the Atlantic - Pacific water mass boundary in the Arctic Ocean [4, 20]. These shifts hint at the possibility of a transient or permanent shift in Arctic Ocean circulation patterns with consequences for fresh water export. However, the reasons for the differences among models are widely unknown. It is important to identify numerical and conceptual differences that lead to different responses. Model intercomparisons can be used to determine the necessary resolution and the effect of different parameterizations and numerical representations in different models.

Whether the recent changes in the Arctic climate present a permanent shift or a phase of a long term oscillation is an important question for the role of the Arctic Ocean in large scale climate developments. It is natural to focus on the NAO as the dominant interannual and interdecadal forcing pattern for the long term variability in the area. Model results do not indicate a unique relationship between the NAO and fresh water outflow through Fram Strait. Some models show a switch-on behaviour of a cyclonic current loop along the rim of the whole Nordic Seas - Arctic Ocean domain that reacts with high outflow to high NAO index forcing. On the other hand, propagating salinity anomalies with travel times between 3 and 10 years between source region and Fram Strait [10,16] seem to prevent such a close relationship. However, travel times close to the periodicity of the forcing can enhance the fresh water flux signal in the outflow. Important questions are whether different modes of variability exist in the Arctic Ocean - as suggested by the model results - how they depend on the local atmospheric forcing, upstream conditions in the Atlantic, and what phase relationship exists between these events and the forcing. It is not sufficient to concentrate on the interior Arctic Ocean because the inflow of Atlantic water into the Arctic Ocean varies and the pathways of this inflow - Barents Sea or Fram Strait - vary with consequences for the fate of the anomalies carried by the branches.

Many of the dynamical relationships between forcing and transport fluctuations at individual passages are poorly known; the sensitivity and feedbacks need quantification before reliable estimate for long-term developments can be made. Comprehensive numerical models of the climate system or its subsystems are an important means to achieve this goal. Here we have highlighted the variance of different numerical results and possible causes. While numerical properties and differences in resolution and parameterizations can certainly not be neglected, it appears that many of the differences actually reflect different modes of the system.

546

Acknowledgements

I thank Jinlun Zhang for providing the oceanic fresh water flux data from his model, Cornelia Köberle and Michael Karcher for making available new model results, Markus Harder for providing his ice model code, and Frank Röske for the ECMWF data. This work was in part funded by the BMBF through grant 07 VKV01 and by the European Commission Mast III Programme through contract MAS3-CT96-0070 (VEINS). This is contribution No. 1566 of the Alfred-Wegener-Institut für Polar- und Meeresforschung, Bremerhaven.

References

1. Carmack, E.C., Macdonald, R.W., Perkin, R.G., McLaughlin, F.A., and Pearson, R.J. (1995) Evidence for warming of Atlantic Water in the southern Canadian Basin of the Arctic Ocean: results from the Larsen-93 expedition, Geophysical Research Letters, **22**, 1061-1064
2. Morison, J., Steele, M., and Anderson, R. (1998) Hydrography of the upper Arctic Ocean measured from the nuclear submarine USS Pargo, Deep-Sea Research, **45**(1), 15-33
3. Carmack, E.C., Aagaard, K., Swift, J.H., Macdonald, R.W., McLaughlin, F.A., Jones, E.P., Perkin, R.G., Smith, J.N., Ellis, K.M., and Killius, L.R. (1997) Changes in temperature and tracer distributions within the Arctic Ocean: results from the 1994 Arctic Ocean section, Deep-Sea Research, **44**(8), 1487-1502
4. McLaughlin, F.A., Carmack, E.C., Macdonald, R.W., and Bishop, J.K.B. (1996) Physical and geochemical properties across the Atlantic/Pacific water mass front in the southern Canadian Basin, J.Geophys.Res. **101**(C1), 1183-1197
5. Steele, M. and Boyd, T. (1998) Retreat of the cold halocline layer in the Arctic Ocean, J.Geophys.Res., **103** (C5), 10419-10435
6. Grotefendt, K., Logemann, K., Quadfasel, D., and Ronski, S. (1998) Is the Arctic Ocean warming? J.Geophys.Res., **103** (C12), 27679-27687
7. Schauer, U., Rudels, B., Jones, E.P., Anderson, L.G., Muench, R.D., Björk, G., Swift, J.H., Ivanov, V., and Larsson, A.-M. (1998) Water mass distribution in the eastern Eurasian Basin and connections with the Canadian Basin: Results from ACSYS-96, J.Geophys.Res. (submitted)
8. Rudels, B., Jones, E.P., and Anderson, L.G. (1994) On the Intermediate Depth Waters of the Arctic Ocean, The Polar Oceans and their Role in Shaping the Global Environment (Johannessen, O.M., Muench, R.D., Overland, J.E.), American Geophys. Union, Washington, 85, 33-46
9. Schauer, U., Muench, R.D., Rudels, B., and Timokhov, L. (1997) Impact of eastern Arctic shelf waters on the Nansen Basin intermediate layers, J.Geophys.Res. **102**, 3371-3382
10. Gerdes, R. and Köberle, C. (1998) Numerical simulation of salinity anomaly propagation in the Nordic Seas and the Arctic Ocean, Polar Research (submitted)
11. Dickson, R.R., Osborn, T.J.,Hurrell, J.W., Meincke, J., Blindheim, J., Adlandsvik, B., Vinje, T., Alekseev, G., Maslowski, W., and Cattle, H. (1998) The Arctic Ocean Response to the North Atlantic Oscillation, J.Climate (in press)
12. Steele, M. and Flato, G.M. (1999) Sea ice melt, growth, and modeling: A review (this volume)
13. Vinje T., Nordlund, N., and Kvambekk, A. (1998) Monitoring ice thickness in Fram Strait, J.Geophys.Res., **103**(C5), 10437-10449
14. Zhang, J., Rothrock, D.A., and Steele, M. (1998), Warming of the Arctic Ocean by a strengthened Atlantic Inflow: Model results, Geophysical Res. Letters, **25**(10), 1745-1748
15. Walsh, J.E., Chapman, W.L., and Shy, T.L. (1996) Recent decrease in sea level pressure in the central Arctic, J.Climate, **9**, 480-486
16. Häkkinen, S. (1993) An Arctic source for the Great Salinity Anomaly: A Simulation of the Arctic ice-ocean system for 1955-1975, J.Geophys.Res., **98** (C3), 16397-16410
17. Dickson, R.R., Meincke, J., Malmberg, S.-A., and Lee, A.J. (1988) The "Great Salinity Anomaly" in the northern North Atlantic 1968-1982, Prog.Oceanogr., **20**, 103-151
18. Gerdes, R. and Schauer, U. (1997) Large-scale circulation and water mass distribution in the Arctic Ocean from model results and observations, J.Geophys.Res., **102**(C4), 8467-8483
19. Aagaard, K. and Carmack, E.C. (1989) The role of sea ice and other fresh water in the arctic circulation. J.Geophys.Res., **94**, 14485-14498

20. Jones, E.P., Anderson, L.G., and Swift, J.H. (1998) Distribution of Atlantic and Pacific waters in the upper Arctic Ocean: Implications for circulation, Geophysical Research Letters, **25**(6), 765-768

21. Hurrell, J.W. (1995) Decadal trends in the North Atlantic Oscillation: Regional temperatures and precipitation, Science, **269**, 676-679

22. Proshutinsky, A.Y. and Johnson, M.A. (1997) Two circulation regimes of the wind driven Arctic Ocean, J.Geophys.Res., **102** (C6), 12493-12514

23. Swift, J.H., Jones, E.P., Aagaard, K., Carmack, E.C., Hingston, M., Macdonald, R.W., McLaughlin, F.A., and Perkin, R.G. (1997) Waters of the Makarov and Canada basins, Deep-Sea Research, **44**(8), 1503-1529

24. Rudels. B., Schauer, U., and Muench, R.D. (1997) Deep saline mixed layer and temporal variations in the Eurasian Basin water column. Proceedings of the Conference on Polar Processes and Global Climate, Rosario, USA, ACSYS Office, Oslo, 220-222

25. Simonsen, K. and Haugan, P.M. (1996) Heat budgets of the Arctic Mediterranean and sea surface heat flux parameterizations for the Nordic Seas, J.Geophys.Res., **101**(C3), 6553-6576

26. Rudels, B., and Friedrich, H.J. (1999) The transformations of Atlantic Water in the Arctic Ocean and their significance for the freshwater budget (this volume)

27. Aukrust, T. and Oberhuber, J.M. (1995) Modeling of the Greenland, Iceland, and Norwegian Seas with a coupled sea ice-mixed layer-isopycnal ocean model, J.Geophys.Res., **100** (C3), 4771-4789

28. Cattle, H. (1999) The Arctic Ocean freshwater budget of a climate general circulation model (this volume)

29. Holland, D.M., Mysak, L.A., and Oberhuber, J.M. (1996) Simulation of the mixed layer circulation in the Arctic Ocean, J.Geophys.Res., **101** (C1), 1111-1128

30. Oberhuber, J.M. (1993) Simulation of the Atlantic circulation with a coupled sea ice-mixed layer-isopycnal model, I, Model description, J.Phys.Oceanogr., **23**, 808-829

31. Zhang, J., Hibler, W.D., Steele, M., and Rothrock, D.A. (1998) Arctic ice-ocean modeling with and without climate restoring, J.Phys.Oceanogr., **28**, 191-217

32. Maslowski, W., Parsons, A.R., Zhang, Y., and Semtner, A.J. (1998) High resolution Arctic Ocean and sea ice simulations. Part I: Ocean model design and early results, J.Geophys.Res. (in press)

33. Nazarenko, L., Sou, T., Eby, M., and Holloway, G. (1997) Arctic Ocean-ice system studied by contamination modelling. International symposium on representation of the cryosphere in climate and hydrological models, Victoria, BC, edited by J.E. Walsh, 17-21

34. Mellor, G.L. and Häkkinen, S. (1994) A review of coupled ice-ocean models, The Polar Oceans and their Role in Shaping the Global Environment (Johannessen, O.M., Muench, R.D., Overland, J.E.), American Geophys. Union, Washington, **85**, 33-46, 1994.

35. Maslanik, J.A., Serreze, M.C., and Barry, R.G. (1996) Recent decrease in Arctic summer ice cover and linkages to atmospheric circulation anomalies, Geophys. Res. Letters, **23**, 1677-1680

SEA ICE GROWTH, MELT, AND MODELING: A SURVEY

MICHAEL STEELE[1] AND GREGORY M. FLATO[2]
[1]Polar Science Center, Applied Physics Laboratory, University of Washington, Seattle, WA, 98105 USA
[2]Canadian Centre for Climate Modelling and Analysis, Atmospheric Environment Service, Victoria, BC, Canada

1. Introduction

The freshwater budget of the Arctic as a whole can be stated quite simply as: inflows minus outflows equals net change within the region. If the region is defined to include all ice-covered seas, then the growth and melt of sea ice is an internal term that has no net effect. This is illustrated schematically for the ice-ocean system in **Figure 1**, where net ice growth in the Arctic Ocean is balanced by net ice melt in the Greenland-Iceland-Norwegian (GIN) Seas.

Yet the phase of freshwater in the Arctic system *does* concern us. Why? Several answers come to mind. First, growth and melt patterns create areas of net salinification and net freshening of the underlying ocean (due to the expulsion of brine by freezing ice), which in turn affects mixed layer properties and water mass formation. The timing and location of these phase changes can affect the variability and ultimate fate of freshwater fluxes. An example is provided by the freshwater transport through Fram Strait, which may influence deepwater formation in the GIN Sea (Aagaard and Carmack [1]). **Figure 2** indicates that the interannual variability of the sea ice flux may be much larger than the oceanic flux (owing to the strong influence of local winds on sea ice drift). Second, ice and liquid water have very different dynamic and thermodynamic properties which affect not just the freshwater but also the momentum and heat budgets of the Arctic regions and their sensitivity to climate change. An example is the "ice-albedo feedback", wherein solar heating decreases sea ice coverage by melting, reducing the albedo which leads to further heating. This feedback contributes to the polar amplification of global warming simulated by coupled climate models, although the details of this feedback and its representation in such models remain somewhat uncertain. Finally, examination of growth and melt patterns can be a useful tool in diagnosing the performance of a numerical sea ice model. Even when observations are scarce, much can be learned by intercomparison of model-generated fields of these fundamental quantities.

E.L. Lewis et al. (eds.), The Freshwater Budget of the Arctic Ocean, 549-587.
© 2000 *Kluwer Academic Publishers.*

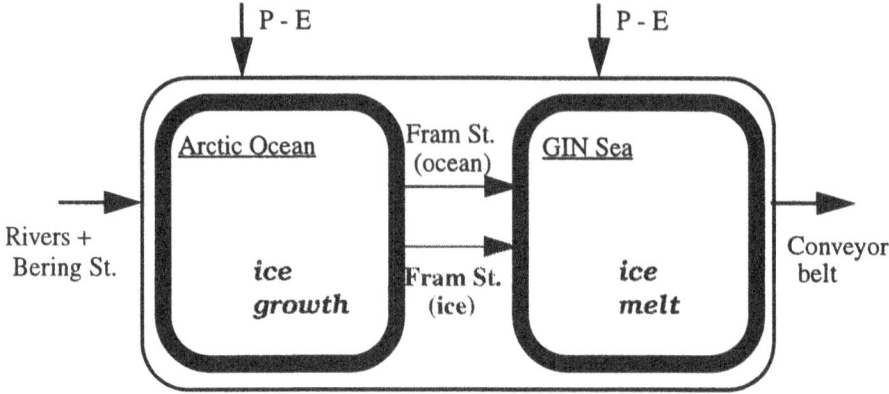

Figure 1. The freshwater budget of the Arctic Ocean and Greenland-Iceland-Norwegian (GIN) Sea. Inputs are from rivers, net precipitation minus evaporation (P-E), and the relatively fresh water that flows northward through Bering Strait from the north Pacific Ocean. The output is denoted "conveyor belt", representing the global thermohaline circulation that brings freshwater southward in the Atlantic Ocean and northward in the Pacific Ocean, as described for example by Wijffels et al. [2]. Net ice growth in the Arctic Ocean is balanced by export southward through Fram Strait and net ice melt in the GIN Sea. Thus growth and melt is an internal term if the "system" is defined to include all of the ice covered area.

Here we present a survey of ice thermodynamics and their role in the simulation of growth/melt cycles in one-dimensional and two-dimensional numerical models. These models typically involve many interacting processes, both dynamic and thermodynamic. We focus here on the latter. Further, our emphasis is on just one component of the ice mass balance, namely, the seasonal cycle of growth and melt and its spatial variability over the Arctic regions. Thus we are not directly concerned with the models' representations of typical state variables such as mean ice drift or even mean ice thickness, except as they relate to growth and melt. Our focus is on the large spatial scales (i.e., the Arctic Ocean) and long time scales (seasonal to interannual) relevant to global climate. Topics not discussed or mentioned only briefly include small-scale (molecular, crystal) thermodynamics, biological and sediment influences, and Southern Hemisphere issues.

The mass balance of sea ice is conveniently described in terms of the thickness distribution function, $g(h)$, which evolves according to (Thorndike et al. [3]; Hibler [4]):

$$\frac{\partial g}{\partial t} = -\nabla \bullet (u g) + \psi - \frac{\partial (fg)}{\partial h} + F_L \tag{1}$$

where $\int_{h_1}^{h_2} g \, dh$ is the fraction of the ocean surface in a given region that is covered by ice with thickness h between h_1 and h_2. The units of each term in Eq. 1 are m^{-1} s^{-1}. The total areal fraction of ice is $g_I = \lim_{\varepsilon \to 0} \int_\varepsilon^\infty g \, dh$, while $\int_0^\infty g \, dh = 1$. With this definition, $g(0)$ represents the areal fraction of open water. The first moment of $g(h)$ yields the mean ice thickness, i.e. $\int_0^\infty h g \, dh = \bar{h}$. A typical $g(h)$ is shown in **Figure 3**. This represents an

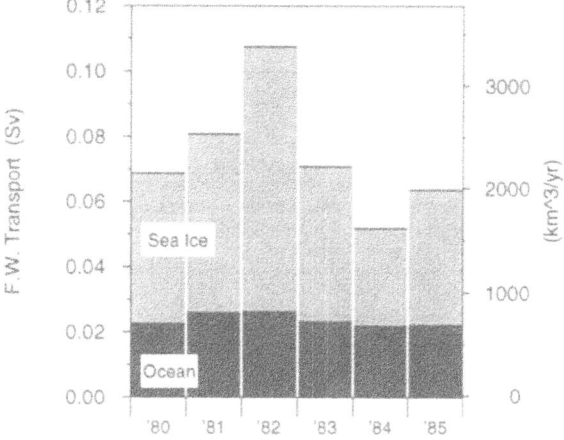

Figure 2. Annual average freshwater fluxes (southward) through Fram Strait during the early 1980's, as simulated by a simple ice-ocean model using a reference salinity of 34.8 (from Steele et al. [5]). The interannual variability of the solid (sea ice) component is much greater than the liquid (oceanic) component, owing to the high correlation between local winds and ice drift.

"end of summer" distribution, in which there is a large peak representing multi-year ice and a smaller peak representing open water. An "end of winter" distribution typically has another peak between the two above, representing first-year ice.

The first two terms on the right hand side of Eq. 1 represent the influence of ice dynamics, where **u** is the vector ice velocity and ψ is a mechanical redistribution term describing the rearrangement of thickness by ridging. The final two terms on the right hand side of Eq. 1 represent thermodynamic processes which are the focus of this study. The parameter f (units of m s^{-1}) is the vertical growth rate (which is negative for melt), while F_L is the lateral melt rate (units of m^{-1} s^{-1} = (m s^{-1}) m^{-2}, i.e., lateral melt rate per unit horizontal area).

One can integrate the first moment of Eq. 1 to obtain an expression for the volume (or mass, assuming constant density) balance of the Arctic Ocean:

$$\frac{d}{dt}\langle V \rangle = \langle G \rangle - \langle O \rangle \qquad (2)$$

where V is ice volume, G is thermodynamic growth, and O is outflow predominantly through the Fram Strait (although the Bering Strait, the Canadian Archipelago and exchanges with the Barents Sea also contribute). Angled brackets indicate an average over the Arctic Ocean (**Figure 4**). Alternatively, if the domain of interest also includes

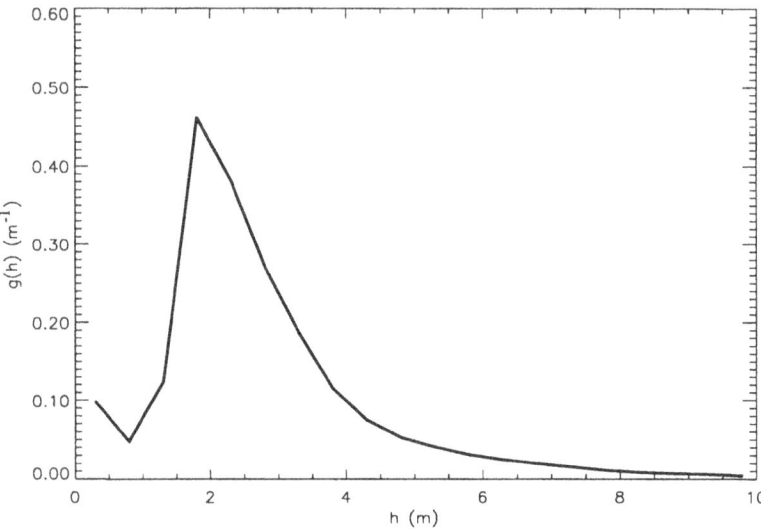

Figure 3. Thickness distribution obtained by the submarine USS Queenfish at the North Pole in August, 1970 (data provided by J. Walsh and described in McLaren et al. [6]).

the GIN Sea where ice is transported out of the Arctic Ocean and melted, then the last term in Eq. 2 is zero as implied by Figure 1.

Observations are insufficient to estimate the time variation of terms in Eq. 2, but one can obtain estimates of the two right-hand-side terms as follows: Koerner [7], based on measurements along an Arctic transect, estimated that overall there is roughly 1.1 m ice growth and 0.6 m melt per year, for a balance of about 0.5 m net growth over an annual cycle (i.e. 0.5 m is an estimate of $\langle G \rangle$ per unit area); Vinje et al. [8] provide an updated estimate of Fram Strait outflow of 2850 km^3 yr^{-1} which, given an area of the Arctic Ocean (not including the Barents and Kara Seas) of roughly 7×10^6 km^2, yields an outflow, $\langle O \rangle$, per unit area of about 0.4 m yr^{-1}. Given the uncertainties in both these estimates, the budget is essentially balanced.

Since temporal and spatial variations in V, G, and O are difficult to observe directly, physically-based numerical models can be used to provide some insight into the magnitude and sources of variability and the processes involved. This chapter is divided into two main sections which examine net growth and melt in one-dimensional (1D) and two-dimensional (2D) models. We first present a short primer on ice thermodynamics, then examine the sensitivity of the seasonal cycle of growth and melt to choices in model physics and forcing, and also review the relevant observational studies. We close with a summary and discussion of outstanding issues and trends in thermodynamic sea ice modeling.

2. Growth and Melt in the "Central Arctic": Theory, Observations, and 1D Models

In this section we briefly summarize the theory of sea ice thermodynamics. The focus is primarily on vertical transfer between top and bottom surfaces, although lateral heat transfer is also discussed. We then review observations and models of the "central Arctic", a somewhat ambiguously defined region that is usually taken as the climatological end-of-summer sea ice extent (Figure 4). This region contains a mixture of perennial (multi-year) and seasonal (first-year) sea ice that evolves according to both dynamic and thermodynamic processes.

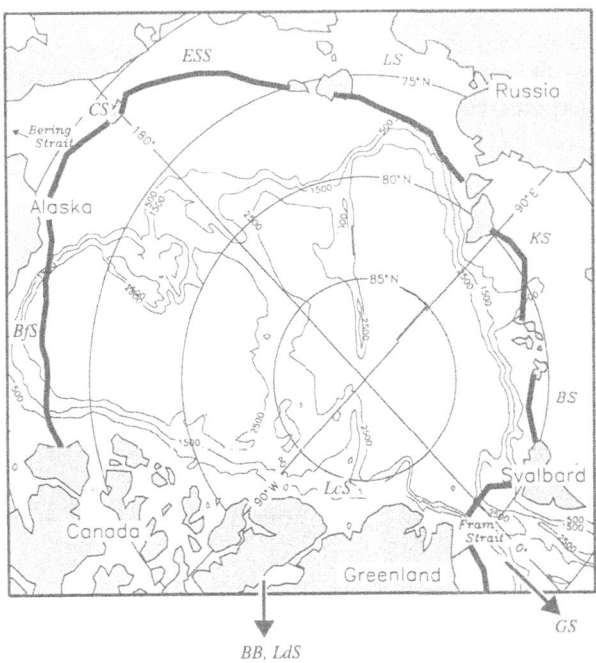

Figure 4. Arctic bathymetry (m) and regional seas. The heavy solid line approximates the climatological minimum sea ice cover (usually observed in August/September) which is the region very roughly referred to as the "central Arctic" in many one-dimensional studies. Also shown are the following peripheral seas that contain seasonal sea ice (i.e., ice that nearly or completely melts away during many summers): Barents Sea (BS), Kara Sea (KS), Laptev Sea (LS), East Siberian Sea (ESS), Chukchi Sea (CS), and Beaufort Sea (BfS). The Lincoln Sea (LcS) contains perennial sea ice that is generally thicker than other areas of the "central Arctic." The peripheral seas plus the central Arctic together comprise the Arctic Ocean. Other regions referred to in the text are the Greenland-Iceland-Norwegian Sea (GS) which lies south of Fram Strait, and Baffin Bay (BB) and the Labrador Sea (LdS) which lie to the west of Greenland.

2.1 THEORY

The observed sea ice pack is a complex jumble of smooth and deformed ice. In thermodynamic modeling this is usually implicitly simplified to a uniform slab or an ensemble of floes with varying thickness. These ice floes are assumed to be in contact with the atmosphere on the top surfaces and with the ocean on their bottom and lateral surfaces (with some lateral exposure to the atmosphere depending on their freeboard). Most theoretical effort so far has been devoted to the vertical heat transfer problem assuming horizontal homogeneity within the ice. However, many models do treat lateral melting as well.

2.1.1 *Vertical melt and growth*
Sea ice grows and melts vertically in response to imbalances between net heat fluxes at the boundaries and internal conduction of heat. The one-dimensional unsteady heat conduction equation, including internal heat sources due to penetrating radiation, can be written as

$$\rho c_p \frac{\partial T}{\partial t} = \frac{\partial}{\partial z}\left(k\frac{\partial T}{\partial z}\right) + F_{sw}I_0(1-\alpha)\kappa e^{-\kappa z}, \tag{3}$$

where $T(z,t)$ is the temperature, t is time, and z is depth measured positive downward from the upper surface (which can be ice or snow). In the interest of simplicity we hereafter drop the explicit time dependence and write the temperature at a given depth using a subscript, e.g., the surface temperature is T_0. Also, ρ is the density, c_p is the specific heat capacity, k is the thermal conductivity, α is the albedo, and κ is the bulk extinction coefficient. The downwelling shortwave radiative energy flux is F_{sw}, while the fraction of this flux that penetrates the surface is I_0. Equation (3) is subject to the following boundary condition at the ice underside:

$$T_h = T_f, \tag{4}$$

where h is the total thickness of ice plus any overlying snow and T_f is the freezing temperature of seawater (i.e., the ice underside is always at the freezing point). At the upper surface, the boundary condition is slightly more complicated, namely

$$
\begin{aligned}
T_o &= T_m & F_o > 0, \; T_o' \ge T_m \\
k\frac{\partial T}{\partial z}\bigg|_{z=0} &= F_o & \text{otherwise}
\end{aligned}
\tag{5}
$$

where T_m is the melting temperature of fresh water (since surface melt is either snow or relatively fresh ice) and F_o is the net heat flux into the surface. Equation (5) states that the upper surface boundary condition is the net surface heat flux, obtained from a surface energy budget calculation, except when the surface is melting. When melt occurs (i.e., when there is net heating of the surface $F_o > 0$ and the surface energy balance predicts a preliminary surface temperature $T_o' \geq T_m$), the boundary condition is forced back to the melting temperature, with the absorbed heat accounted for by the latent heat of fusion of melting ice.

Growth and melt at the ice underside are computed from the difference between the conductive heat flux away from the ice underside and the heat flux out of the upper ocean; therefore

$$L_f \, f_h = k\frac{\partial T}{\partial z}\bigg|_{z=h} - F_h \; , \tag{6}$$

where L_f is the volumetric heat of fusion, f_h is the growth rate (negative for melt rate) at the bottom, and F_h is the heat transferred from the upper ocean to the ice underside (positive upward). Melt at the upper surface is likewise computed from the difference between the conductive flux and the net surface flux; any snow is melted first and the remaining heat is used to melt ice, i.e.,

$$L_f \, f_o = k\frac{\partial T}{\partial z}\bigg|_{z=0} - F_o \; . \tag{7}$$

where f_o is the melt rate at the upper surface. The sum, $f=f_o+f_h$, is the overall melt rate which appears in Eq. 1.

In general, the temperature profile through the ice and snow is nonlinear, representing the storage of sensible heat (or heat deficit). Thermodynamic calculations based on Eqs. 3 - 7 are complicated by the nonlinear dependence of ρ, c_p, and k on this (temperature) profile. Theoretical and empirical relationships between these thermal properties and temperature (and salinity), which arise because of internal phase transitions between pockets of concentrated liquid brine and the surrounding ice, have been discussed by Schwerdtfeger [9], Untersteiner [10] and Ono [11], and their role in the energy partition between internal and surface melt has been addressed recently by Bitz and Lipscomb [12]. More details can be found in these references or in Maykut and Untersteiner [13]. These brine pockets and their parameterized physics represents the storage of latent heat (or latent heat deficit).

There is a further nonlinearity due to the dependence of the net surface energy flux, F_o, on surface temperature, T_o. The surface energy budget equation is

$$F_o = F_{lw} - \varepsilon\sigma T_o^4 + (1-\alpha)(1-I_o)F_{sw} + F_{lat} + F_{sens} \; , \tag{8}$$

where ε is the surface emissivity, σ is the Stefan-Boltzmann constant, F_{sw} and F_{lw} are the downward shortwave and longwave radiative fluxes respectively, and F_{lat} and F_{sens} are the downward latent and sensible heat fluxes respectively. All of the terms in Eq. 8, except for the downward longwave flux, depend explicitly or implicitly on surface temperature, generally in a nonlinear way. A detailed review of the ice surface energy balance is provided by Maykut [14].

Because of the nonlinearities mentioned above, thermodynamic growth and melt calculations are generally performed with numerical models. Although earlier simplified models had been proposed (as far back as Stefan [15]), the first comprehensive one-dimensional thermodynamic sea ice model based on a numerical solution of Eqs. 3 - 8 was presented by Maykut and Untersteiner [13]. Subsequently, Semtner [16] simplified the numerical scheme in various ways, resulting in an efficient hierarchy of models still widely used today. More recent examples of strictly one-dimensional models are provided by Gabison [17], Flato and Brown [18], and Lindsay [19]. Thermodynamic calculations for a region of sea ice containing a variety of ice thicknesses, based on the 1-D version of Eq. 1, are described by Maykut [20], and more recently by Björk [21], Schramm et al. [22], and Holland et al. [23]. In addition, a more sophisticated treatment of shortwave energy interaction with the surface, including melt pond and surface albedo parameterizations, was presented by Ebert and Curry [24].

2.1.2 *Lateral melt*

Lateral melt arises from the transfer of heat absorbed by open water to the sides of ice floes. Several mechanisms may influence lateral melting in ways that are very different from bottom melting. First, buoyancy tends to flush fresh melt water away from lateral floe surfaces, while it tends to "pool" melt water at the ice underside, inhibiting turbulent transfer. Second, surface waves can act to enhance lateral heat transfer by increasing turbulent mixing. Finally, due to strong stratification, the water in leads can be much warmer than that under adjacent ice, and the extent of mixing between lead and under-ice waters is poorly understood.

The seminal paper by Maykut and Perovich [25] presented observations and a simple model describing the melting process on all three floe surfaces (top, bottom, and lateral). They observed lateral melt rates that were comparable to bottom melt rates. Further, they estimated that in the central Arctic, top, bottom and lateral melting would all contribute substantially to the total volume melt. The exact partition depended in their scheme on lead width. The lateral melt parameterization that gave the best fit to observations was a rather severe modification of that obtained in laboratory experiments by Josberger and Martin [26].

Steele [27] derived equations for the mass balance of a field of ice floes that depended on the mean diameter of floes. He found that lateral melting was significant only for floes less than several hundred meters in diameter, and noted the lack of observed floe diameter data with which one might assess the significance of lateral melting in the central pack.

Most models allow for some partition between lateral and bottom melting. Some also allow for lateral growth in leads representing wind-blown frazil accumulation, although this is perhaps a minor effect. In general, the lateral melt rate is specified in one of two ways: as a function of either the energy absorbed by the open ocean, or the elevation of the ocean surface temperature above freezing. The former choice is usually used when assuming that the ocean surface remains at the freezing point in the presence of sea ice.

The latter choice explicitly recognizes that the ocean surface mixed layer generally warms slightly (often to several tenths of a degree above freezing) in the vicinity of a sea ice cover (although by continuity it remains at the freezing point immediately adjacent to the sea ice). Mellor and Kantha [28] and Harvey [29] both show that simple models can be quite sensitive to various bottom and lateral melt rate parameterizations. The model of Ebert et al. [30] indicates that the fraction of summertime solar energy used for bottom melting is about twice that for lateral melting.

2.2 OBSERVATIONS (ONE-DIMENSIONAL)

In this section we briefly survey one-dimensional observations of growth and melt, meaning observations on a single floe or on a variety of floes and open water without explicit regard for horizontal variability. The first such observations taken on a single ice floe over a full seasonal cycle were made by Untersteiner [10] in the Beaufort Sea. Data were collected for two summer melt seasons (1957 and 1958) and the intervening winter growth season (**Table 1**). The mean ablation values for the two summers are shown in Table 1. In fact, while bottom melt was very similar in the two years (22 cm in 1957 and 24 cm in 1958), ice melting on the top surface was very different in 1957 (19 cm) and in 1958 (41 cm). Nonetheless, the mean mass balance shows a near-equilibrium cycle.

TABLE 1. Observed (*obs*) and modeled (*mod*) annual sea ice growth and melt (cm) in the central Arctic Ocean. The first two studies considered the mass balance on a single ice floe with an annual average thickness of 3 m. The last four considered more complex ice thickness distributions. Growth and melt can occur on top (*top*), bottom (*bot*), and/or lateral (*lat*) floe surfaces.

(cm)	..obs.. Untersteiner [10] 3 m floe	..mod.. Maykut & Untersteiner [13] 3 m floe	..obs.. Koerner [7] g(h)	..mod.. Maykut [20] g(h)	..mod.. Holland et al. [23] g(h)	..mod.. Ebert & Curry [24] 2-level
Growth	57	45	110	130	114 107* 7 (bot) (lat)	115 66 49 (bot) (lat)
Melt	53 30 23 (top) (bot)	45 40 5 (top) (bot)	60	41 (94†)	72 21 30 21 (top) (bot) (lat)	74 48 10 16 (top) (bot) (lat)
Net	4 growth	0	50 growth	89 (36†) growth	42 growth	41 growth

* Includes both congelation and frazil ice growth.

† Includes additional mass losses related to melt ponds, lateral melting, and pressure ridge keels.

Mass balance observations along a transect of varying ice thickness were described by Koerner [7]. Accounting for thin ice approximately doubles the annual mean ice growth estimate over that for 3 m ice alone (Table 1). It should be noted that ice mass balance studies have also been performed for time scales shorter than one year (e.g., McPhee and Untersteiner [31]; Wettlaufer [32]).

New data on the mass balance of a variety of ice floe thicknesses and types (e.g., deformed vs. undeformed ice) should be available soon from a variety of sources. Foremost among these is the SHEBA (Surface HEat Budget of the Arctic Ocean) project, in which over a year of such data has just been collected in the Beaufort and Chukchi Seas. Also, formerly classified Soviet data from drifting ice camps will hopefully be released and processed soon.

2.3 MODELS (ONE-DIMENSIONAL)

How well do the models reproduce the observations of growth and melt, and how do these results depend on model physics? These two issues are examined in this section. First, we show the results of Maykut [20] as an example of a model simulation of the seasonal mass balance. We then discuss model sensitivity. Following the literature, we focus on how growth and melt are affected by the inclusion or absence of physical processes. In particular, we examine how growth and melt depend on (i) the thickness distribution, (ii) thermal inertia in the snow and ice, (iii) snowfall, (iv) albedo and shortwave energy absorption, and (v) parameterized ice dynamics.

2.3.1 *Models: The mean seasonal cycle*
Maykut [20] examined the seasonal mass balance using a one-dimensional model that resolved the thickness distribution. **Figure 5a** (Maykut's Figure 3a) shows the annual ice production for 8 thickness categories. The growth rate is a highly nonlinear function of ice thickness, which leads to annual growth of roughly 30 cm for each of three very different thickness categories: 0-20 cm, 20-80 cm, and 80-300 cm. **Figure 5b** (Maykut's Figure 3b) shows the corresponding seasonal cycle of growth and melt. Ice grows during much of the year in the central Arctic, with a short but intense melt period in summer. The annual total growth in this model is 130 cm, while the annual total melt is 41 cm (or 94 cm when accounting for melt ponds, lateral melting, and ridge keels). In steady state, the balance would be exported southward through Fram Strait into the North Atlantic Ocean, where it would melt. Table 1 shows a comparison of these results with the observations of Koerner [7].

2.3.2 *Model sensitivity: Thickness distribution*
Figure 5b also shows the mass balance of a simple "slab" model in which $g(h)$ is approximated by a single thickness category. This model was first introduced in the work of Maykut and Untersteiner [13], and as Table 1 shows, it provides reasonable agreement

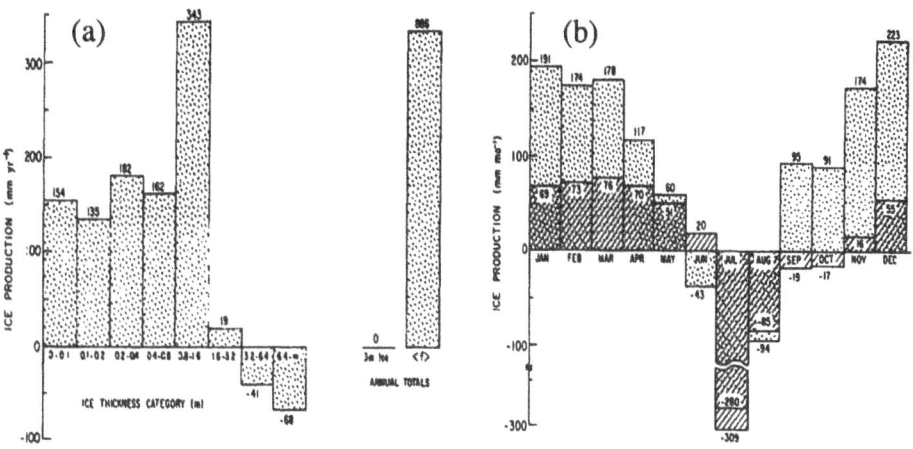

Figure 5. Ice growth and melt in the one-dimensional model of Maykut [20]. (a) Annual totals for 8 thickness categories. (b) The seasonal cycle for the full model (stippled) and the "slab" model (cross-hatched). See text for more details.

with the "slab" observations of Untersteiner [10]. Of course it severely underestimates the total winter growth since it contains no open water or thin ice. The simplest scheme that addresses this problem is the so-called "two-level" model, in which $g(h)$ is approximated by two categories, thin and thick ice, where the former includes open water. Ebert and Curry [24] used this scheme in a one-dimensional model, obtaining results similar to the observations of Koerner [7] (Table 1). Recent studies (e.g., Holland et al. [23]) have extended these results to include a multi-category representation of the thickness distribution (Table 1). **Figure 6** is an updated version of Figure 7a from Schramm et al. [22] (J. Schramm, Univ. of Colorado, personal communication, 1999) which shows that improved resolution of $g(h)$ tends to increase the annual mean thickness but has little effect on the amplitude of the seasonal cycle. Simulations over a greater range of thickness categories show larger differences in annual mean thickness (R. Lindsay, Univ. of Washington, personal communication, 1999), with two-level models producing annual mean ice thickness about 50 cm smaller than shown in Figure 6. Bitz **[33]** showed a similar result for a simulation of the years 1979-1995 (see her Figure 5.19). Better resolution of $g(h)$ means better resolution of thin ice thermodynamics. Thin ice grows much faster than thick ice, but its melt rate is fairly similar. Therefore in a one-dimensional model at equilibrium, the enhanced thin ice growth rate can only be offset by an overall decrease in conductive heat flux, i.e. in an increase in the mean annual thickness.

2.3.3 *Model sensitivity: Thermal inertia*

The vertical profile of temperature is in general non-linear due to the heat capacity or 'thermal inertia' of the ice and snow (as described in section 2.1). **Figure 7**, from Maykut

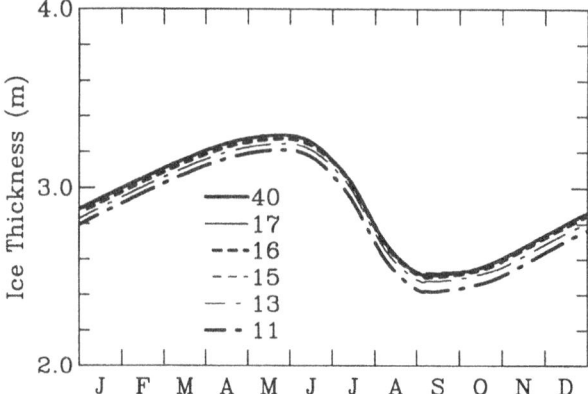

Figure 6. The sensitivity of the seasonal cycle of mean ice thickness in an updated version of the model of Schramm et al. [22] to the number of thickness categories used to resolve the thickness distribution *g(h)* (J. Schramm, Univ. of Colorado, personal communication, 1999). Similar two-category simulations produce annual mean thickness about 50 cm smaller than shown here.

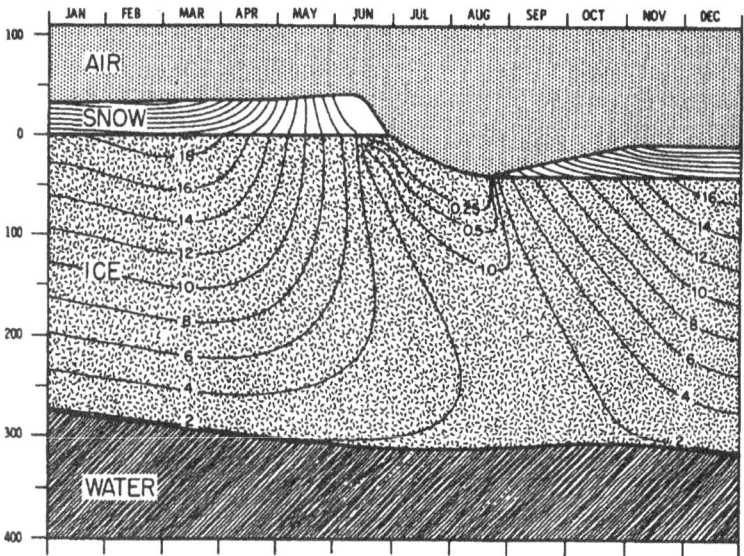

Figure 7. Temperature contours through a climatological seasonal cycle of snow and thick ice according to the model of Maykut and Untersteiner [13].

and Untersteiner [13] (their Figure 2), shows the temperature profile through snow and thick ice when resolved with a model grid spacing of 10 cm. The figure illustrates that the thermal inertia effect is most pronounced in the transition seasons of fall and spring. During fall, for example, it takes some time for the bottom of thick ice to react to the rapidly cooling surface. At this time, thin ice often grows while thick ice is still melting. Note, however, that a linear temperature profile, which is obtained by assuming zero heat capacity, is in fact quite a reasonable approximation during much of the winter and summer. It is also a good approximation throughout the annual cycle for ice up to about 80 cm thick (Maykut [20]).

Semtner [16] explored simplified thermodynamic schemes for use in large-scale climate models. **Figure 8** (his Figure 9) shows how well a model with only three resolved

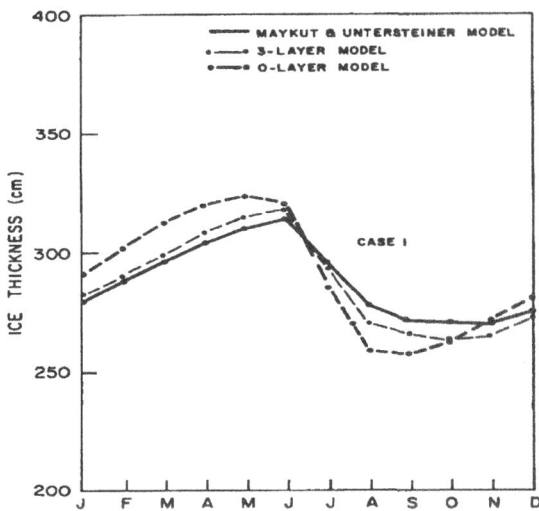

Figure 8. The seasonal cycle of mean ice thickness as a function of how well the temperature profile through snow and ice is resolved, from Semtner [16]. The full Maykut and Untersteiner [13] model is compared with the 3-layer and 0-layer models of Semtner [16].

temperature points in the snow and ice reproduces the seasonal cycle of ice thickness produced by the higher-resolution model of Maykut and Untersteiner [13]. Figure 8 shows a comparison with the full Maykut and Untersteiner model as well as with a "0 layer" model. The latter assumes a linear temperature profile (i.e. assumes zero heat capacity), with the albedo and thermal conductivities adjusted to best match the annual mean thickness of the full model. Even with this tuning, the amplitude of the seasonal cycle is too large when thermal inertia is crudely resolved (3-layer) or neglected (0-layer). Also, the phase of melt and growth onset is incorrect in these simulations, which is not surprising in light of the previous discussion (Figure 7). Further, Bitz et al. [33] have demonstrated the somewhat counterintuitive result that reducing the resolution of the thermal inertia tends to reduce simulated interannual variability in a one-dimensional climate model. It is important to note here that most two-dimensional models currently employ the 0-layer approximation (see Table 3 below).

562

2.3.4 *Model sensitivity: Snow*

Snow has competing effects on the annual mean ice thickness. On the one hand, it is a good insulator and so more snow means less ice growth. On the other hand, it has a high albedo so that more snow means a delayed melt season. Maykut and Untersteiner [13] and Ledley [34] summarized these and other more minor processes. The net result is that the equilibrium ice thickness is relatively insensitive to snowfall variations of less than about 50% (Ebert and Curry [24]). (Outside of this range the albedo effect dominates and thin (thick) snow cover leads to thin (thick) ice.) However, as **Figure 9** shows, the seasonal cycle of growth and melt is quite sensitive to snowfall variations (Semtner **[35]**), as is the net growth of first-year ice (Flato and Brown [18]). It should also be noted that the observed snow cover on the sea ice pack can be highly inhomogeneous and transitory, processes generally not well modeled at present.

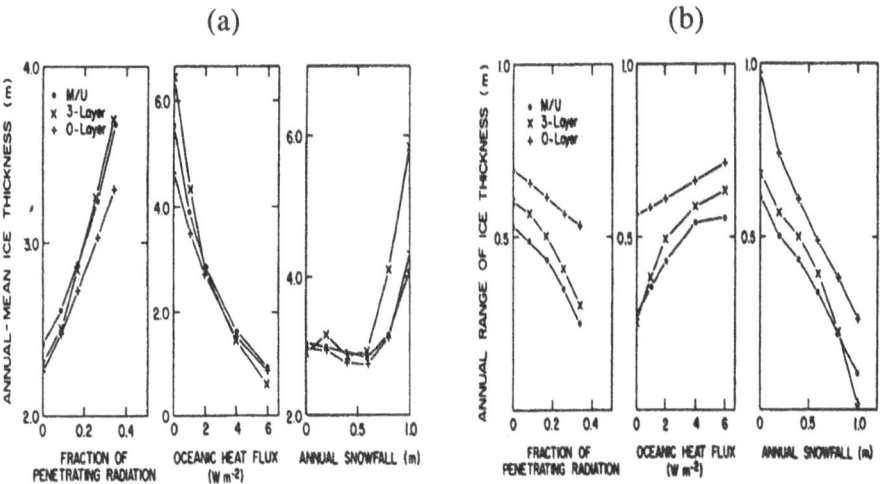

Figure 9. The influence of penetrating radiation through the snow and ice, upward oceanic heat flux, and annual snowfall on (a) annual mean ice thickness and (b) the amplitude of the seasonal cycle, from Semtner [35]. Three models are compared with different treatments of thermal inertia: Maykut and Untersteiner [13], Semtner's [16] 3-layer model, and Semtner's [16] 0-layer model.

2.3.5 *Model sensitivity: Albedo and shortwave energy absorption*

The amount and timing of summer melt is dominated by absorbed solar radiation which in turn is directly related to albedo and its dependence on surface properties like temperature, snow depth and age, melt ponds, etc. It is the dependence of albedo on surface conditions which leads to a suite of ice-albedo feedbacks which were explored in some detail by Curry et al. **[36]**. They found that the sensitivity of modeled ice thickness or surface temperature to forcing perturbations was much higher when the Ebert and Curry [24] scheme (which includes an explicit parameterization of the role of melt ponds in reducing albedo) was used compared to simpler schemes like those used in the models of Maykut and Untersteiner [13] or Semtner [16]. Similarly, Semtner [35] found that neglect of penetrating radiation creates overly thin ice and an amplified seasonal cycle of growth and melt (Figure 9). It is unfortunately still the case that, despite the number of direct observations of surface albedo, its area average (necessary for models) and its dependence on surface properties is still rather uncertain. One of the goals the SHEBA field program is to improve this understanding (Randall et al. [37]).

2.3.6 *Model sensitivity: Ice dynamics*

Ice dynamics can be simply parameterized in a one-dimensional model by adding a constant source of open water area (e.g., Mellor and Kantha [28]) that accounts for the large-scale divergence observed in the Arctic Ocean ice pack. Another option is to specify a time-dependent deformation (e.g., Maykut [20]) using relatively rare data sets such as that collected over one year during the Arctic Ice Dynamics Joint Experiment (AIDJEX). The creation of open water throughout the year enhances both winter growth and summer melt. **Figure 10** is from Mellor and Kantha [28] and shows how more divergence leads to larger seasonal cycles. High divergence also leads to thinner ice, which means that the additional summer melt tends to dominate over winter growth (see also Ebert and Curry [24]).

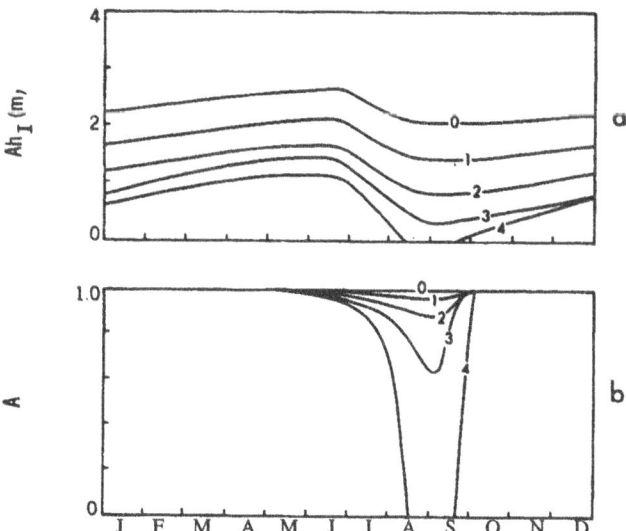

Figure 10. The (a) area-averaged ice thickness and (b) ice concentration for cases in which the constant value for ice divergence was varied from $0\text{-}4 \times 10^{-9}$ s^{-1}, from Mellor and Kantha [28].

2.3.7 *Summary*

A qualitative summary of sections 2.3.2-4 is presented in **Table 2**. Reducing the resolution of the thickness distribution provides a poor representation of thin ice, which leads to an underestimate of the mean thickness (Schramm et al., [22]; Bitz **[38]**), but seems to have little effect on the amplitude of the seasonal cycle. Neglecting thermal inertia in the snow and ice can, with the proper tuning of thermal conductivity and albedo, have little effect on the mean ice thickness, but will inevitably affect the amplitude and phase of growth and melt (Semtner [16]). Snow has little effect on the mean ice thickness (for snow thickness variations ~50% about the observed values), but does have a significant influence on the seasonal cycle of multi-year ice (Maykut and Untersteiner [13]), and interannual variability of landfast ice (Flato and Brown [18]). Neglecting the penetration of solar energy through the ice increases top surface melting more than it decreases bottom surface melting, leading to thinner ice and an increased seasonal cycle (Semtner [35]). Finally, neglecting open water produced by ice divergence underestimates the seasonal cycle of open water growth and the larger effect of melting induced by lead heating, which then leads to an overestimate of the annual mean ice thickness (Mellor and Kantha [28]; Ebert and Curry [24]).

Table 2 indicates that the annual average ice thickness may have very different sensitivity to model physics and/or forcing relative to the amplitude of the seasonal cycle of growth and melt. Although the mean thickness is an important diagnostic, the amplitude of its seasonal cycle is perhaps equally important, as discussed in the Introduction.

3. Spatial Patterns of Growth and Melt in the Arctic Seas: Observations and 2D Models

Net annual growth or melt provide, respectively, a sink or source of freshwater at the ocean surface, and are therefore useful diagnostics of the role of sea ice in the Arctic freshwater budget. Direct observations of these quantities (in the two-dimensional spatial sense) are rare; however, good proxy information regarding the accumulated net growth or melt can be obtained using oceanic chemical tracer data from the upper hundred meters or so. These results are discussed in Section 3.1. To be consistent with some of the existing literature, and to avoid confusion with the implicit sign of the terms 'growth' and 'melt', we will refer to the annual total of growth minus melt (i.e. the integral, at a particular location, of fg in Eq. 1 over all thickness and over an annual cycle) as the **net freezing rate**, whose units are m yr^{-1}. Net freezing rate is typically estimated by averaging over several annual cycles. Where positive it implies a net salt flux at the ocean surface, and where negative it implies a net freshwater flux. Section 3.2 provides a review of net freezing rate patterns in two-dimensional numerical models. We examine regions

TABLE 2. How annual mean ice thickness and the seasonal amplitude of growth and melt respond to simplifications of model physics in one-dimensional models. Little change is indicated by "ε."

Simplification	Mean Thickness	Seasonal Cycle Amplitude	References
reduce resolution of g(h)	reduced	ε	Schramm et al. [22] Bitz [38]
neglect thermal inertia	ε (with proper tuning)	increased (& distorted phase)	Semtner [16] Bitz et al. [33]
neglect snow	ε (MY)* increased (FY)*	increased	Maykut & Untersteiner [13] Semtner [16]; Ledley [34] Ebert & Curry [24] Flato & Brown [18]
neglect penetrating radiation	reduced	increased	Semtner [35] Curry et al. [36]
neglect effect of ice divergence	increased	reduced	Mellor & Kantha [28] Ebert & Curry [24]

* MY = multiyear ice; FY = first year ice.

where the models tend to agree as well as where they disagree and seek explanations for both. We also provide illustrations of sensitivity to model physics and forcing choices.

3.1 OBSERVATIONS (TWO DIMENSIONAL)

There are essentially no direct observations of the spatial variability of growth and melt or net freezing rate in the Arctic seas. In fact, even mean ice thickness is only qualitatively known from submarine sonar data (e.g., Bourke and Garrett [39]), although the quality and quantity of these data are improving (e.g., Keigwin and Johnson [40]). Thus a better picture of the spatial distribution of ice thickness and its seasonal and interannual variability might be on the near horizon. Yet Eq. 1 shows that even if the mean annual cycle of ice thickness was known, it is insufficient to infer growth and melt without additional information about ice transport.

Winter ice growth in coastal polynyas has been studied by Martin and Cavalieri [41] and Cavalieri and Martin [42]. They calculated the area of open water using passive microwave data from satellites and performed a surface energy balance to estimate the ice growth. Mean ice growth in Eastern Arctic polynyas (Martin and Cavalieri [41]) was about 10 m yr^{-1} with a 30% uncertainty. Mean values for Western Arctic polynyas were generally higher (Cavalieri and Martin [42]) but might have been biased by the inability of the microwave sensor to distinguish between open water and thin ice (S. Martin, Univ. of Washington, personal communication, 1999). Higher resolution models in use today might be able to just barely resolve these polynyas, which generally have open water areas of order 2000 km^2 or less.

Recently, oceanic tracer studies using oxygen isotope ratios have been used to address the net freezing rate problem. Combining these with salinity data allows a partition of freshwater between that contributed by sea ice melting and that contributed by river inflows. It is very important to note that this type of analysis is fundamentally Lagrangian, i.e., it determines the amount of sea ice melt water (or river water) that has accumulated over the life history of a water parcel. The picture becomes even more complex when vertical column inventories are taken since individual water layers may move quite independently.

Bauch et al. [43] examined data from a series of cruises in the Eurasian sector of the Arctic Ocean. **Figure 11** shows the result for a sea ice meltwater inventory, where positive values indicate accumulated net melt and negative values accumulated net growth. The zero contour lies roughly on the eastern flank of the transpolar drift stream (e.g., Colony and Thorndike [44]). Accumulated net growth in this region is of order 4-5 m. Given a residence time of 8-10 years in the upper 100 m (Jones et al. [45]; Wallace et al. [46]) this indicates a net freezing rate of about 0.5 m yr^{-1}, in agreement with the limited observations and with models of the central Arctic (Table 1). Further south, accumulated net freezing is positive (i.e. net melt) with a magnitude of about 1 m, forced by warm air advection from the south (Serreze and Barry [47]) as well as upwelling of water in the warm West Spitsbergen Current (Aagaard et al. [48]). Ekwurzel [49] has recently expanded this analysis to include more stations within the Arctic Ocean.

As noted previously, this type of plot contains Lagrangian information and cannot by itself be compared rigorously to an Eulerian plot of net freezing rate that a numerical model might produce. However, there is no reason why such a model could not create, via numerical tracers, a similar plot describing the simulated life history of freshwater which could then be directly compared to isotope-derived data. Such a study has not yet been performed.

Oxygen isotope analysis has also been applied to Canadian Arctic data by Melling and Moore [50] and Macdonald et al. [51]. Most of this work has focused on the shelf and shelf break within several hundred kilometers of the Mackenzie River delta. These studies reveal a net freezing rate of 1.5-2 m yr^{-1} in this region, i.e., 3-4 times the value in the central Arctic. The Lagrangian assumption here implies that this estimate is probably applicable to a larger area such as the Beaufort Sea. As shown in the next section, some models do predict enhanced ice growth in this region, but others predict weak or even strong net melting!

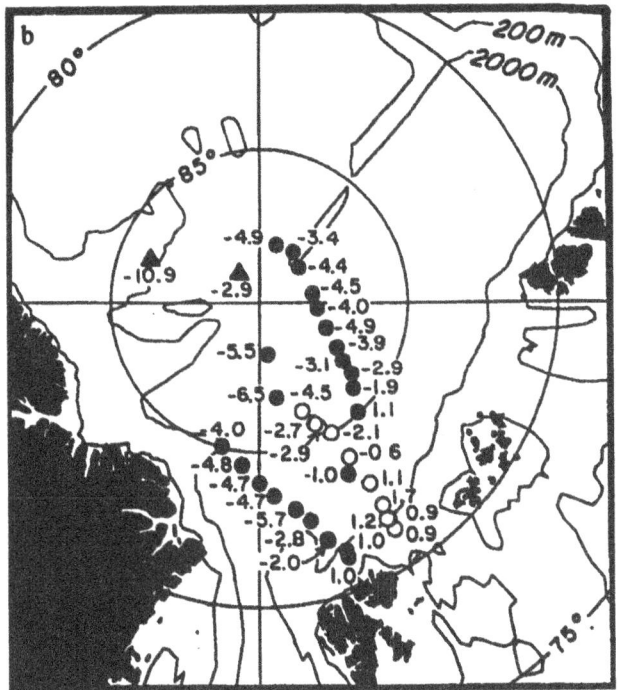

Figure 11. The integrated sea ice melt water thickness (m) at each site from a variety of cruises and ice camps, from Bauch et al. [43]. The calculation is performed using both salinity and oxygen isotope ratio $^{18}O/^{16}O$. Positive values indicate net melting and negative values indicate net growth.

3.2 MODELED NET FREEZING RATE

Many modeling papers include a figure showing the spatial distribution of net freezing rate: these will be surveyed here. These net freezing rate fields do not provide any information about the seasonal cycle of growth and melt, and indeed are largely a measure of net ice divergence or convergence in a region (as can be seen from a local version of Eq. 2). Nevertheless, net freezing rate does reflect the annual input of salt (or equivalently the annual loss of freshwater) at the ocean surface, and so it is of central importance to the Arctic freshwater budget. We start with a very recent study which we use as a basis for comparison. When comparing models, we attempt to distinguish where the choices of model physics, numerics, and/or forcing are likely causing differences in the prediction of net freezing rate. **Table 3** lists various model attributes, while **Table 4** summarizes annual net freezing rate for selected regions. Since our focus is on the Arctic Ocean, results from two-dimensional model studies of Antarctic sea ice are not included; however the

TABLE 3. Details of the two-dimensional sea ice models discussed in Section 3, in chronological order.

Author(s)	coupled?	g(h)	thermal inertia?	grid (horizontal spacing)	atmospheric forcing	ice-ocean heat flux
Hibler [52]	ice only	2-level (slab)*	yes (table)†	125 km	1962-63 8-day winds * thermo table	fixed constant 2 W m^{-2} (specified from thermo table)
Hibler [4]	ice only	explicit g(h)	no	125 km	1962-63 8-day winds * monthly climat. T_{air}	fixed constant 2 W m^{-2}
Walsh et al. [53]	ice only	2-level (uniform)*	no	222 km	1951-80 daily winds * monthly climat. T_{air}	fixed constant 2 W m^{-2}
Hakkinen & Mellor [54]	ice-ocean	2-level (uniform)	yes (no advection)	variable (28-150 km)	monthly climol. & daily (1987) winds * monthly climat. T_{air}	variable (given by interactive ocean model)
Flato & Hibler [55]	ice only	2-level (slab)	no	160 km	1981-1983 daily winds * monthly climat. T_{air}	spatially variable (specified from an off-line ice-ocean run)
Chapman et al. [56]	ice only	2-level (uniform)	no	125 km	1960-1988 daily winds * monthly climat. T_{air}	fixed constant 0-5 W m^{-2}
Thomas et al. [57]	ice only	explicit g(h)	yes (table)	7 linked regions	1979-1985 ice buoy motion * thermo table	fixed constant 2 W m^{-2} (specified from thermo table)
Fichefet & Maqueda [58]	ice-upper ocean	2-level (uniform)	yes (with advection)	3° global	monthly climat. winds * monthly climat. T_{air}	variable (given by interactive upper ocean model)
Zhang et al. [59]	ice-ocean	2-level (uniform)	no	40 km	1979-1985 daily winds * monthly climat. T_{air}	variable (given by interactive ocean model)
Hilmer et al. [60]	ice only	2-level (uniform)	no	110 km	1958-1996 daily winds * 1958-1996 daily T_{air}	spatially variable (specified from an off-line ice-ocean run)

* "slab" denotes model thermodynamics that treats sea ice as a slab with a single thickness; "uniform" denotes model thermodynamics that treats sea ice as a uniform distribution of thicknesses.

† "table" denotes model thermodynamics obtained via the table presented in Thorndike et al. [3].

TABLE 4. Regional net freezing rate in the two-dimensional sea ice models discussed in Section 3, in chronological order. Net growth (positive) is in **bold** font, while net melt (negative) is in regular font. Dark shaded cell headings denote regions with net growth in most simulations, while light shaded cell headings denote regions with net melt in most simulations. The Beaufort and Lincoln Seas' cell headings are unshaded since they are areas of disagreement between simulations, with some predicting net growth and others predicting net melt.

Author(s)	Central Arctic (m yr⁻¹)	Laptev Sea (m yr⁻¹)	Beaufort Sea (m yr⁻¹)	Lincoln Sea (m yr⁻¹)	Chukchi Sea (m yr⁻¹)	GIN Sea (m yr⁻¹)	Barents Sea (m yr⁻¹)
Hibler [52], [4]	**<0.4** '79 -0.6 '80	**<2.8** '79 **<3.0** '80	**<2.4** '79 -1.2 to **1.2** '80	-0.2 '79 ~ -0.6 '80	**<0.4** '79 -0.5 '80	NA	NA
Walsh et al. [53]	**<0.5**	~0.5	**<0.5**	~0	-0.5 to **0.5**	<-2.0	<-1.0
Hakkinen & Mellor [54]	-0.4 *mo** ~0.3 *da**	**<0.8** *mo* **<2.0** *da*	**<0.8** *mo* **<1.5** *da*	**<0.2** *mo* -0.3 to **0.3** *da*	~ **0.2** *mo* -0.3 to **0.8** *da*	<-0.8 *mo* -0.5 to **0.5** *da*	-0.3 to **0.2**
Flato & Hibler [55]	**0.5** to **1.0**	**<2.0**	<-1.0 *VP*† <-2.0 *CF*†	-0.5 to **0.5**	<-2.0	<-2.0	-1.0 to **0.5**
Chapman et al. [56]	~0.5	**<2.0**	-0.3 to **2.0**	~0	<-0.5	-1.0 to **2.0**	-1.0 to **1.0**
Thomas et al. [57]	**0.30** ± 0.08	**0.32** ± 0.11	**0.15** ± 0.41	NA	-0.12 ± 0.28	NA	NA
Fichefet & Maqueda [58]	~0.5	~1.0	-0.5 to **1.0**	~ -0.25	<-0.5	<-4.0	<-2.0
Zhang et al. [59]	~0.5	**<2.0**	-2.0 to **2.0**	~0.5	<-2.0	<-3.0	-3.0 to **1.0**
Hilmer et al. [60]	~0.5	**<1.0**	~0.5	~0.5	<-1.0	<-4.0	<-2.0

*"mo" denotes monthly average wind forcing; "da" denotes daily winds.

† "VP" denotes viscous-plastic rheology; "CF" denotes cavitating fluid rheology.

interested reader is referred to Lemke et al. [61], Owens and Lemke [62], Stössel et al. [63], and Stössel et al. [64]. Finally, note that we refer here to synoptic sea ice models as "two-dimensional" since vertical processes are generally neglected or parameterized (e.g., ridge building and vertical heat transfer).

A recent example of net freezing rate was provided by Hilmer et al. [60], where a 110 km resolution sea ice model (Harder et al. [65]) was integrated over the period 1958-1997. Daily wind and air temperatures, produced by the National Center for Environmental Research / National Center for Atmospheric Research (NCEP/NCAR) reanalysis, were used as forcing. The ocean was not explicitly modeled, but was instead represented by geostrophic currents and upward heat fluxes derived from a coupled model run separately. The ice model is based on the formulation of Hibler [52], but includes an additional prognostic equation for deformed ice (similar to that of Flato and Hibler [66]), and an improved numerical scheme (Zhang and Hibler [67]). Thermodynamic calculations are based on the model of Parkinson and Washington [68] with some improvements including a prognostic snow layer and an assumed linear temperature profile through the ice and snow applied to seven thickness categories as first discussed in Walsh et al. [53]. **Figure 12** shows the net freezing rate obtained by Hilmer et al. [60], adapted from their Figure 1 (M. Hilmer, Inst. für Meereskunde an der Universität Kiel, personal communication, 1998).

The figure is consistent with the observations and models of the central Arctic (Table 1) in the prediction of a positive net freezing rate (net growth) of about 0.5 m yr^{-1}. In fact, most of the Arctic Ocean in this model is a net ice producer of about this magnitude. Higher net freezing rates are seen in the Kara and Laptev Seas, forced mostly by persistent offshore winds. The same effect is observed in northern Baffin Bay, which may be partially a result of the artificial barrier within the Canadian Archipelago. Negative net freezing rate regions are the Chukchi, Barents, Greenland, Iceland, and Labrador Seas. The strongest melt occurs near the ice edge in the northeastern Atlantic Ocean, where the water is especially warm as it advects northward in an extension of the subpolar gyre, yielding rates approaching -4 m yr^{-1}.

A comparison of regional net freezing rates in this and other models is shown in chronological order in Table 4. The values in this table were obtained by subjective interpretation of the appropriate figures in each paper, and are meant only as a qualitative point of comparison. Where a model produced a strong gradient in net freezing rate, the extreme value in the region is indicated. In the central Arctic, there is generally good agreement among models, with all producing a net freezing rate close to 0.5 m yr^{-1}. Similarly, there is general agreement that freezing rates in the Laptev Sea are larger than in the central Arctic, typically by a factor of 2 to 4. The models also generally produce negative net freezing rates (net melt) in the Chukchi, GIN, and Barents Seas, although the magnitude varies considerably. Models with spatially constant ocean heat flux cannot reproduce hydrographic fronts and so do poorly in the GIN Sea, predicting strong net growth in some places. There seems to be no consensus among the models regarding even the sign of net freezing rate in the Beaufort and Lincoln Seas. In the case of the Beaufort Sea, several models produce a 'dipole' structure with freezing rates of the opposite sign in the eastern and western Beaufort, while others produce net growth (or in one case, net melt) over the entire region.

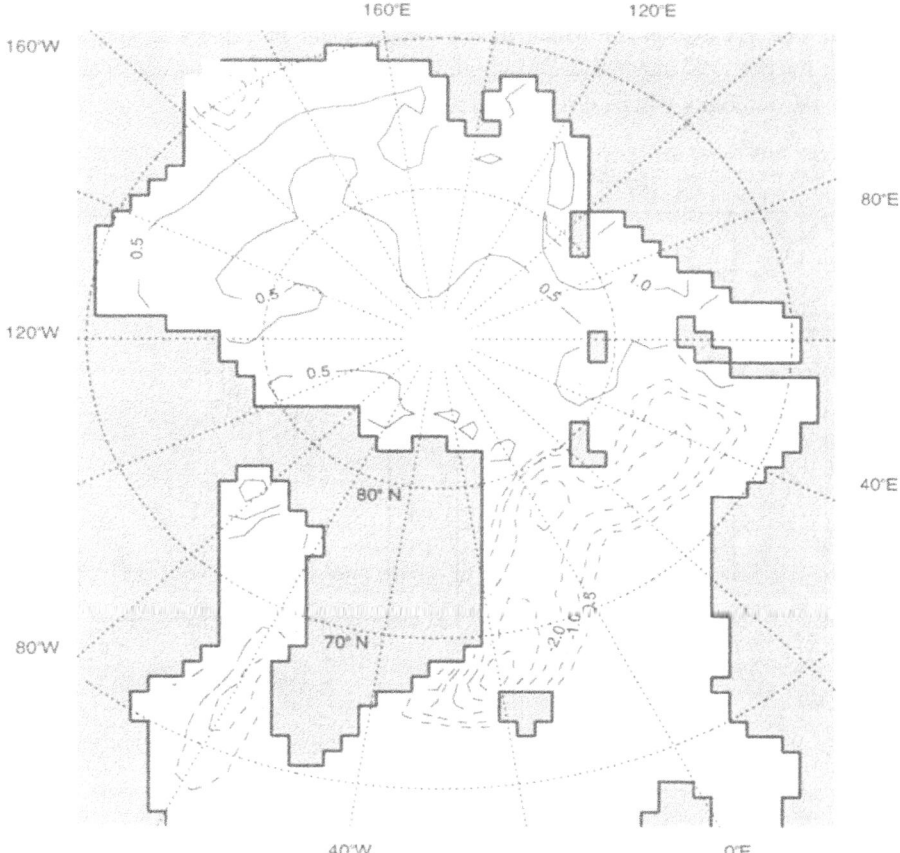

Figure 12. Net freezing rate (m yr^{-1}) for the period 1958-1997, as simulated by Hilmer et al. [60]. Dashed contours indicate regions of net melt while solid contours indicate net growth.

A somewhat more quantitative illustration of regional differences is provided in **Figure 13** by the results of a mass balance study of the Arctic Ocean by Thomas et al. [57]. A companion paper by Steele et al. [5] examined the corresponding freshwater balance in the ice and ocean. This model differs from the others listed in Table 3 in that it does not solve a prognostic equation for ice velocity, but rather diagnoses the mass balance in 7 regions using climatological estimates of growth and melt rates, and assimilation of velocity and concentration data over the period 1979-1985. The crosses

572

in Figure 13a show the mean annual net freezing rate (labelled 'Thermodynamics' in the figure), plotted against the net source or sink due to advection. The diagonal line indicates perfect balance between advection and net freezing rate. In this estimate, only the Chukchi Sea experiences net melt. The ellipses in Figure 13b indicate the 90% confidence interval about the mean, providing a measure of interannual variability. It is noteworthy that the region with largest interannual variability in net freezing rate is the Beaufort Sea, the region for which model disagreement in Table 4 is largest. This indicates the potential importance of sampling variability in estimating net freezing rate. Interannual variability will be discussed in more detail in Section 3.4.

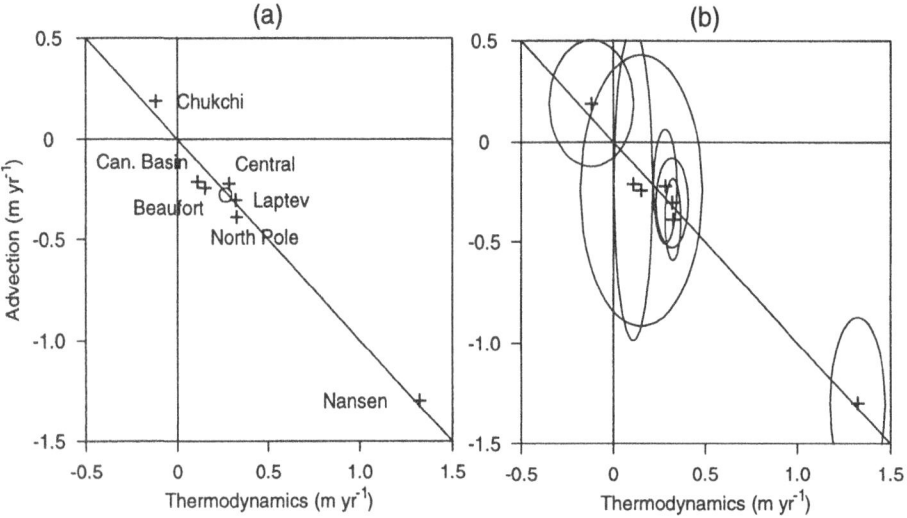

Figure 13. Annual average thickness changes in seven regions (pluses) and in the Arctic Ocean (open circle) from thermodynamics (net freezing rate) and from advection, from Thomas et al. [57]. (a) The mean values over 1979-1985, and (b) the 90% equiprobability ellipse for each mean. All modeled regions except the Chukchi have net growth and export. The region with the most interannual thermodynamic variability is the Beaufort.

Walsh et al. [53] also made a careful examination of the mass balance of the Arctic on a regional basis. They divided their model domain into 13 sub-regions and calculated the monthly mass changes that resulted from net freezing and from transport. **Figure 14** shows two sample plots for the Laptev Sea and the Greenland-Iceland-Norwegian Seas, indicating as before that the former is a region of net ice growth and export, while the latter is a region of net ice melt and import.

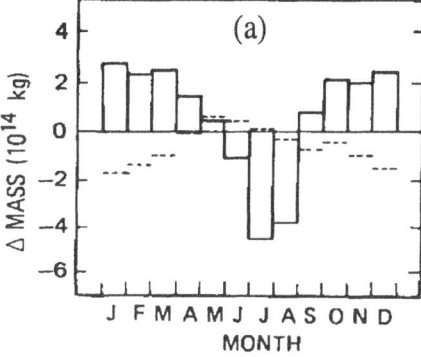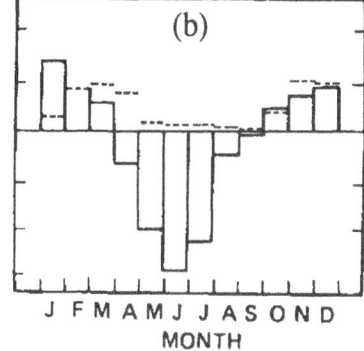

Figure 14. The monthly changes in ice mass from (i) net freezing (histograms) and from (ii) net transport (dashed lines) in (a) the Laptev Sea and (b) the Greenland-Iceland-Norwegian Sea, adapted from Walsh et al. [53].

3.3 SENSITIVITY OF NET FREEZING RATE

The net freezing rate produced by a numerical sea ice model clearly depends on the parameterization of various physical processes, as well as on the atmospheric and oceanic forcing provided to the model. In this section we provide a brief survey of sensitivity experiments which illustrate some of these effects.

Hibler [52] examined the effect of ice dynamics on growth and melt and found (his Figure 23, not shown here) that both growth and melt increase when dynamics are included, relative to a "thermodynamic-only" version. He found that open water produced by dynamics (i.e., divergence and shearing) led to increases in both winter ice growth and summer ice melt. Of course at equilibrium, a thermodynamic-only model produces zero net freezing rate, but this experiment illustrates the role of dynamics in increasing the amplitude of the annual cycle of growth and melt, consistent with the 1D model results summarized in Table 2.

Given the role of sea ice dynamics in both amplifying the annual cycle of growth and melt, and allowing regional net freezing via ice transport, the details of dynamical parameterizations are expected to impact the pattern of net freezing rate. Hibler [52] developed the so-called 'viscous plastic' rheology for sea ice which is still regarded as the standard model and which best reproduces observed buoy drift and North Pole ice thickness (Kreyscher et al. [69]). (Rheology refers to the relationship between stress and deformation in a material.) However, alternative schemes have been proposed. One example is the simpler 'cavitating fluid' rheology, designed for use in global climate models (Flato and Hibler [55]). **Figure 15** compares the annual average net freezing rate obtained using the same forcing (spanning the period 1981-1983), but with models based on either the viscous-plastic or the cavitating fluid rheologies. The figure indicates that the net freezing rate is relatively unaffected by the choice of rheology. (An interesting exception is in the eastern Beaufort Sea, where both models predict net melting, but differ substantially in magnitude. As Table 4 shows, this is one of the main regions of disagreement from model to model.) This conclusion is certainly not definitive, and awaits a more thorough study of how various rheologies affect and interact with model thermodynamics.

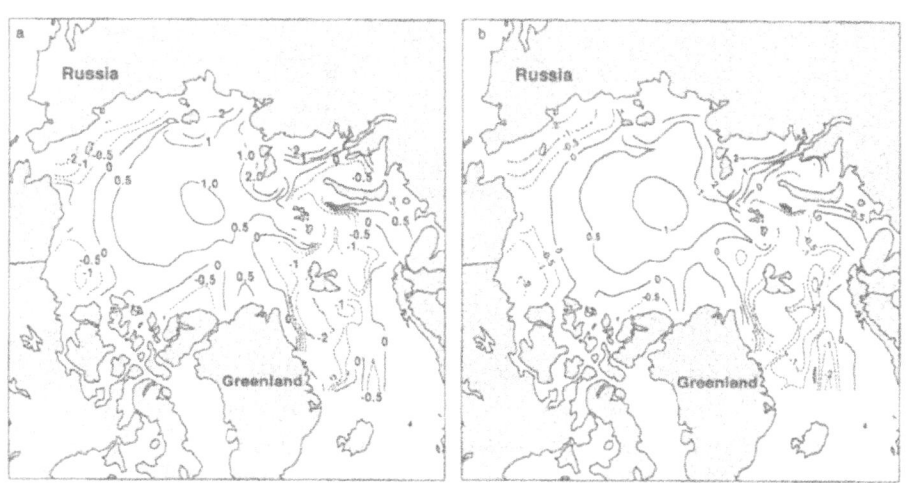

Figure 15. Net freezing rate when using (a) the viscous-plastic rheology, and (b) the cavitating fluid rheology, from Flato and Hibler [55].

Fichefet and Morales Maqueda [58] also performed a variety of interesting sensitivity experiments in which model physics were varied and the effect on the seasonal cycle of total ice volume in each hemisphere (northern and southern) was examined. Seasonal time series of hemispheric-average ice volume are useful since they integrate out dynamic effects (i.e., advection, diffusion, and internal stresses) and can be viewed as the cumulative sum of growth minus melt (see discussion below Eq. 2).

Figure 16a (Figure 7 from Fichefet and Morales Maqueda [58]) shows how the seasonal cycle of ice extent in the Northern Hemisphere changes when thermal inertia (from both sensible and latent heat storage) in the snow and ice are neglected. The main effect is increased melting in summer, when the thinnest first year ice has melted away and the ice cover is dominated by thick ice. This makes sense, since thermal inertia is most obviously important for thick ice (see Section 2.3.3). Fichefet and Morales Maqueda [58] chose not to tune albedos and thermal conductivities to get an annual average ice thickness similar to the standard case, in contrast to the 0-level model of Semtner [16]. Thus **Figure 16b** shows that the mean ice volume is also affected. (In fact,

Semtner [35] does show one example, his Figure 7a, wherein mean thickness behaves similarly to Figure 16b.) Further, dynamics definitely does play a role here; i.e., it moderates the difference between the two experiments relative to the "thermodynamics only" one-dimensional simulation of Semtner [16], especially in terms of the phases of maximum and minimum ice volumes (Fichefet and Morales Maqueda [58]).

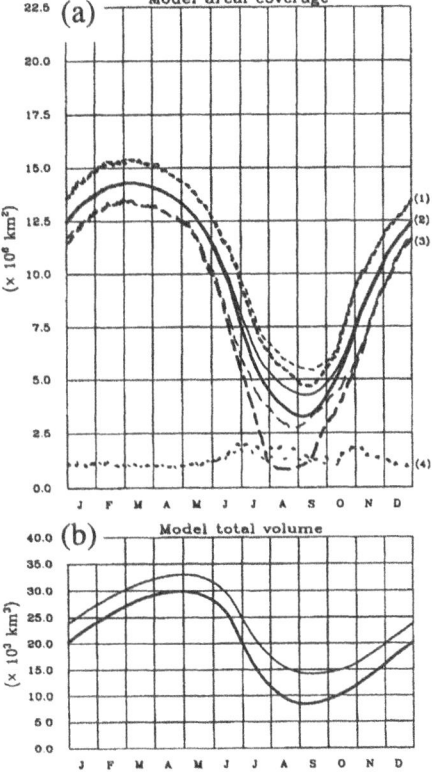

Figure 16. Seasonal cycles of ice area and volume in the simulation of Fichefet and Morales Maqueda [58]. Panel (a) shows the ice extent (curves labeled 1), effective ice area (curves labeled 2), the area enclosed by the 85% concentration contour (curves labeled 3), and the open water area (curves labeled 4). Effective ice area is the sum of grid cell areas containing ice, weighted by ice concentration. The thicker (thinner) curves denote the case with (without) thermal inertia. Panel (b) shows the difference in total domain ice volume in the two cases, which by Eq. 2 provides the net freezing rate over the domain.

Fichefet and Morales Maqueda [58] performed additional sensitivity studies which are summarized here. First, they ran a case wherein sensible heat storage was allowed but latent heat (i.e., the effect of brine pockets) was neglected. The result indicated that latent heat storage via brine pockets is much more important than sensible heat storage. A second sensitivity study was performed in which the uniform thickness distribution parameterization was a two-category scheme in which the ice is treated as either open water or a uniform slab. The result confirmed one-dimensional model results discussed in Section 2.3.2 and summarized in Table 2. A third sensitivity experiment was performed wherein no snow was allowed to accumulate on the sea ice. The model predicted slightly increased winter growth and negligible change in summer melt, leading to a slightly larger ice volume than the standard case, again consistent with one-dimensional studies (Table 2). Finally, a simulation was performed wherein no shortwave energy was allowed to penetrate the snow and ice. The result was reduced ice thickness and slightly more summer melting, in keeping with the results of Semtner's (1984) one-dimensional study. Ignoring penetrating radiation tended to increase the top surface melting, but also reduced the bottom surface melting. The latter effect only partially compensated, so that the net was increased summer melting. Finally, it should be noted that Fichefet and Morales Maqueda [58] found very different sensitivities in Southern Hemisphere sea ice, owing largely to its seasonal nature and thicker snow pack.

We conclude this section with an illustration of the roles of wind and ocean forcing in net freezing rate. **Figure 17** (from Hakkinen and Mellor [54]) shows how the same model responds when daily wind forcing (in this case, from the year 1987) is used instead of monthly means. Comparison of Figure 17a and Figure 17b reveals that the amplitudes of both net growth and net melt in this model have increased under daily wind forcing. The reason is that daily winds have a much larger amplitude than monthly winds, and so the effect of dynamics, discussed above, is damped in the monthly mean forcing case. Smaller monthly winds also affect ice thermodynamics via the turbulent (sensible and latent heat) surface fluxes which are proportional to the square of the wind speed. **Figure 18** (from Zhang et al. [59]) shows the effect in a coupled ice-ocean model of different climate restoring schemes. These schemes are designed to keep the model hydrography from drifting away from known climatologies owing to the accumulation of errors in model physics and/or forcing. The drift can be substantial, which leads to variations in the heat flux from the ocean to the ice, and thus to the substantial variation in net freezing rate shown in Figure 18 (note especially the case with surface restoring).

3.4 TEMPORAL VARIABILITY OF THE ARCTIC ICE MASS BUDGET

So far we have focused primarily on the mean seasonal cycle of the Arctic ice cover. In this section we discuss variability of ice growth, melt and transport in order to illustrate the potential role of sampling variability in explaining some of the differences between models noted in Table 4, and, more importantly, to highlight the role of the Arctic ice cover in driving variability in the North Atlantic Ocean.

We begin with an illustration of interannual variability provided by Hilmer et al. [60] in **Figure 19**. This figure shows the anomaly from the long-term mean (c.f. Figure 12) of net freezing rate for simulated years 1968 and 1995. Also shown is the annual mean sea ice velocity anomaly for each of these years. In the 1968 example (Figure 19a), there is an unusual shift of the transport pattern so as to produce unusually large export into the Barents Sea and then into the central Greenland Sea, with an associated anomaly of ice melt in both regions. The conjecture (Hilmer et al. [60]) is that this anomalous circulation pattern might have played a role in triggering the Great Salinity Anomaly (Dickson et al.

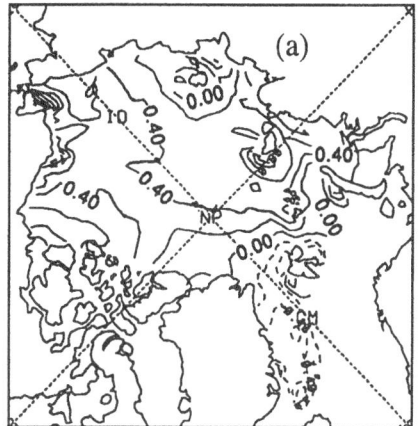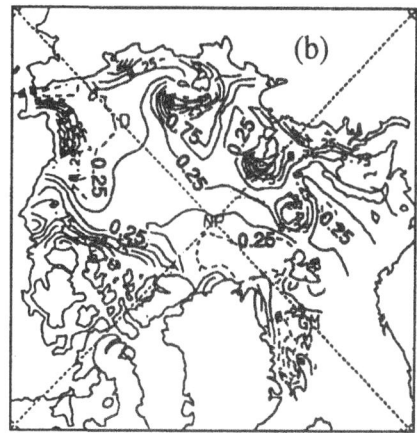

Figure 17. Net freezing rate in the model of Hakkinen and Mellor [54], under (a) monthly climatological wind forcing (contour interval 0.20 m yr^{-1}), and (b) daily wind forcing from the year 1987 (contour interval 0.25 m yr^{-1}).

[70]) of the next few years. The 1995 example, in Figure 19b, shows a more robust southward transport through Fram Strait, leading to enhanced melt rates in the southern Greenland Sea. More subtle differences in net freezing rate are also apparent in the Beaufort Sea, despite qualitatively similar mean velocity fields. In 1968 there is a pattern of anomalously high net freezing rate in the eastern Beaufort and anomalously low net freezing rate in the western Beaufort. This can be contrasted to the 1995 pattern in which the entire Beaufort Sea experiences an anomalously low net freezing rate, sufficient in fact to change the sign (relative to the long-term mean in **Figure** 12) from net growth to net melt. The implication of this comparison is that interannual variability is sufficiently large, particularly in the Beaufort Sea, to offer an explanation for the model discrepancies in Table 4. A similar conclusion was reached by Tremblay and Mysak [71], who found that anomalous winds were the dominant force in creating anomalous modeled summer ice coverage in the Beaufort and Chukchi Seas.

The spatial and temporal variability of Arctic ice thickness was explored by Flato [72], using an updated version of the thickness distribution model of Hibler [4] which was described in Flato and Hibler [73]. The years 1951-1989 were simulated, using daily winds and monthly air temperatures. This study did not present an explicit map of net freezing rate; however, its results on ice thickness variability are of relevance to the subject. Some previously unpublished results from this simulation are also discussed here.

Figure 20 shows 39-year time series of the $\langle V \rangle$, $\langle G \rangle$, and $\langle O \rangle$ terms in Eq. 2. In this case the averaging has been performed over the Arctic Ocean, minus the Barents and Kara

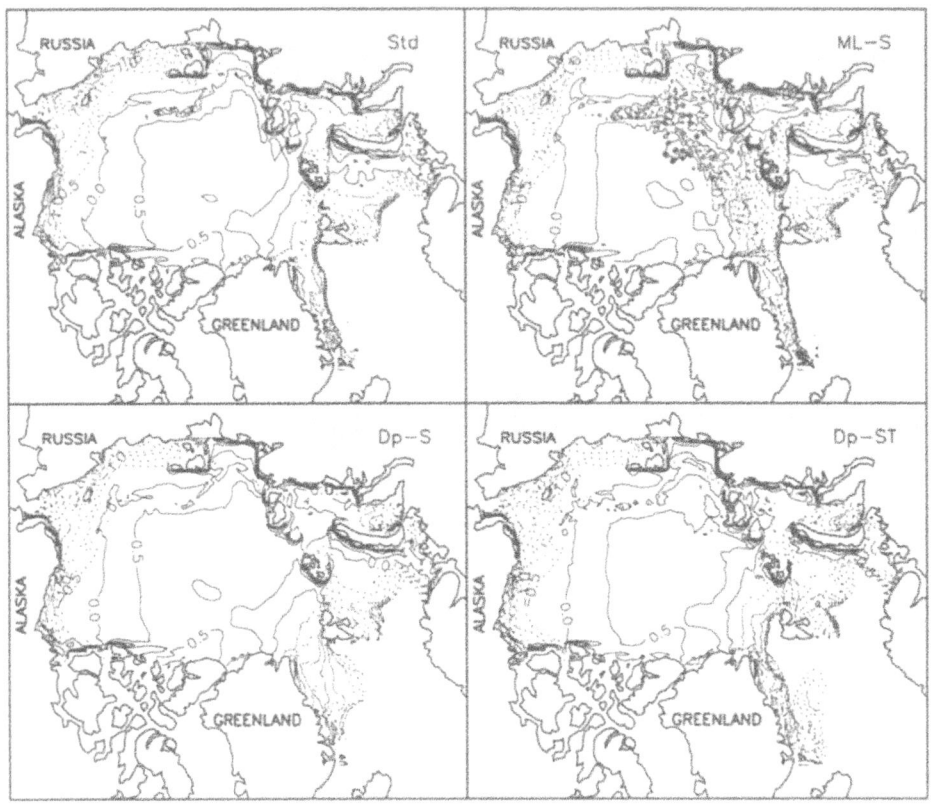

Figure 18. Net freezing rate for the period 1979-1985 as simulated by Zhang et al. [59]. In "Std" no climate restoring was used; in "ML-S" 30-day restoring of the mixed layer salinity was used; and in "Dp-S", "Dp-ST" 5-year restoring of the deep (below the mixed layer) salinity and salinity/temperature was used.

Seas. Ice in- and outflows through the Barents Sea, Bering Strait, and Canadian Archipelago are neglected here in favor of the dominant outflow through Fram Strait.

Figure 20a shows the monthly time series of ice volume, $\langle V \rangle$, as well as its anomaly time series (obtained by subtracting the mean seasonal cycle). A transition from "low volume" to "high volume" occurred in the 1970's, the origin of which is uncertain. Figure 20b shows the corresponding anomalies of the annual mean of $\langle G \rangle$, (which is, by definition, the net freezing rate) and Fram Strait outflow, $\langle O \rangle$, (defined as positive southward). There are several things to note in these figures. First, the modeled interannual variability of the ice growth anomaly is somewhat larger than that of Fram Strait outflow (standard deviations are 647 and 496 km^3 yr^{-1} respectively, which represent roughly 1/3 of their respective mean values). However, this variability is correlated to some extent (correlation coefficient is 0.34, which is significant at the 95% level) such that

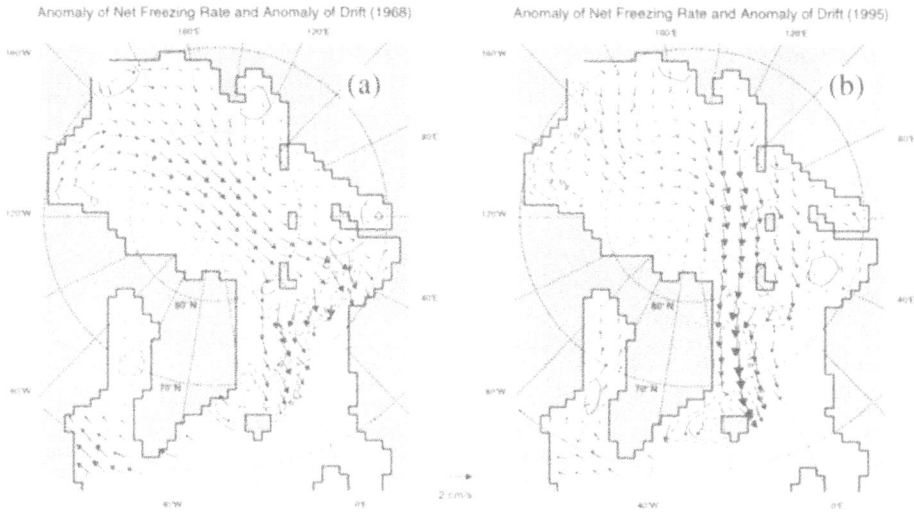

Figure 19. The anomaly (m yr⁻¹) of net freezing rate from the 1958-1997 mean in (a) 1968 and (b) 1995 from Hilmer et al. [60]. Also shown is the anomaly of annual mean ice drift (vectors).

years of high outflow are also years of high growth. This correlation is largely a local effect whereby positive anomalies in the predominantly wind-driven outflow causes anomalous open water and thin ice regions "upstream" of Fram Strait which then enhance local growth. The second point is that the anomalies in growth and outflow are relatively small compared to the predominantly low-frequency variability in the total ice volume. As noted by Flato [72], the variability of ice volume is dominated by the variability of thick, ridged ice which introduces a long time-scale and hence integrates the relatively high-frequency growth and outflow anomalies. However, we note that Bitz et al. [33] illustrated the potential for large thermodynamically-forced variability in a single-column model, and in fact estimated resulting volume anomalies much larger than those shown in Figure 20a.

Figure 21 shows the interannual standard deviation of mean ice thickness for February and August. Highest variability occurs in a band starting in the Greenland Sea and extending north and west across the Lincoln, Beaufort, Chukchi, and East Siberian Seas. Especially high variance is predicted in the Beaufort and Chukchi Seas, which Flato [72] speculated might be from interannual variability in the amount of ridging in these regions (which itself is influenced by the winds). This result reiterates the potential importance of sampling variability in explaining model differences (Table 4), particularly in the Beaufort and Lincoln Seas.

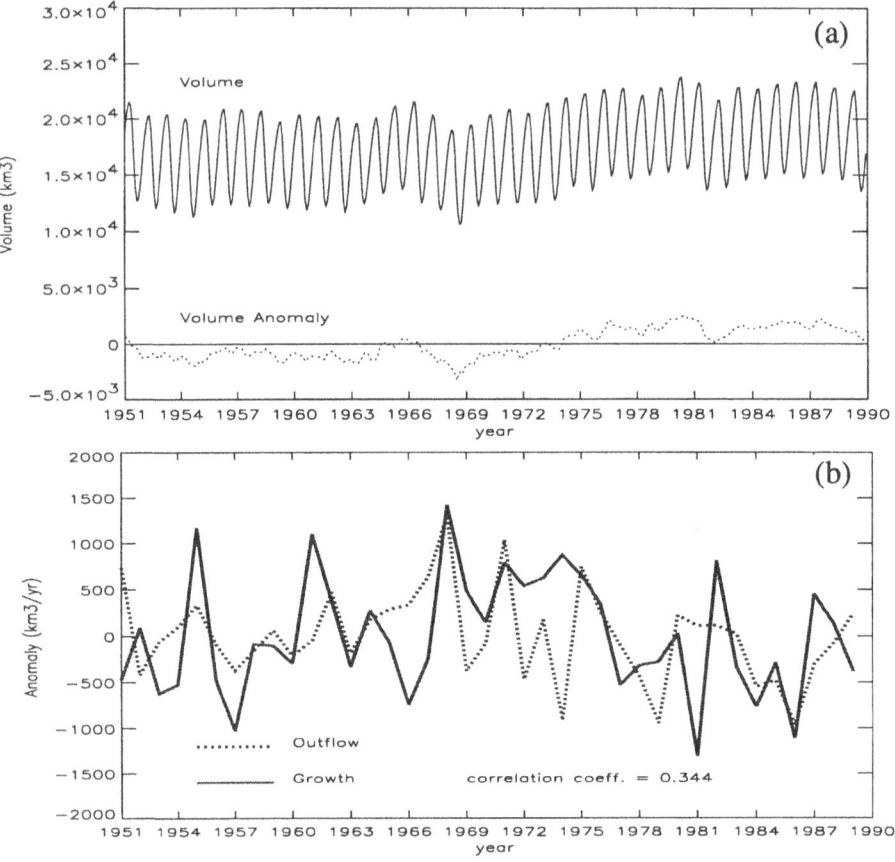

Figure 20. Time series of (a) Arctic ice volume and volume anomaly computed using the model described by Flato [72] and (b) the anomaly of net freezing rate for the Arctic (solid line) and the anomaly of Fram Strait outflow (dashed) from the same model and forcing. To allow easy comparison, the growth rate in (b) has been multiplied by the area of the Arctic Ocean (less the Barents and Kara Seas) to yield units of $km^3 \, yr^{-1}$.

3.5 SUMMARY

Table 5 presents a qualitative summary of the influence of model physics and forcing on the amplitude of the seasonal cycle of growth and melt in two-dimensional sea ice models. One factor that is conspicuously missing is the resolution of the thickness distribution, since there has been nothing published so far on this topic. Also absent is a study of how spatial resolution might affect growth and melt. We expect that most numerical models underestimate growth and melt extrema that occur in polynyas, leads, and river mouths, owing in part to low resolution and in part to a lack of small-scale physical process parameterizations. Finally, we note that daily air temperature data sets are recently

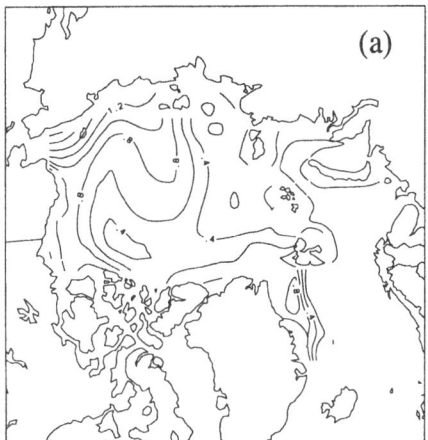 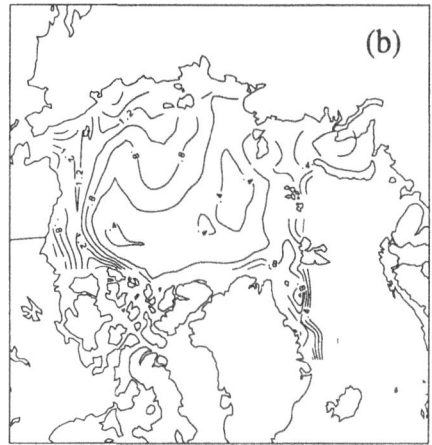

Figure 21. Interannual standard deviation (m) of (a) February and (b) August mean ice thickness over the period 1951-1989, from results discussed in Flato [72].

available (e.g., Martin and Muñoz [74]) that might produce higher growth and melt variability relative to the usual monthly forcing.

It is important to note that model sensitivity may be very different on seasonal and multiyear time scales. For example, ice dynamics has been shown to increase the amplitude of the seasonal cycle of growth and melt in one-dimensional (Mellor and Kantha [28]; Ebert and Curry [24]) and two-dimensional (Hibler [52]) models. Yet dynamics tends to decrease the sensitivity of mean annual ice thickness to forcing perturbations, e.g. changes in air-to-ice or ocean-to-ice heat flux (e.g., Hibler [75]; Flato [76]; Arbetter et al. [77]). As noted in Sections 1 and 2, although the mean annual ice thickness is an important parameter, the seasonal cycle of growth and melt is perhaps equally important in terms of the surface fluxes of heat, salt, and freshwater.

The two-dimensional models described in this section vary in their estimates of net freezing rate by 100-200% in most regions. The best agreement is in the central Arctic, where most models find net growth of about 0.5 m yr^{-1}, a value that agrees well with the admittedly sparse observations. The most glaring differences occur in the Lincoln and Beaufort Seas, where some models predict net annual growth and others predict net annual melt. We stress that the variance between simulations is not really much larger in these regions, only that the sign differences are distressing. Observations in the Beaufort Sea predict strong net growth of about 1.5-2.0 m yr^{-1} (Melling and Moore [50]; Macdonald et al. [51]). Some of this discrepancy might be traced to inadequacies in model physics and/or forcing, but some of it might also derive from sampling variability. Another region of sign disagreement is in the Lincoln Sea, an area filled with very thick ice which might be expected to experience net melting over the year (e.g., Thorndike et al. [3]. Although the mean ice motion is against the coast, there is substantial interannual

582

variability (Figure 19) and perhaps weak divergence in the alongshore component of ice velocity (R. Colony, personal comm.) which might easily produce net growth in this very cold area.

TABLE 5. How the amplitude of the seasonal cycle is affected by simplifications in model physics in two-dimensional models.

Simplification	Seasonal Cycle Amplitude	Reference
no dynamics	reduced	Hibler [52]
monthly average winds	reduced	Hakkinen & Mellor [54]
no thermal inertia	increased	Fichefet & Maqueda [58]
no penetrating radiation	increased	Fichefet & Maqueda [58]

4. Discussion

Sea ice growth and melt, together with horizontal transport, play a central role in the freshwater budget of the Arctic. Because direct observations of growth and melt are sparse, model studies of various kinds have been used to estimate growth and melt, their spatial and temporal variability, and the processes involved. In this chapter we provide a survey of model studies regarding Arctic sea ice growth and melt, and, where possible, compare model results to observations.

Models used in these studies can be classified as one-dimensional or two-dimensional. In the former, spatial variability is not explicitly simulated (although its effects are often parameterized); whereas in the latter, the spatial pattern of growth and melt is simulated. Owing to their simplicity, one-dimensional models have been extensively used to investigate the role of various thermodynamic processes in the seasonal cycle of growth and melt (these are summarized in Table 2). A shortcoming of one-dimensional model studies is that net growth over an annual cycle must be prescribed (by specifying the net divergence).

Two-dimensional models allow explicit representation of local imbalances between growth/melt and transport, and so provide information about the spatial patterns of freshwater fluxes at the ocean surface. A useful diagnostic of this imbalance is the annual total of growth minus melt, termed the 'net freezing rate'. In Table 4 we compare the net freezing rate produced in several regions by various Arctic models. We find, in agreement with the rather sparse observations, that most models produce a net freezing rate of about 0.5 m yr^{-1} in the central Arctic. All the models surveyed produced higher net freezing rates in the Laptev Sea, with values ranging from about 0.5 to 3 m yr^{-1}. The models also generally produced negative freezing rates (i.e. net melt) in the Chukchi, Barents, and GIN Seas, with values ranging from -0.5 to -2 m yr^{-1} in the Chukchi, and values up to -4 m yr^{-1} in certain parts of the GIN and Barents Seas (although there is substantial spatial variability in the latter regions). The Beaufort and Lincoln Seas were regions of disagreement between models, even as to the sign of the net freezing rate. Although observations indicate net growth in the Beaufort Sea, model estimates indicate that this is an area of large interannual variability, which may explain some of the differences. Even when integrated over most of the Arctic Ocean, the available model estimate of interannual variability in net freezing rate (Figure 20) yields a standard deviation of 650 km^3 yr^{-1}, roughly 1/3 of its mean value. However, longer-term simulations with a one-dimensional model imply even higher variability.

Although we have attempted to include the most recent model results, continuing model developments will inevitably render this summary outdated. Particular model improvements currently being implemented include better representation of the sub-grid-scale thickness distribution, better resolution of the vertical temperature profile and associated heat conduction, and better representation of the complex surface processes like snowpack and meltpond evolution which so strongly impact the surface energy balance. In addition, model resolution continues to improve, so that ice transport through the Canadian Archipelago and Bering Strait may soon be resolved. Better resolution would also assist in improving parameterizations of growth and melt processes near river mouths where freshwater enters the Arctic Ocean (Omstedt et al. [78]). Finally, progress in remote-sensing and data assimilation techniques, along with ambitious field programs now underway, offer the promise of improved observational estimates of the ice mass balance.

5. Future Research Priorities

I. Expand the observational data base on sea ice growth and melt. For example, satellite-based studies (using synthetic aperture radar or other new sensors) might help refine our estimates of winter growth. Also, more in situ chemical tracer data would be invaluable in determining freshwater sources, including ice melt and growth. These and other studies should be expanded to cover all high latitude perennial and seasonal sea ice regions. Also, use should be made of recent comprehensive data sets (e.g. NASA Polar Pathfinder, SHEBA annual cycle) to improve the representation of various physical processes affecting growth and melt. Examples include: surface albedo, melt ponds and snow cover, internal heat storage, lateral melt, and surface radiative fluxes.

II. Determine the roles that variability and uncertainty in forcing data play in model estimates of growth and melt. This topic has not been explored in the literature, although Fischer and Lemke [79] did examine how forcing uncertainty affects mean ice thickness.

III. Examine the regional and interannual variability of growth and melt using one or several standard models.

AcknowledgementsWe thank the organizers of the NATO Advanced Research Workshop in Tallinn for a scientifically and culturally rewarding experience. The manuscript benefited from comments by C. Bitz, M. Harder, M. Hilmer, P. Lemke, G. Maykut, H. Stern, and J. Zhang; conversations with S. Martin and B. Semtner also helped to clarify our thinking. We thank M. Hilmer and J. Schramm for access to updated figures and data. M. Steele was supported by an EOS interdisciplinary investigation, POLar Exchange at the Sea surface (POLES), NASA grants NAGW-2407 and NAG5-4375, and by the Surface HEat Budget of the Arctic Ocean (SHEBA) project, NSF grant 0PP9701592. G. Flato is a member of the Canadian EOS interdisciplinary investigation, CRYSYS.

6.   References

1. Aagaard, K. and Carmack, E.C. (1989) The role of sea ice and other freshwater in the Arctic circulation, *J.Geophys. Res.*, **94**, 14,485-14,498.
2. Wijffels, SE., Schmitt, R.W., Bryden, H.L., and Stigebrandt. A. (1992) Transport of freshwater by the oceans, *J. Phys. Oceanogr*, **22**, 155-162.
3. Thorndike, A.S., Rotbrock, DA., Maykut, GA., and Colony, R. (1975) The thickness distribution of sea ice.*J. Geophys. Res.*, **80**, 4501-4513.
4. Hibler, W.D. 111(1980) Modeling a variable thickness sea ice cover, *Mon. Wea. Rev.*, **108**, 1943-1973.
5. Steele, M., Thomas, D., Rothrock, D., and Martin, 5. (1996) A simple model study of the Arctic Ocean freshwater balance, 1979-1985,*J Geophys. Res.*, **101**,20,833-20,848.
6. McLaren. A.S., Bourke, R.H., Walsh, JE., and Weaver, R.L. (1994) Variability in sea-ice thickness overthe North Pole from 1958 to 1992, in *The Polar Oceans and Their Role in Shaping the Global Environment*, edited by O.M. Johannessen et al., pp.363-371, American Geophysical Union.
7. Koemer, R.M. (1973) The mass balance of the sea ice of the Arctic Ocean, J. Glaciol., **12**, 173-1 85.
8. Vinje, T., Nordlund, N., and Kvambekk, A. (1998) Monitoring ice thickness in Fram Strait. *J. Geophys. Res.*, **103**, 10,437-10,449.
9. Scbwerdtfeger, P. (1963) The thermal properties of sea ice. J. Glaciol., **4**, 789-807.
10. Untersteiner, N. (1961) On the mass and heat budget of Arctic sea ice, *Arch. MeteoroL Geophys. Biokumatol.*, A, **12**, 151-182.
11. Ono, N. (1967) Specific heat and heat of fusion of sea ice, in *Physics of Snow and Ice, Volume I*, edited by H. Oura, pp.599-610, Inst. of Low Temperature Sci., Hokkaido, Japan.
12. Bitz, C.M. and Lipscomb, W.R. (1999) An energy-conserving thermodynamic sea ice model, *J. Geophys. Res.*, in review.
13. Maykut, G.A. and Untersteiner, N. (1971) Some results from a time-dependent, thermodynamic model of sea ice,*J. Geophys. Res.*, **76**, 1550-1575.
14. Maykut, G.A. (1986) The surface heat and mass balance, in *The Geophysics Of Sea Ice*, edited by N. Untersteiner, pp.395-463, Plenum.

15. Stefan, J. (1891) Über die Theorie der Eisbildung, insbesondere fiber die Eisbildung im Polarmeere. *Ann. Physik, 3rd Ser*, **42**, 269-286.
16. Semtner, A.J. Jr. (1976) A model for the thermodynamic growth of sea ice in numerical investigations of climate. *J Phys. Oceanogr.*, **6**, 379-389.
17. Gabison, R., (1987) A Thermodynamic model of the formation, growth, and decay of first-year sea ice. *J. Glaciol.*, **33**(113), 105-119.
18. Flato, G.M. and Brown, R.D. (1996) Variability and climate sensitivity of landfast Arctic sea ice. *J. Geophys. Res.*, **101**, 25, 767-25,777.
19. Lindsay, R.W. (1998) Temporal variability of the energy balance of thick Arctic pack ice, *J. Climate*, **11**, 313-333.
20. Maykut, G.A. (1982) Large-scale heat exchange and ice production in the central Arctic, *J. Geophys. Res.*, **87**, 7971-7984.
21. Björk, G. (1992) On the response of the equilibrium thickness distribution of sea ice to ice export, mechanical deformation, and thermal forcing with application to the Arctic Ocean. *J. Geophys. Res.*, **97**(C7), 11, 287-11,298.
22. Scbramm, J.L., Holland, M.M., and Curry, J.A. (1997) Modeling the thermodynamics of a sea ice thickness distribution 1. Sensitivity to ice thickness resolution. *J. Geophys. Res.*, **102**, 23,079-23,091.
23. Holland, M.M., Curry, J.A., and Schramm, J.L. (1997) Modeling the thermodynamics of a sea ice thickness distribution 2. Sea ice/ocean interactions, *J. Geophys. Res.*, **102**, 23,093-23,107.
24. Ebert, E.E. and Curry, J.A. (1993) An intermediate one-dimensional thermodynamic sea ice model for investigating ice-atmosphere interactions. *J. Geophys. Res.*, **98**, 10,085-l0,109.
25. Maykut, G.A. and Perovich, D.K. (1987) The role of shortwave radiation in the summer decay of a sea ice cover, *J. Geophys. Res.*, **92**, 7032-7044.
26. Josberger. E.G. and Martin, 5. (1981) A laboratory and theoretical study of the boundary layer adjacent to a vertical melting ice wall in saltwater ,*J.. Fluid Mech.*, **111**, 439-473.
27. Steele, M. (1992) Sea ice melting and floe geometry in a simple ice-ocean model, *J. Geophys. Res.*, **97**, 17,729-17,738.
28. Mellor, G.L. and Kantha, L. (1989) An ice-ocean coupled model, *J. Geophys. Res.*, **94**, 10,937-10,954.
29. Harvey, L.D.D. (1990) Testing alternative parameterizations of lateral melting and upward basal heat flux in a thermodynamic sea ice model, *J. Geophys. Res.*, **95**, 7359-7365.
30. Ebert, E.E., Scbramm, J.L., and Curry, J.A. (1995) Disposition of solar radiation in sea ice and the upper ocean, *J. Geophys Res.*, **100**, 15,965-15,975.
31. MePhee, M.G. and Untersteiner, N. (1982) Using sea ice to measure vertical beat flux in the ocean, *J. Geophys. Res.*, **87**, 2071-2074.
32. Wettlaufer, J.S., (1991) Heat flux at the ice-ocean interface, *J. Geophys. Res.*, **96**, 7215-7236.
33. Bitz, C.M., Battisti, D.S., Moritz, R.E., and Beesley, J.A. (1996) Low frequency variability in the Arctic atmosphere, sea ice, and upper ocean climate system. *J. Climate*, **9**(2), 394-408.
34. Ledley, T.S. (1991) Snow on sea ice: competing effects in shaping climate. *J. Geophys. Res.*, **96**, 17,195-17,208.
35. Semtner, A.J. Jr. (1984) On modelling the seasonal thermodynamic cycle of sea ice in studies of climatic change, *Climatic Change*, **6**, 27-37.
36. Curry, J.A., Schramm, J.L., and Ebert, E.E. (1995) Sea ice-albedo climate feedback mechanism. *J. Climate*, **8**, 240-147.
37. Randall, D., Curry, J., Battisti, D., Flato, G., Grumbine, R., Hilkinen, S., Martinson, D., Preller, R., Walsh, J., and Weatherly, J. (1998) Status and outlook for large-scale modeling of atmosphere-ice-ocean interactions in the Arctic. *Bull Am. Meteorol. Soc.*, **79**, 197-219.
38. Bitz, C.M. (1997) A model study of natural variability in the Arctic climate, PhD thesis, Univ. of Wash., Seattle, WA, USA.
39. Bourke, R.H. and Garrett, R.P (1987) Sea ice thickness distribution in the Arctic Ocean, Cold Reg. *Sci. Technol*, **13**, 259-280.
40. Keigwin, L.D. and Johnson, G.L. (1992) Using a nuclear submarine for Arctic research, *EOS, Trans. Amer Geophys. Union*, **73**(19).
41. Martin, S. and Cavalieri, D.J. (1989) Contributions of the Siberian shelf polynyas to the Arctic Ocean intermediate and deep water, *J. Geophys. Res.*, **94**, 12,725-12,738.
42. Cavalieri, D.J. and Martin, 5. (1994) The contribution of Alaskan, Siberian, and Canadian coastal polynyas to the cold halocline layer of the Arctic Ocean, *J. Geophys. Res.*, **99**, 18,343-18,362.
43. Bauch, D., Schlosser, P, and Fairbanks, R.G. (1995) Freshwater balance and the sources of deep and bottom waters in the Arctic Ocean inferred from the distribution of $H_2^{18}O$, *Prog. Oceanog.*, **35**, 53-80.
44. Colony, R. and Thorndike, A.S. (1984) An estimate of the mean field of Arctic sea ice motion, *J. Geophys. Res.*, **89**, 10,623-10,629.

586

45. Jones, E.P, Nelson, D.M., and Treguer, P (1990) Chemical oceanography, in *Polar Oceanography, Part B: Chemistry, Biology, and Geology*, 407-476, Academic Press.
46. Wallace, D.W.R., Seblosser, P., Krysell, M., and Bo~m'sch, G. (1992) Halocarbon ratio and tritium/^{3}He dating of water masses in the Nansen Basin, Arctic Ocean, *Deep-Sea Res.*, **39**(S2), S435-S458.
47. Serreze, MC. and Barry, R.G. (1988) Synoptic activity in the Arctic basin, 1979-85, *J.Climate*, **1**, 1276-1295.
48. Aagaard, K., Foldvik, A., and Hillman, S.R. (1987) The West Spitsbergen Current: Disposition and water mass transformation, *J. Geophys. Res.*, **92**, 3778-3784.
49. Ekwurzel, B. (1998) Circulation and mean residence times in the Arctic Ocean derived from tritium, helium, and oxygen-18 tracers, PhD thesis, Columbia Univ., New York, NY, USA.
50. Melling, H. and R.M. Moore (1995) Modification of halocline source waters during freezing on the Beaufort Sea shelf: evidence from oxygen isotopes and dissolved nutrients, *Cont. Shelf Res.*, **15**, 89-113.
51. Macdonald, R.W., Paton, D.W., Carmack, E.C., and Omstedt, A. (1995) The freshwater budget and under-ice spreading of Mackenzie River water in the Canadian Beaufort Sea based on salinity and 180,160 measurements in water and ice, *J. Geophys. Res.*, **100**, 895-919.
52. Hibler, W.D. 111(1979) A dynamic thermodynamic sea ice model, *J. Phys. Oceanogr*, **9**, 815-846.
53. Walsh, J.E., Hibler, W.D. III, and Ross, B. (1985) Numerical simulation of northern hemisphere sea ice variability, 1951-1980, *J. Geophys. Res.*, **90**, 4847-4865.
54. Hlkkinen, S. and Mellor, G.L. (1992) Modeling the seasonal variability of a coupled Arctic ice-ocean system, *J. Geophys. Res.*, **97**, 20,285-20,304.
55. Flato, G.M. and Hibler, W.D. III, (1992) Modeling pack ice as a cavitating fluid, *J. Phys. Oceanogr*, **22**, 626-651.
56. Chapman, W.L., Welch, W.J., Bowman, K.P., Sacks, J., and Walsh, J.E. (1994) Arctic sea ice variability: model sensitivities and a multidecadal simulation, *J. Geophys. Res.*, **99**, 919-935.
57. Thomas, D., Martin, S., Rothrock, D., and Steele, M. (1996) Assimilating satellite concentration data into an Arctic sea ice mass balance model, 1979-1985, *J. Geophys. Res.*, **101**, 20,849-20,868.
58. Fichefet, T. and Morales Maqueda, M.A. (1997) Sensitivity of a global sea ice model to the treatment of ice thermodynamics and dynamics, *J. Geophys. Res.*, **102**, 12,609-12,646.
59. Zhang, J., Hibler, W., Steele, M., and Rothrock, 0. (1998) Arctic ice-ocean modeling with and without climate restoring, *J. Phys. Oceanogr*, **28**, 191-217.
60. Hilmer, M., Harder, M., and Lemke, P. (1998) Sea ice transport: a highly variable link between Arctic and North Atlantic, *Geophys. Res. Lett*, **25**, 3359-3362.
61. Lernke, P., Owens, W.B., and Hibler, W.D. III, (1990) A coupled sea ice-mixed layer-pycnocline model for the Weddell Sea, *J. Geophys. Res.*, **95**, 9513-9525.
62. Owens, W.B. and Lemke, P. (1990) Sensitivity studies with a sea ice-mixed layer-pycnocline model in the Weddell Sea, *J. Geophys. Res.*, **95**, 9527-9538.
63. Södssel, A., Lernke, P, and Owens, W.B. (1990) Coupled sea ice-mixed layer simulations for the southern ocean, *J. Geophys. Res.*, **95**, 9539-9555.
64. Stössel, A., Oberhuber, J.M., and Maler-Reimer, 11(1996) On the representation of sea ice in global ocean general circulation models, *J. Geophys. Res.*, **101**,18,193-18,212.
65. Harder, M., Lenake, P., and Hilmer, M. (1998) Simulation of sea ice transport through Fram Strait: natural variability and sensitivity to forcing, *J. Geophys. Res.*, **103**, 5595-5606.
66. Flato, G.M. and Hibler, W.D. 111(1991) An initial numerical investigation of the extent of sea-ice ridging, *Ann. Glaciol*, **15**, 31-36.
67. Zhang, J. and Hibler, W.D. 111(1997) On an efficient numerical method for modeling sea ice dynamics, *J. Geophys. Res.*, **102**, 8691-8702.
68. Parlanson, C.L. and Washington, W.M. (1979) A large-scale numerical model of sea ice, *J. Geophys. Res.*, **84**, 311-337.
69. Kreyscber. M., Harder, M~, and Lenake, P (1997) First results of the Sea Ice Model Intercomparison Project (SIMIP), *Ann. Glaciol* **25**, 8-Il.
70. Dickson, R.R., Meineke, J., Malmberg, S.-A., and Lee, A.J. (1988) The "Great Salinity Anomaly" in the Northern North Atlantic 1968-1982, *Prog. Oceanog*, **20**, 103-151.
71. Tremblay, L.-B. and Mysak, L.A. (1998) On the origin and evolution of sea-ice anomalies in the Beaufort-Chukchi Sea, *Climate Dynamics*, **14**,451-460.
72. Flato, G.M. (1995) Spatial and temporal variability of Arctic ice thickness, *Ann. Glaciol.*, **21**, 323-329.
73. flato, G.M. and Hibler, W.D. 111(1995) Ridging and strength in modeling the thickness distribution of Arctic sea ice, *J. Geophys. Res.*, **100**, 18, 611-18,626.
74. Martin, S. and Munoz, E.A. (1997) Properties of the Arctic 2-meter air temperature field for 1979 to the present derived from a new gridded dataset. *J. Climate*, **10**, 1428-1440.

75. Hibler, W.D. 111(1984) The role of sea ice dynamics in modeling C02 increases. *Climate Processes and Climate Sensitivity, Geophysical Monograph 29, Am. Geophys. Union,* 238-253.
76. Flato, G.M. (1996) The role of dynamics in warming sensitivity of Arctic sea ice models, *Workshop on Polar Processes in Global Climate, Am. Meteorol. Soc.,* 113-114.
77. Arbetter, T.E., Curry, J.A., Holland, M.M. and Maslanik, J.A. (1997) Response of sea-ice models to perturbations in surface heat flux, *Ann. Glaciol,* **25**, 193-197.
78. Omstedt, A., Carmack, E.C., and Macdonald, R.W. (1994) Modeling the seasonal cycle of salinity in the Mackenzie shelf/estuary, *J. Geophys. Res.,* **99**, 10,011-10,021.
79. Fischer, H. and Leinke, P. (1994) On the required accuracy of atmospheric forcing fields for driving dynamic-thermodynamic sea ice models, in The Polar Oceans and Their Role in Shaping the Global Environment, *Geophys.Monogr Ser,* **85**, 373-381, AGU, Washington, D.C.

FRESH WATER FREEZING/MELTING CYCLE IN THE ARCTIC OCEAN

G.V. ALEKSEEV, L.V. BULATOV. V.F. ZAKHAROV
State Research Center of the Russian Federation
The Arctic and Antarctic Research Institute
Bering St., 38, St. Petersburg, 199397 Russia

1.   Introduction

The Arctic Ocean and its upper layer with sea ice at the surface constitute the main link in the formation of the arctic impact on global climate. Freshwater coming to the Arctic with a polar branch of the global hydrological cycle is accumulated and transformed here. Then it is transported as liquid and solid (sea ice) phases to e North Atlantic. These processes control spreading of a lens of freshened water and sea ice occupying the main portion of the northern polar cap at the planet surface (Fig. 1).

A relatively small salinity and a high content of freshwater, respectively, in a comparatively thin upper layer of the Arctic Ocean prevent its mixing with underlying water. This fact along with strong cooling in the winter provides preservation of drifting ice at the ocean surface at the present time. However. paleoclimatic data show that sea ice did not always form here around 700 kyr BP. The onset of grandiose glacial-interglacial climatic cycles of the Pleistocene epoch belongs approximately to this time.

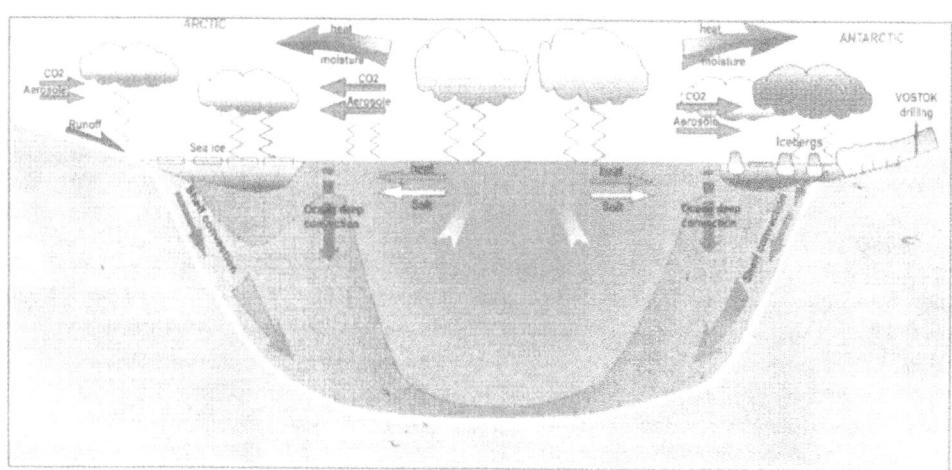

Figure 1. Scheme of interaction of polar and extrapolar processes in the climate system.

(For a colour version of this figure, see p. 620)

E.L. Lewis et al. (eds.), The Freshwater Budget of the Arctic Ocean, 589-608.

590

The latter does not appear a random coincidence testifying to the occurrence of oscillations in the global hydrological cycle with active participation of the processes of freezing and melting of freshwater [1,2]. Historical and current observation data on sea ice extent and water salinity in the upper North Atlantic layer adjoining the Arctic Ocean indicate significant and coordinated fluctuations of this and other climate characteristics in a vast area of high and temperate latitudes with a scale from decades to centuries. Numerical experiments using coupled climate ocean/atmosphere models also simulate such large-scale fluctuations as a climate response to the changes of fresh water balance constituents in the northern polar region. An important role in the development of these fluctuations belongs to the anomalies of spreading of freshened water and sea ice exported from the Arctic Ocean over the North Atlantic. in this respect, it is not clear what are the causes and mechanisms of the anomalous fresh water outflow to the North Atlantic and in particular, the role of the freezing and melting processes.

When the fresh water budget of the Arctic Ocean is considered, one usually means the constituents of the fresh water inflow to the Arctic Ocean and its outflow. However, freshwater transported to the Arctic Ocean undergoes in between these processes multiple phase transformations ("water-ice"), which change the dynamic and thermodynamic properties of the upper ocean layer and the time of residence of incoming fresh water and hence, influence the formation of fresh water outflow. These processes occurring in the ocean constitute a kind of internal arctic hydrological cycle acting together with the external cycle that consists of the processes of fresh water flow to the Arctic, its outflow to the Arctic Ocean and then to the North Atlantic (Fig. 2).

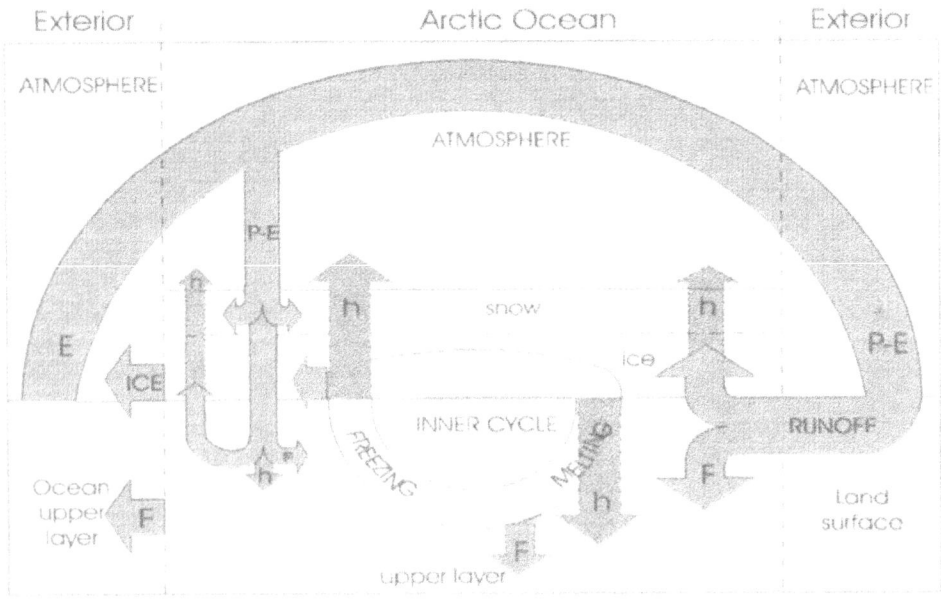

Figure 2. Scheme of fresh water (F) and latent heat (h) cycle in the upper layer of the Arctic Ocean.

(For a colour version of this figure, see p. 620)

One of the consequences of the interaction between the external and internal hydrological cycles in the Arctic is the interseasonal redistribution of freshwater between its liquid and solid phases on which their relative contribution to the fresh water outflow from the Arctic Ocean depends. The empirical and model estimates of these contributions, on average for a year show their considerable variation from a large excess of fresh water outflow in the solid phase over sea ice export to an ice export excess over the outflow. However, most model and empirical calculations yield an estimate close to an equal contribution of both constituents in the resulting fresh water outflow from the Arctic Ocean. The latter signifies that when considering the influence of the Arctic Ocean on global climate, it is necessary to take into account both constituents of the total freshwater outflow from it to the North Atlantic.

2. Freshwater content in the upper layer of the Arctic Ocean

The inflow of fresh water to the Arctic, and its outflow from the Arctic Ocean, comprise only part of the excess of fresh water in its upper layer resulting in the preservation of sea ice. In order to estimate all excess of freshwater, it is necessary to calculate its content relative to some a-priori prescribed salinity. Aagaard and Carmack [3] suggested using a salinity of 34.80 PSU as such a readout salinity. This choice has a deep physical sense since this value is close to the critical salinity value (about 34.82 PSU [4,5]). It separates the regimes of deep and shallow water convection in the central Greenland Sea whose water structure below the upper layer is close to water structure of the Arctic Basin. At water salinity in the upper layer close to critical, the mechanism of low temperature compressibility of sea water can be active, allowing sinking of convective plumes to large depths [3]. From this standpoint, if the content of fresh water in the layer with salinity less than 34.80 PSU were removed, deep convective mixing will be possible and hence, there will be an ice-free regime in the Arctic Ocean.

The calculations of the excess of freshwater in the upper layer from the surface to the depth of the isohaline 34.80 PSU are based on data of 8 winter-spring surveys of the Arctic Ocean by the Soviet "Sever" expeditions in 1955-1979 covering practically all the ocean. Also, data from the Joint U.S. Russian Atlas of the Arctic Ocean [6,7] averaged by decades and over 1950-1989 were employed. The use of water salinity observations in March-May has provided most accurate fresh water content estimates. This is the time when much of the summer fresh water inflow has turned into ice and the thermohaline structure of the upper layer is most homogeneous and its observations at a relatively sparse sequence of levels are most representative.

The mean distribution of freshwater content (FWC) over 1955-1979 in the layer above the isohaline 34.80 PSU (Fig. 3) indicates the maximum in the Beaufort Sea Gyre and a gradual decrease towards the Barents Sea and Fram Strait. The maximum FWC is formed by a combined contribution of strong freshening in the upper 30-50 m layer and the inflow of Pacific water whose influence here is noticeable down to a large depth. The layer where the fresh water content was estimated has a thickness within 120 m in the sub-Atlantic part of the Arctic Basin up to 800 m in the Canadian Basin. The vertical distribution of FWC is also non-uniform in these regions of the Arctic Ocean. In the sub-Atlantic part, its main portion (60%) is contained in the upper 50 m layer whereas in the

Canadian Basin only 40% of the total fresh water amount is in this layer. In general, the largest amount of freshwater is centered in this area under the influence of the anticyclonic circulation with freshening penetrating down to a depth of 800 m. This is, probably, related to its sinking in the center of the anticyclonic gyre [8].

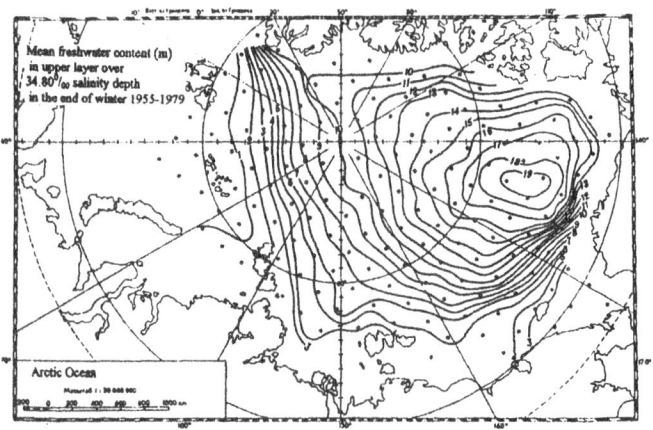

Figure 3. Mean freshwater content (m) in upper layer over 34.80 ‰ salinity depth in the end of
Winter 1955-1979

An analysis of the FWC distribution over 1955-1979 indicates not only its spatial non-homogeneity, but also a significant variability from year-to-year. The interannual FWC changes in the individual flows in the Canadian gyre reached 10 m. An example of the interannual FWC differences is given in Fig. 4 with the 1974 and 1977 distributions. Averaging of the FWC over the Arctic Ocean area (Table 1) shows its decrease from the first to the second half of the 1970s.

This change is even more pronounced in the upper 50 m layer where the "active" part of the FWC is centered participating in the interseasonal phase "water-sea ice" transformation, since this layer is located within the winter convective mixing depth (Fig. 5).

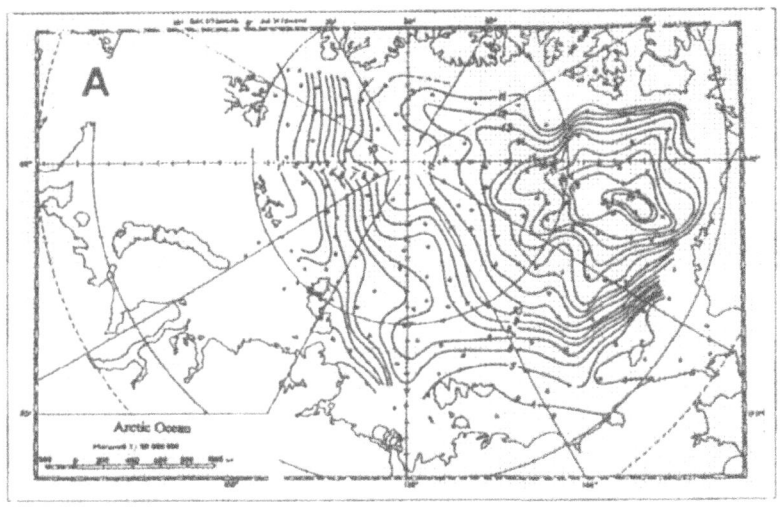

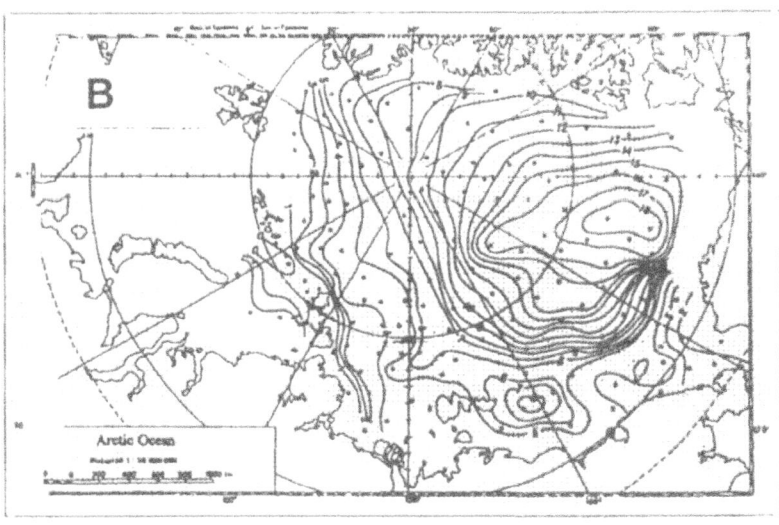

Figure 4. Distribution of freshwater content (FWC) in the layer above the isohaline 34.80 PSU

A) 1974   B) 1977

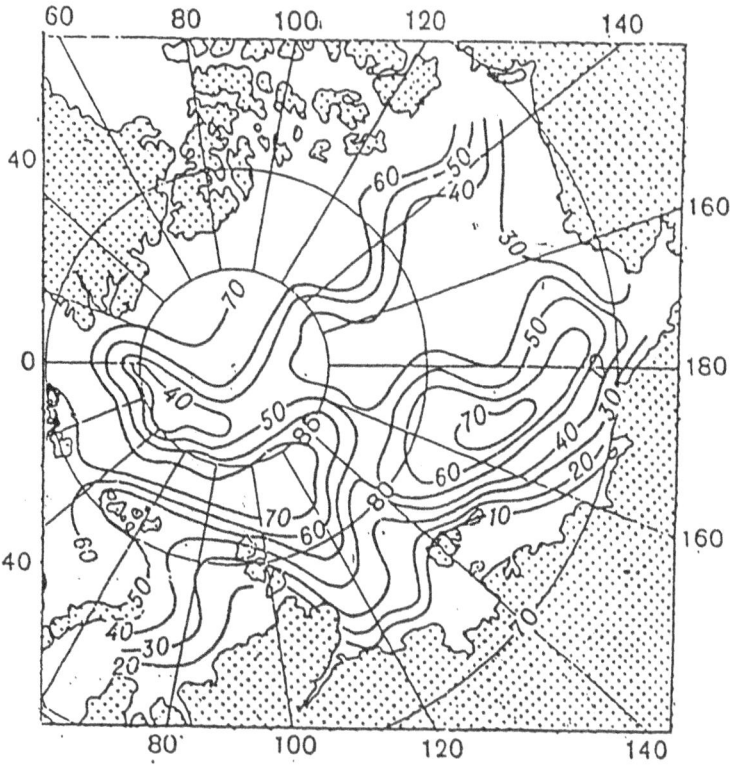

Figure 5. Depth of convection in the Arctic Ocean at the end of winter

    The mean FWC distribution in the upper 50 m layer over 1950-1989 calculated from data [6] is presented in Fig. 6. Unlike the distribution in Fig. 3, two areas of the maximum FWC values in the central Arctic Ocean and a zone of the river runoff influence off the coast of Eurasia are clear. As follows from Table 1, the FWC in this layer has decreased from 1974 to 1977 by 1.3m. This corresponds to a change in the freshwater outflow with the East Greenland Current equaling about 1000km^3 per year (at flow width of about 150 km and a speed of 15 cm/s).

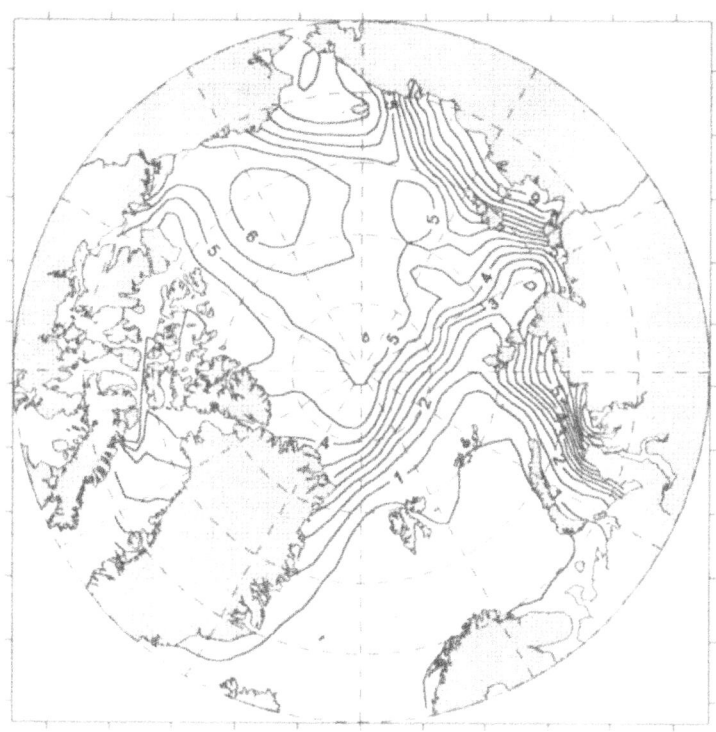

Figure 6. Mean FWC distribution in the upper 50 m layer; 1950-1989

TABLE 1. Mean freshwater content (m) in upper layers of the Arctic Basin: from observations of "5ever" expedition at the end of winter 1955-1977.

Year	0-depth of 34.80 ‰	0-50 m	Number of stations
1955-1956	10.46	4.13	111
1973	11.14	4.88	148
1974	10.47	4.55	137
1975	10.80	4.30	152
1976	9.68	3.69	126
1977	9.91	3.56	113

In the "Sever" expedition data in the area north of Greenland where the East Greenland Current originates there are also significant salinity variations in the upper layer at the end of winter (Table 2), which reflect the changed fresh water content in the upper active layer.

TABLE 2. Mean salinity (‰) at 25 m depth in region to the north of Greenland between 0°E and 45°W from observation of "Sever" expedition

	1955/1956	1973*	1974	1975	1976	1977	1978*	1979	Mean
Salinity, (‰)	31.71	31.10	31.64	31.76	32.27	32.54	32.22	32.54	31.97

* - number of observations were less than other years

Here, a considerable salinity increase from the first half of the period (1955-1975) to the second half (1976-1979) is noticeable equaling 0.84 PSU, on average. The latter corresponds to an FWC difference equal to 1.3 m and using the same parameters of the East Greenland Current (a width of about 150 kin and a speed of about 15 cm/s) to a change in the freshwater outflow by about 1000 km^3 a year.

A significant freshening of the upper layer in the first half of the 1970s as compared with the second half confirms the distribution of mean water salinity differences at a level of 25 m at the end of the winter during the periods 1973-1974 and 1976-1977 (Fig. 7). As is seen, the largest freshening of the upper layer in the early 1970s was in the areas adjoining the Arctic Ocean coast.

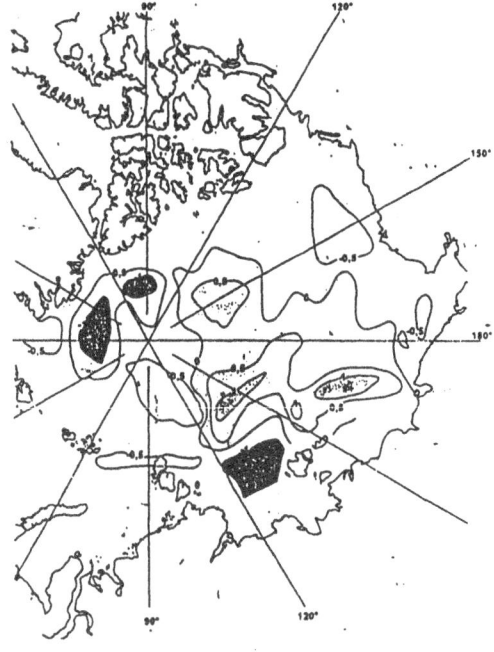

Figure 7. Differences in mean salinity at 25m depth at the end of winter 1973-1974 and 1976-1979 (average for 1973-1974 minus average for 1976-1979).

One more example of a large negative water salinity anomaly in the upper layer of the Arctic Ocean is provided by observations at the drifting station "North Pole-9" along its drift route from the New-Siberian Islands to the North Pole from May 25, 1960 to March 19, 1961 (Fig. 8). As is seen from the transect with the anomalous salinity values relative to the means for 1955-1979 in Fig. 8, they reach 3 PSU indicating an extremely strong freshening of the entire active layer along the drift trajectory of the station.

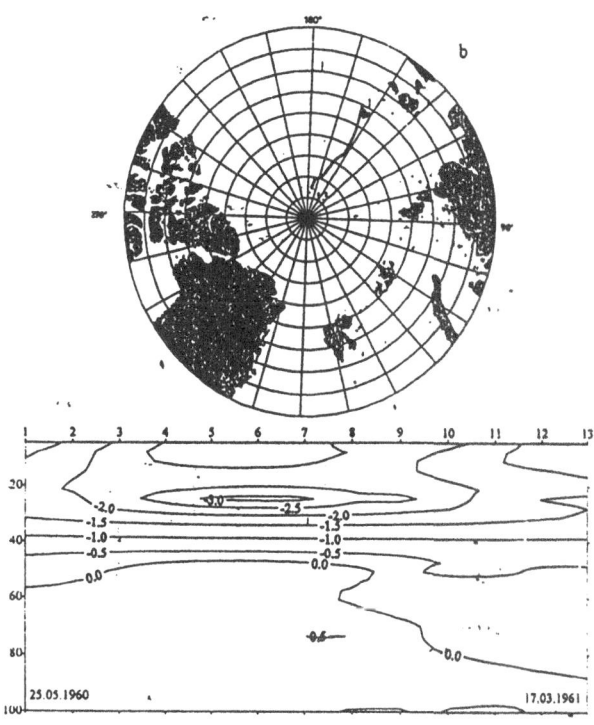

Figure 8. Anomaly of salinity (o/oo) in upper 100 m layer
(a) along the drift trajectory ofNP-9 station (b) in 1960-196 1

3. Contribution of summer melting to fresh water flows to the Arctic Ocean

In connection with such significant salinity variations from year-to-year and correspondingly, freshwater content in the upper layer of the Arctic Ocean, the question arises as to what causes them. The multiyear series of the components of the external freshwater flows to the Arctic Ocean [9,10,11] indicate a certain increase in precipitation and river runoff in the Arctic Ocean in the first half of the 1960s. However, the correlation of these changes with water salinity variations in different regions of the Arctic Ocean and the Nordic Seas is weak [12]. There are probably some other sources of freshening mentioned above and the main candidate could be the increased freshwater inflow in the summer resulting from anomalous ice and snow melting in the Arctic Ocean.

The calculations of freshwater balance constituents indicate that the main portion of fresh water is transported to the Arctic Ocean in the summer (Table 3 and 4). According to the data in these tables, a full inflow during the summer is estimated to be within 6000 to 10000 km3, which is much more than the estimates of the full yearly inflow (Table 3).

TABLE 3. Freshwater inflows to upper layer of the Arctic Ocean

Components	Summer inflow (June-August), km^3	Annual inflow, km3	Author
River runoff	1850	2850	Ivanov, [13]
P-E	1300	1300	Serreze et al, [14]
Sea ice melting	4250	-	Shpaikher, [15]
Total	7400	4150	

TABLE 4. Estimates of the summer inflow of fresh water to the upper layer of the Arctic Ocean

Method	Volume, km^3	Author
Calculation of freshwater budget	7400	Ivanov [13], Serreze et al. [14], Shpaiker [15]
Calculation of freshwater local budget in the Arctic Basin	6600-6700	Alekseev, Buzuyev [16]
Calculations of sea ice melting only in the Arctic Basin	3960-4580	Golovin et al. [17]
Calculation of salinity change in the upper layer from winter to summer.	5800-10,000	Treshnikov [18], Alekseev [19]

The discrepancy is eliminated during the winter by transformation of a considerable portion of freshwater to sea ice. However, the changed conditions of summer melting and to a lesser extent of winter ice formation can lead to a disturbed balance in the "freezing-melting" cycle and correspondingly to freshening or salination of the upper layer in some years.

Let us consider in more detail the features of development of the processes of summer melting and the freshwater inflow. Mean multiyear (1937-1975) interseasonal salinity changes in the upper layer in the Arctic Ocean are characterized [20] by the distribution of mean water salinity over the winter (January-May) and summer (July-October) at a depth of 5 m (Fig.9). The largest seasonal salinity change occurs in the zone of influence of the continental outflow and Bering water and the least in the central Arctic Ocean.

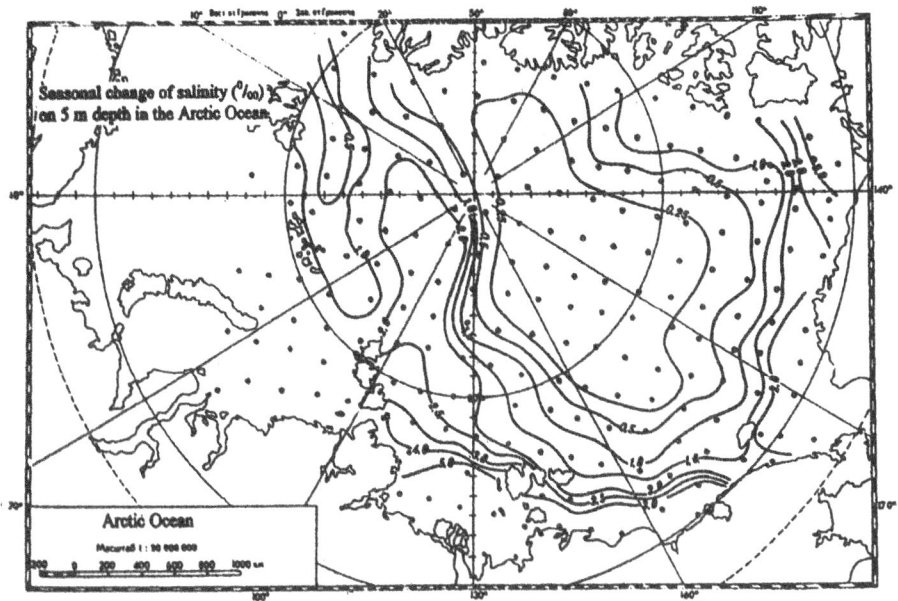

Figure 9. Seasonal change of salinity (‰) on 5m. depth in the Arctic Ocean.

The quantitative representation of the components of the summer freshwater flow to the upper layer in this region of the Arctic Basin is given in Table 5. It contains estimates of these components in the observation data of the drifting station "North Pole-16" in 1970-1971 in the region near 85^{0}N 140^{0}W. The main contribution to the summer inflow here belonged to ice and snow melting (the contribution to the fresh water layer is 24 and 20 cm, respectively).

Melting of snow at the ice surface together with summer liquid precipitation produced 26 cm, which exceeds the mean multiyear estimates for this region [21]. The deviation from the norm is attributed to significant interannual variations of the amount of precipitation, primarily of snow whose depth before the onset of melting in the Arctic Basin changes from 25 to 85 cm from observations at the drifting stations [22]. The conditions of summer melting of snow also vary from year-to-year as follows from Fig. 10, which presents the duration of the snow-free period at the drifting ice surface in the Arctic Basin (after complete snow melting). It is interesting to note that in the years from 1958 to 1966 there was complete snow melting in different regions of the Arctic Basin.

According to data of Table 5, the contribution of ice melting from the top is only 2 cm, which is much less than the mean value of ice melting from the top over the Arctic Basin. From observation data at the drifting stations in 1951-1981, this value was about 40 cm, on average, which changed depending on the ice surface shape (Table 6).

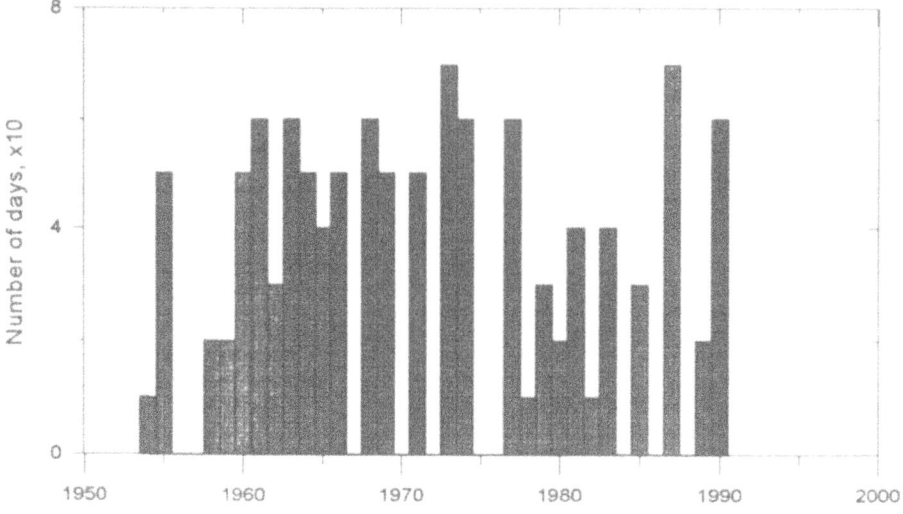

Figure 10. Duration of the snow-free period at the drifting ice surface in the Arctic Basin.

TABLE 5. Components of local summer inflows of fresh water to the upper layer of the Arctic Basin from observations at the drifting station NP-16 [16]

Component	Direct Estimate	Indirect estimate	
	Inflow, cm	Freshening layer thickness,	Inflow, cm
Ice melting from the top	2.0	0-3	21.0
Snow melting	20.0	3-5	12.0
Ice melting from the	7.0	5-10	12.0
Lateral melting	15.0	10-25	4.0
Liquid precipitation	6.0	In puddles on the ice	4.0
Total	50.0		53.0

TABLE 6. Ablation of sea ice (cm) from different shape of sea ice surface [23]

Shape of surface	Number of observations	Ablation, cm	
		Mean	Maximum
Hummock top	38	30.3	50
Hummock slope	65	30.0	74
Level ice	46	27.9	87
Trough without a snow puddle	34	35.9	87
Snow puddle	47	57.6	104

Melting of ice from the bottom corresponding to a 7 cm layer of fresh water in Table 5 is also less than the known estimates of ice melting from the bottom according to data of underwater observations at the drifting stations [24]. These estimates vary depending on the bottom ice surface relief indicating the largest melting (up to 40 cm for the summer) under the snow puddles and the smallest (20-25 cm) under the level ice.

The contribution of lateral ice melting in the data of Table 5 was the largest as compared with ice melting from the top and bottom surfaces. As shown by observations at the drifting stations, at a comparatively small area of fractures between the ice floes in the summer, practically all heat coming to the surface of fractures is lost due to lateral ice melting [25] and the lateral ice floe surface changes its profile under the effect of melting (Fig.11).

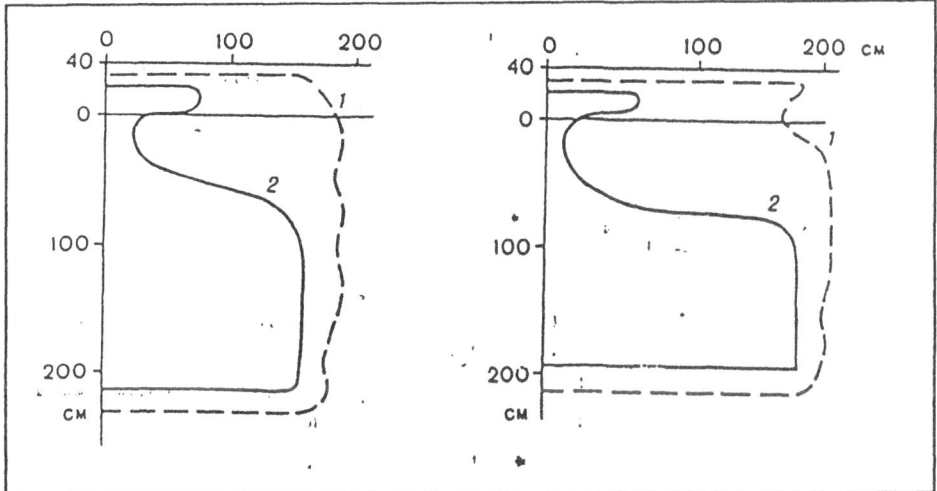

Figure 11. Lateral profiles of ice floe in the start (1) and the end (2) of melting
(observations on NP-16 in 1970-1971, [25]

Resulting from intense summer melting of snow and ice, a thin 1-2 m layer of fresh water is formed at the surface of fractures between the ice floes (Fig.12). By the end of summer up to 1000 km^3 of fresh water can concentrate within the Arctic Basin (at a drifting ice concentration of 90-95%).

602

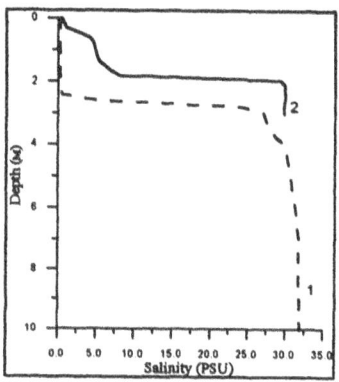

Figure 12. Typical summer salinity profiles in upper layer of the Arctic Ocean in sammer (in leads): 1.North eastern part of the Fram strait (Polarshtern 1993) 2.The Beaufort sea (ice drifting station "North Pole"-31, 1989)

4. Winter ice growth and the annual freezing/melting cycle balance

During the winter period (from October to May) the main mass of freshwater from summer melting freezes predominantly due to the formation of young ice, which turns into hummocked ice due to intense deformations of the drifting ice. The measurements of multiyear ice growth in the wintertime together with estimates of salination of the upper layer indicate that the process of freshwater freezing is related to a great extent to the growth of young ice, rather that to the increase in the old ice thickness. The ice growth at the bottom surface of multiyear ice in the underwater observation data at the drifting station "North Pole-22" during the winter was 20-30 cm [24]. The mean value of growth in the residual ice of the Arctic Basin is estimated to be 50 cm over the winter period [8, 27]. At the same time according to data of observations of the winter salination of water in the upper layer of the Arctic Basin the total ice growth in the winter is estimated to be 180-200 cm [18, 19].

The calculation of mean sea ice volume at the surface of the whole of the Arctic Ocean (Table 7) allows us to estimate the volume of melted *(Vmelt)*, frozen over *(Vfreeze)* and exported sea ice *(Vexp)* and the corresponding thickness of fresh water layer (h) based on the simple balance ratios and data from Table 7:

TABLE 7. Mean volume of sea ice (km3) in the Arctic Ocean in the middle of the month

	IX	X	XI	XII	I	II	III	IV	V	VI	VII	VIII
Volume	15520	15802	16661	18424	20180	21991	23485	24703	25406	24218	21114	17279

$$V_5 = V_9 + V_{melt} + V_{exp} \tag{1}$$

$$V_5 - V_9 = V_{melt} + V_{exp} = V_{freeze} = 9886 \ km^3 \tag{2}$$

$$V_{out} = V_{freeze} - V_{exp} = 7039 \ km^3 \tag{3}$$

$$h_{freeze} = 110 \ cm \tag{4}$$

$$h_{melt} = 80 \ cm \tag{5}$$

$$h_{exp} = 30 \ cm \tag{6}$$

The calculations of mean volumes of melting and freezing over in the annual sea ice cycle performed in this work (Table 8) yield close estimates of the volumes of melted and frozen over ice.

TABLE 8. Mean volume (kin³) of freeze and melt ice during year cycle

	IX	X	XI	XII	I	II	III	IV	V	VI	VII	VIII	Year
Freeze and melt on open water	173	734	940	941	881	706	558	383	-310	-1082	-1485	-1110	1329
Freeze and melt on multi-year ice	0	502	646	609	559	442	262	204	128	-1360	-1113	-574	305
Freeze on free area after export of ice	109	105	155	193	176	143	129	80	30	0	0	0	1167
Sum	282	1388	1741	1743	1616	1191	949	667	-152	-2442	-2598	-1684	2801

For calculation of values in Table 8 the following parameter values were used (Table 9).

TABLE 9. Parameter values used for calculation of mean freeze and melt cycle

No	Parameter	Value
1	Arctic Ocean area	8042 x10^3 km^2
2	Open water area in June-August	204 x 10^3 km^2
3	Sea ice area at the end of August	5838 x 10^3 km^2
4	Mean ice thickness (without ridges, at the end of August)	2.54 m
5	Reduced ice thickness in August: April :	1.85 m 2.59 m
6	Ice growth in open water from September to May	1.84 m
7	Ablation of multiyear ice	44 cm
8	Growth of multiyear ice	50 cm
9	Ablation of first-year and second-year ice	77 cm

The volumes of frozen over and melted ice based on data in Table 8 are estimated as follows:

$$V_{freeze} = 9677 \ km^3 \tag{7}$$

$$V_{melt} = 6876 \ km^3 \tag{8}$$

whose difference gives the volume of exported ice V_{exp} = 2801 km^3.

The increase in the ice mass in the winter is accompanied by the release of a large amount of heat, which composes the main part of its sink from the ocean surface during the winter. The estimates based on the observations at the drifting stations in the Arctic Basin indicate that the heat flux connected with freezing of freshwater comprises about 12 W/m^2 during the autumn-winter period. Of this value, about 7 W/m^2 is provided by freezing of fresh water from local sources (local summer melting) and 5 W/m^2 is due to "advective" freshwater [30]. Recalculated to the mean annual value, this flux is equal to 3 W/m^2, which is close to the estimate of full total flux at the surface of the Central Arctic Basin [31].

5. Possible causes of variations in the freshwater flow from the Arctic to the North Atlantic

The most vivid result of the increased outflow of freshwater from the Arctic Ocean was strong freshening in the upper layer of the North Atlantic in 1963-1973 called the great salinity anomaly [32]. All investigators who considered this phenomenon point to the sources of this anomaly in the Arctic Ocean (for example [33,34]. However, they define differently the processes responsible for the formation of anomalies — from changes in the atmospheric circulation over the Arctic and thus the sea ice drift and current speed in the upper layer up to the increased runoff of the Mackenzie River in the 1960s. All these factors contributed in fact to the formation of changes in the Arctic climatic system in the 1960s. However, in our opinion, they all manifested a large-scale primarily summer warming in the Arctic from the end of the 1950s to the middle of the 1960s. Increased snow and ice melting and the increased outflow from the surrounding continents and changes in the atmospheric circulation over the Arctic Ocean accompanied it.

The development of the salinity anomaly in the 1960s included, first, decreased water salinity in the upper water layer north of Iceland in the summer of 1963 [35]. Second, in the data of [36] the largest salinity decrease at a level of 50 m was observed here in 1969.

At this time according to [37] the lowest water salinity and temperature were observed off the western coast of Greenland in the region of Fylla Bank in the area of the West Greenland Current. This fact points to a practically simultaneous development of freshening in both regions suggesting two ways of freshwater transport from the Arctic to the North Atlantic: through Fram and Denmark Straits and through Baffin Bay and Davis Strait from the west.

As mentioned above, the main source for the anomalous freshening of the upper layer of the Arctic Ocean could be a strong summer melting of snow and ice over the Arctic Ocean area and at the surface of the adjoining land. An analysis of air temperature changes in the Arctic in 1948-1996 indicates that in 1957-1962 there were unusually high air temperatures in the summer near the coast of the Arctic Ocean and especially in the region of West Greenland, Baffin Bay and the adjoining part of the Canadian archipelago (Fig. 13, 14). Mean multiyear air temperature in this area in the summer is the lowest in the entire Arctic and here the largest amount of snow and ice accumulate in the winter. Large positive air temperature anomalies here contributed to intense summer melting and fresh water outflow to the Arctic Ocean, Canadian Straits, Baffin Bay and Hudson Bay. As is seen in Fig. 13, mean air temperature in the summer in the Arctic has again significantly increased at the end of the 1980s- beginning of the 1990s, which can be a signal of the formation in the Arctic of a new freshening impulse for the North Atlantic.

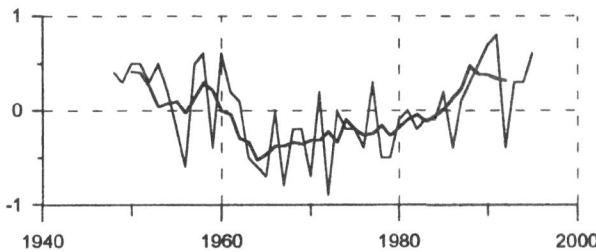

Figure 13. Anomalies of mean air temperature in summer (June-August) in the Arctic for 1948-1996

606

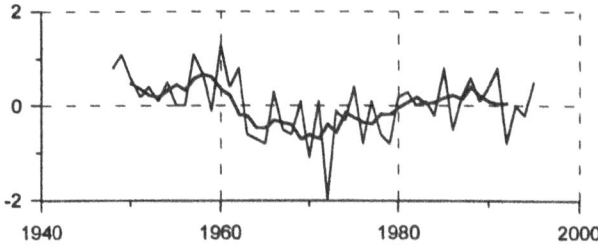

Figure 14. Anomalies of mean air temperature in summer (June-August) in Baffin Bay region for 1948-1996

6. Conclusion

In the upper layer of the Arctic Ocean fresh water transported with the polar branch of the global hydrological cycle accumulates at its surface with sea ice and its outflow to the North Atlantic is formed. Between these two processes in the Arctic Ocean, repeated phase "water-ice" transformations occur. They constitute the internal Arctic hydrological cycle changing the dynamic and thermodynamic properties of the upper layer, and the time of residence of freshwater and thus influence the formation of its outflow.

The fresh water flows to the Arctic and its outflow from the Arctic Ocean comprise an active part of a full excess of fresh water in it resulting in the preservation of sea ice. The total content of the fresh water "excess" in the Arctic Ocean estimated relative to the salinity of 34.80 PSU is non-uniformly distributed over its area changing from year-to-year. The climatic FWC maximum is located in the Canadian Basin and is formed by the fresh water flows with precipitation, river runoff and Bering water. The active FWC portion is centered in the upper 50 m layer within the winter convective mixing depth in the Arctic Ocean. The interannual changes of the mean FWC over the area reach here 1.3 m of the equivalent layer of freshwater. A similar value of FWC changes was obtained from water salinity changes in the region of sources of the East Greenland Current north of Greenland. The interannual FWC changes are characterized by increased values up to the mid-1970s and decreased FWC after 1975. The differences between them are most noticeable at the periphery of the Arctic Ocean.

The main cause of the significant interannual FWC changes is connected with the summer flows of fresh water to the Arctic Ocean. According to different estimates, they are within 6000 to 10000 km^3, which exceeds by several times the net annual inflow. The changes in the conditions of summer melting lead to a disturbed balance in the "freezing-melting" cycle and as a result to freshening or salination of the upper layer of the Arctic Ocean. The summer fresh water flows are formed due to snow melting (whose depth before melting is 36 cm, on average, for the Arctic Basin varying within 25 to 85 cm from data of observations at some drifting stations) and ice melting from the top and the bottom and lateral ice melting. The average value of ice melting determined from data of observations at the drifting stations for multiyear ice comprises 44 cm for melting from the top and 20 to 40 cm for melting from the bottom. A contribution from lateral ice melting is also significant.

The volume of winter ice growth in the Arctic Basin was estimated from climatic data as 9677-9886 km^3 and the volume of summer melting as 6876-7086 km^3, which corresponds to the prescribed volume of the exported ice from the Arctic Ocean of 2800÷2850 km^3.

The excess of the winter ice formation over summer melting serves as the main source of heat released in the winter period from the surface of the Arctic Ocean to the atmosphere.

The anomalous water freshening in the upper layer of the North Atlantic observed in the 1960s-early 1970s is primarily related to the increased flow of freshwater to the Arctic Ocean due to intense snow and ice melting in the Arctic, especially in 1958-1963 in the region of the Canadian archipelago, Baffin Bay and West Greenland. The increased summer air temperature in the Arctic, which began at the end of the 1980s can be an indication of the formation of a new impulse of freshening in the Arctic Ocean.

Acknowledgement

The study was prepared with the support of the Russian Foundation of Basic Researches, grant 97-05-65933.

References

1. Zakharov V.F. (1978) *World Ocean and ice stage of the Pleistocene,* Gidrometeoizdat, Leningrad (in Russian).
2. Zakharov V.F. (1996) *Sea ice in the climate system,* Gidrometeoizdat, St. Petersburg, (in Russian).
3. Aagaard K.A. and Carmack EC. (1989) The role of sea ice and other freshwater in the Arctic circulation. *J. Geophys. Res.,* **94**, *14485-14498.*
4. Alekseev G.V., Ivanov V.V., and Korablev NA (1994) Interannual variability of the thermohaline structure in the convective gyre of the Greenland Sea, in OM. Johannessen, RD. Muench, JE. Overland (eds.). *The Polar Ocean and their role in shaping the global environment,* Nansen Centennial Volume, American Geophysical Union, Washington, 485-496.
5. Alekseev G V , Ivanov V V , Korablev AX, (1995) Interannual variability of the Greenland sea Deep Convection, *Oceanology, 35, 45.52* (in Russian).
6. Joint U.S. Russian Atlas of the Arctic Ocean (1997) *Oceanography Atlas for winter period,* L.A. Timokliov and F. Tannis (eds.), CD-ROM, NSIDS/CIRES.
7. Joint U.S. Russian Atlas of the Arctic Ocean (1998) *Oceanography Atlas for summer period,* L.A. Timokhov and F. Tannis (eds.), CD-ROM, NSIDS/CIRES.
8. Treslmikov A.F., Baranov GI. (1972) *Water circulation structure in the Arctic Basin,* Gidrometeoizdat, Leningrad (in Russian).
9. Lettenmajer D. and Bowing L (1999) Hydrology of the Arctic Drainge Basin, in EL. Lewis (ed.), *The freshwater budget a/the Arctic Ocean,* Kiuwer Academic Publishers, Dordrecht.
10. Walsh J. (1999) Global atmospheric circulation patterns and their effect on AFW fluxes, in EL. Lewis (ed.), *The Freshwater Budget 0/the Arctic Ocean,* Kluwer Academic Publishers, Dorderecht.
11. Shildomanov I., Shildomanov A., Lammers R. (1999) The dynamics of river water inflow to the Arctic Ocean, in EL. Lewis (ed.), *The Freshwater Budget 0/the Arctic Ocean,* Kiuwer Academic Publishers, Dordrecht.
12. Alekseev G.V., Zakharov V.F., Smirnov NP., Smirnov AN. (1999) Multiyear variations of ice conditions and atmospheric circulation in the Arctic and Northern Atlantic, *Meteorology and Hydrology,* (in press).
13. Ivanov V.V. (1976) Freshwater balance of the Arctic Ocean, *Proc./A4RJ, 323,* 138-147 (in Russian).
14. Serreze MC., Barry R.G., Walsh i.E. (1996) Aerological estimation of precipitation minus evaporation over Arctic, *WCRP-93, 14M0/TD,* **739,** 71-74.
15. Shpaikher A.O. (1976) Freshwater content in sea ice of polar areas, *Proc./A4RJ,* 323, 168-177 (in Russian).
16. Alekseev G.V., Busuyev A.Ya. (1973) Evolution of the "sea ice-upper layer" system in the vicinity of the drifling station "NP-16", *Problems o/the Arctic and the Antarctic,* 42, 37-43 (in Russian).
17. Golovin P.N., Kochetov 5.V., Timokliov L.A. (1995) Freshening of under ice !ayer by sea ice me!ting, *Oceanology, 35, 525-530* (in Russian).
18. Treshnikov AT. (1959) Surface water in the Arctic Basin, *Problems o/the Arctic and the Antarctic, 7,* 5-14 (in Russian).

19. Alekseev G.V. (1976) Sea ice/atmospheric circulation interaction in the Central Arctic, *Proc./AARJ*, 332, 109-112 (in Russian).
20. Gorshkov *5.0.* (1980) Atlas of World Ocean, *the Northern Ice Ocean*, Ministry of Defence of the USSR, Moscow.
21 .Khrol V.P. ed. (1996) Atlas *0/water balance a/the Northern Polar Region*, Gidrometeoizdat, St. Petersburg (in Russian).
22. Radionov V.F., Bryazgin NI., Aleksandrov Ye.I. (1996) *Snow cover in the Arctic Basin*, Gidrometeoizdat, St. Petersburg (in Russian).
23. Buzuyev A.Ya., Gorbunov Yu. A., GudeovichZ. M., Losev SM., Mironov EU. (1998), Study of dynamics and morphology of the Arctic Basin ice *coverProblems a/the Arctic andAntarctic*, **71**, 102-124 (in Russian).
24. Grischenko V.D., (1981) Peculiarities of underwater melting and growth of sea ice in the Arctic Basin, *Proc. AARJ*, 372, 123-128 (in Russian).
25. Aiekseev G.V., Buzuyev AYa. (1973) Lateral ice melting in leads, *Proc. AARJ*, 307, 169-178 (in Russian).
26. Galbraith P.S., Ingram R.G. (1994) Hydrology beneatb the *iceReports onPolarRes.*, **145**, 85-105.
27. Maykut GA, Thomdike A. 5. (1973) An approach to coupling the dynamics and thennodynamics of Arctic sea *ice,AIDEX Bulletin*, 21, 23-29.
28. Lebedev A.A., Uralov N.S. (1981) Assessment of the balance of sea ice volume in the northern hemisphere of the Earth, *Proc. AARJ* **384**, 6 1-77) (in Russian).
29. Vinje T., Norlund N. and Kvambekk A. (1998) Monitoring ice thickness in Fram Strait, *J. Geoph. Res.*, 103, 10437-10450.
30. AJekseev G.V., Makshtas AP. (1998) Study of processes of ocean/atmosphere interaction in the Arctic Basin, *Problems 0/the Arctic andAntarctic*, 71.
31. Khrol V.P., ed. *(1993)Atlas o/the heat balance o/ the Northern PolarRegion*, Gidrometeoizdat, St. Petersburg.
32. Dickson R.R., Osbom I.J., Hurrell J.W., Meincke J., Blindheim J., Adladsvik B., Vinje T., Alekseev G., Maslowski W. and Cattle H. (1997) The Arctic Ocean response to the North Atlantic Oscillation, *Proceedings Conference on Polar Processes and Global Climate*, Rosario, Orcas Island, Washington, USA, 3-6 November 1997, 46-47.
33. Mysak L.A., Manak D.K., and Marsden R.F. (1990) Sea ice anomalies observed in the Greenland and Labrador Seas during 1901-1984 and their relation to an interdecadel Arctic climate cycle. *Climate Dynamics.* 5, 111-132.
34. Hakkjnen 5. (1993) An Arctic Source for the Great Salinity Anomaly: A simulation of the Arctic ice-ocean system for 1955-1975,.]. *Geophys. Res.*, 98, 16397-16410.
35. Stefansson U. (1969) Temperature variations in the North Icelandic coastal area during recent decades, *Jakull*, **19**, 18-28.
36. Malmberg S-A., and Blindheim J. (1994) Climate, cod, and capelin in northern waters, *ICES Marine Science Symposia*, **198**, 297-3 10.
37. Drinkwater K.F. (1994) Climate and oceanographic variability in the Northwest Atlantic during 1980s and early 1990s, *NAFO SCR Doc.*, **94/7** 1.

Colour Plates

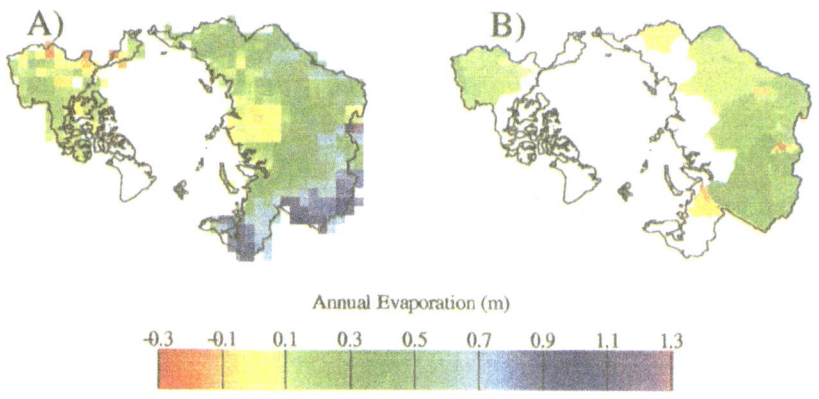

Annual Evaporation (m)

-0.3   -0.1   0.1   0.3   0.5   0.7   0.9   1.1   1.3

Figure 8. Average annual evaporation computed from
A) atmospheric water balance and B) basin water balance

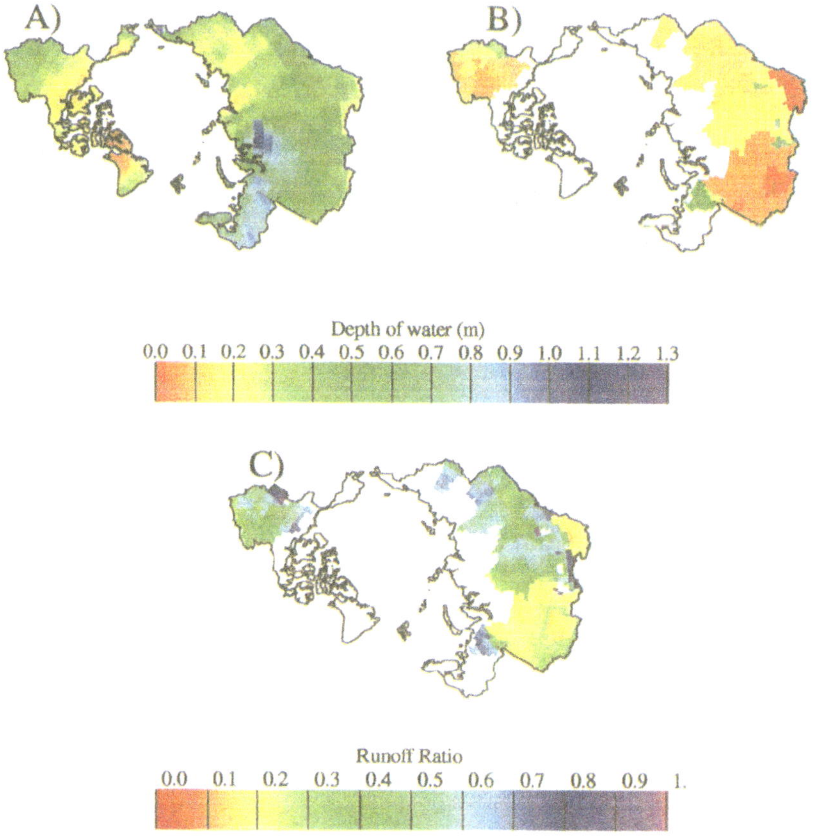

Depth of water (m)

0.0 0.1 0.2 0.3 0.4 0.5 0.6 0.7 0.8 0.9 1.0 1.1 1.2 1.3

Runoff Ratio

0.0   0.1   0.2   0.3   0.4   0.5   0.6   0.7   0.8   0.9   1.

Figure 9. A) Average annual precipitation (1971–1994), B) Average
annual discharge (variable records), C) Average annual runoff ratio

(See also p. 72)

612

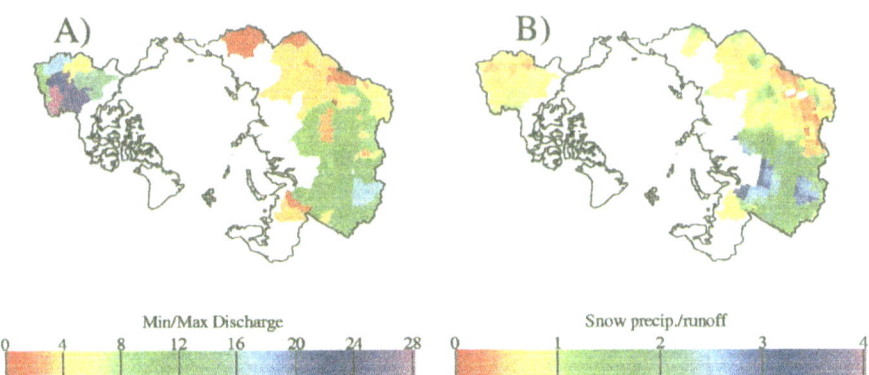

Figure 10: A) Mean ratio of minimum monthly discharge to maximum monthly discharge, and
B) Ratio of snow precipitation to runoff

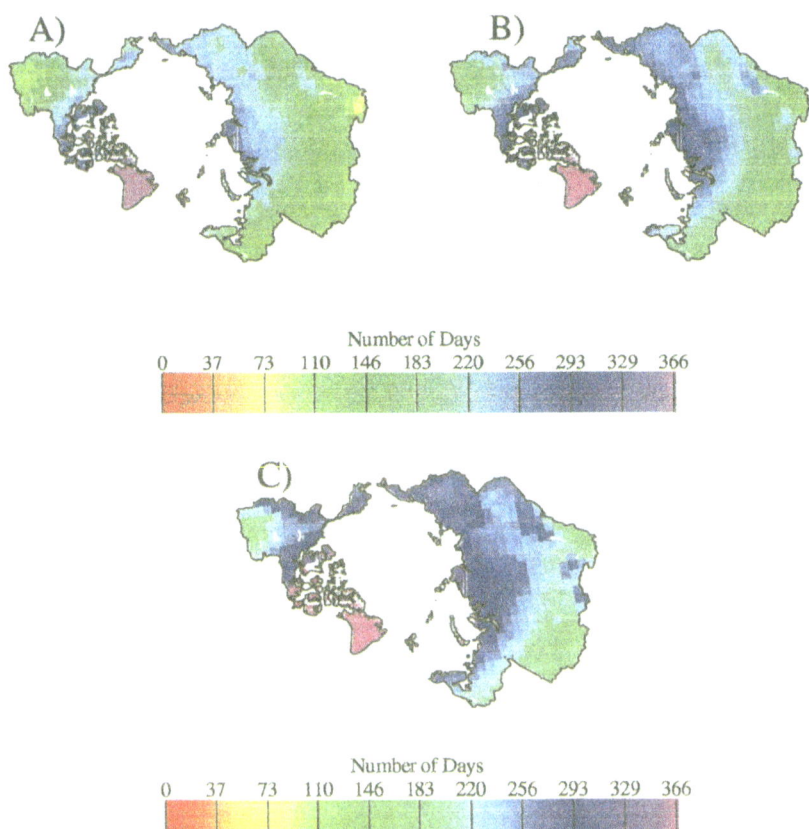

Figure 11. Number of days with snowcover (1971 –1994) A) Minimum
B) Average and C) Maximum

(See also p. 74)

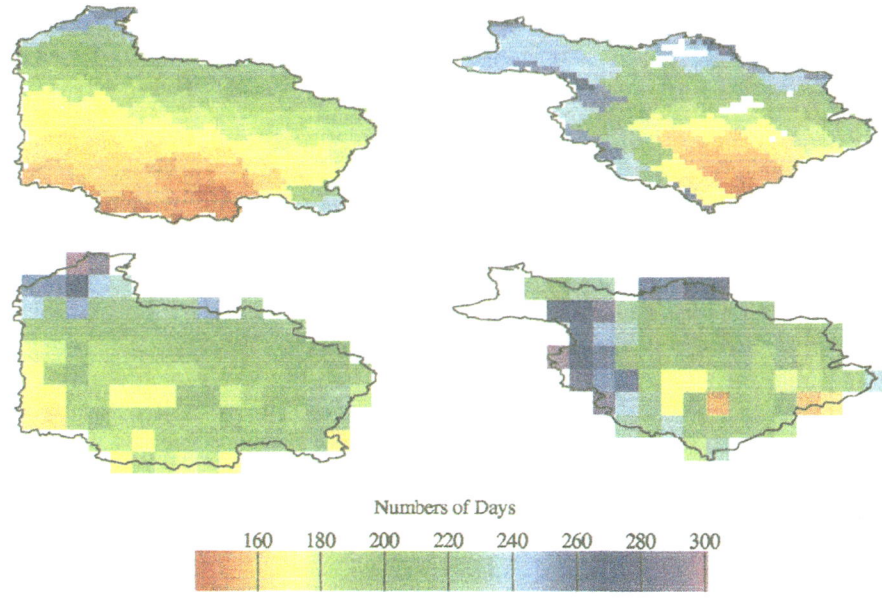

Figure 16. Remotely sensed (top) and simulated (bottom) number of days with snowcover. A) Ob River and B) Mackenzie River

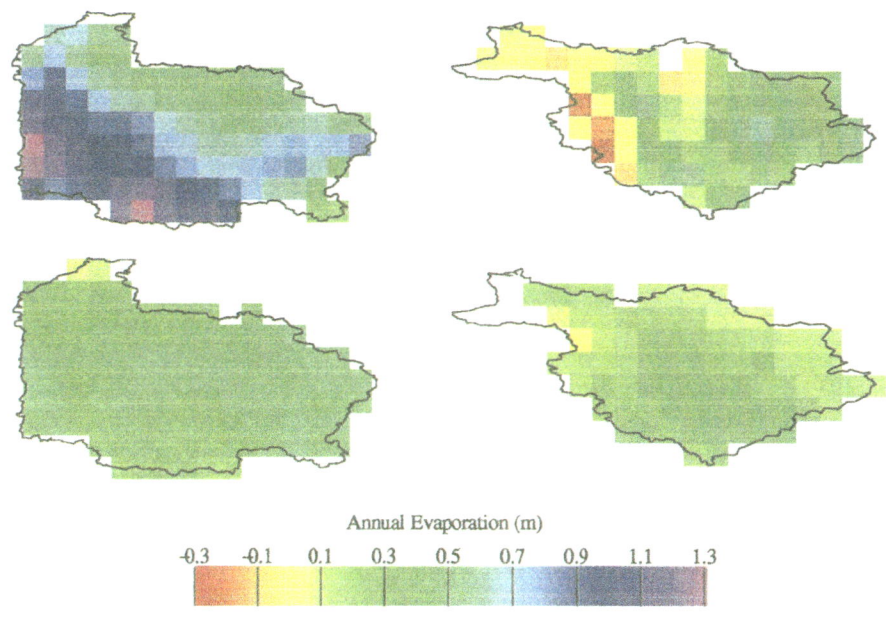

Figure 17. Annual evaporation derived from DAO reanalysis fields (top) and simulated (bottom). A) Ob River and B) Mackenzie River

(See also p. 82)

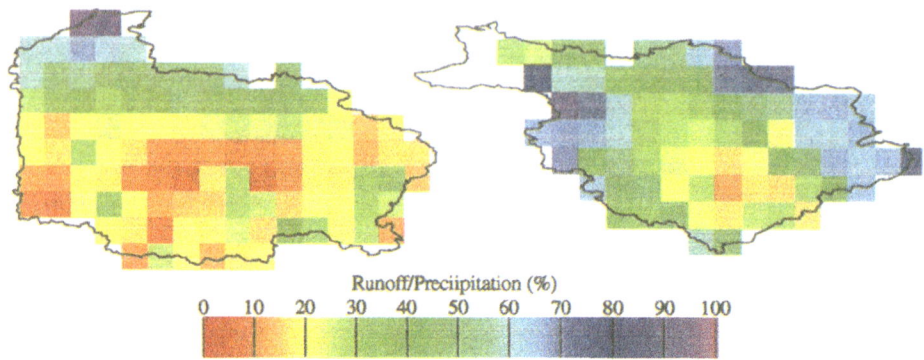

A) Annual Runoff/Annual Precipitation: Ob River (left) and Mackenzie River (right)

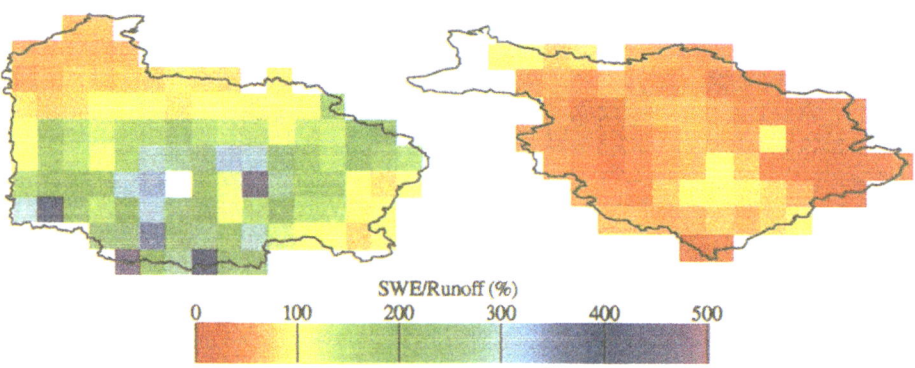

B) Maximum Snow Water Equivalent (SWE)/Annual Runoff: Ob River (left) and Mackenzie River (right)

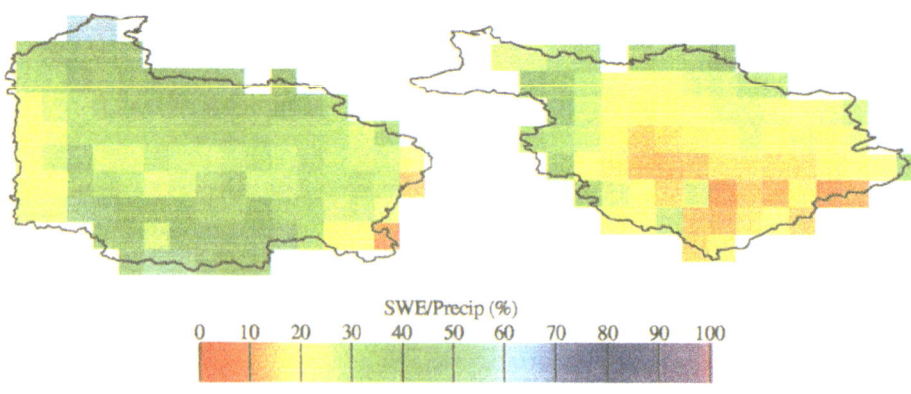

C) Maximum SWE/Annual Precip: Ob River (left) and Mackenzie River (right)

Figure 18: Simulated Water Balance Components

(See also p. 84)

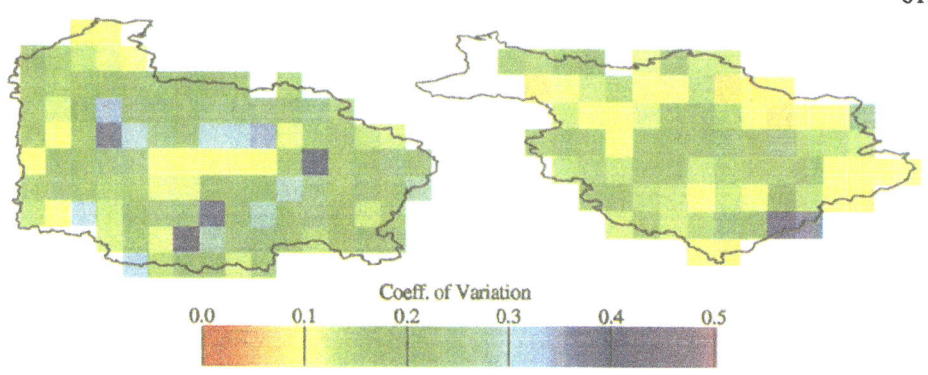

A) Runoff ratio: Ob River (left) and Mackenzie River (right)

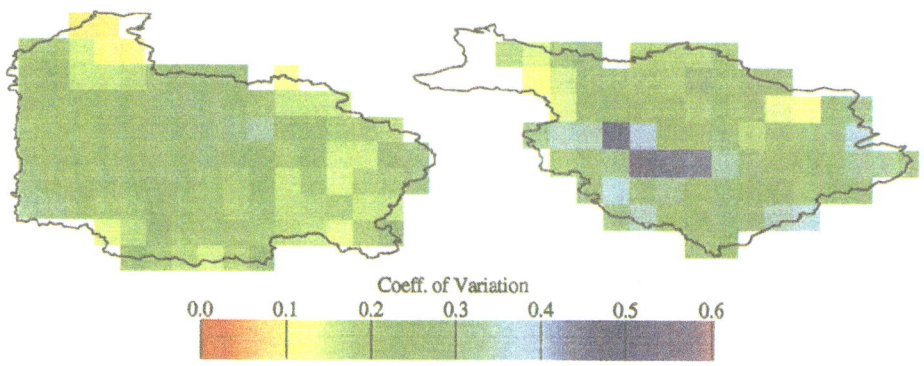

B) Maximum Snow Water Equivalent (SWE): Ob River (left) and Mackenzie River (right)

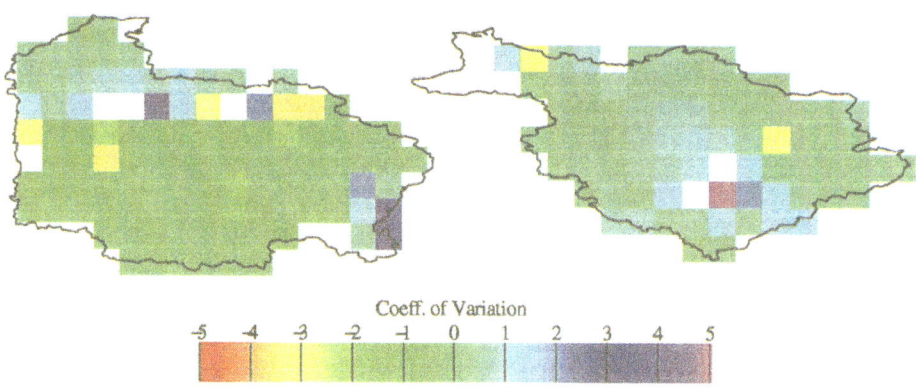

C) Annual Runoff –Maximum SWE: Ob River (left) and Mackenzie River (right)

Figure 19: Coefficient of Variation of Simulated Water Balance Components

(See also p. 85)

616

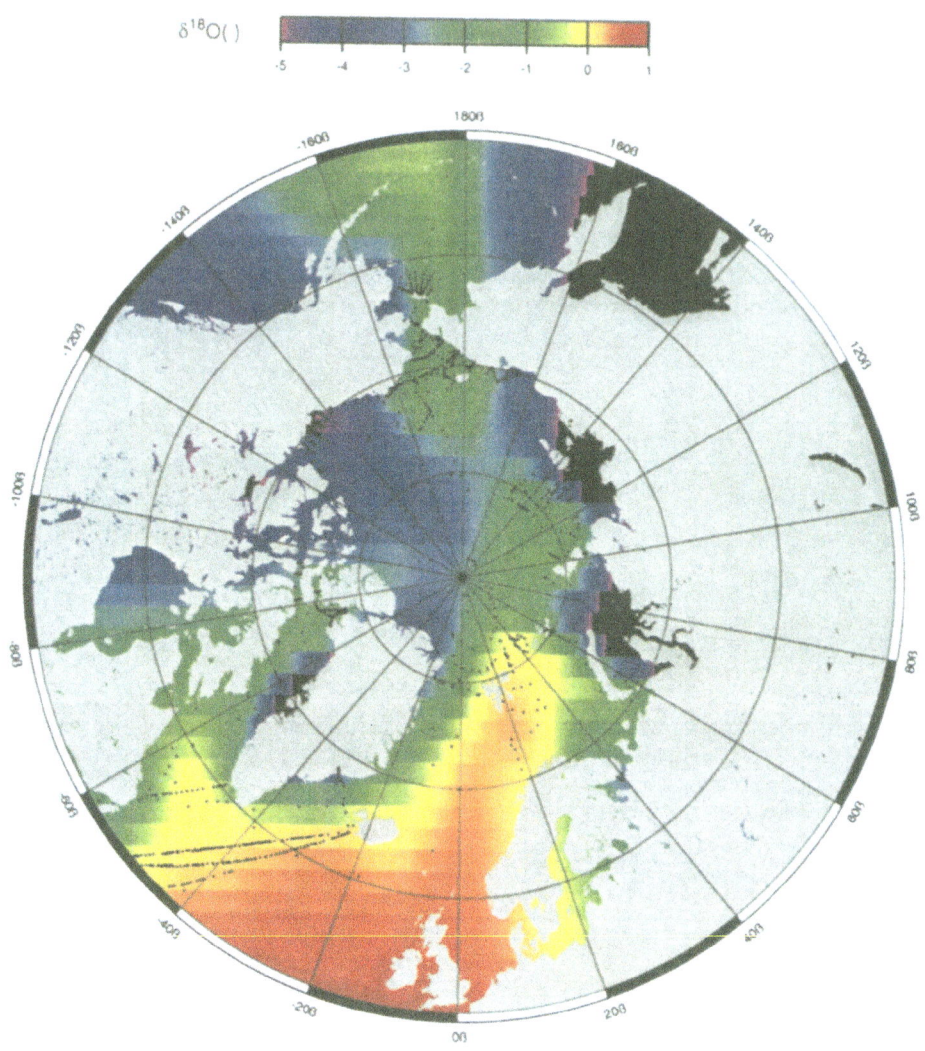

Figure 5a: Distribution of $\delta^{18}O$ in the top 15 meters of the water column. For discussion, see text. Black shading near the coast of Greenland and Siberia indicates values < -5‰.

(See also p. 466)

Salinity (↑15m)

20 21 22 23 24 25 26 27 28 29 30 31 32 33 34 35 36

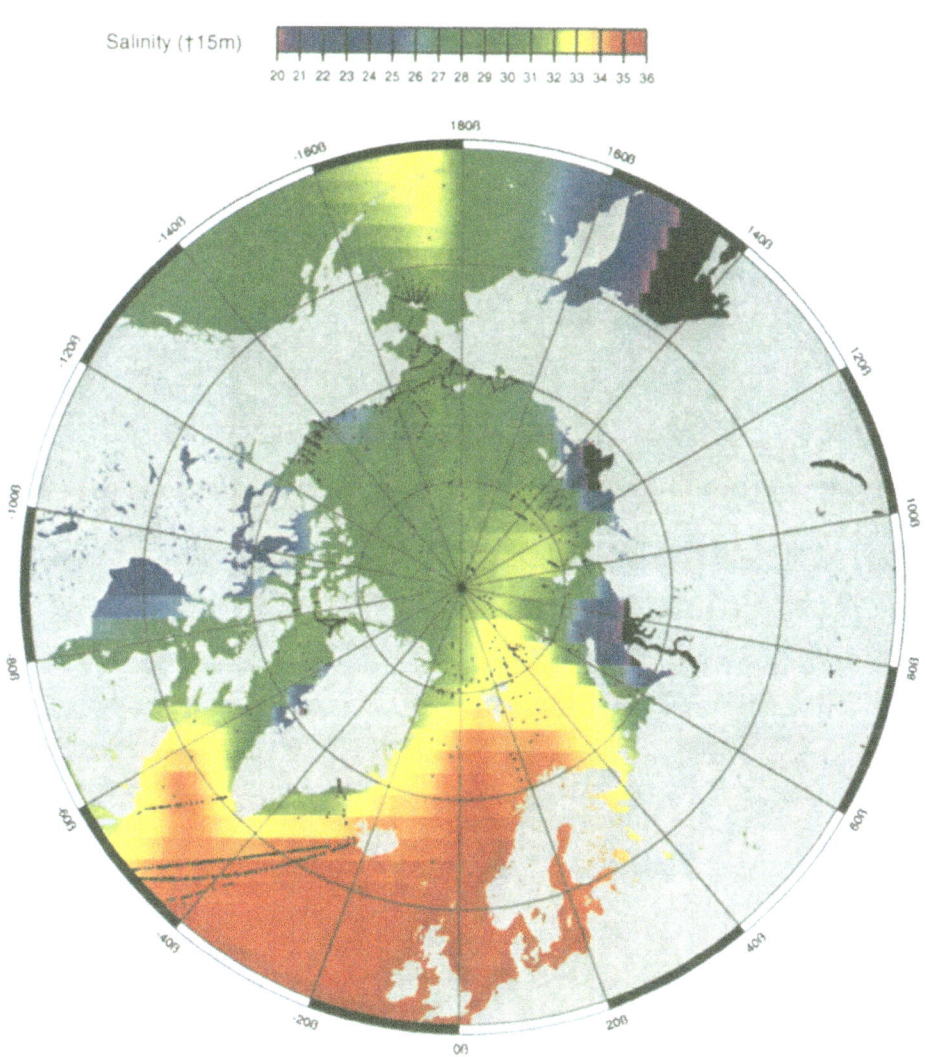

Figure 5b: Map of the salinity of the top 15 meters of the water column using the same stations as in Fig. 5a. For discussion, see text.

(See also p. 468)

618

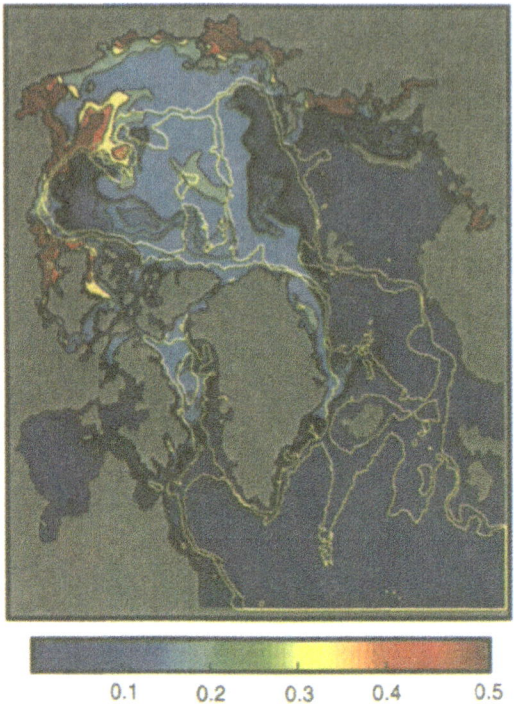

Figure 6a: Results of the simulation of the Arctic freshwater distribution in a high-resolution GCM. Displayed are the fractions of freshwater.

(See also p. 471)

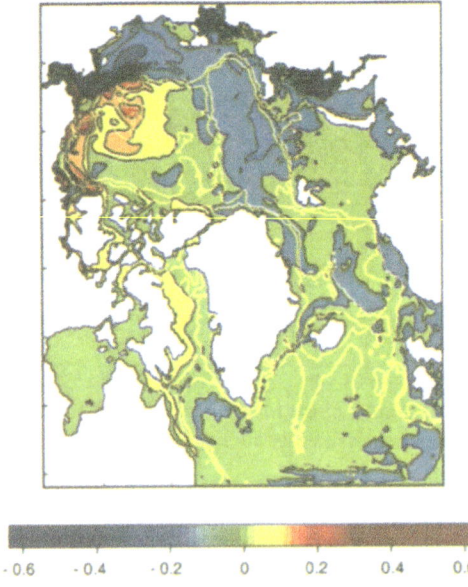

Figure 6b: Difference in freshwater fraction distribution between 1993 and 1979, as simulated by a high-resolution GCM.

(See also p. 472)

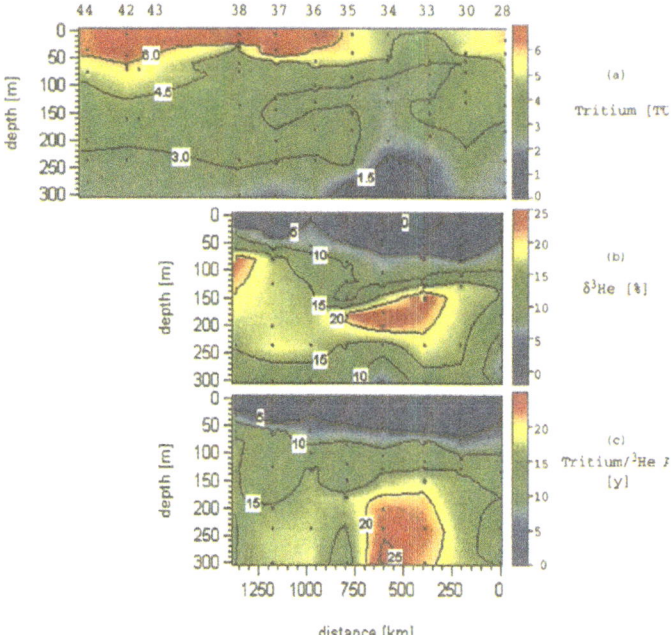

Figure 8: (a) Tritium section along a portion of the cruise track of the US nuclear submarine POGY (1996). (b) same as (a) except for δ^3He. (c) same as (a) except for the tritium/^{3}He age.

(See also p. 474)

Figure 1. Scheme of interaction of polar and extrapolar processes in the climate system.

(See also p. 589)

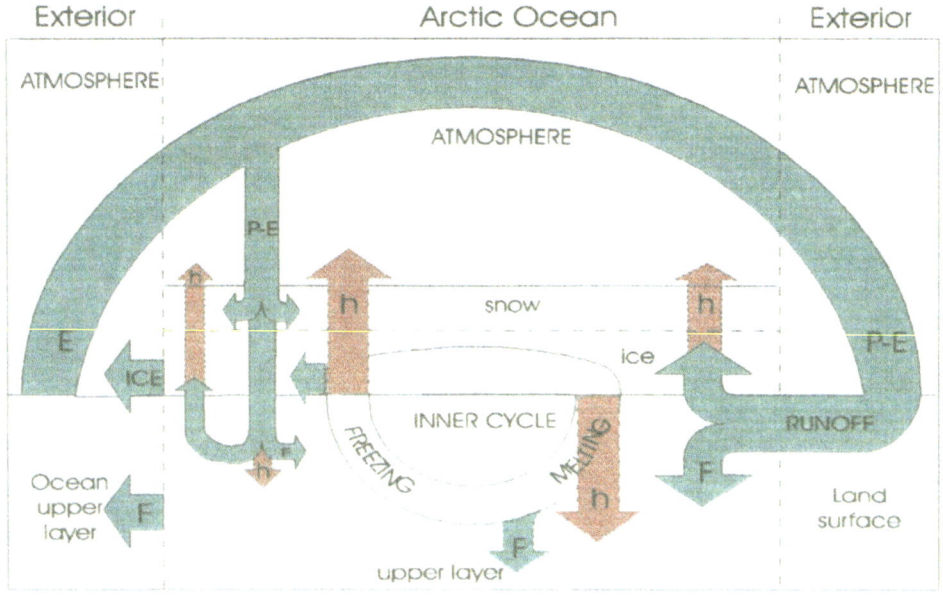

Figure 2. Scheme of fresh water (F) and latent heat (h) cycle in the upper layer of the Arctic Ocean.

(See also p. 590)

SUBJECT INDEX

The manufacturer's authorised representative in the EU is Springer
Nature Customer Service Centre GmbH, Europaplatz 3, 69115 Heidelberg,
Germany. If you have any concerns regarding our products, please
contact ProductSafety@springernature.com

Printed and bound by CPI Group (UK) Ltd, Croydon, CR0 4YY
30/04/2026
02100146-0001